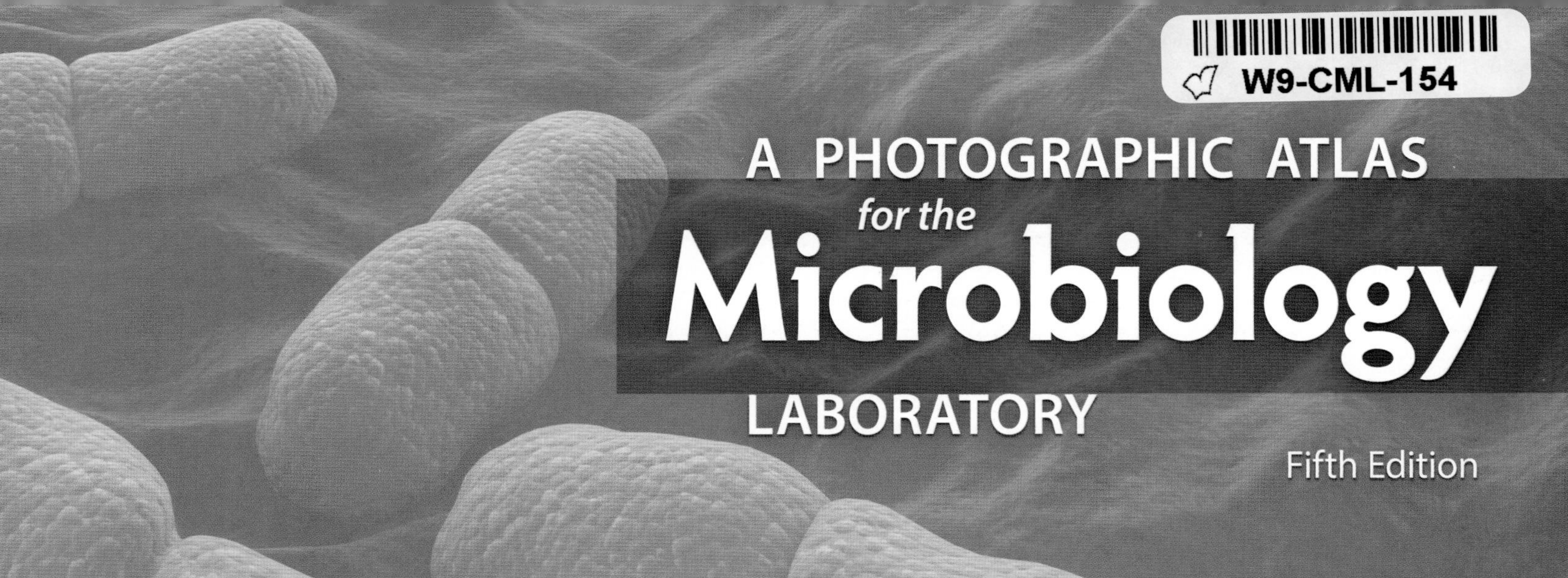

A PHOTOGRAPHIC ATLAS for the Microbiology LABORATORY

Fifth Edition

Michael J. Leboffe
San Diego City College

Burton E. Pierce

925 W. Kenyon Avenue, Unit 12
Englewood, CO 80110
morton-pub.com

Book Team

President and CEO	David M. Ferguson
Senior Acquisitions Editor	Marta R. Pentecost
Supervising Editor	Adam W. Jones
Editorial Project Managers	Rayna S. Bailey, Trina Lambert
Production Manager	Will Kelley
Production Associate	Joanne Saliger
Indexer	Teri Jurgens
Illustrations	Imagineeringart.com, Inc.
Cover Design	Will Kelley
Cover Image	*Listeria* bacteria
Cover Credit	Science Picture Co/Science Source

Dedication

This book is dedicated to Dr. Oliver A. Ryder, Kleberg Endowed Director of Conservation Genetics at the San Diego Zoo Global Institute of Conservation Research. In 1976, I was a second-year graduate student at San Diego State University trying to make the best of a research project that simply didn't resonate with me. I mentioned my struggles to a zoology faculty member at SDSU and he introduced me to Ollie, who had himself recently taken a postdoctoral position in the San Diego Zoo Research Department at a time when the Frozen Zoo[1] was in its infancy. Much to my surprise, he agreed to take me into his lab, where I was assigned the project of characterizing six lemur species based on serum proteins in an effort to clarify their degrees of relatedness. It wasn't until many years later that I realized I was his first graduate student. I was so blindly lucky!

Research wasn't easy for me (it's rarely "easy" for anyone), but I completed the project and while I didn't save the world with my research, my experiences working with Ollie helped make me a better teacher of the scientific process. In my four decades of teaching biology—everything from botany to human anatomy, to microbiology, and more—there hasn't been a class where my lessons learned from Ollie weren't woven in.

Ollie continues his important work in conservation genetics. Check out these links[2,3] and I guarantee you'll be amazed at what is being done to address the global crisis that is extinction.

Thank you, Ollie, for taking me into your lab in 1976 and for the work you've done in your marvelous career. It was an honor working with you and to be around during the Frozen Zoo's early days. No professor had a greater impact on my science education than you did.

Mike

[1] Frozen Zoo: **https://www.smithsonianmag.com/science-nature/san-diegos-frozen-zoo-180971276/**

[2] Dr. Ryder's work: **https://institute.sandiegozoo.org/oliver-ryder-phd**

[3] Extinction Vortex: **https://www.youtube.com/watch?v=EVzhs1WjzGg**

Printed in the United States of America

10 9 8 7 6 5 4 3 2 1

ISBN-10: 1-61731-903-1
ISBN-13: 978-1-61731-903-7

Library of Congress Control Number: 2020941853

Preface

Microbiology may be the intended general topic of *A Photographic Atlas for the Microbiology Laboratory*, but surprisingly, silver emerged as a subplot of it. For some (many?), it came as a change in hair color due to the length of time required to produce the finished product. For others, it was silver's ancient association with the moon, symbolic of the late hours spent bringing it to completion. But the real reason is that the publication year ended up being 2021, 25 years after the first edition was published. Ladies and Gentlemen, welcome to the Silver Anniversary edition of *A Photographic Atlas for the Microbiology Laboratory*! (But we're certain you already knew that because of the silver-embossed commemorative seal on the cover . . . huh? What? Oh . . .)

On to business. The previous edition of this book experienced a complete make-over in terms of organization, and this edition continues that general organization. After the Introduction, the next 8 sections follow the process used by a working microbiologist: isolation of an unknown (Section 2); examination of growth characteristics (Section 3); microscopy and staining (Sections 4, 5, and 6); and then identification by traditional biochemical tests (Section 7), serological tests (Section 8), and modern molecular techniques (Section 9). We then move on to the microorganisms themselves. Viruses (Section 10), Domain Bacteria and Bacterial pathogens (Sections 11 and 12), and Domain Archaea (Section 13). Domain Eukarya is spread over Sections 14 through 17, starting with simple eukaryotes, followed by Fungi of Clinical Importance, Protozoans of Clinical Importance, and Parasitic Helminths. The remaining sections cover a variety of topics. Section 18 is devoted to quantitative techniques. Section 19 covers a potpourri of topics under the general headings of medical, environmental, and food microbiology. Section 20 deals with a superficial treatment of host defenses. Biochemical pathways are relegated to the Appendix, where they can be most easily referenced.

Though the organization is unchanged from the 4th edition, every section was thoroughly scrutinized and updated. Following is a list of some global changes.

- This edition has more than 300 photographs and almost 400 light micrographs. Very few images from 25 years ago remain, but you're welcome to hunt for them; some are still present.
- Some older photos and micrographs were edited to improve contrast or color where it was not possible to replace them with new ones. They did it to the first Star Wars movie, and Ted Turner colorized classic black-and-white movies, so we don't feel badly about doing it, too. That's an advantage of the digital world we live in.
- Time and effort were spent in the San Diego State University Electron Microscope Facility producing transmission and scanning electron micrographs to enhance the sections on bacterial structure.
- The organismal sections take a phylogenetic approach. Relevant taxa are included in the photo captions to assist the reader in maintaining their bearings in the great sea of taxonomy.
- Lastly, an issue of cell sizes in micrographs has been nagging at us for years. Without going into the details of why this wasn't done sooner, we have attempted to print bacterial micrographs with the cells on the page approximating their size as seen through a 1000× oil immersion lens. Bucket list item, this.

Here's a section-by-section synopsis of changes we made this time around.

- **Section 1 Introduction** The treatment of phylogeny with Domains was updated and images of lab safety equipment were added.
- **Section 2 Isolation Techniques and Selective Media** Most photos were replaced; Vogel-Johnson agar replaced tellurite-glycine agar.
- **Section 3 Bacterial Growth** More than a dozen new colony morphology photos were added and the entire section on colony morphology was organized by colony features.
- **Section 4 Microscopy** Included are new photos of a light photomicroscope, transmission electron microscope, and scanning electron microscope, as well as new TEM and SEM images.
- **Section 5 Bacterial Cellular Morphology, and Simple and Negative Stains** There are some miscellaneous new light micrographs and a couple of new TEMs showing binary fission.

- **Section 6 Bacterial Cell Structures and Differential Stains** The most noticeable additions are five new TEMs. A handful of light micrographs were also replaced.
- **Section 7 Differential Tests and Multiple-Test Systems** New artwork for O-F medium and the effect of SXT on folate synthesis have been added. Other artwork has been clarified and several new photos replace older ones. Tryptone water was added to the indole test and a treatment of the Vitek 2 Compact system was added.
- **Section 8 Serology** The Lancefield group agglutination test was added and the ELISA test was updated.
- **Section 9 Molecular Techniques** We updated to next generation DNA sequencing technique, added quantitative PCR, gene chip hybridization, and MALDI-TOF MS. While these are not particularly photogenic topics, there are several new photos and artwork to illustrate them.
- **Section 10 Viruses** In keeping with the phylogenetic theme running through the organismal sections of the *Atlas*, we included a treatment of the Baltimore viral classification system. CPE micrographs were enhanced for contrast and new photos of an inverted microscope were added.
- **Section 11 Domain Bacteria** Using *Bergey's Manual of Systematic Bacteriology*, 2nd ed., as an organizational skeleton, we rewrote and reorganized this entire section. At latest count, there are 31 new micrographs included alongside another handful of replacements. But binary fission being what it is, who really knows what the count is now . . . .
- **Section 12 Bacterial Pathogens** We abandoned the phylogenetic organization of the other organismal sections for an alphabetical organization in this section, figuring that students would have an easier time finding the organisms by name rather than by taxon. This section was also almost completely rewritten, and approximately 30 micrographs are either new or replacements of former ones.
- **Section 13 Domain Archaea** This section was also heavily revised and enhanced with electron micrographs of *Sulfolobus* and *Thermoplasma*, halophile samples collected from the Great Salt Lake, and methanogen samples from sludge digesters at a wastewater treatment plant in San Diego, CA. Yet, the cow remains . . . .
- **Section 14 Domain Eukarya: Simple Eukaryotes** The taxonomy was updated, and more than 20 new or replacement photos are in this section. In many instances, this section serves as a background in the basic biology of protozoan and fungal pathogens covered in Sections 15 and 16.
- **Section 15 Fungi of Clinical Importance** This section was trimmed down for a number of reasons, primarily because other sections grew and we had a page limit. Further, most introductory microbiology courses don't handle fungi, especially pathogenic ones. We did expand our treatment of *Candida albicans* and added the opportunistic pathogen *Rhodotorula mucilaginosa*. Nine of 22 images are either new or replacements of old ones.
- **Section 16 Protozoans of Clinical Importance** Not much new here, but we did spend time improving the quality of most of the micrographs. There is one differential interference micrograph of *Cryptosporidium*.
- **Section 17 Parasitic Helminths** Ten photos out of nearly 60 are either new to the book or are replacements of previous ones. In addition, the majority of micrographs were edited to improve contrast, because many of the original specimens were old when we photographed them more than 20 years ago. There was also an issue of a blue filter in a borrowed photomicroscope that we were unaware of when photographing wet mounts, which made the images look blue. We've adjusted the colors to what the stained specimens actually looked like. Lastly, we've added more labels to the worm specimens.
- **Section 18 Quantitative Microbiology** Eleven out of 12 figures are either new to this edition or are replacements of previous ones. We've expanded coverage of the urine streak by including HardyCHROM UTI media.
- **Section 19 Medical, Environmental, and Food Microbiology** We've greatly expanded the topic of water treatment and added a section on yogurt. Eighteen of 44 figures are either new or replacements of previous ones.
- **Section 20 Host Defenses** Eighteen of 25 micrographs are either new to this edition or are replacements of previous ones. A new overview of immunology was written.
- **Appendix Biochemical Pathways** Here is where we put the biochemical pathways mentioned in the text. All have been modified to some extent. Two new figures have been added. One shows how catabolism and anabolism of amino acids, nucleic acids, polysaccharides, and lipids tie into glycolysis, the entry step, and citric acid cycle. The other shows various fermentation pathways, inspired by a subway map.

Acknowledgments

Well, here we go again: thanking the people at Morton Publishing. (Let the record show that this is the 17th acknowledgments Leboffe and Pierce have written.) But, one can never thank people enough who go out of their way to make one's work easier by being supportive, understanding, and excelling in what they do.

First on the Morton appreciation list must always be the late **Doug Morton**, who founded Morton Publishing in 1977 with a vision of publishing high quality lab books at a reasonable price, and that vision continues to this day. When he signed Mike Leboffe and Burt Pierce to write the first edition of *A Photographic Atlas for the Microbiology Laboratory* in 1995, we had no clue about what kind of reception it would get. And truth be told, you could truncate that previous sentence right after the word *clue*. As an example, when asked about a timeline, we (Mike) told him we could complete the book in seven weeks. We hadn't even gotten the long-neglected photomicroscope into working condition at seven weeks! (Relevant aside: In the last 25 years, Morton Publishing has impressively improved the quality and diversity of its products, as well as signing many capable authors. On the other hand, one of us [Mike] has become no better at predicting timelines.) But back to the point, Doug was a kind and generous man, and his willingness to take a chance on two unproven and unknown authors was life changing for us. We will forever be grateful to him for that.

Today, Morton Publishing has a corporate structure but is still a small company where everyone in the office is a short walk from everyone else. Three key people keep Morton on track and the company is in good hands with this leadership group.

Currently tasked with overseeing the entire operation is President and CEO **David Ferguson**. He began at Morton as a sales representative and now look at him! He's living proof that hard work coupled with opportunity can result in great things. He has many good qualities, but most important to us is that he has promised not to interrupt our nap time in retirement.

Vice President of Operations **Chrissy DeMier**, has the distinction of having discovered Leboffe and Pierce at San Diego City College in 1995 while passing through as a Morton sales representative. She didn't get an adoption from us, but she did create the first link between Morton and Leboffe and Pierce, a collaboration that has been beneficial for all of us. We should add the element of good timing to hard work and opportunity for Chrissy!

We've worked with **Carter Fenton**, Vice President of Sales and Marketing, since 2006 and he truly is one of a kind. We spent some time a while back at a conference discussing marketing and his philosophy of selling, a successful philosophy that is simple and honest: a low-key approach and customer centered. If Leboffe and Pierce ever go into sales, we want to grow up to be Carter Fenton.

Morton Editorial Team. We are grateful to **Marta Pentecost**, Senior Acquisitions Editor, who over the years has assisted us by creating surveys and collating the results to guide us in developing our revisions. We also have appreciated the all-too-infrequent conversations with her about, well, everything. **Adam Jones**, Supervising Editor, and the team of Customs editors also deserve recognition and our gratitude because they are responsible for customizing Morton titles for adopters who don't want (and wouldn't use) the whole book. We also want to express our sincere gratitude to our Project Editor, **Rayna Bailey**, who worked every bit as hard on this book as we did and takes justifiable personal pride in the end-product. Her persistence, attention to detail, and professionalism has elevated the quality of all our work with which she has been associated over the years.

Morton Production Team. Will Kelley, Production Manager, is responsible for the overall design, layout, and typesetting of our books. Morton's books have a distinctive style, and this is due, in large part, to Will's talents. He also is responsible for creating some of the art we use. Thanks, Will, for your outstanding contributions. Another talented graphic designer, **Joanne Saliger**, also deserves mention because she was responsible for producing our books from the beginning. Now a semiretired Morton employee, she has certainly contributed to our books' successes over the years. Thank you, Joanne, for your years of dedication and support for our projects.

Morton Sales Team. Many thanks to **Lee Brooks**, Sales Manager, and all the sales representatives who spread the word about our products. Without them, Leboffe and Pierce would be known only to our families and each other. We meet with publisher's representatives on campus and we know what a difficult job it is.

Thanks again to **Imagineeringart** of Toronto, Canada, for their beautiful artistic renderings. We've worked with them for more than 10 years and their products are extraordinary.

Polly Parks was added to the Leboffe and Pierce team for this fifth edition of the *Atlas*. Polly was a teaching intern for one of us (Mike) at San Diego City College and impressed the whole department with her work ethic, good nature, openness to innovation, and desire to become involved in college functions beyond the classroom. After the internship, she was hired as an adjunct at City College, and then in 2017 was hired full time at El Camino College in Torrance, California. Polly was responsible for bringing a fresh set of eyes to the *Atlas*. She compiled the sections in rough form, suggested and developed several parts of the molecular biology section, and assisted in photography. Of equal importance, she was a discipline expert with whom ideas could be exchanged, and the outcome is a better product as a result of our collaboration. Thanks, Polly!

We are indebted to others at San Diego City College. Lab technicians **Laura Steininger**, **John Tolli**, and **Erica Zhang** willingly assisted us in locating deeply hidden lab equipment and autoclaving all the biohazardous waste we created. Thanks also go to microbiology faculty **Dave Singer**, **Sarah Hawkins**, **Sabine Kurz-Sherman**, and **Puja Saluja** who offered helpful suggestions along the way and tolerated our commandeering of a refrigerator and incubator almost to the point of ownership. And to the other Life Sciences faculty—**Jennifer Chambers**, **Kevin Jagnandan**, **Roya Lahijani**, **Erin McConnell**, **Heather McGray**, **Erin Rempala**, and **Gary Wisehart**—thanks for sharing your space with us. Thanks also to Dean **Randy Barnes** and Senior Secretary **Zenia Torres** for their assistance in getting approval of the Civic Center Contract necessary for us to use San Diego Community College District property for personal business.

One constant throughout the 25-year history of writing *A Photographic Atlas for the Microbiology Laboratory* (and our other microbiology books) has been the extraordinary willingness of total strangers to help Leboffe and Pierce by granting us access to the materials, equipment, organisms, etc., that we don't ordinarily have access to, or sharing their particular expertise that dwarfs ours by comparison, or putting us in contact with someone who knows when they don't. So here, in no particular order, we express our gratitude to the following:

- **Wdee Thienphrapa**, MLS, is a good friend of Polly's and supported her efforts in this project by informing her about the work done by 21st century Medical Lab Scientists. She also provided firsthand explanations of sophisticated modern equipment, such as MALDI-TOF MS. It was comforting having a frontline MLS as a resource. Many thanks, Wdee.
- **Ingrid Reynolds Niesman**, Director of San Diego State University's Electron Microscope Facility. We didn't know each other, but Ingrid graciously agreed to help with producing TEMs and SEMs of bacterial samples. One of us (Mike) had some EM training, but after more than a decade, the skills were rusty. Ingrid did the majority of sample preparation and retrained Mike on the equipment. Together we created some wonderful images (Sections 4, 5, and 6) and developed a friendship as a bonus.
- **Steve Barlow**, former Director of San Diego State University's Electron Microscope Facility. We valued his friendship over the years and for assisting us in previous editions of the *Atlas* and other books. We are grateful to him for supplying the image of phage lambda for this edition (Fig. 10.5).
- **Rick Bizzoco**, Biology Professor Emeritus at San Diego State University. Rick went to great lengths, digging deeply into his photo files to locate and make new prints of *Sulfolobus* and *Thermoplasma*. You can see his work in Figures 13.2, 13.3, and 13.11.
- **Bonnie Baxter-Clark**, Director of the Great Salt Lake Institute (GSLI) and Professor at Westminster College, and **Jaimi Butler** GSLI Coordinator. The Baxter-Clark/Butler team was helpful beyond belief. Bonnie generously supplied us with written resources and answered multiple rounds of questions about halophiles. Jaimi spent a day with us while we collected Great Salt Lake water samples in the rain. She also acted as tour guide and natural history expert. Further, Bonnie took time to help identify microbes in our water sample micrographs (see Figs. 11.45 and 13.7), and when she ran into a stumper, she immediately put us in contact with a colleague and friend of hers, **Juergen Polle**, Biology Professor at Brooklyn College-CUNY. We sent him the photos in question on November 26 and heard back from him on November 28, Thanksgiving Day. His email began with an apology about not responding more quickly! Juergen kindly provided the information we lacked (see Fig. 14.28). Thank you, one and all! You are truly amazing.
- **National Park Service at Yellowstone National Park**; **Rachel Cudmore**, Special Use Permits Coordinator. Rachel provided us with stock photos of Octopus Springs and Mushroom Pool (Figs. 11.1 and 11.5, respectively), which are not open to public viewing. In addition, she put us in contact with **Annie Carlson**, Research Coordinator, who in turn put us in contact with **Eric Boyd**, Associate Professor of Microbiology and Immunology at Montana State University. Eric was kind enough to examine our wish list of organisms to photograph and suggest locations around the park where we might find them. He also verified their identities (to the extent possible from photographs alone) upon our return. Yellowstone photos are found in Sections 11, 13, and 14. The book is so much better because of all of your help.
- **San Diego County Public Health Laboratory**. We have had a relationship with them since 1997 and many of the parasite, virus, fungus, and bacterial pathogen photos are due to their willingness to allow us to photograph specimens in their laboratory. Moving to the present, **Syreeta Steele**, Assistant Laboratory Director, graciously welcomed us back and allowed us to photograph state-of-the-art equipment and shared her laboratorians with us. These included **Anthony Aziz**, **Tracy Basler**, **Robin Ellison Delgado**, and **Paul Temprendola**. Paul also took the micrographs used in Figures 12.31, 12.36, 12.43, and 12.45.

- **City of San Diego's Water Department, Alvarado Water Treatment Plant.** The plant welcomed us and allowed us into their lab to photograph tests and equipment. Thanks to **Dan Silvagio,** Senior Biologist, and the others in the department who also provided assistance, including **Jan Rust, Jeff Noller,** and **Jed Gordon.** Special thanks to **Roy Tamanaha,** who spent considerable time with us performing the Colilert (Figs. 19.15 and 19.16) and Quantitray (Fig. 19.25) tests and provided micrographs of *Giardia* (Fig. 4.2D) and *Cryptosporidium* (Fig. 16.14B). Special thanks also go to **Heather Rerecich** who demonstrated operation of the Vitek 2 machine (Figs. 7.107 through 7.110).
- **City of San Diego Environmental Monitoring Lab at NTC.** They also welcomed us and allowed us to photograph equipment and media they use in testing seawater and other recreational water. Thanks again to **Dan Silvagio** (same biologist as at Alvarado, but different location). Special thanks to **Lara Asato,** Biologist III, who spent time walking us through the water testing processes they perform, and to her staff, including **Angela Entera, Justin Fabi, Mark Frilles, Aaron Russell,** and **Bryan Santos,** who performed those tests flawlessly under the pressure of our unblinking camera lens. See Figures 19.17 through 19.20 for their work.
- **City of San Diego Wastewater Treatment Plant, Point Loma.** We wanted photos of a wastewater treatment plant and, hopefully, access to some sludge to photograph, but we had great difficulty connecting with them. We asked **Jan Rust** (at Alvarado) how we might be able to contact them, and she marched us over to **Elvira Mercado's** office. One phone call later, we had an appointment to visit the Point Loma Wastewater Treatment Plant. Thanks, ladies! Plant Superintendent **David Marlow** generously spent a couple of hours showing us around and even parted with some much-appreciated treated sludge that we could photograph. See Figs. 13.5 and 13.6.
- **Sanjay Kumar** and **Patrick Weaver** of **ThermoFisher Scientific.** Thank you for providing the microarray chip shown in Fig. 9.14. You made a difficult acquisition easy.
- **Craig Rappaport** and **Kerry Israel** of **Olympus America.** In a last-minute request, they supplied photographs of a compound microscope (Fig. 4.1) and an inverted microscope (Fig. 10.12A). It is fitting that Olympus America came to our rescue at the last minute, because 25 years ago they were the first to help us get started by loaning us a state-of-the-art photomicroscope when ours failed us. We love the symmetry.
- **Steve Joens,** Southwest Regional Sales Manager for **Hitachi High-Tech America, Inc.** Thanks for providing the transmission electron microscope photo (Fig. 4.6), as well as some lessons in transmission electron microscopy.
- **Dr. Karin Bayha** of **Carl Zeiss Microscopy GmbH.** Thanks for providing the scanning electron microscope photo (Fig. 4.12).
- **Brian Sykes** of the Prepared Slides Department, **Carolina Biological Supply Company.** We are grateful for the information Bryan provided about the *Agrobacterium tumefacians* slide (Fig. 11.34).
- **Paige Leboffe.** Thank you, Paige, for being so willing to publicly share your home pregnancy test results with us (Fig. 8.20A), announcing the arrival of Cole Matthew Leboffe. Thanks also to **Eric Leboffe** for finding such a wonderful life partner and for being such a great dad and son.

Although retired, much of the credit for this book still belongs to my friend and coauthor, **Burton Pierce.** My father worked for Consolidated Aircraft in San Diego during World War II. Their motto was, "Nothing short of right is right." Burt, whose father also worked for Consolidated Aircraft, personified that motto more than anyone I have known (other than my dad). Burt labored over details and wording until he created the exact blend of rigor and clarity that represented "right" for him. I admit that I was influenced by my dad's work ethic, but that's not nearly as cool as working with a colleague and friend motivated by the same standards. So, thanks Burt for all you've done to make our projects better. Your contributions still constitute a significant part of our books.

Over the last quarter-century, many people have kindly supported our efforts with their expertise and resources. You have been acknowledged personally in previous editions. Please accept our continued global appreciation of you all. We haven't forgotten you.

Mike

La Mesa, CA
June 2020

Photo Credits

All photos and micrographs provided by Michael J. Leboffe and Burton E. Pierce unless noted here.

Section 4

- Figure 4.1: Photograph courtesy of Olympus America Inc.
- Figure 4.2D: Micrograph courtesy of Roy Tamanaha, Alvarado Water Treatment Plant, San Diego, CA
- Figure 4.6: Photograph courtesy of Hitachi High-Tech America, Inc.
- Figure 4.12: Photograph courtesy of Carl Zeiss NTS GmbH
- Figures 4.7 and 4.13: Specimens prepared by Ingrid Niesman, Director of the San Diego State University EM Imaging Facility

Section 5

- Figures 5.9A and B: Specimens prepared by Ingrid Niesman, Director of the San Diego State University EM Imaging Facility

Section 6

- Figure 6.4, 6.5, 6.24, 6.25, and 6.40: Specimens prepared by Ingrid Niesman, Director of the San Diego State University EM Imaging Facility

Section 7

- Figures 7.99 and 7.100: Photographs courtesy of bioMérieux
- Figure 7.106A and B: Code data from EnteroPluri–*Test* reprinted with permission. Liofilchem S.r.l.–*Microbiology Products*

Section 10

- Figure 10.5: Specimen preparation and micrograph by Steve Barlow, former Director of the San Diego State University EM Imaging Facility
- Figure 10.8: Photograph courtesy of Dr. Rachel Schrier and Dr. Clayton Wiley
- Figure 10.12A: Photograph courtesy of Olympus America, Inc.

Section 11

- Figure 11.1: Photograph by Bob Lindstrom, courtesy of the National Park Service
- Figure 11.3: Culture for photograph provided by Karsten Engler, University of California, San Diego Bioengineering Department
- Figure 11.5: Photograph courtesy of the National Park Service
- Figure 11.18A: Photograph courtesy of Ian and Todd Molloy, Crikey Adventure Tours

Section 12

- Figures 12.36 and 12.43: Micrographs courtesy of Paul Temprendola, Microbiologist at San Diego Public Health Services Laboratory
- Figure 12.54: Micrograph by Billie Bird, courtesy of the CDC Public Health Image Library
- Figure 12.76: CDC Public Health Image Library, photograph by Larry Stauffer, Oregon State Public Health Laboratory

Section 13

- Figures 13.2A and B, 13.3, and 13.11: Electron micrographs provided by Rick Bizzoco, San Diego State University

Section 16

- Figure 16.14B: Micrograph courtesy of Roy Tamanaha, Alvarado Water Treatment Plant, San Diego, CA

Section 18

- Figure 18.7: Micrograph taken at the San Diego State University EM Imaging Facility

Contents

Section 1
Introduction

Section 2
Isolation Techniques and Selective Media

Section 3
Bacterial Growth

Section 4
Microscopy

Section 5
Bacterial Cellular Morphology, and Simple and Negative Stains

Section 6
Bacterial Cell Structures and Differential Stains

Section 7 Differential Tests and Multiple-Test Systems 73

Section 8 Serology 129

Section 9 Molecular Techniques 141

Section 10 Viruses 153

Section 11
Domain Bacteria 163

Section 12
Bacterial Pathogens 183

Section 13
Domain Archaea 223

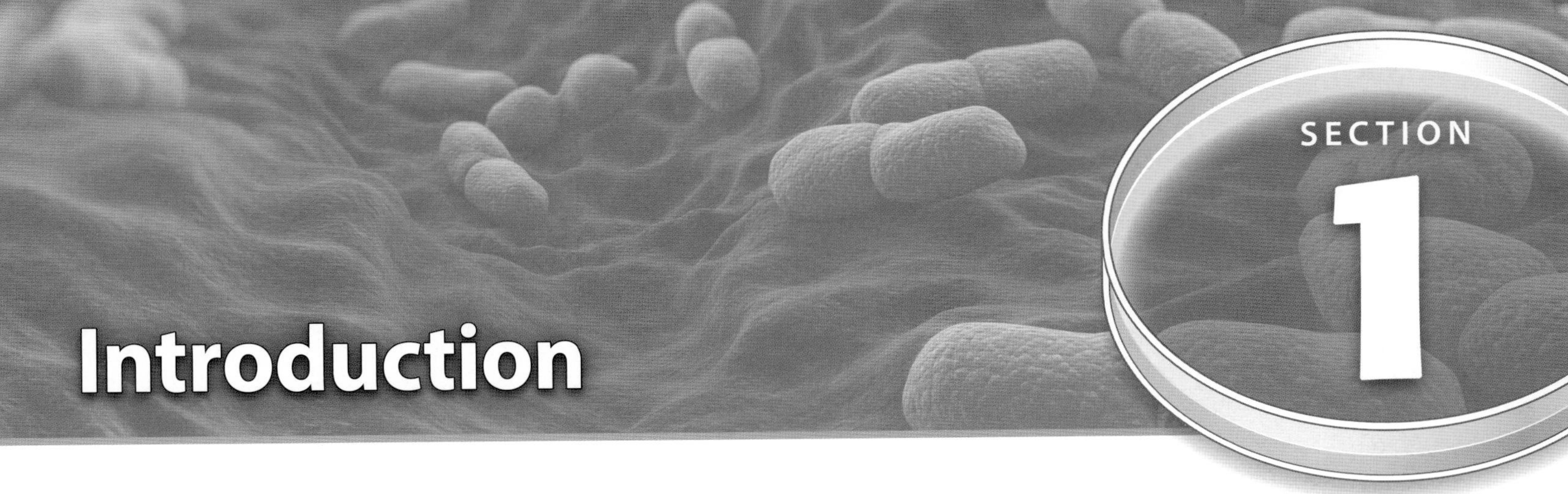

Microbial and Eukaryotic Systematics

> ***Student:*** *Dr. Einstein, aren't these the same questions as last year's (physics) final exam?*
>
> ***Dr. Einstein:*** *Yes, but this year the answers are different.*

We honestly don't know if that final exam story is true or not, but even if it isn't, who better than Albert Einstein (or some imposter using his name) to remind us that nature changes, which it does . . . but that wasn't his point. He was actually saying our *understanding* of nature changes, which can be frustrating to someone studying for a final exam (and it sounds like someone who was also *repeating* the course!), but it is what keeps career scientists interested in their profession.

As we learn more, even more that we don't know is revealed to us. (Albert Einstein, again, "The more I learn, the more I realize how much I don't know."). Deeper and novel understanding of our natural world is exciting. However, change in our understanding of nature also comes in the form of a totally revolutionary replacement of long-held ideas (based on experimental evidence, of course), and this can be difficult to accept because scientists are first and foremost humans and this is something humans often find hard to do. Still, human nature aside, good scientists remain flexible enough to change their views even when new evidence crushes them!

All of that brings us to the changes microbiologists have endured over the past several decades with respect to microbial classification (systematics[1]). The authors are of the generation where biologists placed all organisms into two kingdoms: Animal Kingdom and Plant Kingdom. It was easy—anything that wasn't an animal was a plant—so we had such diverse organisms as bacteria, yeast, and mushrooms placed in the same kingdom as roses and pine trees. Oh, and of course the simple nucleated cells (such as *Euglena* and *Amoeba*) were categorized based on their degree of resemblance to plants or animals.

By the early 20th century, biologists had hit a ceiling with the light microscope. Its reliance on making images with light rays (it's a *light* microscope!) put physical limits on the image that it can produce. The issue wasn't about getting greater magnification; it was about *clarity* of the magnified image. (For more about magnification, see pp. 41–43.) But the result was a very limited impression of cell structure: nucleus, chloroplasts, cilia and flagella, vacuoles, cytoplasmic membrane, and the viscous "filler" (cytoplasm) between these were about all we could see.

Then came a big revolution based on information revealed by the electron microscope. It was found that some cells (subsequently called **prokaryotes**) do not have their genetic material encased inside a membranous nucleus. In fact, those that do (subsequently called **eukaryotes**) not only enclose their nucleus in membrane, but also have all kinds of membranous compartments within their cytoplasm. Endoplasmic reticulum, Golgi bodies, mitochondria, and many other structures were seen for the first time. And those that we already knew about were revealed in detail we could never have previously imagined.

Equally surprising was the realization that prokaryotic cells have a similar, but qualitatively different construction than eukaryotes; their interior is very simple by comparison. Figure 1.1 illustrates an obvious difference between prokaryotic and eukaryotic cells visible with the light microscope: size. Recognition of their fundamental differences led to a restructuring of biology at the kingdom level based on the presence or absence of a nucleus, mode of nutrition (photosynthetic or not), and degree of complexity.

[1] Taxonomy is simply a classification based on convenient criteria. In biology, systematics is a taxonomy based on natural relationships. That is, species with a recent common ancestor will be placed in taxa (categories) that reflect their close relatedness—perhaps they'll be in the same genus or family. More distantly related species will be placed in taxa that put them farther apart—perhaps in different phyla or classes.

All prokaryotes (bacteria and others) were placed in the Kingdom Monera. The eukaryotes were divided into four kingdoms: Fungi, Metaphyta (for the plants), Metazoa (for the animals), and Protista (for all the simple eukaryotes that didn't "fit" in the other three kingdoms). This system served us well over the second half of the 20th century.

New technology allowing us to compare organisms at the molecular level led to another big revolution—including the addition of a taxonomic category more inclusive than Kingdom: **Domain**. Based on comparisons of ribosomal RNA (rRNA), DNA, and proteins, the biological world is now seen to be composed of three Domains: Bacteria (comprising all bacteria), Archaea (formerly the archaebacteria), and Eukarya (comprising all eukaryotes).

A short list of defining characteristics for each domain is given in Table 1-1; Figure 1.2 illustrates their evolutionary history (**phylogeny**) as currently perceived by many biologists[2]. Examples are shown in Figures 1.3 through 1.6.

[2] Other interpretations for the relationships between the three domains have been made. No consensus has been reached on which, if any, is most correct.

In past editions we ignored this new system, in large part because our emphasis has been on medical bacteriology. In the fourth edition, we expanded coverage of the *Atlas* to include more aspects of general microbiology and that demanded addressing the systematics at the time of its writing. We've continued updating systematics in this edition.

History has shown us that even as this edition is being written, microbial systematics itself is being rewritten to be in line with new information largely gathered by molecular microbiologists exploiting all kinds of new techniques previously unavailable to them. And as Dr. Einstein tells us (in so many words), this is the way of science.

A final note relates to the use of the term *prokaryote*. There is a growing sentiment among microbiologists that prokaryote is not a valid designation because it includes representatives (Bacteria and Archaea) on different, ancient lineages (Fig. 1.2). This is an argument to be settled by others more knowledgeable than we. So, until a decision is reached, we will continue to use prokaryote because it is such a convenient term!

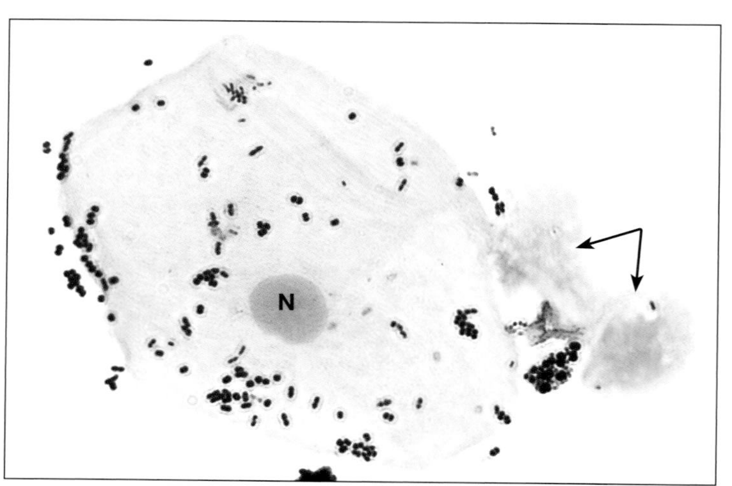

1.1 Comparison of Bacterial and Eukaryotic Cell Size • One difference between prokaryotic (non-nucleated) cells and eukaryotic (nucleated) cells is size. The sample for this micrograph was taken from the base of a molar tooth below the gumline (in the gingival pocket). The large, pink cell is an epithelial cell from the donor's gingival pocket. They are easily scraped away; no blood was shed. The tiny purple and red cells are bacteria, normal inhabitants of the donor's gingival pocket. The epithelial cell is approximately 70 µm in length. Remember that 1 mm equals 1000 µm, or stated another way, 1 µm is 0.001 (1/1000) mm. So, this cell is 0.07 mm long. (Break out that old metric ruler and get a feeling for that size!) The small, dark-purple dots are about 1 µm in diameter, or 0.001 mm. (Put your ruler down and don't even bother . . .) They, along with the other purple and small red cells of various shapes and sizes are bacteria, and it is cells such as these that comprise the primary subjects of this book. While you're here, be sure to notice the obvious nucleus in the epithelial cell (**N**) and the two nucleated, but squashed cells at the right (**arrows**). Those are phagocytic cells (neutrophils) that were patrolling the gingival pocket at the time of sampling.

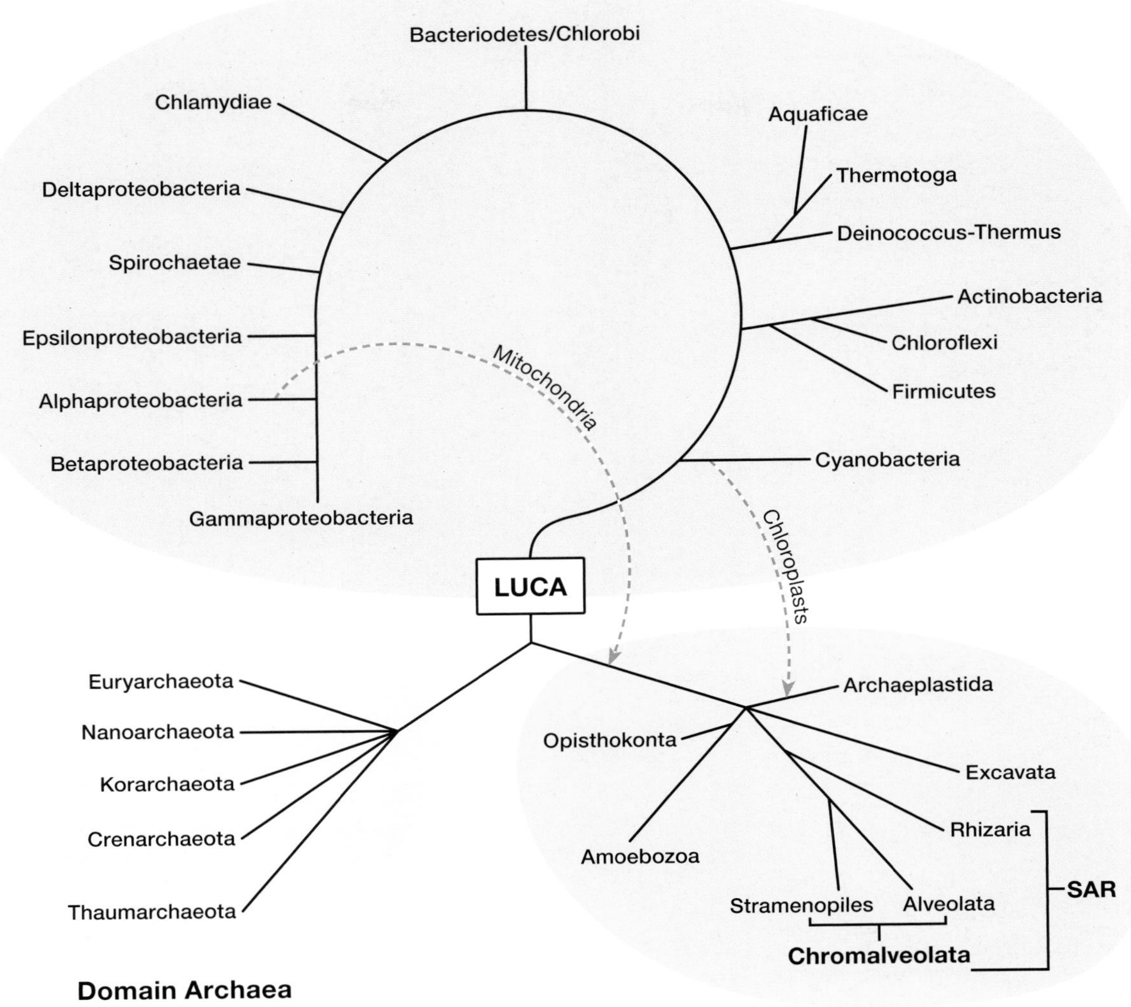

1.2 Three Domains • LUCA stands for "Last Universal Common Ancestor," an unknown entity from which scientists assume all life has descended. Domains are based on rRNA sequencing results and other molecular comparisons. The Archaea and Bacteria comprise an as yet undetermined number of kingdoms. Domain Eukarya includes all eukaryotic organisms and is divided into so-called supergroups, an informal category in which to temporarily place eukaryotes until enough evidence has been accumulated to make more informed decisions about their relationships and placement into kingdoms. The six supergroups are Rhizaria, Chromalveolata (subdivided into Stramenopiles and Alveolata), Excavata, Opisthokonta, Amoebozoa, and Archaeplastida. Some authors combine the **S**tramenopiles and **A**lveolata with the **R**hizaria to form the SAR (the acronym of the three groups) supergroup, resulting in four supergroups.

TABLE **1-1** Comparison of the Three Domains

Characteristic (Present in at Least Some Species)	Bacteria	Archaea	Eukarya
Cell Structure			
Membrane-bound organelles	No	No	Yes
Cell wall of peptidoglycan	Most	No	No
Ester-linked, straight-chained fatty acids in membrane	Yes	No	Yes
Ether-linked, branched aliphatic chains in membrane	No	Yes	No
Ribosome type	70S	70S	80S
Genetic			
Genome surrounded by nuclear membrane (envelope)	No	No	Yes
DNA molecule covalently bonded into circular form	Yes	Yes	No
Introns common	No	No	Yes
Operons present	Yes	Yes	No
Plasmids present	Yes	Yes	Rare
RNA polymerase	One	Several	Three
Initiator tRNA	Formylmethionine	Methionine	Methionine
mRNA processing (poly-A tail and capping)	No	No	Yes
DNA associated with histone proteins	No	Some	Yes
Metabolic			
Oxygenic photosynthesis	Yes	No	Yes
Anoxygenic photosynthesis	Yes	No	No
Methanogenesis	No	Yes	No
Nitrogen fixation	Yes	Yes	No
Chemolithotrophic metabolism	Yes	Yes	No
Ribosomal susceptibility to streptomycin, kanamycin, and chloramphenicol	Yes	No	No

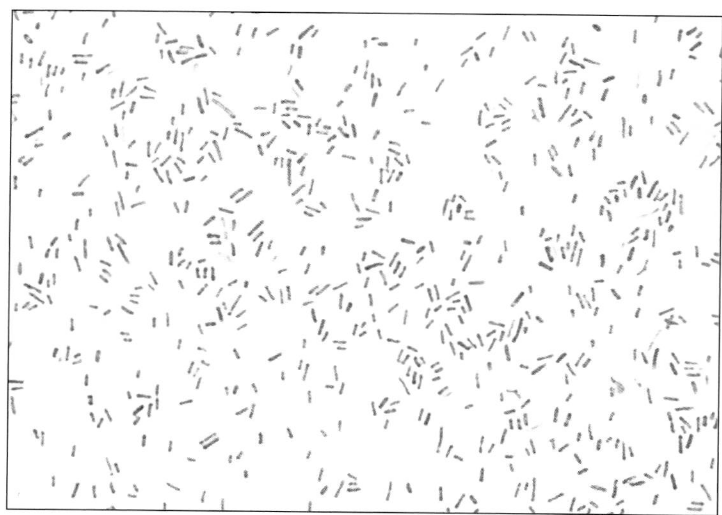

1.3 ***Escherichia coli*** **— A Bacterium** • *E. coli* is the most studied and most well-known of the Bacteria. It is a natural inhabitant of mammalian colons and is an opportunistic pathogen. Cells are about 1.0 μm wide by 2.0 to 4.0 μm long.

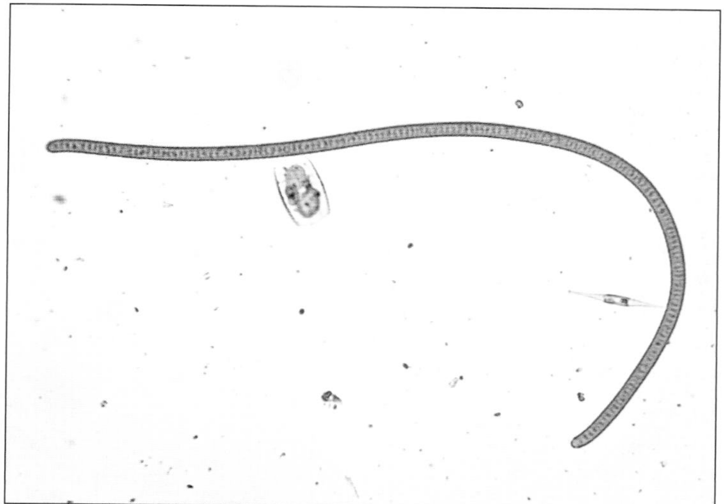

1.4 ***Oscillatoria*** **— A Cyanobacterium** • *Oscillatoria* is a filamentous cyanobacterium. That is, the organism is made of disk-shaped cells stacked together. This specimen is about 170 cells long, with individual cells about 10 μm wide. These Bacteria are capable of photosynthesis that produces oxygen, just like plants. The large, barrel-shaped cell is a diatom, a eukaryote.

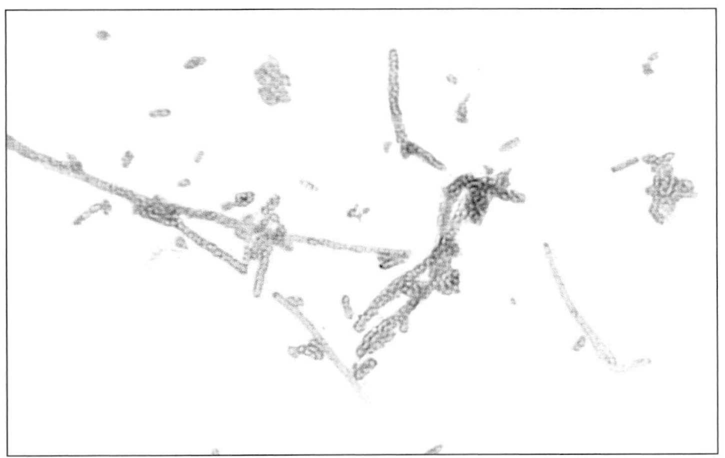

1.5 ***Halobacterium*** **— An Archaeon** • *Halobacterium* is an **extreme halophile**, which means it grows in environments with salt concentrations approaching 25%. Gas vacuoles in the cells provide buoyancy that allows it to get near the water's surface where oxygen and light are more abundant.

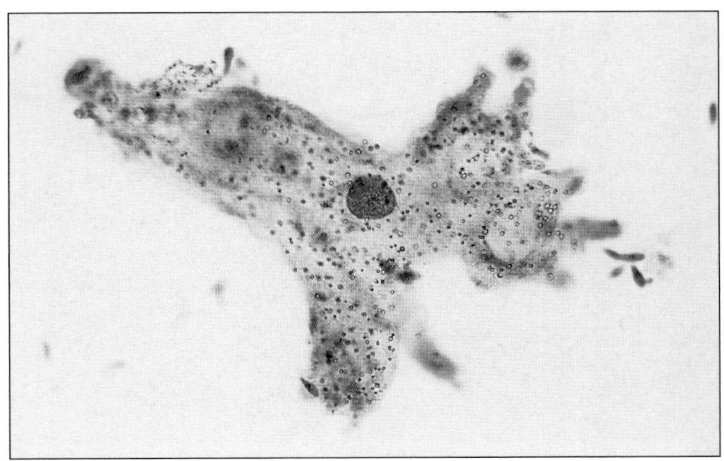

1.6 ***Amoeba*** **— A Eukaryote** • Visible are the numerous pseudopods, used both for movement as well as engulfing prey. Notice the nucleus within the cell.
(Commercially Prepared Slide.)

The Working Microbiologist

Traditionally, the role of microbiologists has been to isolate a microbe from a mixed culture, grow the isolate in a pure culture, and then identify the isolate. A **culture** is composed of one or more kinds of organisms grown under artificial, but suitable conditions. If more than one species is present, it is said to be a **mixed culture**. If only one species is present, it is said to be a **pure culture**.

Cultures may be grown in or on different kinds of **media**. A medium is the material that contains essential resources for growth (a carbon source, a nitrogen source, a sulfur source, and so on). Gases, such as oxygen, carbon dioxide, or diatomic nitrogen are generally not supplied by the medium, but must be made available in the culture container. A medium may be solid with nutrients suspended in the solidifying agent (usually **agar**) or it may be a liquid **broth**. Examples are shown in Figures 1.7 and 1.8.

Identification requires a pure culture because most microbes are not identified based on appearance or possession of unique physical structures (Fig. 1.9). Rather, the identification (**diagnostic**) process involves running biochemical tests using the **isolate** and recording the results. When enough relevant tests are run, the results are compared to a standard database of test results. The best fit leads to a provisional identification of the isolate.

One can now see that if tests are run on a mixed culture, *the results will be a composite of all organisms' positive results and will most likely lead to misidentification!* Even as more molecular diagnostic tests are being used for microbial identification, the demand for pure cultures and the basic idea of collecting test results about the isolate and comparing them to a database remain unchanged. The famous German physician and microbiologist Robert Koch said, "The pure culture is the foundation for all research on infectious disease." He was right—then and now.

This approach taken by microbiologists provides some structure to this *Atlas*. Earlier sections are devoted primarily to bacteriology and progress as a working microbiologist would as he or she tries to identify an isolate. Section 2 addresses methods by which bacteria can be isolated. Once isolation is achieved, Section 3 provides information on how species can be differentiated based on macroscopic features.

Sections 4, 5, and 6 carry this preliminary identification process on to include microscopic features. Section 7 presents differential physiological tests commonly used during the identification process. Section 8 does the same for diagnostic serological tests. This is followed by Molecular Techniques (Section 9), which includes an introduction to some nucleic acid techniques used in identification that are largely replacing the conventional physiological tests.

From here, we depart from the process of diagnostics to the subjects: the microbes themselves. Section 10 covers viruses, Sections 11 and 12 are devoted to Bacteria, and Section 13 addresses Archaea. Eukaryotic microbes, which are identified in large part by microscopic structural features, are covered in Sections 14 through 17. The *Atlas* concludes with three sections devoted to miscellaneous, albeit important, topics: Quantitative Microbiology (Section 18), Medical, Environmental, and Food Microbiology (Section 19), and Host Defenses (Section 20).

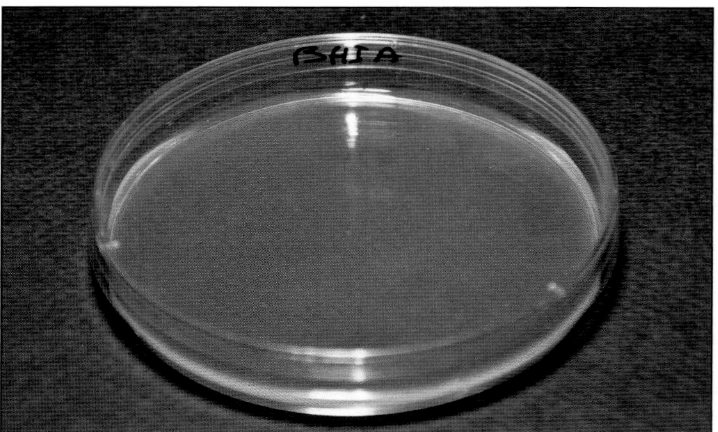

1.7 Agar Plate • Plated media are often used for isolating individual species from a mixed culture (Section 2) or for counting the number of cells in a diluted sample (Section 18). Some differential tests also use plated media (e.g., milk agar, DNase agar, starch agar—Section 7). Shown is a BHIA plate, which stands for "brain-heart infusion agar." The 1.5%–2% agar in the medium acts as a solidifying agent to suspend the nutrients: extracts of brain and heart tissues. Brain and heart provide carbon, nitrogen, and other essential nutrients for growth, as well as energy.

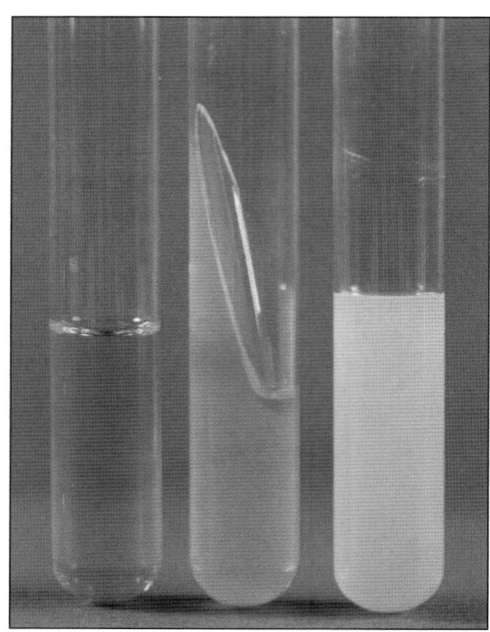

1.8 Tubed Media • From left to right: a broth, an agar slant, and an agar deep tube. The solid media are liquid when they are removed from the autoclave (used for sterilization). Agar deeps are allowed to cool and solidify in an upright position, whereas agar slants are cooled and solidified on an angle. Broths are often used to produce fresh cultures, slants are used to maintain stock cultures, and agar deeps are used for many biochemical tests requiring low oxygen levels.

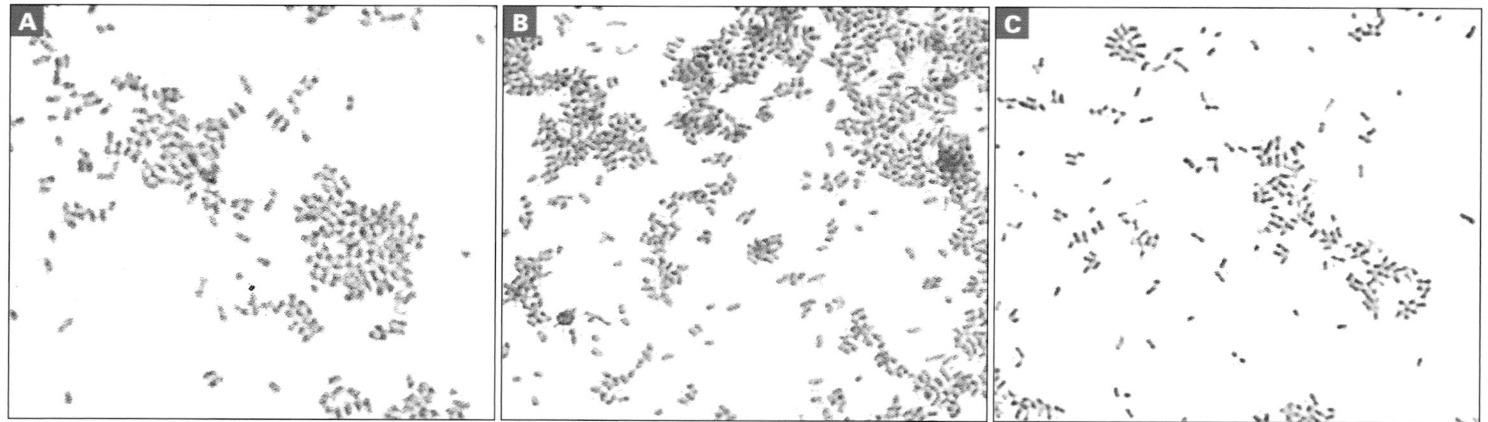

1.9 Comparison of Cell Structure • Bacteria are typically too simple to identify based on structural features. Imagine trying to identify these three species on appearance alone. Instead, bacteria are identified using physiological (Section 7), serological (Section 8) and/or nucleic acid tests (Section 9). Still, microscopic examination combined with special stains can get the identification process started. (**A**) *Klebsiella pneumoniae;* (**B**) *Citrobacter koseri;* (**C**) *Salmonella typhimurium.*

A Word About Laboratory Safety

Because microorganisms present varying degrees of risk to laboratory personnel (oneself, other students, technicians, and faculty), people outside the laboratory, and the environment, microbial cultures must be handled safely. Classifying microbes into four biosafety levels (BSLs) provides a set of minimum standards for laboratory practices, facilities, and equipment to be used when handling organisms at each level.

These biosafety levels, defined in the U.S. government publication, *Biosafety in Microbiological and Biomedical Laboratories,* 5th edition (2009), are summarized in the following text. For complete information, readers are referred to the original document.

BSL-1: Organisms do not typically cause disease in healthy individuals and present a minimal threat to the environment and lab personnel. Standard microbiological practices are adequate. These microbes may be handled in the open, and no special containment equipment is required. Examples include *Bacillus subtilis*, *Escherichia coli* (most strains), *Rhodospirillum rubrum*, and *Lactobacillus acidophilus*.

BSL-2: Organisms are commonly encountered in the community and present a moderate environmental and/or health hazard. These organisms are associated with a variety of human diseases, most of which can be successfully treated if identified in a timely manner. The infection routes of primary concern are ingestion, inhalation, or penetration of the skin (percutaneous). Individuals performing work prone to splashes or aerosol generation (even though these organisms are not generally known to be transmitted by aerosols) should work in a biological safety cabinet (BSC, Fig. 1.10). Otherwise, laboratory work may be done using standard microbiological practices. Examples include *Salmonella*, *Staphylococcus aureus*, *Clostridium difficile*, and *Borrelia burgdorferi*.

BSL-3: Organisms are of local or exotic origin and are associated with respiratory transmission and serious or lethal diseases where treatment and/or vaccines may or may not be available. Special ventilation systems are used to prevent aerosol transmission out of the laboratory, and access to the lab is restricted. Specially trained personnel handle microbes in a Class II or III biosafety cabinet (BSC), not on the open bench. Examples include *Bacillus anthracis*, *Mycobacterium tuberculosis*, and West Nile virus. (Imagine Robert Koch working with *B. anthracis*–anthrax, and *M. tuberculosis*–tuberculosis, in his laboratory in the 1870s with none of this protection!)

BSL-4: Organisms have a great potential for lethal infection. Inhalation of infectious aerosols, exposure to infectious droplets, and autoinoculation are of primary concern. The lab is isolated from other facilities, and access is strictly controlled. Ventilation and waste management are under rigid control to prevent release of the microbial agents to the environment. Specially trained personnel perform transfers in Class III BSCs. Class II BSCs may be used as long as personnel wear positive pressure, one-piece body suits with a life-support system. Examples include agents causing hemorrhagic diseases, such as Ebola, Marburg, and Lassa fever viruses.

Biohazardous waste disposal differs in details from lab to lab, but there are general guidelines. Broken glass or any items that would puncture an autoclave bag are disposed of in a sharps container (Fig. 1.11A). Nonreusable, contaminated items are disposed of in an autoclave bag, which may be supported by a stand (Fig. 1.11B) or lining a step-on waste can. Sterilization of fresh media, glassware, and many instruments, and decontamination of biohazardous materials is done in an autoclave (Fig. 1.12).

Biosafety Cabinets

Biosafety cabinets are classified into three categories, all of which draw air into the cabinet to minimize microbial contamination back into the room.

Class I BSCs resemble chemical fume hoods with a protective glass in the front, but have a HEPA filter along the exhaust path to prevent microbes from entering the environment.

Class II BSCs use laminar airflow, which minimizes turbulence, and HEPA filters to protect the user and materials within the cabinet from contamination. The exhaust passes through a HEPA filter before its release into the environment.

Class III BSCs are completely sealed and gas tight with a fixed viewing window. Incoming air passes through a HEPA filter, whereas materials to be handled are passed through a double-door system where the intermediate compartment is an autoclave. (Other methods may also be used, but accomplish the same purpose.) Exhaust air passes through two HEPA filters or a HEPA filter and an air incinerator. Materials are handled with arm-length, heavy-duty gloves built into the wall of the cabinet.

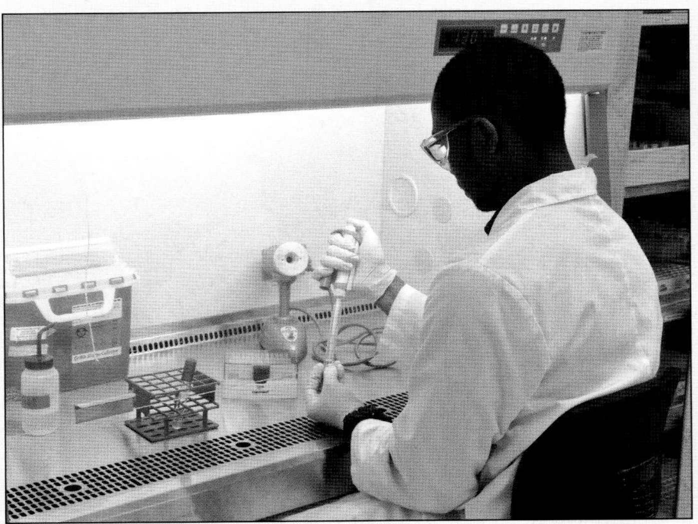

1.10 Biological Safety Cabinet (BSC) in a Teaching Laboratory • In this Class II BSC, air is drawn in from the room and is passed through a HEPA filter prior to release into the environment. This airflow pattern is designed to keep aerosolized microbes from escaping the cabinet. The microbiologist is pipetting a culture. When the BSC is not in use at the end of the day, an ultraviolet light is turned on to sterilize the air and the work surface.

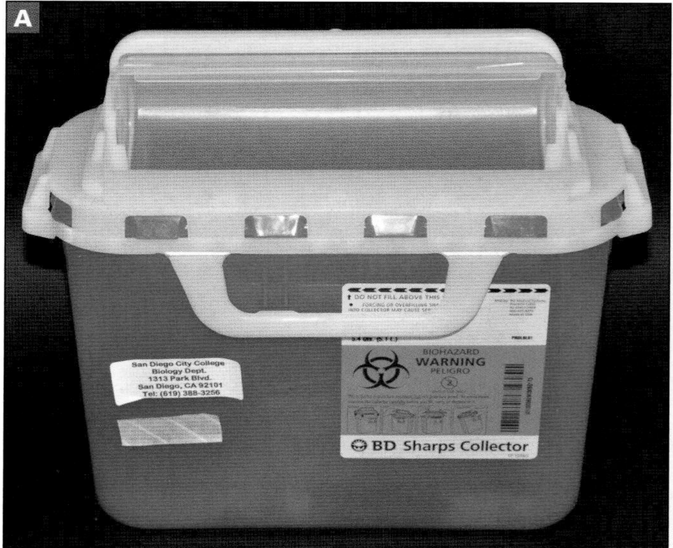

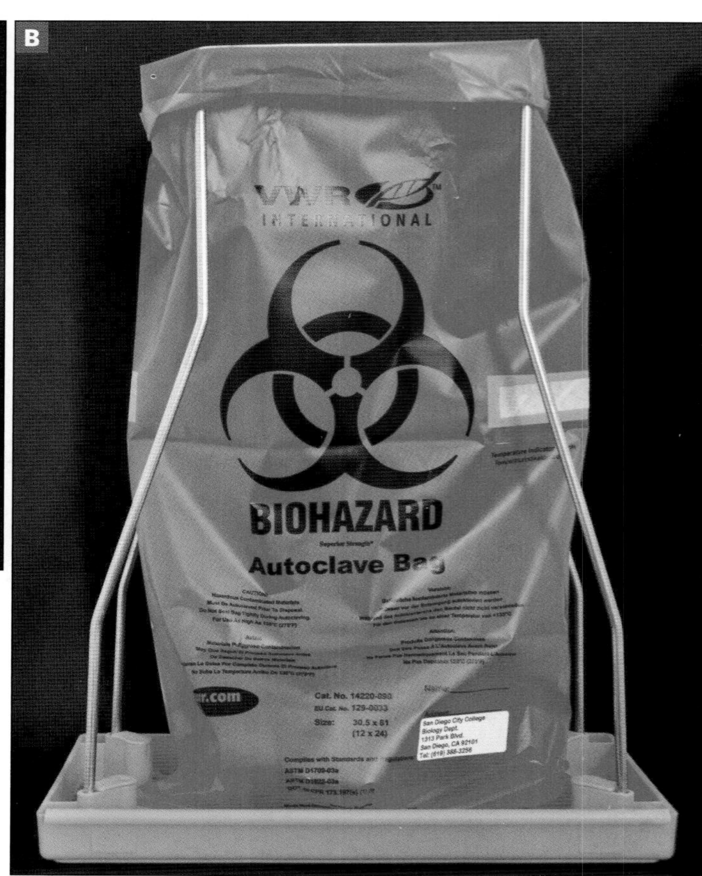

1.11 Disposal Containers ● (A) Needles, broken glass, and other contaminated items that can penetrate the skin or an autoclave bag are disposed of in a sharps container. The dashed black line indicates the fill level. The white stripes on the autoclave tape (lower left) will turn black after proper autoclaving. Above the autoclave tape is the address of the institution that produced the biohazardous waste. (B) Nonreusable items (such as plastic Petri dishes) are placed in an autoclave bag for decontamination. Notice the autoclave tape at the middle right and the institution's address that produced the biohazardous waste at the lower right.

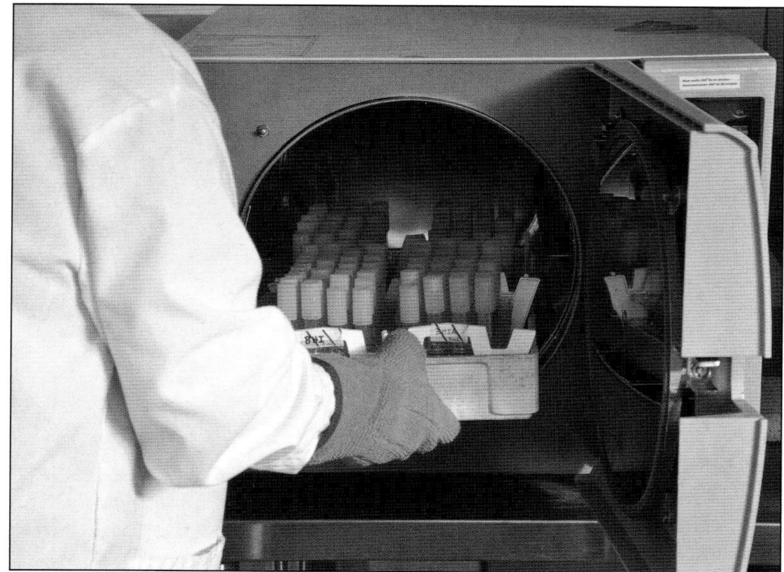

1.12 Autoclave ● Fresh media (as shown here), equipment, and biohazardous materials (e.g., old cultures) are sterilized in an autoclave. Steam heat at a temperature of 121°C (produced at atmospheric pressure plus 15 psi) for 15 minutes is effective at killing even bacterial spores. Some items that cannot withstand the heat, or have irregular surfaces that prevent uniform contact with the steam, are sterilized by other means.

SECTION 2

Isolation Techniques and Selective Media

Streak Plate Methods of Isolation

Purpose

The identification process of an unknown microbe relies on obtaining a pure culture of that organism. The streak plate method is designed to produce individual colonies on an agar plate. A portion of an isolated colony then may be transferred to a sterile medium to start a pure culture.

Principle

A microbial culture consisting of two or more species is said to be a **mixed culture**, whereas a **pure culture** contains only a single species. Obtaining isolation of individual species from a mixed sample is generally the first step in identifying an organism. A commonly used **isolation technique** is the **streak plate**.

In the streak plate method of isolation, a bacterial sample (always assumed to be a mixed culture) is inoculated over the surface of a plated agar medium (Fig. 2.1). During streaking, the cell density decreases, eventually leading to individual cells being deposited separately on the agar surface.

Cells that have been sufficiently isolated will grow into **colonies** consisting only of the original cell type. Because some colonies form from individual cells and others from pairs, chains, or clusters of cells, the term **colony-forming unit** (**CFU**) is a more correct description of the colony origin.

Several patterns are used in streaking an agar plate, the choice of which depends on the source of inoculum and microbiologist's preference. Although streak patterns range from simple to more complex, all are designed to separate deposited cells (CFUs) on the agar surface so individual cells (CFUs) grow into isolated colonies.

A **simple zig-zag** pattern (Fig. 2.2) is generally used with samples suspected of low cell density, whereas a **quadrant streak** pattern is used for samples containing higher cell densities. The **quadrant method** uses the four-streak pattern shown in Figures 2.3 and 2.4.

Streaking for isolation is frequently performed on **selective media** designed to encourage growth of certain organismal types while inhibiting growth of others. The selective media considered in this section are used specifically to isolate pathogenic bacteria or yeast from human or environmental samples containing a mixture of organisms.

Some selective media contain indicators that expose differences between organisms. Such media are considered to be selective and **differential**. See Tables 2-1, 2-2, and 2-3 for summaries of terms related to media, organisms, and roles of common ingredients found in selective media.

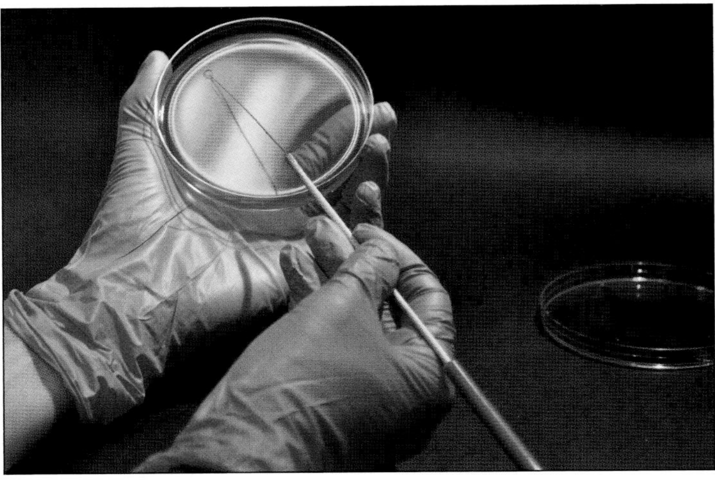

2.1 Streaking a Plate • Some microbiologists prefer to hold the Petri dish in the air when performing a streak plate. To do this, place the plate lid down on the table and lift the base from it, holding the plate comfortably on an angle. Streak across the agar surface using the edge of the loop's face.

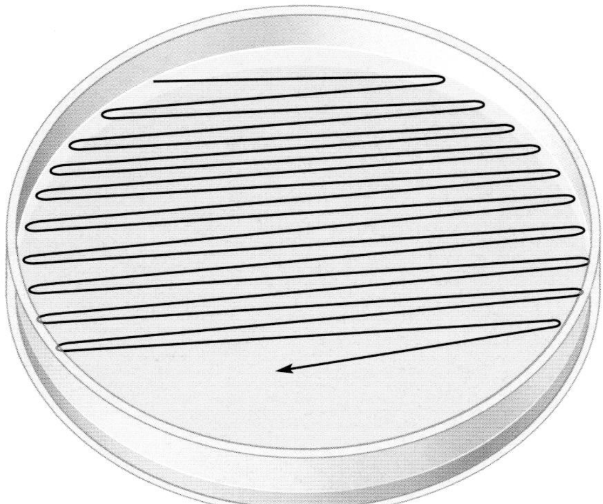

2.2 Simple (Zig-Zag) Streak Pattern on an Agar Plate ● Streak the inoculum in a zig-zag pattern across the agar's surface using the edge of the loop's face. Sometimes this is done with a cotton swab, in which case roll the swab as you streak across the agar's surface. This streak pattern is used to obtain colony isolation from samples with an anticipated low cell density.

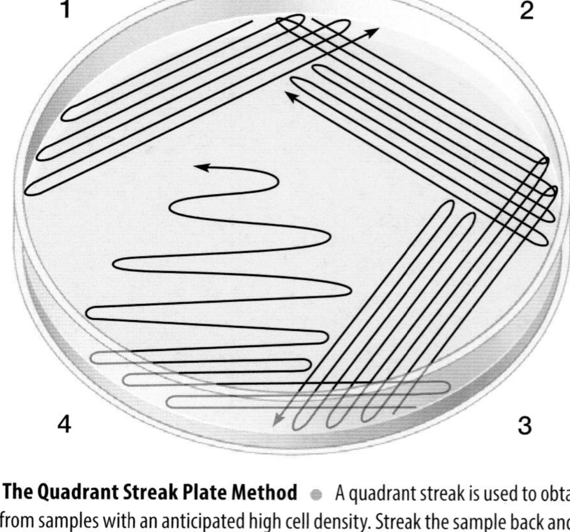

2.3 The Quadrant Streak Plate Method ● A quadrant streak is used to obtain colony isolation from samples with an anticipated high cell density. Streak the sample back and forth in one quadrant of the agar plate. Stay close to the plate's edge and make the streaks long, using the edge of the loop's face. Be careful not to cut the agar with the loop. Flame the loop, and then proceed, streaking into the second quadrant by pulling inoculum from the first quadrant, as shown. Repeat for the third and fourth quadrants.

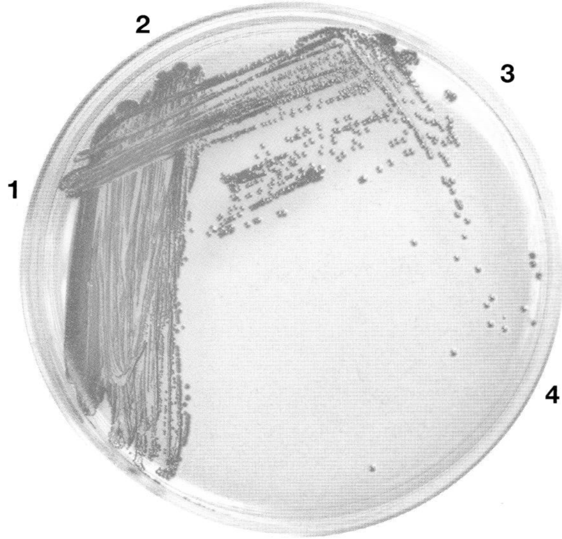

2.4 Streak Plate of *Serratia marcescens* ● Note the decreasing density of growth in the four streak patterns (indicated by numerals). On this plate, isolation is first achieved at the edge of the first streak, but it's in the second streak, and better yet in the third streak. However, the microbiologist would not know this at the time of streaking, so all four streaks are performed in the hope that isolation will occur in at least one of them. Cells from an isolated colony (one that is not touching another colony) can be transferred to a sterile medium to start a pure culture.

TABLE **2-1** Terms Related to Media

Term	Definition
Defined medium	A medium in which the chemical identity (e.g., glucose) and exact amounts (e.g., 1.0 g) of all ingredients are known
Undefined (complex) medium	A medium in which the amount and/or chemical identity of at least one ingredient is not known; examples of ingredients that make a medium undefined often have infusion, extract, or digest in their name
Selective medium	A medium that contains an inhibitor to prevent or slow the growth of undesired organisms
Differential medium	A medium that is formulated in such a way that differences in the biochemistry/physiology between organisms can be detected

TABLE **2-2** Terms Related to Organisms

Term	Definition
Non-fastidious	Refers to an organism whose nutrient requirements are relatively simple because it can make most of its own biochemicals with minimal environmental support; for example, an organism that can make all of its biochemicals from glucose is categorized as non-fastidious
Fastidious	Refers to an organism whose nutrient requirements are complex, very specific, or both because it is incapable of synthesizing some portion of its own biochemicals; for example, an organism that requires all 20 amino acids from the environment (because it can't make its own) is categorized as fastidious. Microorganisms exhibit a wide range of fastidiousness
Enteric	Refers to any gut bacterium, but usually to members of the *Enterobacteriaceae*, which are Gram-negative rods that ferment glucose to acid end-products and share many other features in common
Coliform	A member of the *Enterobacteriaceae* that produces acid (and gas) from lactose fermentation (**Note** that this shared ability is useful for identification purposes, but is not of taxonomic significance)
Noncoliform	A member of the *Enterobacteriaceae* that does not ferment lactose

TABLE **2-3** Common Ingredients in Media and Their Roles

Ingredient	Role
Carbon, hydrogen, oxygen, nitrogen, phosphorous, sulfur, etc.	These elements are required by all organisms and must be supplied by the medium or the culture's environment; they can be supplied by specific compounds (e.g., glucose for carbon) or from digests of animal or plant material (e.g., beef extract)
pH indicator	Plays a major role in making a medium differential; it detects acid or base production, depending on the medium
Bile salts (oxgall)	Used to select against organisms incapable of surviving passage through the gut, especially Gram positives
Lactose	Used as the fermentable carbohydrate in media that differentiate between coliforms and noncoliforms
Thiosulfate (in some form)	Used as an electron acceptor by organisms capable of reducing sulfur to H_2S
Ferric ion	Used as an indicator of sulfur reduction by reacting with H_2S to form a black precipitate

Bacteroides Bile Esculin (BBE) Agar

Purpose

Bacteroides bile esculin (BBE) agar is used for isolation and presumptive identification of *Bacteroides fragilis* and its close relatives in the *B. fragilis* group (see p. 186). *B. fragilis* is the most abundant bacterium found in the human colon, reaching densities of 10^{11} cells per gram of feces! It also is the most common anaerobic human pathogen.

Principle

BBE agar is a selective and differential medium with nutrition supplied by a base medium of tryptic soy agar, which includes digests of casein (milk protein) and soybean meal. Other anaerobes in the sample are inhibited by oxgall (bile). Facultative anaerobes, also abundant in feces, compete with obligate anaerobes when grown anaerobically. These are inhibited by the antibiotic gentamicin.

The medium also includes esculin, which *B. fragilis* is capable of hydrolyzing to produce esculetin. Esculetin in turn reacts with ferric ions in the medium to produce a brown coloration around *B. fragilis* growth. Presumptive identification of *B. fragilis* is based on its ability to grow on BBE and darken the medium (Fig. 2.5).

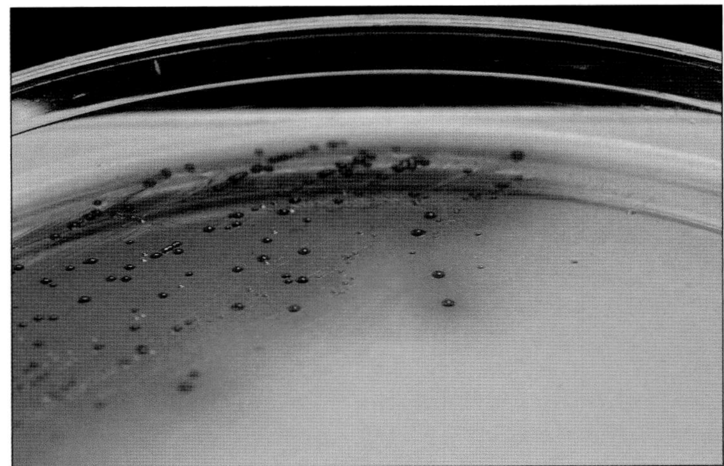

2.5 ***Bacteroides* Bile Esculin Agar** • This photo shows a fecal specimen streaked on BBE. The larger colonies within the darkened medium are presumptively identified as *B. fragilis* group. The smaller, lighter colonies not producing darkening of the medium are something other than *B. fragilis*.

Bismuth Sulfite Glucose Glycine Yeast (BiGGY) Agar

Purpose

BiGGY agar is used to isolate species of the yeast *Candida*. Presumptive identification of *Candida* spp. is also possible because of the differential results. *Candida albicans* is a common inhabitant of the normal microbiota of the oral cavity, gastrointestinal tract, and vagina, but it is also an opportunistic pathogen, especially in immunocompromised individuals.

Principle

BiGGY agar (Fig. 2.6) is a selective medium with carbon, nitrogen, and other nutrients supplied by yeast extract and dextrose. Glycine provides an additional nutrient source, but also inhibits bacterial growth.

During autoclaving, sodium sulfite and bismuth ammonium citrate react to form bismuth sulfite, which is the medium's primary bacterial inhibitory agent, but does not affect *Candida* species. *Candida* species reduce the bismuth sulfite (at slightly acidic or neutral pH) and produce a brown pigment within, and sometimes around, the colonies.

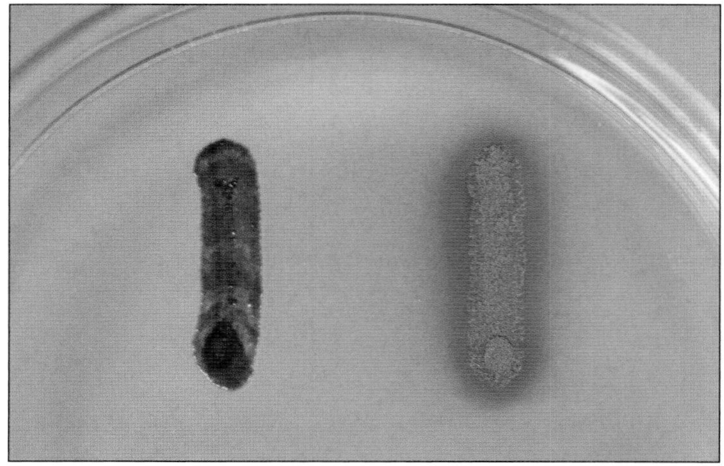

2.6 BiGGY Agar • BiGGY agar is used to isolate and provisionally identify *Candida* species, especially *C. albicans*. Bismuth sulfite inhibits bacterial growth and is also reduced by *Candida* species, making the growth (*Candida albicans* on the left) or the agar (*C. krusei* on the right) shades of brown.

Chocolate II Agar

Purpose

Chocolate II agar is used for isolation and cultivation of fastidious organisms, such as *Campylobacter* (p. 190), *Neisseria* (Figs. 2.7 and 2.8 and p. 206), and *Haemophilus* (Fig. 2.9 and p. 198) species.

Principle

Chocolate II agar is a selective medium made with a blend of casein, peptones, phosphate buffer, corn starch, and bovine hemoglobin. It also contains an enrichment supplement of amino and nucleic acids to encourage growth of *Neisseria* species and provide the **X** and **V blood factors** (protoheme and NAD/NADP, respectively) required by *Haemophilus* species.

This plated medium is typically streaked for isolation and incubated at 37°C in an aerobic environment enriched with carbon dioxide. Subcultures of colonies can then be grown on slanted media and used for diagnostic purposes. Its name comes from the discoloration of hemoglobin as blood is added to the molten mixture during preparation. There is no chocolate in the medium!

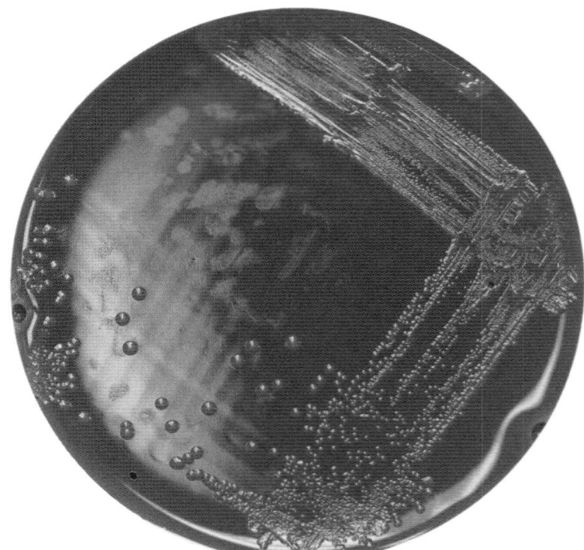

2.7 *Neisseria gonorrhoeae* on Chocolate II Agar • Primary isolates of *N. gonorrhoeae* frequently show variation in colony morphology due to the presence or absence of pili. The *N. gonorrhoeae* on this plate produced smooth, raised, glistening, and colorless colonies typical of piliated cells. *N. gonorrhoeae* colonies are usually smaller than *N. meningitidis* colonies (compare with Fig. 2.8).

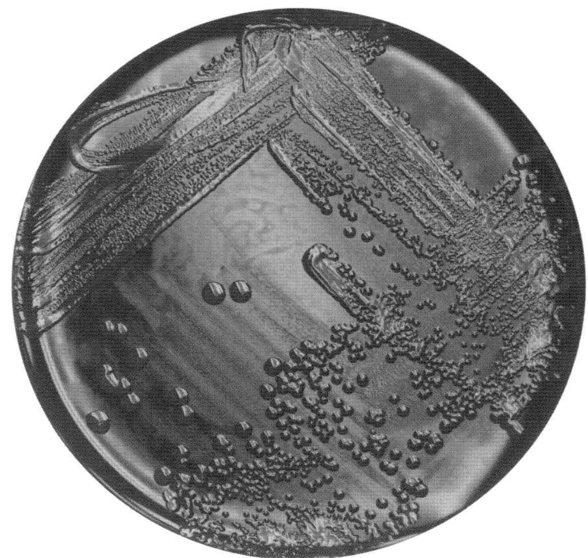

2.8 ***Neisseria meningitidis* on Chocolate II Agar** ● On Chocolate II Agar, *N. meningitidis* colonies are typically large, blue-gray, smooth, convex, and mucoid with an entire edge. Compare colony size with Figure 2.7. *N. meningitidis* colonies are larger than *N. gonorrhoeae* colonies.

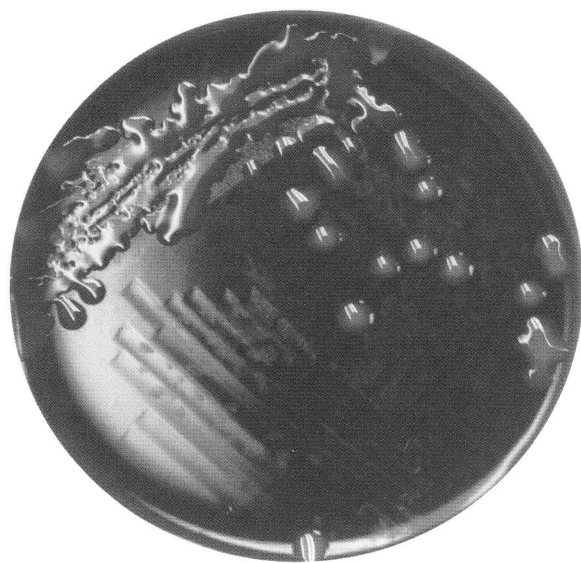

2.9 ***Haemophilus influenzae* on Chocolate II Agar** ● Notice the large, smooth, mucoid colonies, which probably indicate capsule production by this isolate. Cultures often have a musty odor due to the production of indole.

Columbia CNA with 5% Sheep Blood Agar

Purpose

Columbia CNA with 5% sheep blood agar is used to isolate and differentiate staphylococci, streptococci, and enterococci, primarily from clinical specimens.

Principle

Columbia CNA with 5% sheep blood agar is an undefined, differential, and selective medium that allows growth of Gram-positive organisms (especially staphylococci, streptococci, and enterococci) and stops or inhibits growth of most Gram-negative organisms (Fig. 2.10).

Casein, digest of animal tissue, beef extract, yeast extract, corn starch, and sheep blood provide a range of carbon and energy sources to support a wide variety of organisms. In addition, sheep blood supplies the X factor (heme) and yeast extract provides B-vitamins.

The antibiotics colistin and nalidixic acid (CNA) act as selective agents against Gram-negative organisms by affecting membrane integrity and interfering with DNA replication, respectively. They are particularly effective against *Klebsiella, Proteus,* and *Pseudomonas* species. Further, sheep blood makes possible differentiation of Gram-positive organisms based on hemolytic reaction (Figs. 7.8 through 7.10). ●

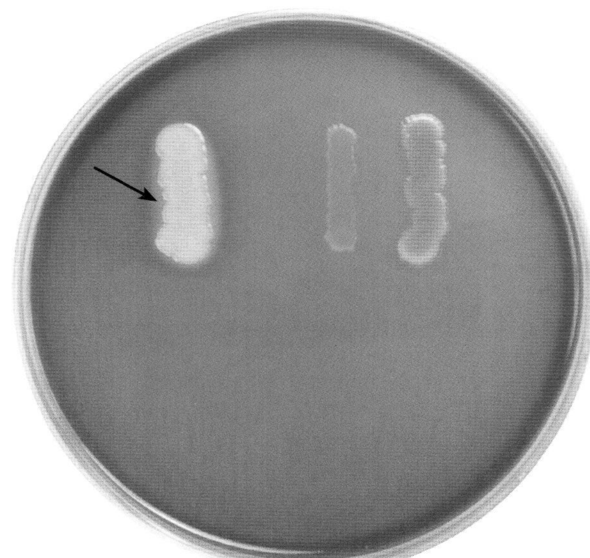

2.10 **Columbia CNA Agar with 5% Sheep Blood** ● This medium is selective for Gram positives and also allows determination of hemolytic reaction (p. 76). From left to right are: *Staphylococcus aureus* (Gram +, β-hemolytic; zone of hemolysis indicated by the **arrow**); *E. coli* (Gram –, no growth), *Streptococcus gallolyticus* (Gram +, nonhemolytic); and *Enterococcus* (Gram +, nonhemolytic). Note that only the Gram-positive organisms grew well on the Columbia CNA agar. *E. coli* didn't grow.

Deoxycholate (DOC) Lactose Agar

Purpose

Deoxycholate lactose (DOC) agar is used for isolation of and differentiation among the *Enterobacteriaceae*. It is also used to detect coliform bacteria in dairy products.

Principle

Deoxycholate lactose agar is a selective and differential medium containing lactose, sodium deoxycholate, and neutral red dye. Lactose is a fermentable carbohydrate, deoxycholate is a Gram-positive inhibitor, and neutral red, which is colorless above pH 6.8 and red below, is added as a pH indicator.

Neutral red turns the bacterial growth red (Fig. 2.11) as acid from lactose fermentation lowers the pH (Fig. 2.12). Lactose non-fermenters will remain their natural color or the color of the medium. Thus, the characteristics to look for on DOC are the quality of growth and color production.

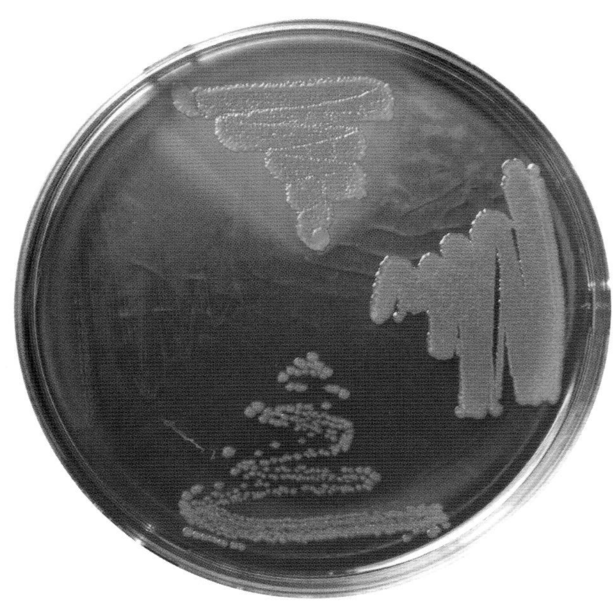

2.11 Deoxycholate Agar • DOC medium inoculated with (clockwise from top): *Escherichia coli, Klebsiella* (*Enterobacter*) *aerogenes, Shigella flexneri,* and *Enterococcus faecalis*. Note the inhibition of the Gram-positive *E. faecalis*. Note also the red coloring of the lactose fermenting coliforms *E. coli* and *K. aerogenes*.

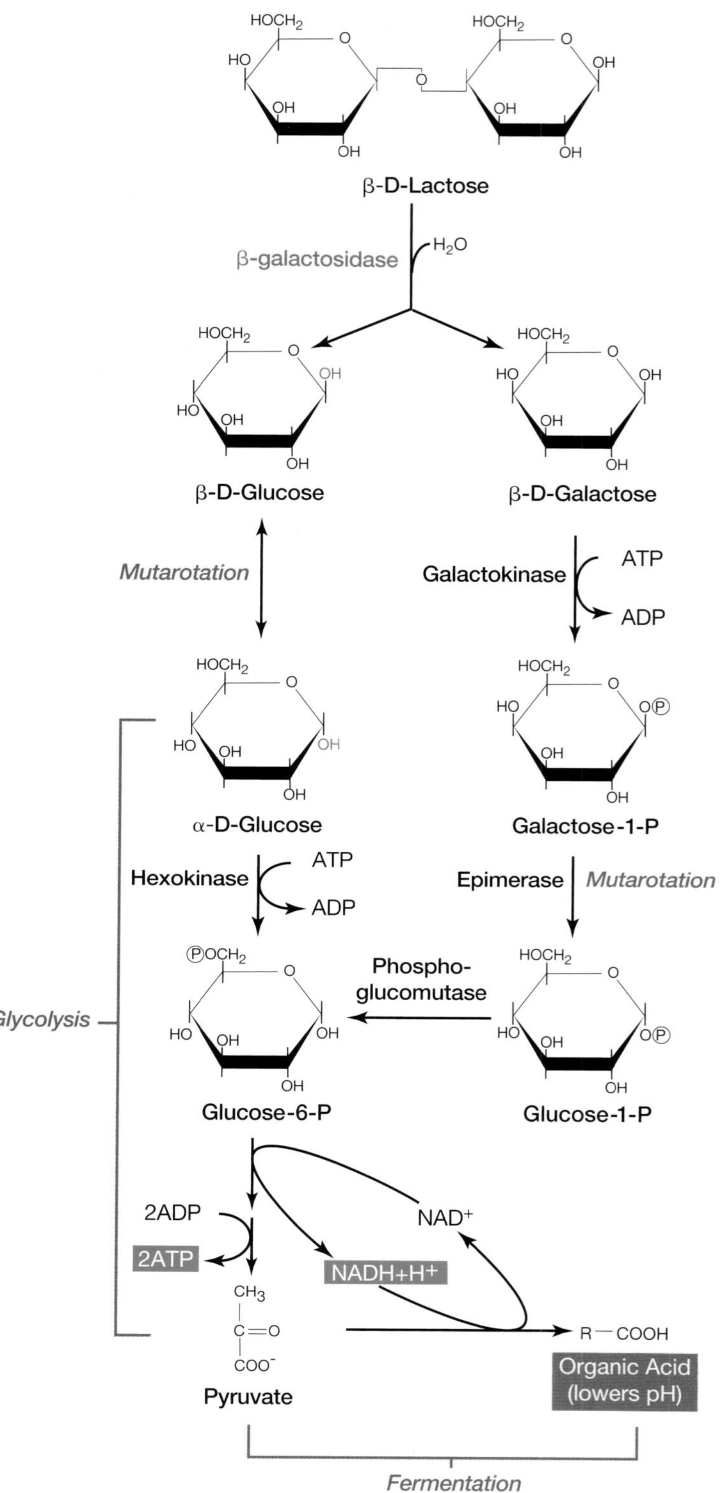

2.12 Lactose Fermentation with Acid End-Products • There are two important points to note here: one, the fermentation provides a mechanism for oxidizing NADH back to NAD^+ so it can be used again in glycolysis; and two, the end-product of the fermentation for coliforms is an acid (and sometimes a gas). For simplicity, only the relevant parts of glycolysis are shown.

m Endo Agar LES

Purpose

m Endo agar LES (Lawrence Experimental Station) is commonly used to detect fecal contamination in water and dairy products. Although its current use is to isolate and identify the presence of enteric lactose fermenters (coliforms), its original use was to isolate and identify *Salmonella typhi*, an enteric lactose non-fermenter (noncoliform).

Principle

m Endo agar LES contains color indicators sodium sulfite and basic fuchsin. These also double as Gram-positive inhibitors. Lactose is included as a fermentable carbohydrate. Coliforms ferment the lactose (Fig. 2.12), appear red, and darken the medium slightly due to the reaction of sodium sulfite with the fermentation intermediate acetaldehyde.

Lactose non-fermenters produce colorless to slightly pink growth (Fig. 2.13). Some lactose fermenters, such as *Escherichia coli* and *Klebsiella pneumoniae* produce large amounts of acid, which gives the colonies a metallic sheen (Fig. 2.14).

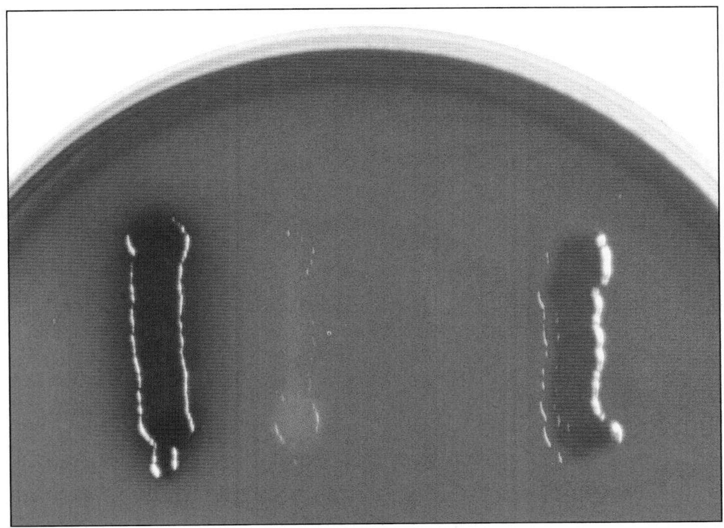

2.13 m Endo Agar LES • m Endo agar LES is a selective and differential medium. Shown is an Endo plate inoculated with (from left to right): *Escherichia coli* (a coliform; note the slight gold metallic sheen along the edges and darkening of the medium), *Klebsiella* (*Enterobacter*) *aerogenes* (a coliform; pink because it produces less acid than *E. coli*), *Enterococcus faecalis* (a Gram positive inhibited by the sodium sulfite), and *Shigella flexneri*, (a noncoliform that doesn't ferment lactose).

2.14 Metallic Sheen • Shown is m Endo agar LES streaked with *Escherichia coli* to illustrate the metallic sheen caused by copious acid production from lactose fermentation. The metallic sheen is best seen by looking at the growth from a low angle with reflected light.

Eosin Methylene Blue (EMB) Agar (Levine)

Purpose

Eosin methylene blue (EMB) agar (Levine) is used for isolation of fecal coliforms. EMB agar can be streaked for isolation or used in the Membrane Filter Technique as discussed on page 300.

Principle

Eosin methylene blue agar is a complex (chemically undefined), selective, and differential medium. It contains digest of gelatin, lactose, and the dyes eosin Y and methylene blue. Gelatin provides energy, nitrogen, and organic carbon. Lactose is fermented to acid end-products (Fig. 2.12) by coliforms such as *Escherichia coli* and *Klebsiella* (*Enterobacter*) *aerogenes*, whereas it is not by pathogens such as *Proteus*, *Shigella*, and *Salmonella* species.

The purpose of the dyes is twofold: one, they inhibit growth of most Gram-positive organisms (some *Enterococcus* and *Staphylococcus* species are exceptions), and two, they react with vigorous lactose fermenters whose acid end-products turn the growth dark purple or black. This dark growth is typical of *E. coli* and is often accompanied by a green metallic sheen (Figs. 2.15 and 2.16).

Other less-aggressive lactose fermenters, such as *Enterobacter* species, produce colonies that can range from pink to dark purple on the medium. Lactose non-fermenters typically retain their normal color or take on the coloration of the medium.

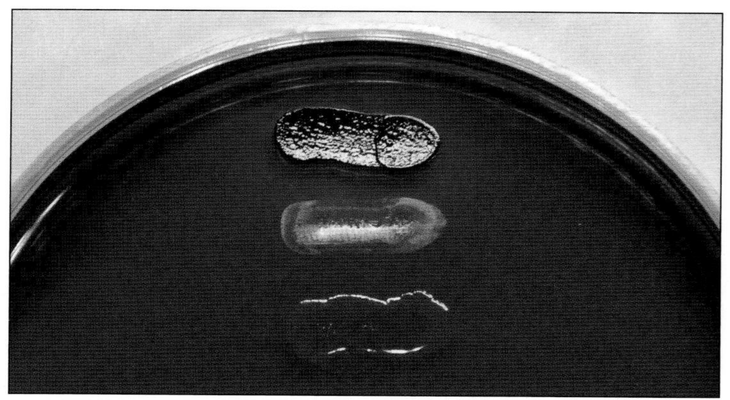

2.15 Eosin Methylene Blue Agar • This EMB plate was inoculated with (from top to bottom): *Escherichia coli* (coliform), *Streptococcus agalactiae* (Gram-positive coccus), and *Salmonella typhimurium* (noncoliform). *S. agalactiae* was inhibited by eosin and methylene blue dyes but still grew enough to form a thin film on the agar. Vigorous lactose fermentation by *E. coli* gave its growth a shiny green appearance. *S. typhimurium* was not inhibited by the dyes but also did not ferment the lactose, and so is a more natural color.

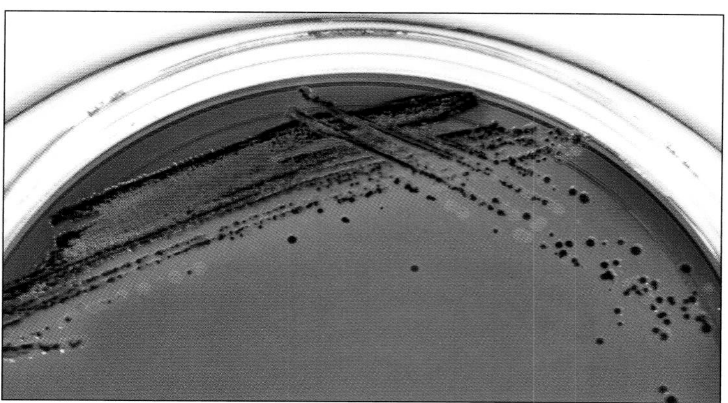

2.16 Eosin Methylene Blue Agar Streaked for Isolation • A mixture of *Escherichia coli* and *Salmonella typhimurium* was streaked for isolation on this EMB agar plate. Note the differences in *E. coli* coloration between regions of dense growth (green metallic sheen) and individual colonies (dark growth) due to differences in the amount of acid produced by lactose fermentation. *S. typhimurium* is a noncoliform and remains its natural beige color.

Hektoen Enteric (HE) Agar

Purpose

Hektoen enteric (HE) agar is used to isolate and differentiate *Salmonella* and *Shigella* species from other Gram-negative enteric organisms.

Principle

Hektoen enteric agar is a complex, differential, and somewhat selective medium designed to isolate *Salmonella* (see p. 211) and *Shigella* (see p. 213) species from other enterics. The test is based on the ability to ferment lactose, sucrose, and/or salicin, and to reduce sulfur to hydrogen sulfide gas (H_2S).

Sodium thiosulfate is included as the source of oxidized sulfur. Ferric ammonium citrate reacts with any sulfur that becomes reduced (H_2S) to form the black precipitate ferrous sulfide (FeS). Bile salts prevent or inhibit growth of Gram-positive organisms. Bile salts also have a moderate inhibitory effect on enterics, so relatively high concentrations of animal tissue and yeast extract are included to offset this situation. Bromothymol blue and acid fuchsin dyes are pH indicators.

Color changes in the colonies and in the agar allow differentiation (Figs. 2.17 and 2.18). Enterics that produce acid from fermentation will produce yellow to salmon-pink colonies. Neither *Salmonella* nor *Shigella* species ferment any of the sugars; instead they break down the animal tissue, which is largely protein. Release of ammonia from protein digestion slightly raises the medium's pH and gives the colonies a blue-green color. Additionally, *Salmonella* species reduce sulfur to H_2S, so the colonies formed also contain FeS, which makes them partially or completely black.

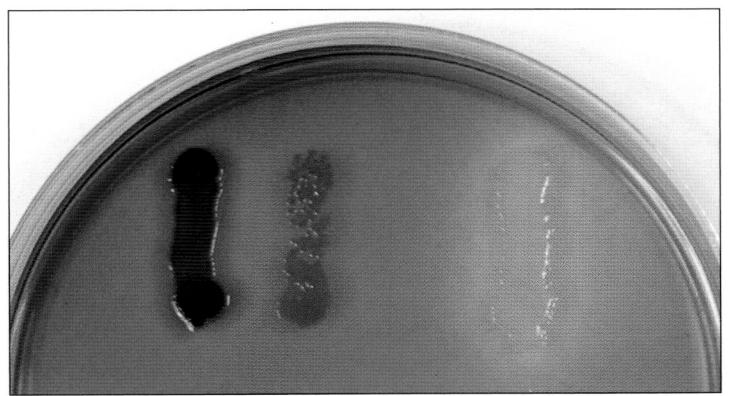

2.17 Hektoen Enteric Agar • This HE plate was inoculated with (from left to right): *Salmonella typhimurium*, *Shigella flexneri*, *Enterococcus faecalis*, and *Escherichia coli*. *E. coli* is a coliform and produced acid from lactose fermentation resulting in the yellow (or sometimes salmon) color of the growth and surrounding medium. *S. typhimurium* and *S. flexneri* are noncoliforms and don't ferment lactose. However, both deaminate protein, resulting in a blue-green color best shown by *S. flexneri*, which is negative for sulfur reduction. The black coloration largely hiding the blue-green color of *S. typhimurium* is due to sulfur reduction. *E. faecalis* is Gram positive and its growth was inhibited by bile salts in the medium.

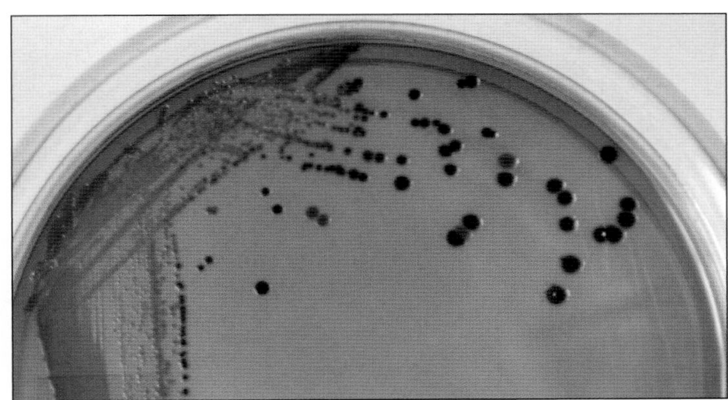

2.18 Hektoen Enteric Agar Streaked for Isolation • This HE agar was streaked with *Salmonella typhimurium* and *Shigella flexneri*. Note the black *Salmonella* colonies due to sulfur reduction and the blue-green *S. flexneri* colonies due to protein deamination.

MacConkey Agar

Purpose

MacConkey agar is used to isolate and differentiate members of the *Enterobacteriaceae* based on the ability to ferment lactose. Variations on the standard medium include MacConkey agar w/o CV (without crystal violet) to allow detection of Gram-positive cocci, or MacConkey agar CS to control swarming bacteria (*Proteus*—see Fig. 3.22) that interfere with other results.

Principle

MacConkey agar is a selective and differential medium containing lactose, bile salts, neutral red, and crystal violet. Bile salts and crystal violet inhibit growth of Gram-positive bacteria. Neutral red dye is a pH indicator that is colorless above a pH of 6.8 and red at a pH below 6.8.

Acid accumulating from lactose fermentation turns the dye red, so lactose fermenters turn a shade of red on MacConkey agar, whereas lactose non-fermenters remain their normal color or the color of the medium (Figs. 2.19 and 2.20). Formulations without crystal violet allow growth of *Enterococcus* and some species of *Staphylococcus*, which ferment the lactose and appear pink on the medium.

2.19 MacConkey Agar • This MacConkey agar plate was inoculated with (from left to right): *Escherichia coli*, *Staphylococcus epidermidis*, and *Shigella flexneri*. *E. coli* is a coliform and fermented lactose to acid end-products and is pinkish. Sometimes a pinkish bile precipitate forms around the growth, as seen here. Only loop marks are visible where *S. epidermidis* was inoculated. Its growth was inhibited by the bile salts and crystal violet. *S. flexneri*, a noncoliform, did not ferment lactose. It appears close to its natural color and has yellowed the medium.

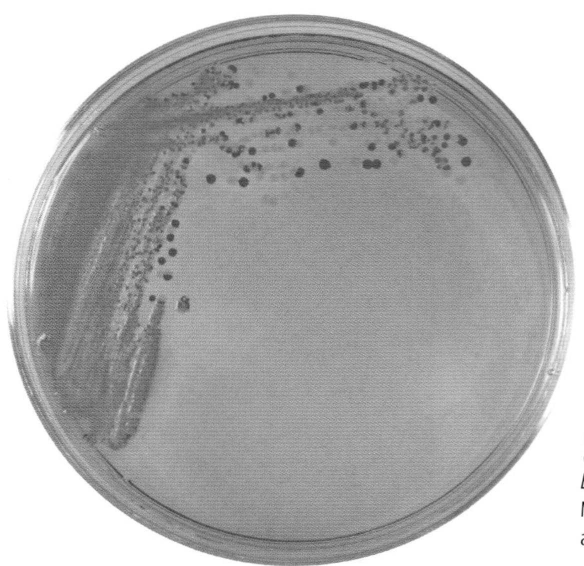

2.20 MacConkey Agar Streaked for Isolation • *Escherichia coli* and *Shigella flexneri* were inoculated on this MacConkey agar plate. Note the absence of precipitated bile salts around this specimen of *E. coli* (compare with *E. coli* in Fig. 2.19).

Mannitol Salt Agar (MSA)

Purpose

Mannitol salt agar (MSA) is used for isolation and differentiation of *Staphylococcus aureus* (see p. 214) from other staphylococci.

Principle

MSA is a selective and differential medium containing the carbohydrate mannitol, 7.5% sodium chloride (NaCl), and the pH indicator phenol red. Phenol red is yellow below pH 6.8, red at pH 7.4 to 8.4, and pink above pH 8.4. Mannitol provides the substrate for fermentation and phenol red indicates acid production by changing color as the pH changes. This makes the medium differential.

Sodium chloride makes this medium selective because its concentration is high enough to dehydrate and kill most bacteria. Staphylococci thrive on the medium, largely because of their adaptation to somewhat salty habitats such as human skin.

Most staphylococci are able to grow on MSA, but do not ferment mannitol, so their growth appears pink or red and the medium remains unchanged. *Staphylococcus aureus* ferments mannitol, which produces acids and lowers the medium's pH (Fig. 2.21). The result is formation of bright yellow colonies usually surrounded by a yellow halo (Fig. 2.22).

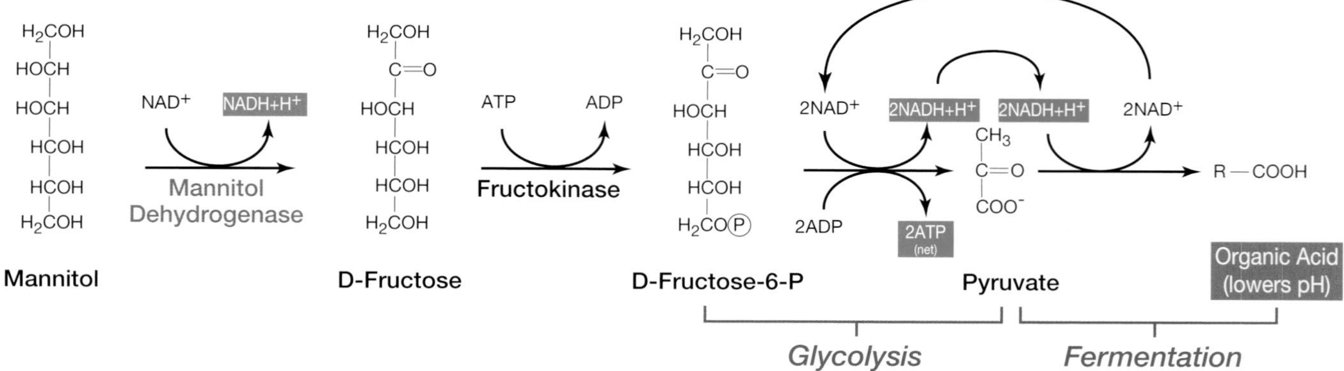

2.21 Mannitol Fermentation with Acid End-Products • There are two important points to note here: one, that this fermentation (as do all fermentations) provides a mechanism for oxidizing NADH back to NAD^+ so it can be used again in glycolysis; and two, that the end-product of the fermentation is an acid—the acid detected by the phenol red pH indicator in MSA. For simplicity, only the relevant parts of glycolysis are shown.

2.22 Mannitol Salt Agar • MSA contains 7.5% NaCl and is selective for members of the genus *Staphylococcus* due to their salt tolerance. It further allows differentiation between *Staphylococcus* species based on the ability to ferment mannitol to acid end-products (e.g., *S. aureus*, left) and those that do not (e.g., *S. epidermidis*, right).

Phenylethyl Alcohol (PEA) Agar

Purpose

Phenylethyl alcohol (PEA) agar is used to isolate staphylococci and streptococci (including *Enterococcus* and *Lactococcus* species) from specimens containing mixtures of bacterial microbiota. Typically, it is used to screen out the common contaminants *Escherichia coli* and *Proteus* species. When prepared with 5% sheep blood, it is used for cultivation of Gram-positive anaerobes.

Principle

Phenylethyl alcohol agar is an undefined, selective medium used to isolate Gram-positive organisms by inhibiting most Gram-negative organisms (Fig. 2.23). It is not a differential medium because it does not distinguish between different organisms that successfully grow on it. Digests of casein and soybean meal provide nutrition while sodium chloride provides a stable osmotic environment suitable for the addition of sheep blood if desired.

Phenylethyl alcohol is the selective agent and may be bacteriostatic or bactericidal depending on concentration. It inhibits Gram-negative organisms by breaking down their cytoplasmic membrane, which affects its function as a permeability barrier. A major consequence of this is inhibition of DNA synthesis due to rapid loss of potassium ions.

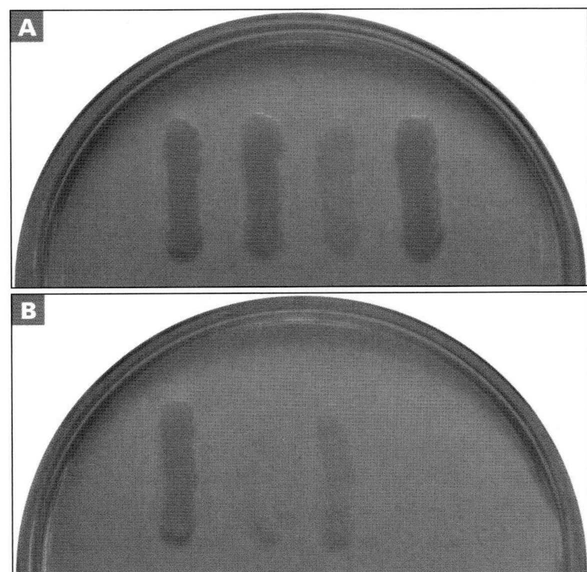

2.23 Trypticase Soy Agar vs. Phenylethyl Alcohol Agar • (**A**) Trypticase soy agar (TSA) is a nutritious, general growth medium. From left to right are: *Staphylococcus aureus*, *Escherichia coli*, *Enterococcus faecium*, and *Klebsiella pneumoniae*. All show excellent growth on TSA. (**B**) Shown are the same organisms inoculated on PEA in the same positions as in (**A**). Notice that *S. aureus* and *E. faecium* grow well on both plates. Phenylethyl alcohol selects against Gram-negative bacteria, but in many cases only slows growth, it doesn't completely stop growth. This is seen here with *E. coli* (partial inhibition) and *K. pneumoniae* (complete inhibition).

Pseudomonas Isolation Agar (PIA)

Purpose

Pseudomonas isolation agar (PIA) is used to isolate non-fermenting, Gram-negative bacteria in clinical samples, especially *Pseudomonas* species (see p. 209). It also allows differentiation of *P. aeruginosa*, a major cause of nosocomial infections (often from contaminated hospital equipment), from other pseudomonads based on its production of the pigment **pyocyanin** (Fig. 2.24).

Principle

PIA is a chemically defined, selective, and differential medium containing the fatty acid synthesis inhibitor, Irgasan[1] (also known as Triclosan), which is inhibitory to many Gram-positive and Gram-negative species. *Pseudomonas* species are not affected by its activity (at its concentration in the medium) due to a membrane efflux pump. Carbon and nitrogen are provided by peptone and glycerol. Pyocyanin is a greenish-blue pigment whose synthesis is enhanced by potassium sulfate, magnesium chloride, and glycerol.

[1] Irgasan is a registered trademark of Ciba-Geigy.

2.24 ***Pseudomonas* Isolation Agar** • *Pseudomonas* isolation agar is selective for pseudomonads and differentiates the majority of species from *P. aeruginosa* based on pyocyanin production. *P. aeruginosa*—note the greenish color from the pigment—is on the right; *P. putida* is on the left.

Sabouraud Dextrose Agar (SDA)

Purpose

Sabouraud dextrose agar (SDA) is an undefined, selective medium designed to grow fungi in general, but especially yeasts and dermatophytes (Fig. 2.25). It is also used in RODAC plates to monitor fungal contamination before and after cleaning of surfaces (p. 297). Substitution of maltose for dextrose has been used to grow *Microsporum* species and *Trichophyton gypseum*.

Principle

The recipe of Sabouraud dextrose agar is relatively simple. Dextrose and digests of casein and animal tissue (in some formulations) act as carbon, nitrogen, and energy sources. The pH is adjusted to 5.8, which is inhibitory to many bacteria. The antibiotics chloramphenicol and gentamicin can be added to improve selection against bacteria.

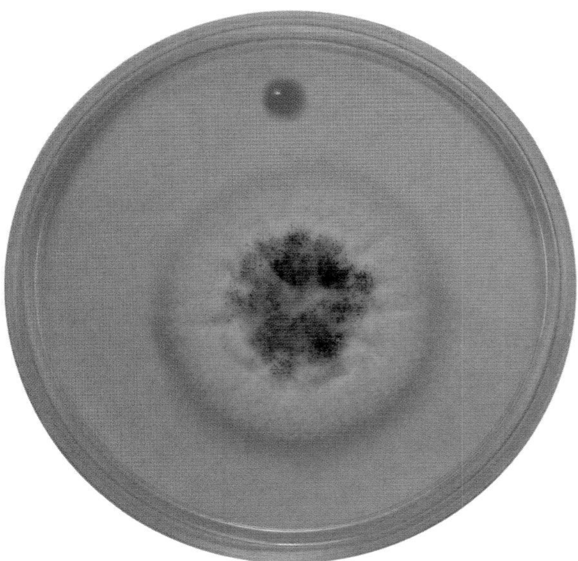

2.25 Sabouraud Agar • Sabouraud agar is selective for fungi. The low pH and optional inclusion of antibiotics make it inhibitory to bacteria (this plate included chloramphenicol). At the top is the yeast *Rhodotorula mucilaginosa* (*R. rubra*). Below is the mold *Aspergillus niger*.

Salmonella–Shigella (SS) Agar

Purpose

Salmonella-Shigella (SS) agar was originally used for the isolation of *Salmonella* (see p. 211) and many *Shigella* (see p. 213) species (i.e., lactose non-fermenting enteric bacteria; noncoliforms) from the lactose fermenting enterics (coliforms). It is no longer recommended for isolation of *Shigella*, because Hektoen (see p. 16) and XLD (see p. 23) agars are more effective, but it is still of use in isolating *Salmonella* species.

Principle

Salmonella-Shigella agar is an undefined, differential, and selective medium with bile salts and brilliant green dye acting as the selective agents against Gram positives and many Gram negatives. Lactose is a fermentable carbohydrate and sodium thiosulfate acts as an electron acceptor for sulfur reducers.

Neutral red is the pH indicator and ferric citrate reacts with H_2S to form a black precipitate, thus indicating sulfur reduction. Lactose fermenters will produce reddish colonies as neutral red changes from colorless to red in the low pH. *Salmonella* and *Shigella* species will be their natural color due to their inability to ferment lactose. *Salmonella* and *Proteus* species typically reduce sulfur, which is indicated by colonies with black centers (Figs. 2.26 and 2.27).

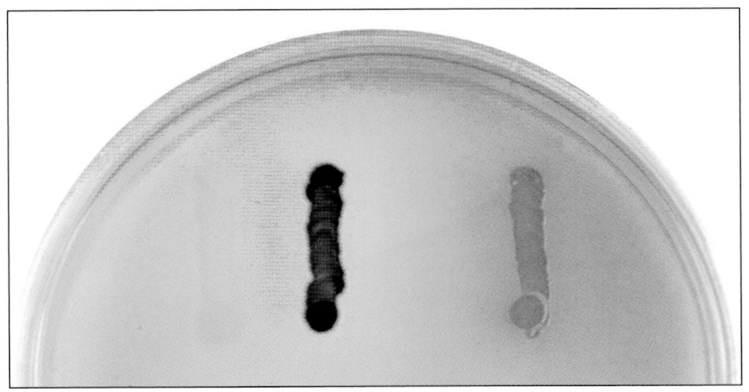

2.26 ***Salmonella-Shigella*** **Agar** • Shown is an SS agar plate inoculated with (from left to right): *Shigella flexneri*, *Salmonella typhimurium*, *Enterococcus faecalis*, and *Escherichia coli*. Note the pink color of *E. coli* due to acid produced by lactose fermentation. *S. typhimurium* reduces thiosulfate to H_2S, which reacts with citrate in the medium to turn the growth black. Both *S. typhimurium* and *S. flexneri* decarboxylate amino acids, which raises the pH and turns the growth (if not covered by the black of sulfur reduction) and the medium yellow. Note the yellow medium surrounding both. They may also remain their natural colors. Growth of *E. faecalis* is inhibited by bile salts and brilliant green dye. Some *Shigella* strains are inhibited by this medium, so Hektoen Enteric agar (see p. 16) or XLD (see p. 23) are recommended for their isolation.

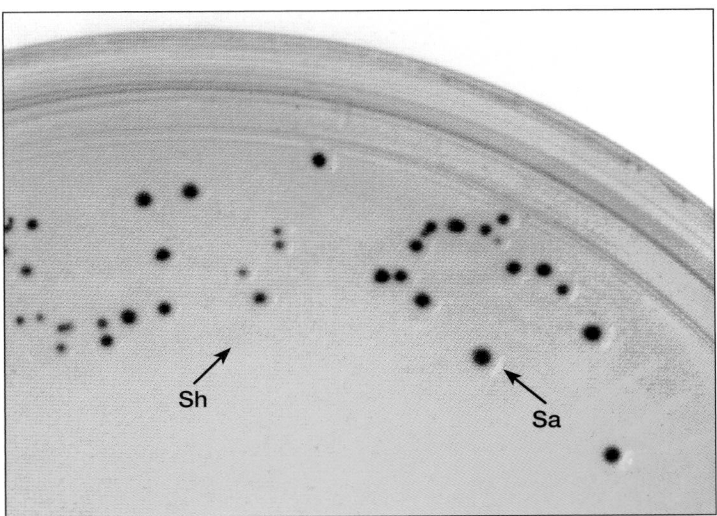

2.27 ***Salmonella-Shigella*** **Agar Streaked for Isolation** • Differentiation between *Salmonella typhimurium* and *Shigella flexneri* colonies is apparent on this SS agar streak plate. Colonies with black centers and clear edges are characteristic of *Salmonella* (Sa) on this medium due to sulfur reduction. *Shigella* (Sh) doesn't reduce sulfur and the colonies are colorless.

Thiosulfate Citrate Bile Sucrose (TCBS) Agar

Purpose

TCBS agar is used for primary isolation of *Vibrio* species. Clinical and nonclinical specimens suspected of fecal contamination are streaked on TCBS in an effort to recover *Vibrio cholerae*, the most important pathogen of the genus.

Principle

TCBS agar is an undefined, selective, and differential medium with an alkaline pH (8.6) that promotes growth of *Vibrio species,* especially *V. cholerae*. Oxgall and sodium cholate are included to inhibit the growth of Gram-positive bacteria. Sucrose is the fermentable carbohydrate and sodium thiosulfate acts as an electron acceptor for sulfur reducers. Bromothymol blue is the pH indicator and ferric ammonium citrate indicates sulfur reduction.

Sucrose fermenters producing acid end-products (such as *V. cholerae*) form yellow colonies (Fig. 2.28) while those of sucrose non-fermenters are blue. Some *Enterococci* ferment sucrose but are partially inhibited by the oxgall. These organisms, as shown in Figure 2.29, produce small yellow colonies. Species able to reduce thiosulfate to H_2S produce black colonies due to the reaction of H_2S with the ferric ion in the medium.

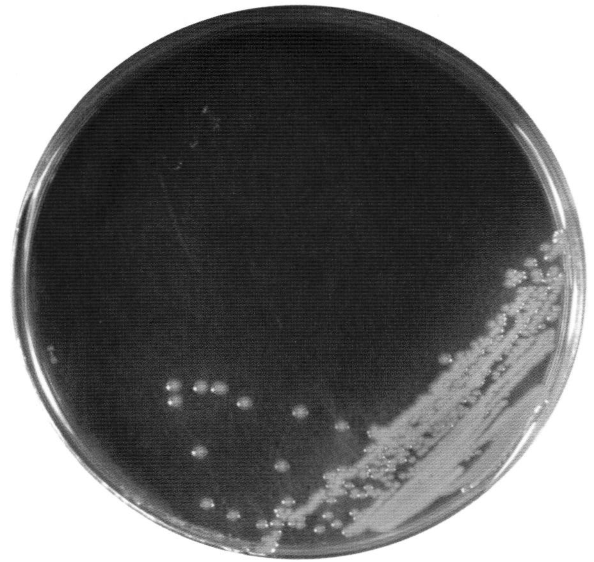

2.28 *Vibrio cholerae* Streaked on TCBS Agar • TCBS agar is primarily used to isolate *Vibrio* species. Large, yellow colonies on TCBS agar are indicative of *Vibrio cholerae,* the most important *Vibrio* pathogen.

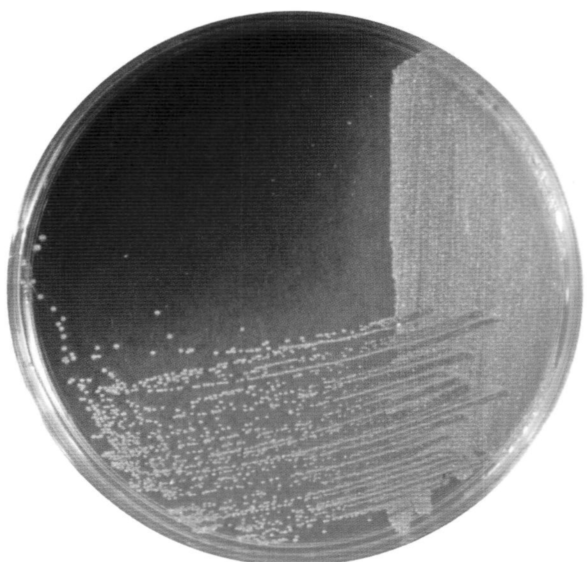

2.29 *Enterococcus faecalis* Streaked on TCBS Agar • This Gram-positive coccus may also be recovered from fecal-contaminated samples; however, its yellow colonies are much smaller than those of *V. cholerae*. Compare the *E. faecalis* colonies with those of *V. cholerae* in Figure 2.28.

Vogel and Johnson (VJ) Agar

Purpose

Vogel and Johnson agar is a modified version of tellurite glycine agar (TGA). Both are used for the rapid isolation of coagulase-positive staphylococci obtained during cosmetics, food, and nutritional supplements testing. The most common and clinically important coagulase-positive staphylococcus is *Staphylococcus aureus* (see p. 214). Because a pH indicator and a higher mannitol concentration are used (than in TGA), VJ agar is also able to detect mannitol fermentation, another characteristic of *S. aureus*.

Principle

Vogel and Johnson agar is an undefined, selective, and differential medium. Among the ingredients in VJ agar are casein and yeast extract, which supply carbon and nitrogen; the fermentable carbohydrate mannitol; and the highly selective agents dipotassium tellurite, glycine, and lithium chloride, all of which inhibit growth of most Gram-negative bacteria and some Gram positives. Phenol red is the pH indicator, which turns yellow under acidic conditions.

Most coagulase-positive *S. aureus* strains are able to ferment mannitol and turn the medium yellow, and reduce tellurite, which produces a black precipitate in the growth. Therefore, an organism that grows well on the medium and produces black colonies on yellow agar is presumptively identified as *S. aureus* and is thus differentiated from coagulase-negative staphylococci and Gram-negative bacteria (Fig. 2.30).

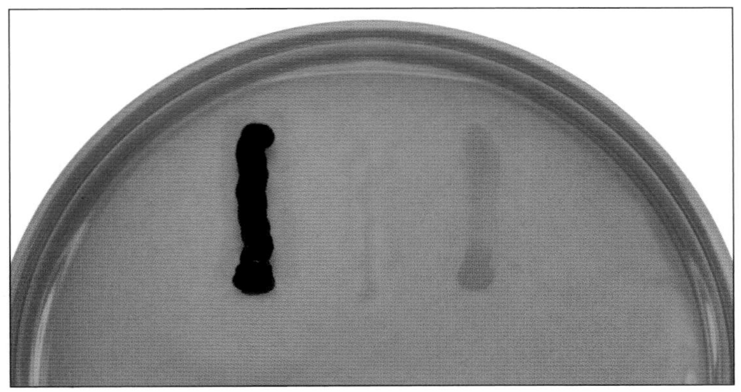

2.30 Vogel and Johnson (VJ) Agar ● VJ agar is used for rapid isolation of *Staphylococcus aureus*. The medium includes three inhibitors, including potassium tellurite, which is reduced by *S. aureus* and produces a black color. *S. aureus* also ferments mannitol to acid end-products, which turns the medium yellow. From left to right: *S. aureus*, *Klebsiella* (*Enterobacter*) *aerogenes* (a Gram negative whose growth was inhibited), and *S. epidermidis*. Some strains of *S. epidermidis* are slightly inhibited by the potassium tellurite but still produce visible growth, and some strains also show slight reduction of tellurite. However, the intense black color of *S. aureus* and the yellow from mannitol fermentation easily distinguishes between the two.

Xylose Lysine Deoxycholate (XLD) Agar

Purpose

Xylose lysine deoxycholate (XLD) agar is used to isolate and identify *Shigella* (see p. 213) and *Providencia* from stool samples.

Principle

Xylose lysine deoxycholate agar is a selective and differential medium containing sodium deoxycholate, xylose, lactose, sucrose, lysine, sodium thiosulfate, phenol red, and ferric ammonium citrate. Deoxycholate is a bile salt that inhibits growth of Gram-positive organisms. Xylose (0.35 g%), lactose (0.75 g%), and sucrose (0.75 g%) are fermentable sugars, and lysine, an amino acid, is provided for decarboxylation (removal of the carboxyl group–COOH, see Fig. 7.26).

Sodium thiosulfate provides oxidized sulfur for organisms capable of reducing it to hydrogen sulfide gas (H_2S). Ferric ammonium citrate performs as an indicator because it releases ferric ions (Fe^{3+}) into the medium that react with H_2S and form ferrous sulfide (FeS), a black precipitate. Phenol red, which is yellow when acidic and red or pink when alkaline, is included as a pH indicator.

Organisms that ferment xylose will acidify the medium and produce yellow colonies. Organisms able to decarboxylate lysine release amines into the medium, which raise the pH, resulting in red colonies. Organisms able to reduce sulfur produce H_2S, which will precipitate with Fe^{3+} and form black colonies.

Shigella and *Providencia* species, which do not ferment xylose, sucrose, or lactose, but decarboxylate lysine, form red colonies on the medium. *Salmonella* species reduce sulfur, ferment xylose (but not lactose or sucrose), and decarboxylate lysine. Xylose fermentation acidifies the medium turning it yellow, but it is in relatively short supply and is consumed during incubation.

The effect of lysine decarboxylation overcomes the acidity as it raises the pH and produces red colonies with black centers from sulfur reduction. Coliform enterics (e.g., *Escherichia, Klebsiella, Enterobacter*) ferment all three sugars and depletion of xylose does not lead to a pH increase from lysine decarboxylation because lactose and sucrose continue to be fermented. These organisms typically produce yellow colonies on the medium (Fig. 2.31). ●

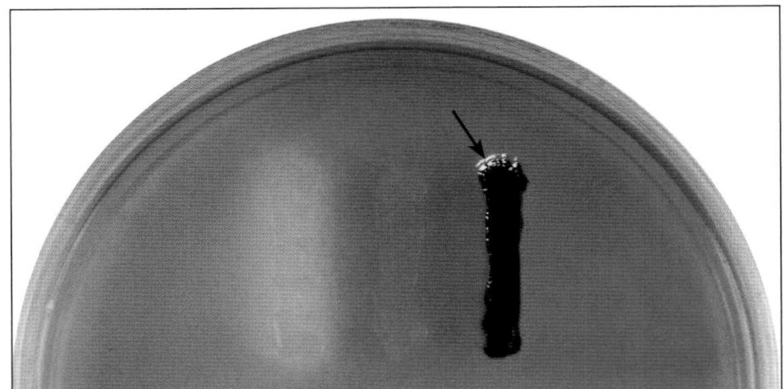

2.31 Xylose Lysine Deoxycholate Agar ● From left to right on this XLD plate is *Escherichia coli*, a xylose, lactose, and/or sucrose fermenter, which has turned the medium (and itself) yellow; *Shigella flexneri*, which is negative for xylose, lactose, and sucrose fermentation and sulfur reduction, but positive for lysine decarboxylation, as evidenced by the red color; and *Salmonella typhimurium*, which ferments xylose, reduces sulfur (making it black), and decarboxylates lysine (note the thin, red border around the edges, most obvious at the top and indicated by the **arrow**). Because xylose fermentation is completed fairly rapidly due to its low concentration, the red of decarboxylation replaces the yellow from fermentation.

SECTION

3 Bacterial Growth

Growth Patterns on Agar

Purpose

Recognizing different bacterial growth morphologies on agar plates is a useful step in the identification process. It is often the first indication that one organism is different from another. Once purity of a colony has been confirmed by an appropriate staining procedure (this is not always done), cells can be transferred to a sterile medium, grown, and maintained as a pure culture, which then acts as a source of that microbe for identification or other purposes. Agar slants are typically used for cultivation of pure cultures. Bacteria also frequently display distinct morphological color and texture on agar slants.

Principle

Agar Plates

When a single bacterial cell is deposited on an appropriate solid nutrient medium, it begins to divide. One cell makes two, two make four, four make eight . . . one million make two million, and so on. Eventually a visible mass of cells—a **colony**—appears where the original cell was deposited.

Color, size, shape, and texture of microbial growth are determined by the genetic makeup of the organism (in many cases by yet unknown mechanisms), but are also greatly influenced by environmental factors, including nutrient availability, temperature, and incubation time.

Colony morphological characteristics may be viewed with the naked eye, a hand lens, a stereo (dissecting) microscope, or a colony counter (Fig. 3.1). The seven basic categories include colony size, whole colony shape, margin (edge), surface, elevation, texture, and optical properties (Fig. 3.2). Following is a list of each category with some representative specific descriptions. It is not an exhaustive list.

1. *Size* is simply a measurement of the colony's dimensions—the diameter if circular, or length and width if shaped otherwise.
2. *Whole colony shape* may be described as **round (circular)**, **erose, (irregular)**, **spindle (tapered at the ends)**, **filamentous**, **rhizoid (branched)**, or **punctiform** (tiny, pinpoint).
3. The *margin* may be **entire** (**smooth**, with no irregularities), **undulate** (wavy), **irregular**, **lobate** (lobed), **filamentous** (unbranched strands), or **rhizoid** (branched like roots).
4. The *surface* may be **smooth, rough, wrinkled (rugose), shiny**, or **dull**.

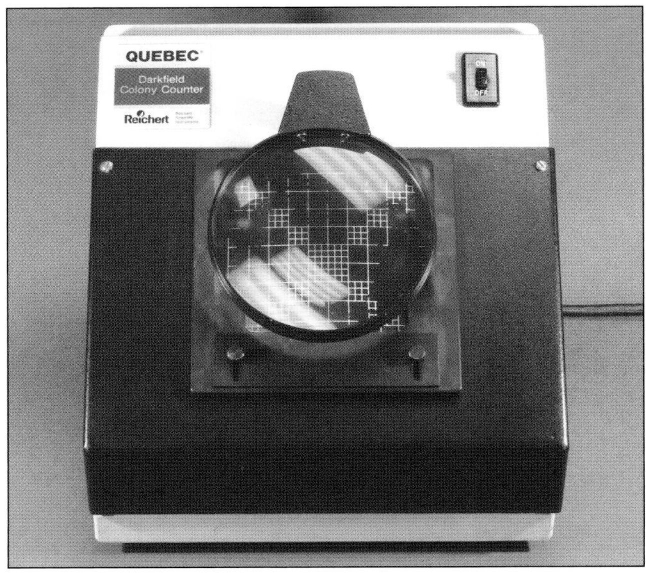

3.1 Colony Counter • Subtle differences in colony shape and margin can best be viewed with magnification, especially if the colony is very small. Magnification can be provided by a colony counter. The **transmitted light** and magnifying glass allow observation of greater detail; however, colony color and many other features are best determined with reflected light. The grid in the background is a counting aid; each big square is 1 square centimeter.

5. The *texture* may be **moist, mucoid** (sticky), **butyrous** (buttery), or **dry.**
6. *Elevations* include **convex, pulvinate** (very convex), **umbonate** (raised in the center), **plateau, flat, and raised, among others.**
7. Other useful features include **color** and optical properties such as **opaque** (light does not pass through) and **translucent** (light passes through).

Features such as colony shape, margin, surface, and color are best viewed by observing from above while holding the plate level with the lid off (if it is safe to do so), but rocking it back and forth slightly so reflected light hits it at different angles. If safe to do so, texture is checked by touching the growth with an inoculating loop or wooden stick.

Elevations are best viewed with the plate tilted slightly at eye level. Opacity and translucence are best viewed by placing the plate on a colony counter or holding it (lid on) so it is illuminated from behind (transmitted light). Colony dimensions are best measured from the plate's base rather than through the lid.

When reporting colony morphology, it is important to include the medium and the incubation time and temperature, all of which can affect a colony's appearance. Refer to Figures 3.3 through 3.26 for examples of colony diversity. Growth patterns of virtually all microorganisms can be influenced by environmental factors. Two important factors influencing bacterial growth are incubation time and temperature.

Figures 3.25 and 3.26 illustrate the effects of incubation time on colony morphology of *Bacillus subtilis* and temperature on pigment production by *Serratia marcescens.*

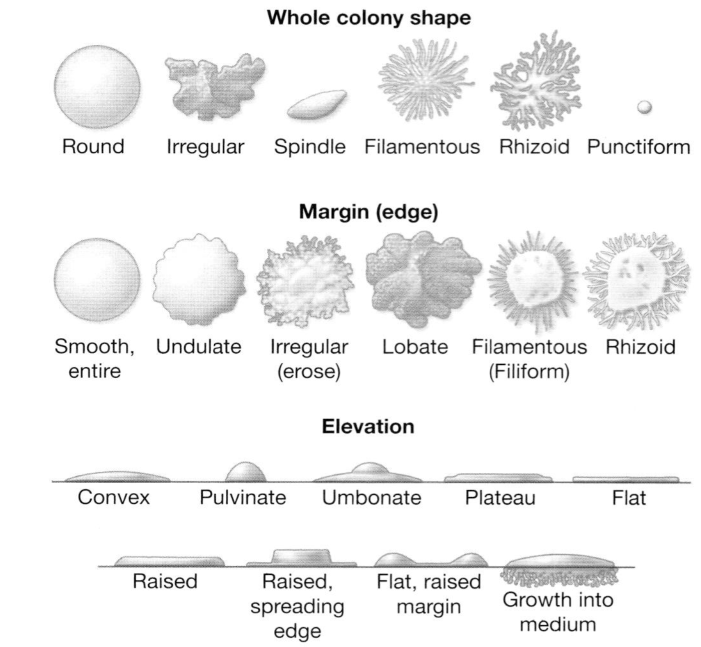

3.2 A Sampling of Bacterial Colony Features ● These terms are used to describe colony morphology. Descriptions also should include color, size, surface characteristics, texture, and optical properties (opaque or translucent). Make up a descriptive name for features you observe that are not covered here! See the text for details.

Diversity of Colony Morphologies in Mixed Cultures (Figures 3.3–3.4)

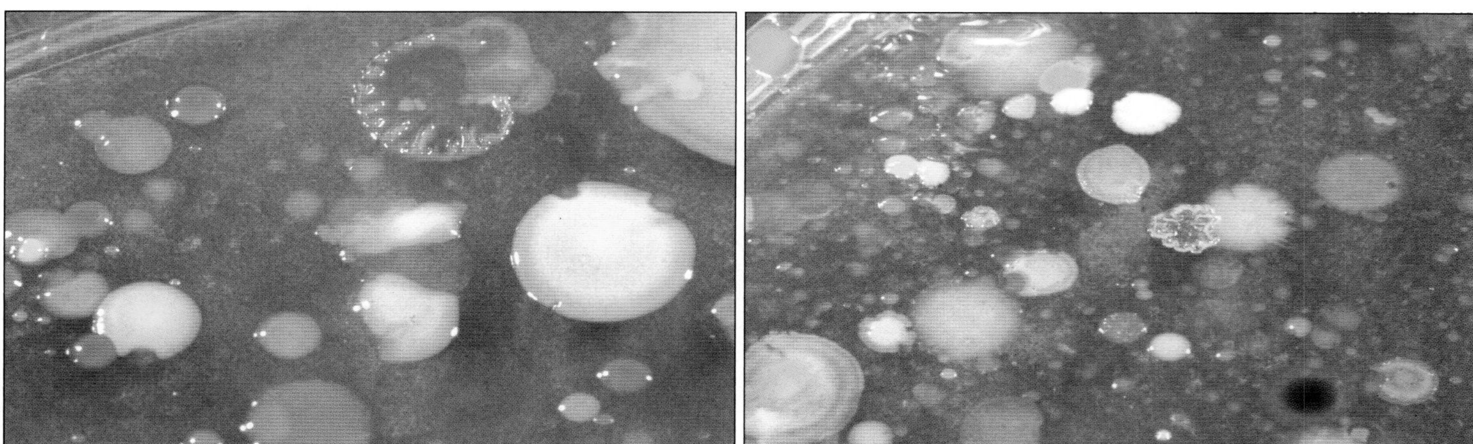

3.3 Two Mixed Soil Cultures Grown on Nutrient Agar ● These plates show the morphological diversity present in two diluted soil samples incubated for 48 hours. If two colonies look different when grown under the same conditions, they most likely are different species. The opposite is not always true, however. Two different species can produce colonies that are virtually identical.

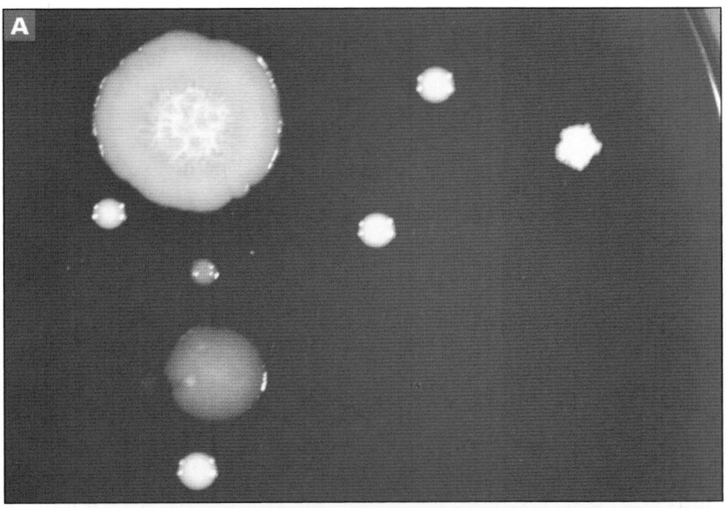

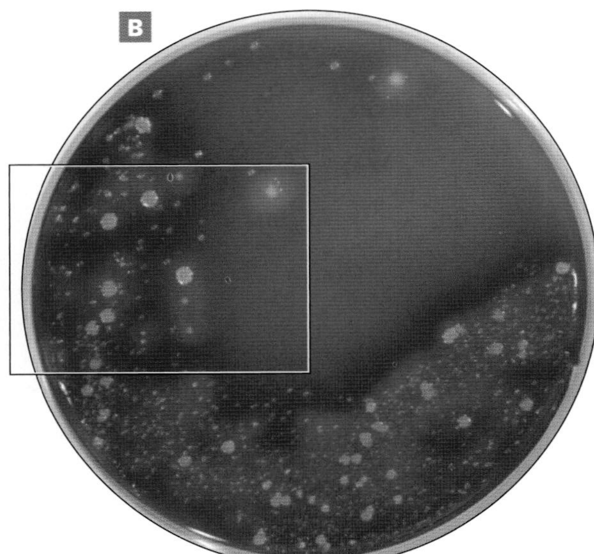

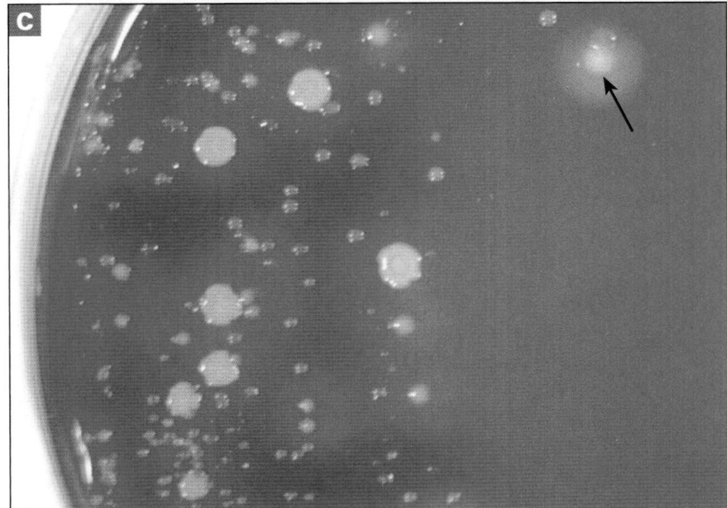

3.4 Throat Cultures Grown on Sheep Blood Agar ● (**A**) Based on morphology, there are probably five different species in this portion of the plate. (**B**) Note the α-hemolysis (darkening of the agar) shown by much of the growth. α-hemolytic organisms are abundant in throat samples and the majority are harmless commensals. (**C**) This is a close-up of the boxed area in (**B**). Note the weak β-hemolysis (clearing of the agar) by the white colony in the upper right (**arrow**). White growth with β-hemolysis is characteristic of *Staphylococcus aureus* (see p. 214). For information about hemolytic reactions on blood agar plates, see page 76.

Common Colony Morphologies (Figures 3.5–3.11)

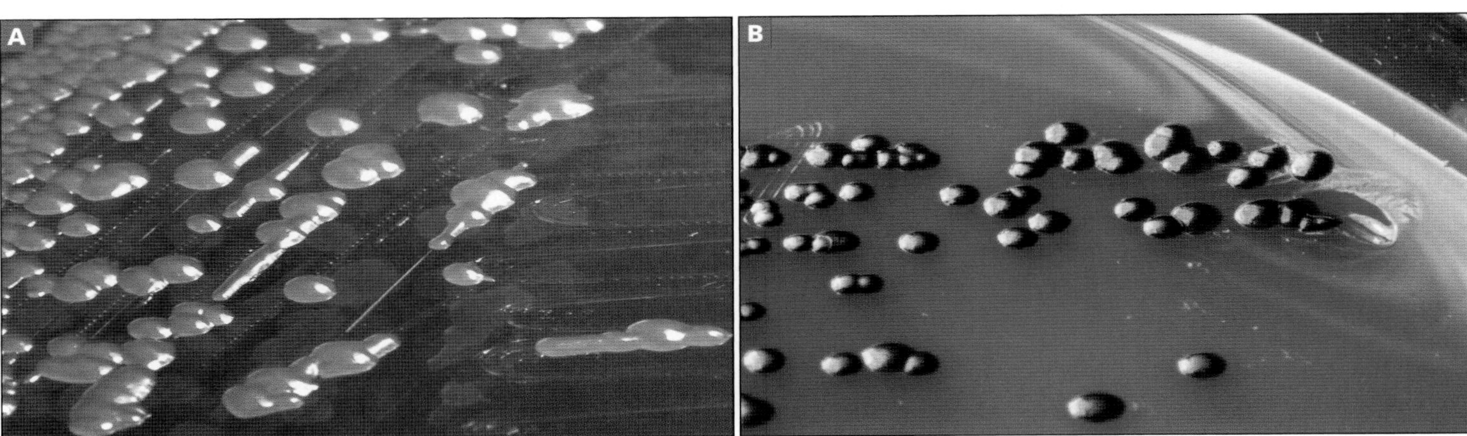

3.5 Round, Shiny, Convex Colonies ● (**A**) These buff-colored colonies of *Providencia stuartii* grew on nutrient agar in 48 hours. *P. stuartii* is a frequent isolate in urine samples obtained from hospitalized and catheterized patients. It is highly resistant to antibiotics. (**B**) The colonies of the soil and water bacterium *Chromobacterium violaceum* are purple (hence "violaceum"). Here they've been grown on sheep blood agar.

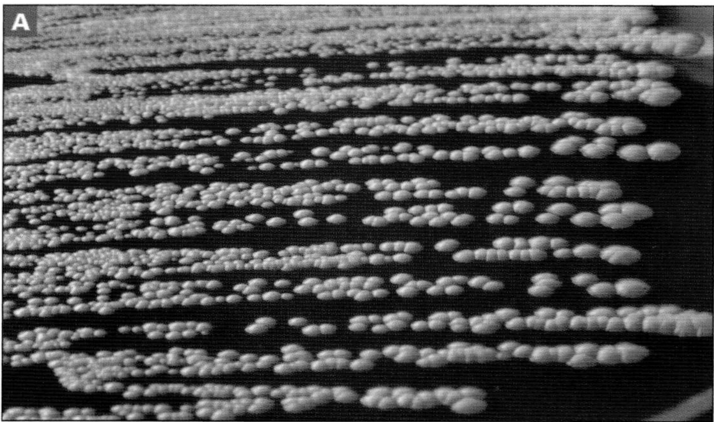

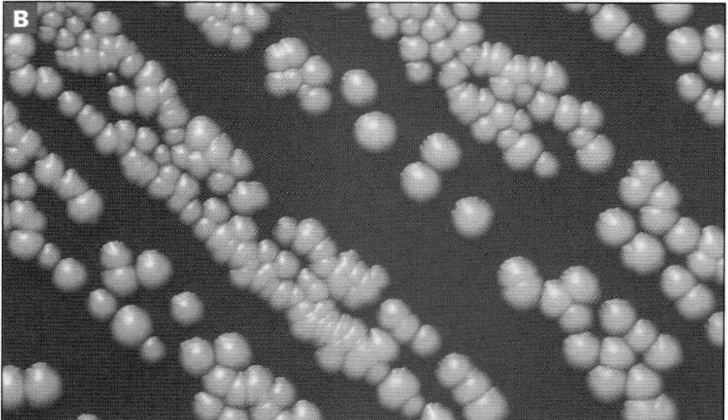

3.6 Round, Dull, Convex Colonies • Dry, buff-colored *Corynebacterium xerosis* (*xero* means "dry") colonies on sheep blood agar photographed from (A) a low angle and (B) from above.

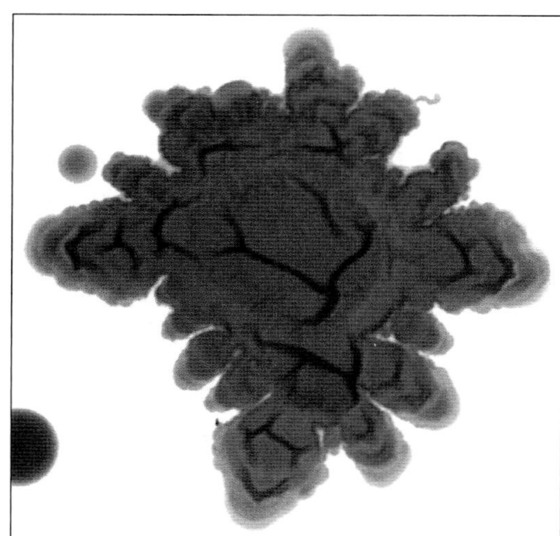

3.7 Irregular Colony Shape • This unidentified contaminant grew on tryptic soy agar after 48 hours and was isolated from a laboratory tabletop. In addition to its irregular shape, it has a lobed margin and wrinkled (rugose) surface. Its widest dimension was approximately 5 mm. This photo was taken through a stereo microscope.

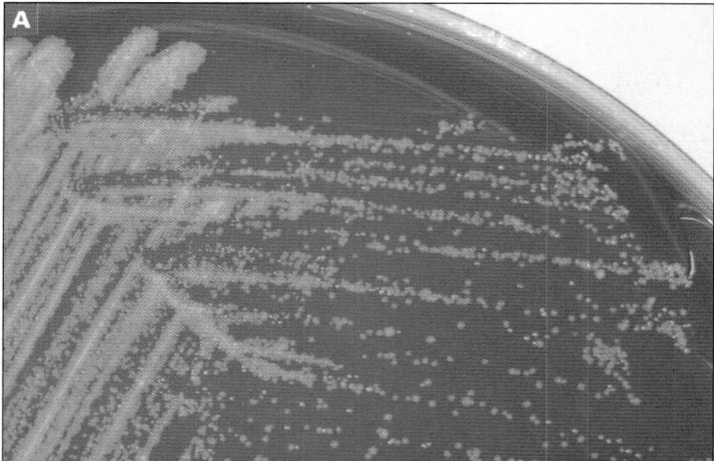

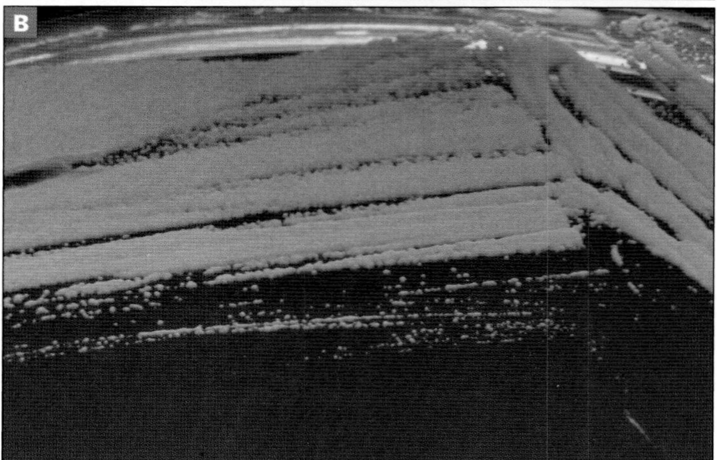

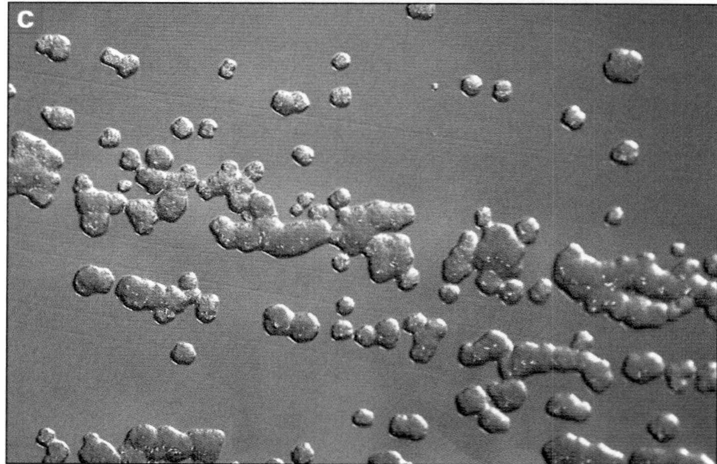

3.8 Punctiform Colonies • (A) Shown are *Mycobacterium smegmatis* colonies grown on sheep blood agar. The colonies of this slow-growing relative of *M. tuberculosis* are less than 1 mm in diameter. (B) These punctiform colonies of the rose-colored *Kocuria rosea* are <0.5 mm in diameter and were grown for 96 hours. (C) These are the same colonies as in (B) but photographed through a stereo microscope. Notice their irregular shape and wrinkled surface. *K. rosea* is an inhabitant of water, dust, and salty foods.

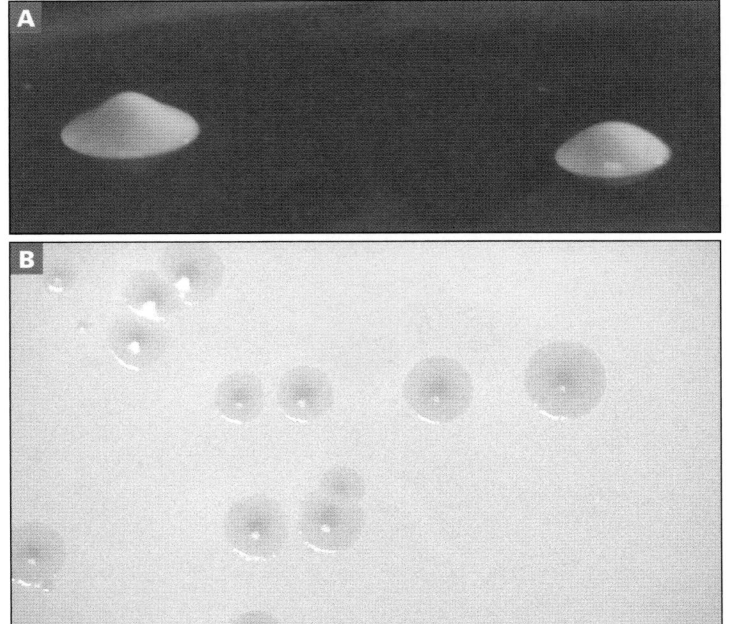

3.9 Umbonate Colonies • (A) The colony of this anaerobic lab contaminant on the left is truly umbonate. The one on the right is getting there. Their diameters are about 3 mm. (B) These *Enterococcus faecium* colonies were grown on tryptic soy agar for 48 hours and were photographed using a stereo microscope. The colonies are 1–2 mm in diameter and are white, circular, and umbonate (note the thicker center) with an entire margin. Notice, though, how flat the colonies are except for the central bump that makes them umbonate. *E. faecium* (formerly known as *Streptococcus faecium*) is found in human and animal feces.

3.10 Flat Colony • This unknown soil isolate is flat only around the edge. The center is concave (but mostly flat, not concave like a bowl) and at the very center is another small, raised ring surrounding another concavity. It's almost like an automobile tire around the wheel and hub. Very unusual! Also note the dull surface.

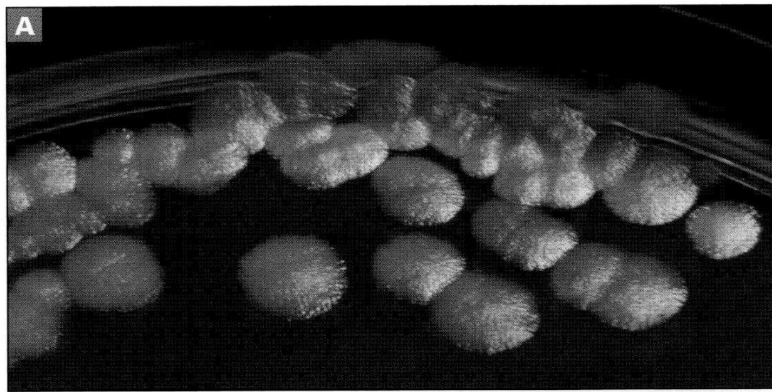

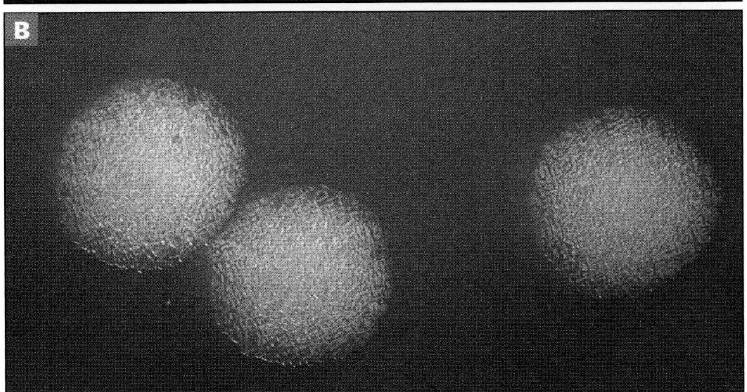

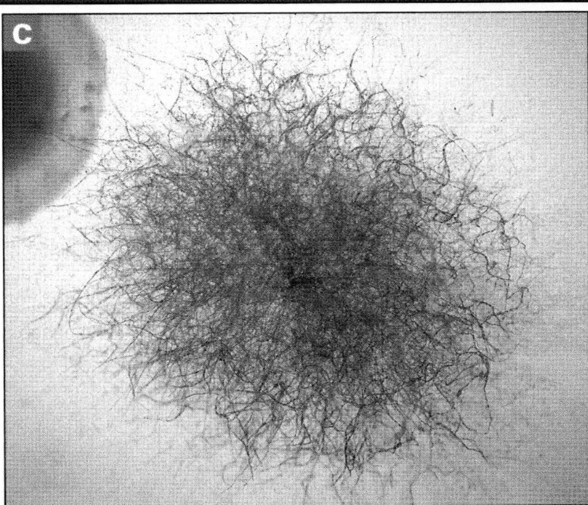

3.11 Filamentous Colonies • (A) These colonies of *Rhizobium leguminosarum* were grown on brain-heart infusion agar and are about 5 mm in diameter. In addition to being filamentous, they are convex, circular, and mucoid with translucent edges. *R. leguminosarum* is capable of causing root nodule formation (*rhiz* means "root") in many legumes and subsequently fixing atmospheric nitrogen. (B) This micrograph of three *R. leguminosarum* colonies on brain-heart infusion agar was taken through a stereo microscope. (C) Fungi other than yeasts grow as filaments called **hyphae**, which collectively are referred to as a mycelium (see p. 240). Shown is an unidentified mold mycelium obtained from the microbiology lab photographed through a stereo microscope. Note the absence of branching in the filaments, which would change the margin description to rhizoid.

Common Colony Margins (Figures 3.12–3.16)

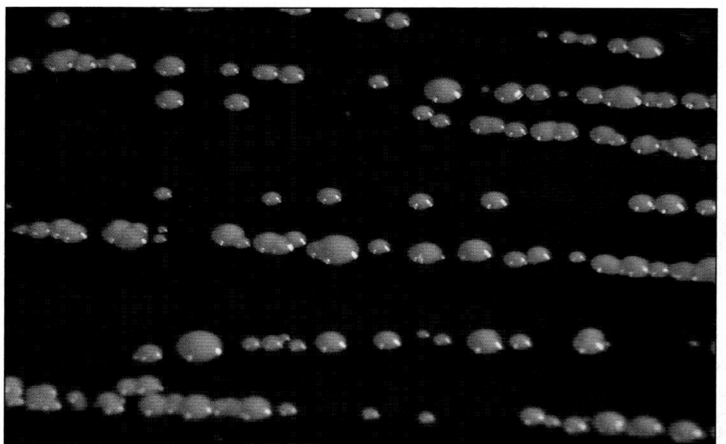

3.12 Entire or Smooth Margin • These circular and convex *Rhodococcus rhodochrous* colonies were grown on brain-heart infusion agar for 48 hours. They are about 1 mm in diameter with a smooth margin and a shiny, pink surface (*rhodo* means "rose colored"). *Rhodococcus* species are soil organisms.

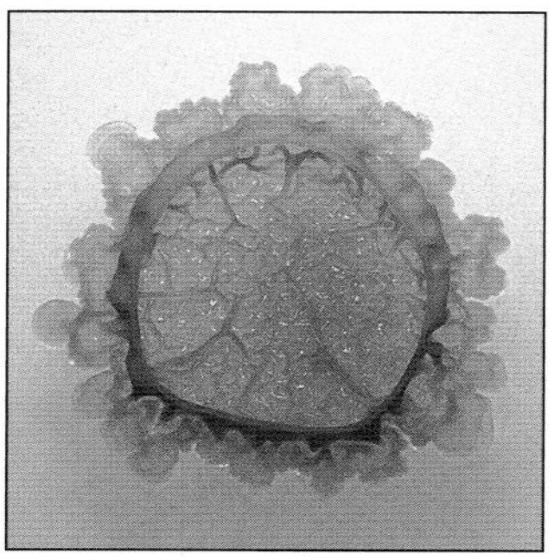

3.13 Lobed Margin • This lab contaminant colony has a lobed margin and a wrinkled surface. Notice that even though the margin is lobed, the overall colony shape is circular. It was approximately 1 cm in diameter after 48 hours of incubation at 35°C.

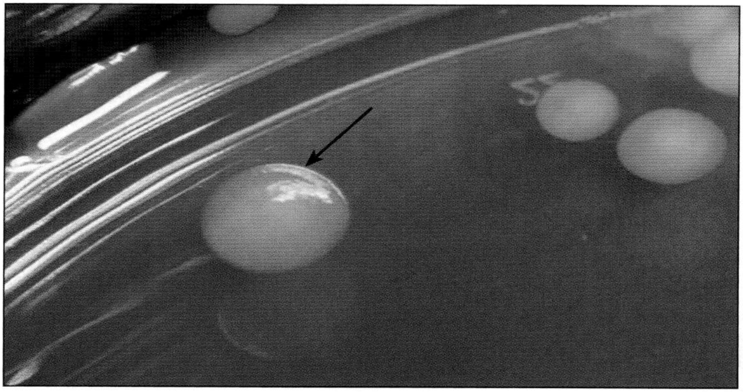

3.14 Spreading Margin • *Erwinia amylovora* colonies show a spreading, sometimes irregular (but not these), margin. Notice that the colony is umbonate and flattens toward the edge, and then gets thinner (**arrow**). These colonies were grown for 72 hours on tryptic soy agar at 25°C. *E. amylovora* is a plant pathogen. Its scientific name literally means "Erwin (Smith's) starch devourer."

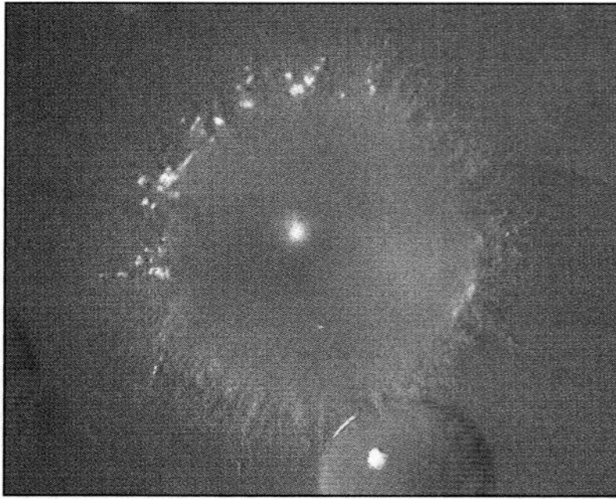

3.15 Filamentous Margin • Viewed with the stereo microscope, the filamentous margin of this unidentified lab contaminant is visible. The colony is also circular, convex, shiny, and about 2 mm in diameter.

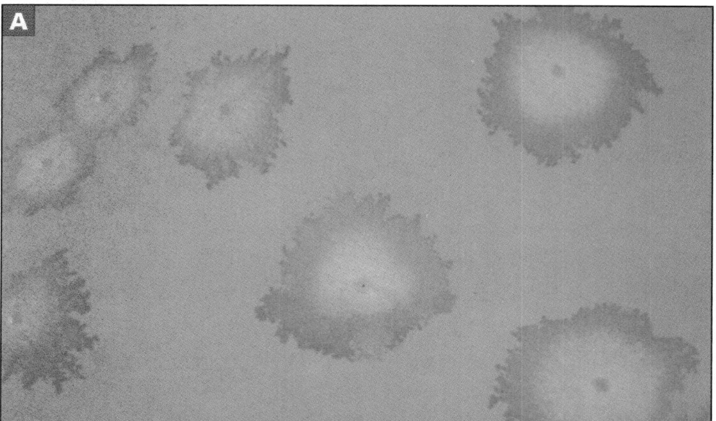

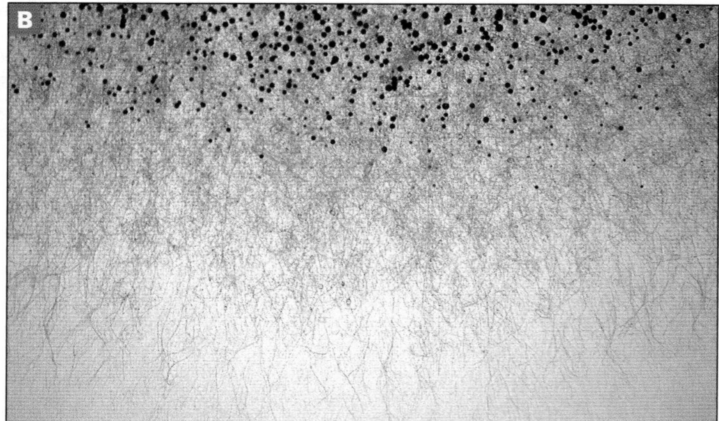

3.16 Rhizoid Margin • (**A**) These irregular colonies of *Clostridium sporogenes* were grown anaerobically on sheep blood agar and were photographed through a stereo microscope. They have a raised center and a flat, spreading edge of branched, tangled filaments. *C. sporogenes* colonies are sometimes referred to as Medusa head colonies because of their resemblance to the mythological creature Medusa, who had snakes for hair! (Colonies of other *Clostridium* species and *Bacillus anthracis* carry the same informal, but memorable, description.) They vary in size from 2 mm to 6 mm. *C. sporogenes* is found in soils worldwide. (**B**) Fungi (other than yeast) are naturally filamentous and sometimes the filaments are branched, as in this lab contaminant (probably *Aspergillus niger*). It takes magnification to see that the margin is rhizoid and not simply filamentous. The black spheres are asexual spores called **conidia**.

Colony Textures (Figures 3.17–3.20)

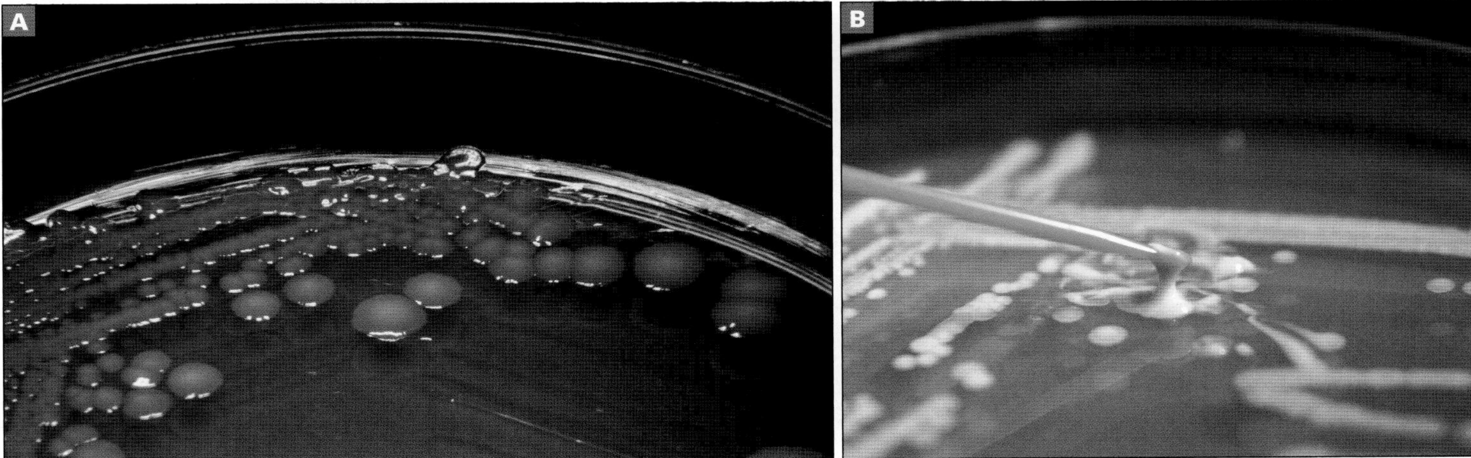

3.17 Mucoid Colonies ● (A) These *Klebsiella pneumoniae* colonies grown on nutrient agar are mucoid, raised, and shiny. While it is a normal inhabitant of the human intestinal tract, it is associated with community-acquired pneumonia and nosocomial urinary tract infections. (B) This *Rhizobium leguminosarum* growth has a thin, mucoid consistency. At times, the growth can become so mucoid that strings a couple of inches long can be formed when picking from a colony.

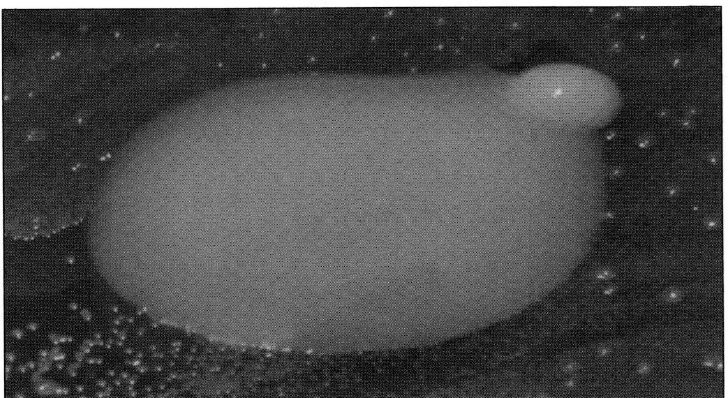

3.18 Butyrous Colony ● This unidentified 12 mm colony was found on a glycerol yeast extract plate inoculated with a diluted soil sample. Butyrous (*butyrum* means "buttery") colonies have the consistency of melted butter. This colony was almost liquid in texture, something that is demonstrated by its "flowing" contact with the yellow colony to its right.

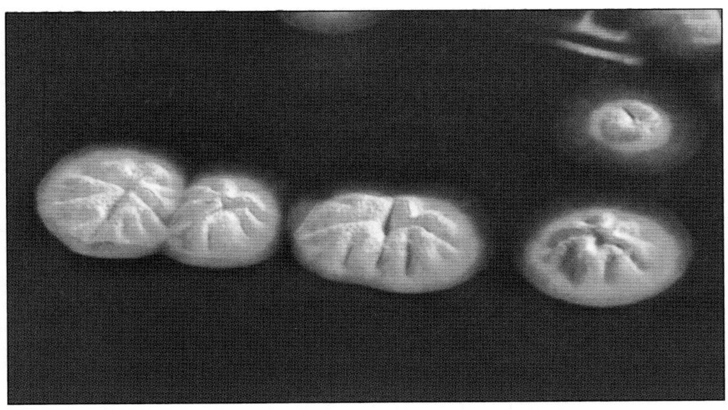

3.19 Granular Colonies ● These colonies of *Streptomyces griseus* grown on brain-heart infusion agar are circular, entire, and granular with a ridged surface. At a later stage of development, they produce yellow reproductive spores. Growth of streptomycetes is associated with an "earthy" smell. This one plate fragranced the entire incubator!

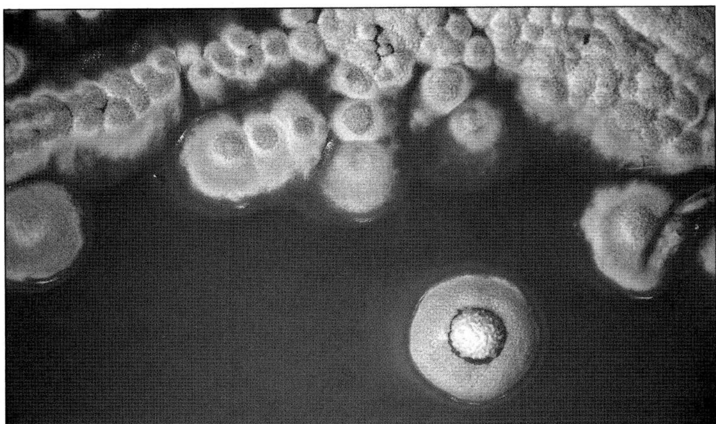

3.20 Dry Colonies ● Shown are colonies of *Streptomyces violaceus* photographed using a stereo microscope after 3 weeks of incubation at 25°C. They are umbonate, granular, and dry. Members of this genus share a growth pattern that resembles fungi (but they are Gram-positive bacteria and the similarities are only coincidental) and some fungal terminology has traditionally been used to describe them. For instance, their filaments are called hyphae and the colony is called a mycelium. They also grow aerial hyphae that produce chains of reproductive spores (different from bacterial endospores). The morphological variation seen in this photo is due to the colonies being in different stages of spore development, with the most advanced one being in the lower right. These colonies were about 4 mm in diameter and adhered tenaciously to the agar surface.

Optical Properties (Figure 3.21)

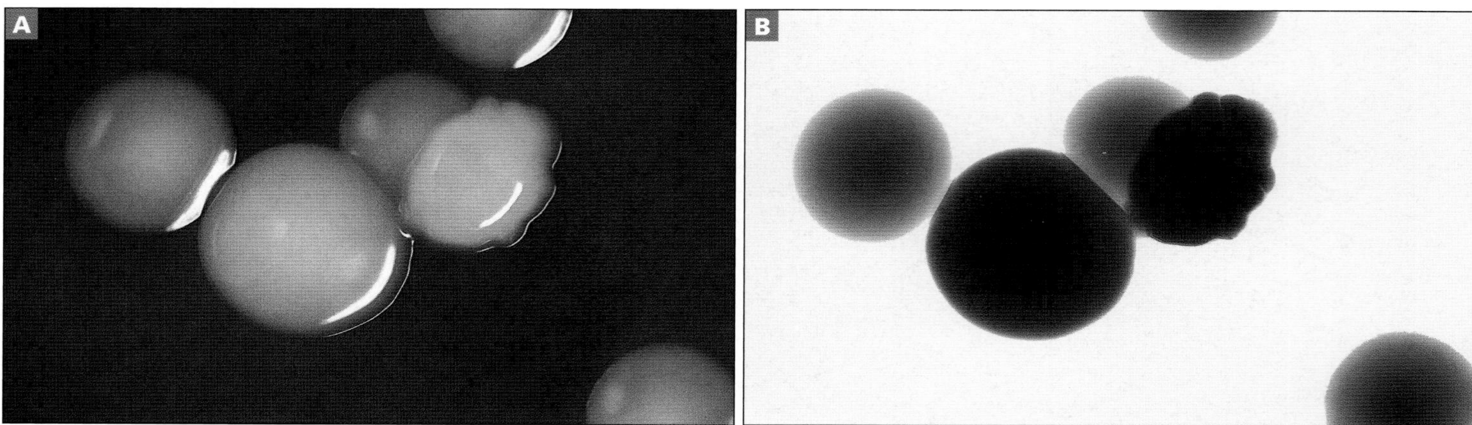

3.21 Opaque and Translucent Colonies ● (A) These colonies were photographed with a stereo microscope using reflected light. There are slight differences in size and color, and one is irregular in shape, whereas the others are circular. But there is another difference not visible with reflected light. (B) Here, the same colonies were photographed with transmitted light and it becomes obvious that the medium-sized round colonies are translucent, whereas the others are opaque.

Other Less Common, But Distinctive, Colony Features (Figures 3.22–3.23)

3.22 Swarming Growth ● Members of the genus *Proteus* will swarm at certain intervals and produce a pattern of concentric rings because of their motility. This photograph demonstrates the swarming behavior of *P. vulgaris* on DNase agar.

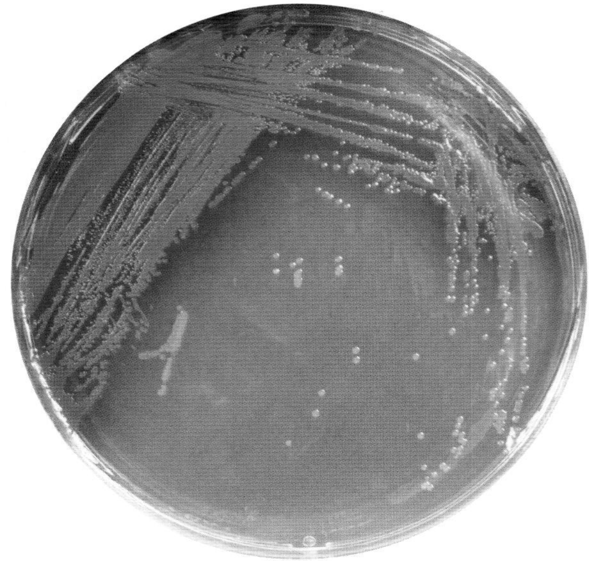

3.23 Diffusible Pigment ● The blue-green pigment **pyocyanin** diffusing from the growth is distinctive of *Pseudomonas aeruginosa*. Here *P. aeruginosa* is growing on tryptic soy agar.

Species Diversity (Figure 3.24)

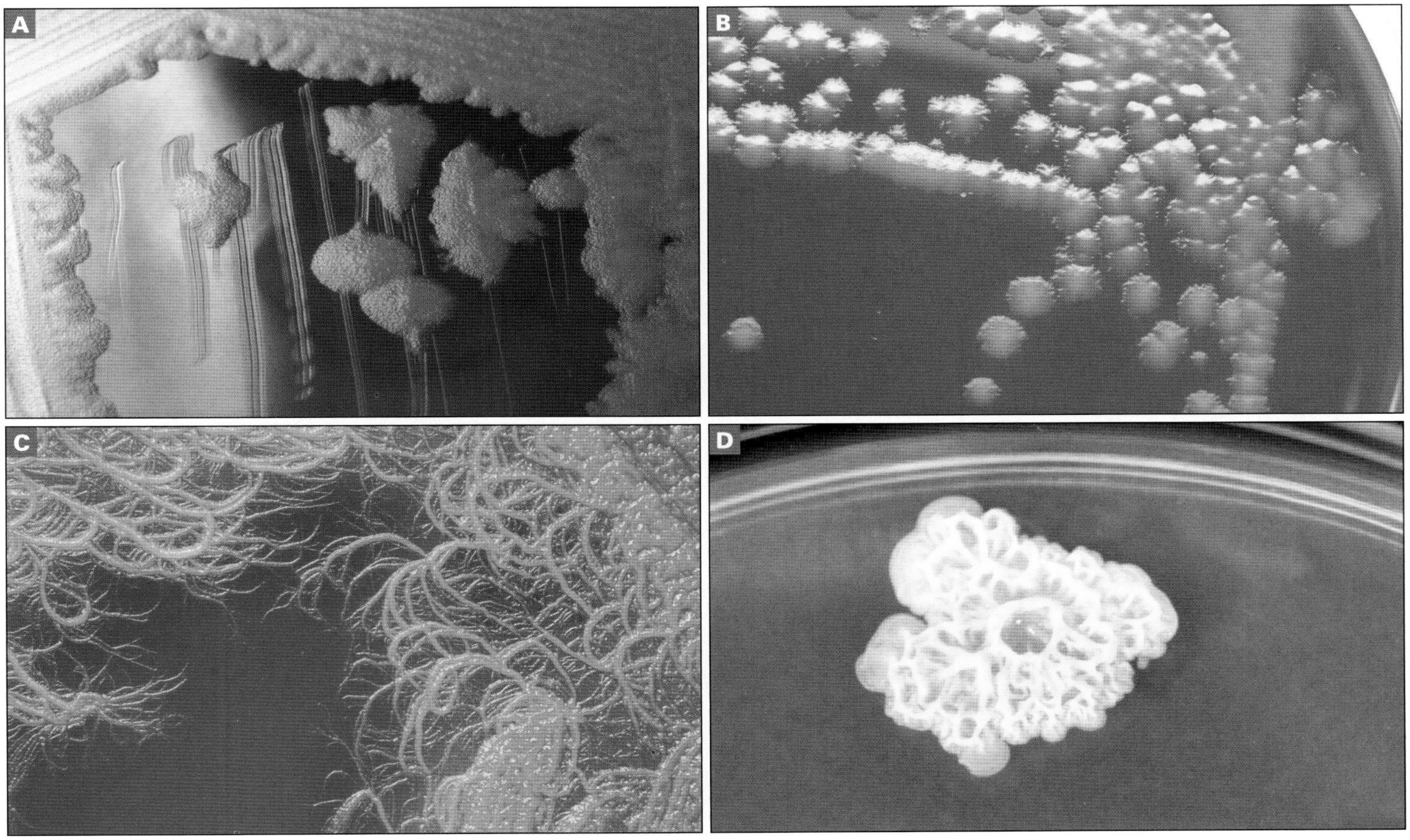

3.24 Comparison of Four *Bacillus* Species Colonies ● (A) *B. cereus* (see p. 185) grown on sheep blood agar (SBA) produces distinctively large (up to 7 mm), gray, granular, irregular colonies. They often produce a "mousy" smell. Also note the distinctive extensions of growth along the streak line. (B) *B. anthracis* (the causative agent of anthrax; see p. 184) colonies on SBA resemble *B. cereus*, but are usually smaller and adhere to the medium more tenaciously. It must be handled at *least* in a BSL-2 laboratory, and sometimes BSL-3 if the cell density is high enough, so it is unlikely you will be testing its tenacity. *B. cereus* and *B. anthracis* also produce Medusa head colonies, but the ones shown in (A) and (B) are either too young to do so or have lost that ability. (C) *B. mycoides* produces distinctive, rapidly spreading, rhizoid (note the branching) colonies. It is a common isolate from soil and is shown here on SBA. (D) This unknown *Bacillus* isolated as a laboratory contaminant produced a wrinkled, irregular colony with an irregular (wavy) margin on tryptic soy agar.

Extrinsic Factors Affecting Colony Morphology (Figures 3.25–3.26)

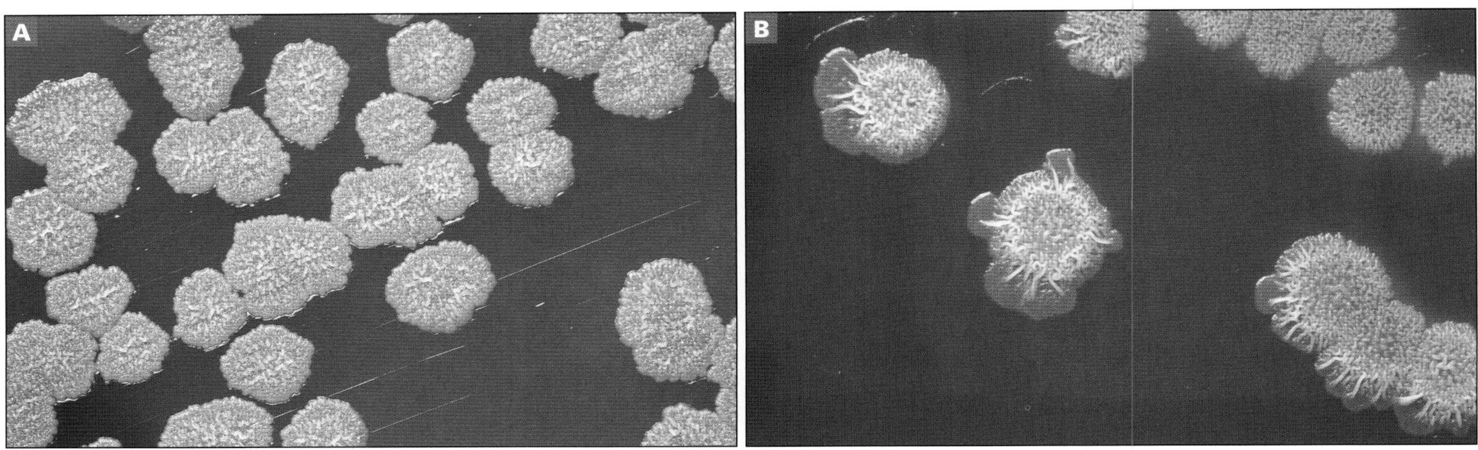

3.25 Effect of Incubation Time on Colony Morphology ● (A) Close-up of *Bacillus subtilis* on sheep blood agar after 24 hours of incubation. (B) Close-up of the same *B. subtilis* culture after 48 hours of growth. Note the wormlike extensions.

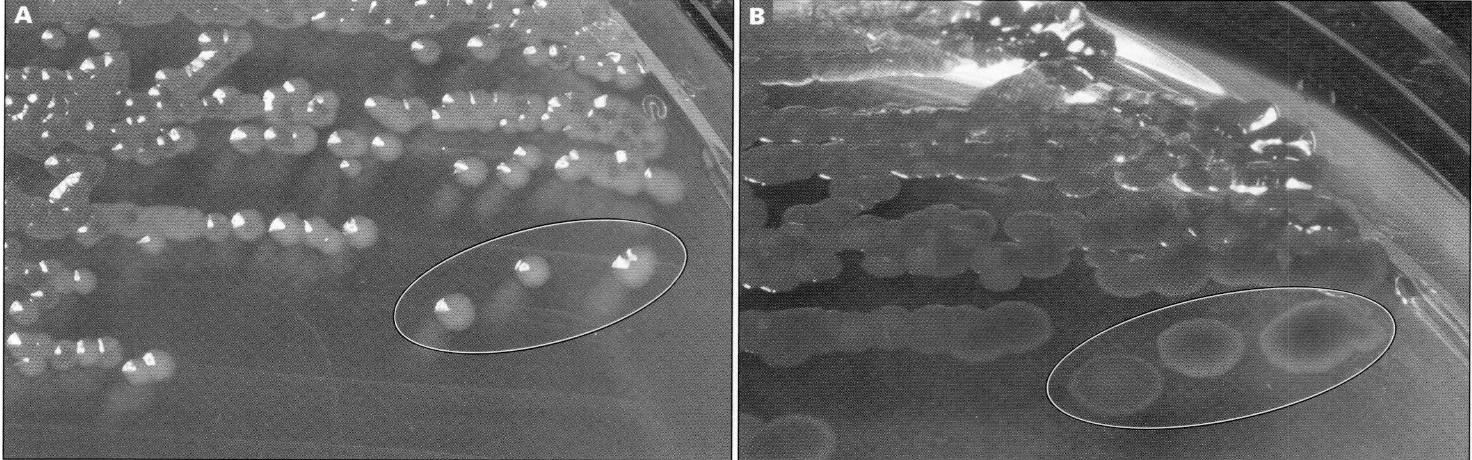

3.26 Effect of Incubation Time on Pigment Production (A) *Serratia marcescens* grown on nutrient agar after 24 hours. (B) The same plate of *S. marcescens* after 48 hours. Note in particular the change in the three colonies in the lower right (circled).

Agar Slants

Agar slants are useful primarily as media for cultivation and maintenance of stock cultures, not for identification. Organisms cultivated on slants, however, do display a variety of growth characteristics. Knowing the characteristic growth pattern of an organism on a slant can come in handy if that pattern looks unusual. That may be an indication of a contaminated pure culture!

Many organisms produce **filiform** growth (dense and opaque with a smooth edge). Growth on slants can vary in texture (e.g., moist or **friable** [dry, crusty]), optical properties (e.g., **opaque, translucent**), and margin (e.g., **smooth, lobed, spreading/effuse, filamentous, rhizoid** [branched], or **echinulate** [spiny]). Margins can be difficult to evaluate or even observe if the slant is not streaked in a straight line or if the organism covers the entire slant with its edge butting up against the glass. Figure 3.27 illustrates some of these features.

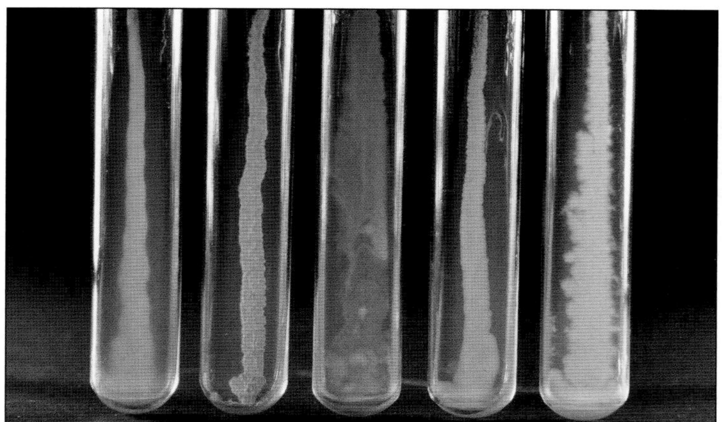

3.27 Growth Patterns on Slants From left to right: *Pseudomonas aeruginosa*, filiform (smooth, even growth) and translucent; *Streptomyces griseus*, friable and opaque; *Erwinia amylovora*, effuse and translucent; *Citrobacter koseri*, echinulate and opaque; and *Bacillus cereus*, rhizoid and opaque.

Color is also a variable feature of growth on slants. All the organisms in Figure 3.28 are naturally **pigmented** (an obvious difference) and all but one are filiform (*Pseudomonas aeruginosa* is effuse).

Growth patterns of virtually all microorganisms can be influenced by environmental factors. Two important factors influencing bacterial growth are incubation time and temperature. Figure 3.29 illustrates the effect of temperature on pigment production by *Serratia marcescens.*

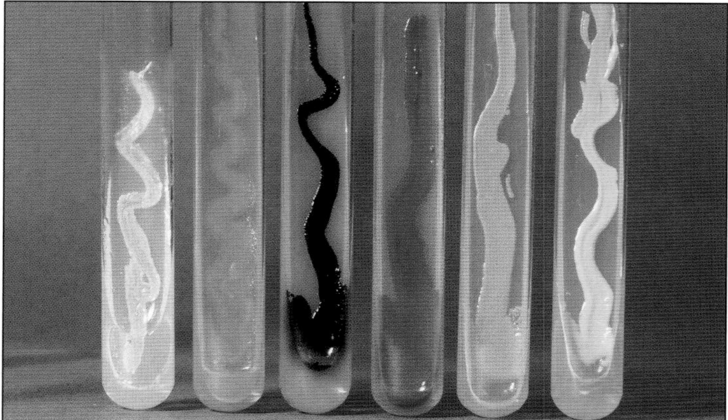

3.28 Pigment Production on Slants All of these organisms naturally produce a pigment. From left to right: *Staphylococcus epidermidis* (white), *Pseudomonas aeruginosa* (green), *Chromobacterium violaceum* (violet), *Serratia marcescens* (red/orange), *Kocuria rosea* (rose), *Micrococcus luteus* (yellow). All but *Pseudomonas*, which is effuse, have an entire (smooth) edge. The waviness you see is due to the inoculation pattern. Bacteria grow where you put them.

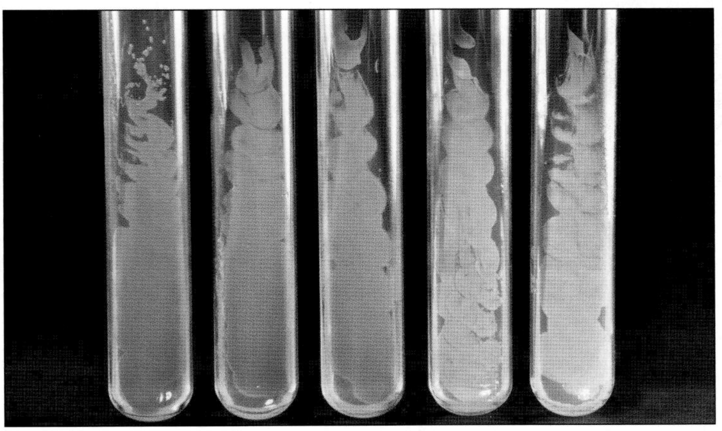

3.29 Influence of Temperature on Pigment Production in *Serratia marcescens* • *Serratia marcescens* was grown for 48 hours on tryptic soy agar slants at five different temperatures. From left to right: 25°C, 30°C, 33°C, 35°C, and 37°C. A difference of 2°C (between 33°C and 35°C) makes the difference between being pigmented or not.

Growth Patterns in Broth

Purpose

Bacterial genera, and frequently different species within a genus, demonstrate characteristic growth patterns in broth that provide useful information when attempting to identify an organism.

Principle

Microorganisms cultivated in broth display a variety of growth characteristics. Some organisms float on top of the medium and produce a type of surface membrane called a **pellicle.** Others sink to the bottom as **sediment.** Some bacteria produce **uniform fine turbidity,** and others appear to clump in what is called **flocculent** growth (like a snow globe). Others show pigmentation or form a ring at the surface. Refer to Figures 3.30 and 3.31. •

3.30 Growth Patterns in Trypticase Soy Broth (TSB) • From left to right: *Klebsiella* (*Enterobacter*) *aerogenes*, uniform fine turbidity—UFT; *Mycobacterium smegmatis*, flocculent growth—notice the flakes within the turbidity; *Bacillus subtilis*, pellicle—a film floating on surface. This pellicle's consistency is illustrated by the portions hanging down into the broth, an artifact of handling during the photo shoot, as is the sediment at the bottom, which fell from the pellicle at the surface; uninoculated TSB for comparison; *Chromobacterium violaceum*, ring at surface, with fine turbidity and a small sediment. Also note the purple coloration of the organism; *Rhizobium leguminosarum*, sediment.

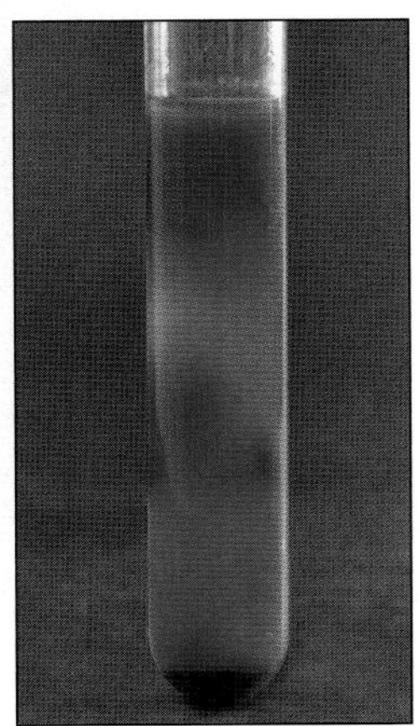

3.31 Pigmentation in Broth • *Rhodospirillum rubrum* has a red color due to carotenoid pigments. It grows as a photoheterotroph in the presence of light and the absence of oxygen.

Aerotolerance

Purpose

Agar deep stabs are a good visual indicator of oxygen tolerance (aerotolerance) in microorganisms.

Principle

Most microorganisms can survive within a range of environmental conditions but, not surprisingly, tend to produce growth with the greatest density in areas where conditions are most favorable. One important resource influencing microbial growth is oxygen. Some organisms require oxygen for their metabolic needs. Some other organisms are not affected by it at all. Still other organisms cannot even survive in its presence. This ability or inability to live in the presence of oxygen is called **aerotolerance.**

Most growth media are sterilized in an autoclave during preparation. This process not only kills unwanted microbes, but also removes most of the free oxygen from the medium as well. After the medium is removed from the autoclave and allowed to cool, oxygen begins to diffuse back in.

In tubed media (both liquid and solid) this process creates a gradient of oxygen concentrations, ranging from **aerobic** at the top, nearest the source of oxygen (air), to **anaerobic** at the bottom. Because of microorganisms' natural tendency to proliferate where the oxygen concentration best suits their metabolic needs, differing degrees of population density will develop in the medium over time that can be used to visually examine their aerotolerance.

Obligate (strict) aerobes, organisms that require oxygen for aerobic respiration, grow at the top where oxygen is most plentiful. Facultative anaerobes grow in the presence or absence of oxygen. When oxygen is available, they respire aerobically. When oxygen is not available, they either respire anaerobically (reducing sulfur or nitrate instead of oxygen) or ferment an available substrate. Refer to the Appendix and Section 7 for more information on anaerobic respiration and fermentation.

Where an oxygen gradient exists, facultative anaerobes grow throughout the medium but appear denser near the top. **Aerotolerant anaerobes,** organisms that don't require oxygen and are not adversely affected by it, grow uniformly throughout the medium. Aerotolerant anaerobes are fermentative even in the presence of free oxygen.

Microaerophiles, as the name suggests, survive only in environments containing lower than atmospheric levels of oxygen because atmospheric levels are toxic. Some microaerophiles, called **capnophiles,** can survive only if carbon dioxide levels are elevated. Microaerophiles will be seen somewhere near the upper middle region of the medium, usually in a fairly narrow band.

Finally, **obligate (strict) anaerobes** are organisms for which even small amounts of oxygen are lethal and, therefore, will be seen only in the lower regions of the medium, depending on how far into the medium the oxygen has diffused.

Agar deep stabs are prepared with tryptic soy agar (TSA) enriched with yeast extract to promote growth of a broad range of organisms. Oxygen, which is removed from the medium during preparation and autoclaving, immediately begins to diffuse back in as the agar cools and solidifies. This process creates a gradient of oxygen concentrations in the medium, ranging from aerobic at the top to anaerobic at the bottom. Agar deeps are stab-inoculated with an inoculating needle to introduce as little air as possible. The location of growth that develops indicates the organism's aerotolerance (Fig. 3.32).

3.32 Agar Deep Stab Tubes Shown are three agar deep tubes inoculated with different organisms. If growth is only seen in the upper couple of centimeters, the organism is an obligate aerobe (*Alcaligenes faecalis*, left). If the organism grows throughout the depth of the stab and better near the surface, it is a facultative anaerobe (*Escherichia coli*, middle tube). Lastly, if the organism only grows in the deeper parts of the agar, it is an obligate anaerobe (*Clostridium butyricum*, right tube).

Cultivation of Anaerobes—Anaerobic Jar

Purpose

Cultivation of obligate anaerobes and microaerophiles requires providing an environment in which oxygen is either absent or considerably reduced, respectively. Various methods have been devised to provide these environments, three of which are covered in the remainder of this section.

The anaerobic jar (Fig. 3.33) is used to grow obligate anaerobes and microaerophiles. Because it is the atmosphere within the jar that is anaerobic, the jar can be placed in a normal incubator alongside aerobically grown cultures.

Principle

Inoculated plates or tubes are placed in the jar and the appropriate gas generating sachet is activated. In the case of the AnaeroGen Gas Generating System by Oxoid, simply opening the packet inside the jar and immediately clamping on the lid is all that is necessary. Ascorbic acid in the packet reacts with free oxygen to make water and in turn releases CO_2 (Fig. 3.34). Within 30 minutes, the atmosphere inside the jar is less than 1% O_2 and between 9 and 13% CO_2.

A methylene blue (or some other) indicator strip is also placed inside the jar. It will turn blue if exposed to air, thus acting as a control to ensure anaerobic conditions have been produced. Figure 3.35 shows two plates inoculated with the same organisms, but one was incubated anaerobically while the other was incubated aerobically.

The Oxoid Campygen sachet works in a similar way, but produces 5% O_2, 10% CO_2, and 85% N_2. It is designed for growing microaerophiles, such as *Campylobacter jejuni*.

$$2\ \text{Ascorbic Acid} + O_2 \longrightarrow 2\ \text{Dehydroascorbic Acid} + 2H_2O$$

3.34 Ascorbic Acid Reaction with Oxygen • Ascorbic acid is the active ingredient within the gas generator packet. When exposed to air in the sealed anaerobic jar, ascorbic acid reacts with oxygen to produce dehydroascorbic acid and water (seen as condensation in Fig. 3.33B), thus making the atmosphere within the jar anaerobic.

3.33 An Anaerobic Jar • (**A**) This anaerobic jar is large enough to hold 12 plates and includes a plate rack, which makes placing them in the jar and retrieving them much easier than if done by hand. A gas generator packet (shown in **B**) that removes oxygen will be activated (see Fig. 3.34), placed inside the jar, and the lid will be clamped securely to make the jar airtight. (**B**) This is the same jar after incubation. The white envelope is the gas generator packet and the small white strip at the top is an oxygen indicator that remains white if oxygen has been removed and changes color if oxygen remains. A less sophisticated indication that oxygen was removed is the condensation on the jar's wall. It should be noted that the "anaerobic" jar is only anaerobic if an anaerobic packet is used. Microaerophiles can be grown in the same jar using a packet that produces an atmosphere suitable for their growth.

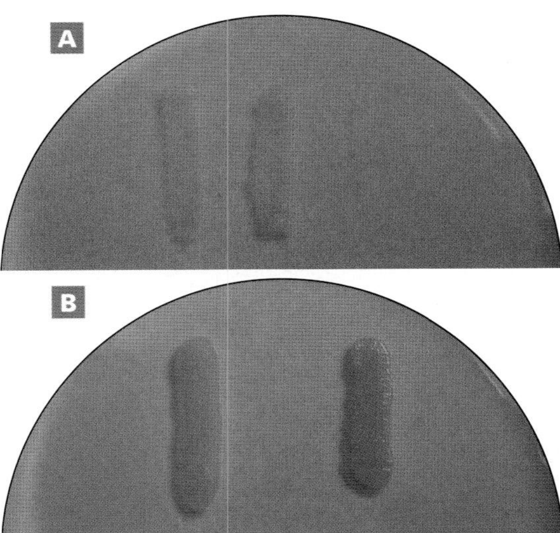

3.35 Plates Incubated Inside and Outside the Anaerobic Jar • Both trypticase agar plates were inoculated with *Staphylococcus aureus* (facultative anaerobe, left), *Alcaligenes faecalis* (obligate aerobe, center), and *Clostridium butyricum* (obligate anaerobe, right). Plate (**A**) was incubated aerobically and plate (**B**) was incubated anaerobically inside an anerobic jar. Compare the relative amounts of growth of the three organisms.

Cultivation of Anaerobes—Fluid Thioglycollate Broth (FTB)

Purpose

Fluid thioglycollate broth (FTB) is a liquid medium designed to grow a wide variety of fastidious microorganisms. It can be used to grow microbes representing all levels of oxygen tolerance; however, it generally is associated with the cultivation of anaerobic and microaerophilic bacteria.

Principle

Fluid thioglycollate broth is prepared as a basic medium or with a variety of supplements, depending on the specific needs of organisms being cultivated. As such, it is appropriate for a broad variety of aerobic and anaerobic, fastidious and non-fastidious organisms. It is particularly well suited for cultivating obligate anaerobes and microaerophiles.

Key components of the medium are yeast extract, pancreatic digest of casein, dextrose, sodium thioglycollate, L-cysteine, and resazurin. Yeast extract and pancreatic digest of casein provide nutrients; sodium thioglycollate and L-cysteine reduce oxygen to water; and resazurin acts as an indicator (pink when oxidized, colorless when reduced). A small amount of agar is included to slow oxygen diffusion.

Oxygen removed during autoclaving will diffuse back into the medium as the tubes cool to room temperature. This produces a gradient of concentrations from fully aerobic at the top to anaerobic at the bottom. Thus, fresh media will appear clear to straw colored with a pink region at the top where the dye has become oxidized (Fig. 3.36). Figure 3.37 demonstrates some basic bacterial growth patterns in the medium as influenced by the oxygen gradient.

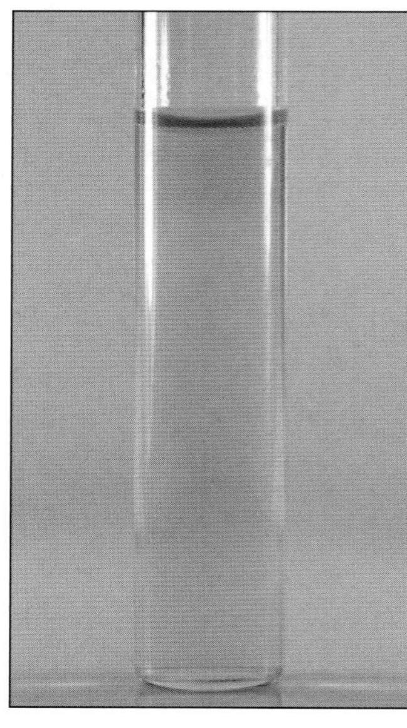

3.36 Aerobic and Anaerobic Zones in Thioglycollate Broth Note the pink region in the upper, aerobic portion of the broth resulting from oxidation of the indicator resazurin. In the lower, anaerobic portion, the dye has been reduced and is colorless, leaving the medium its typical straw color.

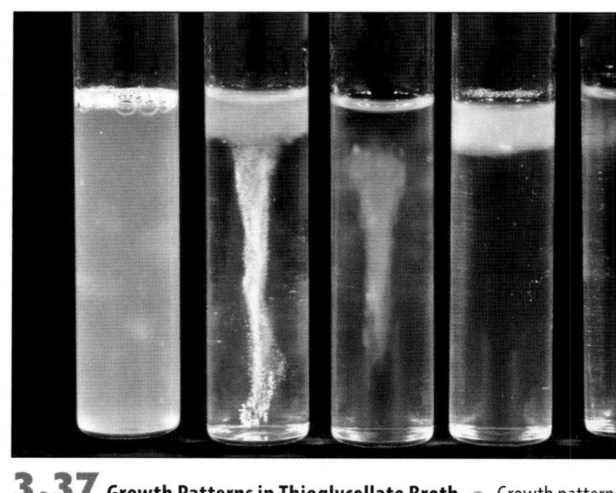

3.37 Growth Patterns in Thioglycollate Broth Growth patterns of a variety of organisms are shown in these fluid thioglycollate broths. Pictured from left to right are: aerotolerant anaerobe, facultative anaerobe, obligate (strict) anaerobe, obligate (strict) aerobe, and microaerophile. Compare these tubes with the uninoculated broth in Figure 3.36.

Cultivation of Anaerobes—Cooked Meat Broth

Purpose

The purpose of cooked meat broth is to grow anaerobes, especially pathogenic clostridia such as *Clostridium perfringens*, *C. tetani*, *C. botulinum*, and *C. difficile* (see pp. 192–195). The medium is also differential and can distinguish between proteolytic and saccharolytic clostridia.

Principle

Cooked meat broth (Fig. 3.38) is a nutrient rich medium, with beef heart, peptone, and dextrose acting as carbon, nitrogen, and energy sources. The beef heart is in the form of meat particles, whereas the other ingredients are dissolved in the liquid. Anaerobic conditions occur as a result of several factors.

One, cardiac muscle contains **glutathione**, a tripeptide that can reduce free molecular oxygen in the medium. Two, the meat is cooked prior to use. This denatures proteins and exposes their sulfhydryl groups, which perform the same function—oxygen reduction. Lastly, the medium with caps loosened is either incubated in an anaerobic jar for 24 hours to remove O_2 or boiled to drive off the O_2.

Caps are immediately tightened to prevent the re-entry of O_2. Because it is the medium that becomes anaerobic, these tubes can be incubated in an aerobic incubator, thus eliminating the need for expensive equipment. Blackening and disintegration of the meat particles indicate proteolytic growth. Acid (not indicated directly) and gas production indicates saccharolytic growth.

3.38 Cooked Meat Broth • The meat particles are visible in each broth. From left to right: *Clostridium butyricum*, uninoculated, *C. sporogenes*. Blackening of the meat particles by *C. sporogenes* is indicative of proteolytic activity. *C. butyricum* grew, but is not proteolytic.

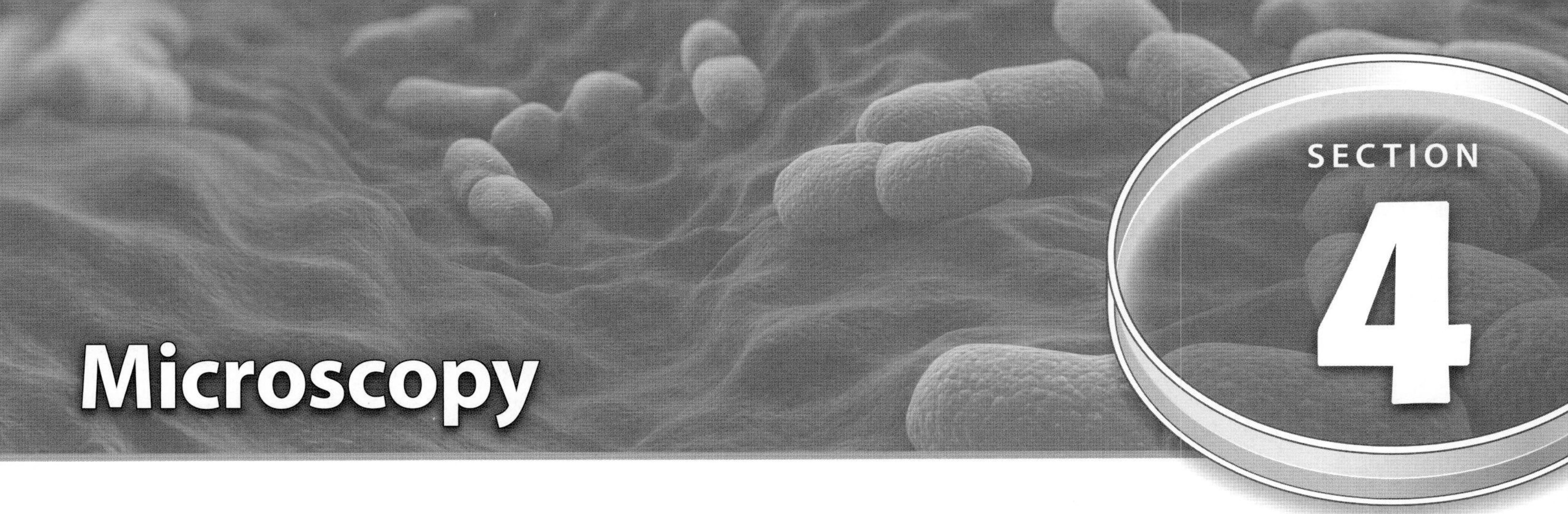

Microscopy

Types of Microscopy

The earliest microscopes used visible light to create images and were little more than magnifying glasses. Today, more sophisticated compound light microscopes (Fig. 4.1) are routinely used in microbiology laboratories. The various types of light microscopy include bright-field, dark-field, fluorescence, and phase contrast microscopy (Fig. 4.2).

Although each method has specific applications and advantages, bright-field microscopy is most commonly used in introductory classes and clinical laboratories. Many research applications use electron microscopy because of its ability to produce higher quality images of greater magnification.

Light Microscopes

Bright-field microscopy produces an image made from light that is transmitted through a specimen (Fig. 4.2A). The specimen restricts light transmission and appears "shadowy" against a bright background (where light enters the microscope unimpeded). Because most biological specimens are transparent, the contrast between the specimen and the background can be improved with the application of stains to the specimen (see Sections 5 and 6). The "price" of improved contrast is that the staining process usually kills cells. This is especially true of bacterial staining protocols.

Image formation begins with light coming from an internal or an external light source (Fig. 4.3). It passes through the **condenser** lens, which concentrates the light

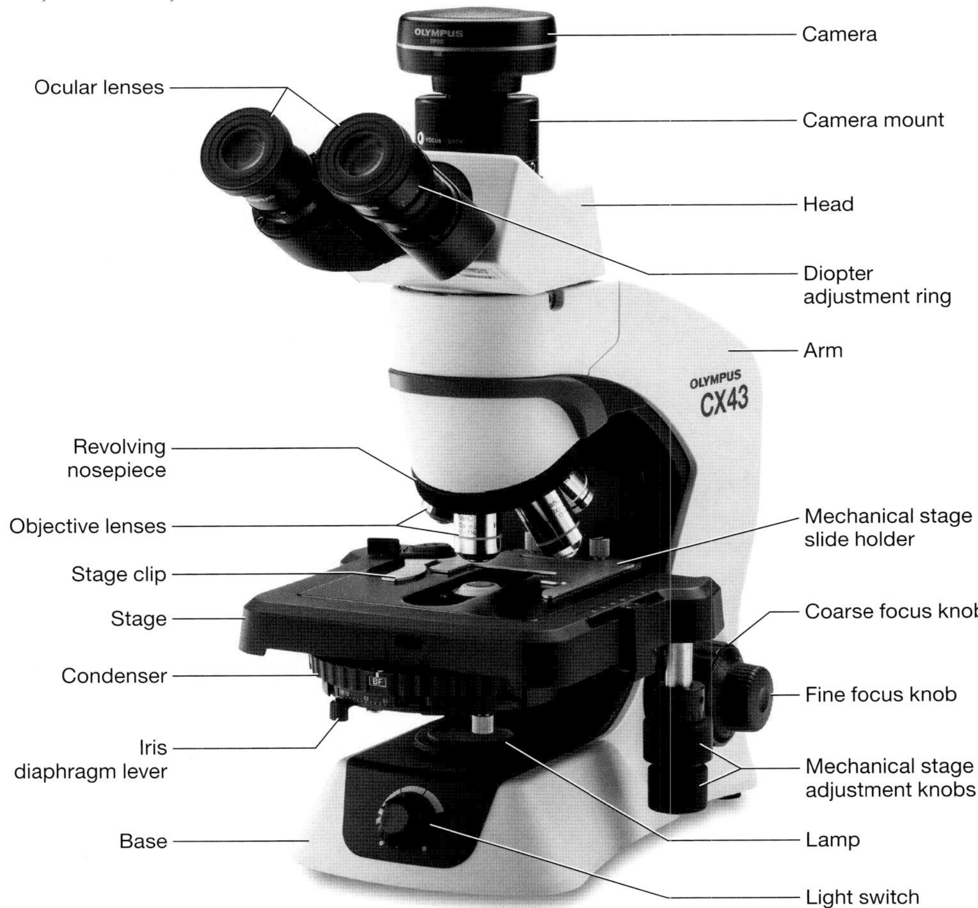

4.1 A Binocular Compound Microscope ● A quality microscope is an essential tool for microbiologists. Most are assembled with exchangeable component parts and can be customized to suit the particular needs of the user. This microscope has a trinocular head, with two oculars and a third to accommodate a camera attachment for photomicrography.

and makes illumination of the specimen more uniform. **Refraction** (bending) of light as it passes through the **objective lens** from the specimen produces a magnified, but inverted, **real image.** This image is magnified again as it passes through the **ocular lens** to produce a **virtual image** that appears below or within the microscope and remains inverted.

The amount of magnification that each lens produces is marked on the lens (Figs. 4.4A through 4.4C). Total magnification of the specimen can be calculated by using the following formula:

$$\text{Total Magnification} = \text{Magnification of the Objective Lens} \times \text{Magnification of the Ocular Lens}$$

The practical limit to magnification with a light microscope is around 1300×. Although higher magnifications are possible, image clarity is more difficult to maintain as the magnification increases. Clarity of an image is called **resolution** (Fig. 4.5). The **limit of resolution** (or **resolving power**) is an actual measurement of how far apart two points must be in order for the microscope to view them as being separate. Notice that resolution improves as the limit of resolution is made smaller.

The best limit of resolution achieved by a light microscope is about 0.2 μm. (That is, at its absolute best, a light microscope cannot distinguish between two points closer together than 0.2 μm.) For a specific microscope, the actual limit of resolution can be calculated with the following formula[1]:

$$D = \frac{\lambda}{NA_{Condenser} + NA_{Objective}}$$

where D is the minimum distance at which two points can be resolved, λ is the wavelength of light used, and $NA_{condenser}$ and $NA_{objective}$ are the numerical apertures of the condenser lens and objective lens, respectively. Because numerical aperture has no units, the units for D are the same as the units for wavelength, which typically are in nanometers (nm).

Numerical aperture is a measure of a lens's ability to "capture" light coming from the specimen and use it to make the image. As with magnification, it is marked on the lens (Figs. 4.4A, 4.4B, and 4.4D). A number of other lens features are also marked on the objective lens and are described in the caption to Fig. 4.4 and Table 4-1.

Using immersion oil between the specimen and the oil objective lens increases its numerical aperture and in turn, makes its limit of resolution smaller. (If desired, oil may also be placed between the condenser lens and the slide.) The result is better resolution.

[1] Different equations have been developed to determine approximate limit of resolution, each made from different assumptions. The one used here assumes the $NA_{Objective} \geq NA_{Condenser}$.

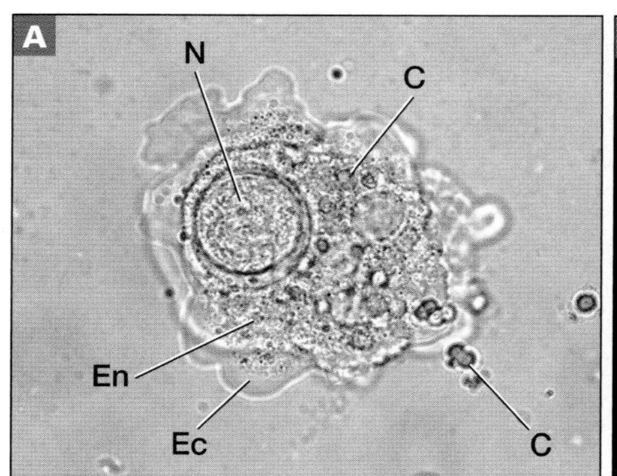

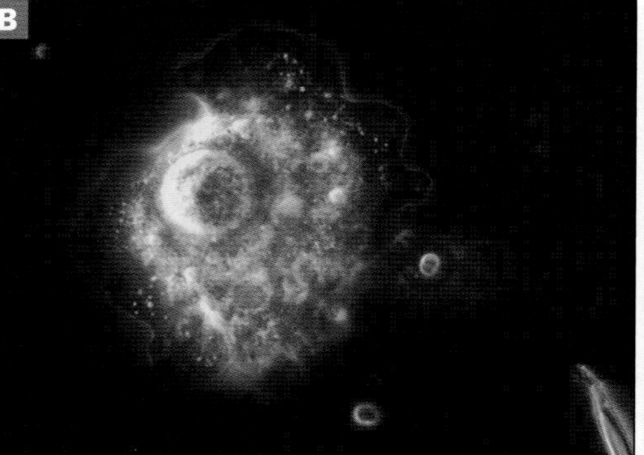

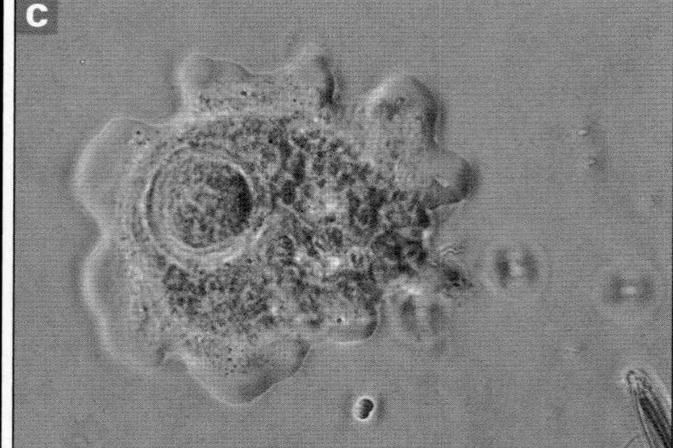

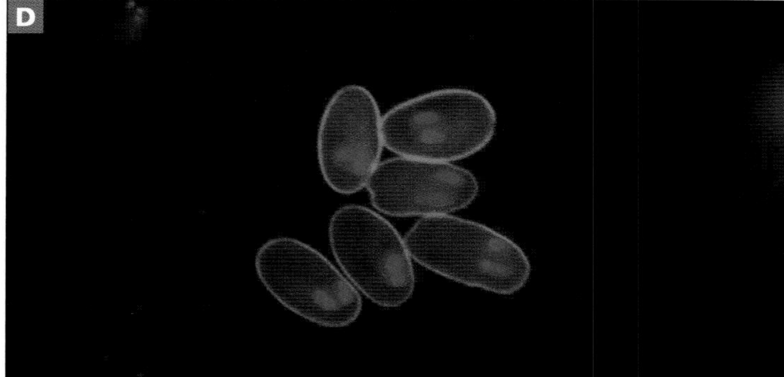

4.2 Types of Light Microscopy • (A) This is a bright-field micrograph of an amoeba (called a **whole mount**). Because of its thickness, the entire organism will not be in focus at once. Continually adjusting the fine focus to clearly observe different levels of the organism will give a sense of its three-dimensional structure. The nucleus (**N**) is obvious, as are the numerous cyanobacteria (**C**), both inside and outside the amoeba. Also visible are the granular endoplasm (**En**) around the nucleus and the ectoplasm (**Ec**) immediately below the cytoplasmic membrane. (B) This is a dark-field micrograph of the same amoeba. Notice the more three-dimensional image and that the ectoplasm is barely visible. (C) This is a phase contrast image of the same amoeba. Different parts of the interior and its detail are visible compared to what is seen in the other two micrographs. (D) This is a fluorescence micrograph of *Giardia lamblia* cysts (see p. 256) stained using two fluorescent dyes. The apple-green color is due to fluorescein isothiocyanate (FITC), which is staining the cyst wall in these cells. The blue dye is 4′,6-diamidino-2-phenylindole (DAPI), which stains DNA.

The light microscope may be modified to improve its ability to produce images with contrast without staining, which often distorts or kills the specimen. Two examples are dark-field and phase contrast microscopy.

In **dark-field microscopy** (Fig. 4.2B), a special condenser is used so only light reflected off the specimen enters the objective. The appearance is of a brightly lit specimen against a dark background, and often with better resolution than that of the bright-field microscope.

Phase contrast microscopy (Fig. 4.2C) uses special optical components to exploit subtle differences in the refractive indices of water and cytoplasmic components to produce contrast. Light waves that are in phase (that is, their peaks and valleys exactly coincide) reinforce one another and their total intensity (because of the summed amplitudes) is high. Light waves that are out of phase by exactly one-half wavelength cancel each other and result in no intensity—that is, darkness.

Wavelengths that are out of phase by any amount will produce some degree of cancellation and result in brightness that is less than maximum but more than darkness. Thus, contrast is provided by differences in light intensity that result from differences in refractive indices in parts of the specimen that put light waves more or less out of phase. As a result, the specimen appears as various levels of "darks" against a bright background.

Fluorescence microscopy (Fig. 4.2D) uses a fluorescent dye that emits fluorescence when illuminated with ultraviolet radiation. In some cases, specimens possess naturally fluorescing chemicals and no dye is needed. Fluorescent dyes can be attached (conjugated) to antibodies that react with specific cellular chemicals, thus providing the ability to specifically stain parts of cells. Using different colored fluorescent dyes attached to different antibodies results in differential staining of the specimen, see page 140.

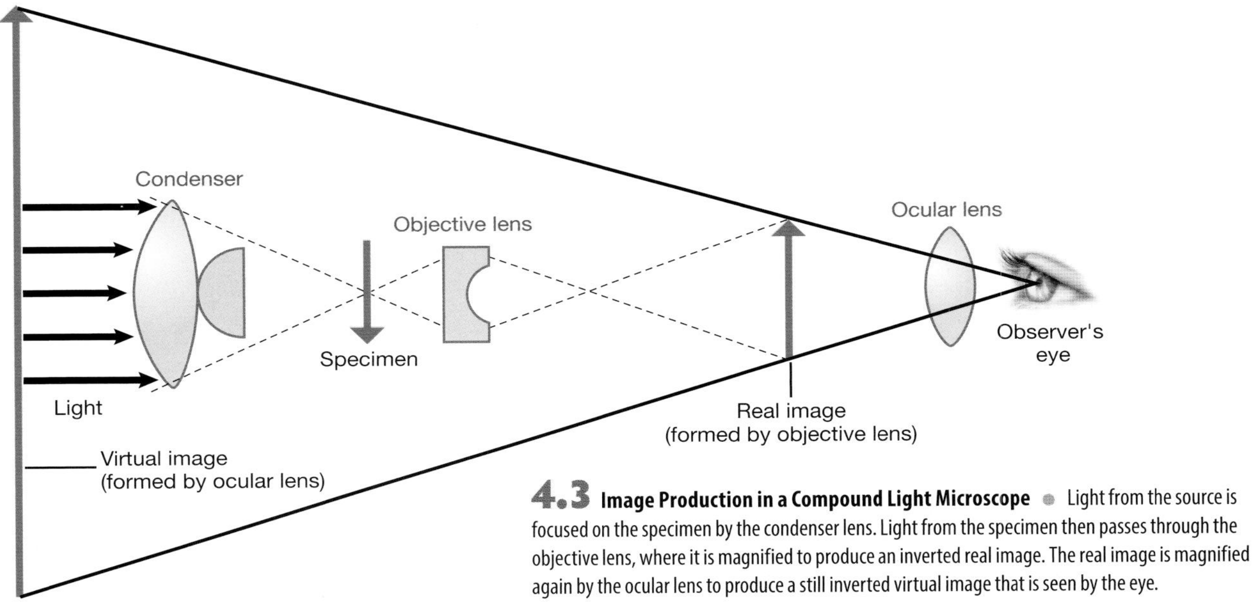

4.3 Image Production in a Compound Light Microscope • Light from the source is focused on the specimen by the condenser lens. Light from the specimen then passes through the objective lens, where it is magnified to produce an inverted real image. The real image is magnified again by the ocular lens to produce a still inverted virtual image that is seen by the eye.

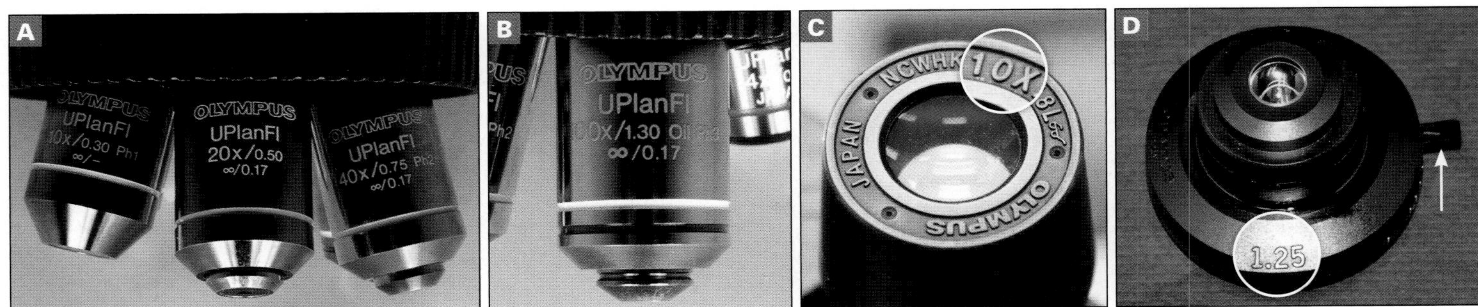

4.4 Markings of Magnification and Numerical Aperture on Microscope Components • Refer to Table 4-1 for an explanation of markings on these microscope components. (**A**) Three plan semi-apochromatic objective lenses are shown. From left to right, the lenses magnify 10×, 20×, and 40×, and have numerical apertures of 0.30, 0.50, and 0.75. Also notice the standard colored rings for each objective: yellow for 10×, green for 20×, and light blue for 40×. (**B**) This is a 100× oil-immersion lens, indicated by the black ring (below the white ring, which is used to indicate a 100× or greater lens). It is the only lens constructed in such a way as not to be damaged by oil and, as such, is the only one with which oil is to be used. The numerical aperture is 1.30. (**C**) A 10× ocular lens. (**D**) A condenser (removed from the microscope) with a numerical aperture of 1.25. The lever at the right (**arrow**) is used to open and close the iris diaphragm and adjust the amount of light entering the specimen.

TABLE **4-1** **Microscope Lens Markings** ● Specific values are given for the microscope shown in Figure 4.4, as are some common alternatives. Your microscope may have different values, but you should be able to interpret them with these for examples.

Item		
Objective Lens	**Numerical Aperture**	**Ring Color**
4× (not shown in Fig. 4.4)	0.13	Red
10×	0.30	Yellow
20× (or 16×)	0.50	Green
40× (or 50×)	0.75	Light Blue
100×	1.30	White (100×) Black (oil immersion)
Other Markings		
Plan	Plan means the lens produces a flat field of view rather than an image in focus at the center but blurring closer to the periphery	
Apo, Fl, or Achro (only Fl is shown in Fig. 4.4)	These indicate the degree of correction for chromatic aberration; **Apo**: An apochromatic lens corrects for red, green, and blue wavelengths; **Fl**: The lens is made of fluorite and makes the lens semiapochromatic, which corrects for chromatic aberration better than an achromatic lens, but not as well as an apochromatic lens; **Achro**: The lens corrects for red and blue wavelengths	
Oil	This indicates the immersion medium to be used; an immersion medium should only be used with a lens intended for immersion; dry lenses are not sealed and the immersion medium can seep into the objective lens and ruin it; glycerol (Gly) is another immersion medium	
Ph1, Ph2, Ph3	These lenses are designed for phase contrast microscopy; the number indicates which phase condenser each is paired with	
Mechanical tube length (∞)	The distance between the nosepiece attachment of each objective and the eyepiece is the mechanical tube length; the infinity symbol indicates that these objectives have been corrected to have an "optically infinite" mechanical tube length; the optics behind this are beyond the scope of this text; however, this allows for insertion of elements into the light path without affecting the focal point of the real image	
Cover glass thickness (0.17)	Lenses are designed to work with a cover glass (slip) of a certain thickness; most lenses and cover glasses are now standardized to 0.17 mm, but variation in a batch of cover slips can occur; to accommodate this, some objectives (not these) may have a correction collar to compensate for variations in thickness	

4.5 Resolution and Limit of Resolution ● The headlights of most automobiles are around 1.5 m apart. As you look at the cars in the foreground of the photo, it is easy to see both headlights as separate objects. The automobiles in the distance appear smaller (but really aren't) as does the apparent distance between the headlights. When the apparent distance between automobile headlights reaches about 0.1 mm, they blur into one because that is the limit of resolution of the human eye.

The Electron Microscope

The **electron microscope** uses an electron beam to create an image, with electromagnets acting as lenses. The limit of resolution is improved by a factor of 1,000 over the light microscope (theoretically down to 0.1 nm, but more realistically down to 2 nm).

The **transmission electron microscope (TEM)** (Fig. 4.6) produces a two-dimensional image of an ultrathin section by capturing electrons that have passed through the specimen. The degree of interaction between the electrons and the heavy metal stain affects the kinetic energy of the electrons, which are collected by a fluorescent plate. The light of varying intensity emitted from the plate is directly proportional to the electron's kinetic energy and is used to produce the image.

The TEM is useful for studying a cell's interior, its **ultrastructure.** A sample transmission electron micrograph is shown in Figure 4.7. Alternative preparation methods, such as freeze-fracture and negative staining, allow production of three-dimensional images by the TEM.

The previous paragraph gave a brief overview of how the TEM works. However, a key to successful transmission electron microscopy is excellent sample preparation. The specimen is fixed by one of various methods (treatment with formaldehyde, glutaraldehyde, or osmium tetroxide) to prevent cell decomposition, stained with an electron dense material (lead citrate, uranyl acetate, or osmium compounds), dehydrated, and embedded in a plastic block (Fig. 4.8).

It is then cut into thin slices using an ultramicrotome (Fig. 4.9A) armed with a glass or diamond blade (Fig. 4.9B).

The slices are captured on a grid (Fig. 4.10A), which is then placed on a specimen rod (Fig. 4.10B), and inserted into the TEM (Fig. 4.10C) so it rests in the electron beam path. Figure 4.11 shows what the microscopist sees when working. This image is also usually displayed on a computer monitor where digital micrographs can be captured.

A **scanning electron microscope (SEM)** (Fig. 4.12) is used to make a three-dimensional image of the specimen's surface. In this technique, a beam of electrons is passed over the stained surface of the specimen. Some electrons are reflected (**backscatter electrons**), whereas other electrons (**secondary electrons**) are emitted from the metallic stain. These electrons are captured and used to produce the three-dimensional image. A sample scanning electron micrograph is shown in Figure 4.13.

As with the TEM, sample preparation for SEM involves fixation, dehydration, and staining (but not sectioning). Once the sample is fixed and dehydrated, it is mounted on a stub (Fig. 4.14) and coated with the "stain" (usually gold) by a process known as "sputter coating." Argon gas is ionized in an electric field within an evacuated chamber. The positively charged argon ions bombard a gold foil, which releases gold atoms that are free to coat the sample. Figure 4.15 shows a sputter coater.

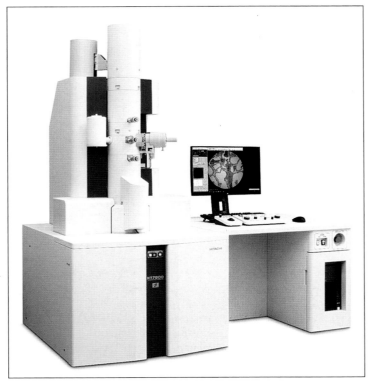

4.6 Transmission Electron Microscope • Shown is a Hitachi HT7800 series Transmission Electron Microscope. Using a standard objective lens mode, magnifications range from ×50 to ×600,000 with a 0.20 nm limit of resolution. Switched to a high-resolution objective lens mode, magnifications range from ×100 to ×1,000,000 with a 0.19 nm limit of resolution. This model differs from traditional TEMs in that there is no fluorescent screen image produced in a viewing chamber (see Fig. 4.11). Rather, a CCD camera captures the image and the digital image produced is viewed directly on the monitor.

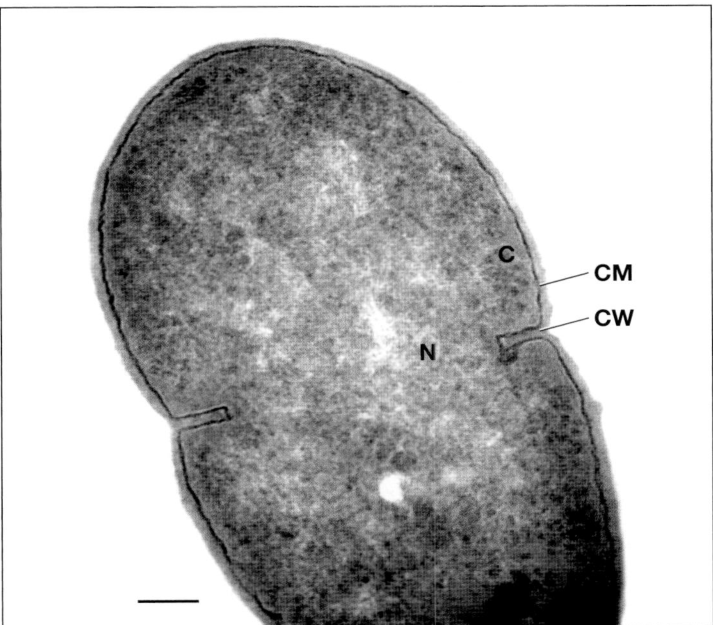

4.7 Transmission Electron Micrograph of *Lactococcus lactis* • Shown is a thin section of a dividing bacterium, *Lactococcus lactis*. *L. lactis* cells are spherical to slightly elongated in shape and are usually singles or pairs. This cell is in the process of dividing by binary fission. Visible in this micrograph are the cell wall (**CW**), cytoplasmic membrane (**CM**), cytoplasm (**C**), and nucleoid (**N**), which is where the cell's DNA is located. Notice the absence of a membrane separating the nucleoid from the cytoplasm. The dark dots in the cytoplasm are ribosomes. (Scale bar = 100 nm)

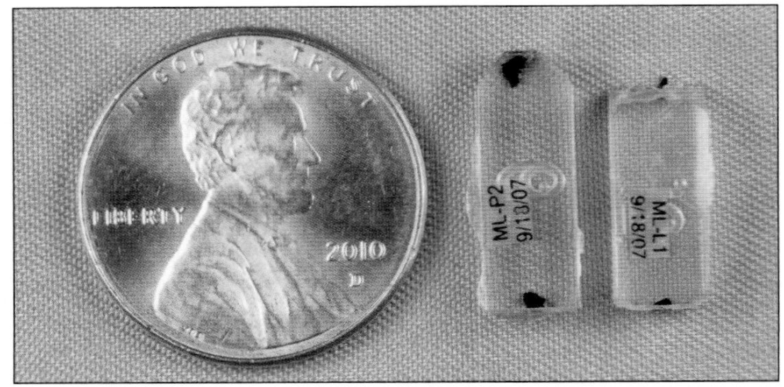

4.8 TEM Specimen Embedded in a Plastic Block • These plastic resin blocks contain specimens, the black spots within the blocks. On the right is a trimmed block that has had excess resin cut away to produce a minute piece of specimen that extends from the block. This is the portion of specimen to be sectioned (Fig. 4.9).

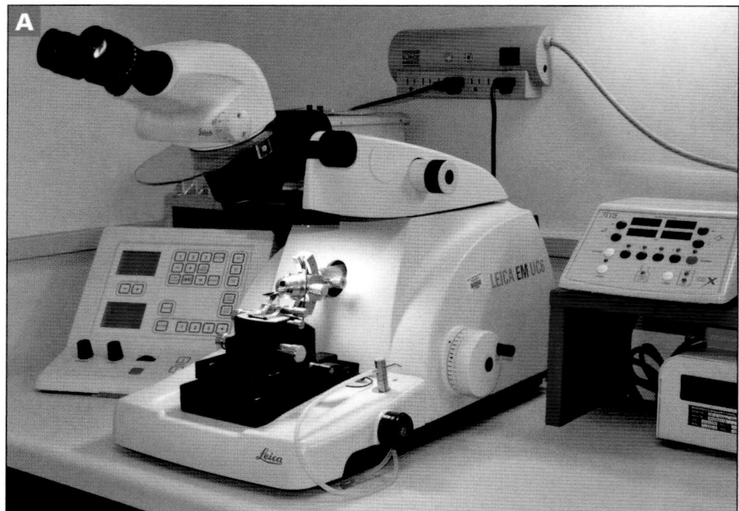

4.9 Ultramicrotome ● (A) This ultramicrotome is capable of producing specimen slices 100 nm in thickness (and less). The arm holding the specimen traces an elliptical path as it approaches and is withdrawn from the blade. In each cycle, it is advanced the distance equal to the desired section thickness, often 100 nm. (B) The specimen block (**S**), with the tiny, trimmed down specimen (**arrow**) facing the blade (**Bl**), is held in the ultra-microtome chuck. As the specimen moves forward and passes by the glass or diamond blade a thin slice is made, which is caught and floated on water in the boat (**Bt**) behind the blade. After passing by the blade, the specimen holder is withdrawn and returns to the starting position, at which point it advances forward by the desired thickness and another cut is made. The process is repeated to produce multiple sections of the same thickness.

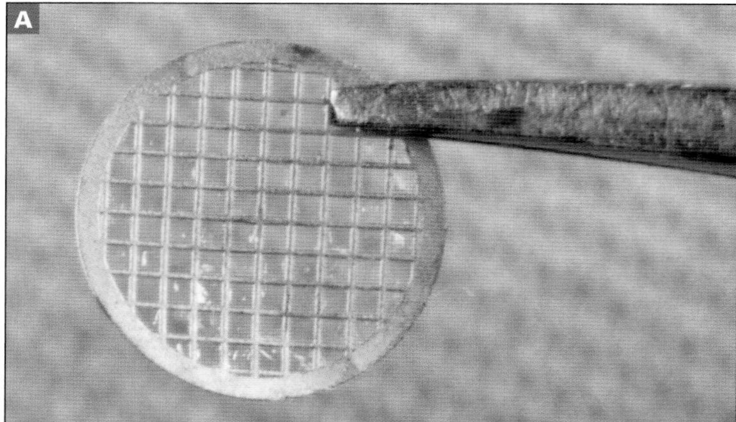

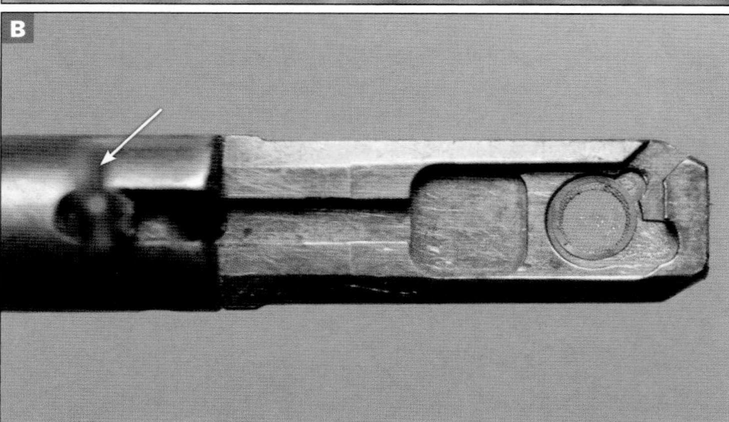

4.10 The Grid and Specimen Rod ● (A) After sectioning, the thin sections are picked up by a grid (shown), which acts as the equivalent of a glass slide in light microscopy. The shiny material between grid bars is a plastic film that fills in the openings and keeps specimens from dropping through. This grid is approximately 3 mm in diameter. (B) For examination, the grid is placed in a specimen rod and held in place by a spring clip, which is out of focus because it is sticking straight up from the rod (**arrow**). It will be carefully lowered to hold the grid in place. (C) Finally, the specimen rod (**arrow**) is inserted into the specimen port in such a way that positions the grid in the electron beam path.

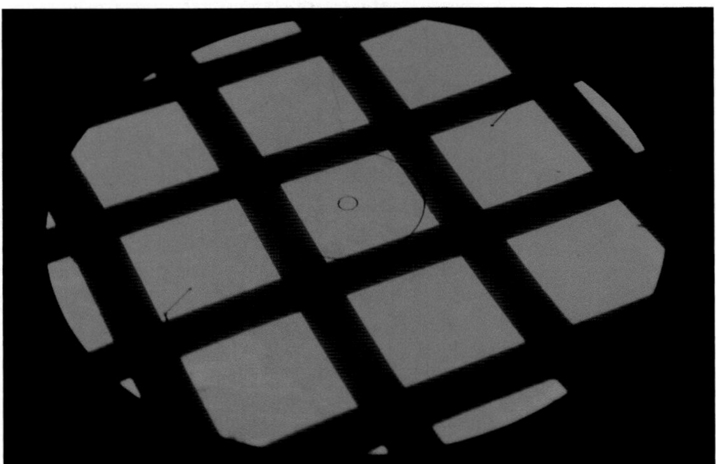

4.11 The Viewing Screen • Electron beams do not produce an image visible to the human eye. In order for the image to be seen, the microscopist views the specimen on this screen coated with a phosphorescent material. The kinetic energy of the electrons hitting the screen is converted to light, which makes the specimen visible. The thick, dark lines are shadows of the grid bars at very low magnification. The image is also captured by a digital camera and viewed on a computer monitor.

4.12 Scanning Electron Microscope • This scanning electron microscope has the ability to magnify from 50× to 2,000,000× with a limit of resolution as low as 1.0 nm.

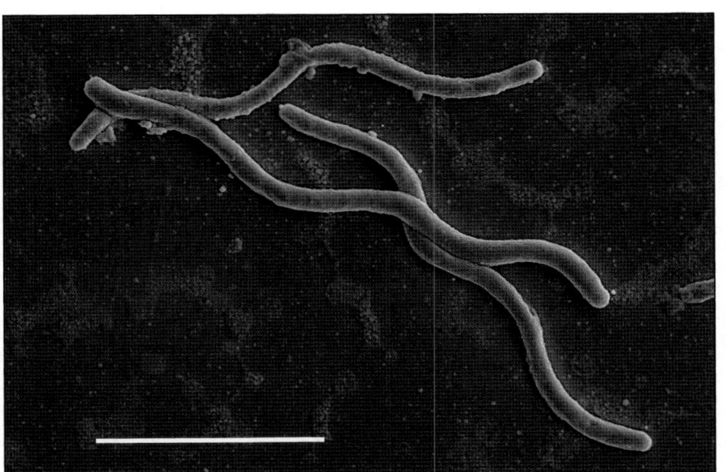

4.13 Scanning Electron Micrograph of *Rhodospirillum rubrum* • Like the TEM, the image produced by the SEM has no color, but it is three-dimensional. This micrograph is of three *Rhodospirillum rubrum* cells. Compare this image with light micrographs of the same organism using the simple and negative staining techniques in Figures 5.3 and 5.6. (Scale bar = 5 µm)

4.14 SEM Specimen Mounted on a Stub • This is a gold-coated pill bug resting on its back upon the platform of the stub. The larger cylinder is a holder for the stub.

4.15 Sputter Coater • Stubs with specimens are placed in the sputter coater chamber, which is then evacuated. Sputtering with gold occurs when ionized argon gas bombards a gold foil to release gold atoms. Two specimens are visible within; the purple is the argon gas.

Bacterial Cellular Morphology, and Simple and Negative Stains

SECTION 5

Simple Stains

Purpose

In Section 4, we introduced you to two of the three important features of a microscope and microscopy: magnification and resolution. A third feature is **contrast**. Cytoplasm is transparent, and while visible, the contrast between the cell and background is generally poor when using bright field (Fig. 5.1). Staining with a dye provides contrast and makes it easier to determine cell morphology, size, and arrangement.

Principle

Stains are solutions consisting of a solvent (usually water or ethanol) and a colored molecule (often a benzene derivative), the **chromogen**. The portion of the chromogen that provides color is the **chromophore**. A chromogen may have multiple chromophores, with each adding intensity to the color. The **auxochrome** is the charged portion of a chromogen. Staining occurs as a result of ionic or covalent bonds between the chromogen (auxochrome) and the cell.

Basic stains[1] (where the auxochrome becomes positively charged as a result of picking up a hydrogen ion or losing a hydroxide ion) are attracted to the negative charges on the surface of most bacterial cells. Thus, the cell becomes colored (Fig. 5.2). Common basic stains include methylene blue, crystal violet, and safranin. Examples of basic stains may be seen in Figures 5.3 and 5.4, and in A Gallery of Bacterial Cell Diversity (Figs. 5.9–5.18, pp. 51–53).

Basic stains are applied to bacterial smears that have been **heat-fixed.** Heat-fixing kills the bacteria, makes them adhere to the slide, and coagulates cytoplasmic proteins to make them more visible. It also distorts the cells to some extent.

[1] Notice that the term *basic* means "alkaline," not "elementary," although coincidentally basic stains can be used for simple staining procedures.

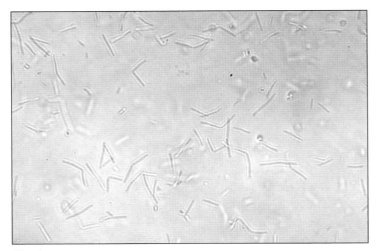

5.1 Unstained Wet Mount ● Shown is an unstained wet mount of *Aeromonas sobria*. There is little contrast between the cells and the water in which they're suspended. However, for this micrograph the iris diaphragm was closed to cut down on excess light and reduce the glare. This enhances what little contrast exists. Also, because the film of water in a wet mount has depth, not all cells will be in focus at the same time. Notice that cells in focus are darker in this micrograph.

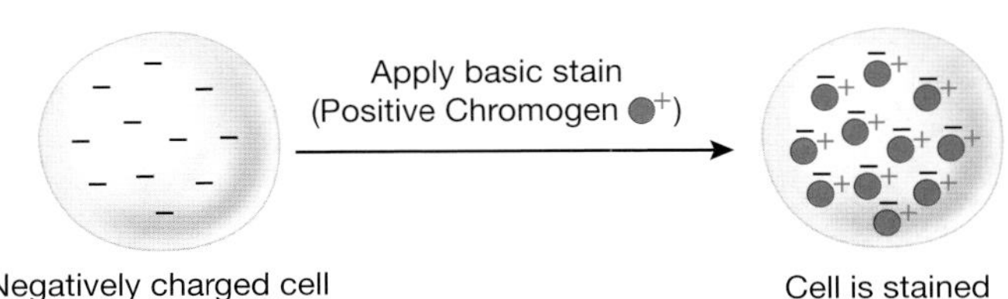

5.2 Chemistry of Basic Stains ● Basic stains have a positively charged chromogen (●+), which forms an ionic bond with the negatively charged bacterial cell, thus colorizing the cell.

5.3 Simple Stain Using Safranin • This is a simple stain using safranin, a basic stain solution. Notice that the stain is associated with the cells and not the background. The organism is *Rhodospirillum rubrum* grown in broth culture and photographed with the oil immersion lens. *R. rubrum* cells range in size from 7–10 µm long by 0.8–1.0 µm in diameter. Compare cell size and morphology in this preparation with the negative stain of the same *R. rubrum* culture in Figure 5.6.

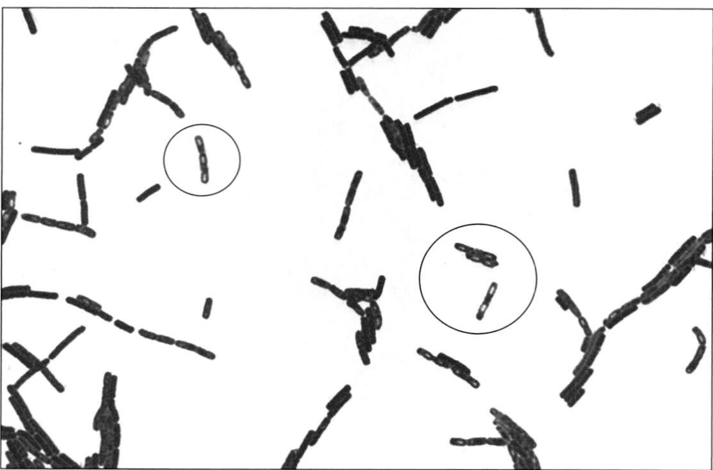

5.4 Simple Stain Using Crystal Violet • These *Bacillus cereus* cells were grown in a broth for 24 hours and stained with the basic stain crystal violet. *B. cereus* grows in chains and is classified as a streptobacillus. Individual cells range in size from about 1 µm wide by 3 to 5 µm in length. Like all members of the genus, *B. cereus* forms endospores (see p. 66–68). The exterior of an endospore is difficult for chemicals to penetrate without applying heat, so in most stained preparations they show up as unstained ovals in the cells (see the circled cells).

Negative Stain

Purpose

The negative staining technique is used to determine morphology and cellular arrangement in bacteria that are too delicate to withstand heat-fixing. A primary example is the spirochete *Treponema*, which is distorted by the heat-fixing of other staining techniques. Also, where determining the accurate size is crucial, a negative stain can be used because it produces minimal cell shrinkage.

Principle

The negative staining technique uses a dye solution in which the chromogen is acidic and carries a negative charge. (An acidic chromogen gives up a hydrogen ion, which leaves it with a negative charge.) The negative charge on the bacterial surface repels the negatively charged chromogen, so the cell remains unstained against a colored background (Fig. 5.5). Examples of acidic staining solutions used in negative stains are Congo red and nigrosin (Figs. 5.6 and 5.7).

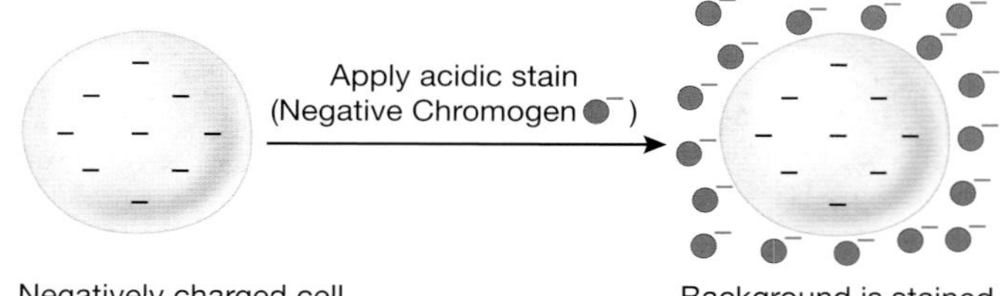

5.5 Chemistry of Acidic Stains • Acidic stains have a negatively charged chromogen (•⁻) that is repelled by negatively charged cells. Thus, the background is colored and the cell remains transparent.

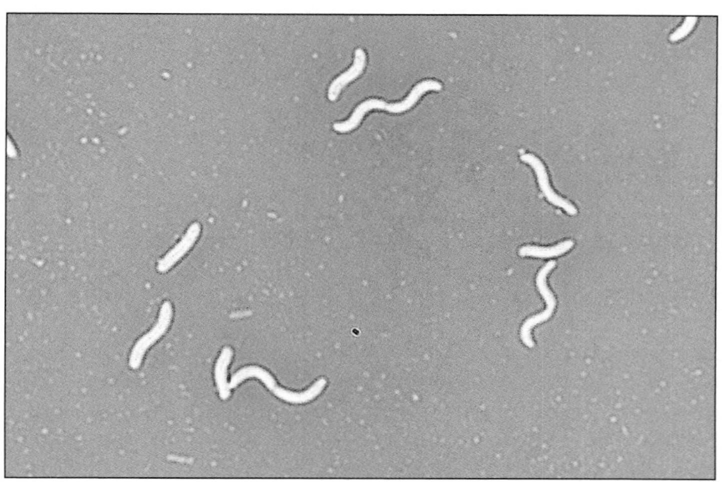

5.6 Negative Stain Using Congo Red • A negative stain procedure is so-called because it is the background that is colorized, not the cells. Heat-fixing is used in most staining procedures and it causes some degree of cell distortion. Because cells in a negative stain are not heat-fixed, their appearance is closer to what they really look like. These *Rhodospirillum rubrum* cells were grown in broth and photographed with the oil-immersion lens. *R. rubrum* cells range in size from 7–10 µm long by 0.8–1.0 µm in diameter. Compare cell size and morphology in this preparation with the simple stain of the same *R. rubrum* culture in Figure 5.3.

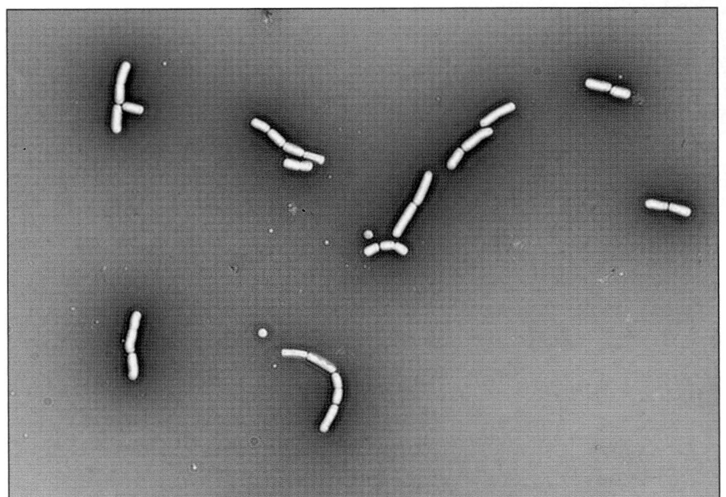

5.7 Negative Stain Using Nigrosin • These *Bacillus cereus* cells came from the same culture used in Figure 5.4, but were stained with nigrosin, an acidic stain. The result is a negative stain of the rod-shaped cells. Compare cell size and morphology in this preparation with the simple stain of the same *B. cereus* culture in Figure 5.4. White spots in this preparation that aren't rods are materials from the broth that repelled the nigrosin. Notice that the endospores are not visible in these cells because the cytoplasm hasn't been stained, so there is no contrast. However, in some *Bacillus* preparations released endospores will be seen as consistently shaped (e.g., elliptical) but unstained objects strewn among the larger cells.

A Gallery of Bacterial Cell Diversity

Bacterial Cell Morphologies

Bacterial cells are much smaller than eukaryotic cells (Fig. 5.8) and come in a variety of **morphologies** (shapes) and **arrangements**. Determining cell morphology is an important first step in identifying a bacterial species. Cells may be spheres (**cocci**, singular **coccus**), rods (**bacilli**, singular **bacillus**), or spirals (**spirilla**, singular **spirillum**).

Variations of these shapes include slightly curved rods (**vibrios**), short rods (**coccobacilli**), and flexible spirals (**spirochetes**). **Pleomorphism** is where a variety of cell shapes—slender, ellipsoidal, or ovoid rods—may be seen in a given sample. Examples of cell morphologies are shown in Figures 5.9 through 5.18.

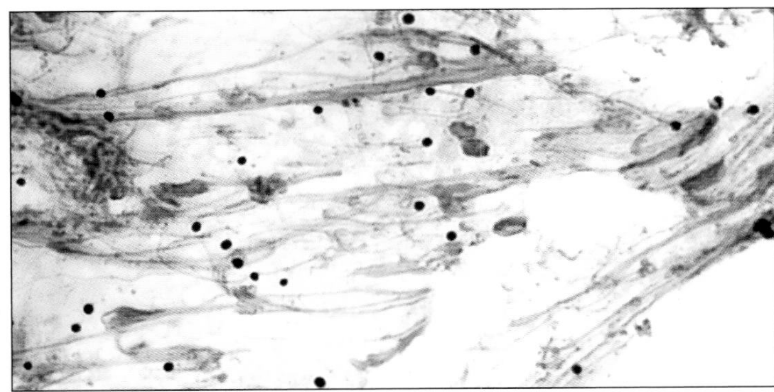

5.9 Single Cocci from a Nasal Swab • This direct smear of a nasal swab illustrates unidentified cocci (dark circles) stained with crystal violet. Each cell is about 1 µm in diameter. The red background material is mostly mucus. (Gram Stain)

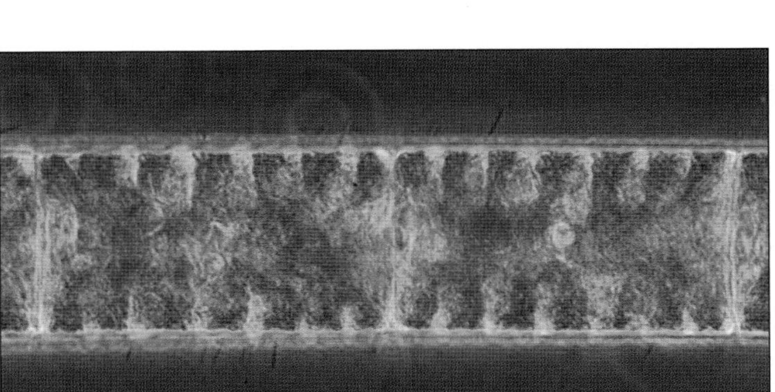

5.8 *Spirogyra* and Cyanobacteria • Bacterial and eukaryotic cells differ in many ways, but this photo emphasizes one that is easily observed: size. The large, green filamentous organism (you can see four different cells) is a green alga named *Spirogyra*. It gets its name from the spiral chloroplasts in its cells. If you look carefully, you can see their helical pattern in this specimen. *Spirogyra* is a eukaryote. The black lines extending from its surface are cyanobacteria. More precisely, *chains* of cyanobacteria. Size difference, indeed! (Phase Contrast)

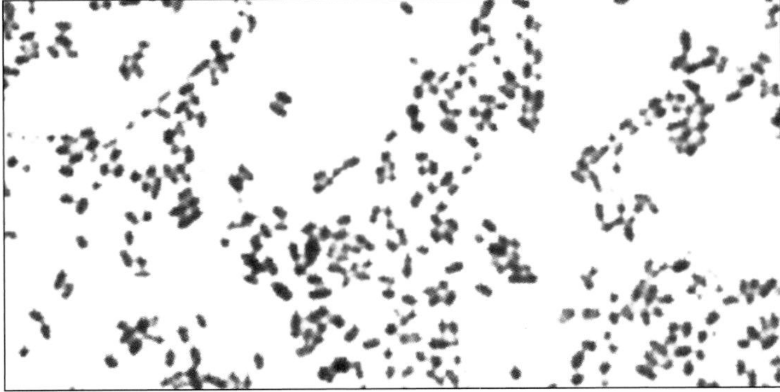

5.10 Ovoid Cocci • *Lactococcus lactis* is an elongated coccus that a beginning microbiologist might confuse with a rod. Notice the slight elongation of the cells, and also that most cells are not more than twice as long as they are wide (see Fig. 5.19A). *L. lactis* is found naturally in raw milk and milk products, but these cells were grown in culture. Cell dimensions are about 1–2 µm long by 1 µm wide. (Gram Stain)

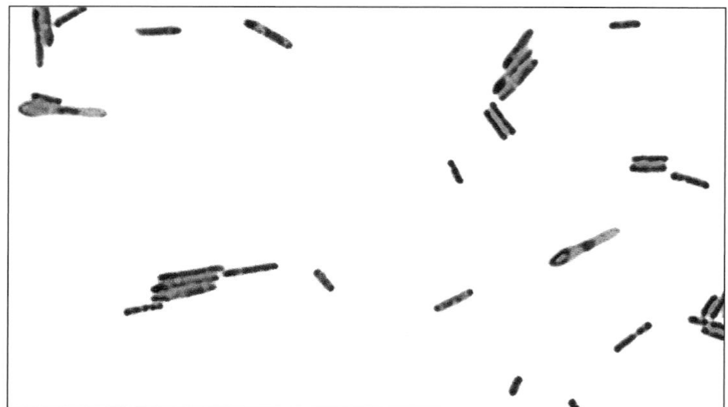

5.11 Single Bacilli • *Clostridium butyricum* generally grows as single bacilli. It is the type species for the genus (meaning that it was the first described *Clostridium* species—back in 1880! It is found in soil and aquatic environments, as well as in human feces and many other habitats. It produces endospores (see p. 66–68), some of which are beginning to form and are visible as bulges toward the ends of some cells. *C. butyricum* cells range in size from 0.2–1.7 µm in diameter by 2.5–7 µm long. (Carbolfuchsin Stain)

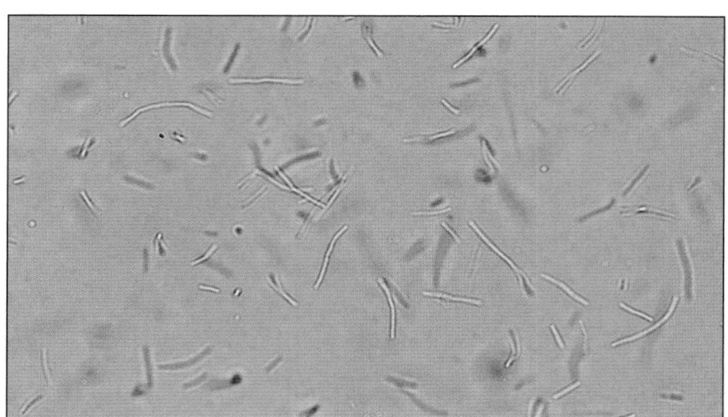

5.12 Long, Thin Bacilli • This *Bacillus subtilis* culture was grown in trypticase soy broth for 24 hours and is shown in a wet mount using high-dry magnification and phase contrast. Notice that the cells are longer and thinner than the *Clostridium butyricum* rods in Figure 5.11 (which was photographed under oil immersion). Cell morphologies of a particular species can vary depending on the medium used (solid vs. liquid) and culture age. The longer "individuals" seen here are actually short chains of two or three cells, as evidenced by transverse constrictions or bends where the cells are attached. Single cells range in size from slightly less than 1 µm in diameter by 2–3 µm. The water layer of any bacterial wet mount is considerably thicker than the cells, resulting in many of the cells being above or below the focal plane and are out of focus. (Phase Contrast Wet Mount)

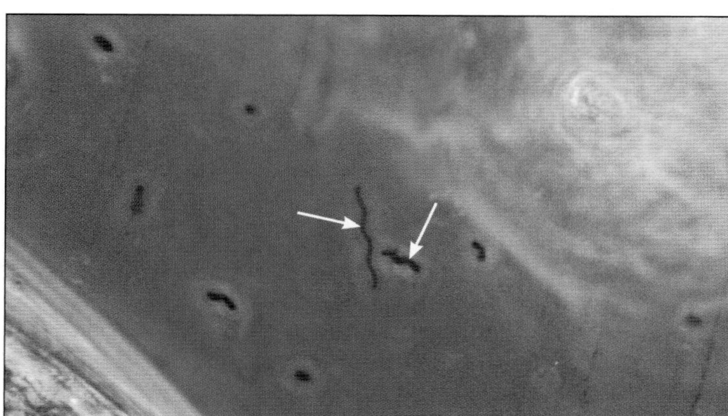

5.13 Two Different Spirilla • These two spirilla (**arrows**) are undoubtedly different species based on their different morphologies: one is long and slender with loose spirals, the other is shorter and fatter with tighter coils. Spirilla are always single cells. Cell dimensions are less than 1 µm wide and up to 35 µm long. (Phase Contrast Wet Mount)

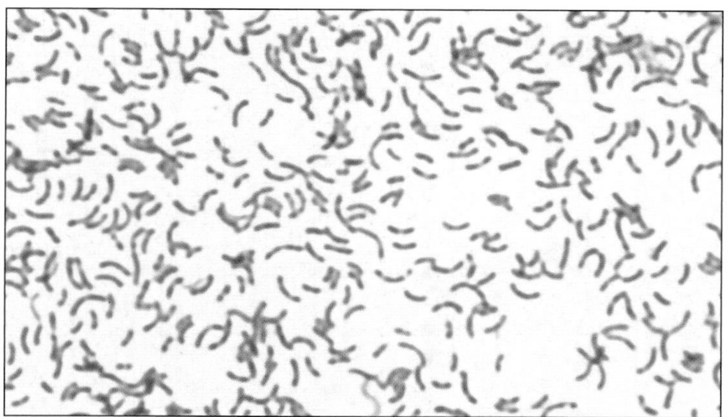

5.14 Spirillum from a Solid Medium • Sometimes the organism's source affects its morphology. Shown is *Rhodospirillum rubrum* grown on an agar slant and stained with carbolfuchsin. Notice that the cells are not spiral shaped and that most are curved rods. Compare their size and shape with Figure 5.15. Cell dimensions are about 1 µm wide by 10 µm long. (Carbolfuchsin Stain)

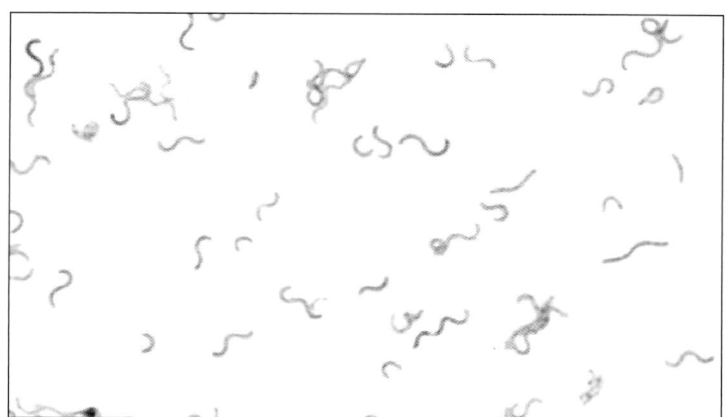

5.15 Spirillum Grown in Broth • Shown is *Rhodospirillum rubrum*, grown in broth and stained with safranin. Compare the size and shape of these cells with those shown in Figure 5.14. (Safranin Stain)

Bacterial Cell Arrangements

Cell division in bacteria usually occurs by a process known as **binary fission.** A cell grows to a size where it is ready to divide and the cell membrane pinches inward to form a **septum** near the center. Concurrently, wall material fills in the space between the facing membranes. Eventually the inwardly growing membranes join and the daughter cells become completely separate cytoplasmic compartments joined by a common wall (Figs. 4.7 and 5.19).

Cell arrangement, determined by the number of planes in which division occurs and whether the cells separate after division, is also useful in identifying bacteria. Spirilla rarely are seen as anything other than single cells, but cocci and bacilli do form multicellular (colonial) associations. Because cocci exhibit the most variety in arrangements, they are used for illustration in Figure 5.20.

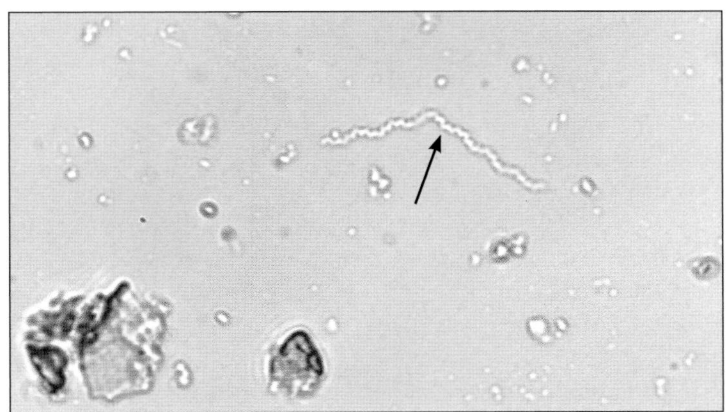

5.16 Spirochaete • This bright field micrograph shows a marine species of *Spirochaeta* (**arrow**). Spirochaetes are tightly coiled, flexible rods. Notice the bend in the center of the cell. Cell dimensions of the genus are less than 1 µm wide by 5–500µm long. (Unstained Wet Mount)

5.17 Vibrios • *Vibrio cholerae* is the causative agent of cholera. Its cells are curved rods, but notice not all the cells are curved. They range in size from 1–3 µm long by less than 1 µm in diameter. (Gram Stain)

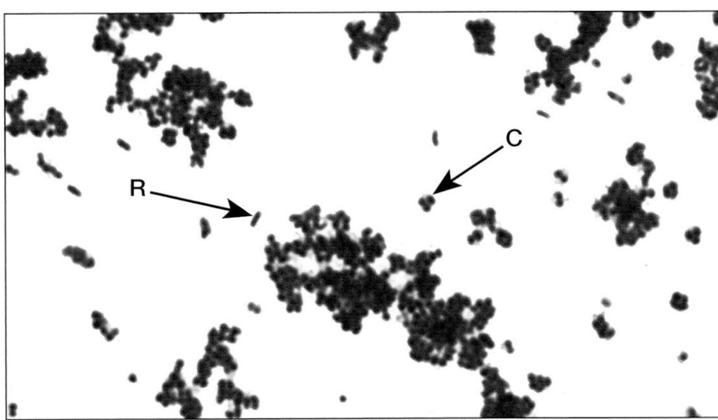

5.18 Bacterial Pleomorphism • *Arthrobacter* and *Corynebacterium* species grow in a variety of shapes and are said to be pleomorphic. Notice that in this specimen the pleomorphic rods of *C. xerosis* range from plump cocci (C) to short, slender rods (R). *C. xerosis* is a normal inhabitant of skin and mucous membranes and is generally not considered to be pathogenic. (Gram Stain)

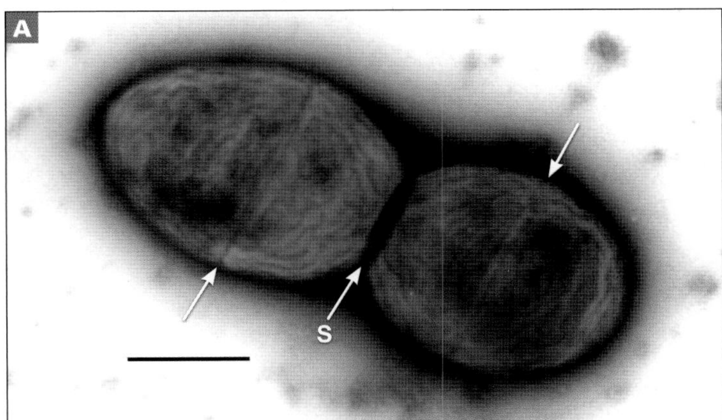

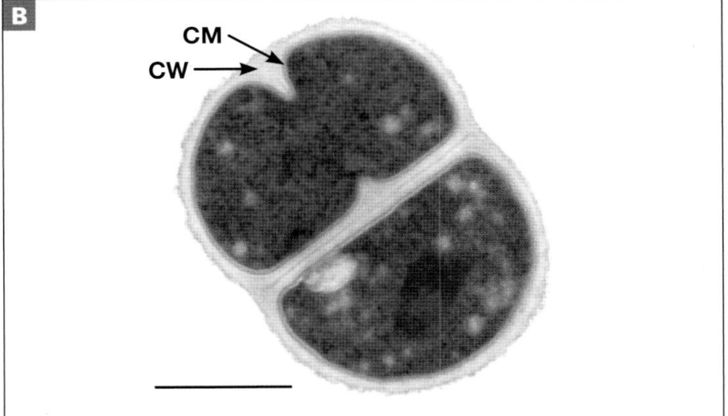

5.19 Coccus Dividing by Binary Fission • Bacterial cell division usually occurs by binary fission. The cell membrane near the center of the cell pinches into the cytoplasm like a tightening belt. When the membranes have completely separated the cytoplasm into two compartments, division is completed. (**A**) These *Lactococcus lactis* cells show evidence of the current division and the previous division. The septum (S) from the previous division is where the cells are attached. Each cell also has a new septum forming (**arrows**). Compare these cells to their morphology as seen in a light micrograph of *L. lactis* in Figure 5.10. (TEM negative stain; scale bar = 500 nm) (**B**) These *Micrococcus luteus* cells are in the process of making a tetrad. The upper cell shows early septum formation with invagination of the cytoplasmic membrane (CM) and wall material (CW) filling in-between. The region between the upper and lower cells shows a completed division. (TEM; scale bar = 500 nm)

If two daughter cells remain attached after a coccus divides, a **diplococcus** is formed. The same process happens in bacilli that produce **diplobacilli**. If the cells continue to divide in the same plane and remain attached, they exhibit a **streptococcus** or **streptobacillus** arrangement.

A second division occurring in a plane perpendicular to the first forms a **tetrad** from a diplococcus. A third division plane perpendicular to the other two produces a cube-shaped arrangement of eight cells called a **sarcina**. Tetrads and sarcinae are seen only in cocci. Division in irregular planes of a coccus produces a cluster of cells called a **staphylococcus**. Figures 5.21 through 5.27 illustrate common cell arrangements.

Some organisms produce more unique arrangements. **Snapping division** in *Corynebacterium* and *Arthrobacter* species produces either a **palisade** or **angular** arrangement of cells (Fig. 5.28). Figure 5.29 illustrates a phenomenon characteristic of virulent strains of *Mycobacterium tuberculosis* in which they grow in parallel chains called **cords**.

Arrangement and morphology are often easier to see when the organisms are grown in a broth rather than on a solid medium (Figs. 5.14 and 5.15), or are observed using a direct smear made from the source rather than a culture. If there is difficulty identifying cell morphology or arrangement, consider transferring the organism (isolate) to a broth culture and trying again.

One last bit of advice: Don't expect nature to conform perfectly to our categories of morphology and cell arrangement. These are convenient descriptive categories that will not easily be applied in all cases (but of course, experience helps). When examining a slide, look for the most common morphology and most complex arrangement. Do not be afraid to report what is seen. For instance, it's okay to say, "Cocci in singles, pairs, and chains."

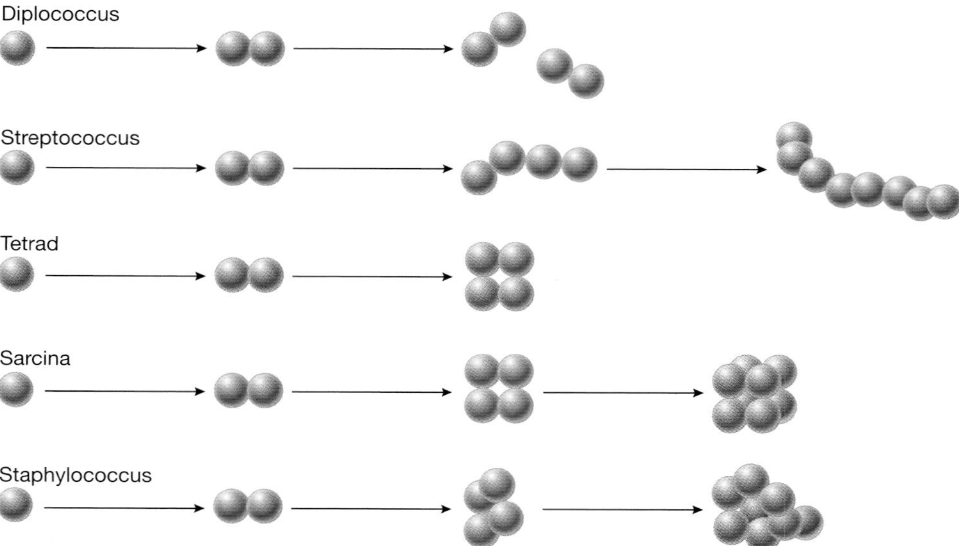

5.20 Division Patterns in Cocci • Diplococci have a single division plane and the cells generally occur in pairs. Streptococci also have a single division plane, but the cells remain attached to form chains of variable length. If there are two perpendicular division planes, the cells form tetrads. Sarcinae have divided in three perpendicular planes to produce a regular cuboidal arrangement of cells. Staphylococci have divided in more than three planes to produce a characteristic grapelike cluster of cells. (***Note*** that rarely will a sample be composed of just one arrangement. Report what you see, and emphasize the most complex arrangement.)

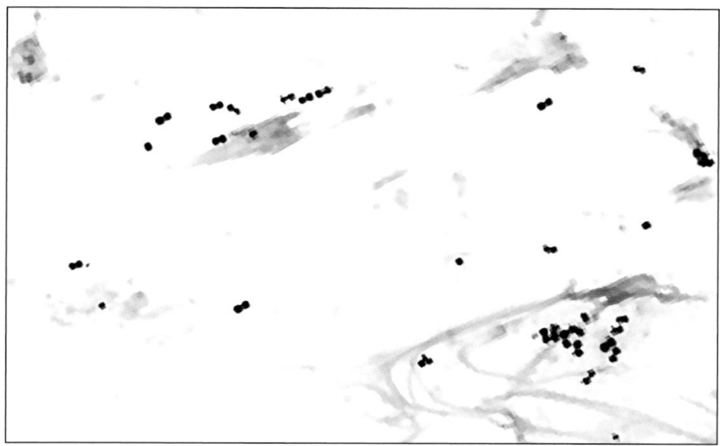

5.21 Diplococcus Arrangement • A few unidentified diplococci are visible in this nasal swab. Single cocci may be the same organism (and haven't divided yet) or a different species. The red background material is host epithelial cells and mucus. (Gram Stain)

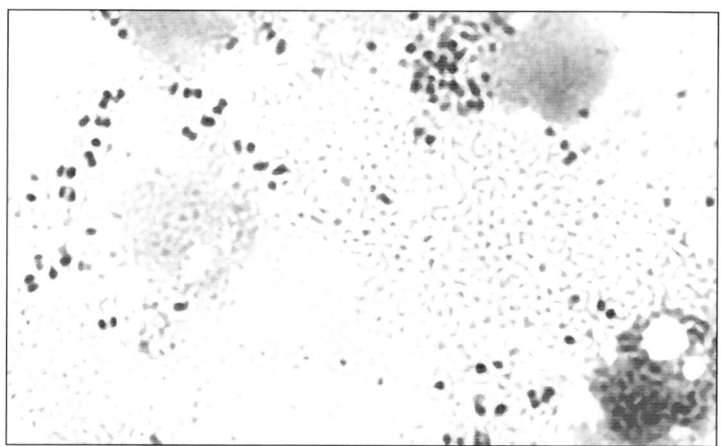

5.22 Another Diplococcus Arrangement • *Neisseria gonorrhoeae* is a diplococcus that causes gonorrhea in humans. Members of this genus produce diplococci with flattened adjacent sides. Their cells range in size from 0.6 μm to 1.9 μm in diameter. (Gram Stain)

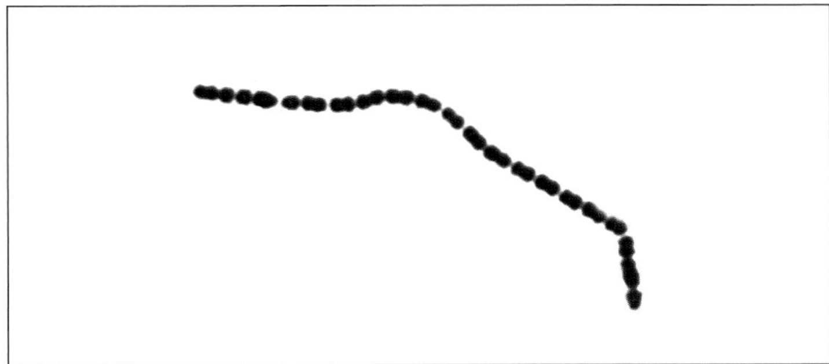

5.23 Streptococcus Arrangement • As its name suggests, *Streptococcus salivarius* forms chains of spherical cells and it inhabits the oral cavity, but primarily the saliva and tongue. This specimen is from a broth culture (which enables the cells to form long chains). Pairs of cells within the chain are in the process of dividing. Individual cells are approximately 1 µm in diameter. (Gram Stain)

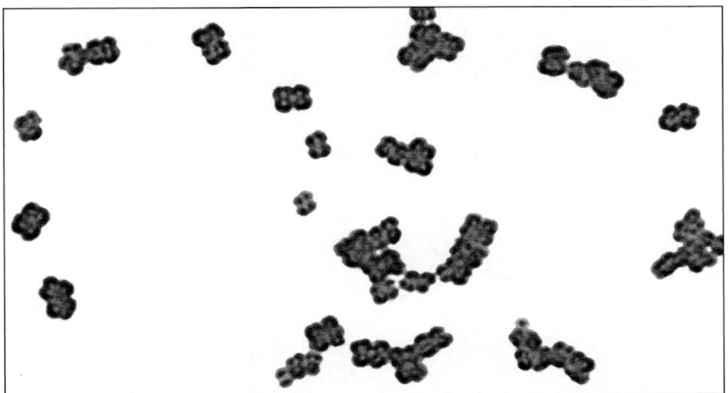

5.24 Tetrad Arrangement • *Neisseria sicca* frequently grows in tetrads, as shown in this Gram stain of a trypticase soy agar culture. Cell arrangement, especially with cocci, is frequently more easily determined if the specimen comes from a broth culture because groups of cells separate from one another in the liquid, but tetrads are pretty easy to spot even when taken from a solid medium. Look at the bigger clumps and you will see tetrads within them. *N. sicca* is a normal inhabitant of the human nasopharynx, saliva, and sputum, and is an opportunistic pathogen. Cells are between 1 and 2 µm in diameter. (Gram Stain)

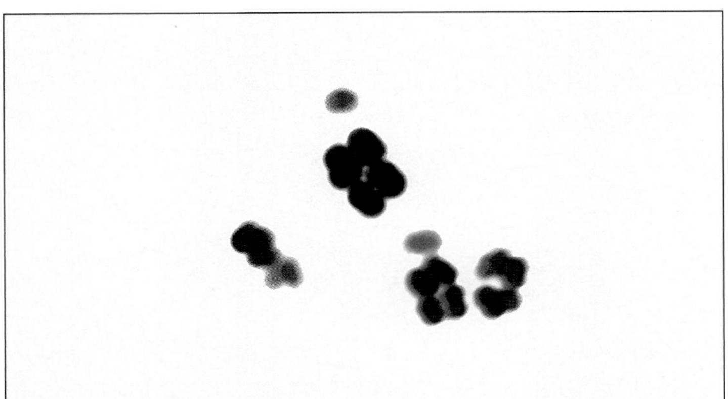

5.25 Sarcina Organization • *Sarcina maxima* exhibits the sarcina organization. The sarcina in the center is clearly a packet of eight cells. However, not all cells in a smear will exhibit the sarcina arrangement—a variety of arrangements leading up to the most complex arrangement is typically seen. Each cell is 2 to 3 µm in diameter, which is large for a coccus and explains the specific epithet: *maxima*. (Crystal Violet Stain)

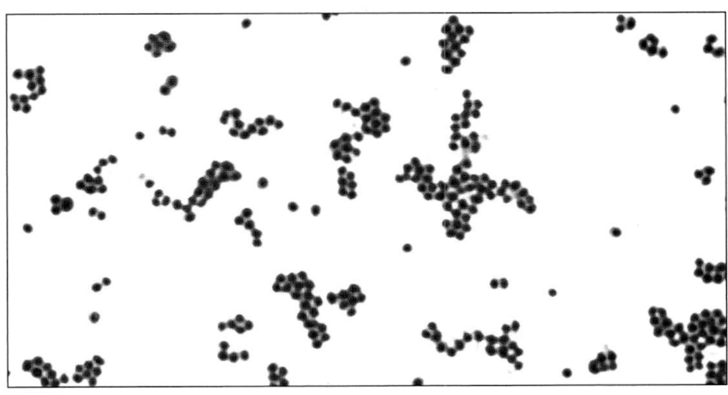

5.26 Staphylococcus Arrangement • *Staphylococcus aureus* cells grown in broth culture are shown. Note that the staphylococcal groups are rarely bigger than a dozen cells or so. Larger clusters are probably two or more groups that happen to be next to each other on the slide. This "bunching" can be especially problematic when making a preparation from cells grown on a solid medium. Be sure to emulsify them thoroughly to spread clusters out as much as possible. *S. aureus* is a common opportunistic pathogen of humans. Cells are approximately 1 µm in diameter. (Crystal Violet Stain)

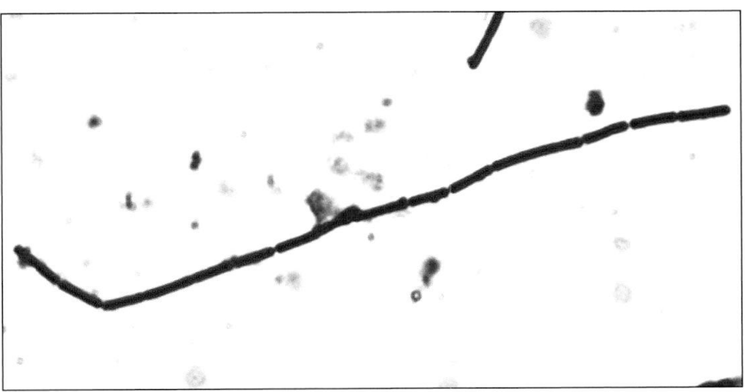

5.27 Streptobacillus Arrangement • *Bacillus subtilis* is a streptobacillus. These cells were obtained from culture and are 0.7 to 0.8 µm wide by 2 to 3 µm long. You can see small indentations where cells join to form the chain. The random stuff in the background is the broth that has taken up stain. (Crystal Violet Stain)

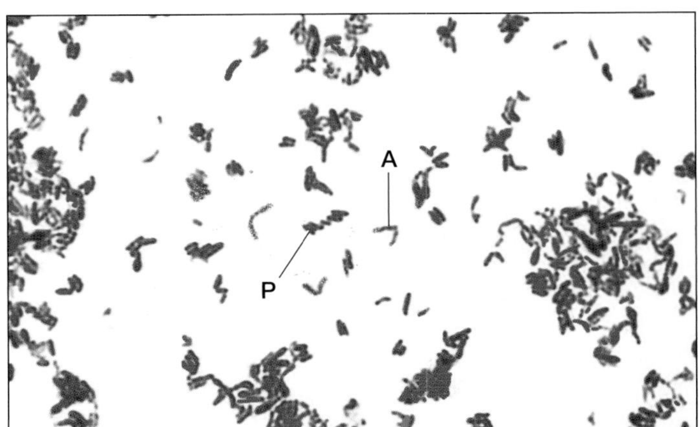

5.28 Angular and Palisade Arrangement • The division plane in most rods is across the cell, but *Arthrobacter* and *Corynebacterium* species divide lengthwise in a process called snapping division. This results in some unique arrangements when the cells remain attached. The palisade arrangement (P) is seen as stacks of rod-shaped cells side-by-side. An angular arrangement (A) is when a pair of rods is bent where the cells join. (Gram Stain)

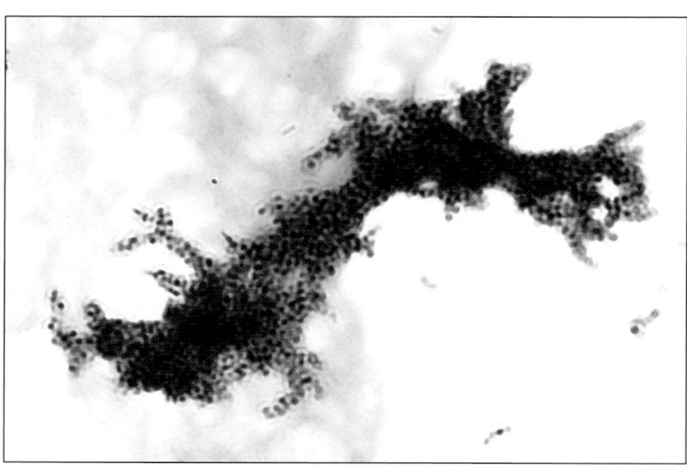

5.29 Cording of *Mycobacterium tuberculosis* • *M. tuberculosis* aggregates in characteristic cords due to adhesion of their waxy, acid-fast cell walls. Cells measure 0.2–0.6 µm wide by 1–10 µm long. Note the characteristic beaded appearance of *Mycobacterium* cells. (Acid-Fast Stain)

SECTION

6

Bacterial Cell Structures and Differential Stains

In Section 5, we introduced you to simple and negative staining techniques, which are used to obtain basic information about cells: their morphology, arrangement, and size. In this section we will introduce you to more sophisticated staining techniques that provide more detailed information about the cells. Stains that have the ability to show differences between cells or cell parts are called **differential stains.** Some differential stains are used to show specific cellular components and may be referred to as **structural stains.**

Gram Stain

Purpose

The Gram stain, used to distinguish between Gram-positive and Gram-negative cells, is the most important and widely used microbiological differential stain. In addition to Gram reaction, this stain allows determination of cell morphology, size, and arrangement. It is typically the first differential test run on a specimen brought into the laboratory for identification. In some cases, a rapid, presumptive identification of the organism or elimination of a particular organism is possible.

Principle

The Gram stain (named after Hans Christian Gram, a Danish physician) is a differential stain in which a **decolorization** step occurs between the application of two basic stains. There are several minor variations of the Gram stain, but they all work in basically the same way (Fig. 6.1). The **primary stain** is crystal violet. Iodine is added as a **mordant** to enhance crystal violet staining by forming a **crystal violet–iodine complex**. Decolorization follows and is the most critical step in the procedure.

Gram-negative cells are decolorized by the solution (generally an alcohol or alcohol/acetone mixture of varying proportions) whereas Gram-positive cells are not. Gram-negative cells can thus be colorized by the **counterstain** safranin, but Gram-positive cells are already violet and cannot. Upon successful completion of a Gram stain, Gram-positive cells appear purple and Gram-negative cells appear reddish-pink (Fig. 6.2).

Electron microscopy and other evidence indicate that the ability to resist decolorization or not is based on the different wall constructions of Gram-positive and Gram-negative cells. Gram-negative cell walls have higher lipid content (because of the outer membrane) and a thinner peptidoglycan layer than Gram-positive cell walls (Figs. 6.3–6.5).

Gram-negative cells
Gram-positive cells
Cells are transparent prior to staining.
Crystal violet stains both Gram-positive and Gram-negative cells. Iodine is used as a mordant.
Decolorization with alcohol or an alcohol/acetone mixture removes crystal violet from Gram-negative cells.
Safranin is used to counterstain Gram-negative cells.

6.1 Gram Stain Overview ● After application of the primary stain (crystal violet), decolorization, and counterstaining with safranin, Gram-positive cells stain violet and Gram-negative cells stain pink/red. Notice that crystal violet and safranin are both basic stains, and it is the decolorization step that makes the Gram-stain differential.

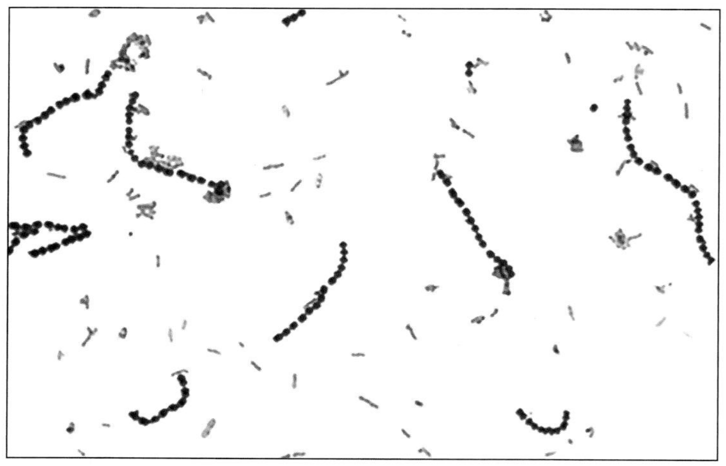

6.2 ***Streptococcus salivarius*** **(+) and** ***Aeromonas hydrophila*** **(–)**
The violet streptococcal cells are Gram positive; the pink rods are Gram negative. Depending on your Gram-stain kit and timing of the counterstain, the Gram-negative cells may stain anywhere from a light pink to a more intense reddish color. In either case, they should be easily distinguishable from the crystal violet color (Gram Stain).

The alcohol/acetone in the decolorizer dissolves the lipid, making the Gram-negative wall more porous and incapable of retaining the crystal violet–iodine complex, thereby decolorizing it. Dehydration of the Gram-positive wall by the decolorizer coupled with its thicker peptidoglycan and greater degree of cross-linking (because of teichoic acids) trap the crystal violet–iodine complex more effectively, making the Gram-positive wall less susceptible to decolorization.

Although some organisms give Gram-variable results, most variable results seen by novice microbiologists are a consequence of poor technique. The decolorization step is the most crucial and most likely source of Gram-stain inconsistency. It is possible to **over-decolorize** by leaving the alcohol on too long and get reddish Gram-*positive* cells.

It also is possible to **under-decolorize** and produce purple Gram-*negative* cells. Neither of these situations changes the actual Gram reaction for the organism being stained. Rather, these are false results because of poor technique.

A second source of poor Gram stains is inconsistency in preparation of the emulsion. A good emulsion is about the size of a dime and dries to a faint haze on the slide. An emulsion that is too heavy or too thin can affect decolorization time.

A third source of Gram-stain inconsistency may be the organisms themselves. Some Gram positives, especially *Bacillus* and *Staphylococcus* species, lose their ability to retain the crystal violet–iodine complex in as little as 24 hours of incubation. Figure 6.6 shows a Gram-stained overnight *Bacillus* culture. Species of the genus *Mycobacterium* can also present Gram-stain challenges, but for a different reason. Their waxy wall (see p. 63) makes Gram staining difficult and they are considered weakly Gram positive. The best practice is to perform your Gram stains on cultures no older than 24 hours.

Until correct results are obtained consistently, it is recommended that control smears of known Gram-positive and Gram-negative organisms be stained along with the organism in question (Fig. 6.7). As an alternative control, a direct smear made from the gumline may be Gram stained (Fig. 6.8) with the expectation that both Gram-positive and Gram-negative organisms will be seen. Under-decolorized and over-decolorized gumline direct smears are shown for comparison (Figs. 6.9 and 6.10). Positive controls also should be run when using new reagent batches.

Potassium hydroxide (KOH) provides an alternative to Gram staining to confirm Gram reaction for particularly difficult species. Part of a colony is emulsified in a drop of KOH for 1 minute, and then the loop is slowly withdrawn. Release of chromosomal material by Gram-negative cells makes the suspension viscous, stringy, and adhesive (Fig. 6.11). Gram positives are unchanged and the emulsion remains watery.

Interpretation of Gram stains can be complicated by nonbacterial elements. For instance, stain crystals from an old or improperly made stain solution can disrupt the field (Fig. 6.12) or stain precipitate may be mistakenly identified as bacteria (Fig. 6.13). In direct smears host cells or noncellular material may be seen (Figs. 6.14–6.17).

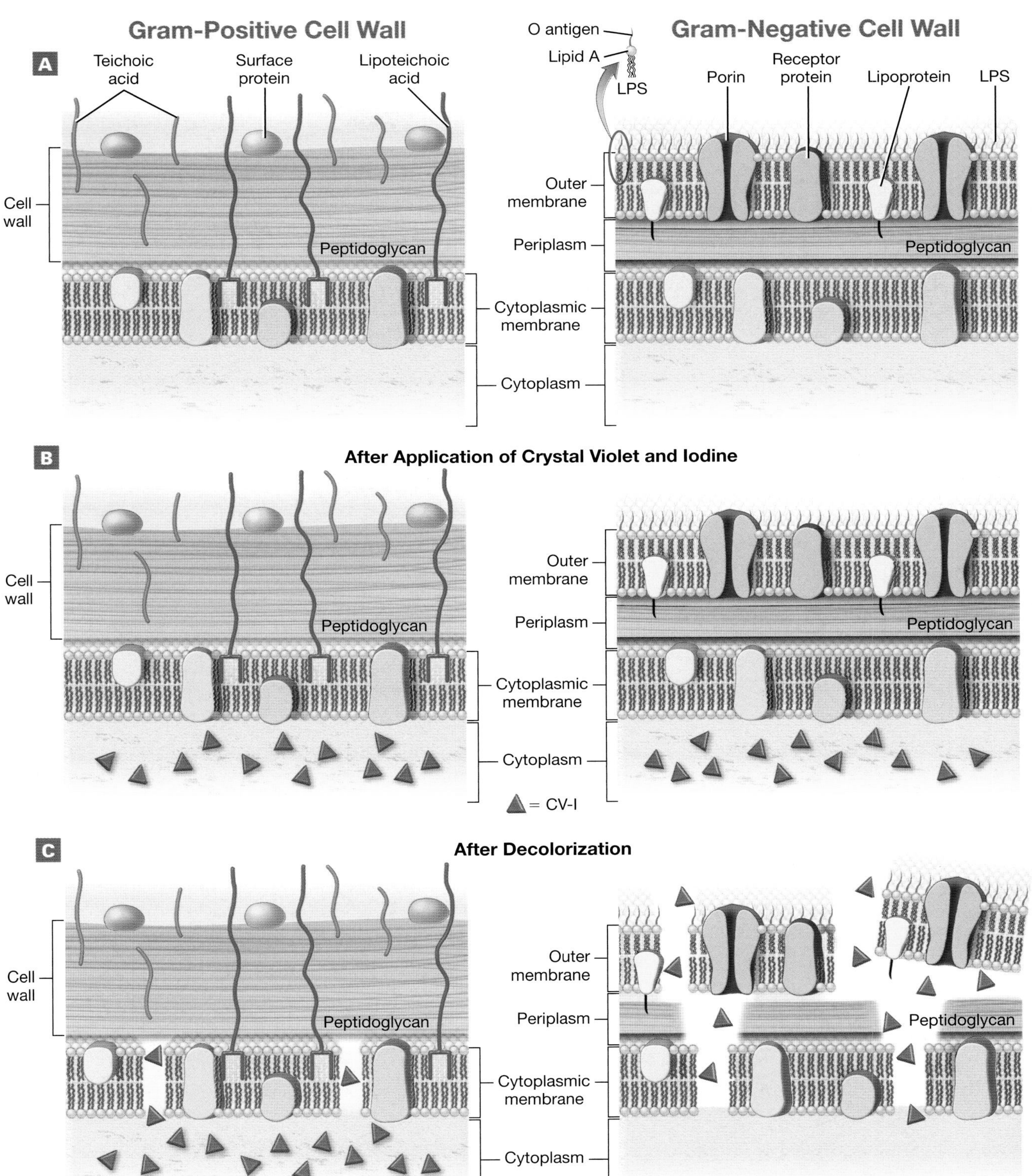

6.3 Bacterial Cell Walls ● (A) The Gram-negative wall (**right**) is composed of less peptidoglycan (as little as a single layer) and more lipid (due to the outer membrane) than the Gram-positive wall (**left**). Though it is shown as a solid layer, the peptidoglycan is actually quite porous, kind of like a three-dimensional chain-link fence. (B) Relatively large crystal violet–iodine complexes (purple triangles) form in the cytoplasm of both Gram-positive and Gram-negative cells. (C) The thicker Gram-positive peptidoglycan layer dehydrates when decolorizer is added, trapping the crystal violet–iodine complexes. The lipid in the Gram-negative cell wall is dissolved by the decolorizer and the thinner peptidoglycan is incapable of preventing extraction of the crystal violet–iodine complex.

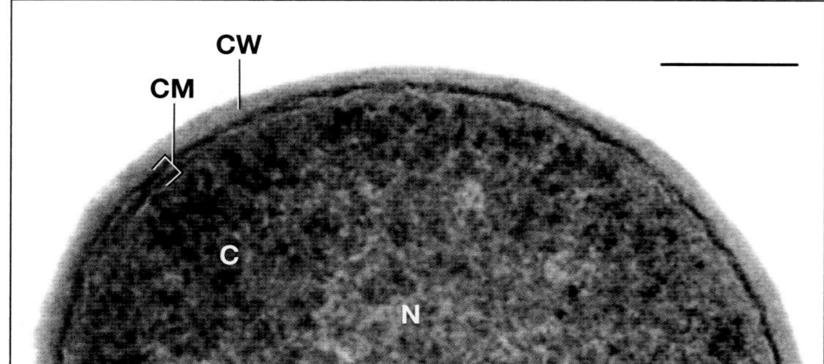

6.4 Gram-Positive Cell Wall • In this TEM of a *Lactococcus lactis* cell, the cytoplasmic membrane (**CM**) is seen as a single dark line in some locations, but as the characteristic two dark lines of a phospholipid bilayer in others. The former is a staining artifact. The thick peptidoglycan (**CW**) is the gray layer external to the cytoplasmic membrane. Inside the cell is darker cytoplasm (**C**) and lighter nucleoid (**N**). (Scale bar = 100 nm) (TEM)

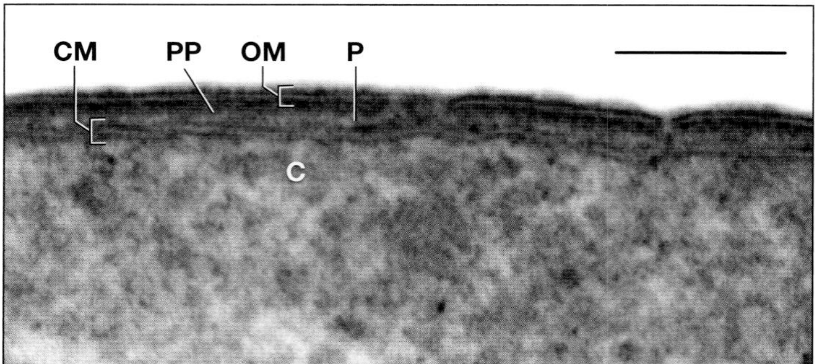

6.5 Gram-Negative Cell Wall • Shown is the *Escherichia coli* cell wall. The cytoplasmic membrane (**CM**) and outer membrane (**OM**) are both visible as double black lines. Each membrane is approximately 7 nm in thickness. Between the two membranes is an extracellular compartment that houses fluid periplasm (**PP**) and the thin peptidoglycan layer (**P**). Periplasm is the site of numerous cellular activities, including hydrolysis of macromolecules, reduction of cytochrome c molecules directly from biochemicals, transport of certain molecules by carrier proteins, and others. Also visible is the cytoplasm (**C**). (Scale bar = 100 nm.) (TEM)

6.6 24-Hour *Bacillus* Culture Illustrating Loss of Gram-Positive Reaction • This *Bacillus* was isolated from aerobic estuarine mud and stained within 24 hours of making the pure culture. Notice some cells are pink and have lost their ability to resist decolorization. This result does not change the fact that *Bacillus* has a Gram-positive wall, though.

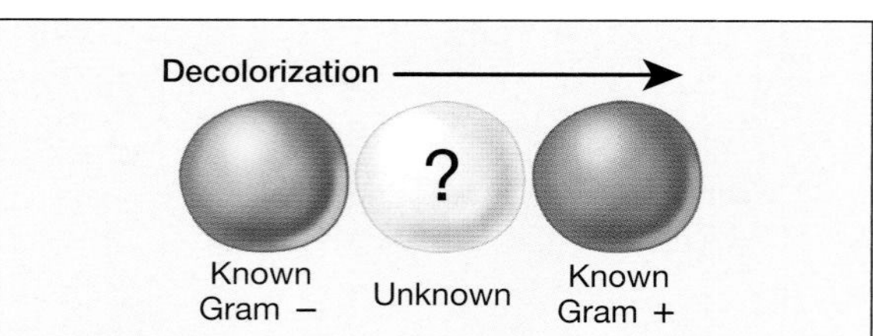

6.7 Positive Controls to Check Your Technique • Staining known Gram-positive and Gram-negative organisms on either side of your unknown organism act as positive controls for your technique. Try to make the emulsions as close to one another as possible. Spreading them out across the slide makes it difficult to stain and decolorize them equally. It is also beneficial to put the Gram-negative control at the end where you will be applying the decolorizer. This will make it easier to see when it is running clear. Putting it at the other end means it will have runoff from the Gram-positive control and the unknown obscuring its runoff.

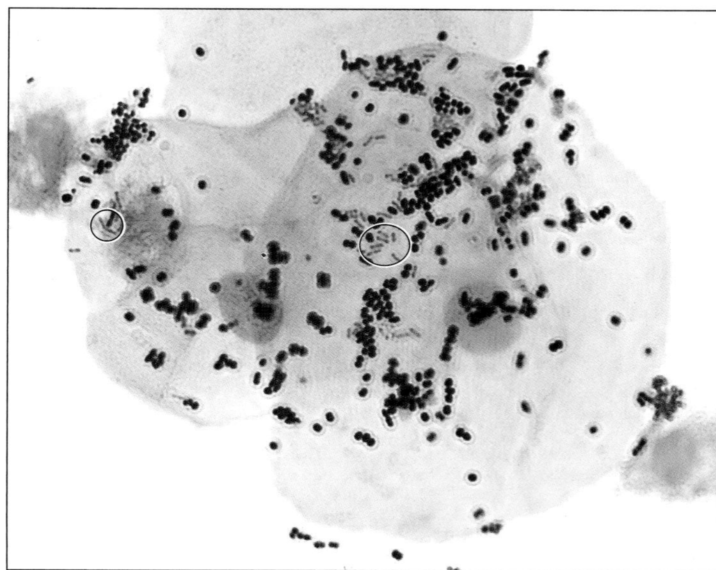

6.8 Direct Smear Positive Control • A direct smear made from the gumline may also be used as a Gram-stain control. Expect numerous Gram-positive bacteria (especially cocci) and some Gram-negative cells, including your own nucleated epithelial cells. In this micrograph, Gram-positive cocci of different sizes predominate, but Gram-negative cells are also present, including two Gram-negative rods (shorter and broader vs. longer and narrower in the two circles). Notice that most bacterial cells are on the epithelial cell's surface, which is where they reside when in the mouth. (Gram Stain)

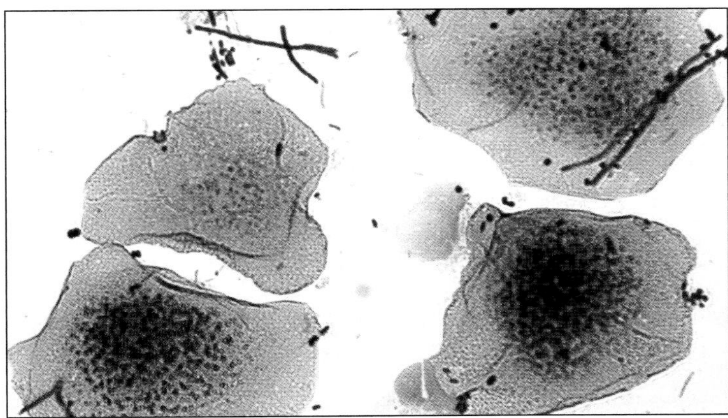

6.9 Under-Decolorized Gram Stain ● This is a direct smear from the gumline. Notice the purple patches of stain on the epithelial cells. Also notice the variable quality of this stain—the epithelial cell to the left of center is stained better than the others.

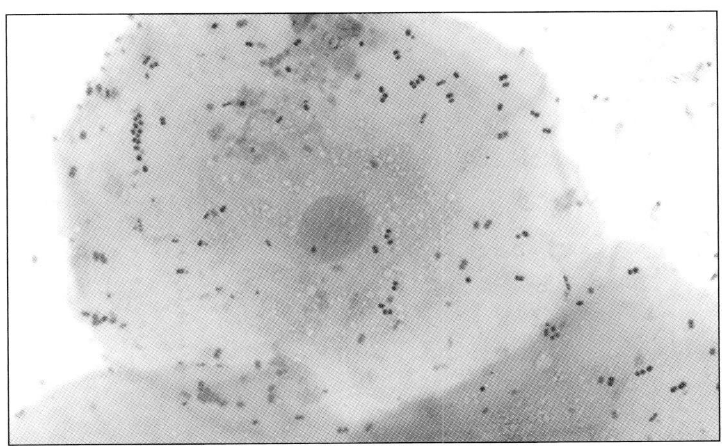

6.10 Over-Decolorized Gram Stain ● This also is a direct smear from the gumline. Notice the virtual absence of any purple cells, a certain indication of over-decolorization.

6.11 KOH Test for Gram Reaction ● This preparation of *Escherichia coli*, a Gram-negative organism, has been emulsified in KOH for 1 minute. The solution has become viscous and stringy due to the release of chromosomal material from the cells. An emulsion of Gram-positive cells does not change in consistency.

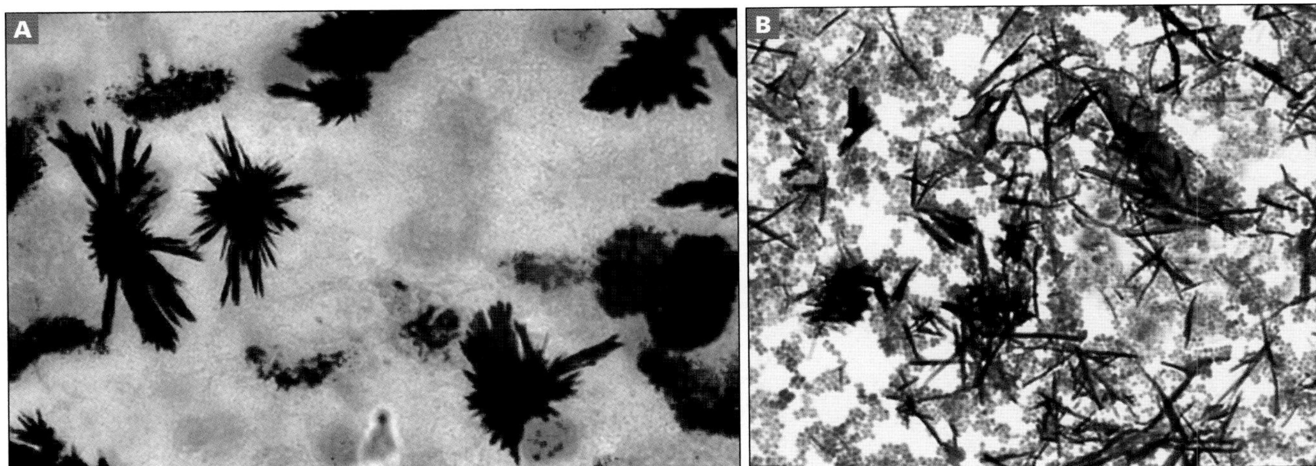

6.12 Crystal Violet Crystals ● If the staining solution is not adequately filtered or is old, crystal violet crystals may appear in the emulsion. Although they are pleasing to the eye, they obstruct the view of the specimen. Crystals from two different Gram stains are shown here. **(A)** This is a gumline direct smear. Epithelial cells and bacterial cells (pink rods) are in the background (this emulsion is way too thick) and the crystals are the large, purple "snowflakes." **(B)** In this Gram stain of *Kocuria rosea* grown in culture, the cells are the lighter purple in the background (can you detect any tetrads?) and the crystals are the darker spikes. (Gram Stain)

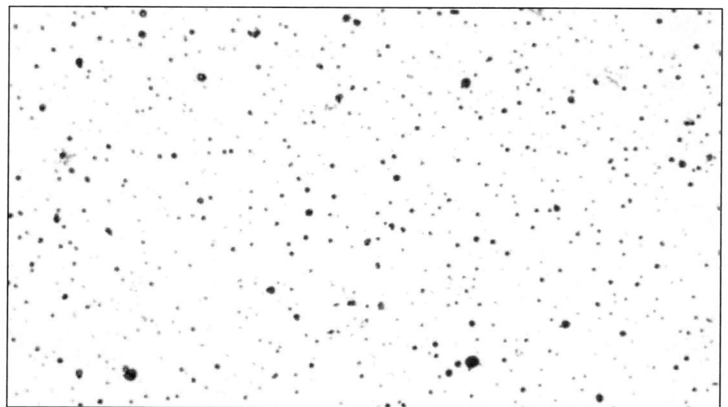

6.13 Stain Precipitate ● If the slide is not rinsed thoroughly or the stain is allowed to dry on the slide, spots of stain precipitate may form and may be confused with bacterial cells. Their variability in size is a clue that they are not bacteria. (Gram Stain)

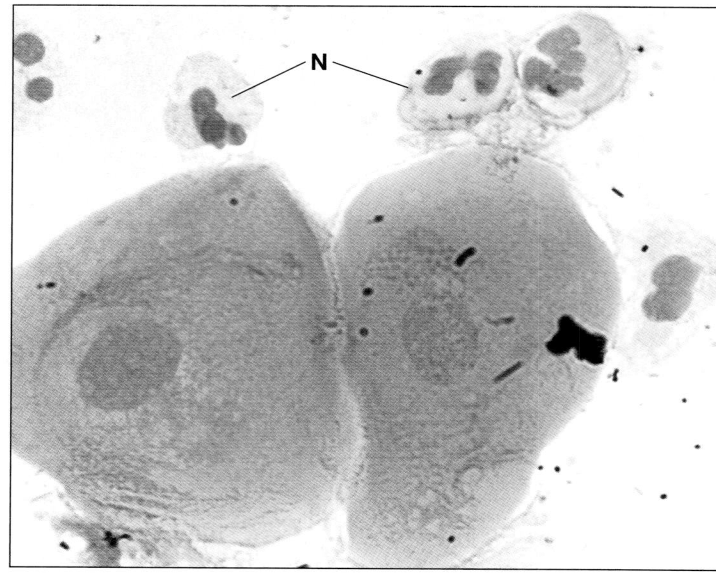

6.14 Neutrophils in a Gumline Direct Smear ● This gumline smear illustrates neutrophils (**N**), cells typically found in inflamed tissue. Notice their size relative to the epithelial cells, their lobed nucleus, and Gram-negative staining reaction. In some preparations, they are very distorted and only the nuclei make them identifiable (see Fig. 6.17). (Gram Stain)

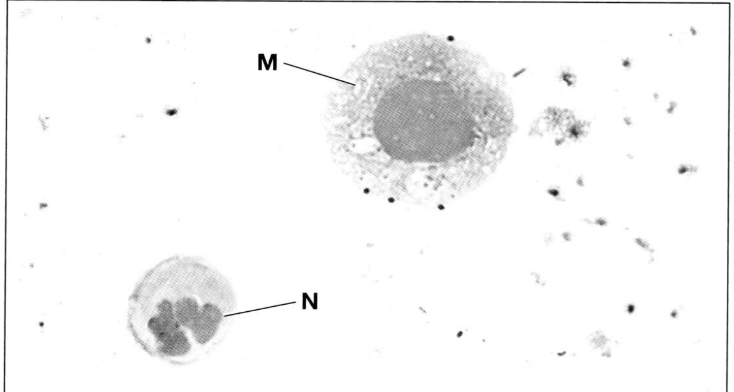

6.15 Macrophage in a Gumline Direct Smear ● In some direct smears, macrophages (**M**) are visible. Compare the size of this macrophage to the single neutrophil (**N**) below it. Notice its spherical nucleus and vacuolated cytoplasm (containing bacteria in the process of digestion). Also notice the Gram-positive cocci on its surface, probably caught in the act of being engulfed. (Gram Stain)

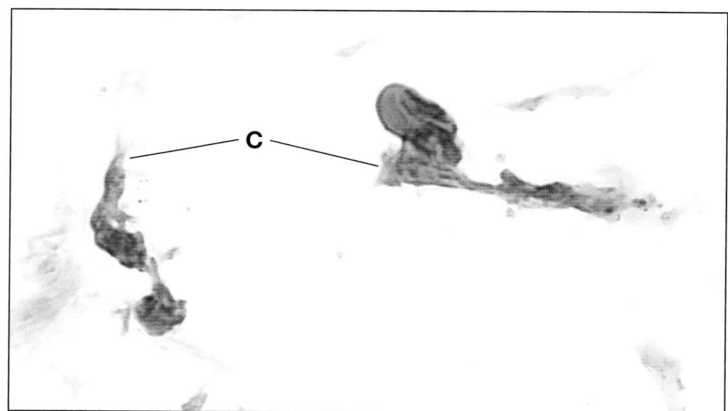

6.16 Respiratory Epithelial Cells in a Nasal Swab Direct Smear ● Two distorted respiratory epithelial cells are seen in this direct smear of a nasal swab. Their columnar and irregular shape, nucleus, and cilia (what are left) provide clues to their identity. (Gram Stain)

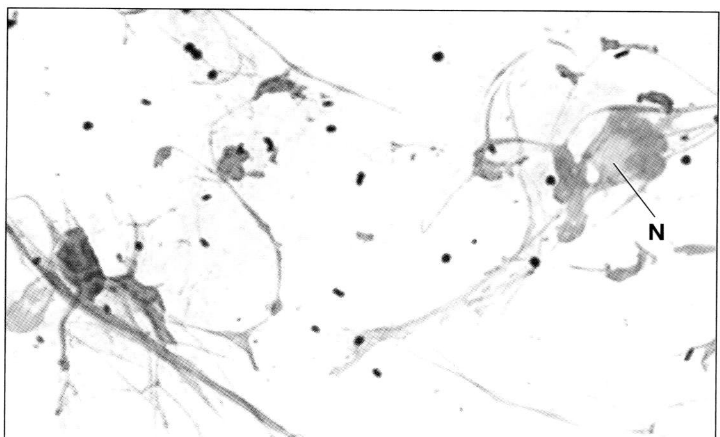

6.17 Mucus in a Nasal Swab Direct Smear ● Mucus strands dominate the field in this Gram-stained nasal swab specimen. Gram-positive cocci and a deteriorating neutrophil (**N**) are also visible. (Gram Stain)

Acid-fast Stains

Purpose

The acid-fast stain is a differential stain used to detect cells capable of retaining a primary stain when treated with an acid alcohol solution. It is an important differential stain used to identify bacteria in the genus *Mycobacterium*, some of which are pathogens (e.g., *M. leprae* and *M. tuberculosis*, causative agents of leprosy and tuberculosis, respectively).

Members of the actinomycete genus *Nocardia* (*N. brasiliensis* and *N. asteroides* are opportunistic pathogens) are partially acid-fast. Oocysts of coccidian parasites, such as *Cryptosporidium* and *Isospora*, are also acid-fast. Because so few organisms are acid-fast, the acid-fast stain is run only when infection by an acid-fast organism is suspected.

Acid-fast stains are useful in identification of **acid-fast bacilli (AFB)** and in rapid, preliminary diagnosis of tuberculosis (with greater than 90% predictive value from sputum samples). It also can be performed on patient samples to track the progress of antibiotic therapy and determine their degree of contagiousness. A prescribed number of microscopic fields are examined and the number of AFB is determined and reported using a standard scoring system (Table 6-1).

Principle

The presence of **mycolic acids** in the cell walls of acid-fast organisms is the cytological basis for the acid-fast differential stain. Mycolic acid is a waxy substance that gives acid-fast cells a higher affinity for the primary stain and resistance to decolorization by an acid alcohol solution (3% HCl in 95% ethanol).

A variety of acid-fast staining procedures are available, two of which are the Ziehl-Neelsen (ZN) method and the Kinyoun (K) method. These differ primarily in that the ZN method uses heat as part of the staining process (this is in addition to heat-fixing the preparation), whereas the K method is a "cold" stain (of a heat-fixed emulsion).

In both protocols the bacterial smear may be prepared in a drop of serum to help the "slippery" acid-fast cells (they are waxy, after all) adhere to the slide. The two methods provide comparable results and the choice comes down to lab conventions and personal preference.

The waxy wall of acid-fast cells repels typical aqueous stains. (As a result, most acid-fast positive bacteria are only weakly Gram positive.) The general sequence of events in an acid-fast stain is shown in Figure 6.18. In both the ZN and K methods, the phenolic compound carbolfuchsin is used as the primary stain, because it is lipid soluble and penetrates the waxy cell wall.

Staining by carbolfuchsin is further enhanced in the ZN method by steam-heating the preparation to melt the wax and allow the stain to move into the wall. The Kinyoun method compensates for not heating the smear by using a more concentrated and lipid-soluble carbolfuchsin as the primary stain. But as a consequence of not heating, the K method is slightly less sensitive than the ZN method.

In both the ZN and K methods, acid alcohol is used to decolorize nonacid-fast cells; acid-fast cells resist this decolorization. A counterstain, such as methylene blue or brilliant

TABLE **6-1** **Acid-fast Smear Reporting Standards***

Number of AFB Seen Fuchsin Stain 1000× Magnification	Number of AFB Seen Fluorochrome Stain 450× Magnification	Number of AFB Seen Fluorochrome Stain 250× Magnification	Reported As
0 AFB per Field	0 AFB per Field	0 AFB per Field	No AFB Seen
1–2 AFB per 300 Fields	1–2 AFB per 70 Fields	1–2 AFB per 30 Fields	Report exact number seen; repeat with another specimen
1–9 AFB per 100 Fields	2–18 AFB per 50 Fields	1–9 AFB per 10 Fields	1+
1–9 AFB per 10 Fields	4–36 AFB per 10 Fields	1–9 AFB per Field	2+
1–9 per Field	4–36 AFB per Field	10–90 AFB per Field	3+
>9 per Field	>36 AFB per Field	> 90 AFB per Field	4+

* Modified from Kent, P.T. and G.P. Kubica. *Public Health Mycobacteriology: A Guide for the Level III Laboratory.* U.S. Department of Health and Human Services, Centers for Disease Control: Atlanta, GA, 1985.

green, is then applied. Acid-fast cells are reddish-purple; nonacid-fast cells are blue or green (Figs. 6.19 and 6.20).

Fluorescent dyes, such as **auramine** (yellow) or **rhodamine** (orange), sometimes in combination, are used in many clinical laboratories and are actually preferable to traditional carbolfuchsin stains for examination of direct smears because of their higher sensitivity. The **fluorochrome** combines specifically with mycolic acid, so no fluorescent antibodies are used (see p. 140).

Acid alcohol is used for decolorization and potassium permanganate is the counterstain. When observed under the microscope with UV illumination, acid-fast cells are yellow and/or orange against a dark background and nonacid-fast cells are not seen (Fig. 6.21). A negative result should be confirmed by a more traditional acid-fast stain.

	Acid-fast	Acid-fast Negative
Cells prior to staining are transparent.		
After staining with carbolfuchsin, cells are reddish-purple. In the ZN stain, steam heat enhances the entry of carbolfuchsin into cells. In the K stain, a higher concentration of a more lipid-soluble carbolfuchsin is used to promote entry into the cells.		
Decolorization with acid alcohol removes stain from acid-fast negative cells.		
Methylene blue or brilliant green is used to counterstain acid-fast negative cells.		

6.18 Acid-fast Stain Overview • Acid-fast cells stain the color of the primary stain, carbolfuchsin, which is reddish-purple. Nonacid-fast cells stain blue with methylene blue or the color of the counterstain if a different one is used.

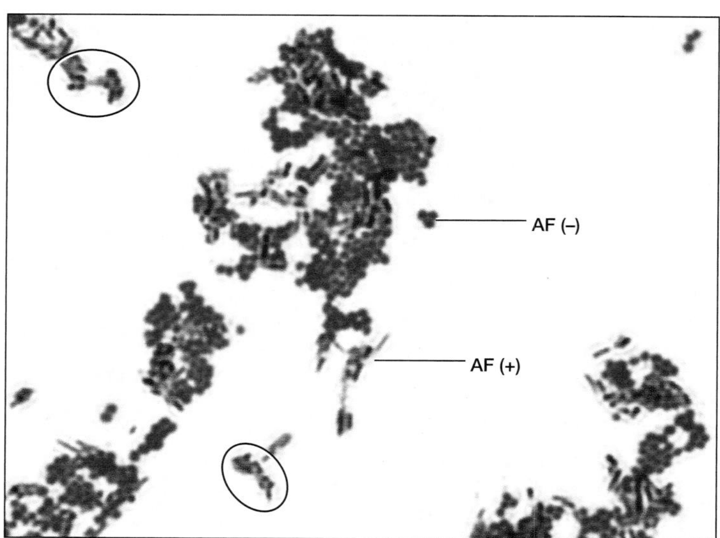

6.19 Acid-fast Stain Using the Ziehl-Neelsen Method • Notice how most of the *Mycobacterium phlei* (**AF+**) cells are in clumps, an unusual state for most rods. They do this because their waxy cell walls make them sticky. To separate them, carefully and gently try mixing them a little longer when preparing the slide than you would for other stains. A few individual cells are visible, however, and they clearly are rods that measure 0.5 µm by 2–3 µm. The *Staphylococcus epidermidis* cells (**AF–**) are also in clumps, but that is because they grow as grapelike clusters. Each cell's diameter is approximately 1 µm. Also notice the characteristic beaded appearance of some AF+ cells (**circled**). Compare this micrograph with Figure 6.20.

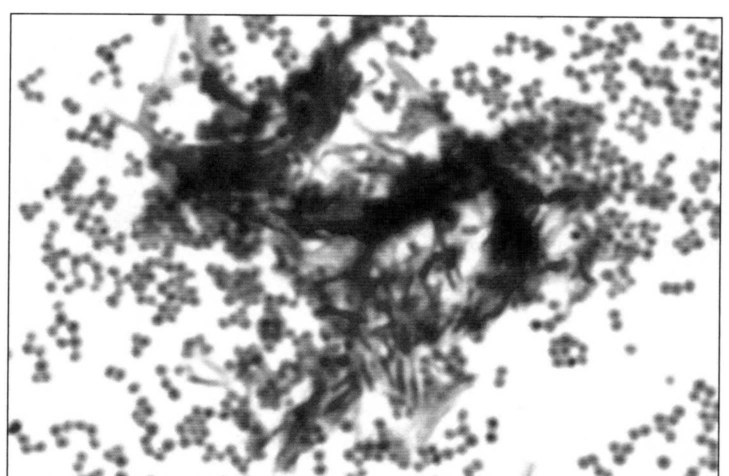

6.20 Acid-fast Stain Using the Kinyoun Method • This is an acid-fast stain of *Mycobacterium smegmatis* (AF+) and *Staphylococcus epidermidis* (AF–) from cultures. Again, notice the clumping of the acid-fast organisms. It is not uncommon for brilliant green dye to stain more of a gray color, but it still contrasts with the carbolfuchsin of acid-fast positive cells. Compare this micrograph with Figure 6.19.

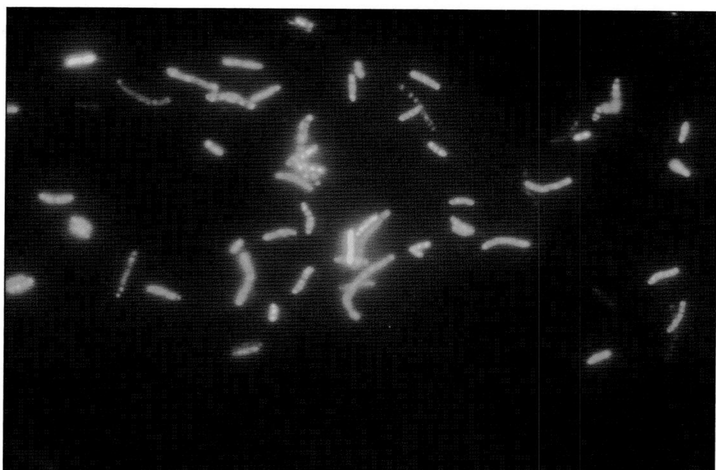

6.21 Fluorochrome Stain of *Mycobacterium kansasii* • Fluorescent staining of AFBs is more sensitive than either of the traditional methods. Notice the characteristic "beaded" appearance and nonuniform staining of these cells. *M. kansasii* causes chronic pulmonary disease. (Auramine O Dye Illuminated with Ultraviolet Light)

Capsule Stain

Purpose

The capsule stain is a differential stain used to detect cells capable of producing an extracellular **glycocalyx** (**capsule**). Capsule production increases virulence in some microbes (such as the anthrax bacillus *Bacillus anthracis* and the pneumococcus *Streptococcus pneumoniae*) by making them less vulnerable to phagocytosis.

Principle

Capsules are composed of mucoid polysaccharides or polypeptides that repel most stains because of their neutral charge. The capsule stain technique takes advantage of this characteristic by staining *around* the cells. Typically, an acidic stain such as Congo red, which stains the background, and a basic stain that colorizes the cell proper are used in combination. The capsule remains unstained and appears as a white halo between the cells and the colored background (Figs. 6.22 and 6.23).

As in a negative stain, cells are spread in a film of an acidic stain and are not heat-fixed, but for a different reason. Heat-fixing causes the cells to shrink and leaves an artifactual white halo around them that might be interpreted as a capsule. In place of heat-fixing, cells may be emulsified in a drop of serum to promote adhering to the glass slide.

In the TEM of *Lactococcus lactis* (Fig. 6.24), a thin capsule is visible as a "fuzzy" layer on the cell's surface.

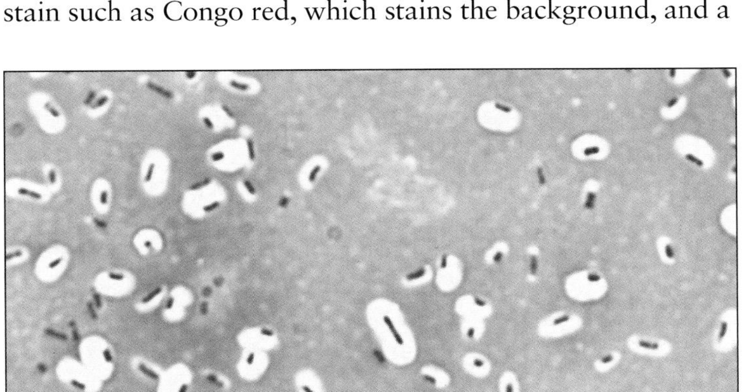

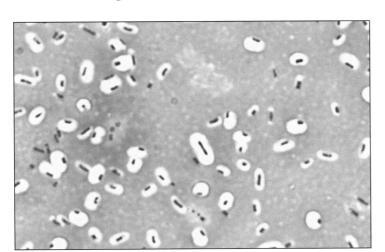

6.22 Capsule Stain of *Klebsiella pneumoniae* • The acidic stain colorizes the background while the basic stain colorizes the cell, leaving the capsules as unstained, white clearings around the cells. Notice the lack of uniform capsule size, and even the absence of a capsule in some cells. Compare this micrograph to Figure 6.23.

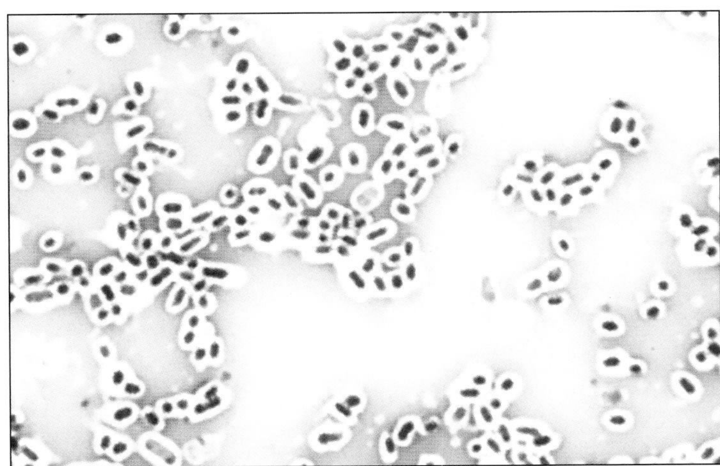

6.23 Alternative Capsule Stain of *Klebsiella pneumoniae* • In this capsule stain, Congo red is the acidic stain and Maneval's is the basic stain. After staining, the Congo red often looks bluish or gray. *K. pneumoniae* is an inhabitant of the intestinal tract of humans and is associated with urinary and respiratory tract infections, but this specimen was grown on an agar slant. The cells are approximately 1 µm wide by 2–4 µm long. Compare this micrograph to Figure 6.24.

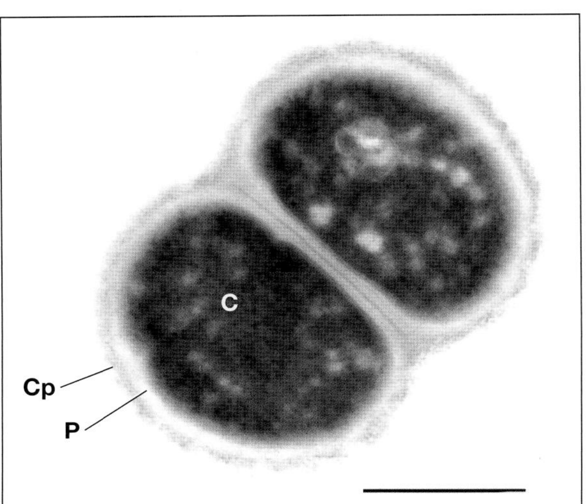

6.24 *Lactococcus lactis* Capsule • Shown is a thin capsule (**Cp**) of *Lactococcus lactis*. Capsules, also known as glycocalyces, are made of mucoid polysaccharides or polypeptides of varying compositions. Also visible are the peptidoglycan (**P**) and cytoplasm (**C**). The cytoplasmic membrane did not stain in this preparation. (Scale bar = 500 nm) (TEM)

Endospore Stain

Purpose

The **endospore** stain is a differential stain used to detect the presence, shape, and location of endospores in bacterial cells. Several genera within the phylum Firmicutes produce endospores, but the two most common are *Bacillus* (more than 100 species) and *Clostridium* (more than 160 species). In addition, several "new" genera formerly classified as *Bacillus* species also produce endospores. These include *Brevibacillus*, *Geobacillus*, *Paenibacillus*, and many others.

Most *Bacillus* species are soil, freshwater, or marine **saprophytes**, but two are pathogens. *B. anthracis* is the causative agent of anthrax and *B. cereus* causes two kinds of food poisoning: emetic and diarrheal (see pp. 184 and 185). Most members of *Clostridium* are soil or aquatic saprophytes or inhabitants of human intestines, but four pathogens are fairly well known: *C. tetani* (tetanus), *C. botulinum* (botulism), *C. perfringens* (gas gangrene), and *C. difficile* (pseudomembranous colitis). These are discussed more fully in Section 12, beginning on page 192.

Principle

Some bacterial species are able to differentiate from active **vegetative cells** (at this point referred to as **spore mother cells**) into dormant endospores when environmental conditions, such as nutrient depletion or high temperatures, are unsuitable for growth. (Note that bacterial endospores are not reproductive cells.) Endospores are highly resistant to heat and chemicals, which allows them to survive in this state for long periods of time. The total absence of ATP within endospores is an indication of how dormant they are.

In addition to nutrient depletion, sporulation has also been shown to be dependent on population density. With increased density, a secreted peptide (called **competence and sporulation factor**, or **CSF**) reaches a critical concentration and results in derepression of sporulation genes, which is a fancy way of saying, "Taking the brakes off" of them.

The electron microscope provides a view of the endospore's fine structure (Fig. 6.25). At the center is the **core.** Unlike cytoplasm, it is almost completely dehydrated and has the consistency of a gel. The most visible core contents are DNA and ribosomes, but also present is **dipicolinic acid,** which assists in dehydrating the core.

Two main layers encircle the core: the **cortex** and the **spore coat.** The cortex is made of modified peptidoglycan, and the spore coat is made of protein layers, including **keratin.** In some species, the spore coat may be enclosed in a membranous **exosporium.** Endospores' resistance is due to a combination of factors, including coat keratin, its dehydrated state, and enzyme inactivation. When conditions become suitable for growth, endospores germinate into metabolically active vegetative cells.

The keratin in the spore coat also resists staining, so extreme measures must be taken to stain an endospore. In the Schaeffer-Fulton method (Fig. 6.26), a primary stain of malachite green is forced into the spore by steaming the bacterial emulsion. Alternatively, malachite green can be left on the slide for 15 minutes or more to stain the spores. Malachite green is water soluble and has a low affinity for cellular material, so **vegetative cells** and **spore mother cells** can be decolorized simply with a water rinse and then counterstained with safranin (Fig. 6.27).

Endospores may be located in the middle of the cell (**central**), at the end of the cell (**terminal**), or between the end and middle of the cell (**subterminal**). Their location in some species is variable and cells may exhibit a combination of terminal and subterminal endospores, for instance. Endospores also may be differentiated based on shape—either **spherical** or **elliptical** (**oval**)—and size relative to the cell, that is, whether or not they cause the cell to look swollen. These structural features are shown in Figures 6.28 through 6.31.

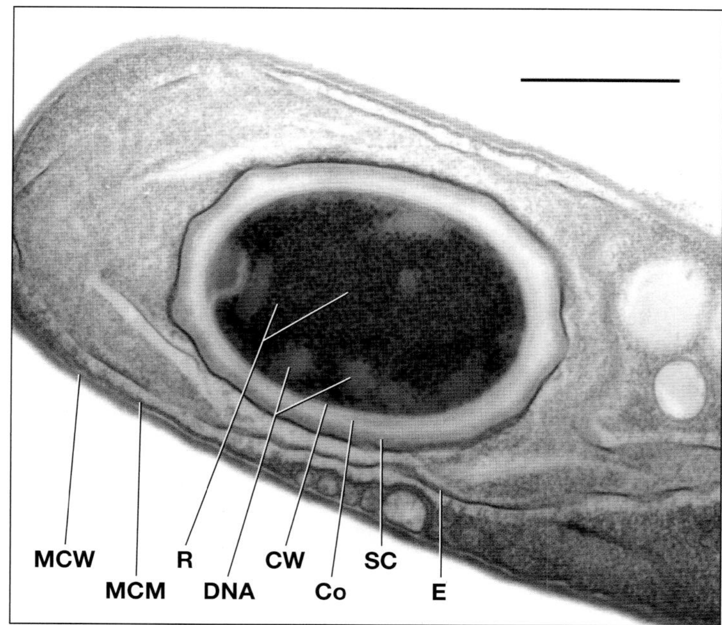

6.25 Endospore • Shown is *Bacillus cereus* in thin section through an endospore still within the mother cell. Occupying the core are ribosomes (**R**) and DNA (**DNA**). Then, moving outward are the core wall (**CW**), cortex (**Co**), spore coat (**SC**), and developing exosporium (**E**). Also shown are the mother cell membrane (**MCM**) and mother cell wall (**MCW**). (Scale bar = 500 nm) (TEM)

A special stain is not required to visualize endospores. Figure 6.32 is a phase contrast micrograph of estuarine mud in which several spore-forming bacteria are visible. Without staining, however, one must be careful not to confuse inclusions, such as sulfur (Fig. 6.33A) or lipid (Fig. 6.33B) granules, with true endospores. The spore stain will definitively identify true endospores. *Corynebacterium* species may also be a source of confusion, because they often have club-shaped swellings (Fig. 6.34).

	Spore producer	Spore nonproducer
Cells and spores prior to staining are transparent.		
After staining with malachite green, cells and spores are green. Heat is used to force the stain into spores, if present.		
Decolorization with water removes stain from cells, but not spores.		
Safranin is used to counterstain cells.		

6.26 Schaeffer-Fulton Endospore Stain Overview • Upon completion, endospores are green, vegetative and spore mother cells are red.

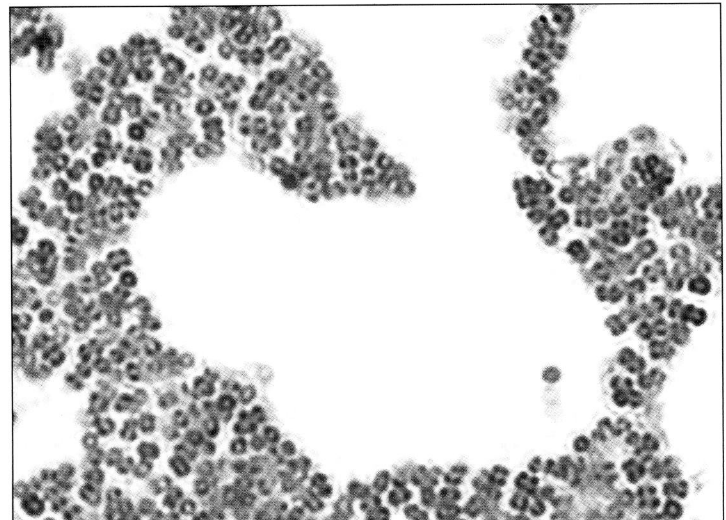

6.28 Spherical Endospores • *Clostridium* and *Bacillus* are the largest and most commonly encountered endospore forming genera, but there are at least a half-dozen others. *Sporosarcina ureae* cells are spherical, as are their endospores. The cells are approximately 2 μm in diameter.

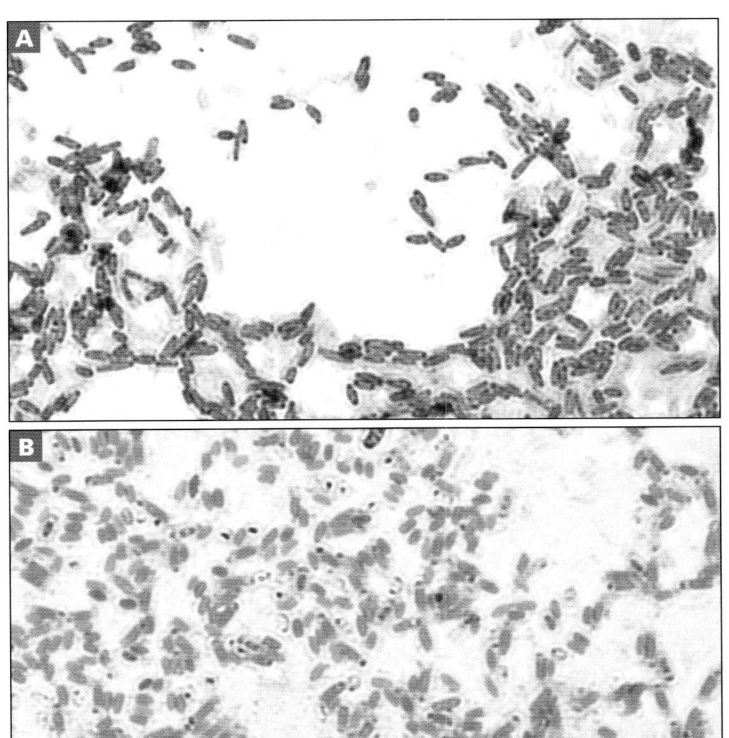

6.27 Culture Age Can Affect Sporulation • Bacteria capable of producing endospores do not do so uniformly during their culture's growth. Sporulation is a cellular response to nutrient depletion, and so is characteristic of older cultures. These two cultures illustrate different degrees of sporulation. (**A**) Most cells in this specimen contain endospores; very few have been released. The light pink rods are probably spore mother cells that have released their spore and have died. (**B**) This specimen consists mostly of released endospores.

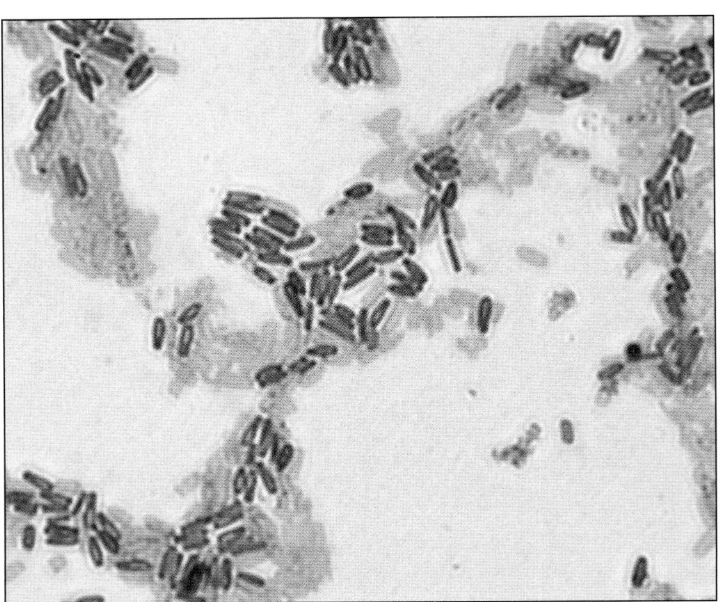

6.29 Central Elliptical Endospores • Most of these *Bacillus megaterium* spores are centrally positioned, but some are subterminal. All are elliptical, but they do not distend the mother cell.

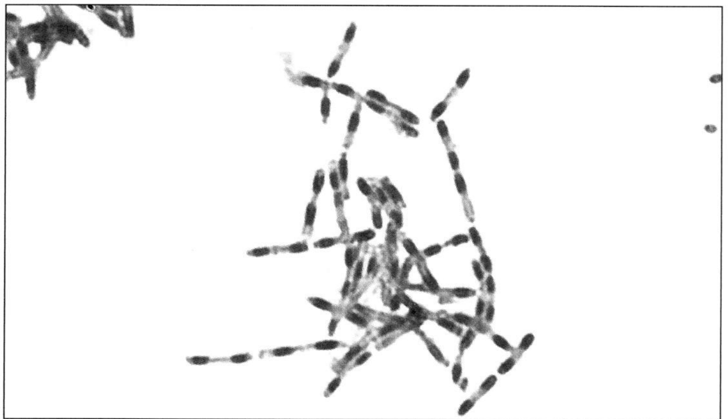

6.30 Subterminal Endospores • These *Bacillus anthracis* endospores are unevenly distributed within the chain of cells, indicating that they're not centrally positioned. Most are subterminal, but a few appear to be terminal. They do not distend the cell.

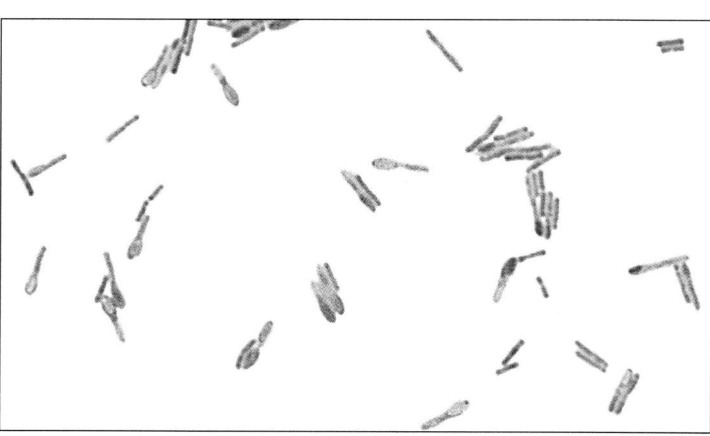

6.31 Elliptical Terminal Endospores • In these *Clostridium butyricum* cells, the green endospores are frequently "capped" by red cytoplasm, but most microbiologists would classify them as terminal. Notice how the spores have caused the ends of the cells to swell.

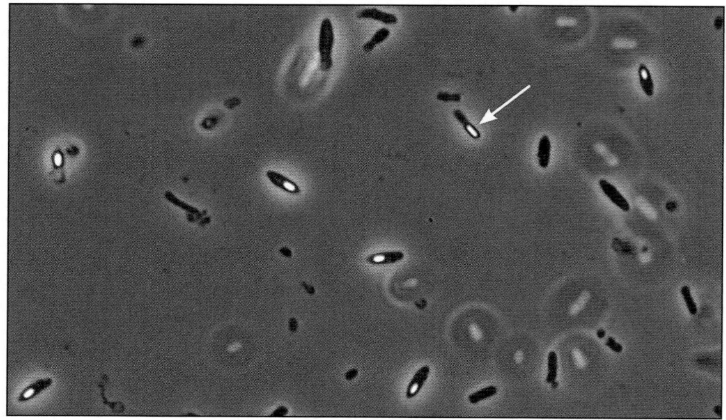

6.32 Endospores as Seen with Phase Contrast Microscopy • The white, elliptical spores are easily seen in these unidentified anaerobic rods found in an estuarine mud sample. Because they are anaerobic, it is likely they are at least two species of *Clostridium* (based on the difference in morphology between the cell marked with the **arrow** and the others).

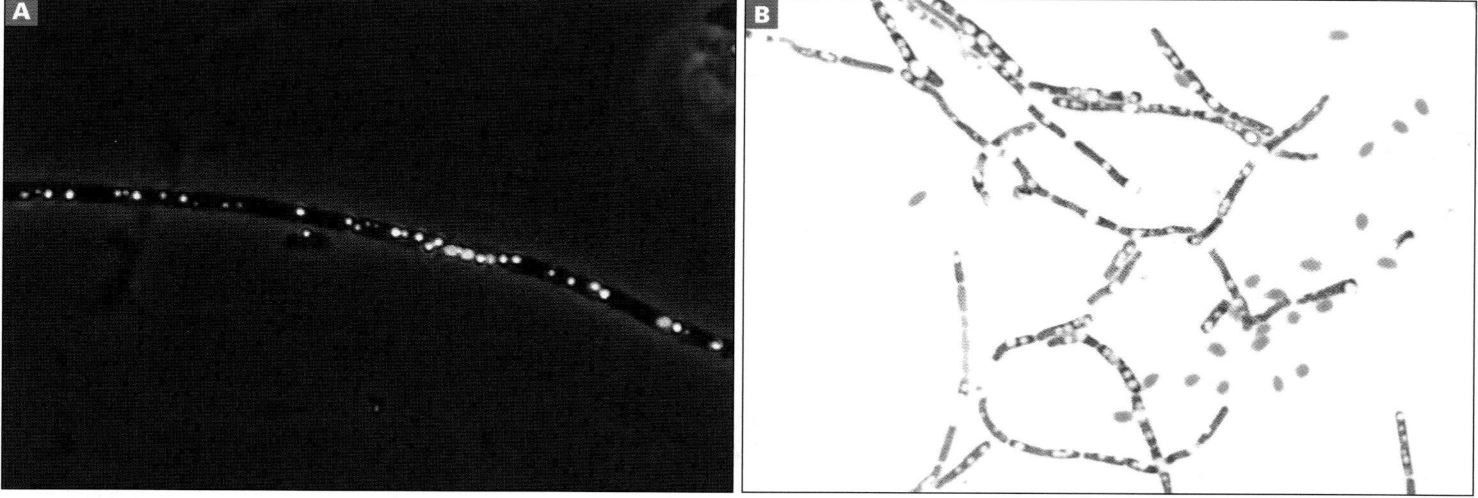

6.33 Endospore Stain Allows Differentiation between True Endospores and Cellular Inclusions • (A) The bright spots in this filamentous bacterium (possibly *Thiothrix*) are sulfur granules, but they might be confused with endospores in this phase contrast micrograph. Their irregular size and the presence of more than one per cell are clues that they are not endospores, but an endospore stain would remove any doubt. (B) This micrograph is an endospore stain of *Bacillus cereus* grown in culture. The spores are green, but notice all the unstained, white spots inside the cells. Some are developing spores similar in size to the stained ones, but the others are lipid granules. Now, imagine looking at this specimen using a simple stain or a Gram stain. Would you be able to identify the spores or would the lipid granules mislead you? For an example of the opposite situation, with stained lipid granules and unstained endospores, see Figure 6.45.

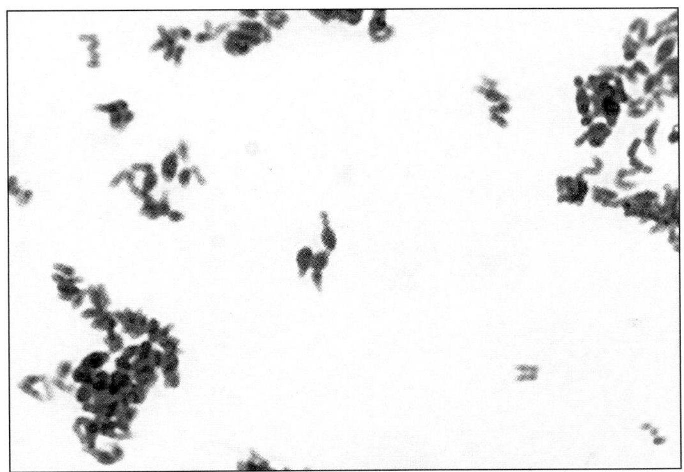

6.34 **Corynebacteria May Have Terminal, Club-Shaped Swellings** ● This is a micrograph of *Corynebacterium diphtheriae* grown in culture. In a simple stain, these swellings might be misinterpreted as spores. One clue that they aren't endospores is that they have taken up the stain.

Flagella Stain

Purpose

The flagella stain allows direct observation of flagella. Presence and arrangement of flagella may be useful in identifying bacterial species.

Principle

Bacterial flagella typically are too thin to be observed with the light microscope and ordinary stains, but there are exceptions. Figure 6.35 shows a wet mount preparation of a spirillum viewed with phase contrast and its flagellar bundles are visible without staining. However, in most instances staining will be necessary.

Unlike other staining procedures, flagella stains do more than add color. Various special flagella stains have been developed that use a **mordant** to assist in coating flagella with stain to a visible thickness. Most require experience and advanced techniques, and generally are not performed in beginning microbiology courses.

The number and arrangement of flagella may be observed with a flagella stain. A single flagellum at the end of a cell is said to be **polar** and the cell has a **monotrichous** arrangement (Fig. 6.36). Other arrangements (shown in Figs. 6.37–6.39) include **amphitrichous,** with flagella at both ends of the cell; **lophotrichous,** with tufts of flagella at the end of the cell; and **peritrichous,** with flagella emerging from the entire cell surface.

Generally speaking, it takes the electron microscope to fully appreciate the structure of bacterial flagella (Fig. 6.40). They are extraordinarily thin (a few nanometers), long (a hundred times, or more, longer than their diameter), flexible, hollow rods of protein. They propel the cell by rotating like a propeller. ●

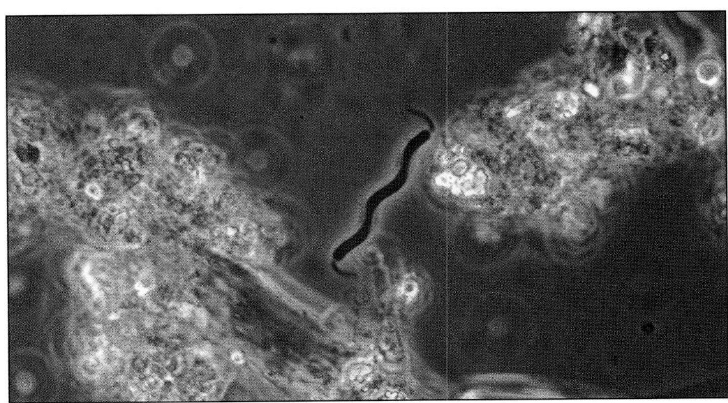

6.35 **Bacterial Flagella Without Staining** ● The flagella of some larger bacteria are visible without staining, as in this wet mount of an unidentified spirillum viewed with phase contrast (they are not nearly as visible with bright field). Each "flagellum" is actually a bundle of several flagella, which is why they are visible without staining. They don't look particularly long because much of their length is out of the focal plane of the objective lens. (Phase Contrast)

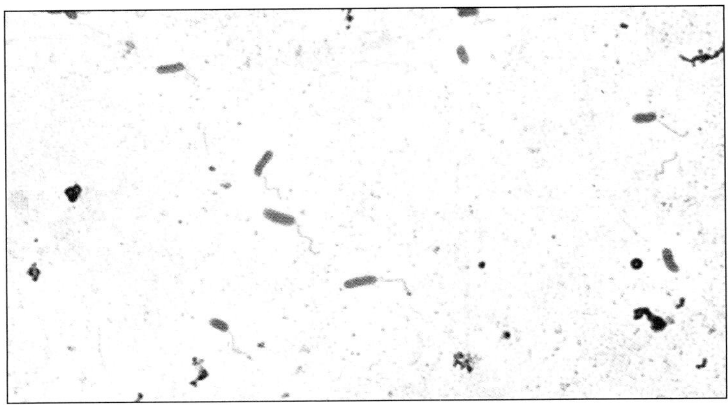

6.36 **Monotrichous (Single Polar) Flagella** ● *Pseudomonas aeruginosa* is often suggested as a positive control for flagella stains, but it is a BSL-2 organism. Notice the single flagellum emerging from the ends of many (but not all) cells. This is due to the fragile nature of flagella, which can be broken from the cells during slide preparation.

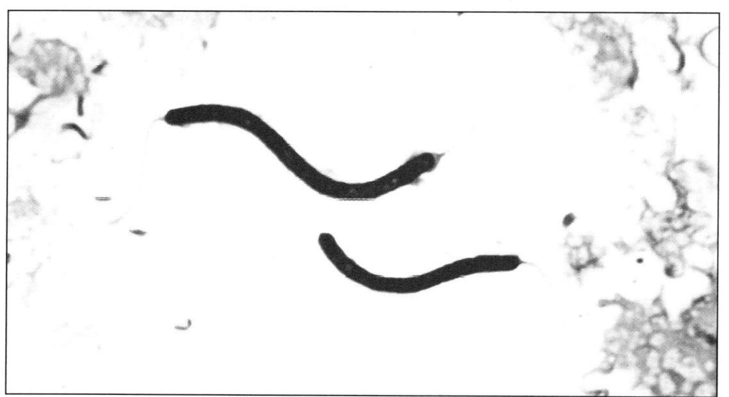

6.37 Amphitrichous Flagella • *Spirillum volutans* has a bundle of flagella at each end, just like the spirillum in Figure 6.35. However, they are more visible when stained and are flattened on a slide so more of the flagellum is in focus.

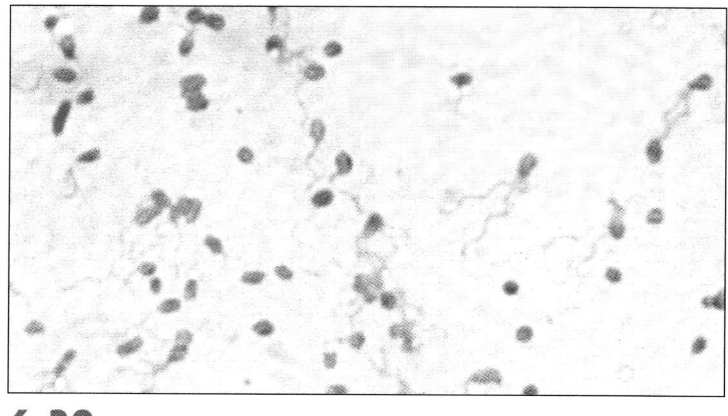

6.38 Lophotrichous Flagella • Several flagella emerge from one end of this *Pseudomonas* species. Again, not all cells have flagella because they were too delicate to stay intact during the staining procedure.

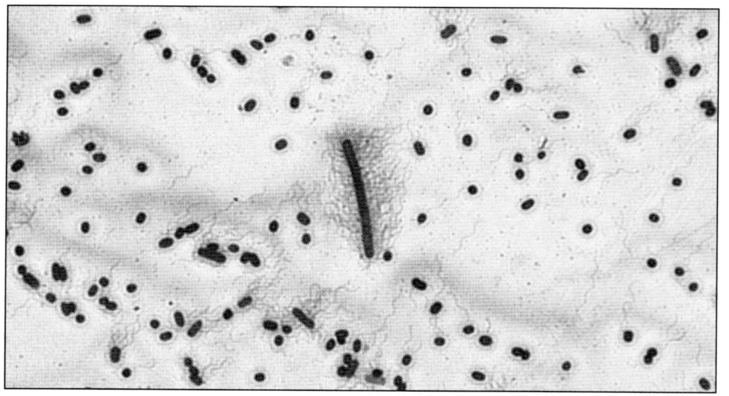

6.39 Peritrichous Flagella • *Proteus vulgaris* is capable of swarming motility (Fig. 3.22), in which the cells spread across the agar surface at specific intervals, then remain in place until the next swarm. The smaller flagellated cells are called swimmers and are the form seen when grown in a liquid medium. They are abundant in this micrograph. When transferred to a solid medium or under certain environmental conditions, swimmers differentiate into swarmers. A swarmer with its peritrichous flagella is seen in the center of this micrograph. Swarmers are larger, contain multiple nucleoids (the site of DNA), and produce an extracellular slime or capsule that assists in swarming. After swarming, they break up into swimmer cells. Swarmers are the more virulent form.

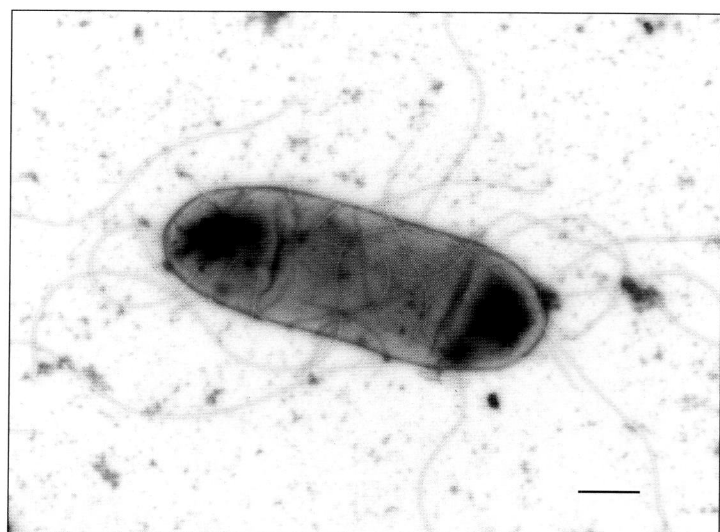

6.40 Flagella • This single *E. coli* cell has peritrichous flagella, which is typical of motile members of the Enterobacteriaceae. Flagella are thin, hollow protein rods that propel the cell by rotating. These flagella are about 5 nm in diameter. Most of them are broken in this micrograph. (Scale bar = 500 nm) (TEM, Negative Stain)

Wet Mount and Hanging Drop Preparations

Purpose

Most bacterial microscopic preparations result in death of the microorganisms due to heat-fixing and staining. Simple **wet mounts** and the **hanging drop technique** allow observation of living cells to determine motility. They also are used to see natural cell size, arrangement, and shape. All of these may be useful characteristics in the identification of a microbe. In addition, observation of dividing cells is possible with the hanging drop preparation.

Principle

A wet mount preparation is made by placing the specimen in a drop of water on a microscope slide and covering it with a cover glass. Because no stain is used and most cells are transparent, viewing is best done with as little illumination as possible by adjusting the iris diaphragm (Fig. 6.41).

Motility often can be observed at low or high dry magnification, but viewing must be done quickly because the preparation will dry out in 15–20 minutes. Swimming bacteria will move independently of one another, either in a straight line or in a zigzag, random path.

A hanging drop preparation allows longer observation of the specimen because it does not dry out as quickly. A thin ring of petroleum jelly is applied around the well of a depression slide (Fig. 6.42). Then a drop of water is placed in the center of a cover glass and living microbes are gently transferred into it. The depression slide is carefully placed over the cover glass in such a way that the drop is received into the depression and is undisturbed. The petroleum jelly causes the cover glass to stick to the slide.

The preparation may then be picked up, carefully inverted so the cover glass is on top, and placed under the microscope for examination. As with the wet mount, viewing is best done with as little illumination as possible, so adjust the iris diaphragm. The petroleum jelly forms an airtight seal that slows drying of the drop, allowing a long period for observation of cell size, shape, binary fission, and motility.

If these techniques are done to determine motility, the observer must be careful to distinguish between true motility and **Brownian motion** created by collisions between cells and water molecules. In the latter, cells will appear to vibrate in place. Nonmotile cells should exhibit Brownian motion. Cells that actually swim will exhibit independent movement over greater distances and the effects of collisions with water molecules will be insignificant.

In addition to Brownian motion, other complications can present themselves when observing wet mounts or hanging drop preparations for motility. Following is a list you can use to troubleshoot if you do not see motility.

- Observation: Cells are in one focal plane and not moving, even by Brownian motion. Interpretation and solution: The cells are stuck to the glass and may or may not be motile if freed (which you won't be able to do). Focus on cells in the water between the coverslip and the slide.
- Observation: Cells are streaming across the field in the same direction. Interpretation and solution: There is a current carrying the cells and they may or may not be motile. The water is either receding due to evaporation or the coverslip was put on too vigorously. Try finding a calmer place to observe the cells or simply wait until the current dissipates. If the water is evaporating, make another slide.
- Observation: Cells are packed together and not moving except, perhaps, by Brownian motion. Interpretation and solution: You have a traffic jam and the cells may or may not be motile. Look around for a part of the slide where the cells are not so densely packed. If you can't find such a region, make a new slide with fewer cells.

One last complication presents itself, and you won't know that it has occurred because most bacterial flagella are too thin to see with the light microscope. Flagella are very delicate and it is possible to damage them when transferring cells to the slide. It is best to check for motility using a broth culture rather than trying to transfer cells from a solid medium to the water drop. Also, do not emulsify the loopful of broth on the slide. Just gently allow the broth to merge with the water drop already on the slide.

6.41 Wet Mount • Shown is an unstained wet mount preparation of *Bacillus megaterium*, a motile Gram-positive rod using the high dry lens. Because of the water's thickness (really a thin film, but size is relative) in the wet mount, cells show up in many different focal planes and are mostly out of focus. To get the best possible image, adjust the condenser height to its proper position and reduce the light intensity with the iris diaphragm. Then, use the fine focus adjustment to scan for motile cells. Just be careful not to hit the cover glass with the objective lens.

6.42 Depression Slide • For short-term viewing of a wet mount, a standard slide and cover glass preparation can be used, but these dry out quickly. For longer-term viewing, a hanging drop slide is prepared with a depression slide. Because the edge of the cover glass over the depression is sealed with petroleum jelly (not shown), the slide doesn't dry out as quickly and can be used to observe cell activities such as division and motility.

Miscellaneous Structures

Bacterial cellular structure is simple compared to eukaryotic cells. However, differentiation of cytoplasmic components is possible to a certain degree. Figures 6.43 through 6.48 illustrate some of these structures as seen with the light microscope.

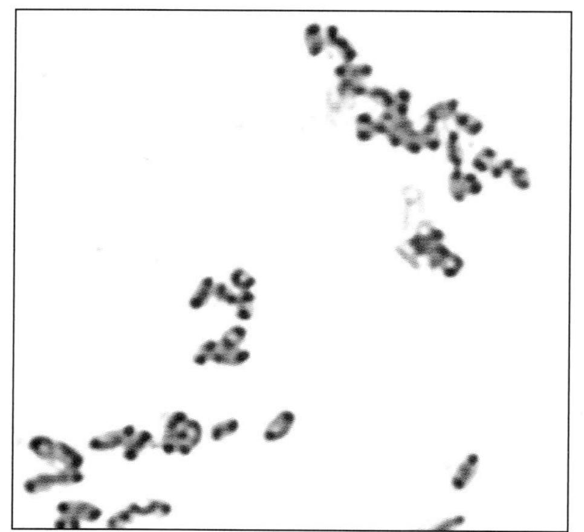

6.43 Bipolar Staining ● Some species in the *Enterobacteriaceae* (Gram-negative rods often found in the large intestine) exhibit bipolar staining, in which the ends of the cells stain darker than the center. These cells are said to be "vacuolated," but they do not have true, membrane-bound vacuoles. This is *E. coli* grown in culture. Bipolar staining is also seen in the *E. coli* TEM, Figure 6.40. (Gram Stain)

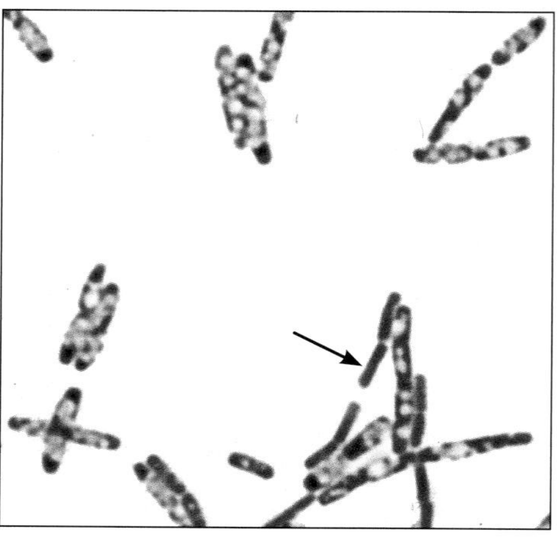

6.44 Vacuolated Cytoplasm ● *Bacillus* cells may stain uniformly (**arrow**) or they may have a foamy, vacuolated appearance (most cells in the field). The large, smooth, white regions within the cells are probably developing endospores, but the more common irregular white regions are probably lipid storage granules. Vacuolated is actually a misnomer, because *vacuus* means "empty," and these are not empty spaces. This is *Bacillus cereus* grown in culture. Compare its appearance in this micrograph with the fat-stained micrograph in Figure 6.45. (Gram Stain)

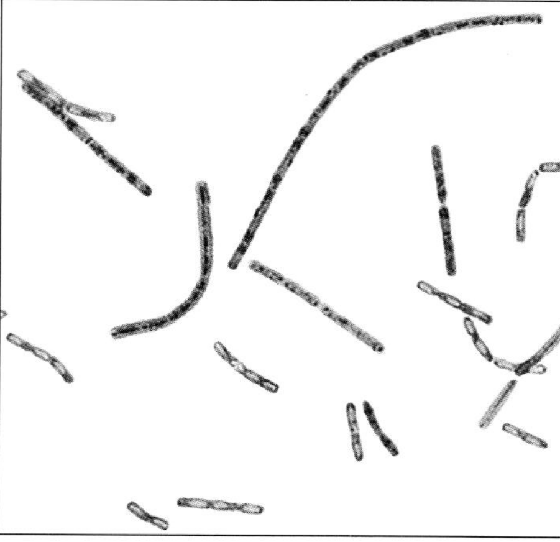

6.45 Organic Cytoplasmic Granules ● Sudan black B stain is specific for lipids, such as poly-β-hydroxybutyrate (PHB) granules. These dark staining PHB granules serve as a carbon and energy reserve. The white ovals are unstained endospores. This specimen is *Bacillus cereus;* other organisms store carbon in the form of starch or glycogen. Most of the intracellular white material in Figure 6.44 would appear black if stained with Sudan black B. (Sudan Black B Stain)

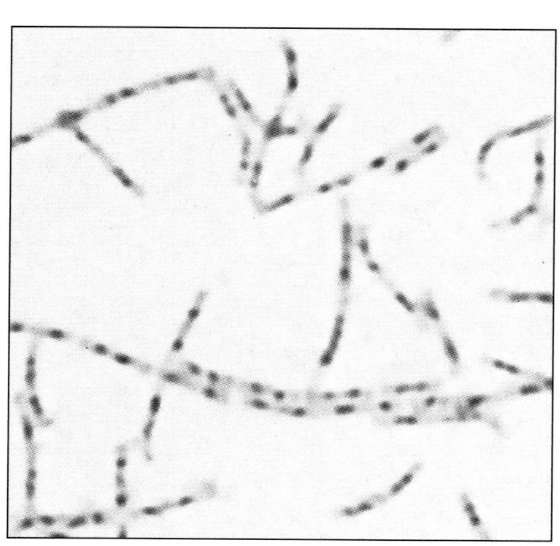

6.46 Nucleoid ● Feulgen and Giemsa stains can be used to stain DNA, with the former producing a red color and the latter staining purple. This micrograph shows the *Bacillus cereus* nucleoid. Remember that what appears to be a nucleus is not. There is no membrane separating the nucleoid from the cytoplasm (see Fig. 4.7).

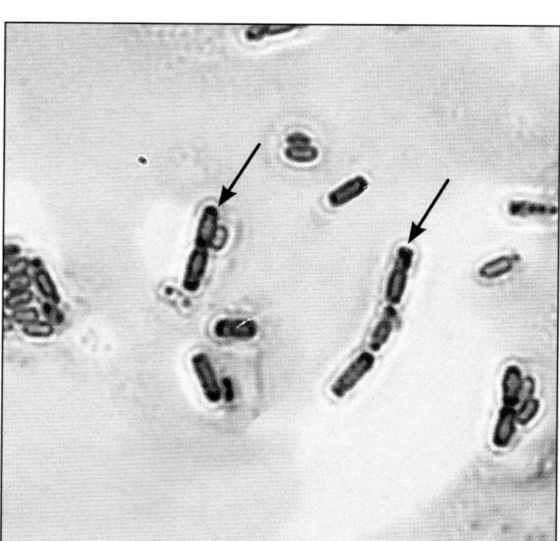

6.47 Parasporal Crystals ● *Bacillus thuringiensis* produces proteinaceous bodies near its spores called parasporal crystals (the dark objects indicated by **arrows**). These crystals kill the larvae of various insect groups (especially Lepidopterans). After ingestion of the crystal, it is activated in the larval gut by a protease enzyme. The result is cytolysis of larval cells and, presumably, a ready nutrient source for the endospore when it germinates. These crystals have been commercially marketed as the insecticide Bt toxin.

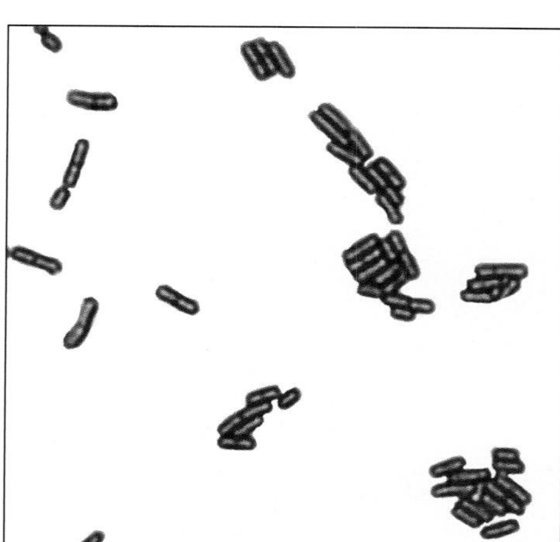

6.48 Cell Wall Stain ● This is a micrograph of *Bacillus coagulans* showing its cell walls. Notice the wall separating cells that form pairs or a chain. This indicates that they are in the process of dividing or have recently completed a division

Differential Tests and Multiple-Test Systems

SECTION 7

Part A: Differential Tests

Bacitracin Susceptibility Test

Purpose

Streptococcus pyogenes, **Lancefield group A** (see p. 136), is differentiated from other **β-hemolytic streptococci** by its susceptibility to the antibiotic bacitracin. It also differentiates *Staphylococcus* species (resistant) from *Micrococcus* and *Rothia mucilaginosa* (*Stomatococcus mucilaginosus*), which are susceptible.

Principle

Antibiotics are antimicrobial substances produced by other microorganisms. **Bacitracin**, produced by *Bacillus licheniformis*, is a powerful peptide antibiotic that inhibits bacterial cell wall synthesis by interfering with transport of peptidoglycan subunits by **undecaprenyl diphosphate** (**bactoprenol**) across the cytoplasmic membrane (Fig. 7.1). Thus, it is effective only on bacteria that have cell walls and are in the process of growing.

The bacitracin test is a simple test performed by placing a bacitracin-impregnated disk (0.04 U) on an agar plate inoculated to produce a bacterial lawn. The bacitracin diffuses into the agar and, where its concentration is sufficient, inhibits growth of susceptible bacteria. Inhibition of bacterial growth will appear as a clearing on the agar plate. Any zone of clearing 10 mm in diameter or greater around the disk is interpreted as bacitracin susceptibility (Table 7-1 and Fig. 7.2).

For more information on antimicrobial susceptibility, refer to the antimicrobial susceptibility (Kirby-Bauer) test, page 291.

$$CH_3\text{-}\overset{CH_3}{\overset{|}{C}}\text{=CH-}CH_2\text{-}[CH_2\text{-}\overset{CH_3}{\overset{|}{C}}\text{=CH-}CH_2]_9\text{-}CH_2\text{-}\overset{CH_3}{\overset{|}{C}}\text{=CH-}CH_2\text{-}Ⓟ\text{-}Ⓟ\text{-NAM-NAG}$$

NAM — D-Ala — D-Glu — L-R_3 — D-Ala — D-Ala

7.1 Undecaprenyl Diphosphate Structure (Bactoprenol) Undecaprenyl diphosphate (UDP) is a lipid soluble, C_{55} molecule derived from 11 isoprene subunits plus one phosphate. It transports the lipid *insoluble* peptidoglycan subunits, which are assembled in the cytoplasm, across the cell membrane during cell wall synthesis. In this illustration, a [NAG-NAM-pentapeptide] subunit is shown attached to the UDP carrier. (Formation of a tetrapeptide crosslink in the wall requires removal of the fifth amino acid, D-Ala. See a general microbiology text for details of peptidoglycan structure.) Bacitracin inhibits removal of a phosphate from UDP after it delivers its peptidoglycan subunit to the cell's exterior, making it incapable of repeating the process and resulting in inhibition of wall growth.

7.2 Bacitracin Susceptibility Test on Sheep Blood Agar • *Staphylococcus aureus* (resistant, R) is above and *Micrococcus luteus* (susceptible, S) is below.

TABLE **7-1** **Bacitracin Test Results and Interpretations (0.04 U Disk)**

Result	Interpretation	Symbol
Zone of clearing 10 mm or greater	Organism is susceptible to bacitracin; probable *Streptococcus pyogenes* if a β-hemolytic streptococcus or *Micrococcus* species if a Gram-positive coccus	S
Zone of clearing less than 10 mm	Organism is resistant to bacitracin; not *Streptococcus pyogenes*; probable *Staphylococcus aureus* if Gram-positive coccus	R

β-Lactamase Test

Purpose

Bacterial resistance to β-lactam antibiotics, penicillins and cephalosporins, and antibiotics in general is an increasing medical issue. Traditional antibiotic resistance tests using disk diffusion (see p. 291) take 24 hours before they can be read, but the β-lactamase test can usually be read within a few minutes.

A positive result indicates that a patient isolate is resistant to penicillins and/or cephalosporins. The test is especially useful in identifying resistant strains of *Neisseria gonorrhoeae*, *Staphylococcus* species, and *Enterococcus* species.

Principle

Penicillins and cephalosporins—called **β-lactam antibiotics** because of the β-lactam ring in their chemical structure—kill bacteria by interfering with cell wall synthesis. The bacterial enzyme **transpeptidase** catalyzes cross-linking between peptidoglycan subunits thus adding rigidity to the wall.

By competing for sites on transpeptidase, β-lactam antibiotics prevent essential cross-linking of the peptidoglycan. Many bacteria have evolved resistance to these antibiotics by producing enzymes called **β-lactamases.** β-lactamases hydrolyze the β-lactam ring, thus destroying the structure of the antibiotic. Partial chemical structures of β-lactam antibiotics are shown in Figure 7.3.

The β-lactamase test is one of many tests used to identify β-lactamase production by measuring resistance to β-lactam antibiotics. In this test, a paper disk containing **nitrocefin** is smeared with the test organism. Nitrocefin is a **cephalosporin** susceptible to most β-lactamases that turns pink when it is hydrolyzed. Therefore, if the test organism produces β-lactamase, it will hydrolyze the nitrocefin and produce a pink spot on the disk (Fig. 7.4).

Penicillin Core

Cephalosporin Core

7.3 β-Lactam Antibiotic Structure • Although penicillins and cephalosporins have unique structures, their bactericidal effects are similar because of their similarities in structure, which includes a β-lactam ring. They interfere with the transpeptidase enzyme that catalyzes peptidoglycan cross-linking, which weakens the cell wall and leads to cell lysis. The **arrows** indicate the site of β-lactamase activity.

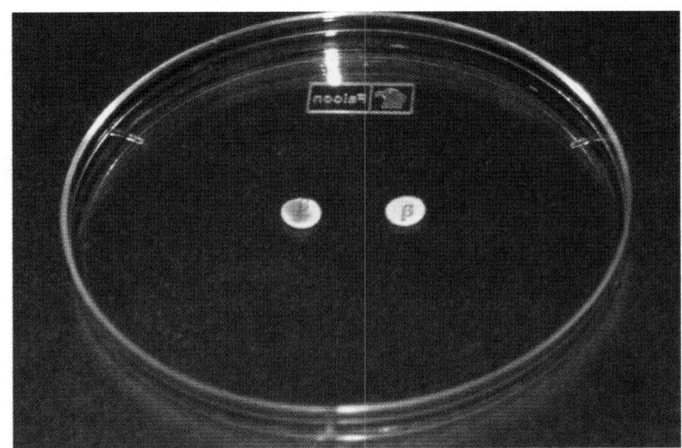

7.4 Cefinase Disks Used in β-Lactamase Testing • A β-lactam resistant strain is on the left and a susceptible strain is on the right.

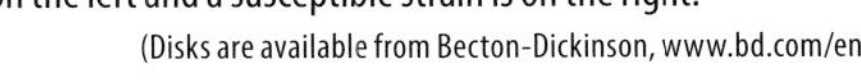
(Disks are available from Becton-Dickinson, www.bd.com/en-us.)

Bile Esculin Agar (Esculinase Test)

Purpose

Bile esculin agar (BEA) is used for the isolation and presumptive identification of bile esculin-positive enterococci (typically *Enterococcus faecalis* and *E. faecium*) and the bovis group of streptococci (**Lancefield group D**: *Streptococcus bovis, S. equinus, S. gallolyticus,* and other streptococci; see Fig. 8.16) from non-group D streptococci. It also can be used to distinguish between the genera *Enterobacter, Klebsiella,* and *Serratia* (bile esculin positive) from other genera in the *Enterobacteriaceae.*

Principle

BEA is an undefined, selective, and differential medium containing beef extract, digest of gelatin, esculin, oxgall (bile), and ferric citrate. **Esculin** extracted from the bark of the horse chestnut tree is a glycoside composed of glucose and **esculetin** joined by a **glycosidic bond** (Fig. 7.5).

Beef extract and gelatin provide nutrients and energy; bile (oxgall) is the selective agent added to separate Group D streptococci (the *Streptococcus bovis* group and enterococci) from other non-Group D streptococci. Ferric citrate is added as a source of oxidized iron to indicate a positive test.

Esculin hydrolysis results in the production of D-glucose and esculetin (Fig. 7.5) and is catalyzed by a β-glucosidase enzyme generically referred to as **esculinase.** Many bacteria possess esculinase, but the number of bacteria that are able to hydrolyze esculin in the presence of bile is much more limited, making this a useful selective and differential medium.

BEA can be prepared as a plated or a slanted medium. If the task is to isolate a streptococcus or enterococcus from a mixed culture, then a BEA plate is streaked for isolation and the differential results are read from that. If testing a pure culture for esculinase activity in the presence of bile, then a slanted medium is used.

When esculin is hydrolyzed in BEA, the resulting esculetin reacts with ferric ions in the medium to produce a dark-brown/black precipitate (Fig. 7.6), which darkens the medium surrounding the growth.

Esculinase → Glycolysis

Glycosidic bond

Esculin + β-D-Glucose + Esculetin

7.5 Hydrolysis of Esculin • Esculinase catalyzes the hydrolysis of esculin at the glycosidic bond joining glucose and esculetin. Many organisms produce esculinase, but the Group D streptococci and enterococci are unique in their ability to do this in the presence of bile salts, which is what this test detects.

When a BEA plate is used, an organism that darkens the medium even slightly is bile esculin positive. An organism that doesn't darken the medium is negative (Fig. 7.7A). A positive result is recorded on BEA slants when more than half of the medium is blackened (Fig. 7.7B). No blackening to less than half-blackened is considered a negative result.

7.6 Bile Esculin Test Indicator Reaction • This test involves the reaction of esculetin, produced during the hydrolysis of esculin, with Fe^{3+}. The result is a dark-brown to black color in the medium.

Esculetin $+ Fe^{3+} \longrightarrow$ Dark Brown Precipitate

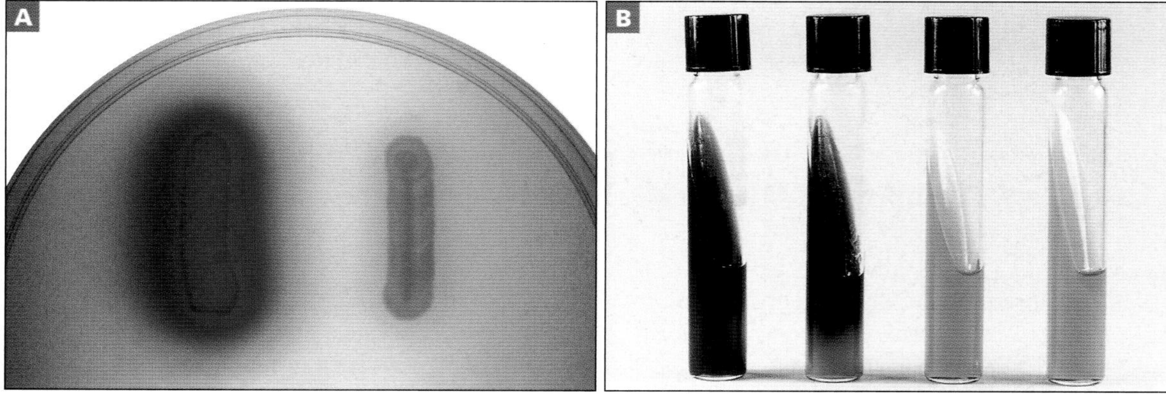

7.7 Bile Esculin Results • **(A)** This plate was inoculated with *Enterococcus faecium* (left) and *Staphylococcus aureus* (right). The darkening of the medium around *E. faecium* indicates a positive result. **(B)** More than half the medium turning black in 48 hours is considered a positive result in the tube test. Less than half is negative. From left to right: *Enterococcus faecium* (+), *Serratia marcescens* (+), *Citrobacter amalonaticus* (–), and an uninoculated control.

Blood Agar (Hemolysis Test)

Purpose

Blood agar is used for isolation and cultivation of many types of fastidious bacteria. It is also used to differentiate bacteria based on their hemolytic characteristics, especially within the genera *Streptococcus*, *Enterococcus*, and *Aerococcus*.

Principle

Several species of Gram-positive cocci produce exotoxins called **hemolysins,** which are able to destroy red blood cells (RBCs) and **hemoglobin.** Blood agar, sometimes called sheep blood agar (SBA) because it includes 5% sheep blood in a tryptic soy agar base, allows differentiation of bacteria based on their ability to hemolyze RBCs.

The three major categories of hemolysis are β-hemolysis, α-hemolysis, and γ-hemolysis. β-**hemolysis**, the complete destruction of red blood cells and hemoglobin, results in a clearing of the medium around the colonies (Fig. 7.8). α-**hemolysis** is the partial destruction of RBCs and produces a greenish discoloration of the agar around the colonies (Fig. 7.9). γ-**hemolysis** is actually no hemolysis and appears as simple growth with no change to the medium (Fig. 7.10).

Hemolysins produced by streptococci are called **streptolysins.** They come in two forms—type O and type S. **Streptolysin O** is oxygen-labile and expresses maximal

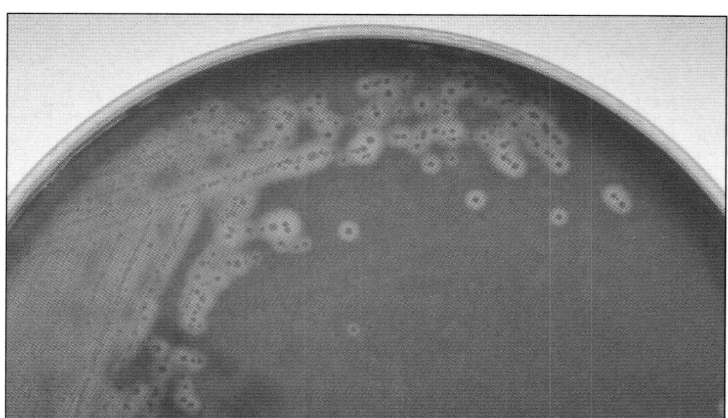

7.8 β-Hemolysis • *Streptococcus pyogenes* demonstrates β-hemolysis. The clearing around the growth is a result of complete lysis of red blood cells. This photograph was taken with transmitted light.

activity under anaerobic conditions. **Streptolysin S** is oxygen-stable but expresses itself optimally under anaerobic conditions as well.

The easiest method of providing an environment favorable for streptolysins on blood agar is what is called the **streak–stab technique.** In this procedure, the blood agar plate is streaked for isolation and then stabbed with a loop. The stabs encourage streptolysin activity because of the reduced oxygen concentration of the subsurface environment (Fig. 7.11).

7.9 **α-Hemolysis** • This is a streak plate of *Streptococcus pneumoniae* demonstrating α-hemolysis. The greenish zone around the colonies results from incomplete lysis of red blood cells. This photograph was taken with transmitted light, which may make the color appear more yellow-green than olive-green.

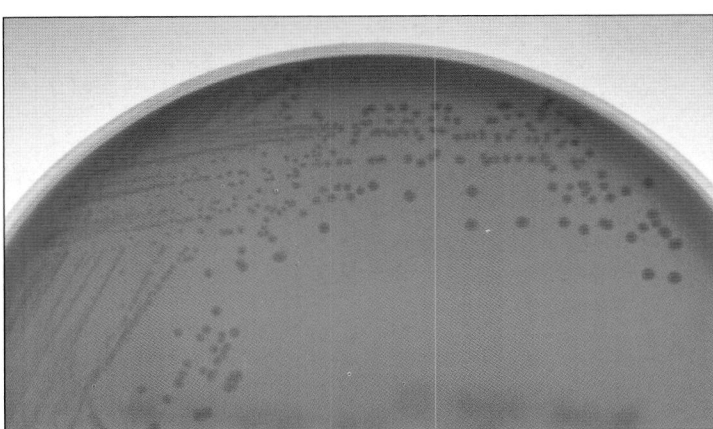

7.10 **γ-Hemolysis** • This streak plate of *Staphylococcus epidermidis* on a sheep blood agar illustrates no hemolysis. This photograph was taken with transmitted light.

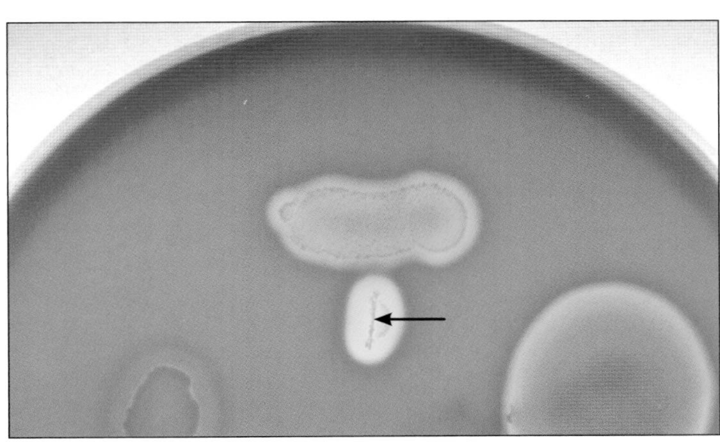

7.11 **Aerobic vs. Anaerobic Hemolysis** • An unidentified throat culture isolate demonstrates less hemolysis when growing on the surface, but β-hemolysis beneath the surface surrounding the stab (**arrow**). This results from production of an oxygen-labile hemolysin, most likely streptolysin S.

CAMP Test

Purpose

The CAMP test (an acronym of the developers of the test—Christie, Atkins, and Munch-Peterson) is used to differentiate **Lancefield group B** (p. 136) *Streptococcus agalactiae* (positive) from other *Streptococcus* species (negative).

Principle

Group B *Streptococcus agalactiae* produces the **CAMP factor**—a hemolytic protein that acts synergistically with the β-hemolysin of *Staphylococcus aureus* subsp. *aureus*. When streaked perpendicularly to an *S. aureus* subsp. *aureus* streak on blood agar (Fig. 7.12), an arrowhead-shaped zone of hemolysis forms and is a positive result (Fig. 7.13).

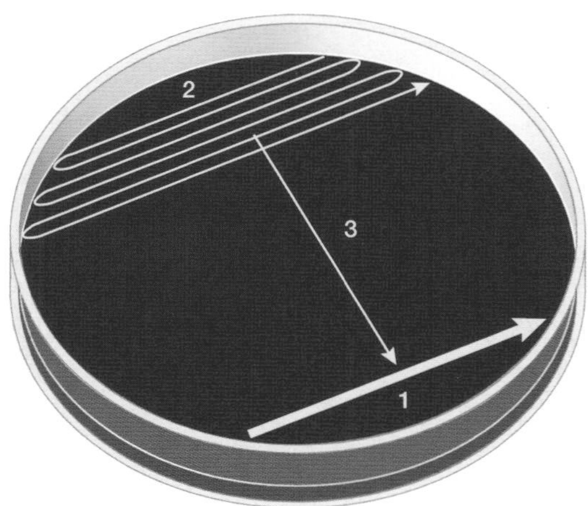

7.12 CAMP Test Inoculation Pattern • Two inoculations are made. First *Staphylococcus aureus* subsp. *aureus* is streaked along one edge of a fresh blood agar plate (**1**). Then the isolate (when testing an unknown organism) is inoculated densely in the other half of the plate opposite *S. aureus* (**2**). Finally, a single streak is made from inside streak 2 toward, but not touching, *S. aureus* (**3**).

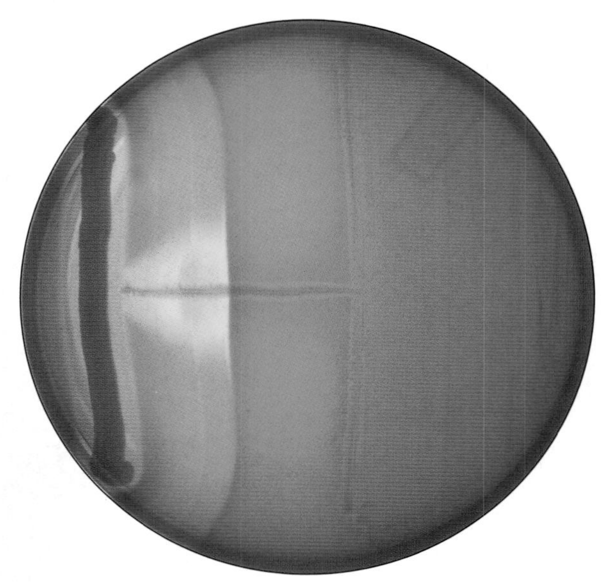

7.13 Positive CAMP Test Result • Note the arrowhead zone of clearing in the region where the CAMP factor of *Streptococcus agalactiae* acts synergistically with the β-hemolysin of *Staphylococcus aureus* subsp. *aureus*.

Casein Hydrolysis (Casease Test)

Purpose

Casein hydrolysis is used for the cultivation and differentiation of bacteria that produce the enzyme casease.

Principle

Many bacteria require proteins as a source of amino acids for synthesis of their own proteins and other compounds. The amino acids can also be used as an energy source. Some bacteria have the ability to produce and secrete enzymes (**exoenzymes**) into the environment that catalyze the hydrolysis of large proteins to smaller peptides or individual amino acids, thus enabling their uptake across the membrane (Fig. 7.14).

Casease is a specific **proteolytic enzyme** that some bacteria produce to hydrolyze the milk protein **casein**, the molecule that gives milk its white color. When broken down into smaller fragments, the ordinarily white casein loses its opacity and becomes clear.

The presence of casease can be detected easily with the test medium milk agar (Fig. 7.15). Milk agar is an undefined medium containing pancreatic digest of casein, yeast extract, dextrose, and powdered milk. When plated milk agar is inoculated with a casease-positive organism, secreted casease will diffuse into the medium around the colonies and create a zone of clearing where the casein has been hydrolyzed. Casease-negative organisms do not secrete casease and, thus, do not produce clear zones around the growth.

$$H_2N-CH(R)-C(=O)-NH-CH(R)-C(=O)\text{-----}NH-CH(R)-C(=O)-OH$$

protein (n amino acids)

$$\xrightarrow[\text{casease}]{H_2O}$$

$$H_2N-CH(R)-C(=O)-OH \quad + \quad H_2N-CH(R)-C(=O)\text{-----}NH-CH(R)-C(=O)-OH$$

amino acid + protein (n-1 amino acids)

7.14 Casein Hydrolysis Reaction • Hydrolysis of any protein occurs by breaking peptide bonds (**red arrow**) between adjacent amino acids to produce short peptides or individual amino acids.

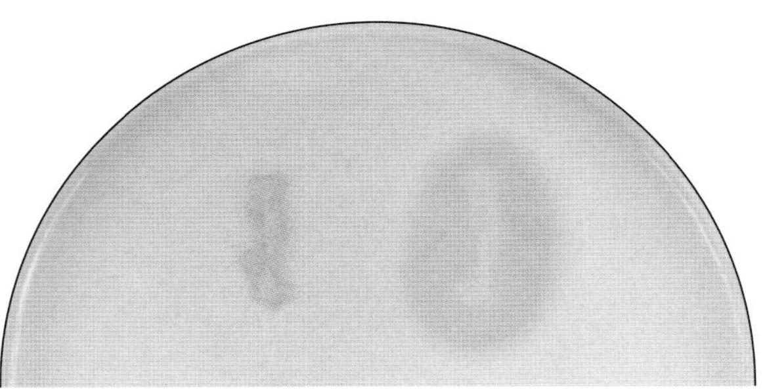

7.15 Casein Hydrolysis Test Results • This milk agar plate was inoculated with casease-negative *Kocuria rosea* (left) and casease-positive *Bacillus megaterium* (right). Clearing around growth is what identifies an organism as casease positive; we are not looking for a color change. *K. rosea* is naturally that color.

Catalase Test

Purpose

The catalase test is used to identify organisms that produce the enzyme **catalase.** It is frequently used to differentiate catalase-positive *Micrococcus* and *Staphylococcus* from catalase-negative *Streptococcus, Enterococcus,* and *Lactococcus*. Variations on this test also may be used in identification of *Mycobacterium* species.

Principle

The **electron transport chains** (**ETC**) of aerobic and facultatively anaerobic bacteria are composed of molecules capable of accepting and donating electrons as conditions dictate. As such, these molecules alternate between their oxidized and reduced states, passing electrons down the chain to a **final electron acceptor** (**FEA**). Energy lost by electrons in this sequential electron transfer is used to perform **oxidative phosphorylation** (i.e., produce ATP from ADP).

In most cases, electrons in the aerobic ETC follow the stepwise path to oxygen, but other paths can be followed and these result in production of toxic forms of reduced oxygen[1]. For instance, one ETC carrier molecule called **flavoprotein** can bypass the next carrier in the chain and transfer electrons directly to oxygen (Fig. 7.16), which produces hydrogen peroxide (H_2O_2), a highly potent cellular toxin.

Reduced **flavin adenine dinucleotide** (**$FADH_2$**) is capable of the same reaction (Fig.7.16). Even if the electrons follow the complete ETC another toxin, the superoxide radical ($O_2^{\bar{\cdot}}$), can be produced in the final step because electrons reduce oxygen one at a time and sometimes it is released before it is completely reduced to H_2O.

[1] As an analogy, a person can walk down stairs one tread at a time and reach the bottom safely. However, the option exists for jumping from the middle tread to the bottom and spraining an ankle. The person arrives at the same place, but damage is done! The potential energy is not released in a slow, controlled way, but in a larger chunk, which has damaging consequences.

Hydrogen peroxide and the superoxide radical are toxic because they oxidize biochemicals and make them nonfunctional. However, organisms that produce them also produce enzymes capable of breaking them down. **Superoxide dismutase** catalyzes conversion of superoxide radicals (the more lethal of the two compounds) to hydrogen peroxide (Fig. 7.16).

Regardless of how hydrogen peroxide is produced, **catalase** converts it into water and gaseous oxygen (Fig. 7.17). In large part (though exceptions exist), the ability to synthesize

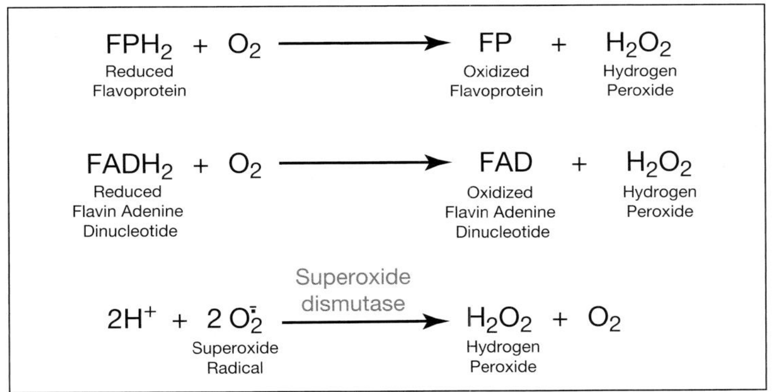

7.16 Microbial Production of Hydrogen Peroxide (H_2O_2) • Hydrogen peroxide (H_2O_2) may be formed through the transfer of electrons from reduced flavoprotein (in an ETC) or flavin adenine dinucleotide ($FADH_2$) to oxygen, or from the action of superoxide dismutase detoxifying the superoxide radical, another destructive form of reduced oxygen.

$$\underset{\text{Hydrogen Peroxide}}{2H_2O_2} \xrightarrow{\text{Catalase}} 2H_2O + O_{2(g)}$$

7.17 Catalase Mediated Conversion of H_2O_2 • Catalase is an enzyme of aerobes, microaerophiles, and facultative anaerobes that converts hydrogen peroxide to water and oxygen gas.

these protective enzymes accounts for an organism's ability to live in the presence of oxygen (Table 7-2).

Bacteria that produce catalase can easily be detected using typical store-grade hydrogen peroxide. When hydrogen peroxide is added to a catalase-positive culture, oxygen gas bubbles form immediately. If no bubbles appear, the organism is catalase negative, though false negatives can occur as a result of culture age. This test can be performed on a microscope slide (Fig. 7.18) or by adding hydrogen peroxide directly to the bacterial growth (Fig. 7.19).

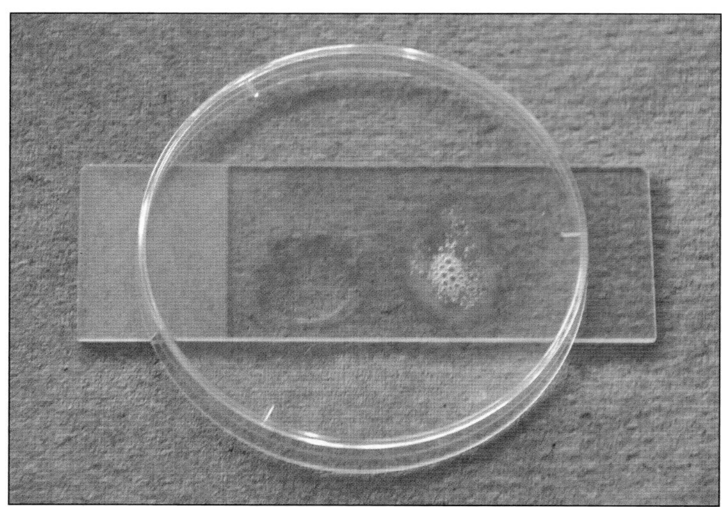

7.18 Catalase Slide Test Results Visible bubble production indicates a positive result in the catalase slide test. *Enterococcus faecium* (–) is on the left, *Staphylococcus aureus* (+) is on the right. It is advisable to cover the slide with a Petri dish lid or base immediately after adding the hydrogen peroxide to prevent splattering of a positive result.

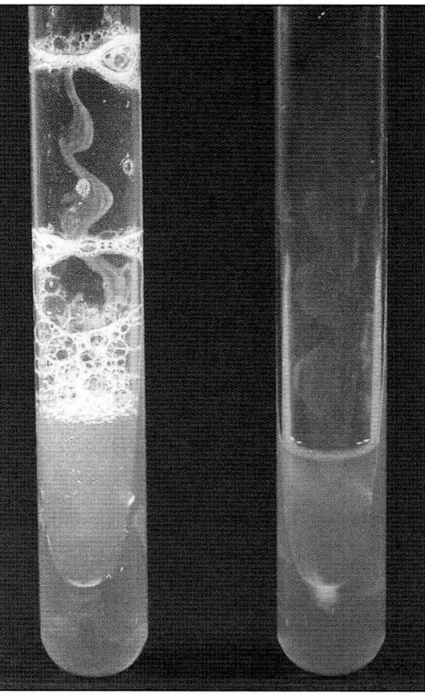

7.19 Catalase Tube Test Results The catalase test may also be performed on an agar slant. *Staphylococcus aureus* (+) is on the left, *Enterococcus faecium* (–) is on the right. Immediately replace the cap after addition of hydrogen peroxide.

TABLE **7-2** **Protective Enzymes in Various Aerotolerance Groups**

Aerotolerance Group	Superoxide Dismutase	Catalase
Obligate aerobe	Present	Present
Facultative anaerobe	Present	Present
Microaerophile	Present	Present in small amounts
Aerotolerant anaerobe	Present	Absent (alternative mechanism)
Obligate anaerobe	Absent	Absent

Citrate Utilization Test

Purpose

The citrate utilization test is used to determine the ability of an organism to use citrate (citric acid) as its sole carbon source. Citrate utilization is one part of a test series referred to as **IMViC** (*I*ndole, *M*ethyl Red, *V*oges-Proskauer and *C*itrate tests) that distinguishes between members of the family *Enterobacteriaceae* and also from other Gram-negative rods.

Principle

The citrate utilization test was designed to differentiate members of the *Enterobacteriaceae*, all of which are facultative anaerobes. That is, they have the ability to ferment carbohydrates and they also have the ability to aerobically respire, which means they have a functional citric acid cycle.

However, the citrate utilization test does not tell us about the citric acid cycle. Instead, it tells about the ability of organisms to use citrate as their sole carbon source and perform citrate fermentation.

Simmons citrate agar is a defined medium. That is, the amount and source of all ingredients are carefully controlled. Because sodium citrate is the only carbon source in the medium[2] it will not support a complex, high-energy yielding respiratory process like the citric acid cycle.

It does, however, provide the means for bacterial species that possess the enzyme **citrate permease** (an **oxaloacetate decarboxylase Na^+ pump**) to transport citrate (actually, oxaloacetate—see Fig. 7.20) into the cell and perform citrate fermentation. These organisms must also be able to survive with ammonium (in the form of ammonium phosphate) as the sole nitrogen source. Bacteria that do not possess citrate permease will not grow on this medium.

Citrate-positive bacteria hydrolyze citrate extracellularly into oxaloacetate and acetate using the enzyme **citrate lyase** (Fig. 7.20). From there, oxaloacetate is decarboxylated to pyruvate, simultaneously using the energy released to pump oxaloacetate/pyruvate into the cell. A variety of products can be formed from pyruvate depending on the cell's pH.

Here is a secret: some bacterial fermentation pathways can make ATP and reducing power (NADH or NADPH), products not made in most fermentations. For instance, *Klebsiella* and *Enterobacter*, both of which are citrate positive, have the ability to make acetyl~CoA[3] from pyruvate (nothing unusual) but then have the ability to convert it to acetyl phosphate, which can then phosphorylate ADP to ATP.

In addition, the same bacteria are capable of using H_2 as an electron source to reduce NAD and/or NADP, the latter of which is used in synthesis reactions (Fig. 7.20).

Bacteria that survive in the medium and utilize citrate also convert ammonium phosphate to ammonia (NH_3) and ammonium hydroxide (NH_4OH), both of which tend to alkalinize the agar. Bromothymol blue dye, which is green at pH 6.9 and blue at pH 7.6, is an indicator, and as the pH goes up the medium changes from green to blue (Fig. 7.21). Thus, conversion of the medium to blue signals a positive citrate test result.

7.21 Citrate Utilization Results • These Simmons citrate slants were inoculated with *Bacillus cereus* (–) on the left and *Citrobacter koseri* (+) on the right. An uninoculated control is in the center.

[2] Of course, there is CO_2 in the air above the agar, but the test is designed primarily to differentiate *Enterobacteriaceae*, none of which are autotrophic.

[3] The symbol "~" within a chemical name or its structural diagram indicates a high-energy bond.

Citrate lyase
Citrate
Oxaloacetate
Acetate
Acetoin
Citrate permease
CO_2
$2CO_2$
NADPH
$NADP^+$
H^+
H_2
Formate
Alkaline pH
Acidic pH
Pyruvate
ATP
ADP
P_i
Acetyl~(P)
Acetyl~CoA
CoA
Acetate
Lactate

7.20 Citrate Utilization Reactions • Citrate lyase hydrolyzes citrate into oxaloacetate and acetate. From there, citrate permease (an oxaloacetate decarboxylase Na^+ pump) transports oxaloacetate into the cell, using the energy released as it is being decarboxylated to pyruvate. Once in the cell, pyruvate can be converted to a variety of products depending on the pH of the environment and the organism's enzymes. One option, the formate/acetyl~CoA pathway shown in detail to the left, produces the useful compounds ATP and NADH or NADPH.

Occasionally a citrate-positive organism will grow on a Simmons citrate slant without producing a change in color. In most cases, this is because of incomplete incubation. In the absence of color change, growth on the slant indicates that citrate is being utilized and is evidence of a positive reaction. To avoid confusion between actual growth and a heavy inoculum, which may be misinterpreted as growth, citrate slants typically are lightly inoculated with an inoculating needle rather than a loop.

Coagulase and Clumping Factor Tests

Purpose

Coagulase and clumping factor tests are used to differentiate *Staphylococcus aureus* (see p. 214) from other staphylococci. Of the 37 *Staphylococcus* species, only a handful are coagulase positive, with *S. aureus* being the most common human pathogen.

Principle

Staphylococcus aureus is an **opportunistic pathogen** that can be highly resistant to the immune response and antimicrobial agents. Its resistance is due, in part, to the production of **coagulase** enzymes. Coagulase works in conjunction with normal plasma components to form protective fibrin barriers around individual bacterial cells or groups of cells, shielding them from phagocytosis and other types of immune defenses.

Coagulase enzymes occur in two forms—**bound coagulase** and **free coagulase**. Bound coagulase, also called **clumping factor**, is attached to the bacterial cell wall and reacts directly with fibrinogen in plasma. The fibrinogen then precipitates causing the cells to clump together in a visible mass.

Free coagulase is an extracellular enzyme (released from the cell) that reacts with a plasma component called **coagulase-reacting factor** (**CRF**). The resulting reaction is similar to the conversion of prothrombin and fibrinogen in the normal clotting mechanism.

Two forms of the coagulase test have been devised to detect these enzymes: the tube test and the slide test. The tube test detects the presence of either bound or free coagulase, while the slide test detects only bound coagulase. Both tests utilize rabbit plasma treated with anticoagulant (usually EDTA) to interrupt normal clotting mechanisms.

The tube test is performed by adding the organism to rabbit plasma in a test tube. Coagulation of the plasma (including any thickening or formation of fibrin threads) within 24 hours indicates a positive reaction (Fig. 7.22). The plasma is typically examined for clotting (without shaking) after about 4 hours because it is possible for coagulation to take place early and revert to liquid within 24 hours.

In the slide test, bacteria are transferred to a glass slide containing a small amount of plasma. Agglutination of the cells on the slide within 1 to 2 minutes indicates the presence of bound coagulase (Fig. 7.23). Equivocal or negative slide test results are typically confirmed using the tube test.

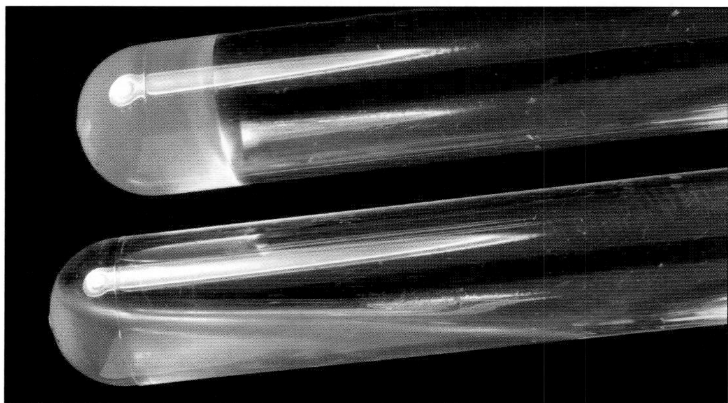

7.22 Coagulase Tube Test Results • The tube test identifies both bound and free coagulase enzymes. These coagulase (rabbit plasma) tubes show *Staphylococcus aureus* (+) above and *S. epidermidis* (–) below after 24 hours of incubation. Tests must be read within 24 hours to avoid reversion of positive tests by staphylococcal fibrinolytic enzyme activity. Coagulase provides protection against phagocytosis and antibodies by surrounding infecting organisms with a fibrin clot.

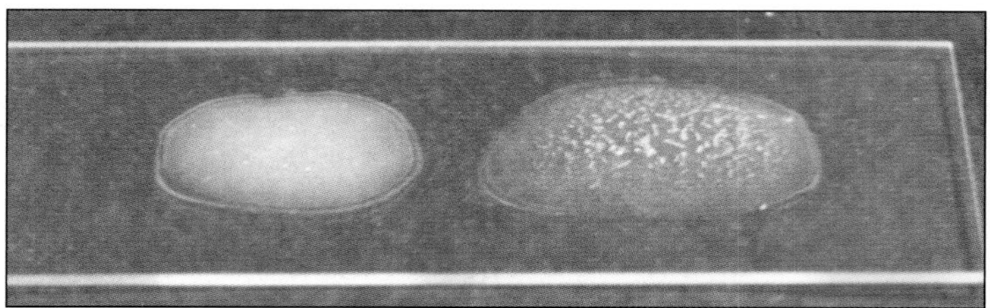

7.23 Coagulase Slide Test Results (Clumping Factor) • The slide test identifies only bound coagulase (clumping factor). Shown are *Staphylococcus epidermidis* (–) on the left and *S. aureus* (+) on the right. Positive results occur within 2 minutes.

Decarboxylase Tests

Purpose

Amino acid decarboxylation can occur with any one of several amino acids. Typically, decarboxylation tests are used to differentiate organisms in the family ***Enterobacteriaceae*** and to distinguish them from other Gram-negative rods. The most frequently used amino acids are arginine, lysine, and ornithine.

Principle

Møller's decarboxylase base medium contains peptone, glucose, the pH indicator bromocresol purple, and the **coenzyme pyridoxal phosphate.** Bromocresol purple is purple at pH 6.8 and above, and yellow below pH 5.2. Base medium can be used with any one of a number of specific amino acid substrates, depending on the decarboxylase to be identified.

After inoculation, an overlay of mineral oil is used to seal the medium from external oxygen and promote fermentation. Glucose fermentation in the anaerobic medium initially turns it yellow due to the accumulation of acid end-products (all *Enterobacteriaceae* ferment glucose to acid end-products). The low pH and presence of the specific amino acid induce **decarboxylase-positive** organisms to produce the enzyme. (That is, the specific decarboxylase gene is "switched on.")

Decarboxylation of the amino acid results in accumulation of **amines,** which are alkaline (Fig. 7.24) and turn the medium purple. If the organism is a glucose fermenter but does not produce the appropriate decarboxylase, the medium will turn yellow and remain so.

If the organism does not ferment glucose the medium will exhibit no color change. Purple color is the only positive result; all others are negative (Fig. 7.25). The three amino acids used most frequently to test for decarboxylase activity are arginine, lysine, and ornithine (Figs. 7.26 through 7.28).

Arginine decarboxylase medium may, in fact, identify a second pathway for arginine catabolism. It is called the **arginine dihydrolase system.** In it, arginine is converted to citrulline and then into ornithine, ATP, CO_2, and two NH_3, which raise the pH and turn the medium purple. If the species is also positive for ornithine decarboxylase, it will produce putrescine and CO_2 from ornithine (Fig. 7.29). However, because the decarboxylase and dihydrolase pathways both produce alkaline end-products, Møller's decarboxylase medium cannot distinguish between them.

R | H₂N—CH—COOH (Amino acid) —Decarboxylase / pyridoxyl phosphate→ R | H₂N—CH₂ (Amine) + CO₂

$$\underset{\text{Amino acid}}{H_2N-\overset{R}{\overset{|}{C}H}-COOH} \xrightarrow[\text{pyridoxyl phosphate}]{\text{Decarboxylase}} \underset{\text{Amine}}{H_2N-\overset{R}{\overset{|}{C}H_2}} + CO_2$$

7.24 Amino Acid Decarboxylation Reaction • Removal of an amino acid's carboxyl group results in the formation of an amine and carbon dioxide.

7.25 Decarboxylation Test Results • Shown are three lysine decarboxylase tubes. *Proteus mirabilis* is on the left, *Klebsiella* (*Enterobacter*) *aerogenes* is on the right, and an uninoculated control is in the middle. *K. aerogenes* (purple) is positive for lysine decarboxylation. *P. mirabilis* (yellow) fermented the glucose to acid end-products, but is negative for lysine decarboxylation. The uninoculated control's color should match a decarboxylase-negative result without fermentation. The color results shown here are representative of all Møller decarboxylase media irrespective of the amino acid used.

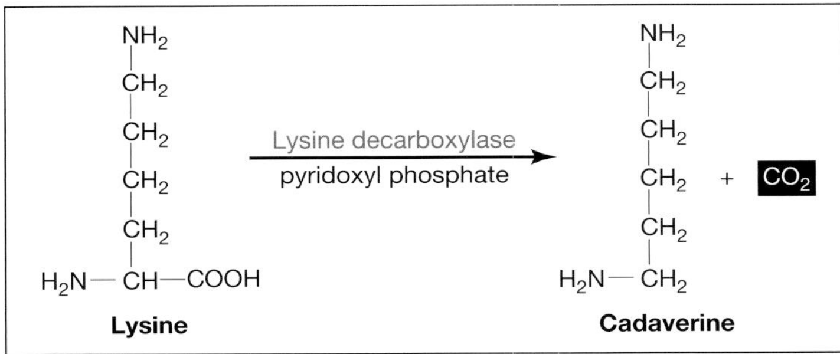

7.26 Lysine Decarboxylation Reaction • Decarboxylation of the amino acid lysine produces cadaverine and CO_2. Note the structural similarity between lysine and ornithine (Fig. 7.27).

$$\underset{\text{Ornithine}}{NH_2-CH_2-CH_2-CH_2-CH(NH_2)-COOH} \xrightarrow[\text{pyridoxyl phosphate}]{\text{Ornithine decarboxylase}} \underset{\text{Putrescine}}{NH_2-CH_2-CH_2-CH_2-CH_2-NH_2} + CO_2$$

7.27 Ornithine Decarboxylation Reaction • Decarboxylation of the amino acid ornithine produces putrescine and CO_2. Note the structural similarity between ornithine and lysine (Fig. 7.26).

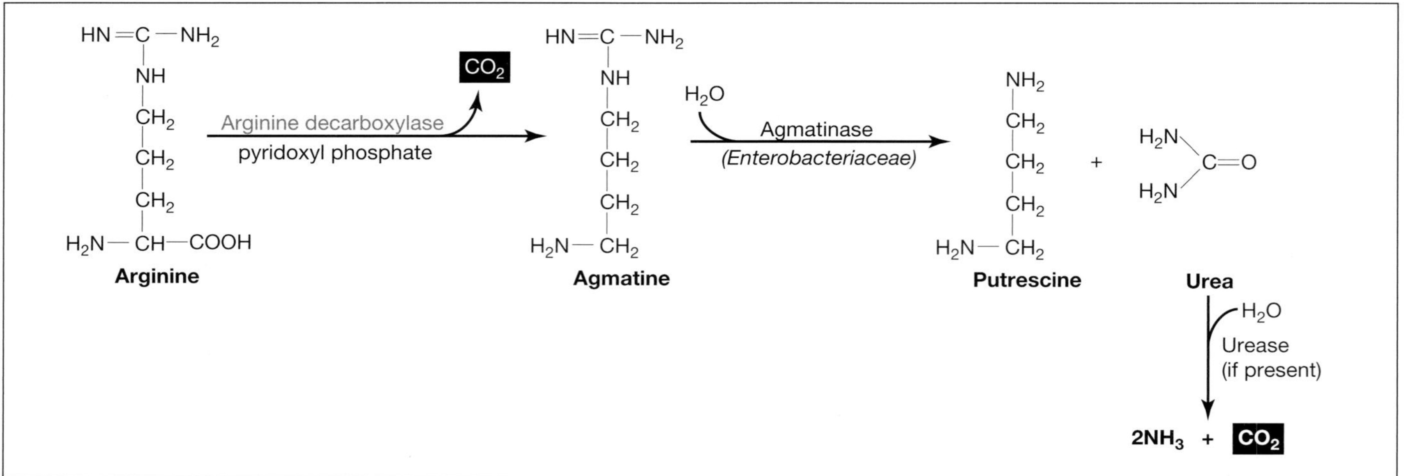

7.28 Arginine Decarboxylation Reaction • Decarboxylation of the amino acid arginine produces the amine agmatine and CO_2. Members of *Enterobacteriaceae* are capable of degrading agmatine into putrescine and urea. Those strains with urease can further break down the urea into ammonia and carbon dioxide. Thus, the end-products of arginine catabolism are carbon dioxide, putrescine, and urea, or (in the presence of urease) carbon dioxide, putrescine, and ammonia.

Arginine — Arginine dihydrolase → NH_3 — Citrulline — Citrulline ureidase (H_3PO_4) → Ornithine — Ornithine decarboxylase, pyridoxyl phosphate → Putrescine + CO_2

Carbamoyl Phosphate — Carbamate Kinase (ADP → ATP) → CO_2 + NH_3

7.29 Arginine Dihydrolase System Reactions • In addition to decarboxylation, arginine may alternatively be catabolized by a dihydrolase enzyme with the production of ornithine, NH_3, and CO_2. In ornithine decarboxylase-positive organisms, ornithine is further degraded into putrescine and CO_2. The example shown here is a dihydrolase system of *Pseudomonas putida* in which ATP synthesis occurs. Regardless of the specifics, the arginine dihydrolase pathway is indistinguishable from the arginine decarboxylase pathway in Møller's decarboxylase medium because both produce alkaline end-products.

DNA Hydrolysis (DNase Test)

Purpose

DNA hydrolysis is used to distinguish *Serratia* species (positive) from *Enterobacter* species, *Moraxella catarrhalis* (negative) from *Neisseria* species (see p. 206), and *Staphylococcus aureus* (positive, see p. 214) from other *Staphylococcus* species.

Principle

An enzyme that catalyzes the **depolymerization** of DNA into small fragments (**oligonucleotides**) or single nucleotides is called a **deoxyribonuclease**, or **DNase** (Fig. 7.30). DNase is an **exoenzyme**; that is, an enzyme that is secreted by a cell and acts on the substrate extracellularly. Extracellular digestion typically allows utilization of a macromolecule too large to be transported into the cell. Ability to produce this enzyme can be determined by culturing and observing an organism on a DNase test agar plate.

One formulation for DNase agar consists of peptides derived from soybean and casein that serve as carbon and nitrogen sources, sodium chloride for osmotic balance, and

DNA as the substrate. DNA also serves as an additional carbon and nitrogen source, or its subunits can be used for DNA synthesis by the growing organism.

After incubation, 1N HCl is added to the plate where it will form a cloudy precipitate with intact DNA, but not with nucleotides. Therefore, clearing around the growth is an indication of DNase activity (DNA hydrolysis) (Fig. 7.31).

A modification of DNase agar includes toluidine blue, which forms a blue complex with intact DNA, but appears pinkish/lavender when complexed with nucleotides. DNase activity is indicated by a pink coloration around the growth (Fig. 7.32). While this medium has the advantage of not using 1N HCl reagent, toluidine blue may inhibit some Gram-positive cocci, so it is recommended for use with *Enterobacteriaceae*.

An alternate modification of DNase test agar contains methyl green dye. The dye forms a complex with polymerized DNA and gives the agar a blue-green color, but no complex is formed with nucleotides. Therefore, clearing that develops around bacterial growth on the medium is an indication of DNase activity and is considered a positive result (Fig. 7.33). The advantage to this medium is that adding 1N HCl is not necessary and it is appropriate for use with both Gram-positive cocci and *Enterobacteriaceae*.

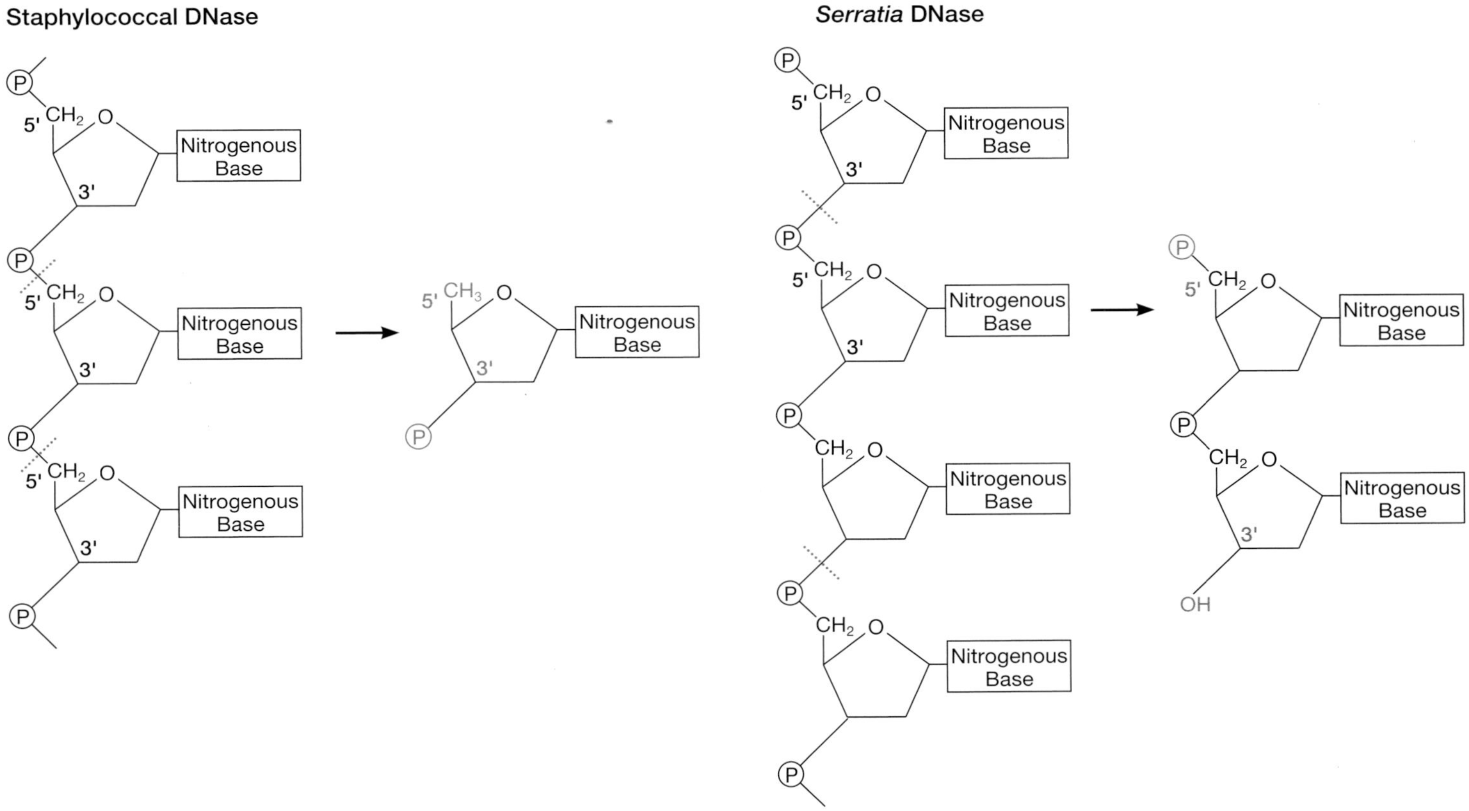

Staphylococcal DNase

Serratia DNase

7.30 DNA Hydrolysis Reactions • *Staphylococcus aureus* DNase and *Serratia* DNase hydrolyze DNA differently. *S. aureus* DNase removes single nucleotides with a 3' phosphate and a 5' methyl group (CH_3). *Serratia* DNase produces fragments of two to four nucleotides (a dinucleotide is shown here) with a 3' hydroxyl and a 5' phosphate.

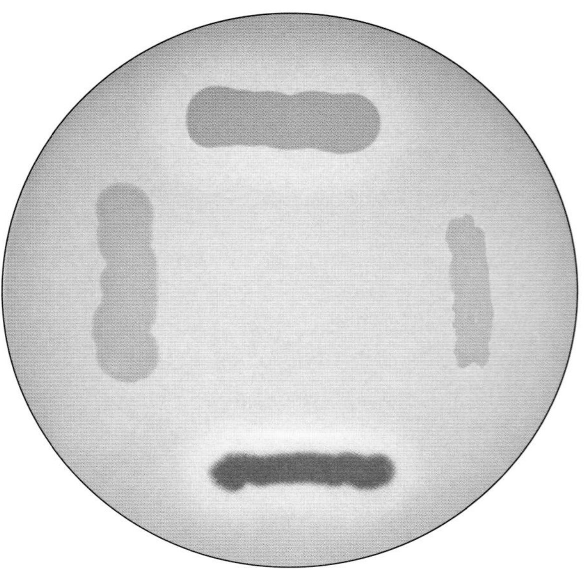

7.31 DNA Hydrolysis Results: DNase Test Agar • After incubation, 1N HCl is added to the medium. HCl causes precipitation of intact DNA, but not of nucleotides or oligonucleotides. Therefore, absence of the precipitate (clearing) around the growth is a positive result. Clockwise from the top: *Staphylococcus aureus* (+), *Staphylococcus epidermidis* (–), *Serratia marcescens* (+), and *Klebsiella* (*Enterobacter*) *aerogenes* (–).

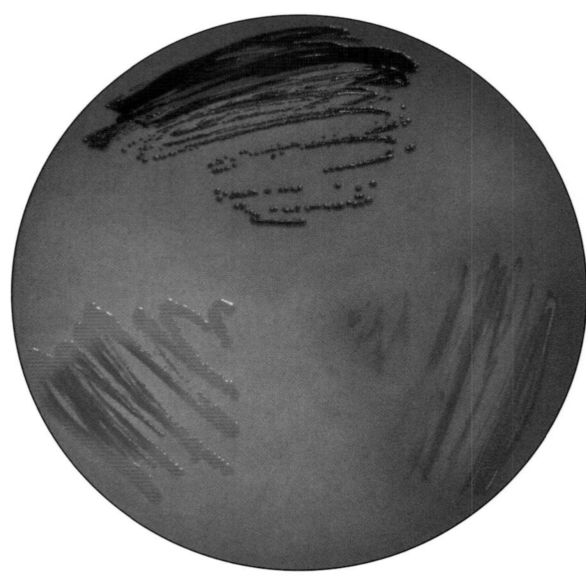

7.32 DNA Hydrolysis Results: DNase Test Agar with Toluidine Blue • Toluidine blue is included in this DNase agar. It forms a pinkish/lavender complex with nucleotides and a blue complex with intact DNA. Clockwise from the top: *Serratia marcescens* (+), *Staphylococcus aureus* (+), and *Klebsiella* (*Enterobacter*) *aerogenes* (–). Note that the pink/lavender color of a positive result is visible, but not very intense in this plate. *S. aureus* strains often grow poorly on this medium. For this reason, DNase plus TB is recommended for use with *Enterobacteriaceae*.

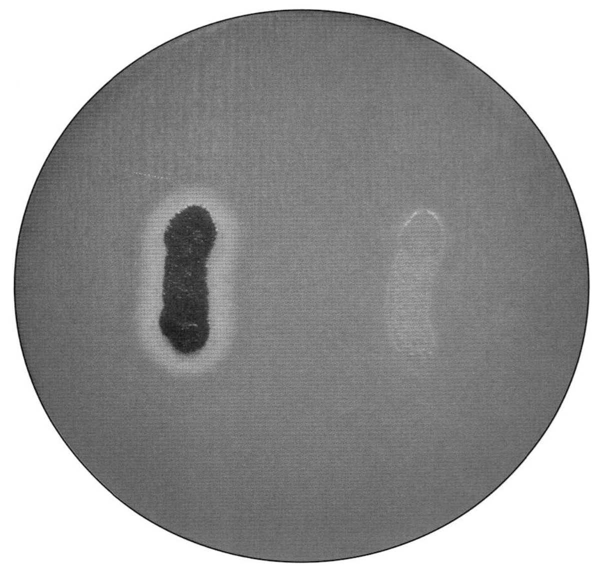

7.33 DNA Hydrolysis Results: DNase Test Agar with Methyl Green • Methyl green is the indicator in this medium. It forms a blue-green complex with intact DNA, but does not form a visible complex with nucleotides and oligonucleotides, so clearing around the growth is a positive result. *Serratia marcescens* (+) is on the left, *E. coli* (–) is on the right.

Fermentation Tests (Purple Broth and Phenol Red Broth)

Purpose

The ability to ferment is an important bacterial differential characteristic. For those that ferment, further differentiation can be made based on which sugars can be fermented and what end-products are produced. More specifically, fermentation tests are used to differentiate species of *Enterobacteriaceae* and to distinguish them from other Gram-negative rods. They are also used to distinguish between Gram-positive fermenters, such as *Streptococcus* and *Lactobacillus* species.

Principle

Carbohydrate fermentation is the metabolic process by which an organic molecule acts as an electron donor (becoming oxidized in the process) and one or more of its organic products act as the **final electron acceptor** (**FEA**), which becomes reduced.

As a rule, if an organism can ferment, it will ferment glucose. However, prior to the actual fermentation, most organisms convert glucose to pyruvate using glycolysis (Appendix, Fig. A.2), although some use alternative pathways (e.g., **Entner-Doudoroff pathway**, Appendix, Fig. A.3).

It is pyruvate or one of its derivatives that actually act as the FEA of the fermentation.

The end-products of fermentation fall into three categories: acids, alcohols or other organic solvents, and gas. The end-products depend on the enzymes possessed by each specific organism (Fig. 7.34 and Appendix, Fig. A.6).[4]

[4] It is customary to include any pathways leading up to the actual reduction of the FEA as part of the fermentation. Glycolysis is included as part of most fermentations, as are any reactions happening prior to glycolysis that produce compounds used in glycolysis. For instance, a lactose fermenter is an organism that splits the disaccharide lactose into the monosaccharides glucose and galactose and then sends these monosaccharides through glycolysis. But, the actual fermentation happens to pyruvate or one of its derivatives.

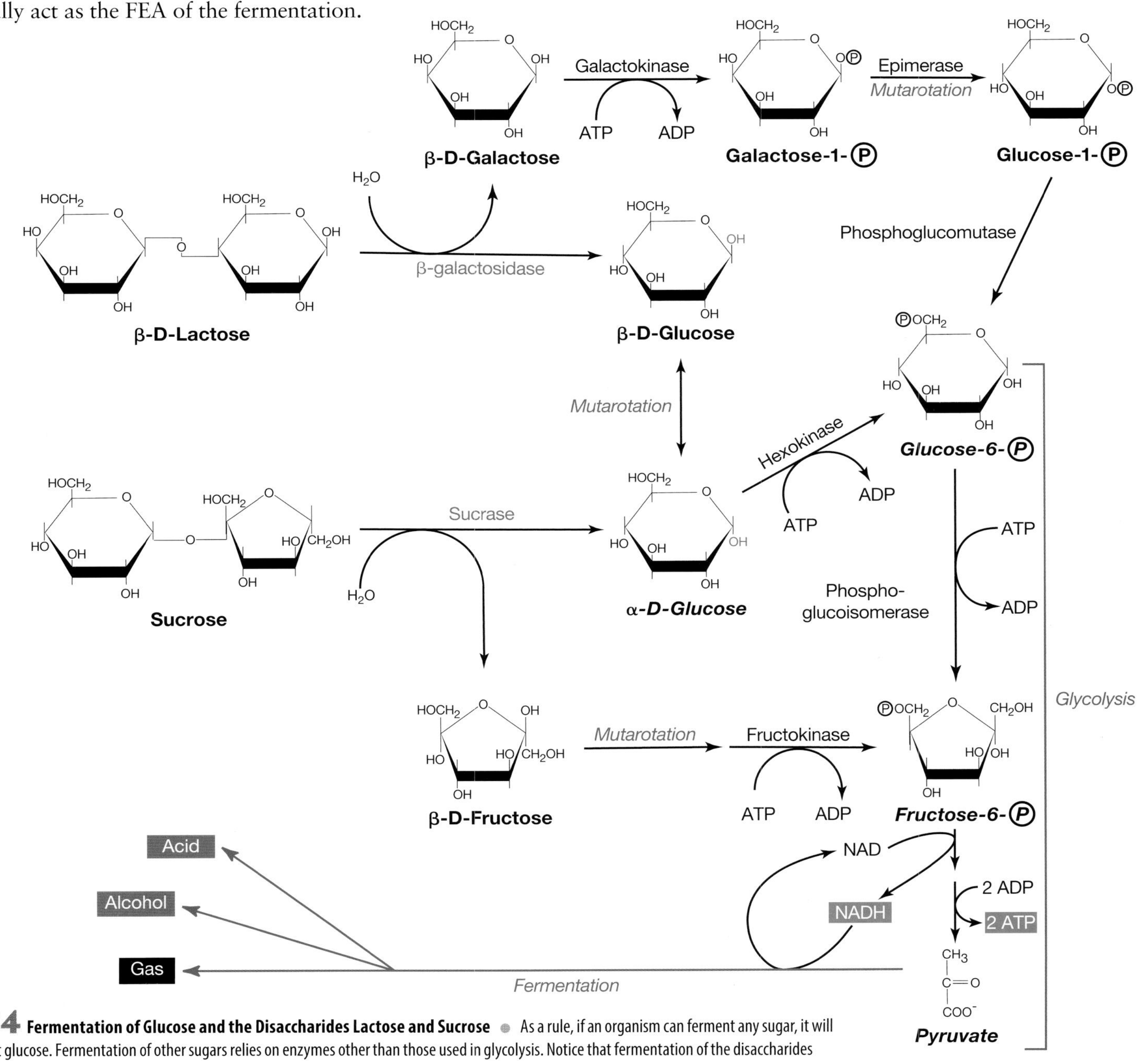

7.34 Fermentation of Glucose and the Disaccharides Lactose and Sucrose ● As a rule, if an organism can ferment any sugar, it will ferment glucose. Fermentation of other sugars relies on enzymes other than those used in glycolysis. Notice that fermentation of the disaccharides lactose and sucrose relies on enzymes to hydrolyze them into two monosaccharides. Once the monosaccharides are formed, they are converted into compounds used in glycolysis and follow the same path to pyruvate that glucose does. Names of glycolysis compounds are italicized.

The principle behind purple broth and phenol red broth is the same. Each **base medium** consists of standard ingredients to which a single fermentable carbohydrate is added. After inoculation and incubation, an organism's ability to ferment that particular carbohydrate can be determined, as can the end-products of its fermentation. Because virtually any carbohydrate can be added to the base, these media are very versatile.

Both base media include protein digests (casein or gelatin) as a carbon and nitrogen source and a pH indicator. Purple broth uses bromocresol purple as the pH indicator (yellow below pH 5.2 and purple above pH 6.8). PR broth includes phenol red (yellow below pH 6.8, pink above pH 7.4, and red in between). During preparation the pH is adjusted to approximately 7 so the broth appears purple or red, respectively.

Any carbohydrate can be added to the base broth, but glucose, lactose, and sucrose are common choices. An inverted Durham tube is placed in each tube as an indicator of gas production.

Acid production from carbohydrate fermentation lowers the pH below the neutral range of the indicator and turns the medium yellow (Figs. 7.35 and 7.36). Gas production, also from fermentation, is indicated by a bubble, or pocket, in the Durham tube where the broth has been displaced.

Though it is not an intended outcome of these fermentation broths, an alkaline reaction can occur if the organism is a non-fermenter or if the organism is an acid-producing fermenter and has been incubated long enough to consume all the carbohydrate. Deamination of amino acids supplied by the protein digest releases ammonia (NH_3), which raises the pH and can turn PR broth from red or yellow to pink and purple broth from yellow to purple.

In the case of a non-fermenter, an alkaline reaction is of no consequence and is scored correctly as a negative. However, in the case of a fermenter, acid that was produced will be neutralized by the alkaline reaction. This is a **reversion reaction** and would lead to a false negative reading. To avoid reversion, cultures should be read after no longer than 24 hours of incubation.

7.35 Purple Lactose Broth Results • Shown are four purple lactose broths. Acid production (A) in purple broth turns the medium yellow. Gas production (G) produces a bubble or pocket within the Durham tube. It is customary to record results using the shorthand of A/G, for acid and gas. A "–" sign indicates that product was not made. From left to right are *Proteus vulgaris* (–/–), an uninoculated control, *Escherichia coli* (A/G), and *Staphylococcus aureus* (A/–).

7.36 Phenol Red Glucose Broth Results • Shown are five PR glucose broths. Acid production (A) in PR broth turns the medium yellow. Gas production (G) produces a bubble or pocket within the Durham tube. An alkaline reaction (K) results in a pink color. The same notation is used for PR results as are used for purple broth. From left to right are *Escherichia coli* (A/G), *Staphylococcus aureus* (A/–), uninoculated control, *Micrococcus luteus* (–/–), and *Alcaligenes faecalis* (K).

Gelatin Hydrolysis (Gelatinase Test)

Purpose

The gelatin hydrolysis test is used to determine a microbe's ability to produce **gelatinases.** *Staphylococcus aureus,* which is gelatinase positive, can be differentiated from *S. epidermidis. Serratia* and *Proteus* species are gelatinase-positive members of *Enterobacteriaceae,* whereas most others in the family are negative. Among Gram-positive rods, *Bacillus anthracis, B. cereus,* and several other *Bacillus* species are gelatinase positive, as are *Clostridium tetani* and *C. perfringens.*

Principle

Gelatin is a protein derived from collagen—a component of vertebrate connective tissue. Gelatinases comprise a family of extracellular enzymes produced and secreted by some microorganisms to hydrolyze gelatin. Subsequently, the cells can absorb individual amino acids and use them for metabolic purposes. Bacterial hydrolysis of gelatin occurs in two sequential reactions, as shown in Figure 7.37.

The presence of gelatinases can be detected using nutrient gelatin, a simple test medium composed of gelatin, peptone, and beef extract. Nutrient gelatin differs from most other solid media in that the solidifying agent (gelatin) is also the substrate for enzymatic activity. Consequently, when a tube of nutrient gelatin is stab inoculated with a gelatinase-positive organism, secreted gelatinase (or gelatinases) will liquefy the medium.

Gelatinase-negative organisms do not secrete the enzyme and do not liquefy the medium (Figs. 7.38 and 7.39). A 7-day incubation period is usually sufficient to see liquefaction of the medium. However, gelatinase activity is very slow in some organisms, so all tubes still negative after 7 days should be incubated an additional 7 days.

A slight disadvantage of nutrient gelatin is that it melts at 28°C (82°F). Therefore, inoculated stabs are typically incubated at 25°C (if the organisms permit) along with an uninoculated control to verify that any liquefaction is not temperature related. If the control liquefies (melts), then all tubes should be put in the refrigerator until the control solidifies. If an inoculated tube remains liquid, the test is recorded as gelatinase positive. If it solidifies, then the test is recorded as gelatinase negative and further incubation may be needed.

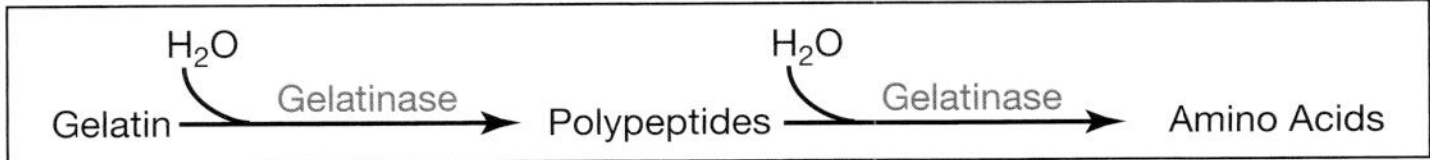

7.37 Gelatin Hydrolysis Reactions • Gelatin is hydrolyzed by the gelatinase family of enzymes into polypeptides and then amino acids.

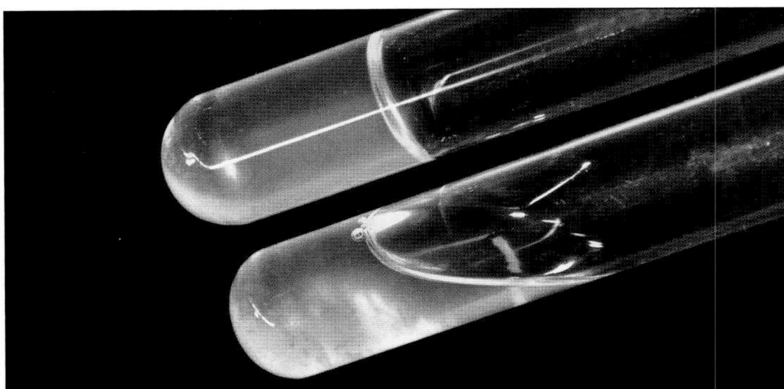

7.38 Gelatin Hydrolysis Results–Nutrient Gelatin • *Escherichia coli* (–) was inoculated in the top nutrient gelatin tube; *Bacillus subtilis* (+) was inoculated in the lower one. Make sure an uninoculated control (not shown) is solid before reading the inoculated tubes to ensure any liquefaction is due to gelatin hydrolysis and not to melting. It is not necessary to tip the tubes very much to see if the gelatin has been liquefied.

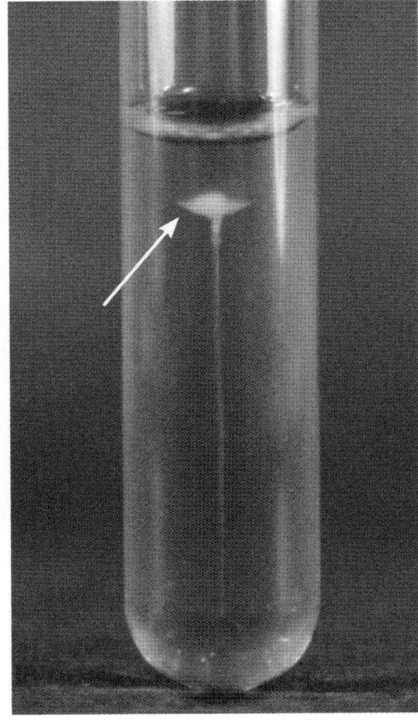

7.39 Crateriform Gelatin Liquefaction • Shown here is *Micrococcus luteus* liquefying the gelatin in the shape of a crater. In addition to being gelatinase positive, this form of liquefaction may be of diagnostic use because not all gelatinase-positive microbes liquefy the gelatin completely.

Indole Production Test

Purpose

The indole test identifies bacteria capable of breaking down **tryptophan** (an amino acid) using the enzyme **tryptophanase**. The test is one component of the **IMViC** battery of tests (**I**ndole, **M**ethyl Red, **V**oges-Proskauer, and **C**itrate) used to differentiate the *Enterobacteriaceae*, especially *Escherichia coli* (indole positive) from *Enterobacter*, *Klebsiella*, *Hafnia*, and *Serratia* (all negative).

Principle

The indole production test can be performed using **SIM medium** or **tryptone water**. (SIM medium also tests for motility and sulfur reduction. SIM is an acronym for Sulfur-Indole-Motility. See p. 99 for motility agar and p. 113 for sulfur reduction.) SIM is a semisolid medium that is formulated with casein and animal tissue as sources of amino acids for the indole test. (Other ingredients are present for sulfur reduction and motility.) Tryptone water is a broth made of tryptone, a digest of casein, which is high in tryptophan.

Indole production in both media is made possible by the presence of tryptophan (an amino acid contained in all proteins, but in these media, specifically casein and animal protein). Bacteria possessing the enzyme tryptophanase can hydrolyze tryptophan to pyruvate, ammonia (by deamination), and indole (Fig. 7.40).

Tryptophan hydrolysis can be detected by the addition of Kovac's reagent after incubation. Kovac's reagent contains ***p*-dimethylaminobenzaldehyde** (DMABA) and HCl dissolved in amyl alcohol. When a few drops of Kovac's reagent are added to the tube, it forms a layer over the medium.

DMABA reacts with any indole present and produces a quinoidal compound that turns the reagent layer red (Fig. 7.41). The formation of red color in the reagent layer in SIM medium or tryptone water indicates a positive reaction and the presence of tryptophanase (Figs. 7.42 and 7.43). No red color is a negative result.

A rapid indole test is available and done by placing bacterial growth on a paper slide impregnated with 5% DMABA (Fig. 7.44). A positive result is production of a pink color on the slide.

Tryptophan + H_2O —tryptophanase→ Indole + NH_3 + Pyruvate → Fermentation / Respiration

7.40 Tryptophan Catabolism in Indole-Positive Organisms • This reaction is a mechanism for getting energy out of the amino acid tryptophan. The enzyme responsible is tryptophanase and the products are indole, ammonia, and pyruvate. Pyruvate can be used in fermentation or cellular respiration.

2 Indole + *p*-dimethylaminobenzaldehyde + HCl + Amyl Alcohol (Kovac's reagent) → Rosindole dye (cherry red)

7.41 Indole Indicator Reaction • Kovac's reagent will form a layer on the medium's surface. Then indole, if present, reacts with the DMABA in Kovac's reagent to produce a red color.

7.42 Indole Test Results–SIM Medium • These SIM tubes were inoculated with *Klebsiella* (*Enterobacter*) *aerogenes* (–) on the left and *Escherichia coli* (+) on the right. Note how little Kovac's reagent is used. Both organisms are also motile, as evidenced by the haziness in the agar (see motility test, p. 99).

7.43 Indole Test Results–Tryptone Water • Even in a broth, Kovac's reagent rests on top where it reacts with indole to produce the red color indicative of a positive result. Shown are *Klebsiella* (*Enterobacter*) *aerogenes* (–) on the left and *Escherichia coli* (+) on the right.

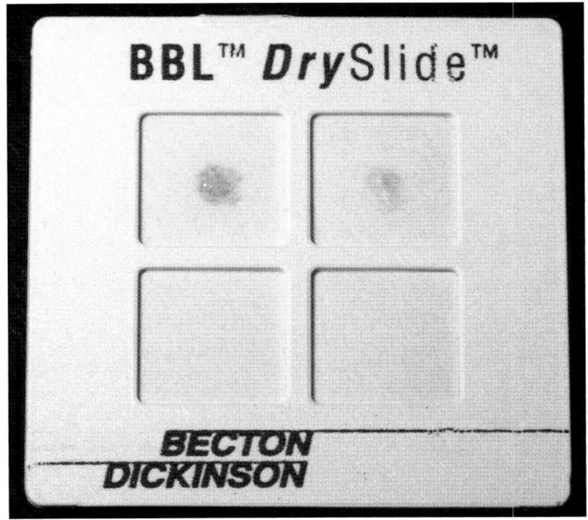

7.44 Indole Test Results—BBL DrySlide Rapid Indole Test • BBL DrySlide (available from Becton-Dickinson, www.bd.com/en-us) provides a means to run the indole test directly from a pure culture and get a quick result. This slide was inoculated with *Escherichia coli* (pink, +) on the left and *Klebsiella* (*Enterobacter*) *aerogenes* (no color change, –) on the right.

Kliger's Iron Agar

Purpose

Kligler's iron agar (KIA) is primarily used to differentiate members of *Enterobacteriaceae* and to distinguish them from other Gram-negative rods such as *Pseudomonas aeruginosa* (see p. 209).

Principle

KIA is a rich medium designed to differentiate bacteria (especially enterics–***Enterobacteriaceae***) on the basis of sulfur reduction, and glucose and lactose fermentation. In addition to the two carbohydrates it includes digests of casein and animal proteins as sources of carbon and nitrogen, and both ferrous sulfate and sodium thiosulfate as sources of oxidized sulfur.

Phenol red is the pH indicator (yellow at pH less than 6.8 and reddish above pH 7.4), and the iron in the ferrous sulfate is the hydrogen sulfide (i.e., sulfur reduction) indicator.

The medium is prepared as a shallow agar slant with a deep butt, thereby providing both aerobic and anaerobic growth environments. It is inoculated by a stab in the agar butt followed by a fishtail streak of the slant. The incubation period is 18 to 24 hours for carbohydrate fermentation and up to 48 hours for hydrogen sulfide reactions. Many reactions in various combinations are possible (Fig. 7.45 and Table 7-3).

When KIA is inoculated with a glucose-only fermenter, acid products lower the pH and turn the entire medium yellow within a few hours. Because glucose is in short supply (0.1%), it will be exhausted within about 12 hours. As the glucose diminishes, ammonia produced from deamination of amino acids by organisms located in the aerobic region (slant) will begin to raise the pH and turn it red.

This process, which takes 18 to 24 hours to complete, is called a **reversion. It only occurs in the slant because the anaerobic conditions in the butt result in slower glucose consumption.** Thus, a KIA with a red slant and yellow butt after a 24-hour incubation period indicates that the organism ferments glucose but not lactose.

Organisms that are able to ferment glucose *and* lactose also turn the medium yellow throughout. However, because the lactose concentration is 10 times higher than that of glucose, more acid is produced for a longer period of time, so both slant and butt will remain yellow after 24 hours.

Therefore, a KIA with a yellow slant and butt at 24 hours indicates that the organism ferments glucose and lactose. Gas produced by carbohydrate fermentation will appear as fissures in the medium or will lift the agar off the bottom of the tube (Fig. 7.45).

Hydrogen sulfide (H_2S) may be produced by the breakdown of cysteine in the peptone or reduction of thiosulfate in the medium (see Figs. 7.85 and 7.86). Ferrous sulfate reacts with the H_2S to form a black precipitate, usually seen in the butt (Fig. 7.45). Acid conditions must exist for thiosulfate reduction; therefore, black precipitate in the medium is an indication of sulfur reduction *and* at the very least, glucose fermentation if the organism is suspected to be an enteric.

If the black precipitate obscures the color of the butt, the color of the slant determines which carbohydrates have been fermented (i.e., red slant = glucose fermentation, yellow slant = glucose and lactose fermentation).

An organism that does not ferment either carbohydrate but deaminates the amino acids will alkalinize the medium (due to the ammonia produced) and turn it red. If the organism can use the peptone aerobically and anaerobically, both the slant and butt will appear red. An obligate aerobe will turn only the slant red (Fig. 7.45). (Note that the difference between a red butt and a butt unchanged by the organism may be subtle; therefore, comparison with an uninoculated control is always recommended.)

Not surprisingly, timing is critical when reading results. An early reading could reveal yellow throughout the medium, leading one to conclude that the organism is a lactose fermenter when it simply may not yet have exhausted the glucose.

A reading after the lactose has been depleted could reveal a yellow butt and red slant leading one to falsely conclude that the organism is a glucose-only fermenter. Tubes that have been interpreted for carbohydrate fermentation can be re-incubated for 24 hours before H_2S determination. Refer to Table 7-3 for information on the correct symbols and method of reporting the various reactions.

TABLE **7-3** **KIA Test Results and Interpretations**

Table of Results		
Result	**Interpretation**	**Symbol**
Yellow slant/yellow butt	Glucose and lactose fermentation with acid accumulation in slant and butt	A/A
Red slant/yellow butt	Glucose fermentation with acid production; amino acids deaminated aerobically (in the slant) with alkaline products (reversion)	K/A
Red slant/red butt	No fermentation; amino acids deaminated aerobically and anaerobically with alkaline products; isolate is not from *Enterobacteriaceae*	K/K
Red slant/no change in butt	No fermentation; amino acids deaminated aerobically with alkaline products; isolate is not from *Enterobacteriaceae*	K/NC
No change in slant/no change in butt	Organism is growing slowly or not at all; isolate is not from *Enterobacteriaceae*	NC/NC
Black precipitate in the agar	Sulfur reduction (an acid condition from fermentation exists in the butt even if the yellow color is obscured by the black precipitate)	H_2S
Cracks in or lifting of agar	Gas production	G

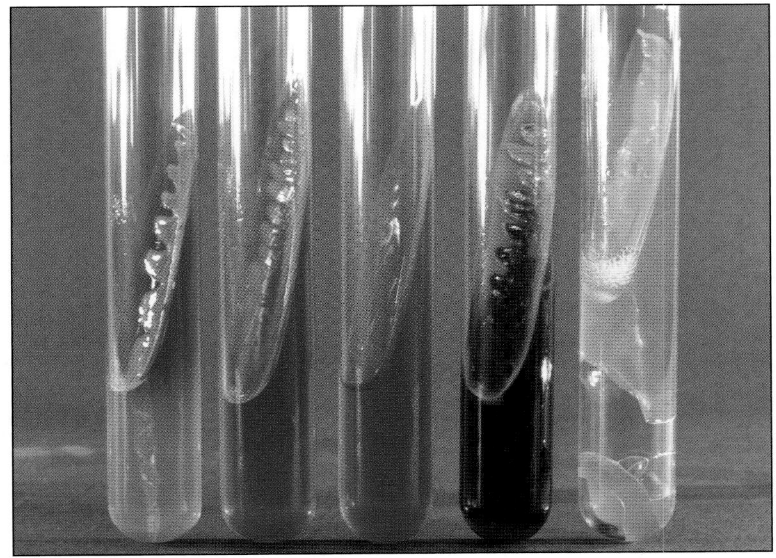

7.45 Kligler's Iron Agar Results • Kligler's iron agar is inoculated by stabbing the agar butt (anaerobic zone) and streaking the slant (aerobic zone), which provides opportunities for many and varied results. It is imperative to read the fermentation results after 18–24 hours of incubation. Once read, further incubation for sulfur reduction results may be necessary. From left to right: *Morganella morganii* (K/A, atypically not producing gas), *Pseudomonas aeruginosa* (K/NC), uninoculated control, *Proteus mirabilis* (K/A, H_2S), and *Escherichia coli* (A/A, G). Refer to Table 7-3 for more details.

Lipid Hydrolysis (Lipase Test)

Purpose

The ability of a microbe to hydrolyze **lipids** can be used to differentiate between species, but it also is used in detecting and quantifying **lipolytic** bacteria, especially in high-fat dairy products. In addition to tributyrin a variety of other lipid substrates, including corn oil, olive oil, and soybean oil, are used to detect differential characteristics among members of *Enterobacteriaceae*, *Clostridium*, *Staphylococcus*, and *Neisseria*. Several fungal species also demonstrate lipolytic ability.

Principle

The word *lipid* is generally used to describe all types of fats, and enzymes that hydrolyze lipids fall into a generic category called **lipases**. Many lipases are **exoenzymes**, that is, they are secreted by the organism to digest lipids outside the cell. Similar in function to other types of exoenzymes, lipases are necessary because intact lipids are macromolecules and too large to be transported into the cell.

Simple fats, known as **triglycerides** or **triacylglycerols**, are composed of glycerol and three long-chain fatty acids. After lipid hydrolysis (lipolysis), glycerol can be phosphorylated and oxidized in its conversion to dihydroxyacetone phosphate, which is an intermediate of glycolysis (Fig. 7.46 and Appendix, Fig. A.1). The process costs one ATP, but reduces one NAD^+ to $NADH + H^+$.

The fatty acids are catabolized by a process called **β-oxidation** in which two carbon fragments are sequentially removed and combined with Coenzyme A to produce acetyl~CoA, one NADH and one $FADH_2$. Acetyl~CoA can then enter the citric acid cycle. Alternatively, glycerol and fatty acids may be used in anabolic pathways.

A variety of simple fats can be used for the lipid hydrolysis (lipase) test, but tributyrin oil is a common choice because it is the simplest triacylglycerol found in natural fats and oils. Tributyrin agar is prepared as an emulsion that makes the agar appear opaque. When the plate is inoculated with a lipase-positive organism, clear zones will appear around the growth as evidence of lipolytic activity (Fig. 7.47). If no clear zones appear, the organism is lipase negative.

Tributyrin agar plates that are more than a week or two old begin to lose their opacity, which can be confused with clearing due to lipase activity or obscure true clearing. Always check the plates for adequate opacity prior to use.

Spirit blue agar is an alternative medium to tributyrin agar. It is prepared as an emulsion with tributyrin oil, but also contains yeast extract and spirit blue dye as a color indicator. The oil and dye form a complex that gives the medium an opaque, light-blue appearance. Lipase-positive bacteria growing on the medium hydrolyze the oil and produce clear halos surrounding the growth. Lightening of the medium such as that produced around *Proteus mirabilis* is not a positive result (Fig. 7.48).

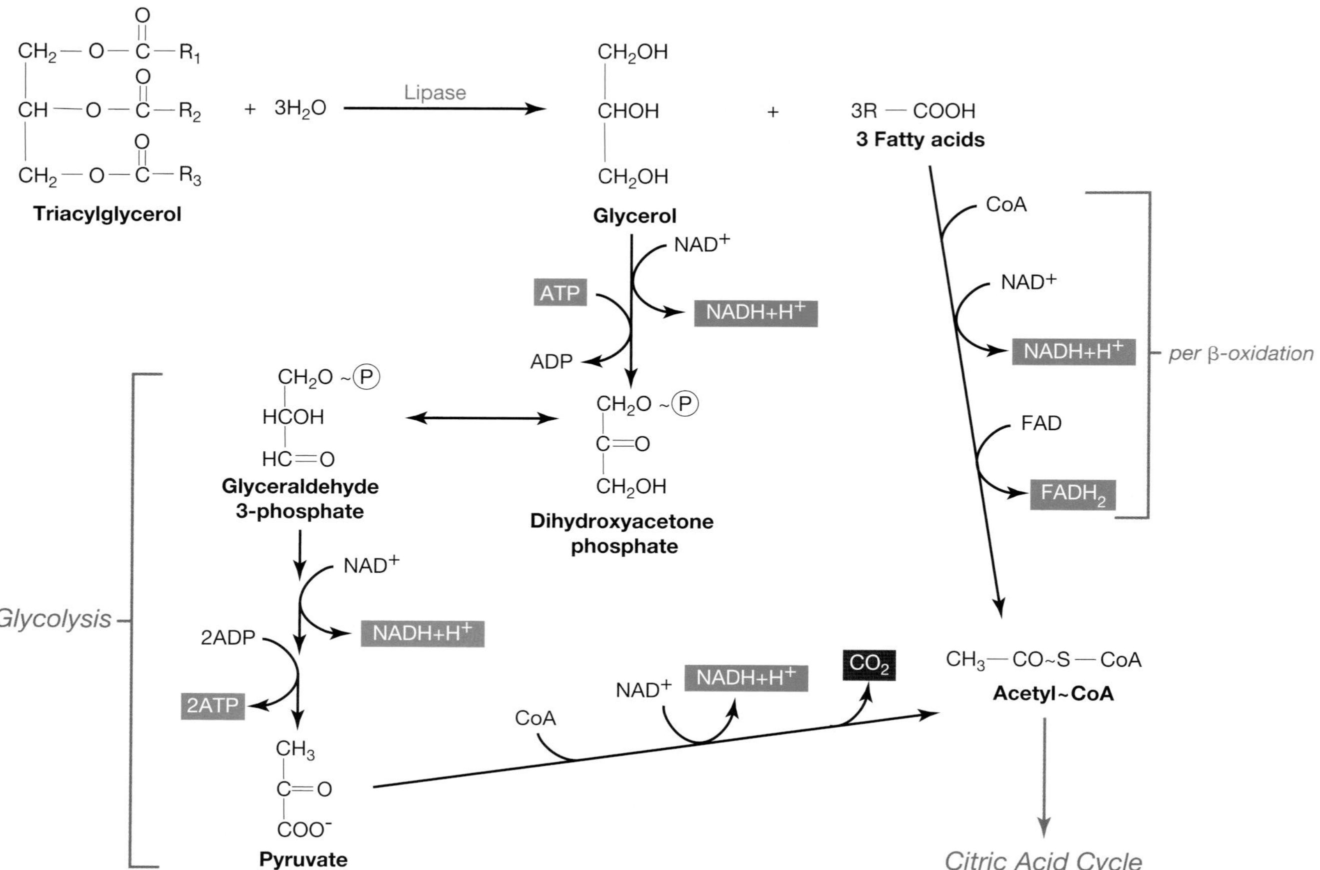

7.46 Lipid Catabolism Reactions ● A triacylglycerol is a simple fat molecule, composed of a three-carbon alcohol (glycerol) bonded to three fatty acid chains (represented by R_1, R_2, and R_3 in this diagram. The products of lipid catabolism can be used in glycolysis (glycerol) and the citric acid cycle (acetyl~CoA).

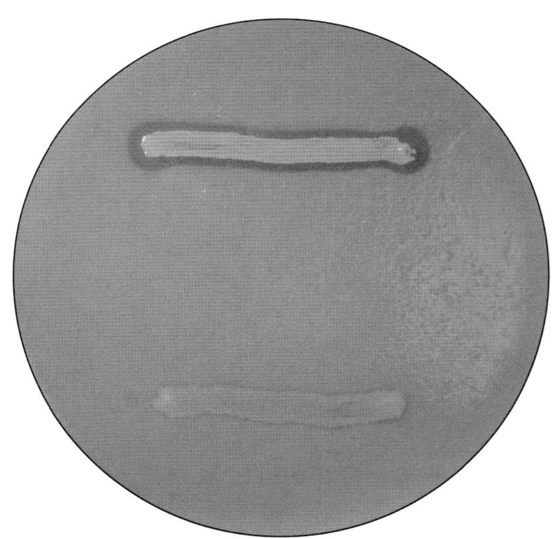

7.47 Lipid Hydrolysis Results–Tributyrin Agar ● *Staphylococcus aureus* (lipase positive, note the clearing in the agar) is above and *Proteus mirabilis* (lipase negative) is below. The yellow color of *S. aureus* growth (*aureus* means "golden") is natural and has nothing to do with the tributyrin agar.

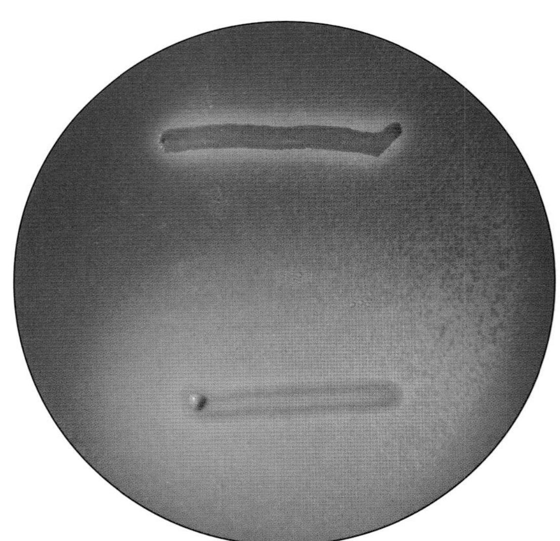

7.48 Lipid Hydrolysis Results–Spirit Blue Agar ● *Staphylococcus aureus* (lipase positive) is above and *Proteus mirabilis* (lipase negative) is below. Clearing, not lightening of the medium (as with *P. mirabilis*) is considered a positive result.

Litmus Milk Medium

Purpose

Litmus milk is used primarily to differentiate members within the genus *Clostridium* (see pp. 192–195). It also differentiates *Enterobacteriaceae* from other Gram-negative bacilli based on the ability of enterics to reduce litmus. Litmus milk can be used to cultivate and maintain cultures of lactic acid bacteria.

Principle

Litmus milk is an undefined medium consisting of skim milk and the pH indicator **azolitmin**. Skim milk provides nutrients for growth, including lactose for fermentation, and protein in the form of casein (which makes milk white). Azolitmin (litmus) is pink at pH 4.5 and blue at pH 8.3. Between these extremes it is light purple.

Four basic reactions occur in litmus milk: lactose fermentation, reduction of litmus, casein coagulation, and casein hydrolysis. In combination these reactions yield a variety of results, each of which can be used to differentiate bacteria. Several possible combinations are described in Table 7-4.

Lactose fermentation acidifies the medium and turns the litmus pink (Fig. 7.49, second tube from the right). This **acid reaction** begins with hydrolysis of the disaccharide into the monosaccharides glucose and galactose by the enzyme **β-galactosidase** (Fig. 7.50). Accumulating acid may cause casein to precipitate and form an **acid clot** (Figs. 7.51 and 7.52).

Acid clots solidify the medium and can appear pink or white with a pink band at the top depending on the oxidation-reduction status of litmus. Reduced litmus is white; oxidized litmus is purple. Acid clots can be dissolved by an alkaline solution. Fissures or cracks in the clot or curd are evidence of **gas** production (Fig. 7.49, third tube from right). Heavy gas production that breaks up the clot is called **stormy fermentation.**

In addition to being a pH indicator, litmus is an $\mathbf{E_h}$ (oxidation-reduction) indicator. As mentioned above, reduced litmus is white. If litmus becomes reduced during lactose fermentation, it will turn the medium white in the lower portion of the tube where the reduction rate is greatest.

Some bacteria produce proteolytic enzymes such as **rennin, pepsin,** or **chymotrypsin** that coagulate casein and produce a **curd** (Figs. 7.49, third tube from the right, and 7.53). A curd differs from an acid clot in that it is not dissolved by an alkaline solution and tends to retract from the sides of the tube revealing a straw-colored fluid called **whey** (Fig. 7.49, tube at far right).

Certain enzymes can digest both acid clots and curds. A **digestion reaction** leaves only a clear to brownish fluid behind (Fig. 7.49, center tube). Bacteria that are able only to partially digest the casein typically produce ammonia (NH_3), which raises the pH of the medium and turns the litmus blue. Formation of a blue or purple ring at the top of the clear fluid or bluing of the entire medium indicates an **alkaline reaction** (Fig. 7.49, tube at far left and third from left).

TABLE **7-4** Litmus Milk Test Results and Interpretations

Table of Results		
Result	**Interpretation**	**Symbol**
Pink color	Acid reaction	A
Pink and solid (white in the lower portion if the litmus is reduced); clot not movable	Acid clot	AC
Fissures in clot	Gas	G
Clot broken apart	Stormy fermentation	S
White color (lower portion of medium)	Reduction of litmus	R
Semisolid and not pink; clear to gray fluid at top	Curd	C
Clarification of medium; loss of "body"	Digestion of peptone; peptonization	P or D
Blue medium or blue band at top	Alkaline reaction	K
No change	None of the above reactions	NC

These results may appear together in a variety of combinations.

7.49 Litmus Milk Test Results ● From left to right: *Alcaligenes faecalis* (K), uninoculated control, *Pseudomonas aeruginosa* (K), *Clostridium sporogenes* (D), *Clostridium acetobutylicum* (AGCR), *Escherichia coli* (A), and *Lactococcus lactis* (ACR). The clear fluid on the surface of the two *Clostridium* cultures is mineral oil used to make the medium anaerobic.

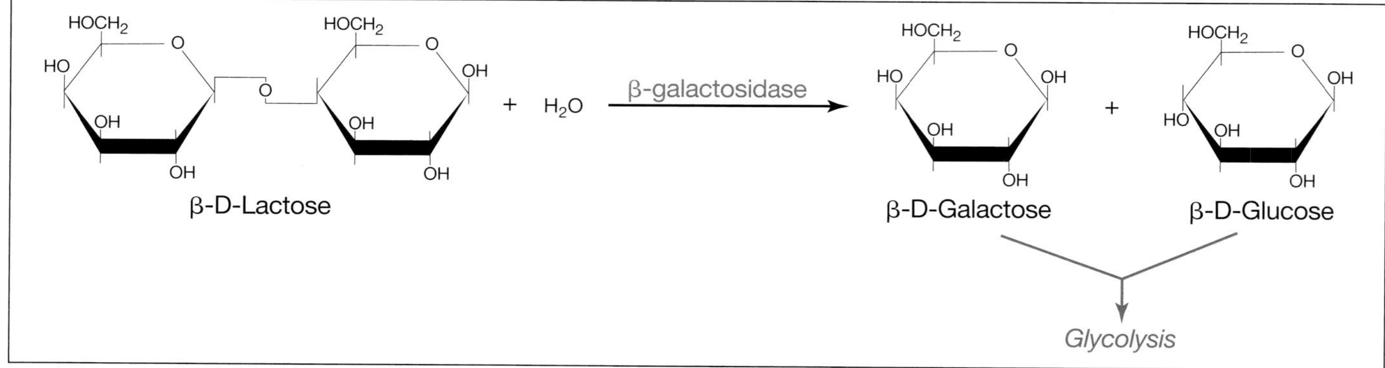

7.50 Lactose Hydrolysis Reaction ● Lactose hydrolysis requires the enzyme β-galactosidase and produces glucose and galactose—two fermentable sugars.

calcium caseinate (soluble salt of casein) —casease, low pH→ caseinogen (acid clot) (insoluble precipitate)

7.51 Acid Clot Reaction ● An acid clot is the result of casease catalyzing the formation of caseinogen, an insoluble precipitate, under acidic conditions.

casein (soluble) —rennin, Ca^{++}→ paracasein (soft curd) (insoluble precipitate)

7.53 Curd Formation Reaction ● Rennin is only one of several enzymes that curdle milk. It converts casein to paracasein to form the curd.

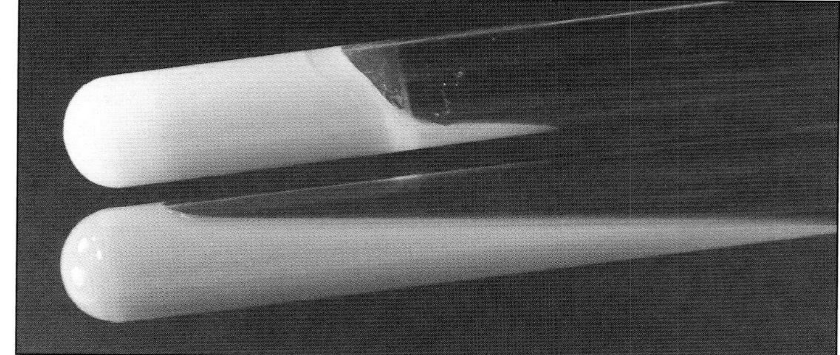

7.52 Acid Clot Formation Results ● An acid clot appears in the top tube; an uninoculated control is below. Note the reduced litmus (white) at the bottom of the clot.

Lysine Iron Agar (LIA)

Purpose

Lysine iron agar (LIA) is used to differentiate enterics based on their ability to decarboxylate or deaminate lysine and produce hydrogen sulfide (H_2S). LIA also is used in combination with triple sugar iron agar to identify members of *Salmonella* (see p. 211) and *Shigella* (see p. 213), because of the fermentation information provided (neither typically ferments lactose or sucrose).

Principle

LIA is a combination medium that detects bacterial ability to **decarboxylate** or **deaminate** lysine and to reduce sulfur. It contains peptone and yeast extract to support growth, the amino acid lysine for deamination and decarboxylation reactions, and sodium thiosulfate—a source of reducible sulfur.

A small amount of glucose (0.1%) is included as a fermentable carbohydrate. Ferric ammonium citrate acts as a sulfur reduction (H_2S) indicator and bromocresol purple is the pH indicator. Bromocresol purple is purple at pH 6.8 and yellow at or below pH 5.2.

LIA is prepared as a slant with a deep butt. This results in an aerobic zone in the slant and an anaerobic zone in the butt. After it is inoculated with two stabs of the butt and a fishtail streak of the slant, the tube is tightly capped and incubated for 18 to 24 hours.

All *Enterobacteriaceae* ferment glucose to acid end-products. If the medium has been inoculated with a **lysine decarboxylase**-positive enteric, acid production from glucose fermentation will induce production of decarboxylase enzymes. The acidic pH will turn the medium yellow, but subsequent lysine decarboxylation will produce the amine **cadaverine** and alkalinize the agar, returning it to purple (Figs. 7.26 and 7.54).

Purple color throughout indicates lysine decarboxylation. Purple color in the slant with a yellow (acidic) butt indicates glucose fermentation, but no lysine decarboxylation took place. In this instance, peptone degradation accounts for alkalinization of the slant.

If the organism produces **lysine deaminase**, the resulting deamination reaction will produce compounds that react with the ferric ammonium citrate and produce a red color. Deamination reactions require the presence of oxygen. Therefore, any evidence of deamination will be seen only in the slant. A red slant with yellow (acidic) butt indicates lysine deamination.

Hydrogen sulfide (H_2S) is produced in LIA by the anaerobic reduction of thiosulfate. Ferric ions in the medium react with the H_2S to form a black precipitate in the butt. Refer to Figure 7.54 and Table 7-5 for the various reactions and symbols used to record them.

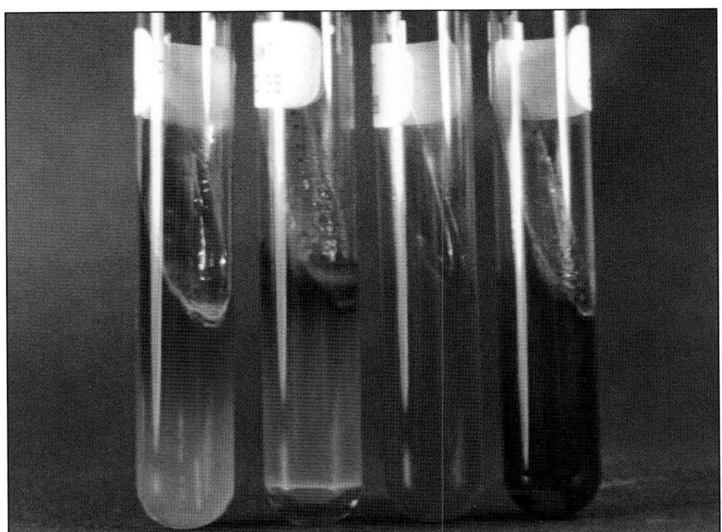

7.54 Lysine Iron Agar Results • Shown are four LIA tubes illustrating typical results. From left to right: *Proteus mirabilis*, (R/A); *Citrobacter freundii*, (K/A, H_2S, note the small amount of black precipitate near the middle and the gas production from glucose fermentation at the base); uninoculated control; and *Salmonella typhimurium* (K/K, obscured by the black precipitate, H_2S).

TABLE **7-5** **LIA Test Results and Interpretations**

Table of Results		
Result	**Interpretation**	**Symbol**
Purple slant/purple butt	Lysine deaminase negative; lysine decarboxylase positive	K/K
Purple slant/yellow butt	Lysine deaminase negative; lysine decarboxylase negative; glucose fermentation	K/A
Red slant/yellow butt	Lysine deaminase positive; lysine decarboxylase negative; glucose fermentation	R/A
Black precipitate	Sulfur reduction	H_2S

Malonate Utilization Test

Purpose

Malonate utilization was originally designed to differentiate between *Escherichia* (see p. 196), which will not grow in the medium, and *Enterobacter*. Its use as a differential medium has now broadened to include other members of *Enterobacteriaceae*.

Principle

One of the many enzymatic reactions of the citric acid cycle, as illustrated in the Appendix (Fig. A.5), is the oxidation of succinate to fumarate. In the reaction, which requires the enzyme **succinate dehydrogenase**, the coenzyme FAD is reduced to $FADH_2$. Refer to the upper reaction in Figure 7.55.

Malonate (malonic acid), which can be added to growth media, is similar enough to succinate in structure to fit in the active site of succinate dehydrogenase (Fig. 7.55, lower reaction) and prevent succinate from doing so. This **competitive inhibition** of succinate dehydrogenase, in combination with the subsequent buildup of succinate in the cell, shuts down the citric acid cycle and will kill the organism unless it can ferment or utilize malonate as its sole remaining carbon source.

Malonate broth is the medium used to make this determination. It contains a high concentration of sodium malonate, yeast extract, and a very small amount of glucose to promote growth of organisms that otherwise are slow to respond. Buffers are added to stabilize the medium at pH 6.7. Bromothymol blue dye, which is green in uninoculated media, is added to indicate any shift in pH.

Malonate-positive organisms utilize sodium malonate as a carbon source and ammonium sulfate as a nitrogen source, which produces sodium hydroxide and raises the pH, turning the medium blue. The concentration of malonate ensures there is enough to compete with succinate and supply malonate-positive organisms with a carbon source.

If an organism cannot utilize malonate but manages to ferment a small amount of glucose, it may turn the medium slightly yellow or produce no color change at all. These are negative results (Fig. 7.56).

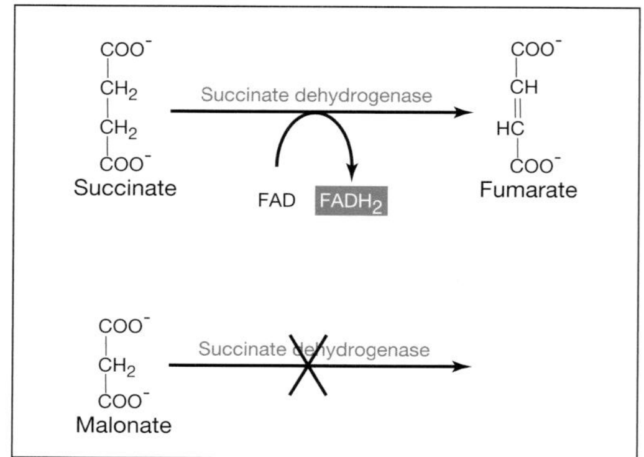

7.55 Competitive Inhibition of Succinate Dehydrogenase • Occupation by malonate of the enzyme succinate dehydrogenase active site prevents attachment of the normal substrate, succinate, and thus, the conversion of succinate to fumarate in the citric acid cycle. This cripples the citric acid cycle.

7.56 Malonate Utilization Results • Malonate broth inoculated with *Klebsiella* (*Enterobacter*) *aerogenes* (malonate positive) is on the left and *Escherichia coli* (malonate negative) is in the center. An uninoculated control used for color comparison is on the right.

Methyl Red Test (MR-VP Broth)

Purpose

The methyl red test is a component of the **IMViC** battery of tests (*I*ndole, *M*ethyl Red, *V*oges-Proskauer, and *C*itrate) used to differentiate the *Enterobacteriaceae* and differentiate them from other Gram-negative rods. Species in the genera *Escherichia*, *Shigella*, and *Salmonella* are MR positive. For more information about these fermentations, refer to Figure A.6 in the Appendix.

Principle

MR-VP broth is a combination medium used for both methyl red (MR) and Voges-Proskauer (VP) tests. (Refer to p. 119 for the VP test.) It is a simple solution containing only peptone, glucose, and a phosphate buffer. The peptone and glucose provide protein (with its nitrogen) and a fermentable carbohydrate, respectively, and the potassium phosphate resists pH changes in the medium.

The MR test is designed to detect organisms capable of performing a **mixed acid fermentation**, which overcomes the phosphate buffer in the medium and lowers the pH (Fig. 7.57 and Appendix, Fig. A.6). Succinate is produced from the addition of CO_2 to phosphoenolpyruvate, whereas the other end-products are derived from the reduction of pyruvate to lactate or its oxidation to acetyl~CoA and formate.

Conversion of acetyl~CoA to acetate results in the formation of one ATP or it can be reduced to ethanol. Formate can be further broken down into H_2 and CO_2 gas. The acids produced by these organisms tend to be stable, whereas acids produced by other organisms may be converted to more neutral products or their fermentation shifts to produce more neutral products in response to the lowered pH.

Mixed acid fermentation is verified by the addition of methyl red indicator dye following incubation. Methyl red is red at pH 4.4 and yellow at pH 6.2. Between these two pH values, it is various shades of orange. Red color is the only true indication of a positive result. Orange is negative or inconclusive. Yellow is negative (Fig. 7.58).

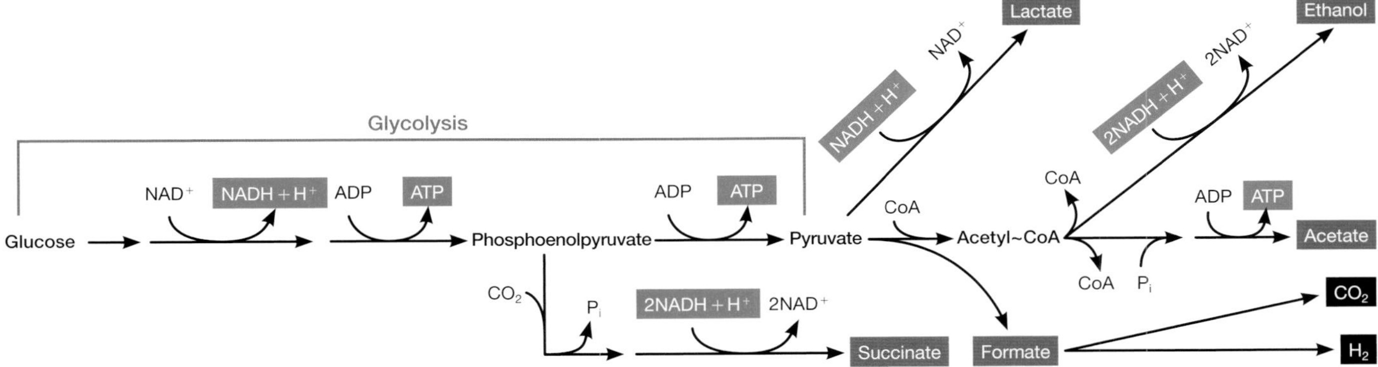

7.57 Mixed Acid Fermentation Reactions of *Escherichia coli* • *E. coli* is a representative methyl red positive organism and is recommended as a positive control for the test. Its mixed acid fermentation produces (in order of abundance) lactate, carbon dioxide, hydrogen gas, ethanol, acetate, succinate, and formate. Most of the formate is converted to H_2 and CO_2 gases. All of the end-products are derived from pyruvate with the exception of succinate, which is produced from phosphoenolpyruvate. *Salmonella* and *Shigella* also are methyl red positive, though some of the specific end-products and amounts differ.

7.58 Methyl Red (MR) Test Results • Methyl red pH indicator turns red at pH 4.4. The products of mixed acid fermentation easily reach that pH, so red is a positive result. *Escherichia coli* (MR positive) is on the left and *Klebsiella* (*Enterobacter*) *aerogenes* (MR negative) is on the right.

Motility Agar

Purpose

Motility agar is used to detect bacterial motility without direct microscopic observation. Motility is an important differential characteristic of *Enterobacteriaceae* and other groups.

Principle

Motility agar is a semisolid medium designed to detect bacterial motility. Its agar concentration is reduced from the typical 1.5% to 0.4%—just enough to maintain its form while allowing movement of motile bacteria. It is inoculated by stabbing with a straight transfer needle. Motility is detectable as diffuse growth radiating from the central stab line (Fig. 7.59).

7.59 Motility Test Results–SIM Medium Without TTC • On the left is *Staphylococcus aureus* (nonmotile) and *Escherichia coli* (motile) is on the right. Notice that motility of *E. coli* is seen only as haziness in the medium. Often, tubes must be compared to an uninoculated control to discriminate between faint haziness and motility. Compare with Figure 7.61.

A tetrazolium salt (TTC) sometimes is added to the medium to make interpretation easier. TTC is used by the bacteria as an electron acceptor. In its oxidized form, TTC is colorless and soluble; when reduced it is red and insoluble (Fig. 7.60). A positive result for motility is indicated when the red (reduced) TTC is seen radiating outward from the central stab. A negative result shows red only along the stab line (Fig. 7.61).

reductase $2H^+$

+ HCl

2,3,5-Triphenyltetrazolium chloride$_{oxidized}$ (TTC_{ox})
colorless and soluble

Formazan$_{reduced}$
red color and insoluble

7.60 Reduction of TTC Reaction • Reduction of **2,3,5-Triphenyltetrazolium chloride (TTC)** by metabolizing bacteria results in its conversion from colorless and soluble to the red and insoluble compound formazan. The location of the growing bacteria can be easily determined by the location of the formazan in the medium.

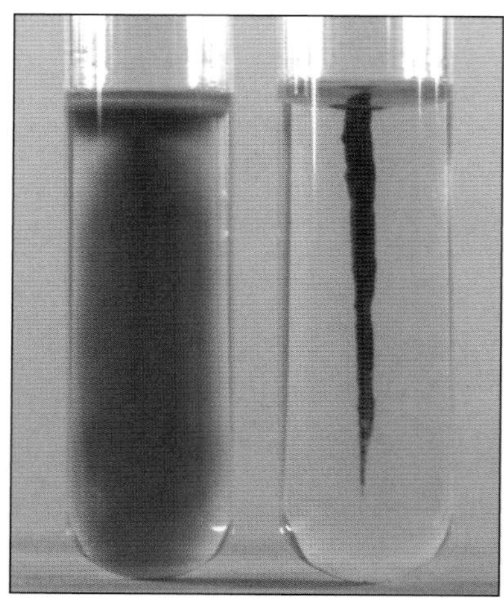

7.61 Motility Test Results–Motility Agar with TTC • Motility test agar tubes containing TTC are easier to read than motility agar without TTC. Shown are two tubes inoculated with *Klebsiella* (*Enterobacter*) *aerogenes* (motile) on the left and *Micrococcus luteus* (nonmotile) on the right. Compare with Figure 7.59.

Nitrate Reduction Test

Purpose

Virtually all members of *Enterobacteriaceae* perform a one-step reduction of nitrate to nitrite. The nitrate test differentiates them from Gram-negative rods that either do not reduce nitrate or reduce it beyond nitrite to N_2 or other compounds.

Principle

Anaerobic respiration involves the reduction of (transfer of electrons to) an inorganic molecule other than oxygen. Nitrate reduction is one such example. Many Gram-negative bacteria (including most *Enterobacteriaceae*) contain the enzyme **nitrate reductase** and perform a single-step reduction of nitrate to nitrite ($NO_3^- \rightarrow NO_2^-$).

Other bacteria, in a multistep process known as **denitrification**, are capable of enzymatically converting nitrate to molecular nitrogen (N_2) or nitrous oxide (N_2O) via nitrite. Some products of nitrate reduction are shown in Figure 7.62.

Nitrate broth is an undefined medium of beef extract, peptone, and potassium nitrate (KNO_3). An inverted Durham tube is placed in each broth to trap a portion of any gas produced. In contrast to many differential media, no color indicators are included. The color reactions obtained in nitrate broth take place as a result of reactions between metabolic products and reagents added after incubation (Fig. 7.63).

Before a broth can be tested for nitrate reductase activity (nitrate reduction to nitrite), it must be examined for evidence of denitrification. This is simply a visual inspection for the presence of gas in the Durham tube (Fig. 7.64). If the Durham tube contains gas the test is complete. Denitrification has taken place. However, gas produced in a nitrate reduction test by an organism capable of fermenting to gas end-products is not determinative because the source of the gas is unknown.

If there is no visual evidence of denitrification (or the test was not determinative), **sulfanilic acid** and **α-naphthylamine** are added to the medium to test for nitrate reduction to nitrite. If present, nitrite will form **nitrous acid** (HNO_2) in the aqueous medium. Nitrous acid reacts with the added reagents to produce a red, water-soluble compound (Fig. 7.65).

Therefore, red color formation after the addition of reagents indicates that the organism reduced nitrate to nitrite. If no color change takes place with the addition of reagents, the nitrate either was not reduced or was reduced to one of the other nitrogenous compounds shown in Figure 7.62. Because it is visually impossible to tell the difference between these two occurrences at this point, another test must be performed.

In this stage of the test, a small amount of powdered zinc is added to the broth to catalyze the reduction of any nitrate (which still may be present as KNO_3) to nitrite. If nitrate is present at the time zinc is added, it will be quickly reduced to nitrite, and the above-described reaction between nitrous acid and reagents will follow and turn the medium red.

In this instance, the red color indicates that nitrate was *not* reduced by the organism (Fig. 7.66). No color change after the addition of zinc indicates that the organism reduced the nitrate to NH_3, NO, N_2O, or some other nongaseous nitrogenous compound.

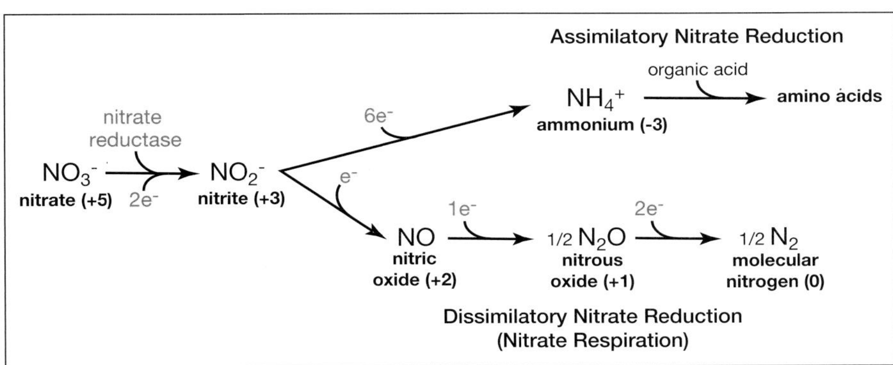

7.62 Possible End-products of Nitrate Reduction • Nitrate reduction is complex. Many different organisms under many different circumstances perform nitrate reduction with many different outcomes. Members of the *Enterobacteriaceae* simply reduce NO_3 to NO_2. Other bacteria, functionally known as "denitrifiers," reduce NO_3 all the way to N_2 via the intermediates shown, and are important ecologically in the nitrogen cycle (see p. 307). Both of these are anaerobic respiration pathways (also known as "nitrate respiration" and "dissimilatory nitrate reduction"). Other organisms are capable of assimilatory nitrate reduction, in which NO_3^- or NO_2^- is reduced to NH_4^+, which can be used in amino acid synthesis. The oxidation state of nitrogen in each compound is shown in parentheses.

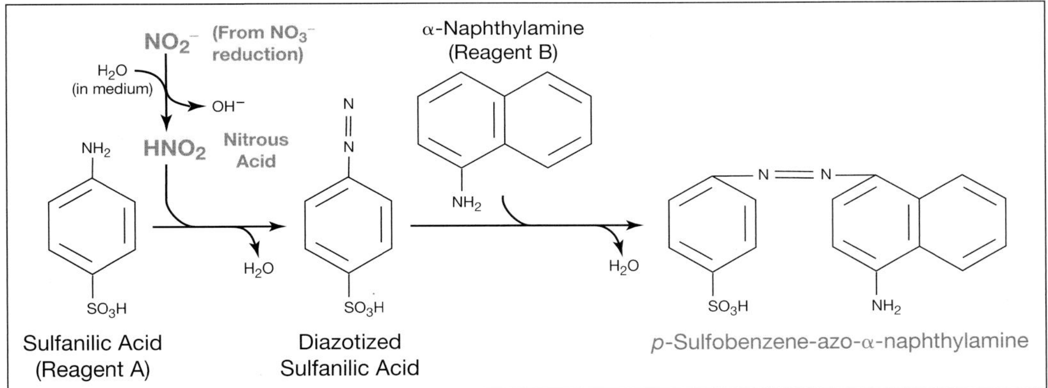

7.63 Indicator Reactions for Nitrate Reduction to Nitrite • If nitrate is reduced to nitrite, nitrous acid (HNO_2) will form in the medium. Nitrous acid then reacts with sulfanilic acid (reagent A) to form diazotized sulfanilic acid, which reacts with the α-naphthylamine (reagent B) to form *p*-sulfobenzene-azo-α-naphthylamine, which is red. Thus, a red color indicates the presence of nitrite and is considered a positive result for nitrate reduction to nitrite.

7.64 Nitrate Reduction Results—Before Addition of Reagents • These are nitrate broths immediately after incubation and prior to addition of reagents. From left to right: *Klebsiella* (*Enterobacter*) *aerogenes*, an uninoculated control, *Enterococcus faecalis*, and two different strains of *Pseudomonas aeruginosa*. Tube 2 is used for color comparison. Note the gas produced in tube 5. (The bubble in the Durham tube is barely visible behind the layer of bubbles on the surface.) It is a known non-fermenter and, therefore, will receive no reagents. The gas produced is an indication of denitrification and a positive result for nitrate reduction. Tubes 1 through 4 will receive reagents. Continue with Figure 7.65.

7.65 Nitrate Reduction Results–After Addition of Reagents ● These are the same tubes as in Figure 7.64 after the addition of sulfanilic acid and α-naphthalamine to the first four tubes. After the addition of reagents, tube 1 shows a positive result. Tubes 3 and 4 are inconclusive because they show no color change. Zinc dust must be added to tubes 2 (control), 3, and 4 to verify the presence or absence of nitrate. Continue with Figure 7.66.

7.66 Nitrate Reduction Results–After Addition of Reagents and Zinc ● These are the same tubes as in Figure 7.65 after the addition of zinc to the middle three tubes. A pinch of zinc was added to these tubes because they have not yet produced a result that could be interpreted. Tube 2 (the control tube) and tube 3 turned red. This is a negative result because it indicates that nitrate is still present in the tube. Tube 4 did not change color, which indicates that the nitrate was reduced by the organism beyond nitrite to some other nitrogenous compound. This is a positive result.

Novobiocin Susceptibility Test

Purpose

Novobiocin susceptibility is used to differentiate coagulase-negative staphylococci. Most frequently it is used to presumptively identify the novobiocin-resistant *Staphylococcus saprophyticus*.

Principle

With the exception of *Staphylococcus saprophyticus*, most clinically important staphylococci are susceptible to the antibiotic **novobiocin**. Novobiocin is an antibiotic, produced by *Streptomyces niveus*. It is related to coumarin and interferes with ATPase activity associated with DNA gyrase, an enzyme necessary during DNA replication.

When agar plates are cultured with a novobiocin-susceptible organism and a novobiocin impregnated disk is placed on it, a large **zone of inhibition** (clearing) around the disk will appear (Fig. 7.67). Conversely, organisms resistant to novobiocin will produce a small zone or no zone at all, depending on several factors.

Factors affecting zone size are the tested strain's susceptibility to novobiocin (not all strains are equally susceptible), the density of the inoculum on the plate, the antibiotic concentration in the agar, and the temperature and duration of incubation. The test organism's density is controlled by diluting it to a 0.5 McFarland turbidity standard (see Fig. 19.3) immediately before inoculation.

The rate and amount of antibiotic diffusion are standardized by using 5 µg novobiocin disks on 5% sheep blood agar and incubating for 24 hours at 35°C. An isolate producing a zone diameter less than16 mm is considered novobiocin-resistant (R). A zone diameter of 16 mm or more indicates susceptibility (S) (Table 7-6).

For more information on antimicrobial susceptibility, refer to the antimicrobial susceptibility (Kirby-Bauer) test, page 291. ●

7.67 Novobiocin Susceptibility Test on Sheep Blood Agar ● The zone diameter cut off for novobiocin susceptibility is 16 mm. The zone diameter of *Staphylococcus saprophyticus* on the left is 7 mm, so this *S. saprophyticus* strain is novobiocin resistant (R). The zone diameter of *Staphylococcus epidermidis* on the right is 38 mm, well above the 16 mm cut off for susceptibility. This *S. epidermidis* is novobiocin susceptible (S).

TABLE **7-6** **Novobiocin Test Results and Interpretations**

Result	Interpretation	Symbol
Zone of clearing 16 mm or greater	Organism is susceptible to novobiocin; not likely *Staphylococcus saprophyticus* if a coagulase-negative staphylococcus	S
Zone of clearing less than 16 mm	Organism is resistant to novobiocin; probable *Staphylococcus saprophyticus* if a coagulase-negative staphylococcus	R

o-Nitrophenyl-β-D-Galactopyranoside (ONPG) Test

Purpose

The ONPG test is used to differentiate late lactose fermenters from lactose non-fermenters in the family *Enterobacteriaceae*. More importantly, it allows rapid determination of the ability to ferment lactose without waiting for the late lactose fermenters to produce a positive result in standard fermentation tests.

Principle

Bacteria that ferment lactose typically produce two enzymes: **β-galactoside permease,** a membrane-bound transport protein, and **β-galactosidase**, an intracellular enzyme that hydrolyzes the disaccharide into β-glucose and β-galactose (see Fig. 7.50).

Bacteria possessing both enzymes are active β-lactose fermenters. Bacteria that cannot produce β-galactosidase cannot ferment β-lactose. Bacteria that possess β-galactosidase but no (or small amounts of) β-galactoside permease are slow to ferment lactose because it does not enter the cell easily. Quickly distinguishing these **late lactose fermenters** from true non-fermenters is made possible by supplying them with a lactose structural analog: ***o*-nitrophenyl-β-D-galactopyranoside (ONPG)**.

When ONPG is made available to bacteria, it freely enters the cells without the aid of a permease. Because of its similarity to β-lactose, it then can become the substrate for any β-galactosidase present. In the reaction that occurs, ONPG is hydrolyzed to β-galactose and ***o*-nitrophenol (ONP)**, which is yellow (Figs. 7.68 and 7.69).

It should be noted that β-galactosidase is an inducible enzyme; it is produced in response to the presence of an appropriate substrate. Therefore, organisms to be tested with ONPG are typically prepared by growing them overnight in a lactose-rich medium, such as Kligler's iron agar or triple sugar iron agar, to ensure that they will be actively producing β-galactosidase, given that capability.

7.68 Conversion of ONPG to β-Galactose and o-Nitrophenol by β-Galactosidase Examine Figure 7.50 as you look at this figure. ONPG is structurally similar enough to the sugar on the left of the disaccharide lactose that β-galactosidase will catalyze its hydrolysis into β-galactose and *o*-nitrophenol. The latter compound is yellow and is indicative of a positive ONPG test.

7.69 ONPG Test Results • The ONPG test allows rapid determination of the ability to ferment lactose even when a late lactose fermenter is being tested. Results occur within a couple of hours of incubation. *Escherichia coli* (ONPG positive) is on the left, and *Proteus vulgaris* (ONPG negative) is on the right.

Optochin Susceptibility Test

Purpose

Optochin susceptibility is used to presumptively differentiate *Streptococcus pneumoniae* (susceptible; see p. 218) from other α-hemolytic streptococci (resistant).

Principle

Optochin is an antibiotic derived from quinine that disrupts **ATP synthase** activity, which results in reduced ATP production in susceptible bacteria. *Streptococcus pneumoniae* is the only streptococcus susceptible to small concentrations of the antibiotic optochin. Therefore, to eliminate the few streptococci that show susceptibility to large concentrations of the antibiotic, the optochin impregnated disks used in this procedure contain a scant 5 µg.

Three or four colonies of the organism to be tested are transferred and streaked on a sheep blood agar plate in such a way as to produce confluent growth over approximately one-half of the surface. The optochin impregnated disk is then placed in the center of the inoculum and the plate is incubated at 35°C for 24 hours in a candle jar or 5% to 7% CO_2.

The antibiotic will diffuse through the agar and inhibit growth of susceptible organisms in the area immediately surrounding the disk. This creates a clearing in the growth or **zone of inhibition** (Fig. 7.70). A zone (14 mm in diameter surrounding a 6 mm disk or 16 mm surrounding a 10 mm disk) is considered presumptive identification of *Streptococcus pneumoniae*. Smaller zones indicate further testing is required (Table 7-7).

For more information on antimicrobial susceptibility, refer to the antimicrobial susceptibility (Kirby-Bauer) test, page 291.

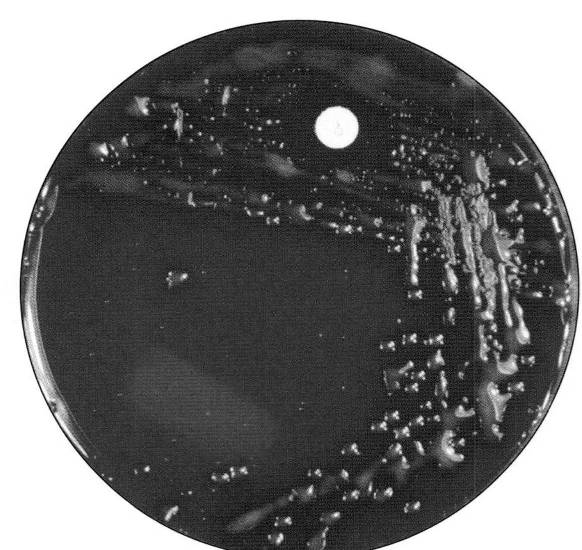

7.70 Optochin Susceptibility Test on Sheep Blood Agar • The zone diameter cut off for optochin susceptibility is 14 mm for a 6 mm disk. The zone of inhibition surrounding the "P" disk on this plate is 19 mm, indicating susceptibility to optochin. The α-hemolysis (green coloration of the agar—see p. 76 for more information) coupled with optochin susceptibility leads to presumptive identification of *Streptococcus pneumoniae*.

TABLE **7-7** **Optochin Test Results and Interpretations**

Result	Interpretation	Symbol
Zone of clearing 14 mm or greater	Organism is susceptible to optochin; probable *Streptococcus pneumoniae* if an α-hemolytic streptococcus	S
Zone of clearing less than 14 mm	Organism is resistant to optochin; not *Streptococcus pneumoniae* if an α-hemolytic streptococcus	R

Oxidase Test

Purpose

The oxidase test is used to identify bacteria containing the respiratory enzyme **cytochrome c oxidase** (or simply "oxidase" for short). Among its many uses is the presumptive identification of oxidase-positive *Neisseria* (see p. 206) and *Moraxella*. It also can be useful in differentiating the oxidase-negative *Enterobacteriaceae* from the oxidase-positive *Pseudomonadaceae*.

Principle

Consider the fate of a glucose molecule in a respiring cell[5]. If it is destined to be used for ATP production, it will become oxidized to 6 CO_2 with the concurrent reduction of **coenzymes:** 10 NAD^+ to 10 $NADH+H^+$ and 2 FAD to 2 $FADH_2$. In addition, 4 ATPs are made by **substrate phosphorylation.** These occur in reactions throughout glycolysis (Appendix, Fig. A.2), the transition step, and the citric acid cycle (Appendix, Fig. A.5).

As you can see, reduced coenzymes have the potential to accumulate rapidly. Therefore, in order to continue oxidizing glucose, these coenzymes must be converted back to their oxidized state. This is done by the **electron transport chain (ETC).**

Aerobes, microaerophiles, most facultative anaerobes, and even obligate anaerobes have ETCs. Electrons from the oxidation of reduced coenzymes are passed down a chain of membrane-bound carrier molecules of the ETC with increasingly positive **reduction potentials** to a final (or terminal) electron acceptor (e.g., ½O_2, NO_3^-, SO_4^{2-}).

A second function of some ETC carriers is to act as a **proton pump**, using the energy lost by the electrons and creating a proton (H^+ ion) gradient across the cytoplasmic membrane that is higher on the outside.

Because of their charge, protons are incapable of diffusing back into the cell through the membrane's phospholipid bilayer. However, the cytoplasmic membrane has channels through which protons *can* diffuse, and the kinetic energy of their diffusion is coupled to **oxidative phosphorylation**—ATP synthesis—catalyzed by membrane-bound **ATP synthases.** So, the energy possessed by electrons stripped away from glucose ultimately is used in ATP synthesis.

[5] Glucose is not completely oxidized to 6 CO_2 in fermenting cells.

There are many different types of electron transport chains (Fig. 7.71), but all share the characteristics listed above. Some organisms use more than one type of ETC, depending on the availability of oxygen or other final electron acceptor(s). *Escherichia coli*, for example, has two ETCs for respiring aerobically and at least one for respiring anaerobically (using nitrate as the FEA). Many bacteria have ETCs resembling mitochondrial ETCs in eukaryotes[6].

The mitochondrial chains contain a series of four electron carriers broadly named Complexes I, II, III, and IV, each of which contains several molecules jointly able to transfer electrons and use the free energy released in the reactions. The last enzyme in the chain, Complex IV, is called **cytochrome c oxidase** because it makes the final electron transfer of the chain from cytochrome c, residing in the periplasm or attached to the periplasmic side of the cytoplasmic membrane of Gram-negative cells, to oxygen inside the cell. In Gram-positive cells, it is membrane-bound.

The oxidase test is designed to identify the presence of cytochrome c oxidase. It is able to do this because cytochrome c oxidase has the unique ability to not only oxidize cytochrome c, but to catalyze the *reduction* of cytochrome c by a **chromogenic reducing agent** called **tetramethyl-*p*-phenylenediamine.** Chromogenic reducing agents are chemicals that develop color as they become oxidized. In the case of tetramethyl-*p*-phenylenediamine, it turns purple/blue (Fig. 7.72).

In the oxidase test, the reducing reagent can be added directly to bacterial growth on solid media (Fig. 7.73), or a bacterial colony can be transferred to filter paper and reagent is added (Fig. 7.74). A color change occurs within seconds if the reducing agent becomes oxidized, thus indicating that cytochrome c oxidase is present. Lack of color change within the allotted time means that cytochrome c oxidase is not present and signifies a negative result.

[6] This is not a surprise, because mitochondria at one time were free living proteobacteria.

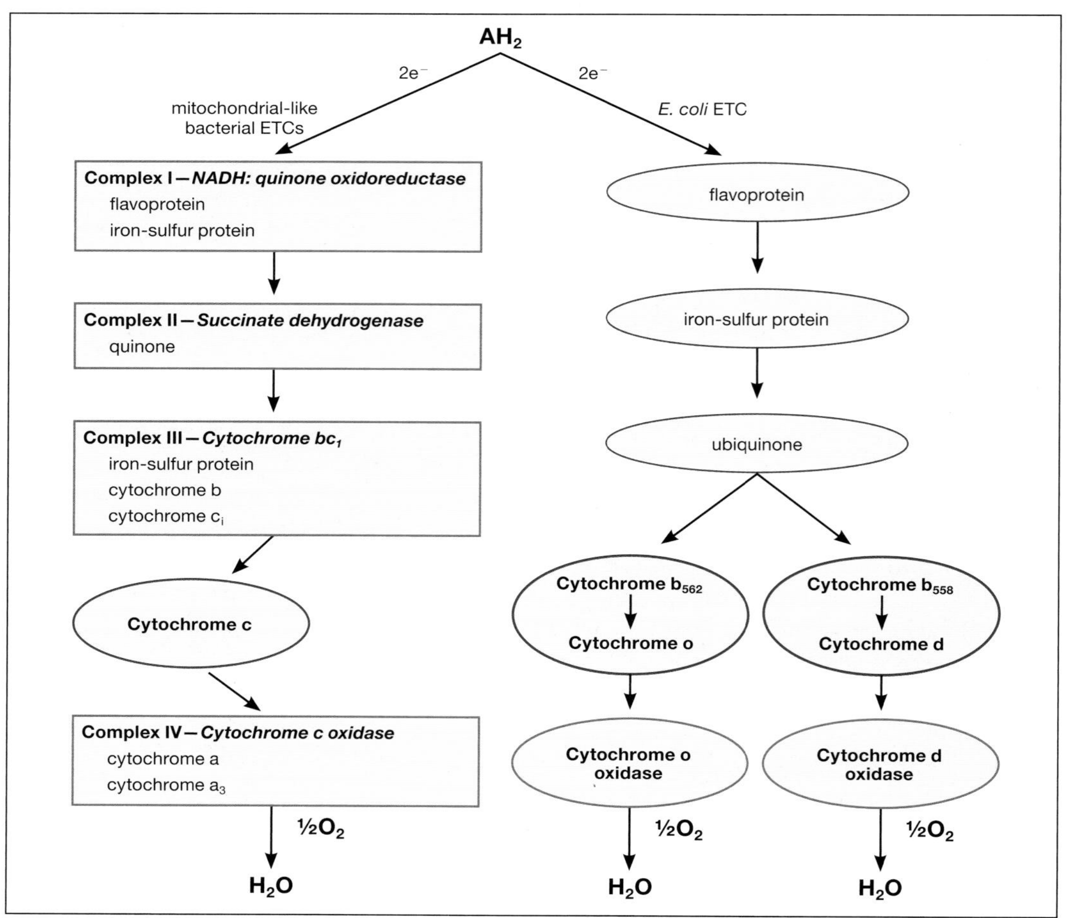

7.71 Aerobic Electron Transport Chains (ETCs) ● Bacterial aerobic electron transport chains (ETCs) show great diversity. However, all begin with flavoproteins that receive electrons from reduced coenzymes, such as NADH or $FADH_2$ (represented by AH_2) and all have pumps that create the proton gradient for ATP synthesis. The part of the ETC relevant to the oxidase test, though, is at the end. Bacteria with chains that resemble the mitochondrial ETC in eukaryotes, shown on the left, contain cytochrome c oxidase (Complex IV), which transfers electrons from cytochrome c to oxygen. These organisms give a positive result for the oxidase test. Other bacteria, such as members of the *Enterobacteriaceae*, are capable of aerobic respiration, but have a different terminal oxidase system and give a negative result for the oxidase test. Both paths shown on the right are found in *E. coli*. The amount of available O_2 determines which pathway is more active. Enzymatic complexes and oxidases are outlined in red.

A

$$2\ \text{Cytochrome } c_{red} + 2H^+ + \tfrac{1}{2}O_2 \xrightarrow{\text{cytochrome c oxidase}} 2\ \text{Cytochrome } c_{ox} + H_2O$$

B

$$\text{Tetramethyl-}p\text{-phenylenediamine}_{red} + 2\ \text{Cytochrome } c_{ox} \xrightarrow{\text{cytochrome c oxidase}} \text{Tetramethyl-}p\text{-phenylenediamine}_{ox} + 2\ \text{Cytochrome } c_{red}$$

Tetramethyl-*p*-phenylenediamine$_{red}$ (colorless)

Tetramethyl-*p*-phenylenediamine$_{ox}$ (deep purple/blue)

7.72 Natural Oxidation of Cytochrome c and Reduction of Cytochrome c in the Indicator Reaction ● **(A)** Under natural conditions, cytochrome c oxidase transfers electrons from cytochrome c to O_2, which produces water. **(B)** When in the oxidized state, cytochrome c is available to be reduced by the phenylenediamine reagent. The phenylenediamine reagent is colorless when reduced and deep purple/blue when oxidized, its condition in a positive oxidase test.

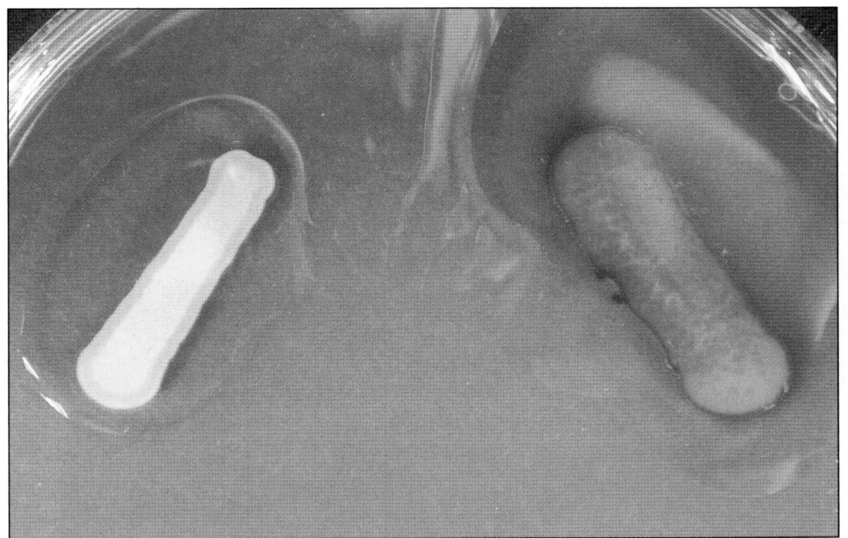

7.73 Oxidase Test Results on Bacterial Growth • Many variations of the oxidase test exist, but in all cases reaction with the phenylenediamine reagent produces a blue or purple color. Reading must be taken within 20 seconds because the reagent can react with moisture in the air and produce a false-positive result. Shown are oxidase-negative *Staphylococcus aureus* (left) and oxidase-positive *Pseudomonas aeruginosa* (right) on an agar plate after addition of the phenylenediamine reagent. *S. aureus* is its natural color; yellow is not the negative result.

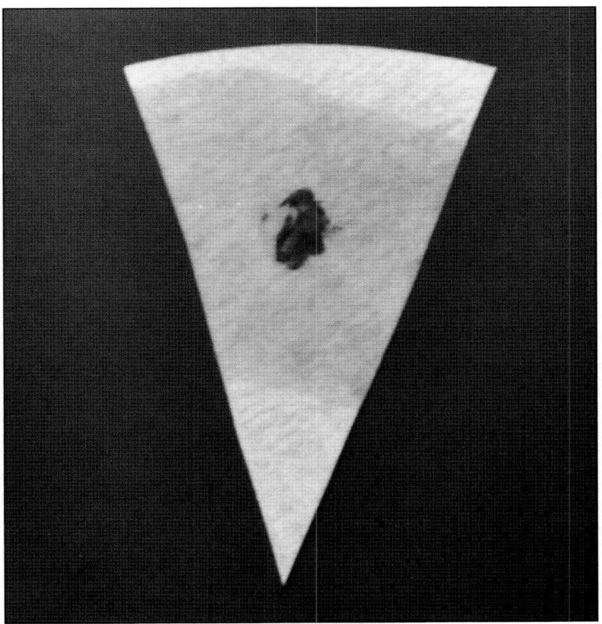

7.74 Oxidase Test Results-Filter Paper • Alternatively, a small amount of growth can be transferred to a piece of filter paper and reagent added. Shown is a positive result produced by *Pseudomonas aeruginosa*.

Oxidation–Fermentation (O-F) Test

Purpose

The oxidation–fermentation (O–F) test is designed to differentiate bacteria on the basis of fermentative or oxidative metabolism of specific carbohydrates (usually glucose). It allows presumptive separation of the oxidative and fermentative *Enterobacteriaceae* from the strictly oxidative *Pseudomonas* and *Bordetella*, and the nonreactive *Alcaligenes* and *Moraxella*.

Principle

In O–F medium, oxidative organisms oxidize the specific carbohydrate (usually glucose) to CO_2 and H_2O with the release of energy, using in the following order: glycolysis, the oxidation of pyruvate (transition step), the citric acid cycle, and finally the electron transport chain (ETC) with oxygen (the final electron acceptor, FEA) being reduced to H_2O.

Similarly, fermentative organisms convert the specific carbohydrate to pyruvate, but because either they *cannot* use O_2 as an FEA or it is *not available*, pyruvate, or one of its derivatives, acts as the FEA and it is reduced to one or more of the familiar fermentation end-products: acid, gas, and alcohol. Consequently, fermenters identified by this test acidify O–F medium to a greater extent than do oxidizers.

Hugh and Leifson's O–F medium includes a high sugar-to-peptone ratio to reduce the possibility that alkaline products from peptone utilization will neutralize weak acids produced by oxidation of the carbohydrate. Bromothymol blue dye, which is yellow at pH 6.0, green at pH 6.9, and blue at pH 7.6, is added as the indicator. A low agar concentration makes it a semisolid medium that also allows determination of motility, evidenced by spreading from the stab lines.

The medium is usually prepared with glucose, but lactose, sucrose, maltose, mannitol, or xylose are sometimes used. The medium is not slanted. Two tubes of the specific sugar medium are stab inoculated several times with the test organism. After inoculation, one tube is sealed with a layer of sterile mineral oil to promote anaerobic growth and fermentation.

The mineral oil creates an environment unsuitable for oxidation because it prevents diffusion of oxygen from the air into the medium. The other tube is left unsealed to allow aerobic growth and oxidation. (Note that the O–F tubes are heated in boiling water and then cooled prior to inoculation. This removes free oxygen from the medium and ensures an anaerobic environment in all tubes. The tubes covered with oil will remain anaerobic, whereas the uncovered medium becomes aerobic as oxygen diffuses back in.)

O–F test results are summarized in Table 7-8 and shown in Figure 7.75. Organisms that are only able to oxidize (O) will turn the unsealed medium yellow (or partially yellow) and leave the sealed medium green or blue. Organisms that are not able to metabolize the sugar (N) will either produce no color change or turn the medium blue because of alkaline products from amino acid degradation (deamination). Organisms that are able to only ferment the carbohydrate or ferment *and* oxidize the carbohydrate will turn the sealed and unsealed media yellow throughout. The inability to differentiate between these occurs because obligate fermenters ferment aerobically and anaerobically, so the result is conservatively reported as "F or O–F" (Fig. 7.76). Slow or weak oxidizers and/or fermenters will turn both tubes slightly yellow at the top and are scored the same way.

7.75 Oxidation–Fermentation (O–F) Test Results These pairs of tubes represent three possible results in the O–F test. Reading from left to right: pair 1, *Pseudomonas aeruginosa* (O); pair 2, uninoculated; pair 3, *Alcaligenes faecalis* (N) and an obligate aerobe; pair 4, *Shigella flexneri* (F or O–F), but only because the test is incapable of distinguishing between the two. *S. flexneri* is actually O–F.

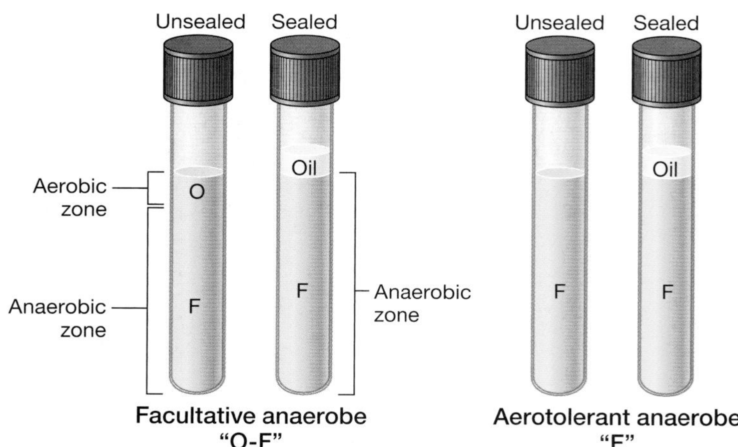

7.76 Ambiguity When Both Tubes are Yellow A facultative anaerobe ferments under anaerobic conditions, but metabolizes oxidatively under aerobic conditions, so both tubes are yellow *and the organism is O–F.* An aerotolerant anaerobe ferments anaerobically AND aerobically, so both tubes end up yellow *and the organism is F.* The O–F test can't distinguish between the two causes of the same result, so if both tubes are yellow the result is recorded as "F or O–F."

TABLE **7-8** **O–F Medium Test Results and Interpretations**

Table of Results			
Sealed (Anaerobic)	**Unsealed (Aerobic)**	**Interpretation**	**Symbol**
Green or blue	Any amount of yellow	Oxidation	O
Green or blue	Green or blue	No sugar metabolism; organism is nonsaccharolytic	N
Yellow throughout	Yellow throughout	Oxidation and fermentation, or fermentation only	O–F or F*
Slightly yellow at the top	Slightly yellow at the top	Slow oxidation and slow fermentation, or slow fermentation only	O–F or F*

* The results of a fermentative organism and one capable of both fermentative and oxidative metabolism (of that specific carbohydrate) look the same in this medium. Therefore, when both tubes are yellow, the result is recorded as (O–F) or (F). See Figure 7.76.

Phenylalanine Deaminase Test

Purpose

Deamination of phenylalanine is used to differentiate the genera *Morganella*, *Proteus*, and *Providencia* (positive) from other members of the *Enterobacteriaceae* (negative).

Principle

One way that amino acids can be catabolized is by removing the carboxyl group in a process called **decarboxylation** (see p. 83). Alternatively, amino acids can be modified by removing their amino group (NH_2) by **deamination** (Fig. 7.77).

Organisms that produce **phenylalanine deaminase** can be identified by their ability to deaminate the amino acid phenylalanine. The reaction, as shown in Figure 7.78, initially removes two hydrogen ions, which combine with oxygen to make water and produce an intermediate acid (not shown). The intermediate is then deaminated to produce ammonia (NH_3) and **phenylpyruvic acid.** Phenylalanine deaminase activity, therefore, is evidenced by the presence of phenylpyruvic acid.

Phenylalanine agar provides a rich source of phenylalanine (2 g/L). A reagent containing ferric chloride ($FeCl_3$) is added to the medium after incubation. The normally colorless phenylpyruvic acid reacts with the ferric chloride and turns a dark-green color almost immediately (Fig. 7.79). Formation of green color indicates the presence of phenylpyruvic acid and, hence, the presence of phenylalanine deaminase. Yellow is negative (Fig. 7.80).

$$\underset{\textbf{Amino Acid}}{H_2N-\overset{\overset{\large R}{|}}{CH}-COOH} \xrightarrow{\text{Deaminase}} \underset{\textbf{Organic Acid}}{H_2\overset{\overset{\large R}{|}}{C}-COOH + NH_3}$$

7.77 Amino Acid Deamination • Deamination is the removal of an amino group (NH_2) from an amino acid. Ammonia and an organic acid are the products.

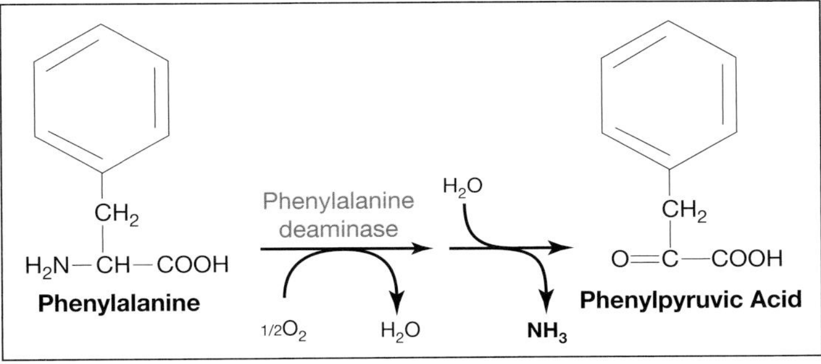

7.78 Deamination of Phenylalanine Reaction • Phenylalanine deamination occurs as a two-step process, but the net result is production of phenylpyruvic acid and ammonia.

$$\text{Phenylpyruvic Acid} + FeCl_3 \longrightarrow \text{Green Color}$$

7.79 Phenylalanine Deaminase Indicator Reaction • Phenylpyruvic acid produced by phenylalanine deamination reacts with $FeCl_3$ to produce a green color, which indicates a positive phenylalanine deaminase test result. The result must be read immediately because the color may fade.

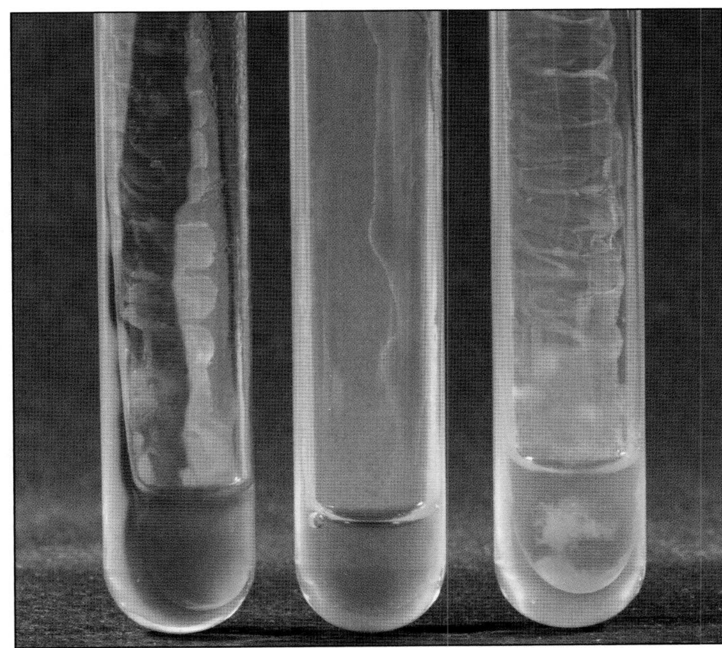

7.80 Phenylalanine Deaminase Test Results • Note the color produced by the stream of ferric chloride in each tube. *Proteus mirabilis* (positive) is on the left, an uninoculated control is in the middle, and *Escherichia coli* (negative) is on the right.

L-Pyrrolidonyl–β-Naphthylamide Hydrolysis (Pyrrolidonyl Arylamidase [PYR] Test)

Purpose

PYR hydrolysis is used for presumptive identification of **Group A streptococci** (*Streptococcus pyogenes*, see p. 219) and enterococci by determining the presence of the enzyme ***L*-pyrrolidonyl arylamidase** (**PYR**).

Principle

Aminopeptidases are enzymes that remove amino acids from the amino end of a protein or peptide. A specific example is L-pyrrolidonyl arylamidase (also known as L- pyrrolidonyl aminopeptidase), which is produced by Group A streptococci (*Streptococcus pyogenes*) and enterococci. This enzyme hydrolyzes L-pyrrolidonyl–β-naphthylamide to produce L-pyrrolidone and β-naphthylamine, all of which are colorless. In the indicator reaction, β-naphthylamine reacts with *p*-dimethylaminocinnamaldehyde to form a red precipitate (Fig. 7.81).

PYR may be performed as an 18-hour agar test, a 4-hour broth test or, as used in this example, a rapid disk test. In each case the medium (or disk) contains L-pyrrolidonyl–β-naphthylamide to which is added a heavy inoculum of the test organism. After the appropriate incubation or waiting period, a 0.01% *p*-dimethylaminocinnamaldehyde solution is added. Formation of a red color within a few minutes is interpreted as PYR positive (Fig. 7.82). Yellow or orange is PYR negative.

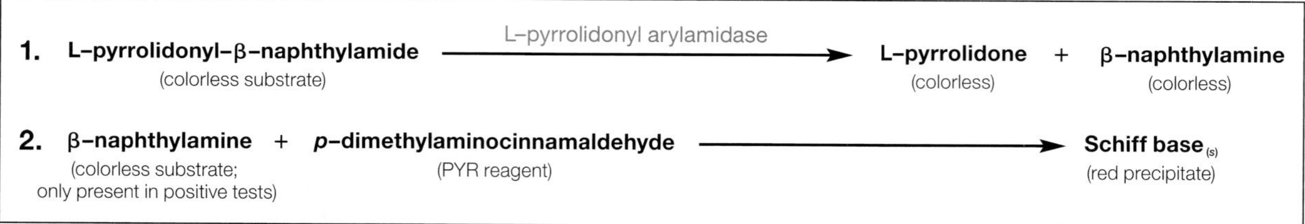

7.81 PYR Test Chemistry • Reaction 1 shows the hydrolytic splitting of the substrate L-pyrrolidonyl-β-naphthylamide included in the medium. (The example shown in Figure 7.82 has this substrate in the filter paper.) Both products of the reaction are colorless. Reaction 2 illustrates the reaction of the PYR reagent (*p*-dimethylaminocinnamaldehyde) and β-naphthylamine (only present in positive tests). In this reaction, the amino group (H_2N) of β-naphthylamine reacts with the aldehyde group (–CHO) of the reagent and produces a Schiff base (–CH=N–), which is red.

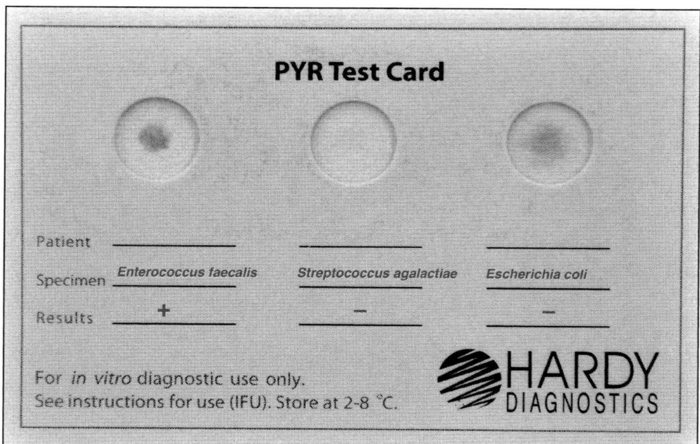

7.82 PYR Disk Test Results • Shown is a Hardy Diagnostics PYR test card. The filter paper in each opening is impregnated with substrate (L-pyrrolidonyl-β-naphthylamide). After inoculation and the appropriate waiting time, the PYR reagent (*p*-dimethylaminocinnamaldehyde) is added to each disk. A pink/red color is considered positive. No color change (center opening) or an orange, salmon, or yellow color is considered negative. Certain *Enterobacteriaceae* (e.g., *Escherichia coli*, as in this example) may produce the blue coloration seen on the right when grown on a high-tryptophan medium. This is also a negative result.

Starch Hydrolysis (Amylase Test)

Purpose

Starch agar originally was designed for cultivating *Neisseria*. It no longer is used for this, but with pH indicators, it is used to isolate and presumptively identify *Gardnerella vaginalis*. Its ability to demonstrate starch hydrolysis aids in differentiating species of the genera *Corynebacterium, Clostridium, Bacillus, Bacteroides, Fusobacterium*, and *Enterococcus*, most of which have amylase-positive and amylase-negative species.

Principle

Starch is a complex polysaccharide compound composed of **α-glucose** polymers in one of two forms—**amylose** (linear) and **amylopectin** (branched)—usually as a mixture with the branched configuration being predominant. The α-glucose molecules in both amylose and amylopectin are joined by **1,4-α-glycosidic** (acetal) **bonds** (Fig. 7.83)[7]. The two forms differ in that amylopectin contains side chains covalently bonded to approximately every 30th glucose in the main chain by a **1,6-α-glycosidic linkage**.

Starch is too large to pass through the bacterial cell membrane. Therefore, to be of metabolic value, it first must be digested extracellularly (or in the periplasmic space, in the case of Gram-negative bacteria) into smaller molecules (Fig. 7.83). There are three enzymes involved in starch digestion: **α-amylase**, **β-amylase**, and **α-1,6-glucosidase**. α-amylase breaks the starch polymer into the monosaccharide α-glucose and the disaccharide α-**maltose**, whereas β-amylase produces β-**maltose**.

Neither amylase can disassemble the glucose subunits around branch points of amylopectin, so small, branched molecules called **limit dextran** remain after amylase activity. **α-(1,6)-glucosidase** breaks these down into short, linear oligosaccharides that α-amylase converts to α-glucose and α-maltose. Because the enzymes are secreted extracellularly, they are referred to as **exoenzymes**.

Starch agar is a simple plated medium of beef extract, soluble starch, and agar. When organisms that produce α-amylase, β-amylase, and oligo-1,6-glucosidase are cultivated on starch agar, they hydrolyze starch in the area surrounding their growth.

Because both starch and its sugar subunits are virtually invisible in the medium, the reagent iodine is used to detect the presence or absence of starch in the vicinity around the bacterial growth. Iodine reacts with starch and produces a blue or dark brown color; therefore, any microbial starch hydrolysis will be revealed as a clear zone surrounding the growth (Fig. 7.84).

[7] The carbons of organic molecules are numbered. A "1,4 linkage" is a covalent bond between the first carbon of one glucose and the fourth carbon of another glucose. A "1,6 linkage" is a covalent bond between the first and sixth carbons of two glucose molecules. The "α" and "β" refer to whether the two glucose molecules have the same orientation (α) or are "flipped" 180° relative to one another (β).

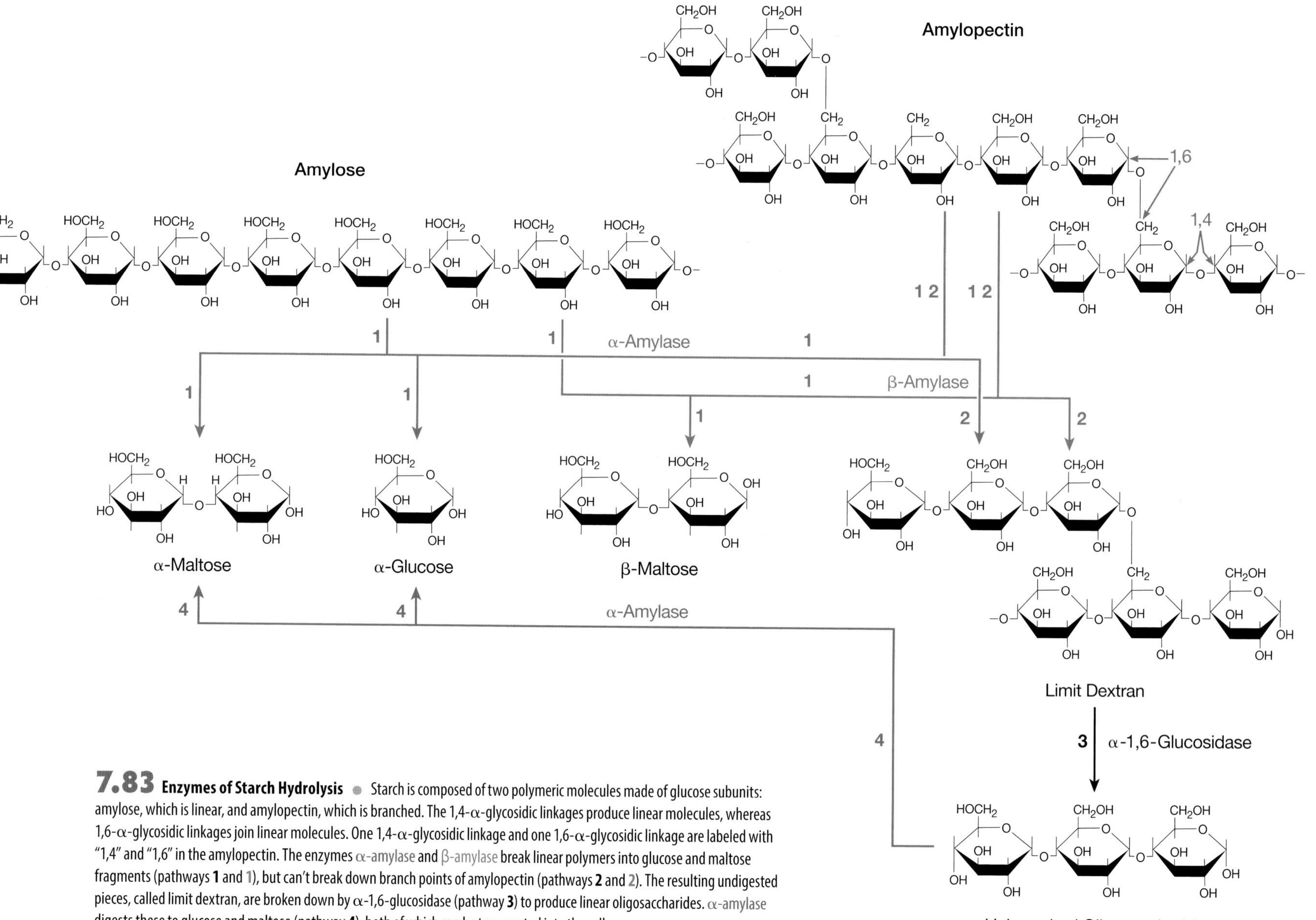

7.83 Enzymes of Starch Hydrolysis • Starch is composed of two polymeric molecules made of glucose subunits: amylose, which is linear, and amylopectin, which is branched. The 1,4-α-glycosidic linkages produce linear molecules, whereas 1,6-α-glycosidic linkages join linear molecules. One 1,4-α-glycosidic linkage and one 1,6-α-glycosidic linkage are labeled with "1,4" and "1,6" in the amylopectin. The enzymes α-amylase and β-amylase break linear polymers into glucose and maltose fragments (pathways **1** and 1), but can't break down branch points of amylopectin (pathways **2** and 2). The resulting undigested pieces, called limit dextran, are broken down by α-1,6-glucosidase (pathway **3**) to produce linear oligosaccharides. α-amylase digests these to glucose and maltose (pathway **4**), both of which can be transported into the cell.

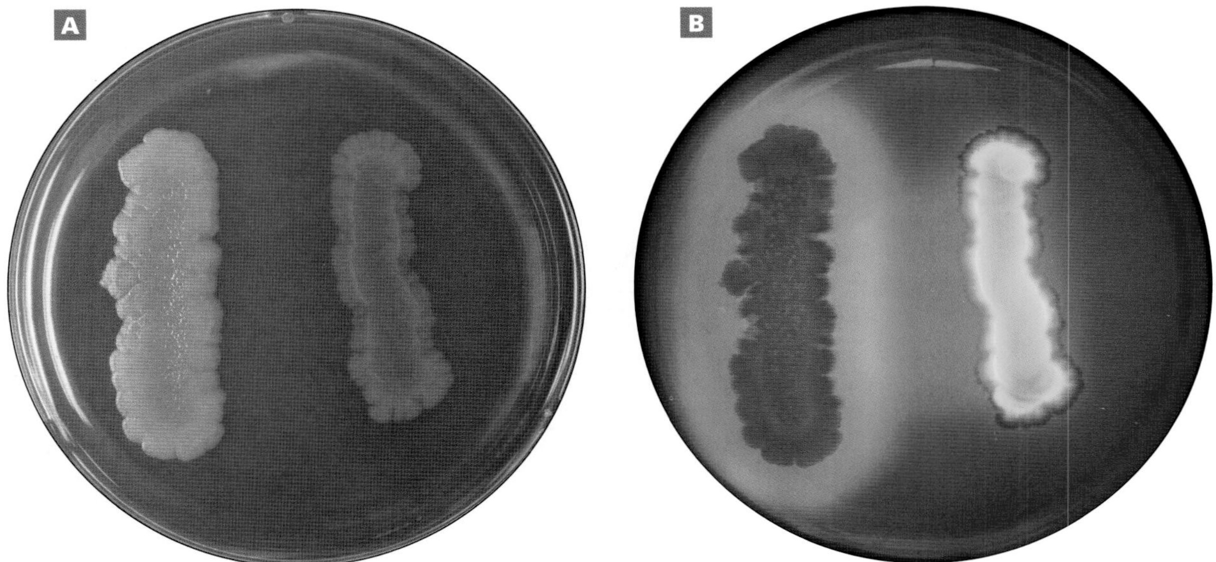

7.84 Starch Hydrolysis Results ● (A) This is a starch agar plate inoculated with *Bacillus subtilis* on the left and *Escherichia coli* on the right before iodine has been added. Notice the wavy margins of both organisms. The photograph was taken against a dark background to show that there is no visible clearing at this point. (B) This is the same plate as in photo A after iodine has been added. The clearing in the medium around *B. subtilis* demonstrates a positive result for starch hydrolysis (amylase positive). *E. coli*, with no clearing, is amylase negative. Note that the wavy margin of *E. coli* growth produced a lighter region on the periphery that might be misinterpreted as clearing. To prevent recording a false positive, it is a good idea to establish the edges of growth of each organism by drawing a line on the plate's base prior to adding iodine.

Sulfur Reduction Test (SIM Medium)

Purpose

Sulfur reduction is used to differentiate members of *Enterobacteriaceae*, especially species of *Salmonella*, *Shigella*, and *Proteus* (positive), from negative *Morganella* and *Providencia* species.

Principle

Sulfur reduction can be evaluated using various media; here we discuss using SIM medium, but also see Kligler's iron agar (p. 91), lysine iron agar (p. 96), and triple sugar iron agar (p. 116). SIM medium also tests for indole production (p. 90) and motility (p. 99).

SIM is a semisolid medium that is formulated with casein and animal tissue as sources of amino acids (most importantly, sulfur-containing cysteine), oxidized sulfur in the form of sodium thiosulfate, and ferrous ammonium sulfate.

Sulfur reduction to H_2S can be accomplished by bacteria in two different ways, depending on the enzymes present.

1. The enzyme **cysteine desulfurase** catalyzes the hydrolysis of the sulfur-containing amino acid cysteine to pyruvate and H_2S during **putrefaction** (Fig. 7.85).
2. In a totally unrelated way, the enzyme **thiosulfate reductase** catalyzes the reduction of sulfur (in the form of sulfate) to H_2S at the end of an anaerobic respiratory electron transport chain (Fig. 7.86).

Both systems produce hydrogen sulfide (H_2S) gas. When either reaction occurs in SIM medium, the H_2S produced combines with iron, in the form of ferrous ammonium sulfate, to form **ferric sulfide** (FeS), a black precipitate (Fig. 7.87). Any blackening of the medium is an indication of sulfur reduction and a positive test. No blackening of the medium indicates no sulfur reduction and a negative reaction (Fig. 7.88). ●

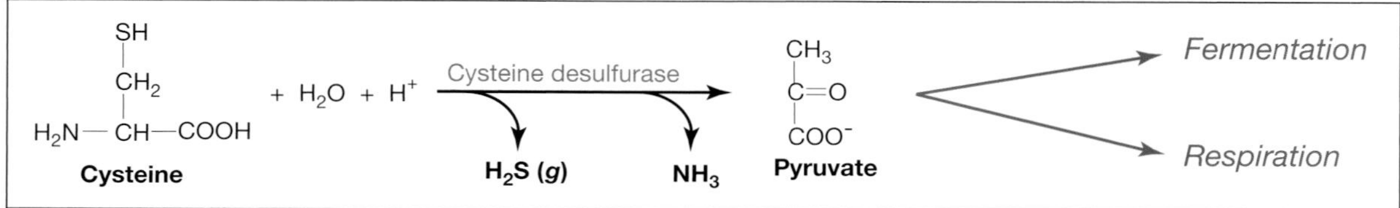

7.85 Putrefaction of Cysteine Reaction • Cysteine desulfurase catalyzes the hydrolysis of the amino acid cysteine with the production of pyruvate, NH_3, and H_2S. This reaction is a mechanism for getting energy out of the amino acid cysteine.

$$3S_2O_3^{2-} + 4H^+ + 4e^- \xrightarrow{\text{Thiosulfate reductase}} 2SO_3^{2-} + 2H_2S\,(g)$$

Thiosulfate

7.86 Thiosulfate Reduction Reaction • Anaerobic respiration with thiosulfate as the final electron acceptor also produces H_2S. Anaerobic respiration produces more ATP per glucose than fermentation, but less than aerobic respiration.

$$H_2S + FeSO_4 \longrightarrow H_2SO_4 + FeS\,(s)$$

Hydrogen sulfide

7.87 H_2S Indicator Reaction • Hydrogen sulfide, a colorless gas, can be detected when it reacts with ferrous ammonium sulfate in the medium to produce the black precipitate ferric sulfide.

7.88 Sulfur Reduction Results–SIM Medium • *Escherichia coli* on the left is H_2S negative and *Proteus mirabilis* on the right is H_2S positive. SIM medium also provides the opportunity to determine motility. Although without a control for comparison it can be difficult to tell, *E. coli* is motile, as evidenced by the haziness in the medium from bacteria swimming away from the stab line. However, when an organism is H_2S positive it is nearly impossible to read motility in this medium.

Sulfamethoxazole-Trimethoprim (SXT) Susceptibility Test

Purpose

The **sulfamethoxazole-trimethoprim** (SXT) susceptibility test is used to differentiate Groups A and B streptococci (SXT resistant) from other β-hemolytic streptococci (SXT susceptible). Used in conjunction with the bacitracin susceptibility test (as in this example) it also differentiates between Group A and B streptococci.

Principle

Folate (actually, tetrahydrofolate) is necessary for purine synthesis as well as synthesis of several amino acids. Without purines nucleic acids can't be made, and without all 20 amino acids proteins can't be made, so its importance can't be overstated.

When combined, sulfamethoxazole and trimethoprim act synergistically to competitively inhibit successive steps in the folate synthesis pathway (Fig. 7.89A). They are able to competitively inhibit because their molecular structures are similar enough to the two enzymes' substrates that they also bind to their active sites and prevent normal function (Figs. 7.89B and 7.89C).

SXT disks typically contain 23.75 μg of sulfamethoxazole and 1.25 μg of trimethoprim, a ratio that maximizes their combined effect. When a disk is placed on the surface of a sheep blood agar plate inoculated to produce confluent growth, a clearing will appear around the disk if the organism is susceptible (S) to the antibiotic mixture. Growth up to the edge of the disk indicates resistance (R).

The synergistic effect of sulfamethoxazole and trimethoprim can be illustrated on an inoculated plate incubated with disks of each antibiotic (rather than both on one SXT disk). See page 294 and Figure 19.7 for a description of this phenomenon.

Performing both SXT and bacitracin (see p. 73) susceptibility tests simultaneously is useful in differentiating β-hemolytic streptococci. The isolate is streaked on a sheep blood agar plate and one of each disk is placed in the region of dense growth (the first streak) at least 4 cm apart (Fig. 7.90). Any clearing around the SXT disk and a zone of 10 mm or more for the bacitracin disk is interpreted as susceptibility[8]. See Table 7-9 for more details.

For more information on antimicrobial susceptibility, refer to the antimicrobial susceptibility (Kirby-Bauer) test, page 291.

[8] Some references say any clearing around the bacitracin disk is interpreted as susceptibility. The disk is 6 mm in diameter, so almost any clearing will approach the 10 mm cutoff.

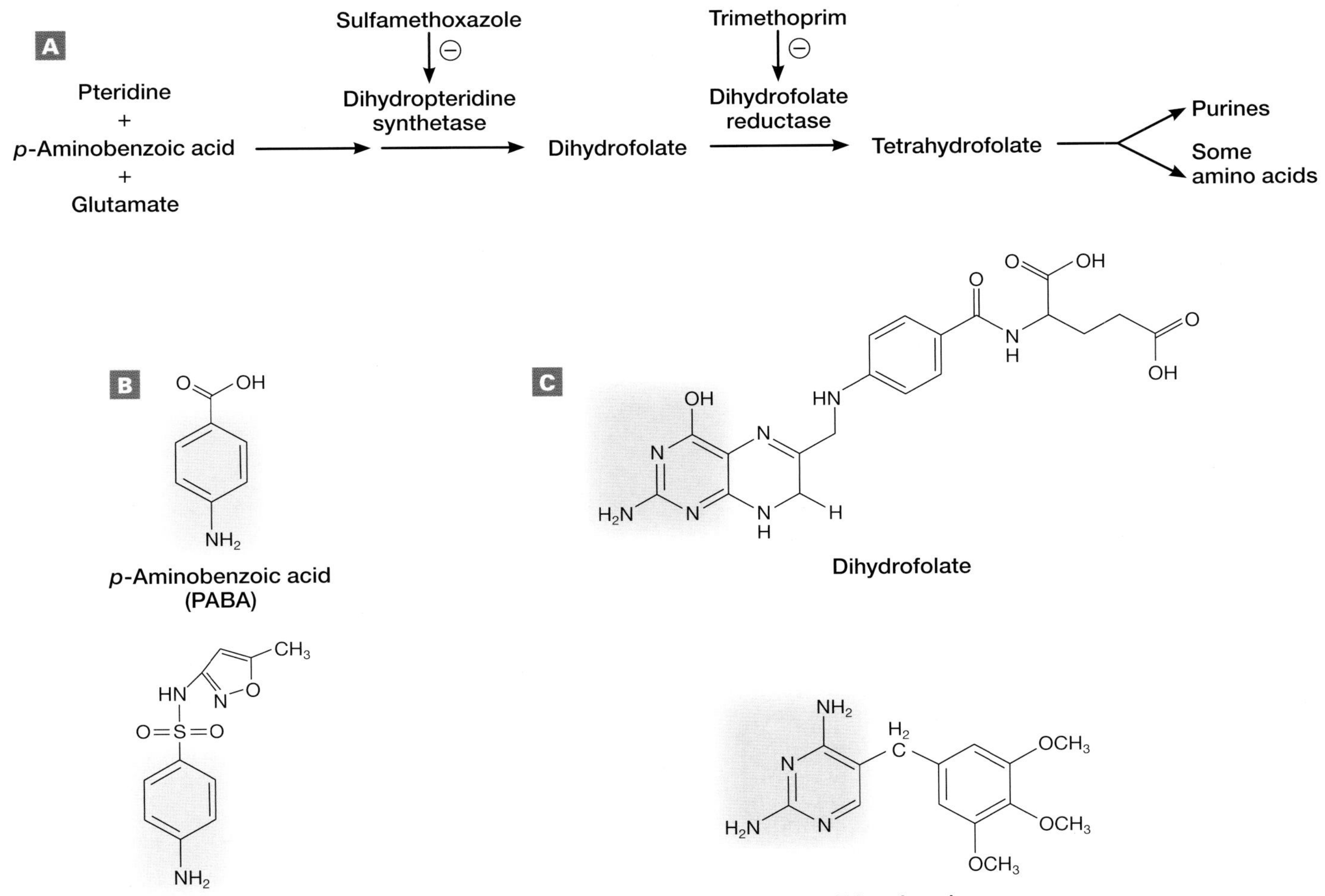

7.89 Effect of SXT on Folate Synthesis (A) Sulfamethoxazole and trimethoprim interfere with tetrahydrofolate synthesis, a coenzyme required for synthesis of purines and some amino acids. Pteridine, *p*-aminobenzoic acid (PABA), and glutamate are assembled into dihydrofolate in a multistep process, with dihydropteridine synthetase being the relevant enzyme because sulfamethoxazole inhibits it. Then, dihydrofolate reductase reduces dihydrofolate to tetrahydrofolate. It is the enzyme inhibited by trimethoprim. (B) The shaded regions of these molecular structures illustrate why sulfamethoxazole can compete with PABA for dihydropteridine synthetase's active site. (C) The shaded regions of these molecular structures illustrate why trimethoprim can compete with dihydrofolate for dihydrofolate reductases's active site.

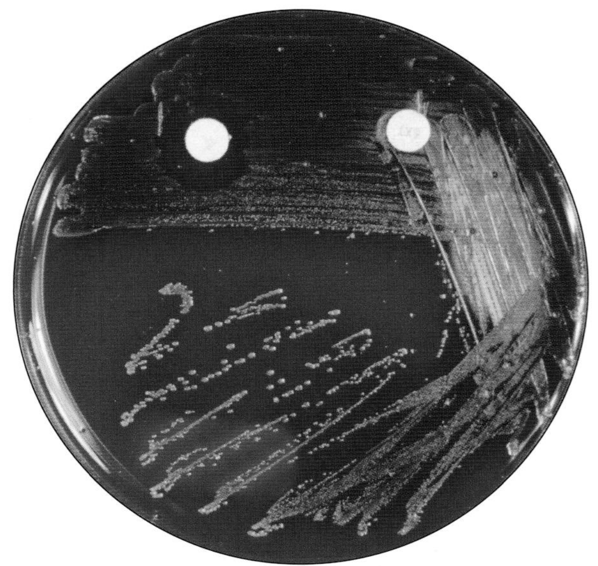

7.90 Bacitracin-SXT Susceptibility on Sheep Blood Agar • This sheep blood agar plate was inoculated with an unknown β-hemolytic streptococcus isolate and incubated with bacitracin (left) and SXT (right) disks. A β-hemolytic streptococcus that is susceptible to bacitracin (clearing) and resistant to SXT (no clearing) is provisionally identified as *Streptococcus pyogenes* (Group A). See Table 7-9 for more details.

TABLE **7-9** **Responses of β-Hemolytic Streptococci to Bacitracin and SXT** •

Table of Results		
β-Hemolytic Streptococcus Group	**Bacitracin**	**SXT**
Group A	S	R
Group B	R	R
Group C, F, and G	S or R	S

Triple Sugar Iron Agar (TSIA)

Purpose

Triple sugar iron agar (TSIA) is primarily used to differentiate members of *Enterobacteriaceae* and to differentiate them from other Gram-negative rods such as *Pseudomonas*.

Principle

TSIA is a rich medium designed to differentiate bacteria (especially enterics–*Enterobacteriaceae*) on the basis of glucose fermentation, lactose fermentation, sucrose fermentation, and sulfur reduction.

In addition to the three carbohydrates, TSIA includes beef extract, yeast extract, and peptone as carbon, nitrogen, and sufur (from the amino acid cysteine) sources. Sodium thiosulfate acts as an alternative source of reducible sulfur. Phenol red is the pH indicator (yellow below pH 6.8, pink above pH 7.4, and red in between) and the iron in ferrous sulfate is the hydrogen sulfide indicator.

The medium is prepared as a shallow agar slant with a deep butt, thereby providing both aerobic and anaerobic growth environments. It is inoculated by a stab in the agar butt followed by a fishtail streak of the slant. The incubation period is 18 to 24 hours for carbohydrate fermentation and up to 48 hours for hydrogen sulfide reactions. Many reactions in various combinations are possible (Fig. 7.91 and Table 7-10).

When TSIA is inoculated with a glucose-only fermenter, acid products lower the pH and turn the entire medium yellow within a few hours. Because glucose is in short supply (0.1%), it will be exhausted within about 12 hours. As the glucose is consumed, amino acid deamination by the organisms located in the aerobic region (slant) produces NH_3 and raises the pH.

This process, which takes 18 to 24 hours to complete, is called a **reversion.** Within the time frame of the test, reversion will only be seen in the slant because glucose consumption is slower in the anaerobic butt. Thus, a TSIA with a red slant and yellow butt after a 24-hour incubation period indicates that the organism ferments glucose but not lactose and/or sucrose.

Organisms that are able to ferment glucose *and* lactose *and/or* sucrose also initially turn the medium yellow throughout (see Fig. 7.34 for the biochemical relationship between these three fermentations). However, because the lactose and sucrose concentrations are 10 times higher than that of glucose, more acid is produced for a longer period of time and both slant and butt remain yellow after 24 hours.

Therefore, a TSIA with a yellow slant and butt at 24 hours indicates that the organism ferments glucose and one or both of the other sugars. Gas produced by fermentation will appear as fissures in the medium or will lift the agar off the bottom of the tube (see the tube second from the right in Fig. 7.91).

Hydrogen sulfide (H_2S) may be produced by the breakdown of cysteine from the protein or by the reduction of thiosulfate in the medium (see Figs. 7.85 and 7.86). Ferrous sulfate in the medium reacts with the H_2S to form a black precipitate, usually seen in the butt.

Acid conditions must exist for thiosulfate reduction; therefore, black precipitate in the medium is an indication of sulfur reduction *and* fermentation. If the black precipitate obscures the color of the butt, the color of the slant determines which carbohydrates have been fermented: a red slant means glucose fermentation; a yellow slant means glucose and lactose and/or sucrose fermentation.

An organism that does not ferment any of the carbohydrates but utilizes peptone and amino acids will alkalinize the medium and turn it red. If the organism can use the peptone aerobically and anaerobically, both the slant and butt will appear red. An obligate aerobe will turn only the slant red. (Note that the difference between a red butt and a butt unchanged by the organism may be subtle. Therefore, comparison with an uninoculated control is always recommended.)

As previously stated, timing is critical in reading TSIA results. An early reading could reveal yellow throughout the medium, leading one to conclude that the organism ferments more than one sugar when it simply may not yet have consumed the glucose. A late reading after the lactose and sucrose have been depleted could reveal a yellow butt and red slant leading one to falsely conclude the organism is a glucose-only fermenter.

Timing for interpreting sulfur reduction is not as critical. Tubes that have been interpreted for carbohydrate fermentation and are negative for sulfur reduction can be re-incubated for 24 hours before H_2S determination. Refer to Table 7-10 for information on the correct symbols and method of reporting the various reactions.

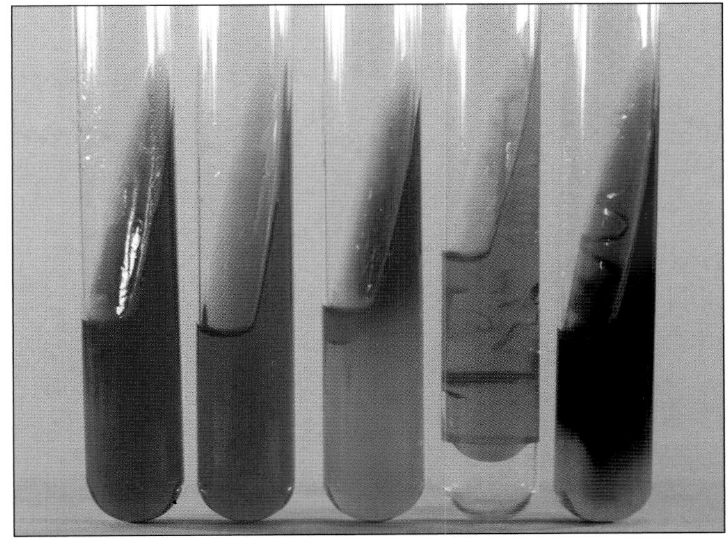

7.91 Triple Sugar Iron Agar Results • TSIA fermentation results need to be read at 24 hours of incubation. Positive sulfur reduction results can be read with the fermentations, but negative results should be reincubated for another 24 hours. From left to right: *Pseudomonas aeruginosa* (K/NC), uninoculated control, *Morganella morganii* (K/A, atypically not producing gas), *Escherichia coli*, (A/A, G), and *Proteus mirabilis* (K/A, H_2S). Notice the acid butt along with the sulfur reduction in the *P. mirabilis* tube. See Table 7-10 for more details.

TABLE **7-10** **TSIA Test Results and Interpretations**

Table of Results		
Result	**Interpretation**	**Symbol**
Yellow slant/yellow butt	Glucose and lactose and/or sucrose fermentation with acid accumulation in slant and butt	A/A
Red slant/yellow butt	Glucose fermentation with acid production; amino acids deaminated aerobically (in the slant) with alkaline products (reversion)	K/A
Red slant/red butt	No fermentation; amino acids deaminated aerobically and anaerobically with alkaline products; not from *Enterobacteriaceae*	K/K
Red slant/no change in butt	No fermentation; amino acids deaminated aerobically with alkaline products; not from *Enterobacteriaceae*	K/NC
No change in slant/no change in butt	Organism is growing slowly or not at all; not from *Enterobacteriaceae*	NC/NC
Black precipitate in the agar	Sulfur reduction; an acid condition, from fermentation of glucose or lactose and/or sucrose, exists in the butt even if the yellow color is obscured by the black precipitate	H_2S
Cracks in or lifting of agar	Gas production	G

Urea Hydrolysis (Urease Tests)

Purpose

Urea hydrolysis is used to differentiate organisms based on their ability to hydrolyze urea with the enzyme **urease**. Urinary tract pathogens from the genus *Proteus* (see p. 208) may be distinguished from other **enteric bacteria** by their rapid urease activity. This medium is also used in identifying *Helicobacter pylori* (see p. 199), which is associated with gastric and duodenal ulcers, as well as stomach cancer. *Streptococcus salivarius*, an oral commensal, also produces urease.

Principle

Decarboxylation of certain amino acids produces urea, which is the primary nitrogenous waste in mammalian urine. It can be hydrolyzed to ammonia and carbon dioxide by bacteria possessing the enzyme **urease** (Fig. 7.92). Urea hydrolysis provides nitrogen in a usable form (ammonia) and it also acts as a virulence factor for some pathogens (e.g., *Helicobacter pylori*) by counteracting acid in the environment.

Many enteric bacteria possess the ability to metabolize urea, and the urease genes are activated when a usable nitrogen source is absent or urea is present. The original urease test used urea agar, which was formulated to differentiate enteric organisms capable of rapid urea hydrolysis ("rapid urease-positive bacteria," such as *Proteus*, *Morganella morganii*, and some *Providencia stuartii* strains) from slower urease-positive and urease-negative bacteria.

Dextrose (glucose) and gelatin are carbon sources in urea agar and the latter also provides nitrogen. Phenol red is the pH indicator used to track the rising pH due to ammonia accumulation. Rapid urease-positive organisms produce a pink color (positive) within a day; slower urease-positive organisms may take several days to produce a positive result (Fig. 7.93).

Urea broth differs from urea agar in two important ways. First, its only nutrient source other than urea is a trace (0.0001%) of yeast extract. Second, it contains buffers strong enough to inhibit alkalinization of the medium by all but the rapid urease-positive organisms mentioned previously. Phenol red is used as the pH indicator. Pink color in the medium in less than 24 hours indicates a rapid urease-positive organism. Orange or yellow is negative (Fig. 7.94). See Table 7-11 for more details on interpretations.

$$(H_2N)_2C{=}O \xrightarrow[\text{Urease}]{H_2O} 2NH_3 + CO_2$$

Urea

7.92 Urea Hydrolysis Reaction • Urea hydrolysis produces ammonia, which raises the pH in the medium and turns the pH indicator pink. Ammonia can be used by some organisms as a nitrogen source. It also can protect the organism against an acidic environment.

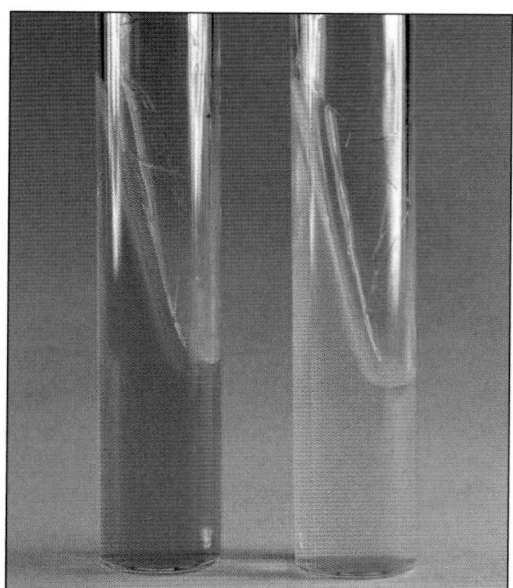

7.93 Urea Hydrolysis Results–Urea Agar • These urea agar tubes were incubated for 24 hours. Shown are *Proteus mirabilis* (left, positive) and *Hafnia alvei* (right, negative).

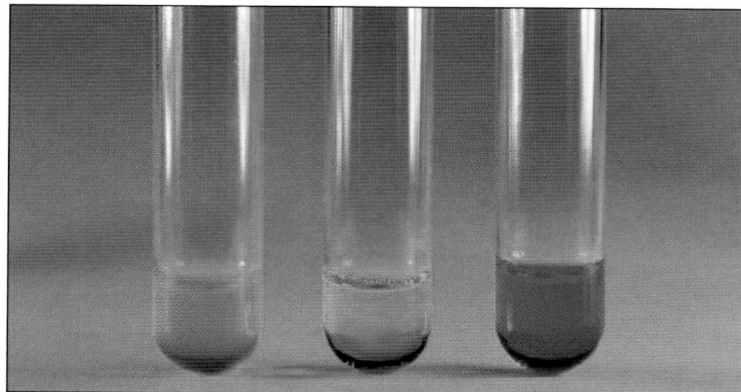

7.94 Urea Hydrolysis Results–Urea Broth • These urea broth tubes were incubated for 24 hours. From left to right: *Shigella flexneri* (negative); an uninoculated control; *Proteus mirabilis* (positive)

TABLE **7-11** **Urease Test Results and Interpretations**

Table of Results			
Agar			
Result			
24 Hours	**24 Hours to 6 Days**	**Interpretation**	**Symbol**
All pink		Rapid urea hydrolysis; strong urease production	+
Partially pink		Slow urea hydrolysis; weak urease production	w +
Orange or yellow	Partially pink	Slow urea hydrolysis; weak urease production	w +
Orange or yellow	Orange or yellow	No urea hydrolysis; urease is absent	–
Broth			
Result			
24 Hours		**Interpretation**	**Symbol**
Pink		Rapid urea hydrolysis; strong urease production	+
Orange or yellow		No urea hydrolysis; organism does not produce urease or cannot live in broth	–

Voges-Proskauer Test

Purpose

The Voges-Proskauer (VP) test is a component of the **IMViC** battery of tests (*I*ndole, *M*ethyl Red, *V*oges-Proskauer, and *C*itrate) used to distinguish between members of the Family ***Enterobacteriaceae*** and differentiate them from other Gram-negative rods. It identifies organisms capable of a **2,3-butanediol fermentation.**

Principle

MR-VP broth is the combination medium used for both Methyl Red (MR) and Voges-Proskauer (VP) tests (see p. 98 for the MR test). It is a simple solution containing only peptone, glucose, and a phosphate buffer. The peptone and glucose provide a nitrogen source and a fermentable carbohydrate, respectively. The potassium phosphate resists pH changes in the medium.

The Voges-Proskauer test was designed to identify organisms that are able to ferment glucose, with the production of acetoin and 2,3-butanediol, which have a neutral pH (Fig. 7.95). VP-positive organisms produce other fermentation end-products prior to making 2,3-butanediol.

They can reduce pyruvate to lactate (which lowers the pH), and make formate (which is broken down into H_2 and CO_2 gases) and acetaldehyde (which is reduced to ethanol). The acetaldehyde path results in no pH change, but formate (at least until its conversion to H_2 and CO_2) and lactate production lower pH. VP-positive organisms compensate for the lowered pH by shifting to 2,3-butanediol production.

Adding VP reagents to the medium oxidizes **acetoin** to **diacetyl**, which in turn reacts with **guanidine nuclei** from peptone to produce a red color (Fig. 7.96). A positive VP result, therefore, is red. No color change (or development of copper color) after the addition of reagents is negative. The copper color is a result of interactions between the reagents and should not be confused with the true red color of a positive result (Fig. 7.97). Use of positive and negative controls for comparison is usually recommended.

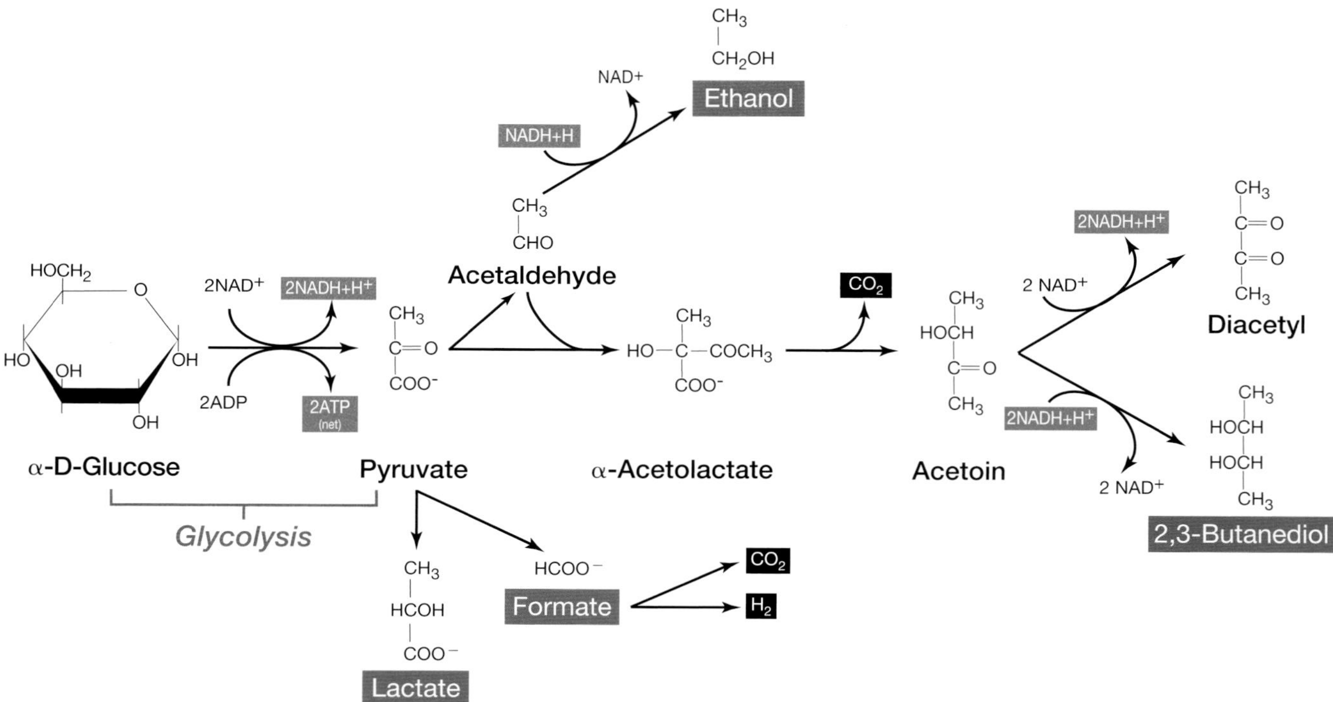

7.95 2,3-Butanediol Fermentation Reactions ● VP-positive organisms have the potential to perform fermentations with ethanol, lactate, formate, 2,3-butanediol, and carbon dioxide and hydrogen gases as end-products. As the pH drops due to acid end-product accumulation, VP-positive organisms switch to a 2,3 butanediol fermentation, which raises the intracellular pH. Acetoin is the last intermediate in this fermentation and its reduction by NADH produces the end-product 2,3-butanediol. Acetoin can also be oxidized to diacetyl with the production of reducing power in the form of NADH+H$^+$. The VP indicator reaction artificially oxidizes acetoin to diacetyl (see Fig. 7.96).

OH
α-Naphthol (VP Reagent A) + Acetoin (CH_3–HOCH–C=O–CH_3) —KOH (VP Reagent B), O_2 → ($2H_2O$) Diacetyl (CH_3–C=O–C=O–CH_3) + Guanidine (in peptone) → **Red color**

7.96 Voges-Proskauer (VP) Test Indicator Reactions ● Reagents A (α-Naphthol) and B (KOH) are added to VP broth after 48 hours of incubation. In a positive VP test, these reagents react with acetoin and oxidize it to diacetyl, which in turn reacts with guanidine (from the peptone in the medium) to produce a red color. No color change occurs with a negative test.

7.97 Voges-Proskauer Test Results ● *Escherichia coli* (VP negative) is on the left and *Klebsiella* (*Enterobacter*) *aerogenes* (VP positive) is on the right.

Part B: Multiple-Test Systems

API 20 E Identification System for *Enterobacteriaceae* and Other Gram-Negative Rods

Purpose

API stands for *A*nalytical *P*rofile *I*ndex. The api 20 E multiple-test system (available from bioMérieux, Inc.) is used clinically for the rapid identification of *Enterobacteriaceae* (more than 5,500 strains) and other Gram-negative rods (more than 2,300 strains). It is one of a series of similar multiple-test systems bioMérieux manufactures that identify Gram-negative nonenteric bacteria, staphylococci and micrococci, streptococci and enterococci, coryneforms, anaerobes, and yeasts.

Principle

The api 20 E system is a plastic strip of 20 microtubes and cupules, partially filled with different dehydrated substrates. Bacterial suspension is added to the microtubes, rehydrating the media and inoculating them at the same time. As with the other biochemical tests in this section, color changes take place in the tubes either during incubation or after addition of reagents. These color changes reveal the presence or absence of chemical action and, thus, a positive or negative result (Fig. 7.98).

After incubation, spontaneous reactions—those that do not require addition of reagents—are evaluated first. Then tests that require addition of reagents are performed and evaluated. Finally, the results are recorded on the api 20 E result screen (Fig. 7.99) available online at https://apiweb.biomerieux.com/login. An oxidase test is performed separately and constitutes the 21st test.

As shown in Figure 7.99, the result screen divides the tests into groups of three, with the members of a group having numerical values of 1, 2, or 4, respectively. These numbers are assigned for positive (+) results only. Negative (–) results have a zero value and are not counted.

The values for positive results in each group are added together to produce a number from 0 to 7, which is entered in the oval below the three tests. The totals from each group are combined sequentially to produce a seven-digit code, which can then be interpreted in the Analytical Profile Index online using APIWEB identification software (Fig. 7.100).

In rare instances, information from the 21 tests (and the seven-digit code) is not enough to identify an organism. When this occurs, the organism is grown and examined on MacConkey agar and supplemental tests are performed for nitrate reduction, oxidation/fermentation of glucose, and motility. The results are entered separately in the supplemental spaces on the APIWEB result screen and used for final identification.

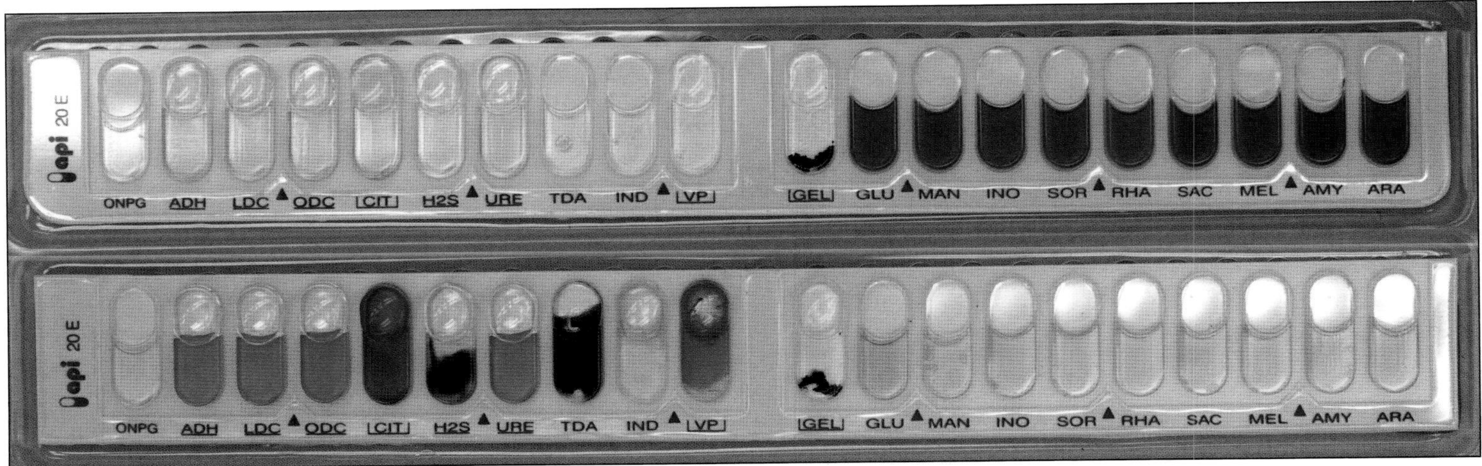

7.98 api 20 E Test Strips • The top strip is uninoculated. The bottom strip (with the exception of GEL) is positive for all results. The 20 tests are, from left to right: **ONPG**=*o*-nitrophenyl-β-D-galactopyranoside; **ADH**=arginine dihydrolase; **LDC**=lysine decarboxylase; **ODC**=ornithine decarboxylase; **CIT**=citrate; **H_2S**=sulfur reduction; **URE**=urease; **TDA**=tryptophan deaminase; **IND**=indole; **VP**=Voges-Proskauer; and **GEL**=gelatinase. The remainder are fermentation tests for the following sugars: **GLU**=D-glucose; **MAN**=D-mannitol; **INO**=inositol; **SOR**=D-sorbitol; **RHA**=L-rhamnose; **SAC**=D-sucrose; **MEL**=D-melibiose; **AMY**=amygdalin; and **ARA**=L-arabinose. Most of these tests have been described individually in this section.

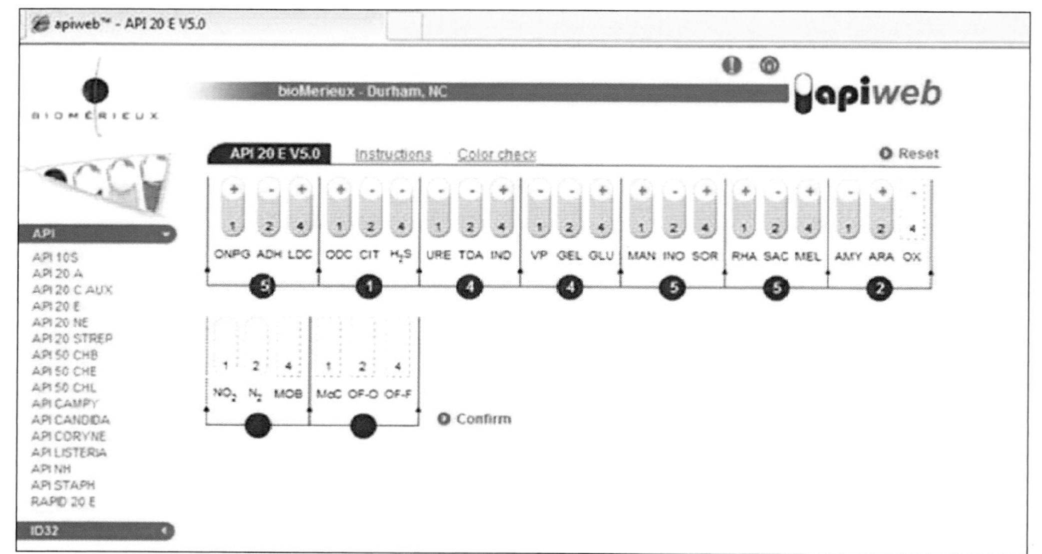

7.99 api 20 E Result Screen • The 21 test results of the api 20 E (oxidase is not part of the strip and is run separately) are divided into groups of three, with each positive result given a numerical value of 1, 2, or 4. In this screen shot from APIWEB, results from a sample have been recorded electronically. The isolate is positive for ONPG, LDC, ODC, IND, GLU, MAN, SOR, RHA, MEL, and ARA. The sum of the numbers assigned to those positive results within each group is shown in the blue circles below each group. For instance, group 1 was positive for ONPG (1) and LDC (4) and a 5 (1 + 4 = 5) is recorded below. Collectively, the blue circles produce a seven-digit number, 5144552, which is matched to the database in APIWEB (see Fig. 7.100). Additional tests, shown below the main tests, can be run if identification is equivocal.

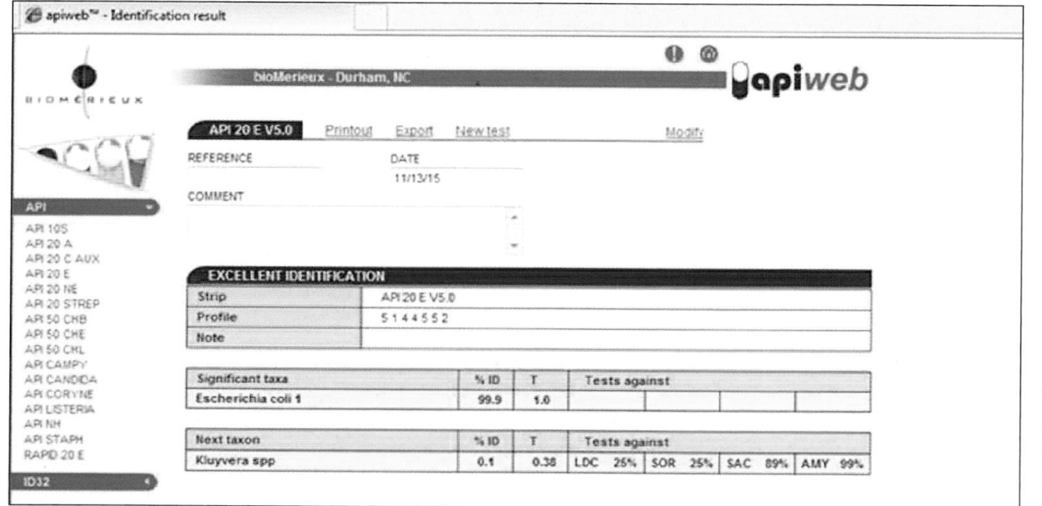

7.100 Identification Using APIWEB • The website APIWEB has an extensive database of test results for *Enterobacteriaceae* and other non-fastidious, Gram-negative rods for which the api 20 E was designed. Shown is a screen shot with the seven-digit number (5144552) from Figure 7.99 entered on the Profile line. The resulting identification was *Escherichia coli* with 99.9% confidence. The next closest match (0.1% confidence) was the genus *Kluyvera*. Tests not conforming to *Kluyvera* are listed at the right under "Tests against."

BBL CRYSTAL Identification System—Enteric/Non-Fermenter (E/NF)

Purpose

The BBL CRYSTAL E/NF (Becton Dickinson Microbiology Systems) is a multiple-test system used to identify aerobic Gram-negative rods of the family ***Enterobacteriaceae***, as well as other Gram-negative bacilli recovered from human samples. Identification is based on the results of 30 biochemical tests.

Principle

The BBL CRYSTAL E/NF (Fig. 7.101) test system panel consists of a base with 30 wells and a lid with 30 plastic tabs, each with a dehydrated substrate on its tip that fits into a different well when the lid is in place. The organism to be identified is first suspended in an inoculum fluid, and then dispensed into the 30 wells of the base. Placement of the lid onto the base immerses the plastic tabs in inoculum and reconstitutes the dehydrated substrates. Separate oxidase and indole tests must also be run.

After 18 to 20 hours incubation, the panel may be read using a light box and comparing the results to a standard color reaction chart. As can be seen in Figure 7.101, there are three rows of 10 tests each (A through J).

Numerical values of 4, 2, or 1 are assigned to any positive results for any test in the top, middle, or bottom rows, respectively, on the score sheet (Fig. 7.102). By adding the values in the 10 columns, a 10-digit BBL CRYSTAL profile number is obtained. The profile number is then matched to a computer data base of approximately 100 taxa for identification. Test scoring is now automated and the 10-digit code is automatically entered into the database to provide identification.

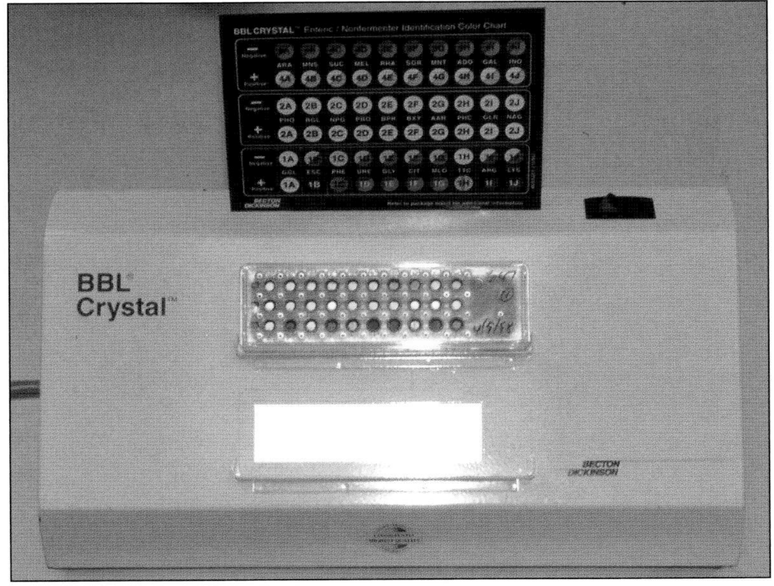

7.101 BBL CRYSTAL E/NF • **First Row:** The top row tests for carbohydrate utilization with acid end-products. Positive tests are indicated by a gold or yellow color and are given a value of 4. The carbohydrates are: **ARA**=arabinose; **MNS**=mannose; **SUC**=sucrose; **MEL**=melibiose; **RHA**=rhamnose; **SOR**=sorbitol; **MNT**=mannitol; **ADO**=adonitol; **GAL**=galactose; and **INO**=inositol.
Second Row: The second row consists of tests that detect the ability to hydrolyze the various substrates with the production of a yellow compound (either *p*-nitrophenol or *p*-nitroaniline). Each positive is assigned a value of 2. The substrates are: **PHO**=*p*-nitrophenyl phosphate; **BGL**=*p*-nitrophenyl α-β-glucoside; **NPG**=*p*-nitrophenyl β-galactoside; **PRO**=proline nitroanilide; **BPH**=*p*-nitrophenyl bisphosphate; **BXY**=*p*-nitrophenyl xyloside; **AAR**=*p*-nitrophenyl α-arabinoside; **PHC**=*p*-nitrophenyl phosphorylcholine; **GLR**=*p*-nitrophenyl β-glucuronide; and **NAG**=*p*-nitrophenyl-N-acetyl glucosamide.
Third Row: The third row consists of various biochemical tests with each positive result being given a value of 1. The tests are as follows with the positive result in parentheses: **GGL**= γ-L-glutamyl *p*-nitroanilide hydrolysis (yellow); **ESC**=Esculin hydrolysis (black); **PHE**=phenylalanine deamination (brown); **URE**=urea hydrolysis (blue); **GLY**=glycine degradation (blue); **CIT**=citrate utilization (blue); **MLO**=malonate utilization (blue); **TTC**=tetrazolium reduction (red); **ARG**=Arginine catabolism (purple); and **LYS**=lysine catabolism (purple).

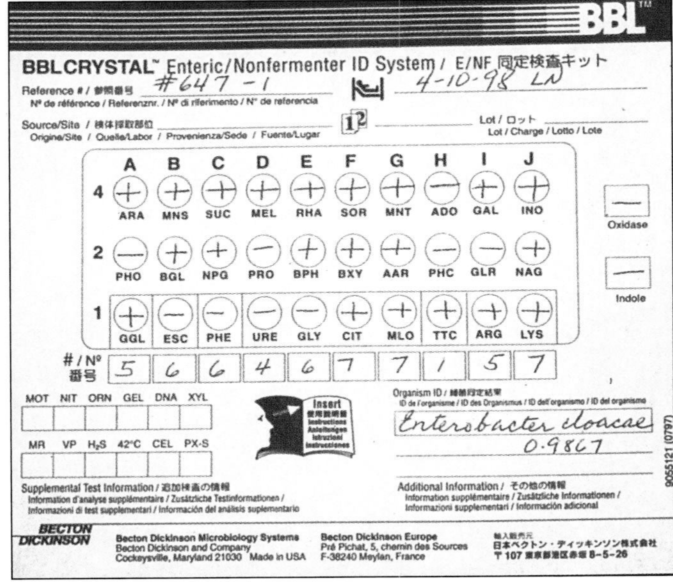
BBL

BBLCRYSTAL™ Enteric/Nonfermenter ID System / E/NF 同定検査キット

Reference # / 参照番号 #647 -1   4-10-98 LN

Source/Site   Lot

	A	B	C	D	E	F	G	H	I	J
4	+ ARA	+ MNS	+ SUC	+ MEL	+ RHA	+ SOR	+ MNT	− ADO	+ GAL	+ INO
2	− PHO	+ BGL	+ NPG	− PRO	+ BPH	+ BXY	+ AAR	− PHC	− GLR	+ NAG
1	+ GGL	− ESC	− PHE	− URE	− GLY	+ CIT	+ MLO	+ TTC	+ ARG	+ LYS
# / N° 番号	5	6	6	4	6	7	7	1	5	7

Oxidase −

Indole −

MOT NIT ORN GEL DNA XYL

MR VP H_2S 42°C CEL PX-S

Organism ID: *Enterobacter cloacae* 0.9867

Supplemental Test Information

Additional Information

BECTON DICKINSON

Becton Dickinson Microbiology Systems, Becton Dickinson and Company, Cockeysville, Maryland 21030 Made in USA

Becton Dickinson Europe, Pré Pichat, 5, chemin des Sources, F-38240 Meylan, France

7.102 BBL CRYSTAL E/NF Score Sheet • Positive test results are assigned a value of 4, 2, or 1, depending on its row. Columns are added and a 10-digit number is made from the sums. This number is then matched to a database containing approximately 100 taxa for identification. The 10-digit number (5664677157) on this score sheet identifies the organism as *Enterobacter cloacae*. This process is now automated, with an AutoReader that can be connected to a computer. The reader scans the test results, records them in the associated software, and produces the 10-digit identification code. A specimen report is generated with the provisional identification.

EnteroPluri-*Test*

Purpose

The EnteroPluri-*Test* is a multiple-test system used for rapid identification of bacteria from the family ***Enterobacteriaceae*** and other Gram-negative, oxidase-negative bacteria.

Principle

The EnteroPluri-*Test* is a multiple-test system designed to identify enteric bacteria based on: glucose acid/gas, adonitol, lactose, arabinose, sorbitol, and dulcitol fermentation; lysine and ornithine decarboxylation; sulfur reduction; indole production; acetoin production from glucose fermentation; phenylalanine deamination; urea hydrolysis; and citrate utilization.

The EnteroPluri-*Test*, as shown in Figure 7.103, is a tube containing 12 individual chambers with the capability of performing 15 tests. Inside the tube, running lengthwise through its center, is a removable wire. After the endcaps are removed (aseptically), one end of the wire is touched to an isolated colony on a streak plate and drawn back through the tube to inoculate the media in each chamber, air holes are opened for tests requiring aerobic conditions, caps are replaced, and the tube is incubated for 18–24 hours.

After incubation, most results can be read as a simple color change (Figs. 7.104 and 7.105). The indole and VP tests require addition of reagents, which is done with a syringe after the first tests are read. All results are recorded on an EnteroPluri-*Test* data chart (Fig. 7.106A). As shown in the figure, the combination of positives entered on the data chart results in a five-digit numeric code. This code is used for identification in the EnteroPluri-*Test* codebook (Fig. 7.106B).

The codebook is a master list of all enterics identifiable by this system and their numeric codes derived from testing. In most cases, the five-digit number applies to a single organism, but when two or more species share the same code, a confirmatory test is performed to further differentiate the organisms.

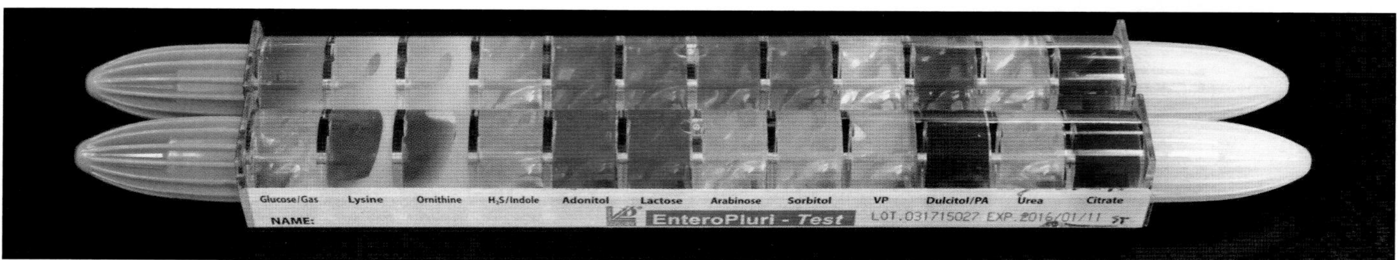

7.103 EnteroPluri-*Test* System ● An inoculated tube is in front; an uninoculated control is behind. Each tube contains 12 compartments, with the ability to run 15 tests. From left to right: acid from glucose/gas from glucose; lysine decarboxylase; ornithine decarboxylase; sulfur reduction/indole production; acid from adonitol; acid from lactose; acid from arabinose; acid from sorbitol; Voges-Proskauer; acid from dulcitol/phenylalanine deaminase; urease; and citrate utilization.

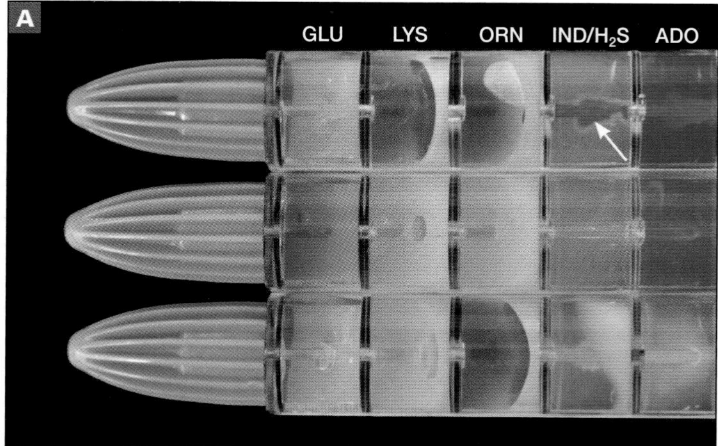

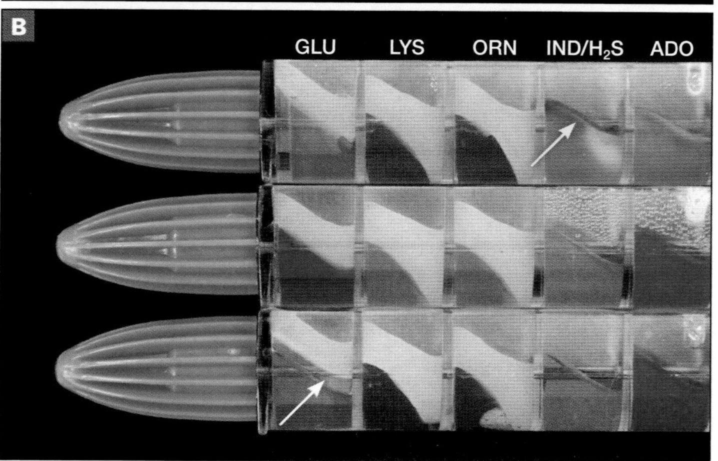

7.104 EnteroPluri-*Test* Results–Part 1 ● **(A)** Compartments viewed from above are (L to R): glucose acid/gas (GLU), lysine decarboxylase (LYS), ornithine decarboxylase (ORN), indole/H_2S (IND/H_2S), and adonitol acid (ADO). The tube in the center is uninoculated for comparison. The other tubes are both positive for acid production from glucose, as they should be if they are enterics. The tube below shows the difference between a negative (yellow LYS) and a positive (purple ORN) decarboxylase result. The tube on top is positive for both decarboxylases and also has reduced sulfur (H_2S positive). Note the black in the center (**arrow**). The indole compartment in the bottom tube is cloudy from addition of Kovac's reagent. The last visible compartment is acid from adonitol. The tube below is positive (yellow); the other tubes are negative (red). Acid from lactose, arabinose, and sorbitol are read the same way and are not shown. **(B)** In this photo, the same compartments are viewed from the side. The center tube is uninoculated, and the others were inoculated with different organisms than shown in Figure **A**. This photo is used to illustrate positive glucose-gas and indole results. Note the space between the white wax and the yellow agar in the lower tube, indicating gas production from glucose (**white arrow**). Compare with the agar/wax interface in the other two tubes. Also notice the pink layer in the fourth (indole) compartment of the upper tube (**yellow arrow**). This is a positive result for indole production and occurred after addition of Kovac's reagent (which had soaked into the agar by the time the photo was taken and produced the white artifact in the medium).

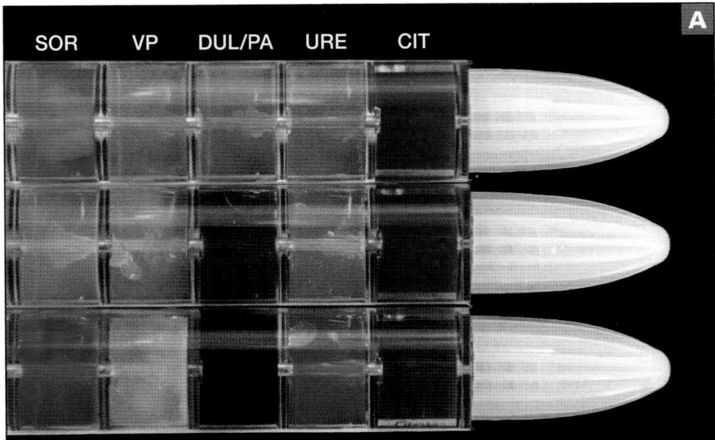

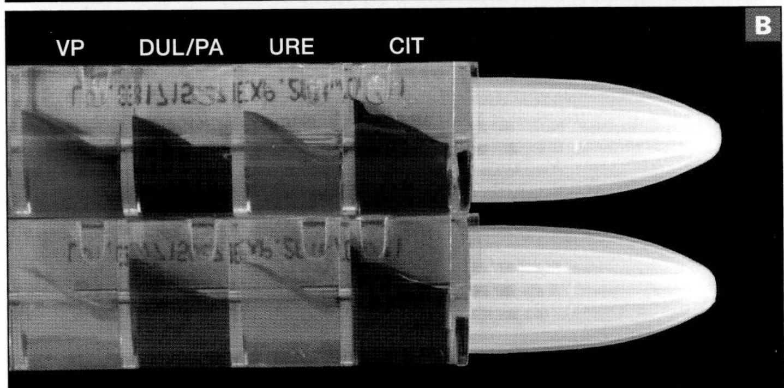

7.105 EnteroPluri-*Test* Results–Part 2 ● (A) All three of these tubes were inoculated and are viewed from above. Compartments from left to right are: sorbitol, VP, dulcitol/phenylalanine deaminase, urea, and citrate. Sorbitol is read the same way as adonitol (Fig. 7.104A): yellow is positive for acid and red is negative. All VP results are negative. The dulcitol/PA compartment will be dark brown if the organism is phenylalanine deaminase positive (below), yellow if dulcitol positive (above) and green if negative for both (center). Conflicting results for these two tests are rare because organisms for which the EnteroPluri-*Test* is designed have an extremely low probability of being positive for both. A positive urease test is pink (above and below) and yellow if negative (center). A positive citrate test is blue (top two) and green if negative (below). (B) Viewed from the side, these tubes show a positive VP (above) and a negative VP (below) result. In this specimen, a positive reaction was seen in about 10 minutes after addition of reagents, but it may take up to 20 minutes to become positive.

A

MODULO DATI / *DATA CHART*

DATA *DATE*	CAMPIONE *SAMPLE*

	GROUP 1			GROUP 2			GROUP 3			GROUP 4			GROUP 5		
Test	Glucose	Gas	Lysine	Ornithine	H_2S	Indole	Adonitol	Lactose	Arabinose	Sorbitol	VP	Dulcitol	PA	Urea	Citrate
Codice di positività *Positivity code*	4	2	1	4	2	1	4	2	1	4	2	1	4	2	1
Risultati *Results*	4	2	1	4	0	0	0	0	1	4	2*	0	0	0	1
Somma dei codici *Code sum*	7			4			1			6			1		

CODICE NUMERICO *NUMERICAL CODE* — MICRORGANISMO *MICROORGANISM* — *VP not visible from this view.

B

ENTEROBACTERIACEAE
ENTEROBACTERIACEAE

Codice numerico *Code number*	Microrganismo *Microorganism*	Test atipici *Atypical tests*
74147	*Serratia liquefaciens*	PA, URE
74150	*Salmonella sp. subsp. choleraesuis*	H_2S, CIT
	Escherichia coli	IND, LAC
74151	*Salmonella sp. subsp. choleraesuis*	H_2S
74160	*Serratia liquefaciens*	CIT
74161	*Serratia liquefaciens*	NONE
74162	*Serratia liquefaciens*	URE, CIT

7.106 EnteroPluri-*Test* Data Chart and Codebook ● (A) The compiled results of the inoculated (front) tube in Figure 7.103 are shown on this data chart, where the tests are arranged into five groups of three. Positive results are assigned their corresponding number (4, 2, or 1) and the sum of the positive results for each group is recorded. The resulting 5-digit number is then found in the EnteroPluri-*Test* Codebook. (B) Shown is a portion of a page from the codebook. The 5-digit code for the unknown scored in Figure A is 74161, which corresponds to *Serratia liquefaciens*. Because of genetic variability, not all *S. liquefaciens* strains will give identical results for the 15 tests. Note that the codes 74147, 74160, and 74162 (and several others not shown in this figure) also code for *S. liquefaciens*. The codebook indicates which results are atypical for the isolate. For instance, 74160 identifies an isolate as *S. liquefaciens* that is atypically citrate-negative. (Can you see how a citrate-negative result would change the code from 74161 to 74160?) Also notice that in some cases a single code may be produced by two organisms (e.g., 74150). In these cases, additional testing would be required to determine the isolate's identity.

Vitek 2–Compact

Purpose

The Vitek 2-Compact instrument (Fig. 7.107; available from bioMérieux, Inc.) is an automated, rapid, and versatile system that can be used to identify clinically and environmentally important organisms. (It also performs antibiotic susceptibility testing, but that is not covered here.) Five batteries of tests are available, their use depending on the circumstances and type of laboratory (environmental vs. clinical).

The tests are: GN for Gram-negative fermenting (*Enterobacteriaceae*) and non-fermenting bacilli; GP for Gram-positive cocci and non-spore-forming bacilli; NH for *Neisseria*, *Haemophilus*, and other fastidious Gram-negative bacteria; ANC for anaerobic and coryneform bacteria; and YST for yeast and yeast-like organisms.

Principle

We will use the GN system as our example of how the machine works. The GN ID card has the capability to identify more than 150 species of *Enterobacteriaceae* and other non-fermenting, Gram-negative bacilli. The card has

64 wells, each loaded with the appropriate substrate for its particular test (Fig. 7.108). Many are the familiar tests covered in this book, such as fermentations of various sugars, decarboxylations, H_2S production, and urea hydrolysis, but there are many more that are not covered here.

While most of the operation is automated, a laboratorian has to prepare the sample. Colonies to be tested are suspended in 3 mL of sterile saline and are adjusted to the prescribed McFarland turbidity level (0.50 to 0.63 for GN; see Fig. 19.3).

Figure 7.108 shows the prepared sample tube and its associated card in a cassette (rack), which can hold up to 10 cards. The cassette is manually placed into the machine's filling station (Fig. 7.109), and the sample is automatically transferred to the card via the blue tube, passing through all the channels and inoculating the 64 wells.

After inoculation, the cassette is manually transferred (within 10 minutes) to the loading station. During incubation, the cassettes are automatically removed at 15-minute intervals and the wells are optically evaluated for color changes that indicate positive test results. All of this information is fed into a computer and final results can be obtained in a few hours. The combination of results (**biopattern**) is compared to a database and a provisional identification is displayed by the computer (Fig. 7.110).

Confidence levels are reported as "excellent" (96–99%), "very good" (93–95%), "acceptable" (89–92%), "low discrimination" (85–88%), and "unidentified organism" (<85%). "Low discrimination" means two or three organisms have the same biopattern and is accompanied by additional tests that can discriminate between them. "Unidentified" either means more than three organisms share the same biopattern or the biopattern is not in the database.

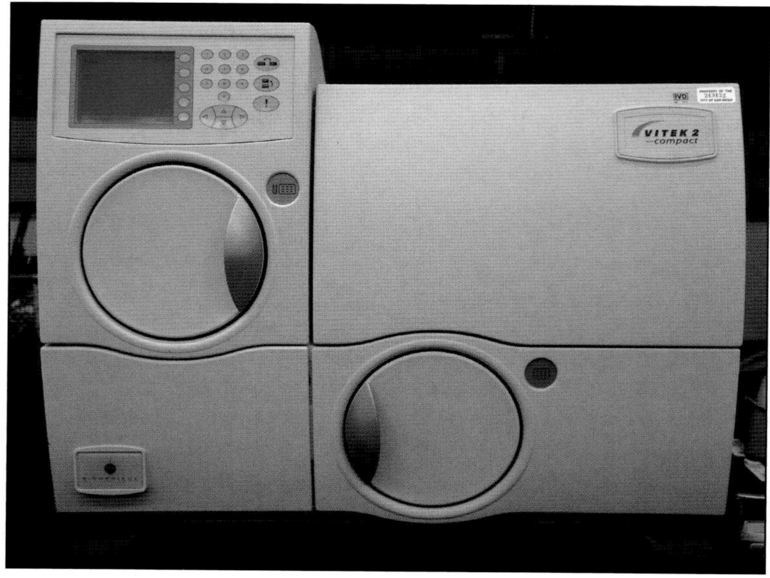

7.107 Vitek 2–Compact Machine • The Vitek 2 Compact is a small table top apparatus used for identification of multiple groups of organisms as well as antibiotic susceptibility testing. The circular compartment on the left ("filling station") prepares the test cards by adding the microbial suspension to the cards' wells. The circular compartment on the right ("loading station") is where the cards are incubated and results are read. Identification results are provided in as little as 5 hours. Other than getting isolation, preparing the sample to a MacFarland standard, loading the machine, and working the computer, the process is automated. This minimizes culture handling by the laboratorian and eliminates manual inoculation of 64 media!

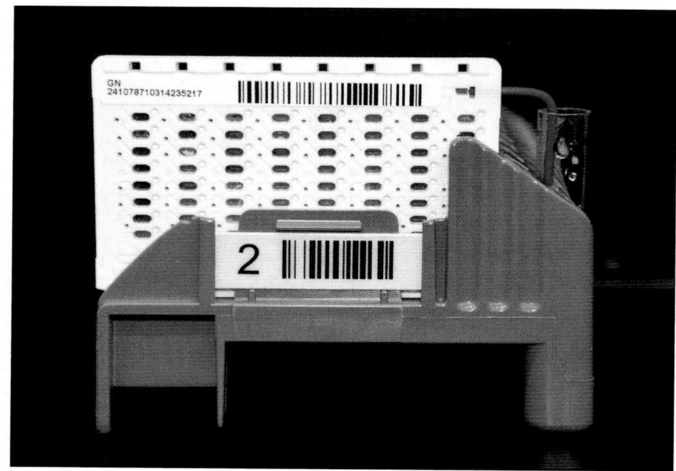

7.108 Vitek 2–Compact GN ID Card in a Cassette with its Sample Tube • Shown is a GN ID card with 64 wells in which tests for identification of Gram-negative enterics and non-fermenting, Gram-negative rods are located. The prepared sample of an environmental isolate is in the tube on the right. The card is identified by a bar code and is linked to its associated tube to allow for sample tracking. This cassette (the gray rack) can hold up to 10 cards.

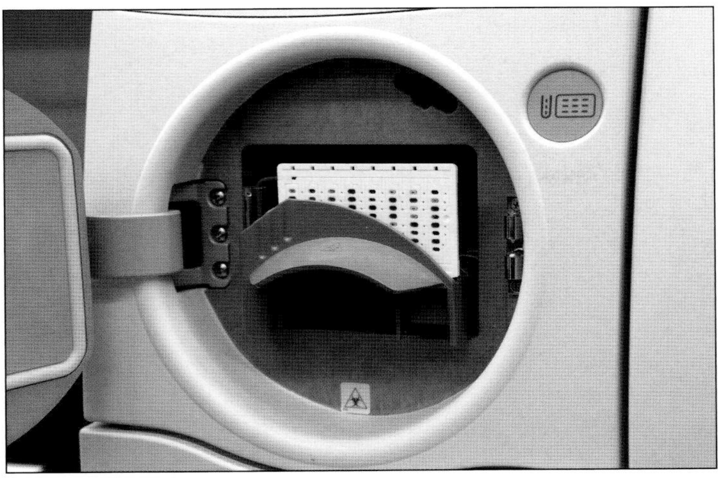

7.109 Vitek 2–Compact Vacuum Filling Station ● The laboratorian has loaded the cassette carrying the card into the filling station. Inoculation of the wells is an automated process where each sample is transferred through the blue tube to a system of channels leading to the card's 64 wells. When inoculation is completed, the cassette is manually moved to the loading station where the blue transfer tube is cut and the card is sealed. The identification process commences and the 64 wells are optically examined at 15-minute intervals for color changes that indicate positive results. The compiled data are loaded onto a computer, which examines a database for a best fit identification.

Identification Information	Card:	GN	Lot Number:	2410787103	Expires:	Jan 22, 2020 12:00 GMT-08:00
	Completed:	Apr 17, 2019 14:37 GMT-08:00	Status:	Final	Analysis Time:	5.00 hours
Selected Organism	98% Probability		Leclercia adecarboxylata			
	Bionumber:	4625730542130011			Confidence:	Excellent identification

7.110 Vitek 2–Compact Results Report ● Comparing the test results with a database resulted in identification of the sample shown in Figure 7.108 as *Leclercia adecarboxylata* with 98% confidence. It took only 5 hours. *L. adecarboxylata* (formerly known as *Escherichia adecarboxylata*) is an opportunistic pathogen of immunocompromised patients. It is also found in environmental samples (this was a water sample) and foods. Its specific epithet refers to its inability to decarboxylate arginine, lysine, and ornithine.

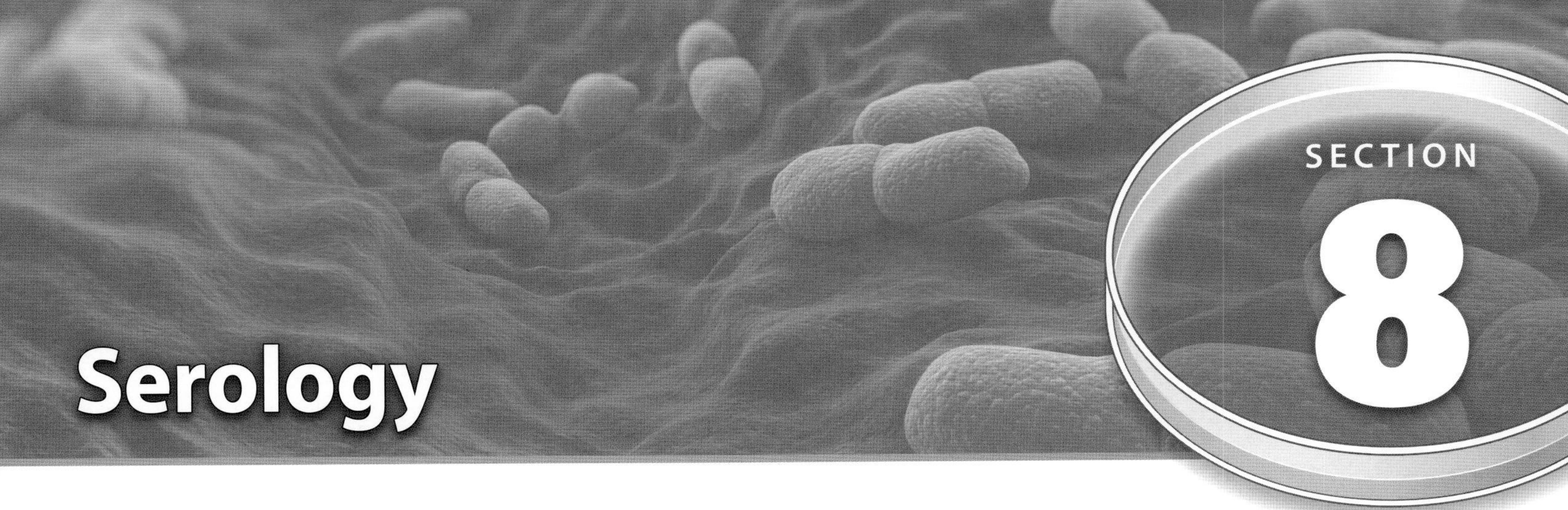

Introduction to Antigens and Antibodies

This section presents the in vitro use of antigens and antibodies as diagnostic tools to identify the presence of one or the other in a patient's sample. This field of microbiology is called **serology** because the earliest applications—even before microbiologists knew exactly what was happening at the molecular level—involved using the **serum** component of blood (which is where antibodies were first found) to identify the presence of antigens in samples. In order to fully appreciate these serological tests, an introduction to antigens and antibodies is necessary.

Antigens are high molecular weight molecules with a definite three-dimensional shape that the body recognizes as being "foreign." Not all foreign molecules are recognized as being foreign by the immune system and that is why antigens are special. Antigens stimulate the production of specific **antibodies** (or **immune cells**[1]) and will react specifically with them once produced. In general, proteins make the best antigens, followed by polysaccharides. Lipids and nucleic acids do not make good antigens.

Antibodies are glycoproteins (**immunoglobins–Ig**) produced by specialized lymphocytes (B-cells/plasma cells) and each is composed of two identical long polypeptides called **heavy chains** and two identical shorter polypeptides called **light chains** (Fig. 8.1A). Functionally, each antibody molecule behaves as if it has the shape of the letter "Y."

There are two **antigen binding sites**, one at the end of each arm of the "Y." This part of the antibody is called the **F_{ab} portion** (for **"fragment of antigen binding"**). The stem of the "Y" (the **F_c portion**; for **"crystallizable fragment"**) is formed by the remainder of the two heavy chains and performs **effector functions**, such as binding to phagocytic cells or activating the complement system.

Obviously, there must be great antibody diversity in order to protect against all the antigens the world presents. On a gross level, antibody diversity begins with the type of heavy chain present. The five basic immunoglobin classes (with heavy chains in parentheses) are: IgG (γ), IgM (μ), IgA (α), IgD (δ), and IgE (ε) (Fig. 8.1B).

Additional diversity occurs because light chains come in two forms: kappa (κ) and lambda (λ), which are able to combine with any of the heavy chains. Further, antibodies may be joined into more complex arrangements, such as in IgM, where five Y-shaped antibodies form a pentamer, and in IgA that exists as a dimer in mucus and other secretions (Fig. 8.1B).

The majority of antibody diversity occurs as a result of each lymphocyte's ability to produce its own version of its heavy and light chains. Each heavy and light chain has an amino acid sequence along most of its length that is more or less characteristic of its type. This is called the **constant region**.

However, each heavy and light chain also contains a **variable region**, which differs in amino acid sequence from those produced by all other lymphocytes and from each other. When assembled, the variable regions combine to form a unique three-dimensional pocket that will bind an antigen with a complementary shape.

Molecules that are good antigens have a three-dimensional shape and a surface topography. Because antigens have a relatively unchanging three-dimensional shape with a varied surface topography, different parts of an antigen can stimulate production of different antibodies, each reacting with a different surface feature. These surface features are called **epitopes** (or **antigenic determinants**), and their presence makes an immune response to any antigen molecule more complex than just one antibody for one antigen (Fig. 8.2).

[1] While immune cells are essential to the immune system in vivo, our topics in Section 8 deal only with the behavior of antigens and antibodies in vitro, so there will be no further reference to immune cells.

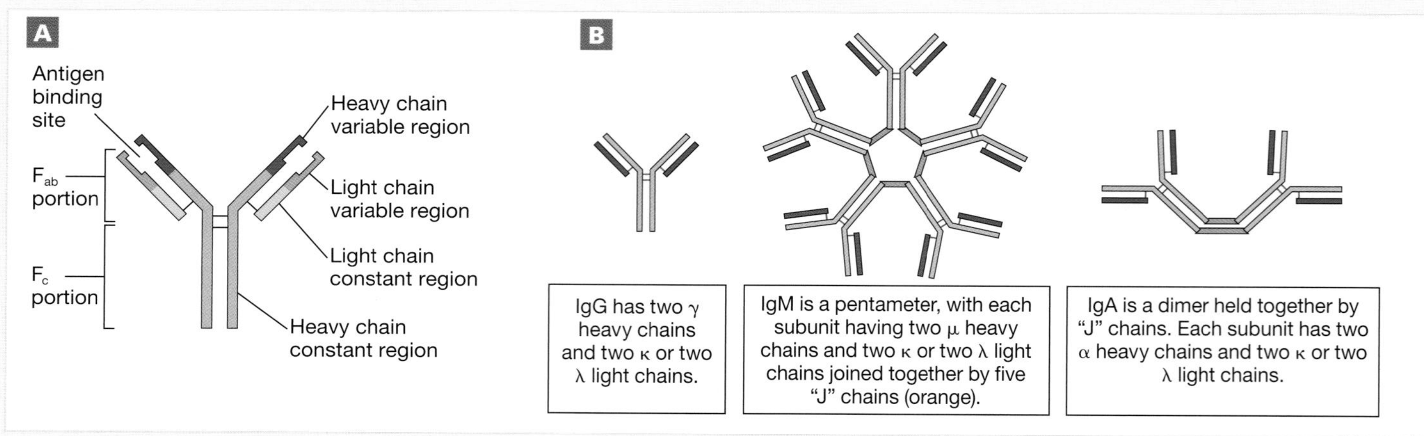
A
Antigen binding site
Heavy chain variable region
F_{ab} portion
Light chain variable region
Light chain constant region
F_c portion
Heavy chain constant region
B
IgG has two γ heavy chains and two κ or two λ light chains.
IgM is a pentameter, with each subunit having two μ heavy chains and two κ or two λ light chains joined together by five "J" chains (orange).
IgA is a dimer held together by "J" chains. Each subunit has two α heavy chains and two κ or two λ light chains.

8.1 Antibody Structure • (A) The fundamental antibody structure is two identical heavy chains and two identical light chains held together by covalent disulfide bonds. Each chain has a constant region and a variable region composed of "standard" and unique amino acid sequences, respectively. The variable regions are unique to each B lymphocyte and join together to form the antigen binding site, thus accounting for the ability of each B-cell to produce antibodies that react with only a single antigen (epitope). (B) There are five structural antibody classes identified by the heavy chain in each. IgG, IgD, IgE, and some IgA have the basic "Y" structure, but IgM and most IgA combine to form more complex arrangements.

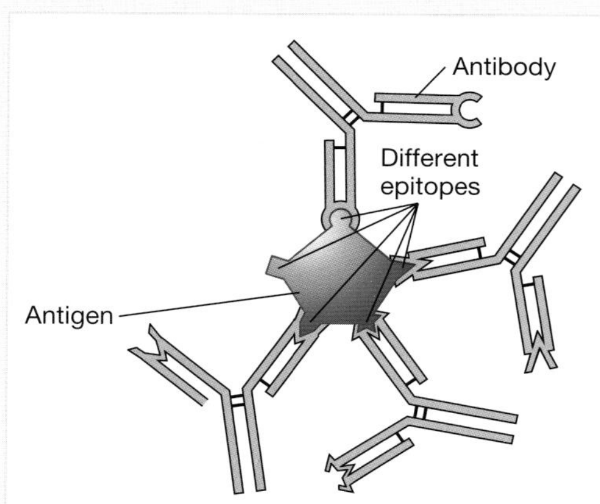
Antibody
Different epitopes
Antigen

8.2 Epitopes • Shown is an artist's representation of an antigen molecule with multiple epitopes. Each epitope has the ability to stimulate lymphocytes to produce antibodies that will bind them because of the complementary shapes. Often biologists refer to the whole molecule (and sometimes even the whole cell or virus particle) as an "antigen," but in reality, the immune system "sees" only epitopes. In this illustration, a single molecule has resulted in the production of four antibodies targeting it. The "square" surface feature on the left didn't stimulate antibody production and is therefore not an epitope. However, in another individual that same "square" might act as an epitope. Every individual makes their own arsenal of antibodies.

Serological Reactions

Antigen–antibody reactions are highly specific and occur in vitro as well as in vivo. Serology is the discipline that exploits this specificity as an in vitro diagnostic tool. All serological tests can be designed to identify either antigen or antibody in a sample. The choice for any particular test largely depends on the circumstances of the disease process; that is, whether it is easier to identify antigens or antibodies in a patient sample.

All serological tests also fundamentally rely on the reaction between antigen and its **homologous antibody**, but they differ in the way a positive reaction is seen. You will see several examples in the upcoming topics of how different indicator reactions reveal positive results.

Precipitation Reactions

Purpose

Precipitation reactions can be used to detect the presence of either antigen or antibody in a sample. They have mostly been replaced by more sensitive serological techniques for diagnosis, but are still useful for a simple demonstration of serological reactions. Double-gel immunodiffusion is used to check samples for identical, related, or unrelated antigens.

Principle

Soluble antigens may combine with **homologous antibodies** to produce a visible **precipitate.** Precipitate formation in vivo serves to inactivate toxins or enzymes produced by infectious agents. The precipitates can subsequently be cleared by cells such as eosinophils. Precipitate formation in vitro can serve as evidence of antigen–antibody reaction in a diagnostic test and is considered to be a positive result.

Precipitation is produced because each antibody has (at least) two **antigen binding sites** and many antigens have multiple copies of **epitopes on their surface.** This results in the formation of an interconnected **antigen–antibody complex** and produces the visible precipitate—a positive result.

As shown in Figure 8.3, if either antibody or antigen is found in too high a concentration relative to the other, no visible precipitate will be formed even though both are present, and produces a false negative result. (Of course, the microbiologist would record a negative result and would be unaware that it is truly positive and that the test has produced a false negative result.) **Optimum proportions** of antibody and antigen are necessary to form precipitate, and they occur in the **zone of equivalence.**

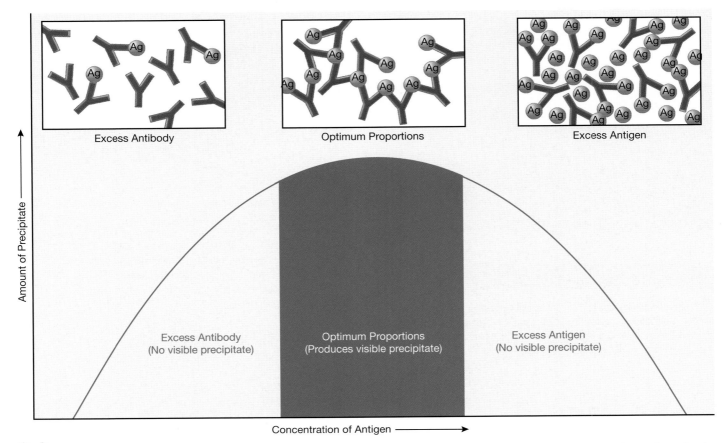

8.3 Precipitation Reactions ● Precipitation occurs between soluble antigens and homologous antibodies where they are found in optimal proportions (at the zone of equivalence) to produce a cross-linked antigen-antibody complex. Excess antigen or excess antibody prevents substantial cross-linking, so no visible precipitate is seen—even though both antigen and antibody are present. This is a critical factor that must be addressed in developing serological tests. To determine optimum proportions (as shown in this graph), antibody concentration is kept constant as antigen concentration is adjusted.

Several styles of precipitation tests have been developed. The **precipitin ring test** is performed in a small test tube or capillary tube. **Antiserum** (containing antibodies homologous to the antigen being looked for) is placed in the bottom of the tube. The sample with the suspected antigen is layered on the surface of the antiserum in such a way that the two solutions have a sharp interface.

In a positive test, as the antigens and antibodies diffuse past each other, precipitation occurs where optimum proportions of antibody and antigen are found (Fig. 8.4). In a negative test, there is no antigen-antibody binding. This test can also be used to identify antibody in a sample by using known homologous antigen.

In its simplest form, **gel immunodiffusion** involves two wells formed in a saline agar plate. In one well is the antiserum; in the other is a sample of unknown antigen composition. The two diffuse out of their respective wells toward each other.

If the sample has an antigen that will react with the antibody, then a precipitation line will form at the region of optimal proportions as their diffusion paths pass. If no precipitation line forms, the test is considered negative. That is, the sample has no antigen that will react with the antibodies. Gel immunodiffusion can also identify the presence of antibody in a sample by using a known homologous antigen.

Gel immunodiffusion is also used to determine antigen relatedness, that is, how many shared epitopes antigens have. The more shared epitopes, the more closely related they are (Fig. 8.5), with results being recorded as **identity** (all epitopes are the same), **partial identity** (some epitopes are the same), and **nonidentity** (no shared epitopes).

The test is performed using a saline agar plate with several wells placed around a center well (Fig. 8.6). Antiserum containing a mixture of antibodies that bind specifically with the different epitopes on the antigen molecules is placed in the center well. Samples of known or unknown antigen (epitope) composition are put in the surrounding wells. The diffusion path out of a well is not linear. Rather, it is radial in all directions. Because of this, precipitation lines may form between the center well and any or all of the surrounding wells, depending on their epitope composition.

The precipitation line pattern between neighboring wells is indicative of antigen relatedness in those wells (Fig. 8.7). A single, smooth, curved precipitation line indicates the two antigens in neighboring wells share identical epitopes (**identity**) for the epitopes being tested.

Two spurs indicate unrelated antigens that share no epitopes (that the test can detect) and is considered **nonidentity** because two completely separate precipitation lines formed. A single spur indicates the antigens share some, but not all, epitopes (**partial identity**) because two lines formed—a line of identity and a line of nonidentity (Fig. 8.8).

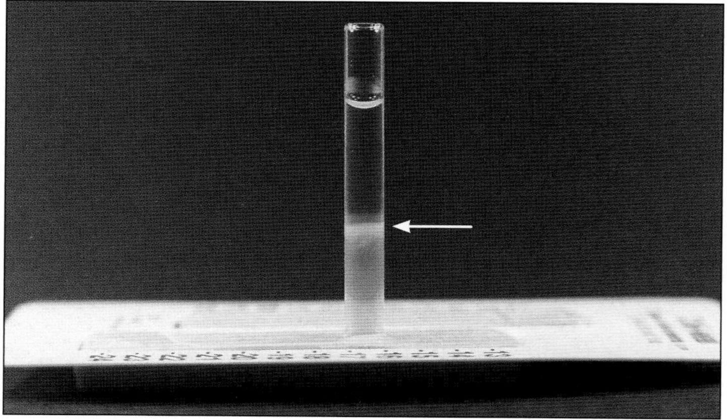

8.4 Positive Precipitin Ring Test • A sample of antigen has been layered over an antiserum containing antibodies. Assuming both are present, the antigens and antibodies will diffuse past one another, and at some point a white precipitation ring (**arrow**) will form at the site of optimum proportions. If no ring appears, the result is negative.

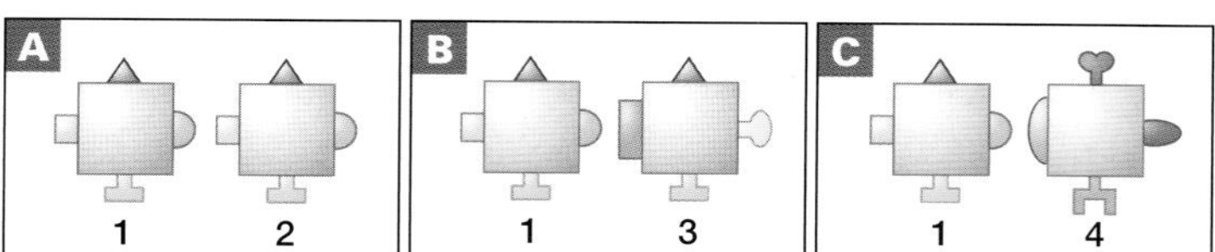

8.5 Degrees of Antigen Relatedness • Shown are three pairs of antigens, each with four epitopes, which could be placed in adjacent wells of a double gel immunodiffusion test. In each pair, the antigen on the left is the same. Compare these results with Figure 8.7 as you read. (**A**) In this pair, antigens 1 and 2 have identical epitopes and would produce a line of identity (so really both are Antigen 1 or both are Antigen 2). (**B**) Antigens 1 and 3 share two epitopes but differ in the other two. They would show a pattern of partial identity. (**C**) Antigens 1 and 4 have no epitopes in common and would produce a pattern of nonidentity.

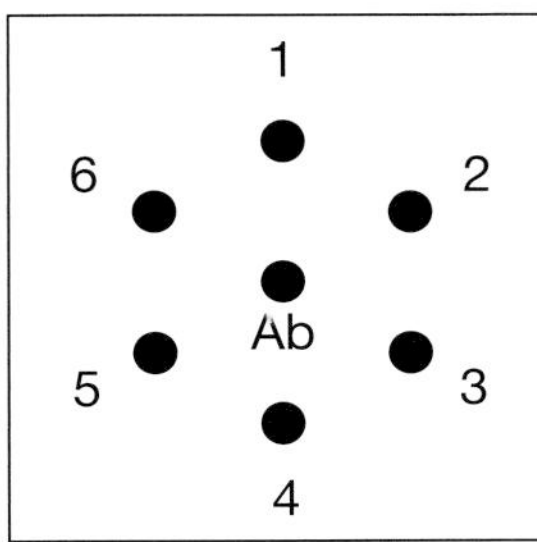

8.6 Double Gel Immunodiffusion Well Placement • Wells are punched out of a saline agar plate in this basic pattern: a central well surrounded by outer wells. Antiserum with multiple antibodies is placed in the center well. The outer wells are loaded with different antigen samples.

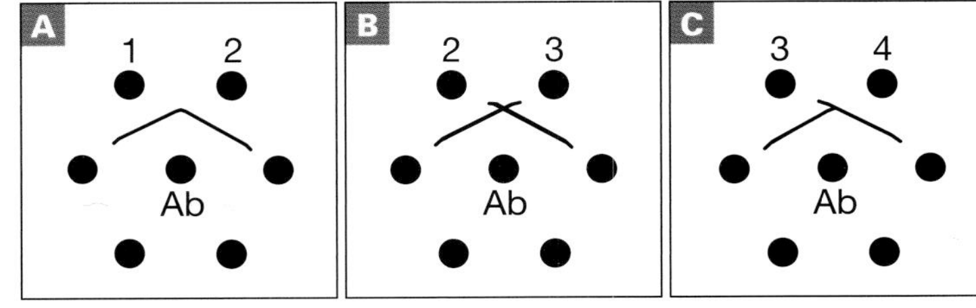

8.7 Double Gel Immunodiffusion Precipitation • This diagram shows three possible precipitation patterns formed between the antibodies in the center well and antigens in pairs of outer wells. Compare these results with Figures 8.5 and 8.8 as you read. **(A)** Antigens 1 and 2 demonstrate the precipitation pattern of identity. This indicates that these antigens have identical epitopes (at least for the epitopes being tested). **(B)** The precipitation pattern demonstrated by antigens 2 and 3 is nonidentity, which means that the antigens are not related and share no epitopes being tested. **(C)** The precipitation pattern produced by antigens 3 and 4 shows partial identity; the antigens share some, but not all, epitopes.

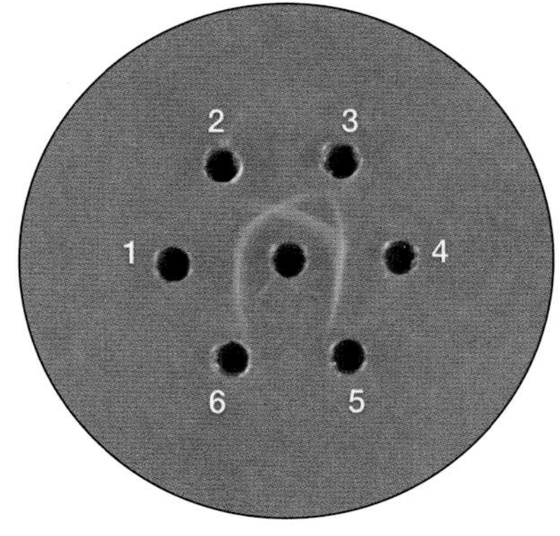

8.8 Double Gel Immunodiffusion Results • Antibodies were put in the center well and different antigens were put in the outer wells. Precipitation lines without spurs formed between the central well and wells 1 and 2. They illustrate identity of the antigens in wells 1 and 2. The antigens in wells 2 and 3 illustrate nonidentity, as evidenced by the two spurs at the intersection of their precipitation lines. The lines formed between the central well and wells 3 and 4 illustrate partial identity. The antigens share one or more epitopes (curved precipitation line), but well 4 has at least one additional epitope absent in well 3, as evidenced by the single spur. No reaction occurred between the antibodies and antigens in wells 5 and 6.

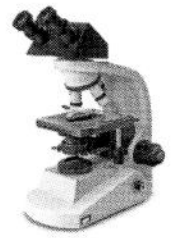

Agglutination Reactions

Purpose

Agglutination reactions may be used to detect the presence of either antigen or antibody in a sample. **Direct agglutination** reactions are used to diagnose some diseases, determine if a patient has been exposed to a certain pathogen, and are involved in blood typing. **Indirect agglutination** is used in some pregnancy tests as well as in diagnosing disease.

A quick sampling of diagnostic indirect agglutination test suppliers showed tests for bacteria (*Escherichia coli* O157, *Haemophilus influenzae*, *Legionella*, *Listeria*, *Neisseria gonorrhoeae*, *N. meningitidis*, *Salmonella*, *Staphylococcus aureus*, *Streptococcus pneumoniae*, streptolysin O antibodies, and *Treponema pallidum*); viruses/viral diseases (infectious mononucleosis, rubella virus, and rotavirus); fungi (*Cryptococcus*); protozoans (toxoplasmosis); and autoimmune disease (rheumatoid factor, cardiolipin). Clearly, agglutination tests are effective and popular!

Principle

Particulate antigens (such as whole cells) may combine with homologous antibodies to form visible clumps called **agglutinates**. In vivo, these agglutinates immobilize the cells until they can be removed by phagocytic cells. In vitro, **agglutination** can serve as evidence of antigen–antibody reaction and is considered a positive result. Agglutination reactions are highly sensitive and may be used to detect either the presence of antigen or antibody in a sample.

There are many variations of agglutination tests (Fig. 8.9). **Direct agglutination** relies on the combination of antibodies and naturally particulate antigens. **Indirect agglutination** relies on artificially constructed systems in which agglutination will occur. These involve coating particles (such as RBCs, or more commonly, latex microspheres) with either antibody or antigen, depending on what is being looked for in the sample.

Addition of the homologous antigen or antibody will then result in clumping of the artificially constructed particles and turns what would have naturally been a less-sensitive precipitation reaction into a more-sensitive agglutination.

Agglutination tests are done on a microscope slide or a card. Shown in Figure 8.10 is a slide agglutination of *Salmonella* antigens. Currently, only two *Salmonella* species are recognized: *S. enterica* (with six subspecies) and *S. bongorii*. **Serotyping,** that is, characterizing strains based on possession of identifiable and unique antigen combinations, has identified more than 2,500 *Salmonella* **serotypes** using three antigens.

The **O antigen** is polysaccharide from the outer membrane's **LPS layer,** the **H antigen** is **flagellin** protein, and the **Vi antigen** is a polysaccharide in the capsule. Identifying *Salmonella* serotypes is of public health importance because different serotypes often produce different diseases (e.g., *Salmonella* serotype Typhi causes typhoid fever and *Salmonella* serotype Typhimurium causes salmonellosis) or are associated with different modes of transmission, information that is useful to public health agencies in identifying outbreaks.

Hemagglutination is a general term applied to any agglutination test in which clumping of red blood cells indicates a positive reaction. Blood tests as well as a number of indirect diagnostic serological tests are hemagglutinations.

The most common form of blood typing detects the presence of **A** and/or **B antigens** on the surface of red blood cells. An individual with type A blood has RBCs with the A antigen and produces anti-B antibodies. Conversely, an individual with type B blood has RBCs with the B antigen and produces anti-A antibodies. People with type AB blood have *both* A and B antigens on their RBCs and lack anti-A and anti-B antibodies. Type O individuals lack A and B antigens but produce *both* anti-A and anti-B antibodies.

While genetically determined, the A and B antigens are not direct gene products, but rather are produced by enzymatic reactions that modify the O antigen, a short (5-sugar) polysaccharide, found on all RBC membranes (Fig. 8.11). People with type A blood produce an enzyme that adds the sugar N-acetylgalactosamine to the end of the O antigen to make the A antigen.

In like fashion, people with type B blood have a different enzyme that adds the sugar galactose to the end of the O antigen. People with type AB blood make both enzymes and modify the O antigen in both ways, and people with type O blood can only synthesize the O antigen, which remains unmodified. (Think about this: The antigenic difference between type A and type B blood is a single sugar at the end of the same molecule!)

ABO blood type is ascertained by adding a patient's blood separately to anti-A and anti-B antiserum and observing for any signs of agglutination (Fig. 8.12). Agglutination with

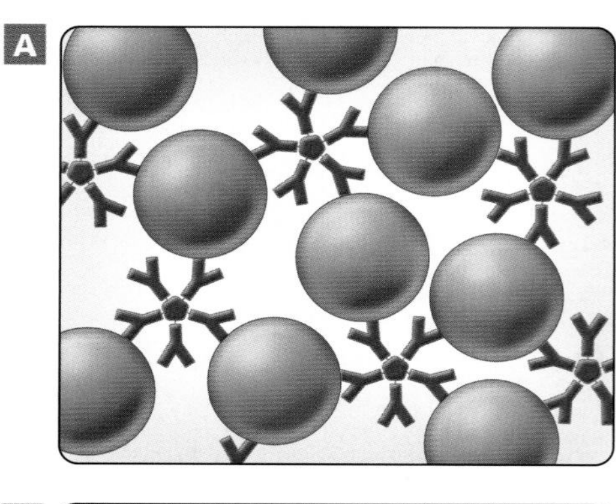

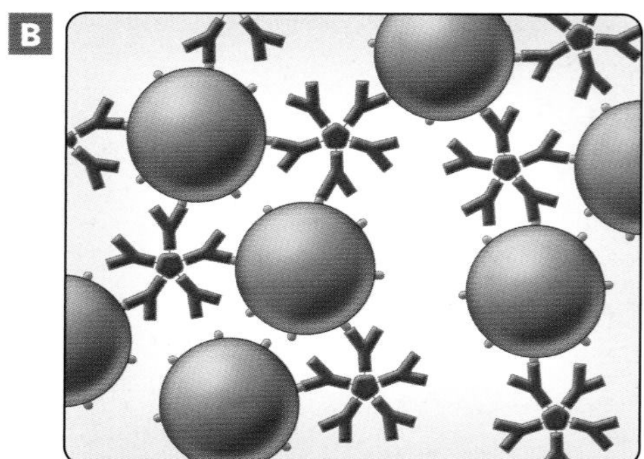

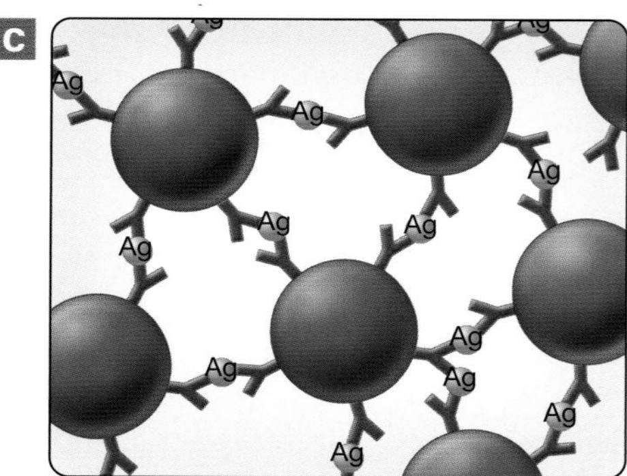

8.9 Direct and Indirect Agglutination Reactions • (A) Direct agglutinations involve naturally particulate antigens (blue spheres) reacting with antibodies (purple snowflakes–IgM). The antigen is naturally particulate and reacts with the homologous antibody to produce an agglutinate. (B) Indirect agglutinations require preparation of artificial particles. In this illustration, the soluble antigens (light blue dots) have been affixed to latex microspheres so they become part of a particle. When they react with the homologous antibody, a visible agglutinate forms. This style of indirect agglutination is used to identify antibody in a patient's sample. (C) Alternatively, antibodies (IgG) can be attached to the latex particle. When they react with homologous antigen, agglutination occurs. This style of indirect agglutination is used to identify antigen in a patient's sample.

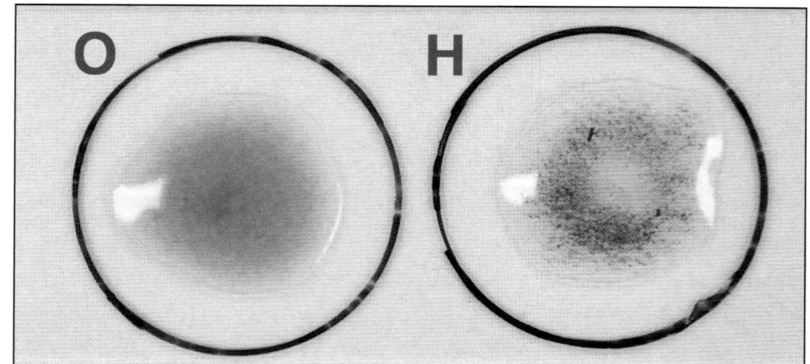

8.10 Salmonella H and O Antigen Slide Agglutination • *Salmonella* O and H antigens are used to serotype *Salmonella* isolates. Shown is a simple slide agglutination test used as a demonstration of anti-H antiserum reactivity with O antigen (left) and H antigen (right). Note the granular agglutinate that has formed in the mixture of H antigen and anti-H antibodies, but not with O antigen and anti-H antibodies. To be of use, all serological tests need to have high specificity. That is, a positive reaction should only occur when antigen and its homologous antibody are mixed together. False positives should occur as infrequently as possible.

anti-A antiserum indicates the presence of the A antigen and type A blood. Agglutination with anti-B antiserum indicates the presence of the B antigen and type B blood. If both agglutinate, the individual has type AB blood; lack of agglutination occurs in individuals with type O blood.

A similar test is used to determine the presence or absence of the **Rh factor**, an erythrocyte membrane polypeptide. There are actually three Rh antigens (C, D, and E), and possession of any one of them makes the person Rh positive, but the one usually detected in blood testing is the **D antigen**.

If clumping of the patient's blood occurs when mixed with anti-D antiserum, the patient is Rh positive (Fig. 8.13). The ABO and Rh blood factors are genetically unrelated, so any combination of the two is possible (e.g., "O–positive" or "O–negative").

Indirect hemagglutination may be used to detect the presence of either antigens or antibodies in a sample. The example shown in Figure 8.14 uses sheep RBCs coated with surface antigens extracted from *Treponema pallidum* (the causative agent of syphilis; see p. 220) to produce the particulate antigen.

When added to patient serum containing anti-*T. pallidum* antibodies, agglutination occurs and is evidenced by a smooth mat formed on the bottom of the well. This method and indirect agglutination methods using gelatin or latex particles are slightly less sensitive than fluorescent antibody tests (see

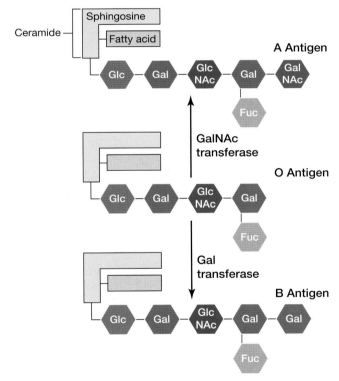

8.11 A and B Blood Antigens • All humans produce the short oligosaccharide that produces type O blood. However, if a person possesses at least one allele (either genotype I^AI^A **or** I^Ai) to produce the enzyme *N*-acetylgalactosamine transferase (GalNAc transferase) they add the sugar *N*-acetylgalactosamine to the O oligosaccharide and have type A blood. If a person has at least one allele (either genotype I^BI^B **or** I^Bi) to produce the enzyme galactose transferase (Gal transferase) they add the sugar galactose to the O oligosaccharide and have type B blood. The presence of both alleles (genotype I^AI^B) results in both reactions occurring and produces type AB blood. The absence of both alleles (genotype ***ii***) leaves the O oligosaccharide unmodified and the person has type O blood. Because the O oligosaccharide is present in all humans, it is not detected as an antigen. Key: glucose (Glc), galactose (Gal), *N*-acetylglucosamine (GlcNAc) *N*-acetylgalactos-amine (GalNAc), and fucose (Fuc).

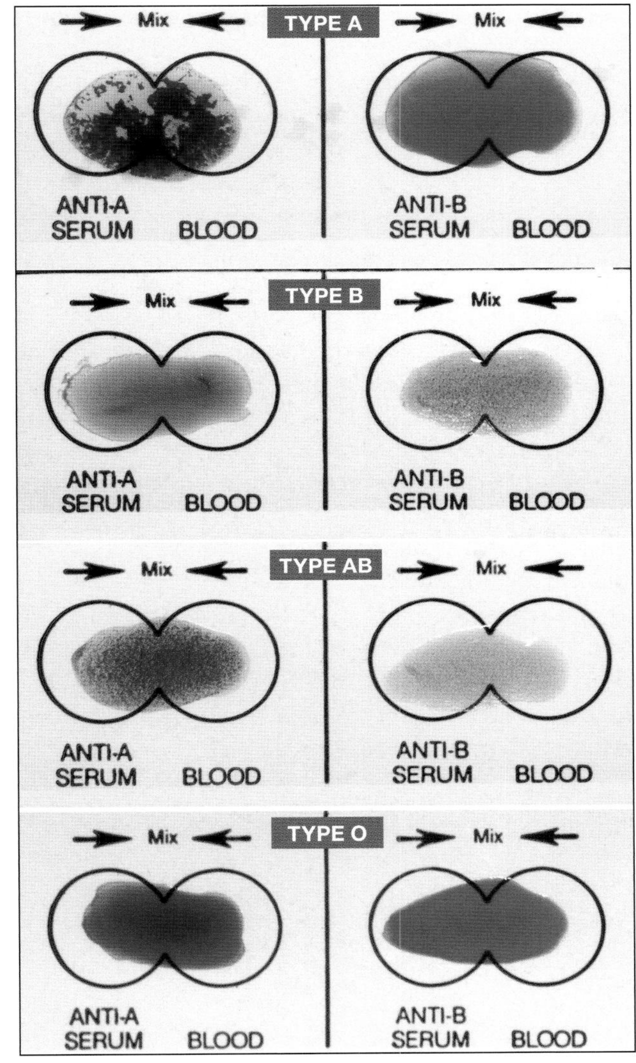

8.12 ABO Blood Group Test • Blood typing relies on agglutination of RBCs by anti-A and/or anti-B antisera. The blood types are as shown. Note that the degree of agglutination is not equal in all samples.

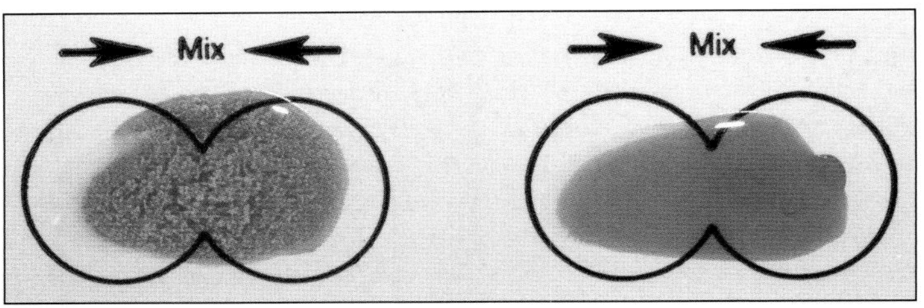

8.13 Rh Blood Group Test • Rh blood type is determined by agglutination with anti-Rh (anti-D) antibody. Rh-positive blood is on the left, Rh-negative is on the right.

p. 140), and the hemagglutination is the least reliable of the three, but is still good enough to be used.

The **rapid plasma reagin test** (**RPR**) for "antibodies" to *T. pallidum* is actually a **flocculation** reaction, but the visible outcome is similar to an agglutination test, so it is covered here. **Reagin** is an antibody-like molecule produced by people with syphilis.

During infection, treponemal-damaged cells release the phospholipids **cardiolipin** and **lecithin**, which react with reagin. The RPR test reagent contains cardiolipin- and lecithin-coated particles mixed with charcoal particles. In a positive RPR test, there is clumping of the particles by reagin.

Charcoal enhances the visibility of clumping, making the test easier to read (Fig. 8.15). Because it does not test for *T. pallidum*, it is considered a **nontreponemal test** (NTT), but is often the first test run in syphilis screening. Positive results are then subjected to a **treponemal test** (TT), such as the hemagglutination or latex agglutination tests described previously.

Characterization of clearly defined species in the genus *Streptococcus* has been challenging, though modern molecular techniques are helping to sort them out. Fortunately, the most clinically significant species can frequently be identified by a combination of **hemolysis reaction** (see p. 76) and serological reaction.

The serological reactions identify surface carbohydrates and proteins that act as antigens. Rebecca Lancefield identified 20 antigens over many years of work during the 20th century. The organisms are assigned to a **Lancefield group** based on their antigen(s). While Lancefield's work used precipitation reactions, modern serotyping utilizes latex agglutination kits (Fig. 8.16).

8.14 Microhemagglutination Test for *Treponema pallidum* (MHA-TP) ● Coating RBCs with *Treponema pallidum* antigens is the basis for this hemagglutination test. Because the antigens do not naturally cause agglutination, this test is an example of an indirect agglutination. The top row consists of serially diluted standards to act as positive (reactive) controls. A positive result is evidenced by a smooth mat of cells in the well (as in well A1). A negative result is a button of cells (as in well A12). Patient samples are in rows B and C. Patients B1, B3, B4, and B5 test positive for syphilis antibodies; patient B2 is negative. Samples in row C correspond to samples in row B and are negative (nonreactive) controls. In each case, the patient's serum is mixed with unsensitized RBCs to assure that agglutination (in Row B) is actually due to reaction with *Treponema* antibodies and not the RBCs themselves.

Examples of clinically important Lancefield groups are group A streptococci (GAS), which is mainly β-hemolytic *S. pyogenes* (see p. 219), the causative agent of streptococcal pharyngitis and scarlet fever; group B streptococci (GBS), which is mainly β-hemolytic *S. agalactiae* (see p. 216), the causative agent of neonatal sepsis; and group D streptococci, which includes *Enterococcus faecium*[1] (varied hemolytic reactions and diseases) and *S. bovis* (varied hemolytic reactions and diseases). ●

[1] The genus *Enterococcus* was formed in 1984 and removed a handful of species from the genus *Streptococcus* that were informally referred to as enterococci. At the same time, the lactococci were placed in the new genus *Lactococcus*.

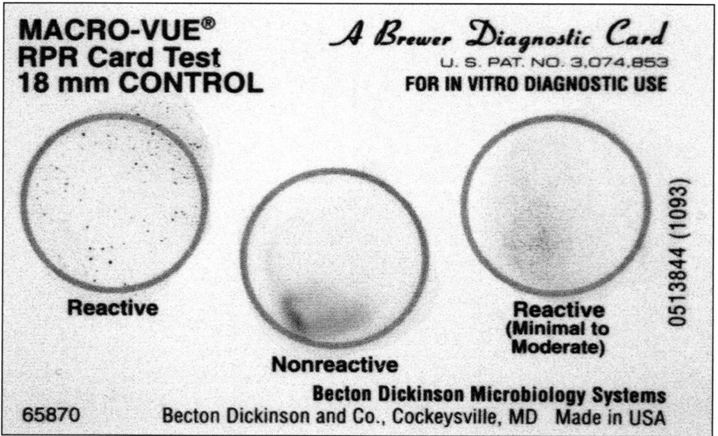

8.15 Rapid Plasma Reagin (RPR) Test for Syphilis ● The rapid plasma reagin (RPR) test for syphilis relies on antigen-coated particles interacting with reagin (an antibody-like substance) present in an infected individual's plasma. This control card shows the range of possible reactions and must be run with every set of tests.

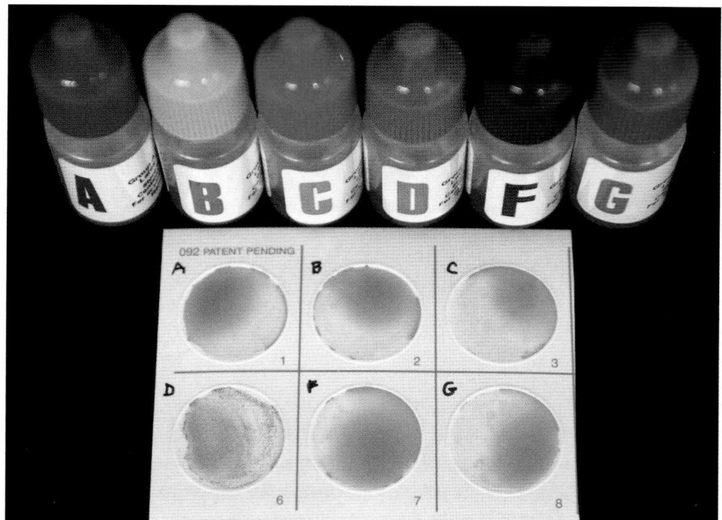

8.16 Lancefield Group Latex Agglutination Test for *Streptococcus* Species ● Shown is a Lancefield group latex agglutination testing kit. Only antisera for the six medically most important groups are included. The α-hemolytic isolate tested negative for all but Group D antigen and was provisionally identified as *Enterococcus* or *Streptococcus bovis*. Further testing was not done. The D antigen is a teichoic acid located between the cell membrane and cell wall. Testing kit is available from Hardy Diagnostics, https://catalog.hardydiagnostics.com/cp_prod/Hardy_Part_Description.aspx

Enzyme Immunoassay (EIA)/ Enzyme-Linked Immunosorbent Assay (ELISA)

Purpose

EIA and ELISA are two names for essentially the same procedure and can be used to detect the presence and amount of either antigen or antibody in a sample. In an antibody capture EIA, the test is used to screen patients for the presence of HIV antibodies (4 to 8 weeks after exposure), rubella virus antibodies, West Nile virus antibodies, and others.

An antigen capture EIA is used to detect hormones (such as HCG in some pregnancy tests and LH in ovulation tests), drugs, and viral (e.g., HIV p24 antigen within 4–8 weeks after exposure) and bacterial antigens.

Principle

As with other serological tests, EIAs can be used to detect antigen in a sample or antibody in a sample (Fig. 8.17). If antigen in a sample is to be identified, the system must have purified antibody to that antigen (**homologous antibody**). If the goal is to identify antibody in the sample, the system must have purified **homologous antigen**. In both systems, a **conjugate** (secondary) antibody with an attached enzyme acts as an indicator of antigen–antibody reaction.

The **direct EIA** detects the presence of antigen in a sample. A microtiter well[2] is first coated with homologous antibody specific to the antigen being sought. This is followed by coating the well with a blocking agent to prevent nonspecific binding of the following components to the well, especially the conjugate antibody.

The sample being assayed is added to the well. If the antigen is present, it will react with the antibody coating the well; if none is present, no reaction occurs. In this form of EIA, the conjugate antibody is also specific for the antigen. When added to the well, it binds to the antigen, if present.

After allowing time for the conjugate antibody to react with the antigen, the well is washed to remove any unbound conjugate antibody (which would produce a false positive when the substrate is added in a negative test). Lastly, addition of the enzyme's substrate produces a color change if the enzyme/conjugate antibody/patient antigen complex is in the well, and it is seen as a positive result.

The **indirect EIA** detects the presence of antibody in a sample. In this type of EIA, the microtiter well is lined with antigen specific to the antibody being sought and a blocking agent. The sample being assayed is added to the well and, if the antibody is present, it will react with the antigen coating the well; if none is present, no reaction occurs.

In this case, the conjugate antibody is an ***anti-human immunoglobin antibody***—its antigen is actually an antibody! When added to the well, it binds to the antibody in the sample, if present. As with the direct EIA, unbound conjugate antibody must be washed away to prevent a potential false-positive result. Enzyme substrate is added and a color change indicates a positive test.

Figures 8.18 and 8.19 illustrate a quantitative form of indirect ELISA, i.e., it is used to determine both the presence of antibody in a sample and the amount. Figure 8.20 illustrates a different version of the ELISA technique—a home pregnancy test—which screens for the presence of **human chorionic gonadotropin** found in pregnant women.

[2] A microtiter plate with 96 wells is shown in Figure 8.18.

Direct EIA (ELISA) method

Step 1. The direct EIA detects the presence of *antigen* in a sample. Antibody (antiserum) specific for the antigen is coated (adsorbed) onto the wall of a microtiter well.

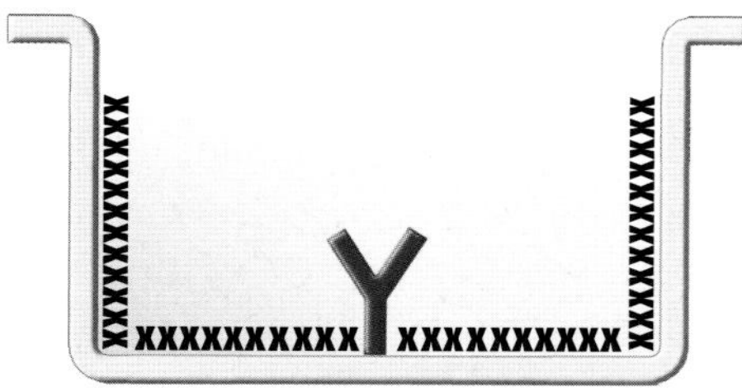

Step 2. A blocking agent, such as gelatin or albumin, is added to coat the well's surface to prevent its interacting nonspecifically with any test reagents.

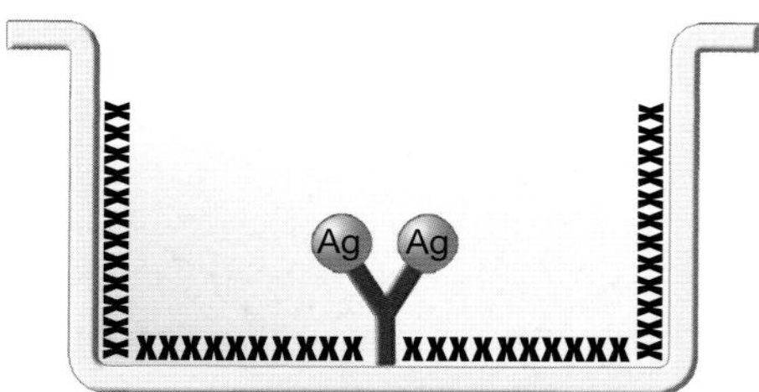

Step 3. The sample is added. If the *antigen* is present, it will bind to the antibody coating the well.

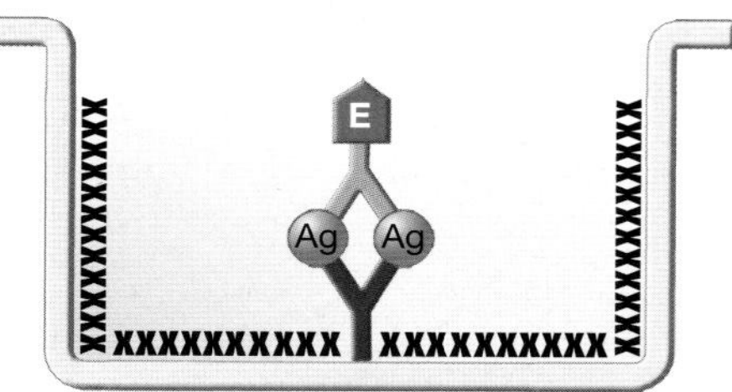

Step 4. A second (conjugate) antibody with an attached enzyme specific for the same antigen is added. Unbound conjugate antibody is washed away.

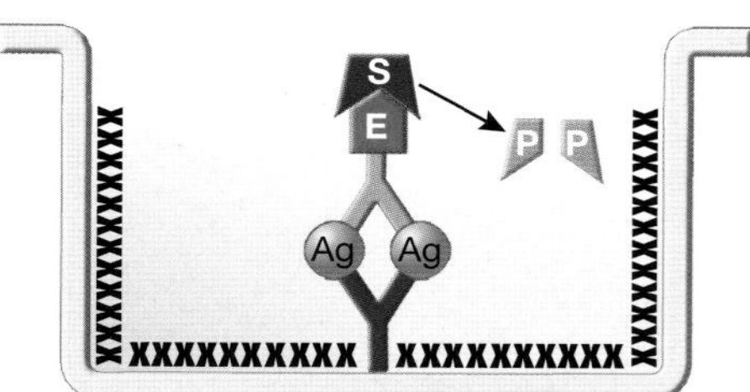

Step 5. Substrate for the enzyme is added. Conversion of substrate to product is evidenced by a color change. A color change means the sample has the antigen; no color change is a negative result.

Indirect EIA (ELISA) method

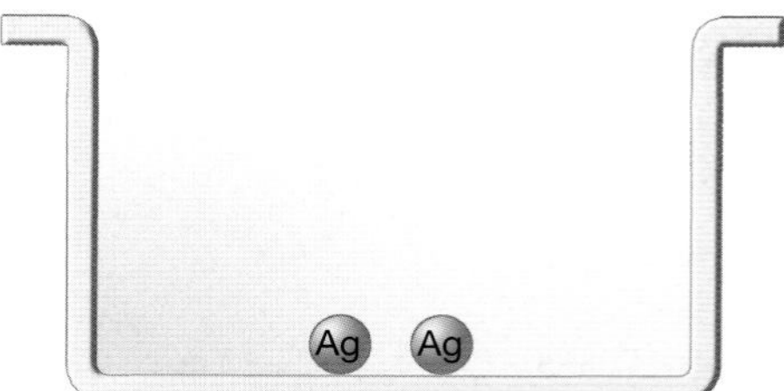

Step 1. The indirect EIA detects the presence of *antibody* in a sample. The antigen specific for that antibody is coated (adsorbed) onto the wall of a microtiter well.

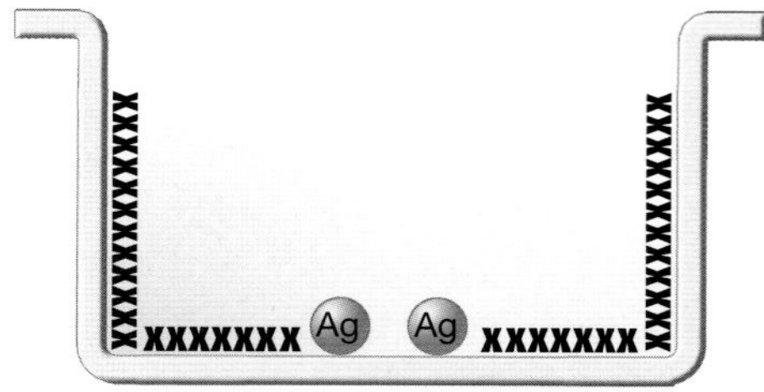

Step 2. A blocking agent, such as gelatin or albumin, is added to coat the well's surface to prevent its interacting nonspecifically with any test reagents.

Step 3. The sample is added. If the *antibody* is present, it will bind to the antigen coating the well.

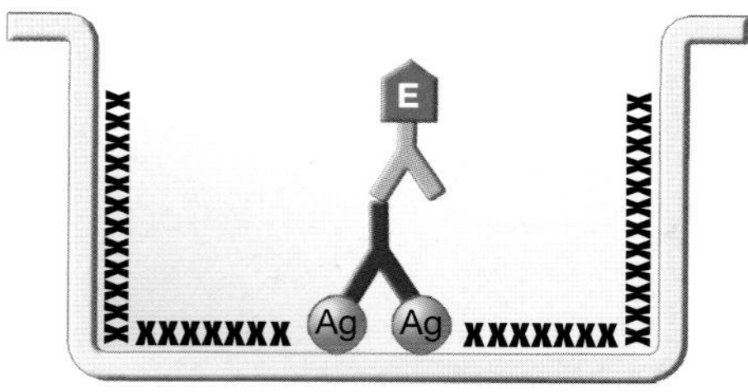

Step 4. An anti-human immunoglobin (IgG) antibody with an attached enzyme is added. Unbound conjugate antibody is washed away.

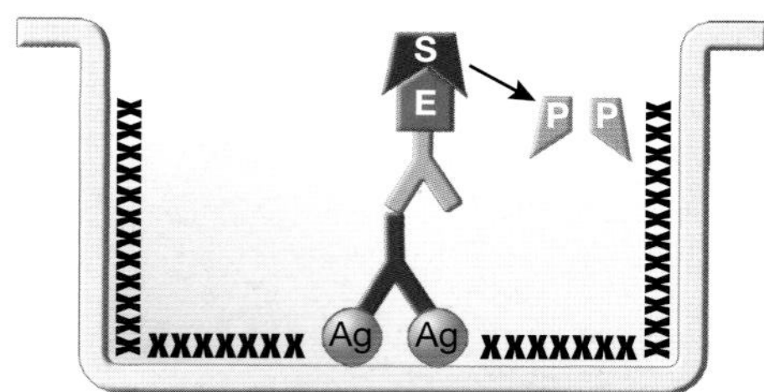

Step 5. Substrate for the enzyme is added. Conversion of substrate to product is evidenced by a color change. A color change means the sample has the antibody; no color change is a negative result.

8.17 Direct and Indirect EIAs ● A direct EIA is used to identify antigen in a sample. An indirect EIA identifies antibody in a sample. In both examples, a positive test is illustrated.

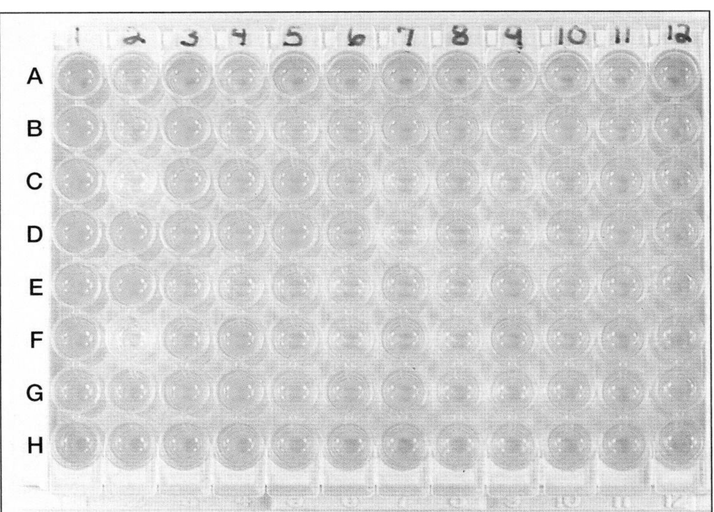

8.18 Quantitative Indirect ELISA for HIV Antibodies ● In this ELISA, a dark-yellow color indicates a negative reaction. The lighter the color, the higher the antibody titer. Color is read by a photometer, which is much more sensitive than the human eye (Fig. 8.19), and results are fed into a computer. A variety of controls are also used. Serially diluted antibody samples of known concentration are in column 1, rows A through H and column 2, rows A through C. Absorbance values from these are used to develop a standard curve correlating antibody titer with absorbance. Patient samples are in the other wells. Each patient's antibody titer can then be determined by comparison with the standard curve.

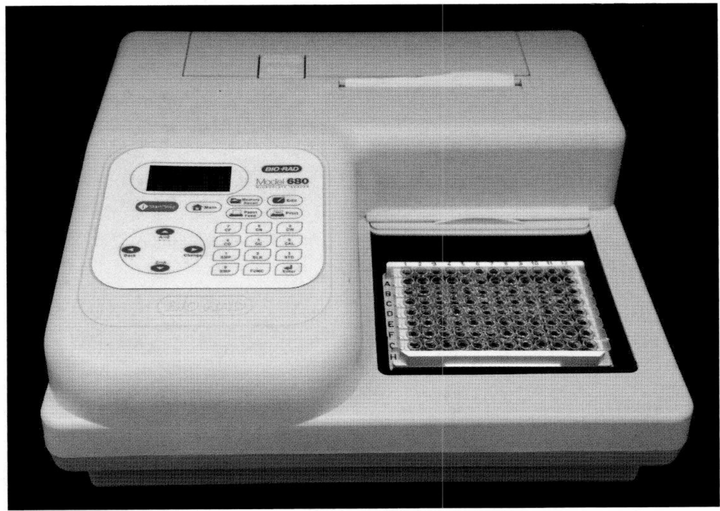

8.19 Microplate Reader ● Samples and reagents are placed in the wells of a microplate and allowed to react. Modern molecular techniques require very small volumes and one microplate replaces 96 conventional test tubes. The plate is placed on a tray (as shown) and is then drawn to the left into the machine (like a CD or DVD). Once inside, absorbance readings can be taken and recorded.

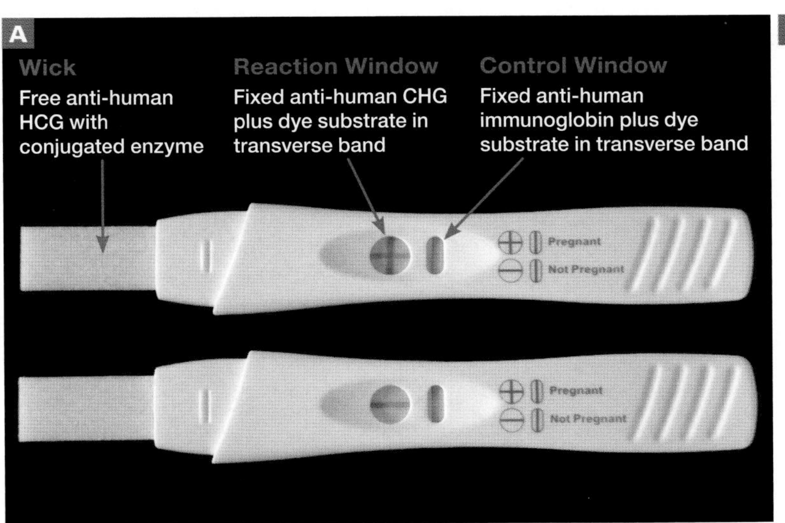

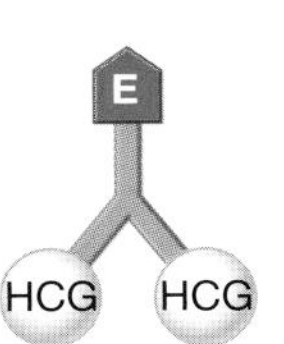

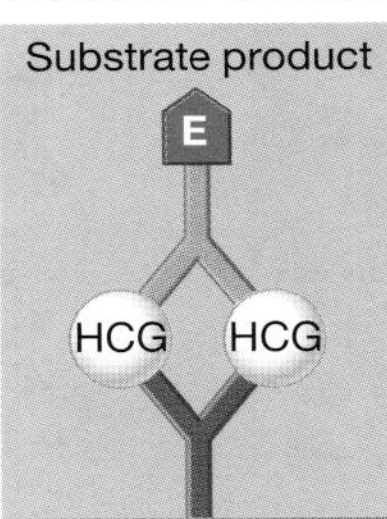

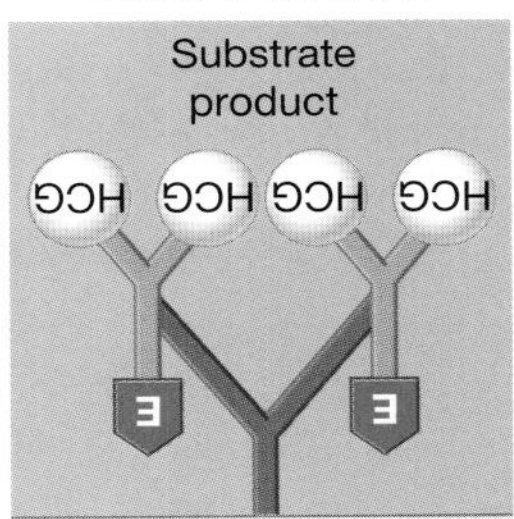

8.20 Membrane ELISA–A Home Pregnancy Test ● Fixing of antibodies to a membrane (rather than a microtiter dish well) and other advances have made ELISAs easier to perform and available to home consumers. (**A**) Shown is a home pregnancy test that identifies human chorionic gonadotropin (HCG), an antigen present in the urine of pregnant women. A "+" sign shows up in the reaction window in a positive test (above, who happens to be Cole Matthew Leboffe checking in). A "–" appears in the reaction window for a negative result (below). Positive or negative, a single line should appear in the control window to indicate the test is working. (**B**) Three different antibodies are used in this ELISA: free (unattached) anti-human HCG antibody in the wick, fixed anti-human HCG antibody in the reaction window, and fixed anti-human immunoglobin antibody in the control window. Shown are their conditions in a positive reaction. Here's how a positive reaction happens. A urine sample applied to the wick migrates upward by capillary action. In a positive reaction, the woman's urine has HCG and it binds to the free anti-human HCG antibodies in the wick (left), which are then carried to the reaction (round) window. More anti-HCG antibodies are attached to the transverse band (but aren't visible yet) in the reaction window and they bind to the HCG bound to the free antibodies carrying it (middle). This produces antibody sandwiches with HCG in the center. It also brings the enzyme on the free antibodies and their dye substrates together. The indicator reaction occurs, producing a color change and a "+" sign in the window. The woman is pregnant. Not all free antibodies bind to the fixed antibodies in the round window. They continue up the wick and encounter fixed anti-human immunoglobin antibodies and dye. Binding of the free conjugate antibodies (attached to HCG or not) results in a color change in the band and acts as a control (right). Note the color change in both the reaction and control windows. Be aware that different home pregnancy tests use different symbols to indicate positive and negative results. Read the package insert before use.

Fluorescent Antibody (FA) Technique

Purpose

Like most serological tests, the **fluorescent antibody** (FA) technique can be used to identify the presence of either antigen or antibody in a sample. **Direct tests** (**DFA**) identify the presence of antigens; **indirect tests** (**IFA**) detect the presence of antibody in a sample. FAs are useful in diagnosing many viral infections as well as certain parasitic diseases.

Principle

Fluorescent antibodies are conjugated with a fluorescent dye such as **fluorescein isothiocyanate** (**FITC**) **dye**, which fluoresces when illuminated with UV light. In a DFA (Fig. 8.21), a patient sample containing the suspected antigen is fixed to a microscope slide. The fluorescent antibody is added and allowed to react with the antigen.

After rinsing to remove unbound antibody (which could produce a false-positive result), the slide is viewed with a fluorescent microscope with a UV light source. If the suspected antigen is present, the labeled antibodies will have bound to it and will emit an apple green color (Fig. 8.22).

IFAs (Fig. 8.23) are used to detect antibodies in a sample. In this form of the test, the microscope slide comes prepared with the specific antigen fixed to it. Dilutions of the patient's sample are added to several slides and given time to react with the antigen.

The FITC-labeled antibody is an anti-human immunoglobin antibody, so if patient antibody is bound to antigen on the slide, the fluorescent antibody will bind to it. After rinsing to remove any unbound fluorescent antibodies, the slide is viewed under a fluorescent microscope with a UV light source. If the suspected antibody is present, the labeled antibodies will fluoresce and appear apple green (Figs. 8.23 and 10.9).

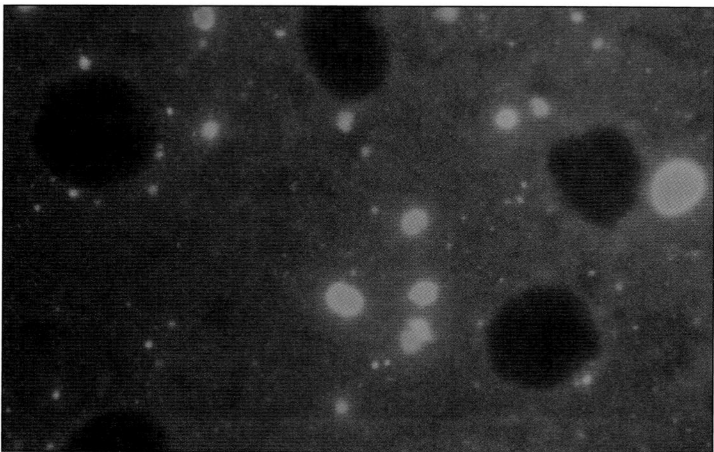

8.22 Fluorescent Micrograph of a Positive DFA for Rabies Virus ● The rabies virus is a negative-sense, single-stranded RNA virus that replicates in host nervous tissue, such as the brainstem. This brainstem preparation was from a dog suspected of having rabies. Antirabies virus antibodies conjugated with FITC dye are specific for rabies ribonucleoprotein, which forms characteristic inclusions in infected cells. When examined with a UV microscope, the inclusions show as light-green ovals in the field. A negative result would look black.

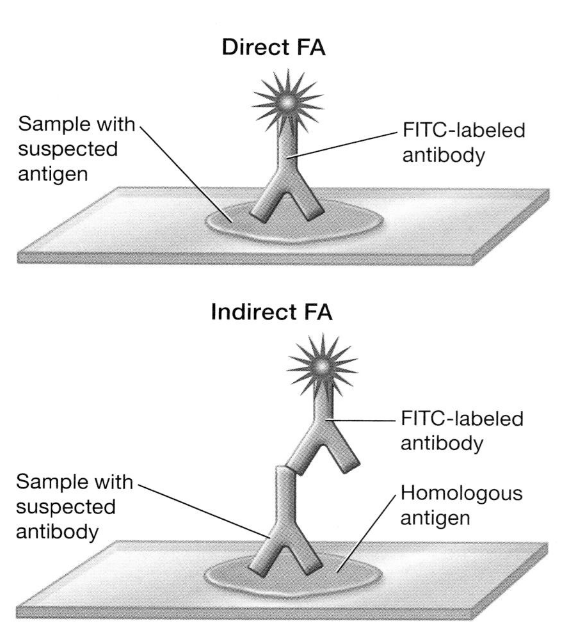

8.21 Direct and Indirect Fluorescent Antibody Techniques ● The DFA and IFA are used to identify antigens and antibodies in a sample, respectively.

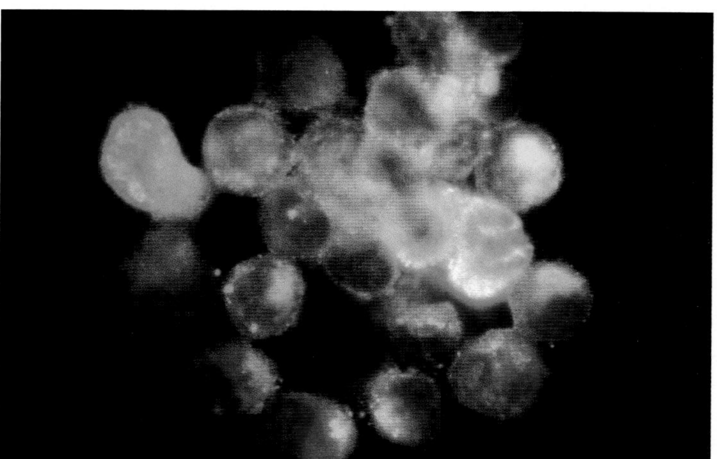

8.23 Positive IFA for Influenza B Virus ● Influenza virus is a negative-sense, single-stranded RNA virus with eight gene segments. Replication is within the nucleus of an infected cell. The influenza IFA comes as a kit with a mixture of influenza-infected cells and uninfected cells on prepared slides. A patient sample suspected of having anti-influenza B antibodies is incubated with the cells. After washing, anti-human IgG antibodies conjugated with FITC are incubated with the cells. If the patient sample has anti-influenza B antibodies, the conjugate antibodies will bind to them and fluoresce an apple-green color under UV illumination. Uninfected cells appear reddish due to counterstaining with Evans blue dye.

SECTION 9

Molecular Techniques

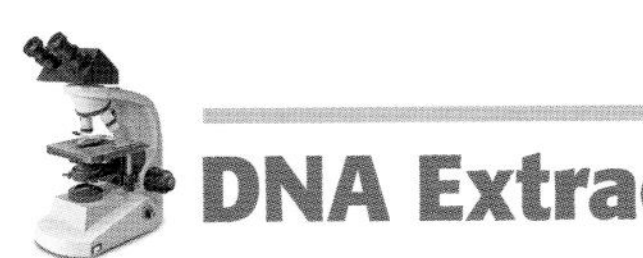

DNA Extraction

Purpose

DNA extraction is the starting point for many lab procedures, including DNA sequencing, species identification, and genetic engineering.

Principle

DNA extraction from cells is surprisingly easy and, classically, occurs in three basic stages.

1. A detergent (e.g., sodium dodecyl sulfate—SDS) is used to lyse cells from a pure culture and release cellular contents, including DNA.
2. This is followed by a heating step (at approximately 65°C–70°C) that denatures proteins (including DNases that would destroy the extracted DNA) and other cell components. Temperatures higher than 80°C will denature DNA, and this is undesirable. A protease and RNase also may be added to remove proteins and RNA, respectively. Other techniques for purification may also be used.
3. Finally, the water-soluble DNA is precipitated in cold alcohol as a whitish, mucoid mass (Fig. 9.1). This mixture is commonly centrifuged to separate the solid DNA from the remaining soluble cell contents.

After extraction, an ultraviolet spectrophotometer (Fig. 9.2) can be used to estimate DNA concentration in the sample by measuring absorbance at 260 nm, the optimum wavelength for DNA absorption. As a rough approximation, 50 μg/mL of double-stranded DNA (dsDNA) produces an absorbance of 1 at 260 nm (A_{260}). To calculate the DNA concentration (X), use an equation of proportions and plug in the sample's A_{260} value:

$$\frac{X\ \mu g/mL\ dsDNA}{A_{260}} = \frac{50\ \mu g/mL\ dsDNA}{1.0}$$

$$X\ \mu g/mL\ dsDNA = 50\ \mu g/mL\ dsDNA \times Abs_{260}$$

Reading absorbance at 280 nm and calculating the following ratio can determine purity of the sample:

$$\text{Sample Purity} = \frac{Abs_{260\,nm}}{Abs_{280\,nm}}$$

If the sample is reasonably pure nucleic acid, the ratio will be about 1.8 (between 1.65 and 1.85). Because protein absorbs maximally at 280 nm, a ratio of less than 1.6 is likely because of protein contamination, whereas higher than 1.8 may indicate RNA contamination. If purity is crucial, the DNA extraction can be repeated. If the ratio is greater than 2.0, the sample is diluted and read again.

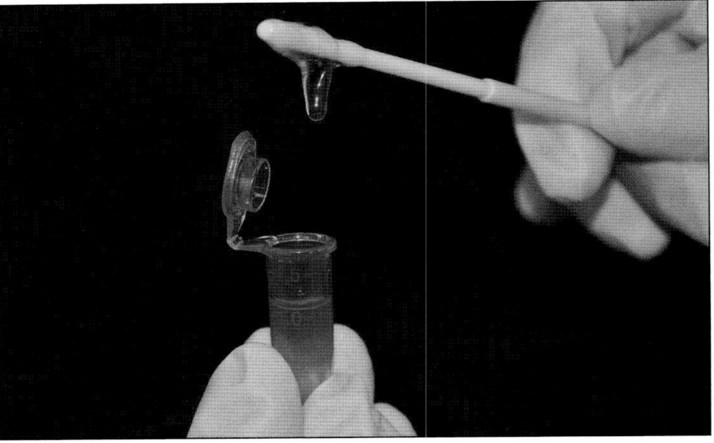

9.1 Precipitated *E. coli* DNA • This *E. coli* DNA has been spun onto the handle of a disposable loop.

Automated DNA extraction and purification systems are now available that perform all steps directly from patient samples using a number of different mechanisms. As in the traditional method, sample preparation begins with cell lysis, deproteinization, and precipitation, but these are automated.

The novel component is how the DNA is separated from the rest of the lysate. A majority of systems utilize magnetic beads made of an iron-oxide core surrounded by silica and a coating of carboxylic acid. The carboxylic acid binds to DNA (or RNA, depending on the bead used and the goal of the extraction) in the lysate. Then, a magnetic field is used to pull the beads out of the solution. At the same time, proteins and other cell components are washed away leaving behind the magnetic beads covered in DNA. Lastly, a buffer is used to release the DNA from the beads.

This technology also provides a means for isolating specific and relevant nucleic acid sequences by coating the beads with complementary ssDNA to which the target DNA can bind. Nucleic acid extraction using magnetic fields is rapid and incredibly precise, remarkably without the need for a pure culture.

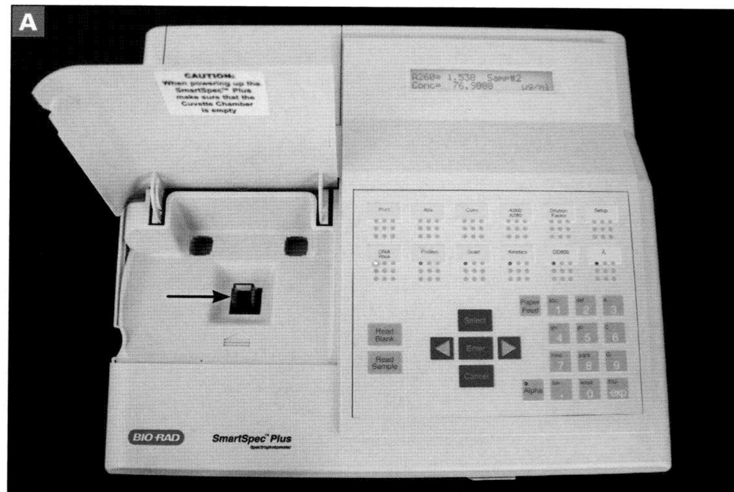

9.2 Ultraviolet Spectrophotometer (**A**) A UV spectrophotometer can be used to determine DNA concentration and purity. A quartz cuvette containing the DNA extract is shown in the sample port (**arrow**). Absorbance readings and other information are displayed in the green window at the upper right. (**B**) In this close-up of the display window we see that this specimen has an absorbance at 260 nm ($A_{260\ nm}$) of 1.538 absorbance units. Because an $A_{260\ nm}$ of 1.0 is equal to 50 μg/mL of dsDNA, this specimen has a concentration of 76.9000 μg/mL. Absorbance also can be used to determine purity of the sample. A relatively pure DNA sample will have an $A_{260\ nm}/A_{280\ nm}$ value of close to 1.8.

DNA Sequencing

Purpose

The key to unlocking DNA's secrets lies in determining its nucleotide sequence. Identifying, then comparing DNA sequences has revealed information about gene functions and what defines nonstructural DNA segments, such as promoters, operators, and introns. Sequencing has also been used to compile **genomic libraries** of known organisms.

Comparing DNA sequences from newly discovered species to sequences in genomic libraries provides information about relatedness and as such is a valuable taxonomic tool. Alternatively, comparison of species specific DNA sequences from patient or environmental isolates to genomic libraries provides a means of rapid identification.

Principle

In vivo DNA synthesis requires a template strand, a primer, DNA polymerase, and a pool of all four deoxyribonucleotide triphosphates[1] (dATP, dCTP, dTTP, and dGTP) (Fig. 9.3). The primer binds to the template, then DNA polymerase adds nucleotides to the 3' hydroxyl end of the primer (and each subsequent nucleotide as the chain grows) in the order determined by base pairing with the template strand. The new bond is formed between the 3' hydroxyl of the replicating strand and the 5' phosphate of the nucleotide being added.

[1] Deoxyribonucleotides are incorporated into the growing DNA chain by removal of the two phosphates to produce the nucleotide structure as it is found in the DNA chain; that is, the sugar-phosphate backbone, with the nitrogenous base extending into the center of the double helix. Removal of the two phosphates provides the energy required for DNA synthesis. The symbol dNTP is used to designate any of the four deoxyribonucleotides in triphosphate form.

In 1980, Frederick Sanger shared the Nobel Prize in Chemistry for developing a method to determine the nucleotide sequence in a DNA molecule. It involves in vitro DNA replication, which uses the basic materials required by in vivo replication as well as other ingredients essential to the process.

The key to understanding Sanger's process is recognizing the importance of the 3' hydroxyl to chain elongation: without it, extension of the replicating strand cannot occur. This fact forms the basis of Sanger's original **dideoxy method** (also known as the **chain termination method**) of DNA sequencing, in which **dideoxyribonucleotides** (ddNTPs) lacking the 3' hydroxyl (Fig. 9.3C) are used along with "normal" dNTPs.

DNA polymerase cannot distinguish between the two and inserts either one (e.g., dATP or ddATP if a thymine nucleotide [T] is in the template strand). But when a ddNTP is added replication stops, resulting in a partially replicated DNA strand ending with a ddNTP. If the mixture of dNTPs and ddNTPs is in proper proportions (about 10:1), replicated strands of every possible length are made because of the random insertion of the ddNTPs.

Using gel electrophoresis (p. 147) these can be separated by size (length). Then, by identifying the ddNTP at the end of each successively longer fragment, the nucleotide sequence of the replicated DNA strand can be determined.

Sanger's original method required that replication occur in four different tubes, each with a supply of one ddNTP, and primers labeled with radioactive phosphates. Replication in each tube produced fragments ending with the same ddNTP and of all lengths dictated by the template. Each sample was then run in its own lane of the electrophoresis gel and fragments were separated by length.

The radioactive primers revealed the location of each fragment in the gel, which was used to determine the nucleotide sequence. While novel and ingenious, this method was labor intensive and involved handling (and disposing of) radioactive chemicals.

Sanger's method has now been automated and is referred to as **next generation sequencing.** Instead of labeling the primer, each of the four ddNTPs is labeled with a differently colored fluorescent molecule. In this method, all four ddNTPs can be mixed in a single tube because the fragments are labeled (and thus identified) with a different color depending on their terminating ddNTP.

Electrophoresis is done in a single lane within a tube through which a laser beam is passed. As each fragment migrates past the laser beam during electrophoresis, it emits the color appropriate to its labeled ddNTP. The color is recorded and the nucleotide sequence is determined from the color sequence (Fig. 9.4).

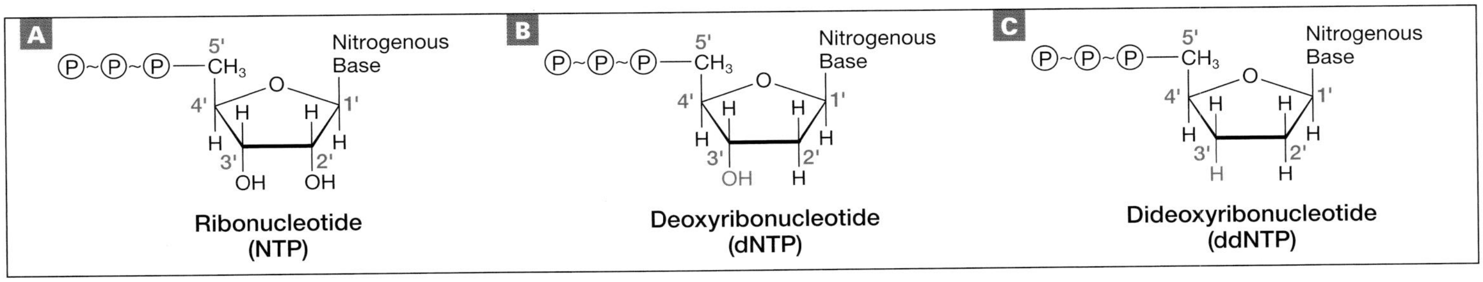

9.3 Nucleotide Structures ● Nucleotides consist of a 5-carbon sugar (either ribose or deoxyribose), one of four nitrogenous bases, and initially three phosphates. **(A)** Synthesis of nucleotides "from scratch" begins with ribulose-5-phosphate produced in the pentose-phosphate pathway (see Appendix, Fig. A.4, p. 328), so initially all nucleotides are ribonucleotides. **(B)** Deoxyribonucleotides (dNTP) are produced by a reduction of the 2' carbon in ribose, which removes an oxygen (hence, "deoxyribose"). During RNA or DNA synthesis, a pyrophosphate (two of the 5' phosphates) is removed and the bond energy released is used to create the bond between the new nucleotide's remaining 5' phosphate and the 3' hydroxyl of the sugar on the 3' end of the growing chain. **(C)** Dideoxyribonucleotides (ddNTP) are produced in vitro by reducing the 3' carbon, which leaves it without the critical hydroxyl group necessary for insertion of the next nucleotide during elongation. During in vitro sequencing replication, insertion of a ddNTP acts as a "cap" and elongation stops.

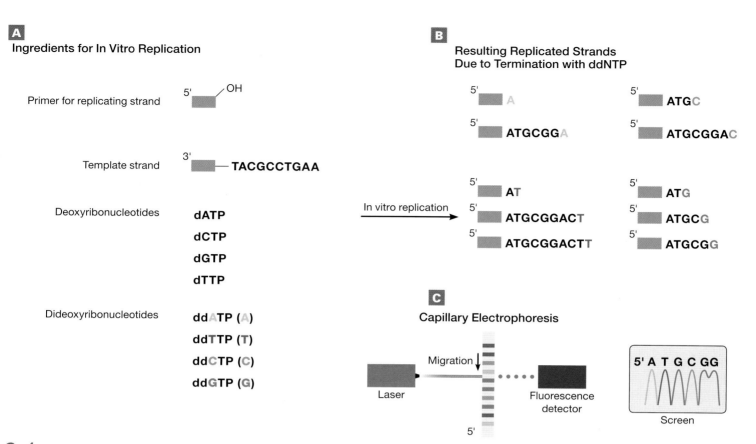

A Ingredients for In Vitro Replication

Primer for replicating strand: 5' ▬ OH

Template strand: 3' ▬ TACGCCTGAA

Deoxyribonucleotides: dATP, dCTP, dGTP, dTTP

Dideoxyribonucleotides: ddATP (A), ddTTP (T), ddCTP (C), ddGTP (G)

In vitro replication →

B Resulting Replicated Strands Due to Termination with ddNTP

5' A
5' ATGC
5' ATGCGGA
5' ATGCGGAC
5' AT
5' ATG
5' ATGCGGACT
5' ATGCG
5' ATGCGGACTT
5' ATGCGG

C Capillary Electrophoresis

Laser
Migration
5'
Fluorescence detector
5' A T G C GG
Screen

9.4 Next Generation DNA Sequencing • (A) The ingredients for next generation sequencing are shown. The reaction mixture includes an RNA primer, template ssDNA, and a supply of dNTPs. Also present are ddNTPs labeled with fluorescent dyes, with each color corresponding to one of the four nitrogenous bases. In this illustration, we've chosen yellow for ddATP, red for ddTTP, blue for ddCTP, and green for ddGTP. This differential labeling has streamlined DNA sequencing from Sanger's original method because it requires a single reaction tube rather than four separate ones. (B) The same principle of chain termination is used (see text) and the partially replicated strands are separated by electrophoresis. (C) However, because the ddNTPs are color coded, the partially replicated fragments can be run in a single electrophoresis lane within a capillary tube rather than the four lanes in Sanger's original method. As the fragments migrate, they pass through a laser beam that causes each dye to fluoresce its color, which is sensed by a detector and displayed on a computer monitor. Reading the colored peaks from left to right provides the sequence of the strand complementary to the template DNA.

Polymerase Chain Reaction

Purpose

Polymerase chain reaction (PCR) is a relatively simple and convenient method of rapidly **amplifying** (making more copies of) DNA. Once multiple DNA copies are made, they can be used in various ways, ranging from research to diagnostics to forensics.

Principle

Polymerase chain reaction (PCR) is a Nobel prize-winning technique conceived in 1983 by Kary Mullis. PCR is used to amplify a desired gene or other short DNA fragment. The process occurs in vitro and requires a dsDNA template, DNA polymerase, two ssDNA primers, a supply of deoxyribonucleotides in triphosphate form[2] (dATP, dCTP, dGTP, and dTTP), and other essential components, such as buffers to maintain the appropriate pH, magnesium as a cofactor for DNA polymerase, and other salts to create an osmotic balance.

In principle, the process is pretty simple and occurs in three general steps: **denaturation** (dsDNA strand separation), **annealing of primers**, and **extension** (DNA replication). In a living cell (in vivo), these activities are supported by a series of stabilizing enzymes. In a tube (in vitro), they are driven by temperature changes.

---

[2] Being in triphosphate form allows nucleotides to supply the energy required for their insertion. dATP is the deoxyribose form of the familiar ATP (which has ribose as its sugar), but it has the same high energy bonds, as do the other dNTPs.

The process is illustrated in Figure 9.5. First, raising the temperature to 92°C–96°C for a few seconds to a few minutes denatures the dsDNA template by breaking hydrogen bonds holding it together. The duration is based on the plastic tube's thickness and the efficiency of the machine performing the heating.

Next, lowering the temperature to 45°C–65°C allows the two primers to base pair with their different target sequences on opposite ssDNA template strands. The actual temperature is determined empirically and depends on primer lengths and their nucleotide composition (G+C percent)[3].

Lastly, the temperature is raised to 72°C for a few seconds to a few minutes, which allows attachment of DNA polymerase to the 3' ends of the primers and strand elongation begins.

The duration of elongation depends on the product's expected nucleotide length. As a general rule, 60 seconds per kilobase (kb) of DNA being synthesized is a good starting point for the protocol. 72°C is the optimal temperature for the commonly used enzyme *Taq* DNA polymerase to function. This enzyme was isolated from the thermophilic bacterium *Thermus aquaticus* (hence, *Taq*), an inhabitant of volcanic hot springs, and it can withstand the high temperatures of PCR. Imagine trying PCR with a DNA polymerase isolated from a mesophile like *E. coli*.

These three steps complete the first cycle of amplification, or round of replication. The process then repeats itself (Fig. 9.5), with each round doubling the single-stranded template molecules. Assuming all the products of one cycle are used as templates for the next cycle, the original two DNA target strands can be amplified a millionfold in a few hours. (20 replication cycles from a single dsDNA template will produce 2^{20} = 1,048,576 double-stranded copies of the original!)

Although Kary Mullis' original protocol involved moving the samples between water baths to achieve the temperature changes at each step, researchers now use a machine called a **thermal cycler** that can be programmed to rapidly perform these temperature shifts (Fig. 9.6).

Initially, PCR was used to simply amplify pieces of DNA so that researchers would have more DNA to use during experiments. The advancing trend of identifying and classifying microbes by their nucleotide sequences has broadened PCR's usefulness. These rely on the ability to identify unique sequences to which specific primers can bind.

For example, PCR of a patient's isolate would use primers specific to the suspected pathogen. Recovery of a PCR product signifies the pathogen is present because the primers bound, leading to replication. If no PCR product is made, then the patient is not infected by that pathogen.

Application of this technology allows identification of difficult-to-grow viral or bacterial agents that don't lend themselves to traditional identification methods (see *Mycobacterium tuberculosis*, p. 204.)

[3] Recall that G-C pairs are held together by three hydrogen bonds and A-T pairs are held together by two. Therefore, annealing temperature is affected by nucleotide composition. The temperature required to separate 50% of the nucleotides within dsDNA is called the melting temperature (T_m). The more G-C pairs, the higher the T_m value.

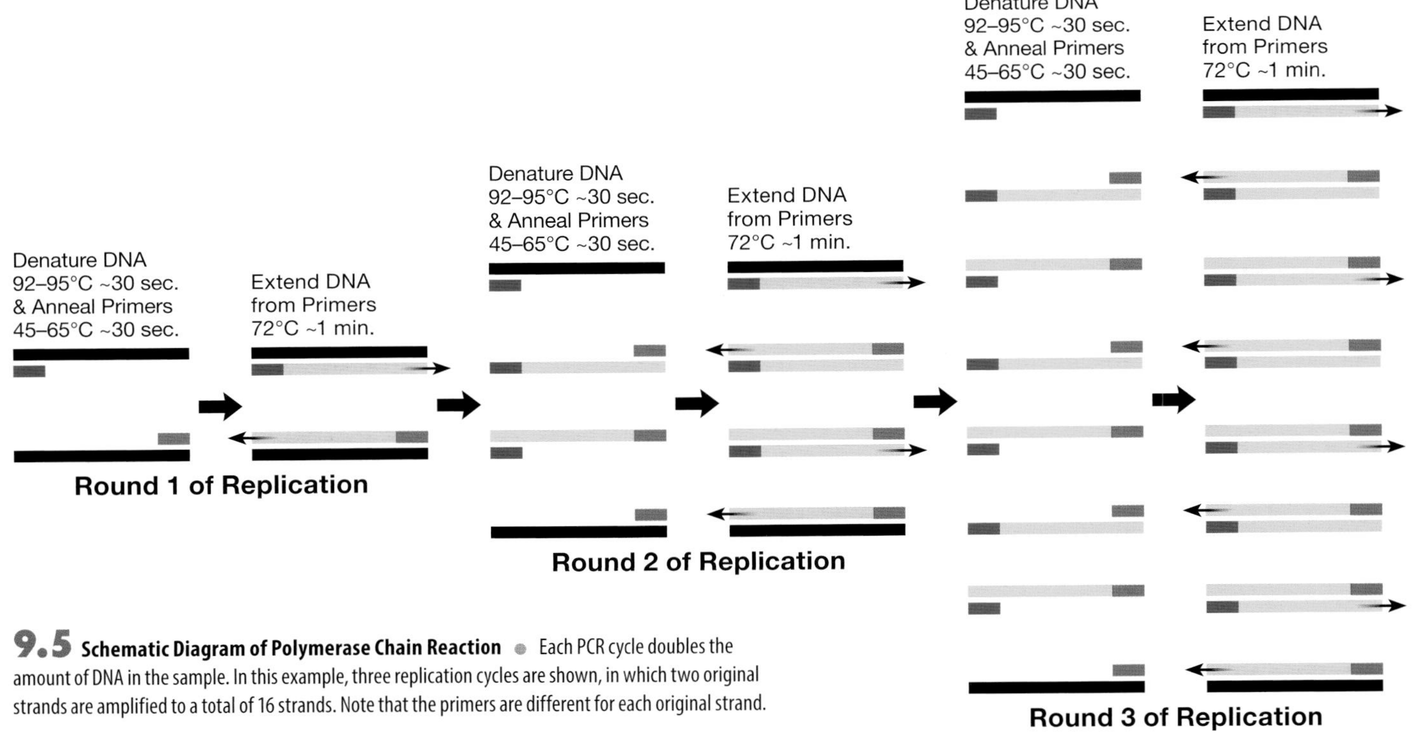

9.5 Schematic Diagram of Polymerase Chain Reaction • Each PCR cycle doubles the amount of DNA in the sample. In this example, three replication cycles are shown, in which two original strands are amplified to a total of 16 strands. Note that the primers are different for each original strand.

Although originally developed for DNA, PCR technology has now been extended to RNA applications, which is especially useful when working with RNA viruses. Reverse transcriptase is an enzyme produced by retroviruses that synthesizes DNA from an RNA template (p. 155).

In vitro, the DNA product is called complementary DNA (cDNA), which can then be amplified by reverse transcriptase PCR (RT-PCR). As an extension of RT-PCR, real time or quantitative PCR (qPCR) quantifies the cDNA by including a fluorescent dye in the reaction mixture that interacts with dsDNA by intercalation (wedging itself into the double helix). A modified thermal cycler with a fluorescence sensor is used (Fig. 9.7).

As a cycle proceeds, the dye intercalates in the dsDNA. A computer collects fluorescence readings after every cycle, with the fluorescence signal increasing in proportion to the amount of dsDNA produced. So, not only do the results (Fig. 9.8) illustrate the presence or absence of a particular gene, comparison to a dsDNA standard curve allows calculation of the sample's original mRNA quantity.

9.6 Thermal Cycler • Temperatures and durations of each step, and number of cycles can be programmed into the thermal cycler to automate PCR.

9.7 Quantitative Thermal Cycler • In qPCR, the reaction mixture used includes the usual ingredients, but also a fluorescent dye that attaches to dsDNA. As replication cycles proceed, the amount of attached dye increases. Fluorescence is monitored at prescribed intervals throughout and a computer compiles a graph like the one in Figure 9.8.

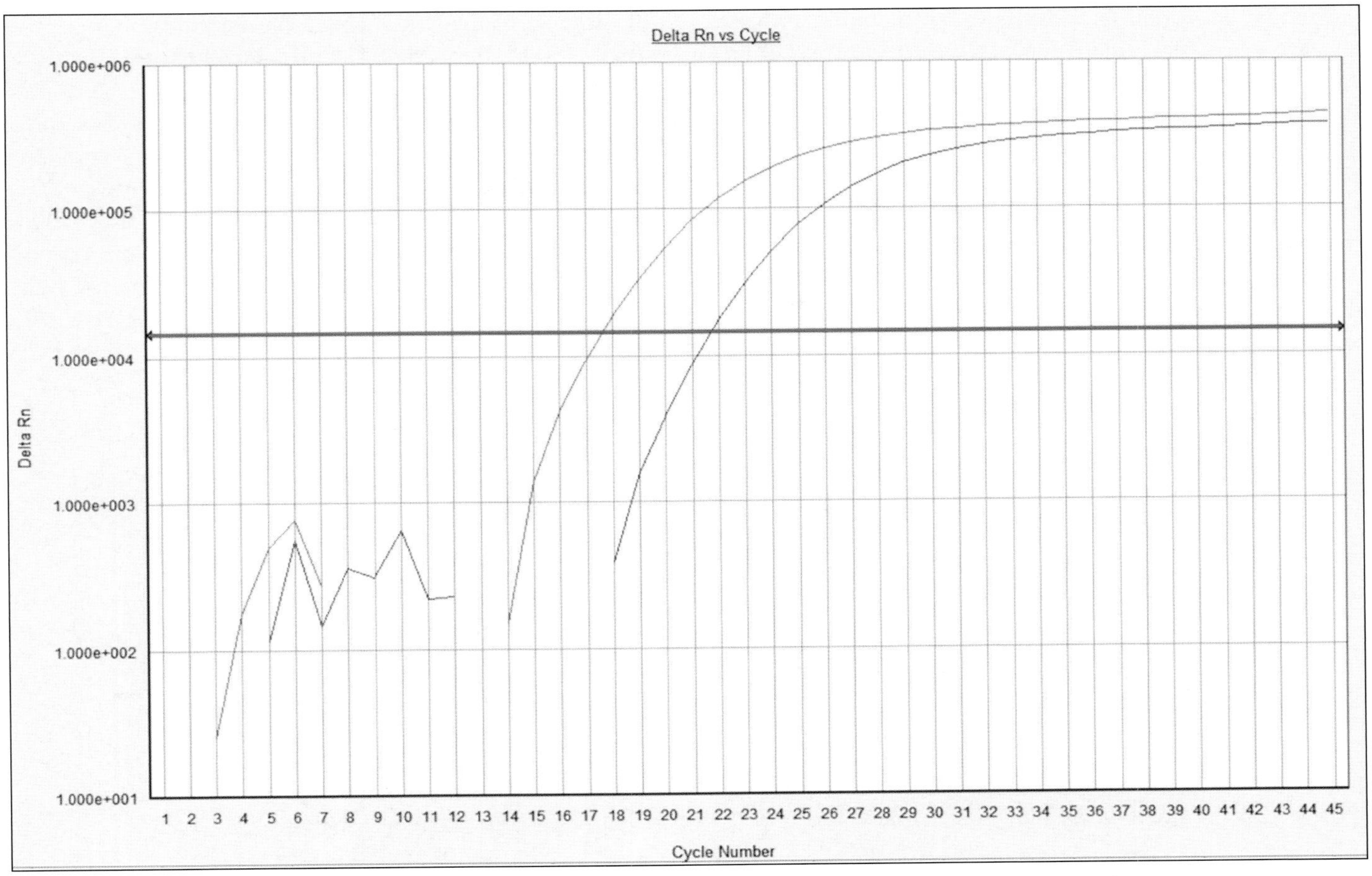

9.8 Representative qPCR Amplification Plot ● In this graph, the PCR cycle number is plotted on the x-axis and logarithmic change in the fluorescent dye is plotted on the y-axis (Delta Rn). The acceptable background fluorescence level is set by the user and is plotted as a horizontal green line across this graph. Only fluorescence values above the line are valid for interpretation. The first few cycles of the PCR reaction have fluctuations in fluorescence, representing some dsDNA production as well as the background fluorescence. This graph illustrates a test for the quantity of Norovirus RNA for two separate proteins in the patient's sample. (Norovirus is a Class IV +ssRNA virus; see p. 155 and Fig. 10.6 for more information.) That the two curves rise higher than the background fluorescence (roughly at cycles 18 and 22) indicates production of two PCR products from the patient's sample and is a positive result for a Norovirus infection. If the patient did not have a Norovirus infection, there would be no curves because there would be no initial templates for PCR in the sample.

Gel Electrophoresis and DNA Hybridization

Purpose

Gel electrophoresis is a technique in which molecules are separated by size and electrical charge in a gel matrix, which acts as a sieve (Fig. 9.9). Once separated, the fragments can be used in a number of ways.

Protein or nucleic acid gel patterns can be compared for taxonomic or identification purposes. Separated molecules can be removed from the gel and used in biochemical or genetic engineering studies. Other techniques, such as DNA fingerprinting and Southern, Western, and Northern blotting, begin with electrophoresis.

Principle

Electrophoresis gels are typically prepared from **agarose** or **polyacrylamide**, depending on the molecules to be separated. Agarose is used for large DNA molecules and polyacrylamide is used for small DNA molecules and proteins.

Gels are cast by melting and then pouring the liquid gel material into a mold where it polymerizes and produces a solid, but porous, gel slab a few millimeters thick. In addition, the gel is prepared with tiny wells at one end to hold the samples.

Once solidified, it is immersed in a buffered solution, such as TAE (Tris-acetate-EDTA) or TBE (Tris-borate-EDTA), to maintain proper electrolyte balance (Fig. 9.10). Precast gels are also available for purchase. Samples to be examined (either nucleic acid or protein) are loaded into the different wells and electrodes are attached to create an electrical field in the gel.

Under the electrical field's influence molecules in the samples migrate through the gel. Because of their negative charges, DNA and most proteins migrate toward the positive pole and they travel various distances due to differences in size and electrical charge.

At the end of the run, the gel is stained to show the location of the separated molecules as bands in each lane. **Coomassie blue** is commonly used for protein (Fig. 9.11), whereas **ethidium bromide** or other, safer fluorescent dyes can be used for nucleic acids (Fig. 9.12).

Beyond simply producing a band pattern in a gel, it is possible to locate the specific nucleic acid band containing a gene of interest using a fluorescently labeled DNA or RNA **probe.** This requires knowing a unique nucleotide sequence within the gene of interest and then creating the probe with the complementary sequence.

The sample is run using ssDNA, which allows binding of the probe to the gene of interest after fragment separation. A wash step removes unbound probe. Shining a UV light on the gel will cause the probe to fluoresce and reveal the band with the gene of interest (Fig. 9.13). For more about hybridization, see the following section.

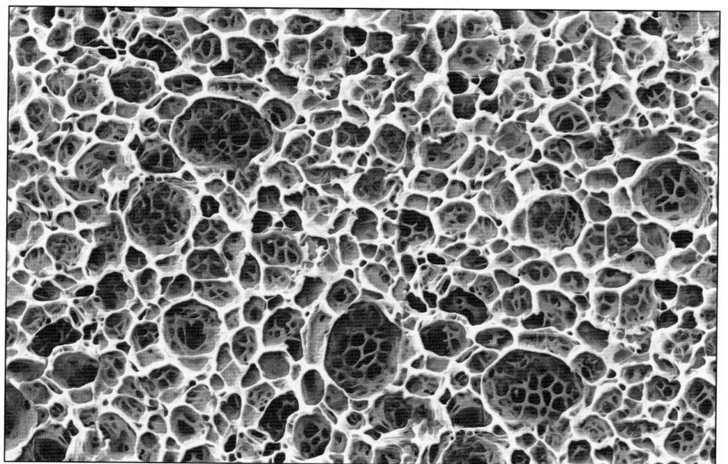

9.9 Artist's Rendition of an Agarose Gel Matrix • Electrophoresis gels are made either of agarose or polyacrylamide, with the former used for nucleic acid separations and the latter for protein and small DNA molecules. Pore size is inversely related to gel concentration and is not uniform, as shown in this illustration. Pore size estimates range from 10^1 to 10^2 nm.

9.10 Electrophoresis Gel Apparatus • This agarose gel is being used to separate DNA fragments. The fragments are running toward the bottom of the photo. The dark blue dye is at the migration front with sample DNA running behind it. It provides a visual indicator of progress through the gel. The run is stopped before the blue dye migrates off the end, which ensures sample won't be lost.

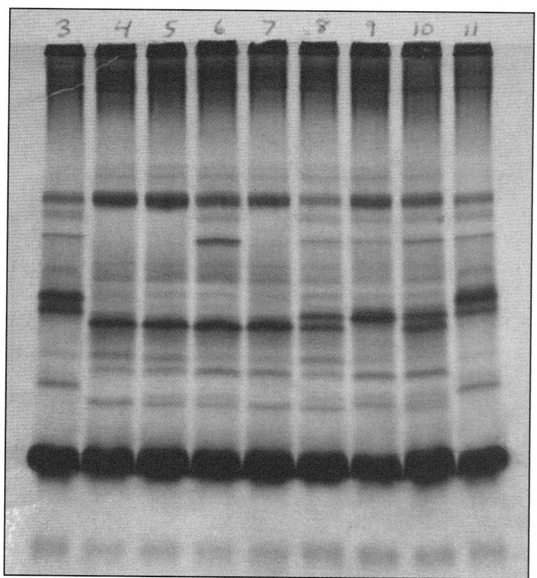

9.11 Polyacrylamide Gel Electrophoresis (PAGE) of Protein • This polyacrylamide gel of serum proteins from several lemur species illustrates staining with Coomassie blue. Migration was from top to bottom. The large band at the bottom is serum albumin.

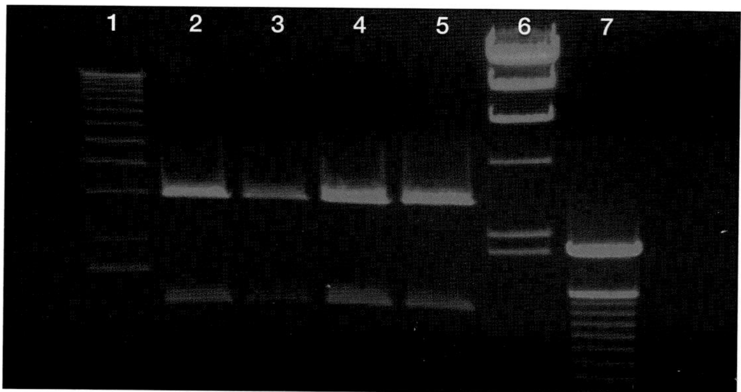

9.12 Agarose Gel Electrophoresis of DNA • This agarose gel was used to separate DNA fragments and to determine their sizes. The first, sixth, and seventh lanes are DNA fragment samples of known sizes. The first lane forms a 1 kilobase (Kb) "ladder" in which the slowest fragment is 12,216 bp (base pairs) in length and the fastest visible fragment is 1,636 bp. (Other smaller fragments are too faint to be seen in this gel.) In lane 6, the fragments range in size from 23,130 bp to 2,027 bp. In lane 7, the range is 2,072 bp to 100 bp. Comparison of migration distance by the sample fragments to those in the ladders provides an estimate of their size. (Ethidium Bromide Stain)

DNA Hybridization and Microarrays

Purpose

DNA hybridization is a versatile technique that tests for the presence of certain DNA sequences[4] in a sample called **target sequences**. The procedure relies on the specificity of base pairing between a single-stranded **probe** and a single-stranded target sequence in a sample's DNA. The information provided by successful hybridization depends on the construction of a particular test.

For instance, a positive result can identify the presence of a particular virus or bacterial species in a patient sample, or it can deepen our understanding of gene expression. It is also used in forensic and commercial genetic testing, among many other applications.

Principle

DNA hybridization tests detect the presence of specific, known nucleotide sequences in a sample using complementary single-stranded DNA (ssDNA) probes. The first step is to amplify sample DNA using PCR (p. 144), after which the dsDNA is converted to ssDNA. Next, a lab synthesized fluorescently labeled ssDNA probe with a nucleotide sequence complementary to the known target sequence in the sample is added. It **hybridizes** (base pairs) with the target, if present, and acts as a molecular beacon signaling the target's presence (Fig. 9.13). If the target is absent, the probe and its label will be washed away.

Frequently, multiple probes, each with a different target and fluorescent label, are used simultaneously. If being used for diagnostics, the combination of results is compared to a genomic library to identify the species based on which genes are present and absent.

A limitation to original DNA probe technology is that only a handful of genes can be tested at once. Scientists advanced this technology into a microscopic scale using either silicon-based computer microarray chips (Fig. 9.14) or polystyrene beads. Attached to each spot on a microarray chip is a specific gene's ssDNA probe (Fig. 9.15). A typical microarray chip can hold up to 40,000 different DNA fragments. Bead technology uses individual beads, each coated with a different probe, attached to a substrate.

To either of these preparations, the sample's fluorescently labeled ssDNA is added and allowed time to hybridize wherever possible. Following a wash step to remove unbound sample DNA, a computer profile is generated, with fluorescence signaling gene presence and darkness gene absence. The fluorescence pattern produced by the sample can be compared to patterns produced by known organisms, with a match leading to identification.

[4] Hybridization can also be applied to RNA target sequences, but to avoid wordiness in describing the process we're only addressing DNA here.

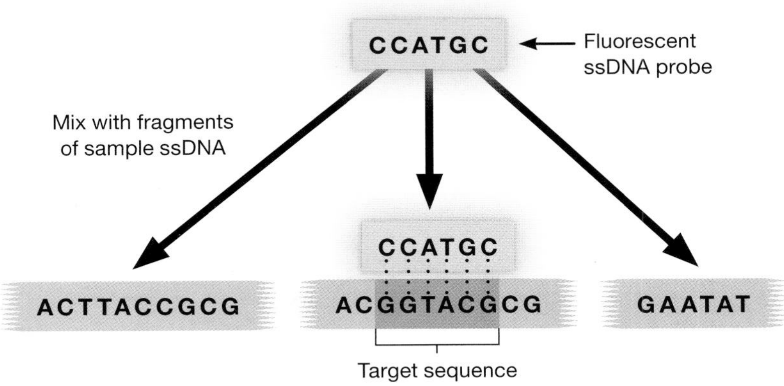

9.13 Diagram of DNA Hybridization • Nucleic acid hybridization allows identification of a specific ssDNA fragment out of a mixture of ssDNA fragments. In this illustration, blue ssDNA represents sample DNA that has been amplified using PCR. The yellow ssDNA represents a lab-synthesized fluorescently labeled probe with a known nucleotide sequence complementary to a target sequence potentially in the sample DNA. As shown, the probe only hybridizes (base pairs) with sample ssDNA carrying the complementary target sequence. Unbound probe is washed away, and fluorescence signals successful hybridization.

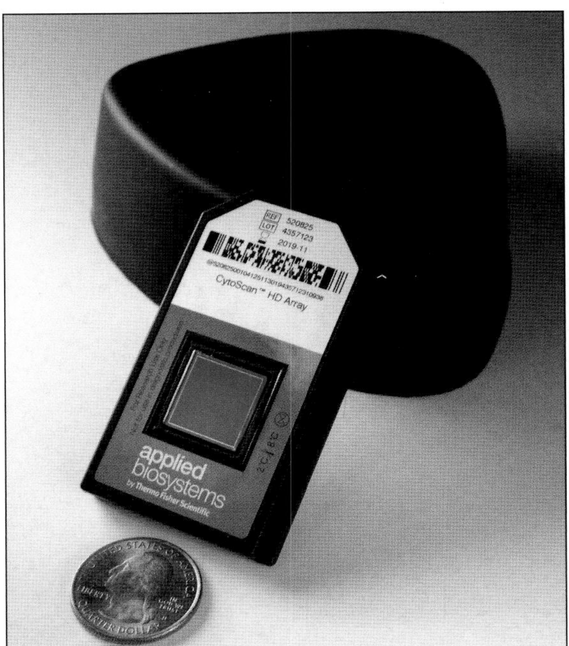

9.14 Gene Chip • Located within the purple region of this GeneChip (available from ThermoFisher Scientific) is a grid with thousands of minute test spots, each with a unique probe. Microchips such as this can be used in a wide range of tests depending on the probes used. Fragmented DNA can be tested for hybridization simultaneously with this system. For more information, see Figure 9.15.

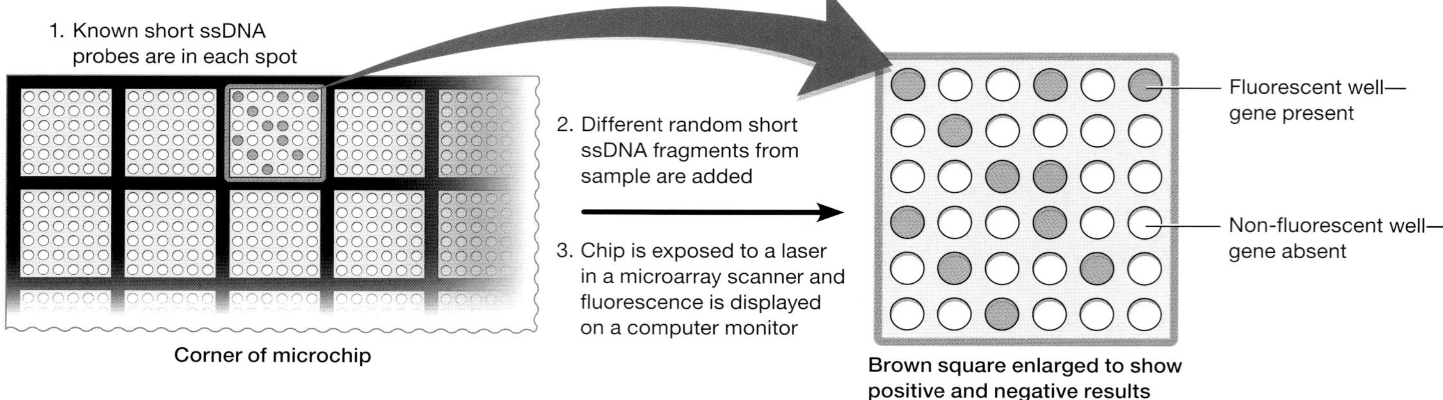
1. Known short ssDNA probes are in each spot
2. Different random short ssDNA fragments from sample are added
3. Chip is exposed to a laser in a microarray scanner and fluorescence is displayed on a computer monitor
Corner of microchip
Fluorescent well—gene present
Non-fluorescent well—gene absent
Brown square enlarged to show positive and negative results

9.15 Diagram of Microarray Analysis ● Microarrays allow running of multiple probes simultaneously. Shown in this illustration is an enlarged view of what a small portion of the chip would look like. Chips are manufactured with different ssDNA probes in each spot. Fluorescently labeled sample ssDNA fragments are added to the microarray chip and allowed time to hybridize, which is followed by a wash step. If a segment of sample DNA is able to bind to one of the probes, it will not be washed away. The chip is then put into a microarray scanner where it is exposed to a laser. Hybridized sample DNA fluoresces and produces a fluorescent spot pattern when displayed on a computer monitor. If used for identification, the pattern is compared to patterns produced by known species for the best match.

Matrix-Assisted Laser Desorption/Ionization Time of Flight (MALDI-TOF) Mass Spectrometry (MS)

Purpose

With advancements in technology, many identification methods have been created based on unique nucleic acid sequences or biochemical capabilities. However, these processes can be costly, time consuming, and require considerable background knowledge about the species or sample source.

Mass spectrometry (MS) is a tool that has been used by chemists for many years to determine the chemical composition of molecules, within seconds and using small quantities of the sample. Newer versions of mass spectrometry are able to rapidly identify species based on protein, carbohydrate, and/or lipid molecules present in cells.

Matrix-Assisted Laser Desorption/Ionization Time of Flight (MALDI-TOF) is a highly sensitive version of mass spectrometry commonly being used to rapidly identify unknown species from pure cultures (Fig. 9.16).

Currently, MADLI-TOF is used to identify groups of bacteria, such as enterics, non-fermenting Gram-negative bacteria, anaerobic bacteria, mycobacteria, *Nocardia*, and actinomycetes. It is also used to identify yeast species. MALDI-TOF also holds the potential for identifying pathogens below the species level (i.e., strains), which could provide the opportunity for more tailored treatment procedures in the future.

Principle

Mass spectrometry is a technique where atoms within molecules become ionized and are then analyzed by a computer based on their **mass-to-charge ratio** (***m/z***). In Matrix-Assisted Laser Desorption/Ionization Time of Flight (MALDI-TOF), a colony from a steak plate is transferred to a metal plate (Figs. 9.17A and 9.17B). A single plate is smaller than a credit card and has 96 target spots for individual specimens. Then a matrix reagent containing low molecular weight organic molecules is pipetted over the sample (Fig. 9.17C).

This mixture is allowed to dry, during which the cells and matrix begin to crystalize. The matrix supplies protons and facilitates the formation of ions from the sample in the next step. Commonly, a number of target spots are dedicated to positive controls usually containing a manufactured **bacterial test standard** (**BTS**) that matches the results for *E. coli*.

After drying, the plate is inserted into a vacuum sealed chamber within the MALDI-TOF mass spectrometer (Fig. 9.18). Each individual target spot is exposed to high-intensity pulses of ultraviolet (UV) laser beams. The matrix molecules absorb the energy and become gaseous. Continued UV exposure causes ionized biomolecules to be released from the cells. This is **desorption**.

The exact mechanism of ionization is not fully understood. The main hypothesis considers that when the matrix is irradiated, matrix molecules transfer hydrogen ions to molecules in the sample. There is also evidence to support the hypothesis that proteins, carbohydrates, and glycoproteins form positive ions, while lipids and glycolipids form negative ions. Different versions of matrix reagents are available and work specifically with the molecules expected to be present in the sample.

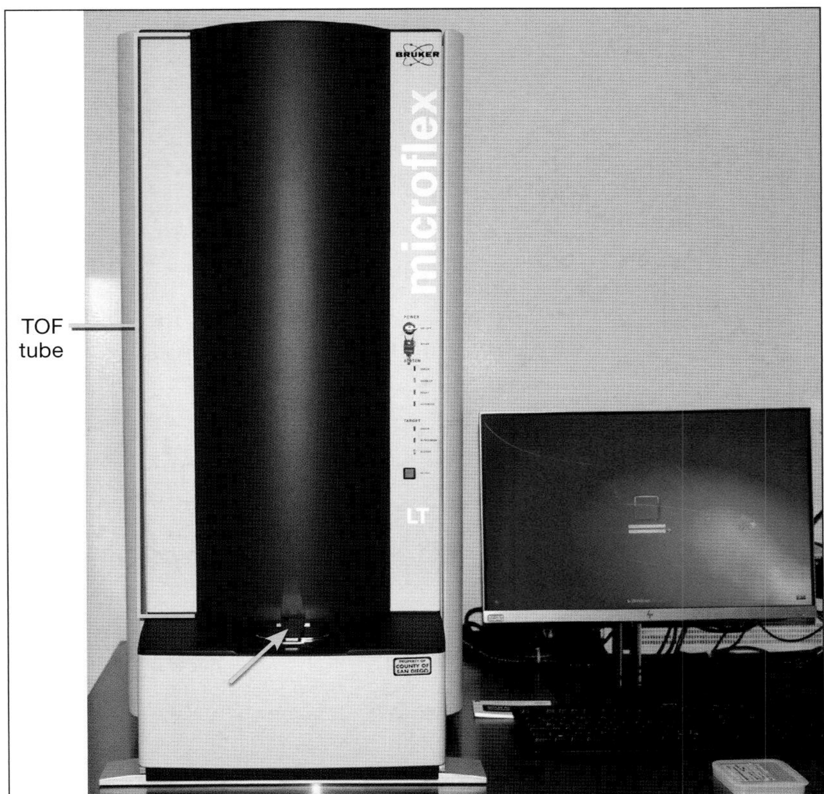

9.16 MALDI-TOF Mass Spectrometer ● The mass spectrometer is on the left. After preparation, a sample card is placed into the port (**arrow**; also Fig. 9.18). The time of flight tube is also shown (TOF). A sensor is at the top of the TOF tube, the data are analyzed, and displayed in graph form on the monitor at the right (see Fig. 9.20).

Next, the **time of flight** (**TOF**) **tube** applies an electrostatic field around the ionized molecules, which provides kinetic energy to them (Fig. 9.19). Then, these ions move into an uncharged area, where they freely drift at their own speed. As the ions drift, they separate based on size, where larger ions take longer to reach the analyzer than smaller ions.

Their travel rates through the uncharged region is measured and used to determine the mass-to-charge (*m/z*) ratio, and a computer graphs the size, charge, and quantity (also called intensity) of each ion present. Identification is made by comparing the unknown sample's graph to a database of known species' graphs (Fig. 9.20).

Cells from different species or strains have unique combinations of molecules. Convenient targets for discriminating between species or strains include storage granules, and cytoplasmic membrane and cell wall components.

For example, many *Pseudomonas* species synthesize poly-β-hydroxybutyrate (PHB) granules while the pathogenic *P. aeruginosa* does not (Fig. 6.45, p. 72 for more information about PHB granules). The fatty acids in membrane phospholipids also contain identifiable differences across species. Similarly, tetrapeptides within peptidoglycan are generally unique to individual species or genera. ●

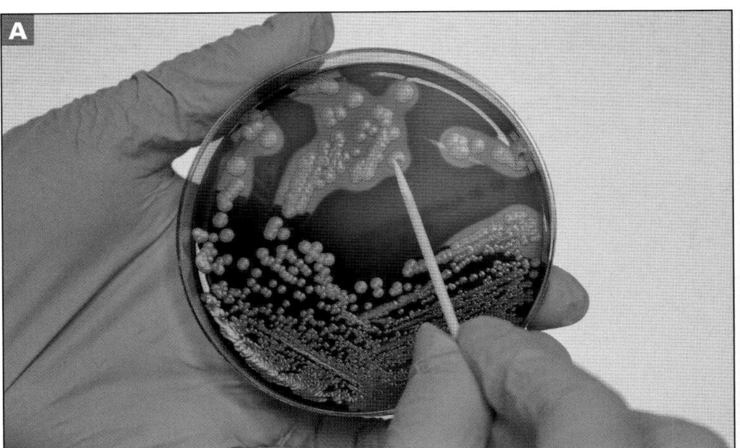

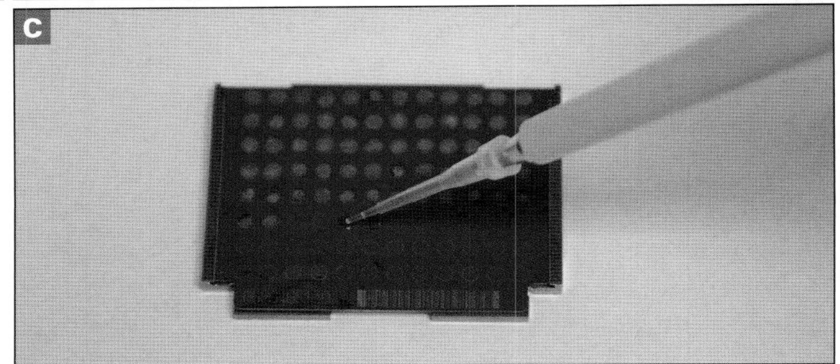

9.17 MALDI-TOF MS–Sample Preparation ● (**A**) The clinical lab scientist uses a sterile toothpick to transfer a small portion of an isolated colony on a streak plate to one target spot on the MALDI-TOF plate (**B**). Multiple samples can be run simultaneously. (**C**) Matrix reagent is added and allowed to dry fully, which takes only a few minutes.

9.18 MALDI-TOF MS–Loading the Sample Plate ● The prepared sample plate is inserted into the sample port (Fig. 9.16, arrow) of the mass spectrometer. After it is in place, a laser beams shoots at the target spots one at a time, producing ionized particles from the sample that travel up the TOF tube (Figs. 9.16 and 9.19). A computer analyzer at the top collects and compiles the molecular information in graphic form (Fig. 9.20).

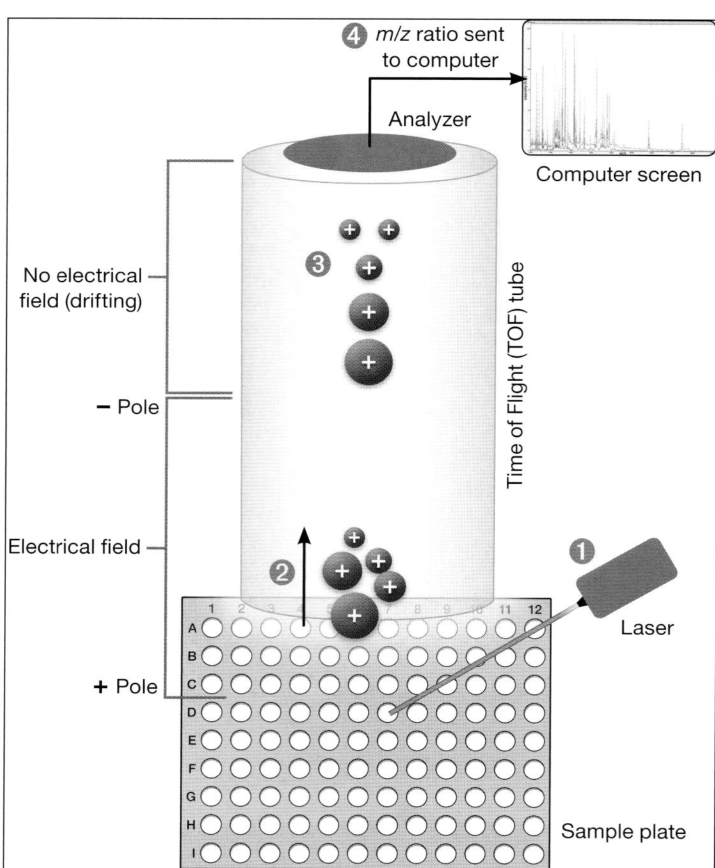

9.19 Diagram of MALDI-TOF MS ● (**1**) The laser beam hits the target specimen causing ionization of microbial molecules. (**2**) These ions begin traveling through the electrical field, being repelled by the positively charged pole toward the negative pole. (Recall that they have picked up hydrogen ions from the matrix material.) The polarity difference creates an accelerative field that propels them into the tube's uncharged region (**3**) where smaller ions drift faster than larger ones. (**4**) As the ions hit the analyzer, their data are collected, mass-to-charge (*m/z*) ratio is calculated, and a computer generates a graph (Fig. 9.20).

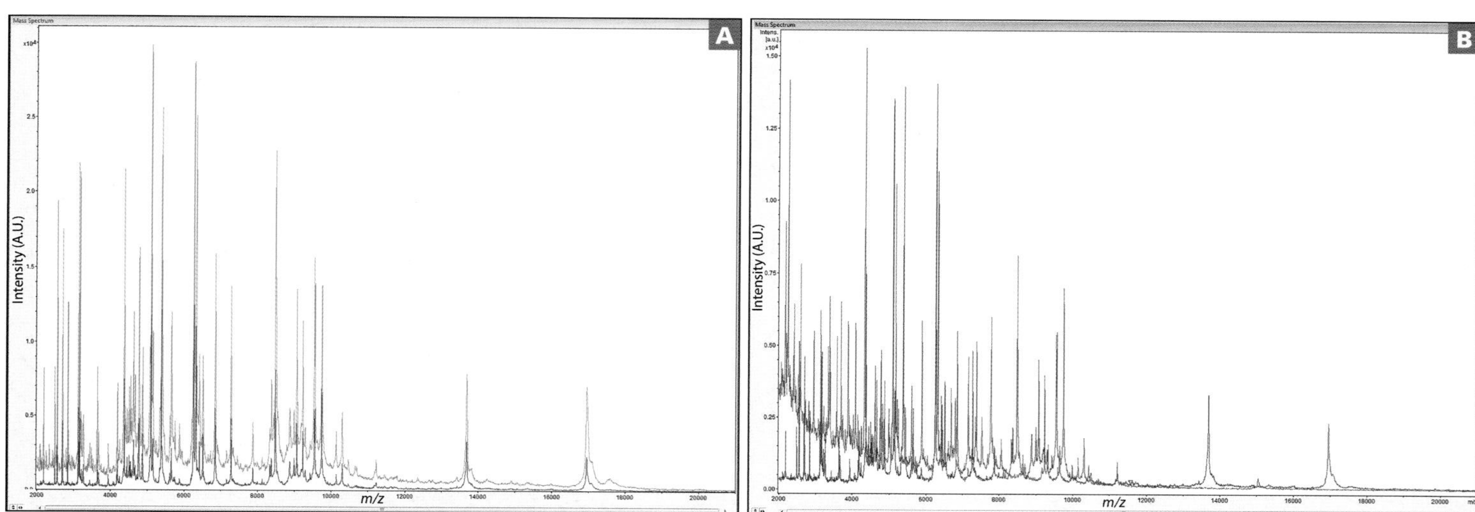

9.20 Representative MALDI-TOF MS Results ● (**A**) Peaks on the graph represent different ions, with mass-to-charge ratio plotted on the x-axis and intensity (amount) plotted on the y-axis. Collectively, the peaks illustrate a profile of the ions produced by the specimen. In this graph, the peaks of the two plots (green and purple) align with one another showing the samples came from the same species. Peak intensities do not need to be equal, as they are affected by the amount of specimen used. (**B**) Shown are results with from two different species as evidenced by the lack of alignment between their peaks.

SECTION

10 Viruses

Introduction to Viruses

In the opinion of most biologists, viruses are not living. This is based primarily on the absence of two major characteristics associated with living things: viruses are not cellular, and they have no metabolism. On the other hand, they do exhibit heredity, are made of biochemicals, and evolve.

Proteins form the structures and receptors of the virus particle (not cell!), and DNA *or* RNA (all cellular organisms use DNA) is the hereditary material. And, whereas cells always possess both DNA and RNA (for protein synthesis), viruses have only one or the other and use it as their genome. They are also very small, usually much smaller than bacteria. They are capable of making more of themselves (replication), but require a cellular host to do the work.

As such, they are **obligate intracellular parasites.** And, unlike cells that remain intact when they reproduce, viruses disassemble during replication, and then the progeny reassemble prior to release from the host—hence "viral replication" rather than "viral reproduction."

A virus particle minimally consists of a protein **capsid** surrounding its **genome.** The capsid is one of two basic geometric shapes and is composed of protein subunits called **capsomeres.** Some viruses have a rod-shaped capsid, in which the capsomeres form a helix and the genome is threaded in the helical grooves (Fig. 10.1). Others have an **icosahedral** capsid with 20 triangular faces, in which capsomeres combine to form the faces (Fig. 10.2).

Many viruses that infect bacteria, called **bacteriophages** or simply **phages,** have a complex structure with a protein tail (Figs. 10.2 and 10.3). Coincidentally, complex viruses frequently have icosahedral capsids. It is not uncommon for a virus to possess enzymes, but they don't have many. Most enzymes that are required for replication are encoded in the genome and are made once inside the host cell.

Some viruses (typically those that infect animals) have an outer **envelope** composed of lipid membrane obtained from the host cell upon release. Envelopes usually have one or more types of protein **spikes** that assist in attachment to host cells to begin the next infection.

Viral replication involves the same basic stages, regardless of the host. These are: **attachment** to the host, **penetration** into the host, **uncoating** of the genome, **genome replication** and **synthesis** of viral proteins, **assembly** of progeny, and **release.** Figure 10.4 illustrates the lytic replicative cycle of bacteriophage T4 in which the host is lysed when the phage progeny are released.

Lambda phages (Fig. 10.5) also infect *E. coli*, but they have two replicative paths. While capable of a lytic path, they more frequently follow a **lysogenic** path and integrate their DNA into the *E. coli* chromosome at one or more

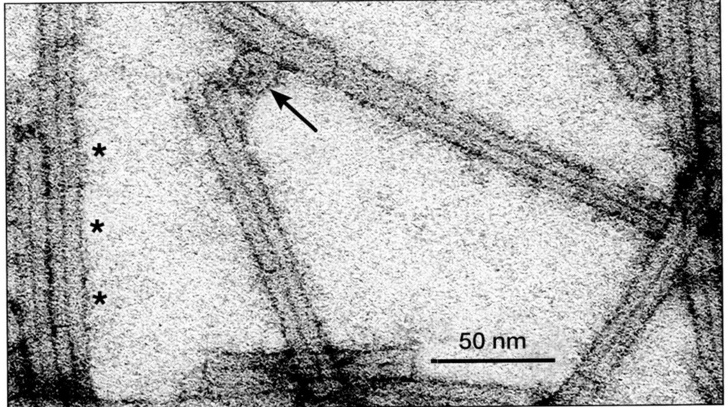

10.1 Tobacco Mosaic Virus (TMV) • TMV is the causative agent of tobacco mosaic disease in plants. The virions are rod shaped and are composed of 2,130 capsomeres helically arranged around a central canal (the dark line in each virion). Striations of the helix are faintly visible at the asterisks (*). These virions are approximately 400 nm long and 15 nm in diameter. Assembly of the TMV capsid involves addition of disk-like subassemblies consisting of two layers of 17 capsomeres each. These assume a "lock washer" shape when associated with the +ssRNA genome and integrate into the growing capsid. An isolated disk of a virion is indicated by the **arrow.** (TEM Negative Stain)

specific sites where they become latent as a **prophage**. When *E. coli* replicates its DNA, it also replicates lambda DNA, resulting in all of its descendants being infected.

At some point, **induction** occurs where the prophage leaves the chromosome and a lytic path that leads to host cell lysis occurs. Induction can occur spontaneously (approximately once per 10^4 host cell divisions) or occur as a result of DNA damage, such as by UV radiation.

On occasion, removal of the prophage during induction takes bacterial DNA and leaves some viral DNA in the chromosome. All the progeny receive an incomplete viral genome. They are still capable of infecting a new host up through penetration, but can't complete the process. However, the bacterial DNA that gets delivered may be able to integrate into the host's DNA and increase its genetic variability. This is called **specialized transduction** because

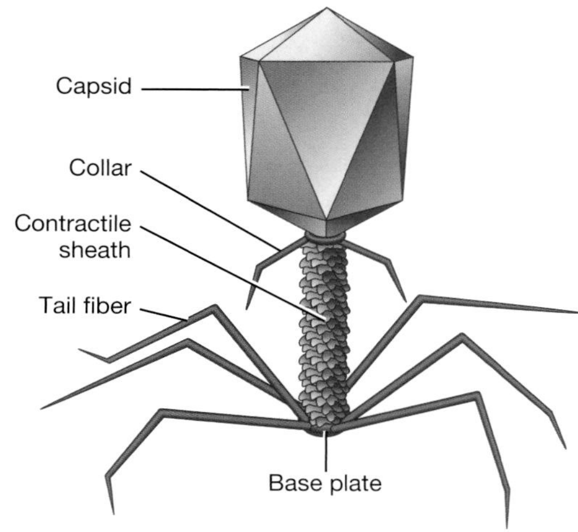

10.2 Artist's Rendition of T4 Coliphage • Bacteriophages frequently have a complex structure that includes the capsid containing the genome and a contractile tail with many parts. T4 has an icosahedral capsid.

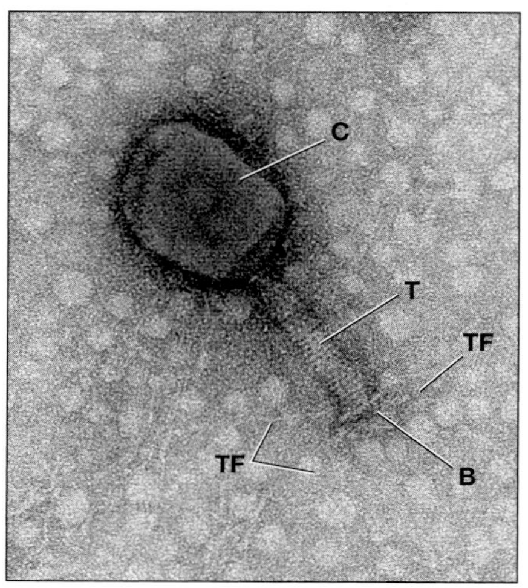

10.3 T4 Coliphage • This is a negative stain of one T4 phage particle. Shown are the capsid (**C**), tail (**T**), base plate (**B**), and tail fibers (**TF**). The length of this phage from base plate to tip of capsid is approximately 180 nm (0.18 µm). (TEM Negative Stain)

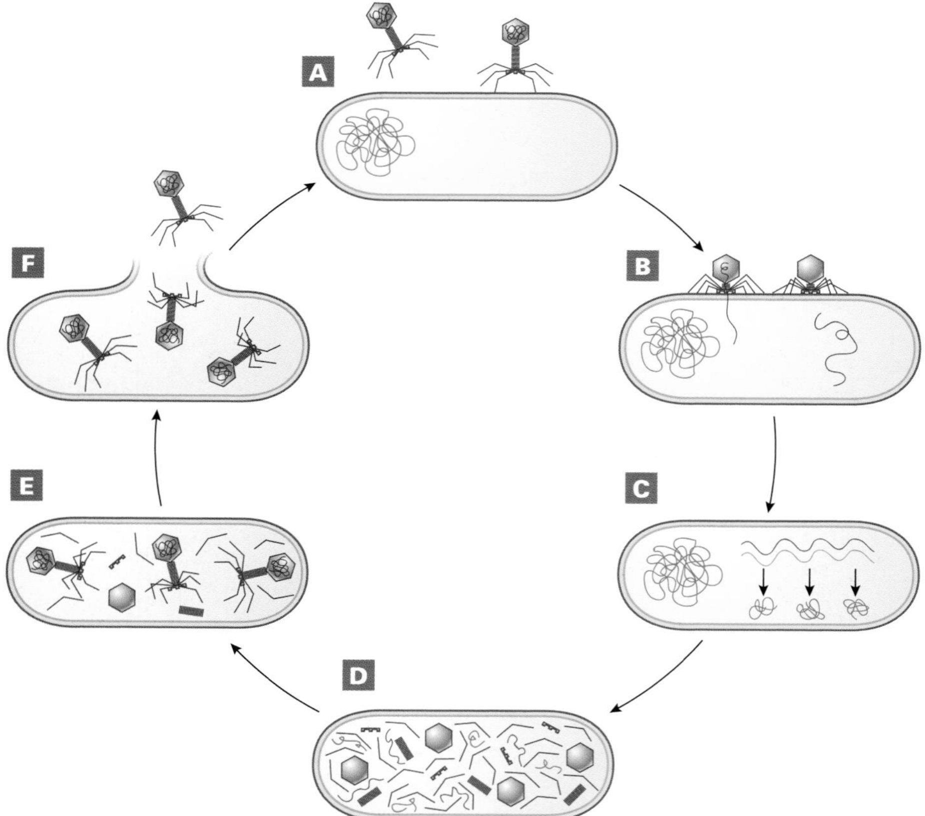

10.4 T4 Phage Lytic Cycle • Shown is a simplified diagram of the T4 coliphage's replicative cycle. (**A**) An infective T4 phage (left) is approaching its host, *Escherichia coli*. On the right, the same phage is shown attached by its tail fibers to specific receptors on *E. coli*. This is the **attachment** phase. The *E. coli* chromosome is the red tangled line at the left of the cell. (**B**) On the left, T4 is in the process of injecting its DNA (purple line) into *E.coli*. This is the **penetration** phase; on the right, the entire phage genome is in the host. Removal of viral genome from its capsid is **uncoating**. (**C**) Phage DNA is in the process of being transcribed into mRNA (**red line**), which is then translated into phage proteins (**squiggly green lines**). This is the **synthesis** phase and is performed by *E. coli* under the direction of the phage DNA. Simultaneously, phage DNA is being replicated, but this is not shown. (**D** and **E**) Synthesis leads to **assembly** of the phage progeny, where the capsid subunits come together to form the capsid and into which the genome is inserted. The tail with all its detailed parts also comes together and attaches to the capsid. Note that host DNA has been degraded, but remains as short, red lines. (This is important because occasionally *E. coli* DNA is randomly put into a viral capsid and is transferred to a new *E. coli* host, providing it with new genes in a process called **generalized transduction**. See text above for a description of specialized transduction with lambda phage.) (**F**) The fully assembled phages are **released** as the cell bursts. Estimates vary, but at 37°C this entire cycle can take between 25 and 60 minutes and release between one- and two-hundred phage progeny, each of which can repeat the process!

only bacterial genes near the site of prophage insertion will be transduced. (See Fig. 10.4 caption for a description of **generalized transduction** with T4 phage.)

The viral genome is either DNA or RNA. Further, some DNA viruses have double-stranded DNA (dsDNA), whereas others have single-stranded DNA (ssDNA), which does not exist in cells. RNA viruses can also have ssRNA or dsRNA (also not found in cells) as their genome. This genome diversity exceeds that seen in cells and viruses have evolved appropriate enzymes to catalyze genome replication and mRNA production.

You are already familiar with cellular enzymes for DNA replication and mRNA transcription: **DNA polymerase** and **RNA polymerase**, respectively. But that nomenclature is inadequate when naming viral enzymes and virologists developed a system based on the template and the product. Using the virologist's system, the viral equivalents of cellular enzymes are **DNA-dependent** (DNA is the template) **DNA polymerase** (DNA is the product) for replication and **DNA-dependent RNA polymerase** for transcription.

Many RNA viruses catalyze genome replication and mRNA production with **RNA-dependent RNA polymerase**. Other RNA viruses have the ability to construct DNA from an RNA template using an **RNA-dependent DNA polymerase** (also known as **reverse transcriptase** or **RT**). Recognize that these are useful enzyme categories, but that there is diversity in each based on the specific virus.

1975 Nobel Laureate (for the discovery of reverse transcriptase) David Baltimore devised a viral taxonomy in 1971 based on viral genome and the steps necessary to get information from the genome to messenger RNA. By convention, mRNA is considered to be " + " sense. A complementary DNA or RNA strand is considered "–" sense. Presently, there are seven categories (Fig. 10.6).

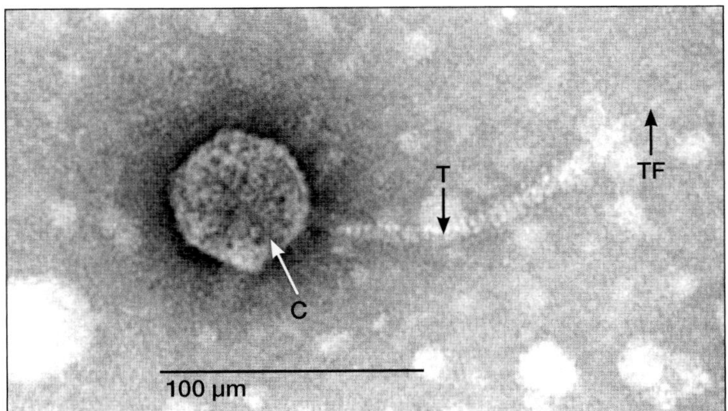

10.5 Lambda Phage • Although there are similarities, lambda phages structurally differ from T4 phages in having only a single tail fiber (**TF**), which is faintly visible sticking straight down from the tail (**T**). Also visible is the capsid (**C**). (TEM Negative Stain)

- **Class I dsDNA viruses.** These viruses make mRNA and replicate their genomes just as cells do, usually using the host's DNA-dependent DNA polymerase and DNA-dependent RNA polymerase enzymes. Examples include herpes simplex viruses, variola virus (small pox), and human papillomaviruses (cervical cancer and genital warts).
- **Class II ssDNA viruses.** These viruses must first replicate their genome to dsDNA using host DNA-dependent DNA polymerase. The resulting dsDNA is transcribed into mRNA by host cell DNA-dependent RNA polymerase. Normal DNA replication provides more ssDNA for the progeny genome and both " + " and "–" strands are distributed equally among progeny. An example is human parvovirus B19 (which causes Fifth disease—a rash, seen mostly in children).
- **Class III dsRNA viruses.** A viral RNA-dependent RNA polymerase uses the negative strand to produce +ssRNA that acts as mRNA. The same enzyme uses both strands as templates for genome replication. Examples include rotavirus (gastroenteritis, mainly in children) and reovirus (mild respiratory and digestive tract symptoms).
- **Class IV +ssRNA viruses.** The genome acts as messenger RNA. A viral RNA-dependent RNA polymerase produces a –ssRNA molecule using the original genome +ssRNA as a template and it, in turn, acts as the template for genome replication. Examples are coronavirus, poliovirus, hepatitis A and C viruses, West Nile virus, and Zika virus.
- **Class V –ssRNA viruses.** These viruses synthesize +ssRNA from the genome that acts as mRNA and as a template for genome replication. Viral RNA-dependent RNA polymerase is responsible for both processes. Some viruses in this class (e.g., influenza) have **segmented genomes**. That is, they have multiple RNA molecules analogous to chromosomes in cells. Examples include rabies virus, Ebola virus, measles virus, mumps virus, and influenza A, B, and C viruses.
- **Class VI +ssRNA retroviruses.** Retroviruses produce an RNA-dependent DNA polymerase (reverse transcriptase) capable of using a +ssRNA template to first make a +RNA/–DNA hybrid molecule. It then replaces the +RNA strand with +DNA to make dsDNA, which is then incorporated into the host genome and becomes latent as a **provirus**. When active, transcription by host DNA-dependent RNA polymerase produces +ssRNA, which can be used as mRNA or genome for the progeny. Examples include human immunodeficiency viruses (HIV—causes AIDS) and human T-lymphotrophic viruses (HTLV—causes T-cell leukemia).

- **Class VII dsDNA viruses with an RNA intermediate.** These viruses have an incomplete dsDNA genome, and replicate via an RNA intermediate (**pregenome**) using viral reverse transcriptase. In a very complex sequence of events, the genome at the time of infection is mostly dsDNA, but there are segments that are only –ssDNA. Thus, there is one complete –ssDNA strand and portions that are double-stranded. Upon entry in the nucleus, the gaps are filled. DNA dependent RNA polymerase is responsible for transcription and production of a + ssRNA that acts as the pregenome. From this template, reverse transcriptase produces a complementary –DNA strand, which then is replicated to make dsDNA copies that act as progeny genomes. The best-known example is hepatitis B virus (HBV), responsible for serum hepatitis.

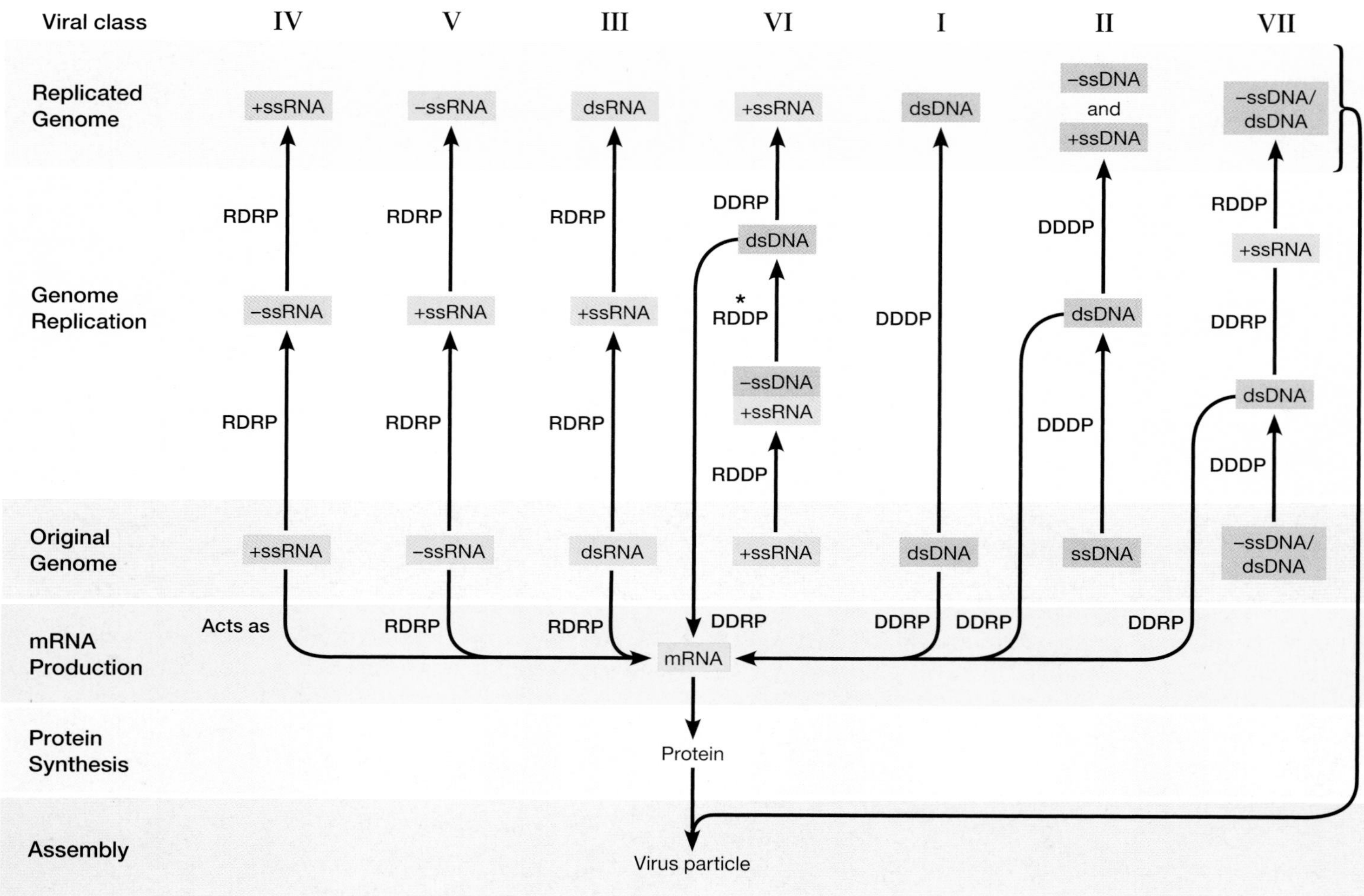

10.6 Baltimore Virus Classification ● There are seven categories in this system, based on viral genome and mechanism for making mRNA, which is always "+" sense. Also shown are simplified paths of mRNA production and genome replication for each class. Enzyme abbreviations: DDDP = DNA-dependent DNA polymerase; DDRP = DNA dependent RNA polymerase; RDRP = RNA-dependent RNA polymerase; RDDP = RNA-dependent DNA polymerase.

* In class IV viruses, the RDDP that synthesizes the second DNA strand is the same enzyme that synthesized the first, but is acting as a DDDP.

Human Immunodeficiency Virus (HIV)

HIV is the causative agent of AIDS (Acquired Immune Deficiency Syndrome). At first, only a single type of HIV was known, but in 1985, a second HIV was isolated. The two forms are now referred to as HIV-1 and HIV-2, respectively.

HIV-1 and HIV-2 are **retroviruses** (class VI: family retroviridae, genus Lentivirus) and have the ability to perform **reverse transcription**; that is, they make DNA from a +ssRNA template, a very unusual process that characterizes the group. The HIV genome consists of only nine structural genes and a handful of regulatory genes.

Morphologically (Figs. 10.7 and 10.8), a phospholipid envelope derived from the host cell membrane encloses the capsid and is lined with **matrix proteins** (p17) that are involved in regulating various stages of the replicative cycle. A protein core made of **capsid proteins** (p24) surrounds two single stranded RNA molecules, each of which is associated with a molecule of reverse transcriptase.

Integrase (IN), which is responsible for incorporation of provirus DNA into the host DNA, and **protease** (PR) enzymes are also present. (Translation of HIV RNA produces three long **polyproteins** that must be separated into their component proteins. Protease does this.) **Glycoprotein spikes** emerging from the envelope are involved in attachment to the host cell. The **surface glycoprotein** (gp120) is involved in attachment and the **transmembrane glycoprotein** (gp41) acts as its anchor.

HIV is transmitted via body fluids such as blood, breast milk, semen, and vaginal secretions. Infection can occur as a result of sexual intercourse with or blood transfusion from an infected individual. Infection may also occur across the placenta during pregnancy or via contaminated needles used for injection of intravenous drugs. An infected mother also may transmit it to a newborn during delivery or nursing. Casual social contact does not appear to be a route of infection. There is no cure for HIV infection, but with antiretroviral therapy its progress can be controlled.

HIV infection can be broken down into three phases: acute, latent, and AIDS[1]. The **acute phase** begins at the time of infection, but symptoms may take 2 to 4 weeks to show. During this time HIV infects cells with CD4 membrane receptors that **T-helper cells** (also called **CD4$^+$ cells**) use during an immune response, but HIV uses for attachment.

Other cells, such as dendritic cells (a type of **antigen presenting cell–APC**), macrophages (also APCs), and monocytes can be infected and play a role in spreading the virus to T-helper cells. During the acute phase, viral load is high and the patient is highly contagious.

Once inside host cells, HIV produces reverse transcriptase that converts its +ssRNA into viral DNA (see Fig. 10.6, class VI), which integrates into the host chromosome as a **provirus** and is largely latent, though some viral replication occurs. After assembly, new virions emerge from the host cell by budding and infect other cells. This is the **clinical latency phase.** During clinical latency, the infection can be controlled using **antiretroviral therapy** (**ART**), which may produce a state of **viral suppression** (<200 HIV per mL of blood) or an **undetectable viral load,** (too low for tests to detect).

Years to decades later, increasing viral load begins to take its toll on T-helper cells. They are essential to the normal operation of the immune system because they promote development of immune cells in both **humoral** and **cell-mediated responses.** A normal range for T-helper cells is between 500/μL (mm^3) and 1200/μL. A patient with a T-helper cell count below 200/μL of blood is considered to have AIDS.

Depletion of T-helper cells cripples the immune system and the patient becomes susceptible to infections by organisms not typically pathogenic. Thus, AIDS is not a single disease, but rather a syndrome of diseases characteristic of patients with HIV infection.

Safe sex practices and using sterile needles if an IV drug user are excellent ways to protect oneself from HIV infection. There is also medication (Truvada) available for people who

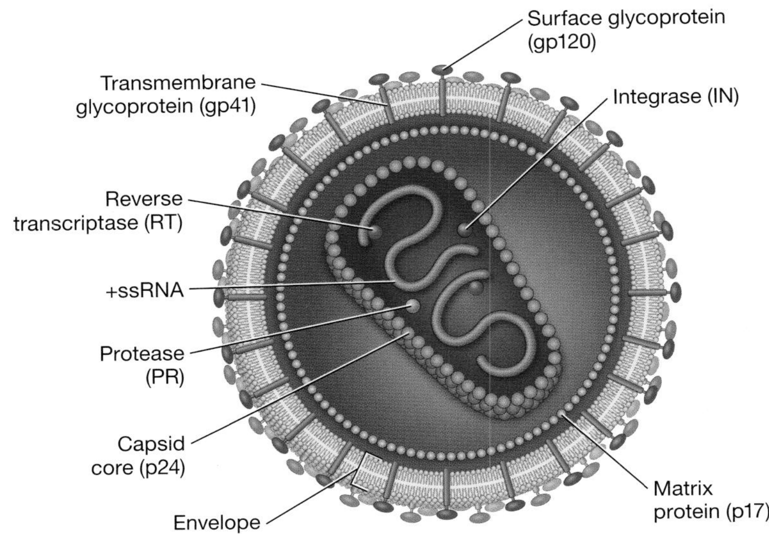

10.7 Artist's Representation of HIV • HIV is an enveloped retrovirus. See text for details.

[1] These descriptions are current as of 2019 and are largely based on the CDC and other reputable websites. Printed books age, but websites are easily updated. Check the CDC website for the most current information on "everything HIV," https://www.cdc.gov/hiv/default.html.

are HIV negative, but are at high risk. This approach is called **preexposure prophylaxis (PrEP)** and should be combined with the previously mentioned behavioral practices.

Other medications for HIV-positive patients target different steps in the replicative process. These include interfering with attachment and reverse transcriptase (nucleoside and nonnucleoside inhibitors), integrase, and protease activities. But there is no cure for HIV infection.

Currently, enzyme-linked immunosorbent assays (ELISAs) are used to determine the presence of anti-HIV antibodies and can discriminate between HIV-1 and HIV-2 (see p. 137 and Fig. 8.18). Antibody ELISAs are often combined with an antigen ELISA that identifies the p24 antigen. Indirect fluorescent antibody tests (Fig. 10.9) are also used for detection of anti-HIV antibodies. Nucleic acid tests (NATs) include reverse transcriptase PCR (RT-PCR) and qualitative-PCR (qPCR) amplification tests for viral load (see p. 146).

According to the World Health Organization (WHO) 37.9 million people globally were living with HIV infection in 2018. Worldwide, 770,000 people died in the same year[2]. Most are in Sub-Saharan Africa.

[2] http://www.who.int/gho/hiv/en/

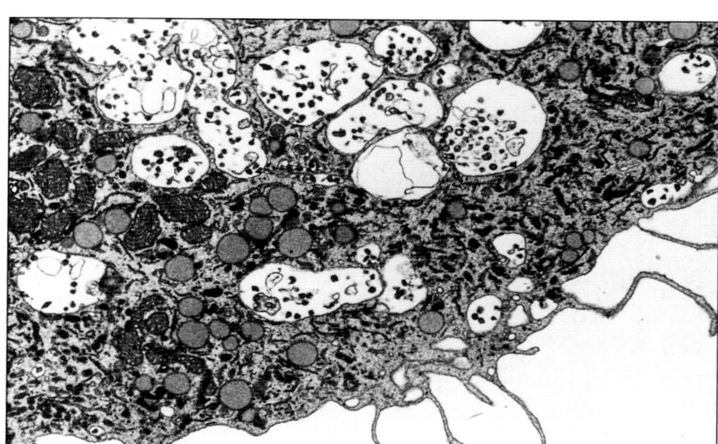

10.8 Electron Micrograph of a Macrophage from an HIV-Infected Person Vacuoles within the macrophage contain viral particles. (TEM)

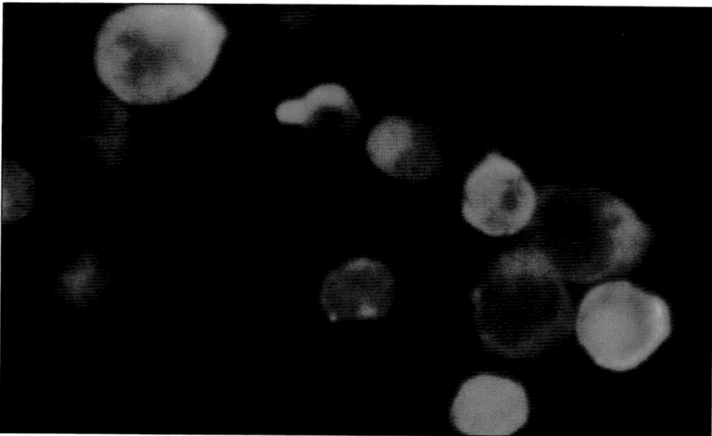

10.9 An Indirect Fluorescent Antibody (IFA) Test Positive for HIV The test system consists of HIV-infected cells attached to the glass slide. When patient serum is added, any anti-HIV antibodies bind to the infected cells. Subsequent addition of fluorescent anti-human immunoglobin antibodies results in binding to patient anti-HIV antibodies, if present. After washing away unbound fluorescent antibodies, the slide is observed under a UV microscope. A bright, apple-green color is a positive result. A negative result would be black.

Viral Cytopathic Effects in Cell Culture

Purpose

This procedure is used for in vitro identification of viruses. Presumptive identification of a virus in a specimen can be made by determining the host cell(s) in which it replicates, how quickly it causes damage, and the type of **cytopathic effect** (damage) it produces. Confirmation can be made using a specific serological test.

Principle

Supplied with the appropriate nutrients and environment, viable virus host cells can be grown in a tube or flat bottle. This is a **cell culture** (Fig. 10.10). Incubation is done in such a way that growth occurs only on one side. The cells divide and produce a characteristic monolayer on the container's inside surface.

Different media are used for cell culture at different stages. These are: **growth medium** and **maintenance medium**. Growth medium is used to begin a cell culture. When the cell layer is confluent, or nearly so, the growth medium is replaced with maintenance medium, which can be adjusted to keep the cells in the proper state to be infected by each specific virus. Both media are supplemented with amino acids, vitamins, growth factors, and calf serum.

To ensure that the serum is free of viral antibodies and certain infectious agents, only fetal, neonatal, or agammaglobulinemic calf serum is used. Antibiotics are also included to inhibit bacterial growth. The accumulation of carbon dioxide from extracellular respiration lowers the medium's pH, which is counteracted by a buffer. A pH indicator (such as phenol red) is used to monitor the effectiveness of the buffer.

A sample suspected of containing virus is introduced into the cell culture. Cultures are grown in an incubator and, if tubes are used, they may be placed in a mechanical roller to keep the medium aerated (Fig. 10.11). As they replicate in the cell culture, viruses inflict damage upon the host cells called a cytopathic effect (CPE), which can be viewed with an inverted microscope (Fig. 10.12).

Depending on the virus and the host cell, CPEs will be evident after as little as 4 days to as much as 4 weeks. Most of the time, they start as small spots (**foci**) in the cell layer, and then spread outward. Common damage to the cells includes rounding (either small or large), a change in texture (either **granular** or **hyaline**—glassy), or formation of a **syncytium** (fusion of infected cells).

Figures 10.13, 10.17, and 10.19 illustrate normal growth of three cell types used in cell culture. Figures 10.14 through 10.16, Figure 10.18, and Figures 10.20 through 10.22 illustrate various CPEs in these three host cell types.

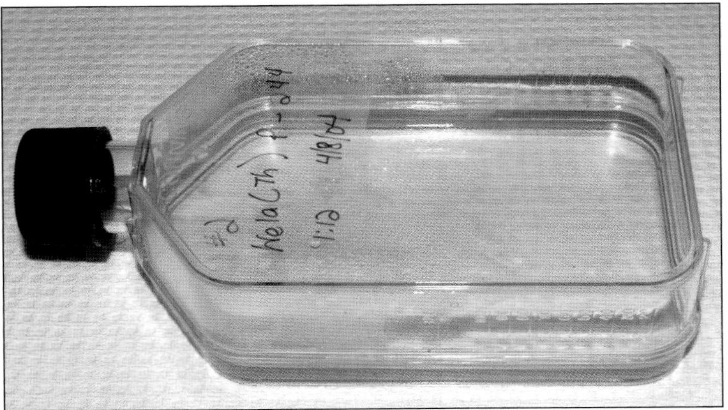

10.10 Cell Culture • Viruses cannot be grown in bacteriological media (such as nutrient agar) because they require living host cells. So, the first challenge in propagating viruses is to grow their hosts in a tube or flat bottle to produce a cell culture. Shown is a culture of HeLa cells. HeLa is an established (continuous) cell line; that is, it has lasted more than 70 passages and continues to maintain its sensitivity to viral infection. In fact, it was the first "immortal cell line" created. The cells were obtained in 1951 from Henrietta Lacks ("HeLa"), who was suffering from cervical cancer. She died soon thereafter, but her cells have been used since then in making many major medical discoveries. This fascinating story has been recounted in *The Immortal Life of Henrietta Lacks*, by Rebecca Skloot, and addresses bioethical issues because her cells were taken without her permission or knowledge.

10.11 Roller Drum • Cultivation of some viruses (e.g., respiratory viruses) requires aeration in order to more closely mimic their natural conditions. Incubation of the culture in a roller drum keeps the medium aerated. In this laboratory, all cell cultures are aerated.

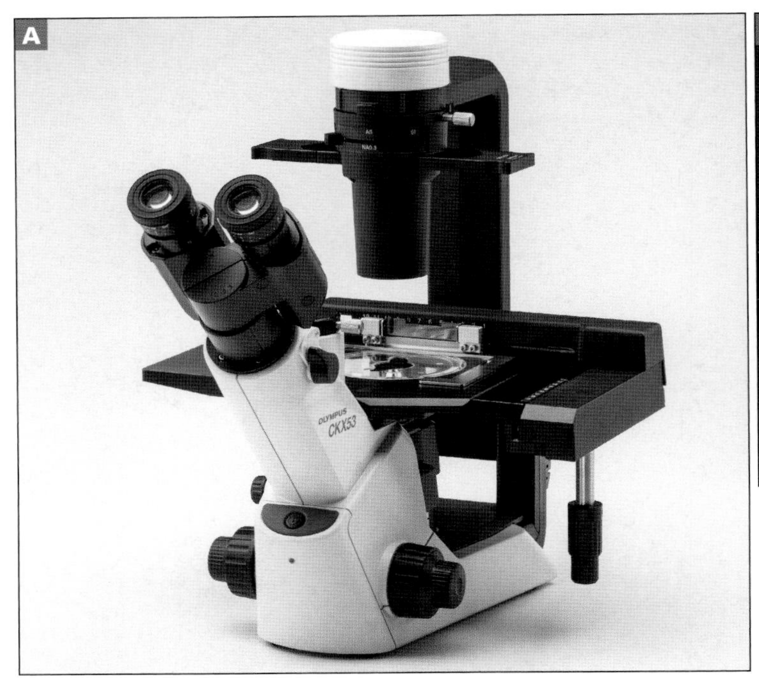

10.12 Inverted Microscope • **(A)** Culture tubes and bottles are more easily viewed with an inverted microscope because the objective lenses are below the stage. This allows viewing of the cell monolayer without looking through the thickness of the flask or tube. **(B)** In this close up, you can see the nosepiece and objectives below the stage and the light source above. The focal lengths of those objectives are too short to view the cells on the bottom of the flask if they were positioned above it.

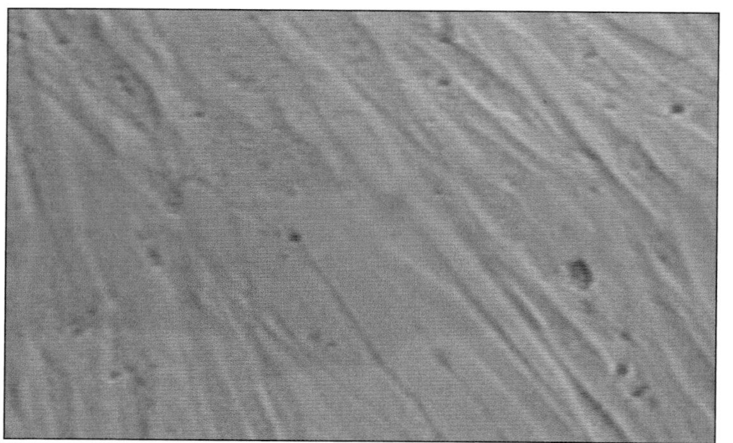

10.13 Normal MRC-5 in Cell Culture • MRC-5 is a cell line of human diploid fibroblasts begun in 1966 from the lung tissue. A cell line has a limited number (about 50) of passages (transfers) before it is no longer useful. MRC-5 is used for isolation of herpes simplex viruses (HSV), varicella-zoster viruses (VZV), cytomegaloviruses (CMV), adenoviruses, enteroviruses, respiratory syncytial viruses (RSV), and rhinoviruses. These cells are uninfected and act as a control to verify that the cell line is still normal.

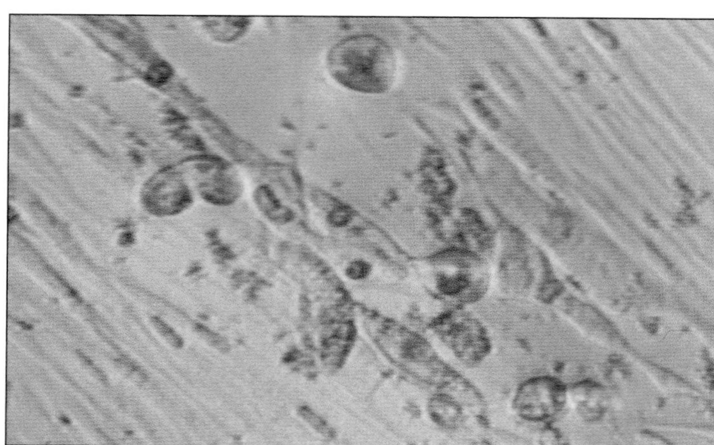

10.14 Cytomegalovirus (CMV) CPE in MRC-5 Cells • CMV grows best in human fibroblasts, such as MRC-5. The infected fibroblasts form a row and are rounded and hyaline (glassy). CMV belongs to the family Herpesviridae, all of which are dsDNA viruses (class I). Though common, its infections are usually asymptomatic. Compare with Figure 10.13.

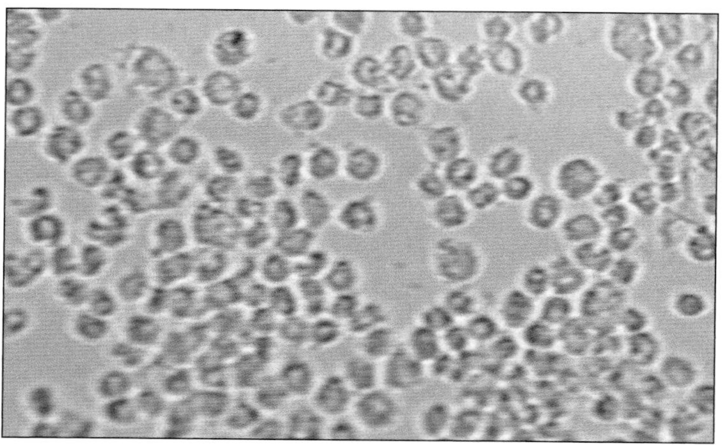

10.15 Enterovirus CPE in MRC-5 Cells • In cell culture, enterovirus infected MRC-5 cells become small (pyknotic) and round. Most enterovirus infections are asymptomatic, but can result in acute diseases such as nonspecific febrile illness, aseptic meningitis, and poliomyelitis. Enteroviruses belong to the family picornaviridae (pico-RNA-viridae—literally "small RNA viruses"). Their genome is +ssRNA (class IV). Compare with Figure 10.13.

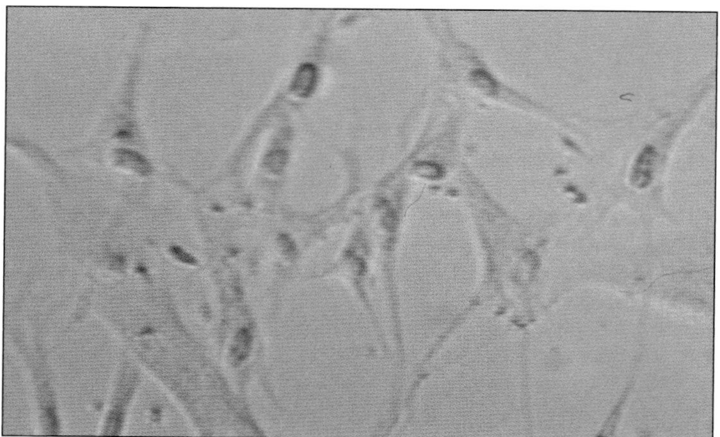

10.16 Varicella-zoster Virus (VZV) CPE in MRC-5 Cells • Enlarged cells with odd shapes characterize the VZV CPE in MRC-5 cells. Varicella-zoster is a dsDNA virus (class I) that belongs to the herpesviridae and is responsible for producing chickenpox (varicella) and shingles (zoster). Compare with Figure 10.13.

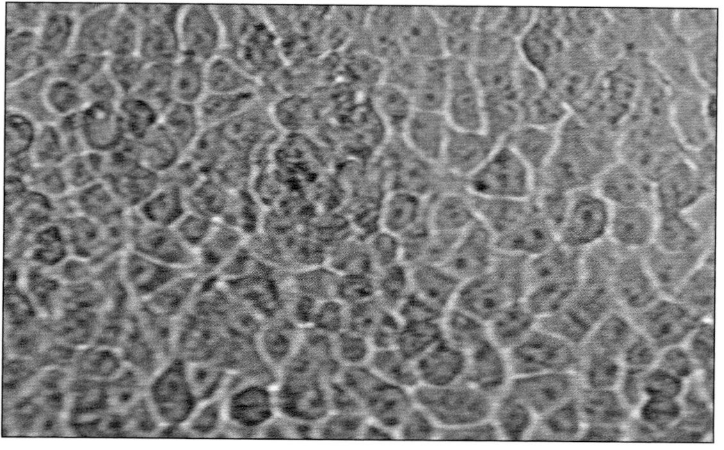

10.17 Normal HeLa Cells in Culture • HeLa is an established cell line used for isolation of poxviruses, respiratory syncytial viruses (RSV), rhinoviruses, and enteroviruses.

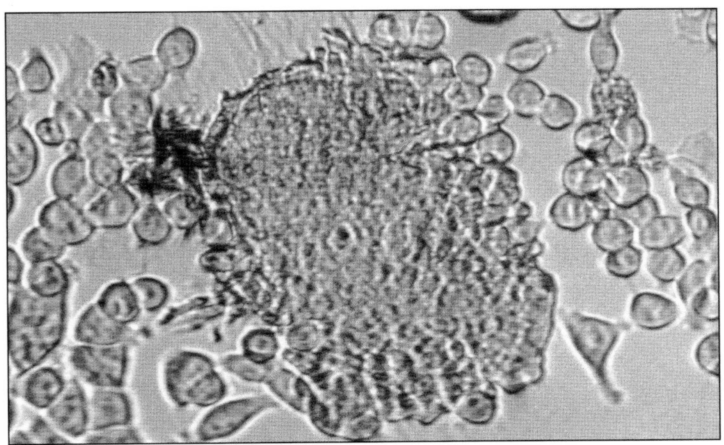

10.18 Human Orthopneumovirus Virus CPE in HeLa Cells • Human orthopneumovirus (formerly respiratory syncytial virus–RSV) in HeLa cell culture produces characteristic syncytia of fused cells. Human orthopneumovirus is a –ssRNA virus (class V) with an unsegmented genome and is a member of the family pneumoviridae. It is responsible for lower respiratory tract infections, especially in infants. Compare with Figure 10.17.

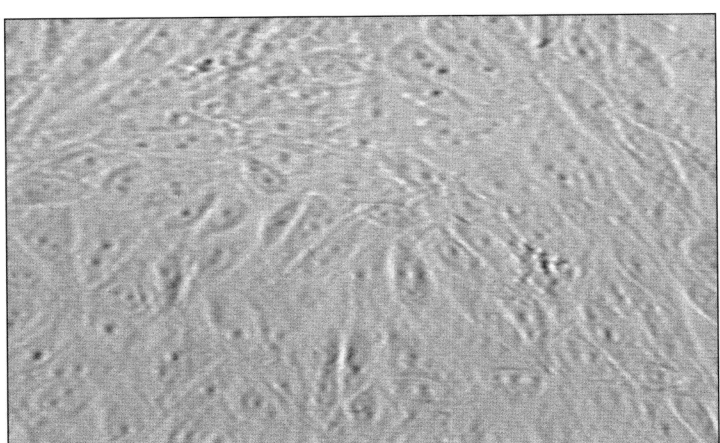

10.19 Normal African Green Monkey Kidney (AGMK) Cell Culture • This cell line is used for isolation of herpes simplex virus (HSV), varicella-zoster virus (VZV), mumps virus, and rubella virus, among others.

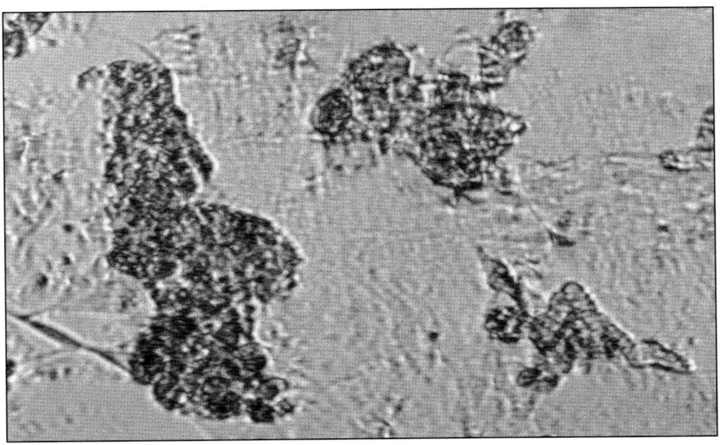

10.20 Measles Virus CPE in AGMK Cells • Syncytia formation is the typical CPE for measles (rubeola) virus in AGMK cells. Measles virus is a –ssRNA virus (class V) with an unsegmented genome and belongs to the family paramyxoviridae. It produces rubeola, also known as seven-day measles. The disease is highly contagious and can lead to serious complications and even death, but a vaccine is available to minimize the likelihood of infection. Compare with Figure 10.19.

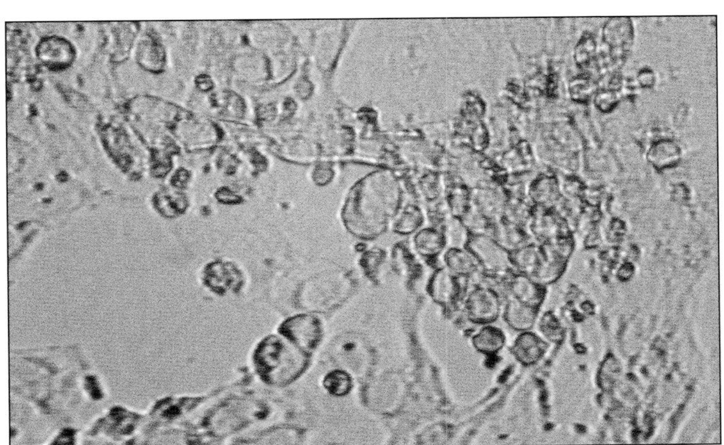

10.21 Influenza A Virus CPE in AGMK Cells • Cell degeneration and syncytia formation characterize the CPEs of influenza viruses in AGMK cells. Influenza virus is a –ssRNA virus (class V) with eight gene segments and is a member of the paramyxoviridae. Influenza is an acute respiratory disease that may reach epidemic and pandemic proportions. Compare with Figure 10.19.

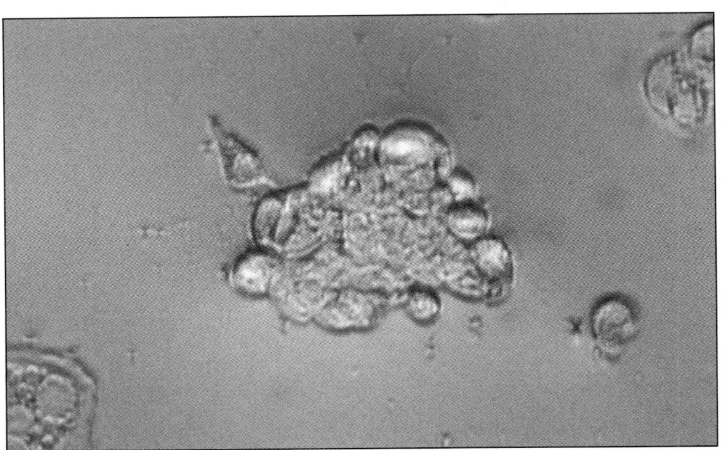

10.22 Adenovirus CPE in AGMK Cells • The typical adenovirus CPE in AGMK cells consists of rounded cells clustered like grapes. Adenoviruses have dsDNA genomes (class I) and cause a variety of upper and lower respiratory diseases, gastroenteritis, meningitis, and encephalitis, among others. Compare with Figure 10.19.

Hemadsorption in Cell Culture

Purpose

Hemadsorption is used for presumptive identification of influenza, parainfluenza, and sometimes mumps virus, because these viruses do not produce much cytopathic effect (CPE) in cell culture. Confirmation is accomplished by serological tests.

Principle

Infection with influenza, parainfluenza, or mumps virus results in viral glycoproteins being present in the infected cell's membrane. When these viruses emerge from the host, they carry the glycoproteins in their envelope (which is actually host cell membrane) and use them for attachment to and penetration of a new host cell. They also have the ability to **adsorb** (attach) to guinea pig RBCs. This property is exploited in the **hemadsorption test**.

Because influenza, parainfluenza, and mumps viruses often produce little or no CPE in cell culture, a hemadsorption test may be run. After incubation of the cell culture inoculated with the patient's sample, guinea pig RBCs are added to the medium. If the virus is present, the infected cells will have viral glycoproteins in their membranes, and the RBCs will adsorb to them (Fig. 10.23).

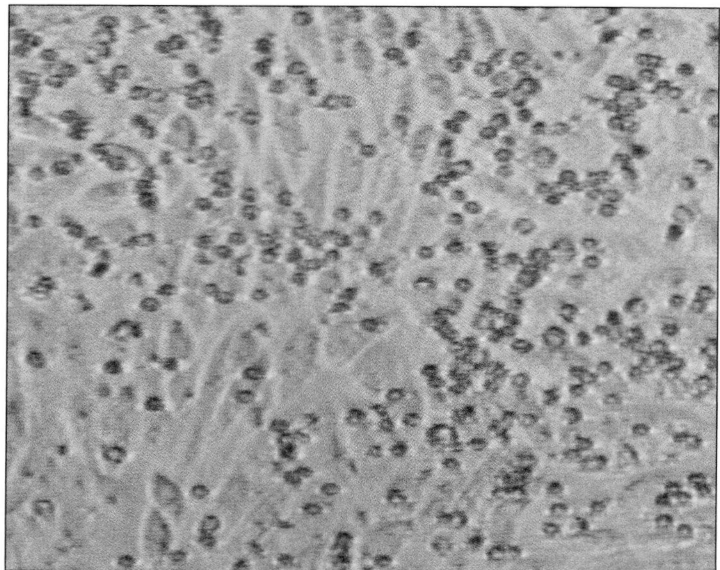

10.23 Hemadsorption Positive Test Result • Shown is hemadsorption of guinea pig RBCs by human diploid fibroblast cells infected with parainfluenza virus. Notice that the RBCs (darker circles) are always associated with the fainter pink fibroblast cells and are not in the spaces between cells. Parainfluenza viruses belong to the family paramyxoviridae and have unsegmented –ssRNA genomes (class V). They cause respiratory diseases, including croup.

Domain Bacteria

SECTION 11

What follows is not an exhaustive treatment of all bacterial groups. Rather, they are among the more commonly encountered Bacteria, or are ecologically important because of unique metabolic capabilities (admittedly, a subjective evaluation on the authors' part). As you read these descriptions, refer to the **cladogram** in Figure 1.2 for a tentative evolutionary perspective. Also, recall the taxonomic hierarchy.

Working from the most inclusive group (domain) downward to species, the taxa are: domain, kingdom, phylum (or division), class, order, family, genus, and species. In some bacterial lineages, a higher taxon has so few species or not enough is known about them that there are no (or few) subdivisions. For instance, the phylum Aquificae, has only one class, one order, and three families.

By convention, many phylum names end with "-ae," most order (or division) names end with "-ales" and those of families end with "-aceae." This helps inform readers familiar with those suffixes where they are reading within the hierarchy.

We have based the organization of this section on *Bergey's Manual of Systematic Bacteriology*, 2nd ed., Volumes 1–5.

Bergey's Manual of Systematic Bacteriology, 2nd ed., Volume 1

Phylum Aquificae

Aquificales is the only order in the phylum. The ancestors giving rise to these modern-day descendants are thought to have diverged from the remainder of Bacteria very early on. All are Gram-negative, chemolithoautotrophic,[1] hyperthermophilic,[2] motile rods. Members of this phylum can perform the reductive Krebs cycle, in which CO_2 is added to the Krebs intermediates by a process called carbon fixation rather than being removed. This ability is also what makes them autotrophs.

Aquifex (Fig. 11.1) is the type genus and it comprises two species. They are facultative anaerobic microaerophiles. Most impressively, they are able to grow between 67°C and 95°C! Electron donors include H_2, S^0, or $S_2O_3^{2-}$; O_2 and NO_3^- are electron acceptors. Metabolic products from sulfur are sulfuric acid and H_2S; NO_3^- produces NO_2 and N_2. The two *Aquifex* species have been recovered from hot springs, sulfur hot springs, and hydrothermal vents.

Sulfurihydrogenibium yellowstonense ("Sulfuri" for short; Fig. 11.2) was first described in 2005 and is found in pH-neutral, sulfur-rich, hot springs between 55°C and 78°C. Its metabolism differs from *Aquifex* in that it is an obligate aerobe, a facultative heterotroph, and is incapable of getting energy from H_2 oxidation. Additionally, arsenic compounds can be metabolized by some species.

[1] Chemolithoautotroph. Yikes! This tongue-twisting, compound word provides information about an organism's energy source, electron source, and carbon source. A chemolithoautotroph gets its energy from chemicals (*chemo*), its electrons from inorganic molecules (*lithos* means "stone"), and is able to make all its organic molecules from carbon dioxide (*auto* means "self"). The polar opposite would be a photoorganoheterotroph that gets its energy from light (*photo*), its electrons from organic compounds (*organo*), and carbon from organic molecules. (*Hetero* means "other," and indicates it is incapable of making its own organic molecules from CO_2; its organic carbon must come from some source *other than themselves*. The microbial world is so metabolically diverse, that all combinations of the three paired terms have been found, though some are more common than others. Finally, *trophos* means "feeder."

[2] *Therm* refers to "heat." A thermophile (*philos* means "loving") grows within a temperature range of 45°C to 80°C, whereas a hyperthermophile (*hyper* means "excessive") grows at temperatures above 80°C up to about 95°C. (Recall that 100°C is the temperature at which water boils.)

The cells form mats composed of streamers that look like fettucine. The streamers have the ability to enhance $CaCO_3$ (calcite) precipitation on and within them, which eventually preserves them as fossils in sculptured layers of travertine (limestone). Astrobiologists are interested in these distinctive formations as potential biomarkers of life on other planets with similar environmental conditions.

11.1 ***Aquifex*** **(Phylum Aquificae; Family Aquificaceae)** • These pink *Aquifex* streamers are seen in the hot runoff channel (about 90°C) of Octopus Springs in Yellowstone National Park. *Aquifex* is a chemolithoautotroph that gets its energy by oxidizing hydrogen gas or reduced sulfur compounds. The cells comprising the streamers are Gram-negative, motile, rods. Their dimensions range from 2–6 µm long by approximately 0.5 µm in diameter.

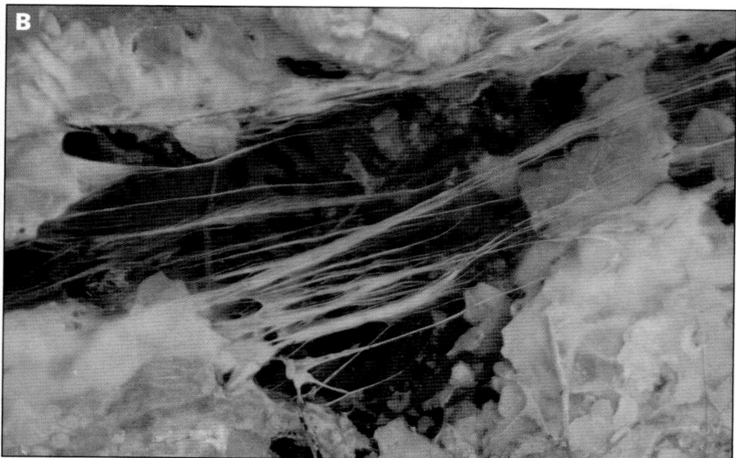

11.2 ***Sulfurihydrogenibium yellowstonense*** **(Phylum Aquificae; Family Hydrogenothermaceae)** • (A) In this photo, the white fringes are composed of *S. yellowstonense* streamers. The cells composing them are approximately 1–3 µm long by 0.6–0.8 µm in diameter. (B) Shown are some streamers growing in a small runoff channel in White Elephant Back Terrace. Note the sheets of extracellular polymeric substance (a mix of various compounds) that serve to strengthen the streamers against the forces of fast-flowing water. The yellow spots are sulfur deposits. (C) Most of the white material is travertine with trapped and fossilized *S. yellowstonense* streamers. You can see them as the gently curving patterns in the three blocks at the center left of the photo. Living streamers would be most abundant at the small waterfall and resulting stream, but they are not easily seen in this view because of the water. All photos were taken in the vicinity of the upper terrace at Mammoth Hot Springs in Yellowstone National Park. Photos (A) and (C) were taken near Canary Spring. Photo (B) was taken by the roadside at the White Elephant Back Terrace.

Phylum Thermotogae

Species of the single order Thermotogales are Gram-negative rods with a distinctive outer sheath ("toga"), which loosely encloses the cell at both ends (Fig. 11.3). All are anaerobic fermentative heterotrophs that produce acetate, lactate, CO_2, and H_2 (among other compounds) from glucose fermentation. H_2 inhibits growth, and to compensate for its production all have the ability to reduce sulfur and sulfate to H_2S. Species are thermophilic, but some are hyperthermophilic and have been recovered from hot springs and hydrothermal vents.

Thermotoga maritima was first isolated in 1986 from a volcanic island near Sicily and grows over a temperature range of 55°C–90°C. Its genome has been sequenced and surprisingly, approximately one-quarter of its genes were obtained by lateral gene transfer with archaeon species. This order, like Aquificae, appears to be an ancient branch within the Domain Bacteria.

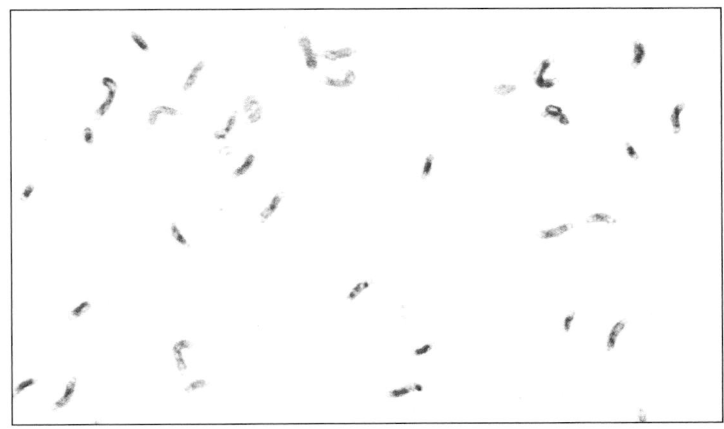

11.3 ***Thermotoga maritima*** **(Phylum Thermotogae; Family Thermotogaceae)** Cell dimensions are approximately 1 µm wide by 1.5–11.0 µm long, though these cells did not reach that maximum length. Note the light, oval regions at the ends of cells. This is where the external sheath (toga) is most prominent. The enclosed space is periplasm. (Safranin Stain)

Phylum Deinococcus-Thermus

Deinococcus species (order Deinococcales) form either spherical or slightly elongated cells (Fig. 11.4). Though they stain Gram positive, their wall has a thin peptidoglycan layer and an outer membrane (as in Gram-negative cells). They are nonmotile, aerobic chemoheterotrophs, and may be mesophilic or thermophilic.

Their genome consists of two different circular chromosomes and two smaller plasmids. *D. radiodurans* is noteworthy for its ability to survive high levels of ionizing and UV radiation, many times greater than the lethal level for humans. Surprisingly, this is due not to its ability to protect DNA from damage, but to its ability to protect the enzymes involved in DNA repair from oxidation (also an outcome of irradiation).

Beyond that, *D. radiodurans* carries multiple copies of repair enzyme genes, providing another mechanism for escaping the effects of DNA damage. Evidence supports the conclusion that this repair system evolved as a protective mechanism against desiccation, which produces similar DNA damage.

Thermus (order Thermales) is a genus of straight Gram-negative rods of variable length. They are thermophilic aerobic respirers (optimum temperature of about 70°C), preferring neutral and slightly alkaline hydrothermal regions (Fig. 11.5). *Thermus aquaticus* is famous for its DNA polymerase (*Taq*1), the enzyme that polymerizes DNA during synthesis. It is stable at high temperatures and is therefore useful in the polymerase chain reaction (PCR) technique of cloning DNA *in vitro* (see p. 144).

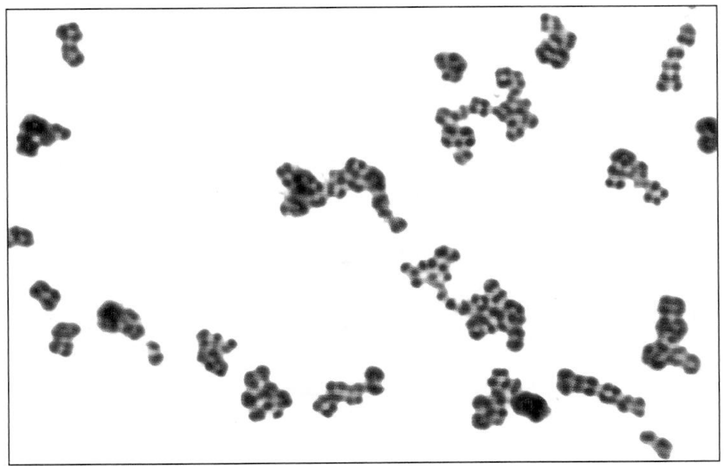

11.4 ***Deinococcus*** **(Phylum Deinococcus-Thermus; Family Deinococcaceae)** *Deinococcus* typically grows in pairs and tetrads (though sometimes as rods). Individual cells are up to 3.5 µm in diameter. They are Gram-positive in spite of an unusually thin peptidoglycan layer in their wall. These specimens were grown in culture. (Gram Stain)

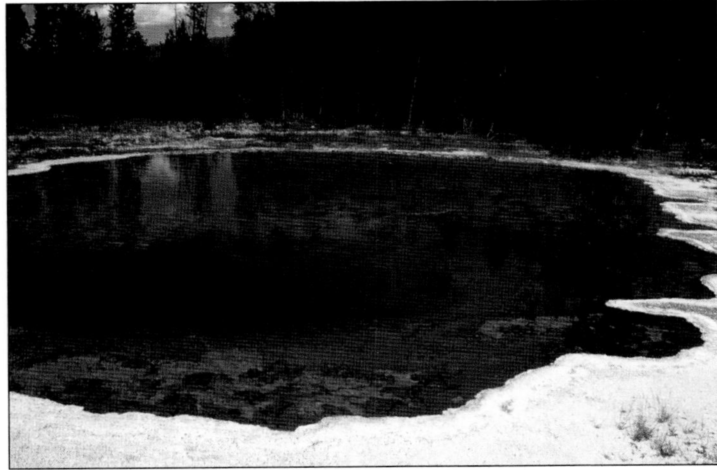

11.5 ***Thermus aquaticus*** **Habitat (Phylum Deinococcus-Thermus; Family Thermaceae)** Mushroom Spring in Yellowstone National Park is the site where Dr. Thomas Brock and Dr. Hudson Freeze first discovered *Thermus aquaticus* in 1969. Mushroom Spring's temperature is near 70°C (give or take a few degrees), which coincides with the optimum temperature for *T. aquaticus* growth. Its temperature range is roughly 50 to 75°C. The heat-stable DNA polymerase from *T. aquaticus* (*Taq*1 polymerase) is used in polymerase chain reaction (PCR).

Phylum Chloroflexi

Members of this phylum are filamentous rods with gliding motility. The phylum is divided into several classes. We focus on the class Chloroflexi, order Chloroflexales, which is composed of metabolically diverse species.

Mainly **photoheterotrophic** when growing anaerobically, they also can perform **anoxygenic photosynthesis** using bacteriochlorophyll *a* but don't use reduced sulfur as an electron source. They were once known as **green nonsulfur bacteria**; compare them to **green sulfur bacteria**, page 169.

Chlorosomes located next to the internal surface of the cytoplasmic membrane hold photosynthetic antenna pigments that transfer absorbed light energy to bacterial chlorophyll *a* located in the cytoplasmic membrane. Chlorosomes are membrane-bound, but not by a typical phospholipid bilayer, and are highly efficient at absorbing light energy allowing chloroflexi to grow in habitats without much light.

Aerobically, many chloroflexi are capable of chemoheterotrophic metabolism. Most are thermophilic. *Chloroflexus* is the type genus (Fig. 11.6). Cells are 0.5–1.0 μm wide by 2–6 μm long and have an atypical Gram-negative wall.

11.6 ***Chloroflexus*** **Habitat (Phylum Chloroflexi; Family Chloroflexaceae)** *Chloroflexus* is the most studied member of the Chloroflexi. In this view of Whirligig Geyser (and other low sulfur hot springs) at Yellowstone National Park, *Chloroflexus* is often responsible for the orange microbial mats in the runoff channel, and is often found in association with cyanobacteria. Its preferred temperature is between 50°C and 60°C. The green algae at the left of the channel grow between 38°C and 56°C, dramatically illustrating the varied habitats of Yellowstone's geological features and the variety of thermophilic organisms that reside in them.

Phylum Cyanobacteria (Cyanophyta)

Cyanobacteria are easily seen without staining because of their combination of photosynthetic pigments, which confer a bluish-green color. In fact, they were formerly known as blue-green algae. All are capable of **oxygenic photosynthesis** (oxygen is a waste product, just like plant photosynthesis[3]) and all are Gram negative, though their peptidoglycan is thicker than most other Gram negatives.

When they are single-celled, they are about the size of bacteria and are best viewed with oil immersion. But when they are found in chains, called **trichomes**, they are easily visible (as a group) at high-dry magnification. Some trichomes have an extracellular **sheath**.

Trichomes often have specialized cells, including **heterocysts**, which are capable of **nitrogen-fixation** (see p. 307 for more about the nitrogen cycle), and **akinetes**, which are resistant spores. Many trichomes are capable of gliding motility, which means movement on a solid surface without the use of flagella. Many different mechanisms have been described, but cyanobacteria secrete a slime and they slide along it. Figures 11.7 to 11.17 illustrate some common forms and cyanobacterial variability. All micrographs are unstained wet mount preparations.

Microbialites are found in aquatic environments and are structures of various shapes formed by an interaction between primary producer(s), sediments, and calcium carbonate. **Stromatolites** are a subset of microbialites that grow in layers. Fossil stromatolites dated to 3.5 billion years before the present contain the earliest known fossils (Fig. 11.18A). In extant microbialites (and those fossils), the most common primary producers are cyanobacteria, hence their inclusion in this section.

One model of extant microbialite growth begins with a cyanobacterial mat and its associated **extracellular polymeric substance** (EPS). The EPS traps organic and inorganic ($CaCO_3$) sediments that add to the developing microbialite's mass. The process continues and is guided by a phototropic response by the cyanobacteria, so the microbialite grows upward toward the sun, ultimately becoming dome-shaped. As the microbialite enlarges, opportunities for occupation by other microbes present themselves, which produces a more complex community. Eventually, the dome collapses (Fig. 11.18B), though the community may survive.

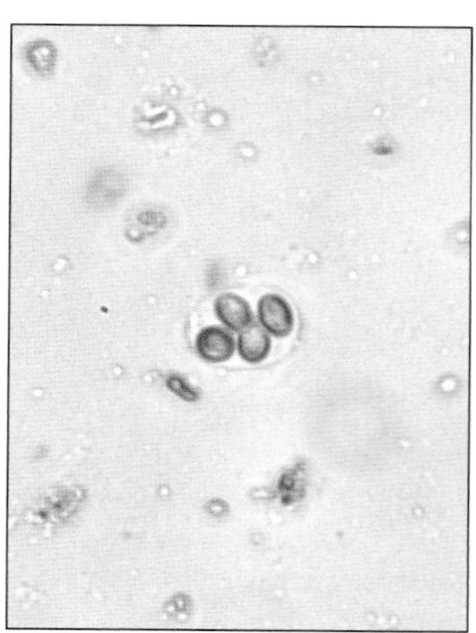

11.7 ***Gloeocapsa*** **(Phylum Cyanobacteria; Order Chroococcales)** This cyanobacterium is distinctive because of its one, two, or four spherical cells held together within a layered sheath. Cell clusters are produced by two successive binary fission divisions from one original cell. Cells are 3–10 μm in diameter. (Light Micrograph)

[3] More correctly we should say that plants do cyanobacterial photosynthesis, because their chloroplasts were once free-living cyanobacteria.

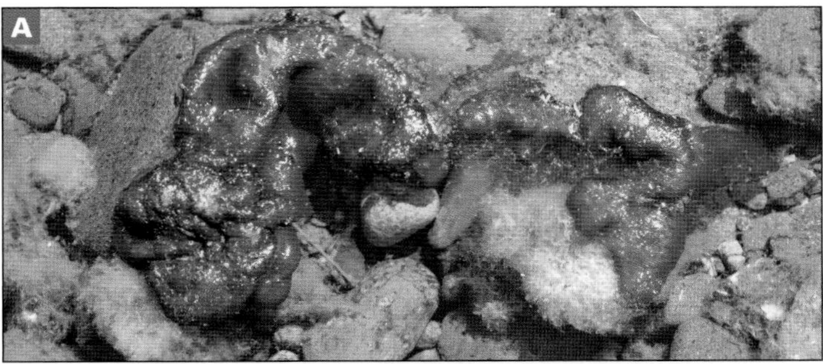

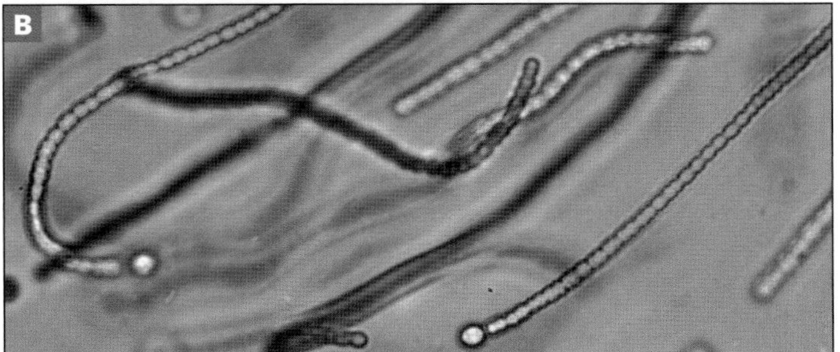

11.8 ***Nostoc* colonies (Phylum Cyanobacteria; Order Nostocales)** • (A) This genus is easily identified macroscopically because of its globular colonies and thick, rubbery mucilage. (B) Trichomes are composed of spherical cells and form a tangled mass within the colony. Note the terminal spherical heterocysts. Trichomes are about 5 µm in width. (Light Micrograph)

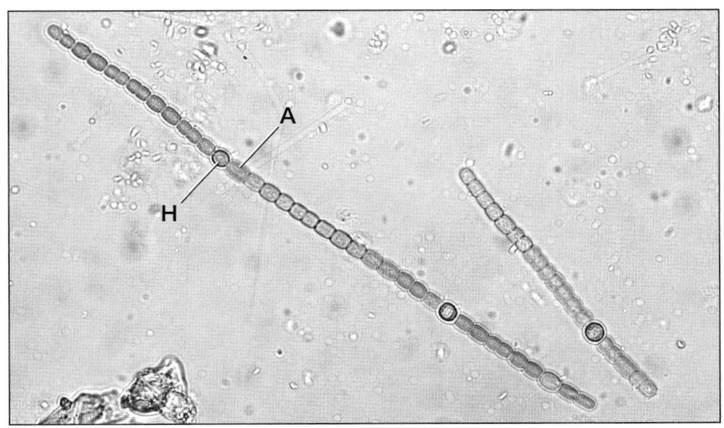

11.9 ***Anabaena* (Phylum Cyanobacteria; Order Nostocales)** • The trichomes of this gliding cyanobacterium may possess thick-walled spores called akinetes (**A**) and specialized, nitrogen-fixing cells called heterocysts (**H**). Trichomes are of variable lengths but are approximately 20 µm in width. Reproduction is by fragmentation of the trichome. (Light Micrograph)

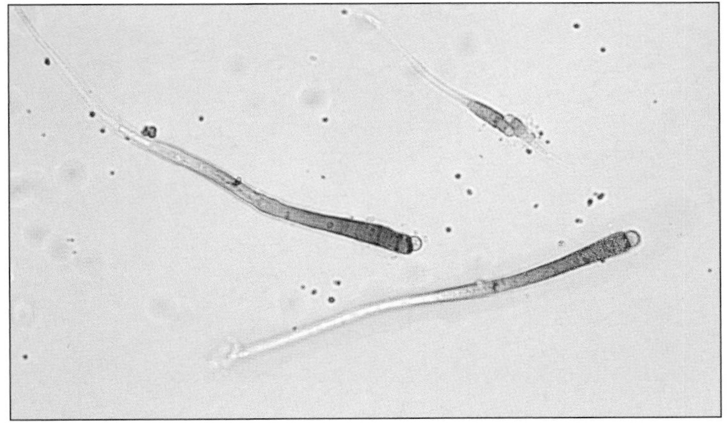

11.10 ***Rivularia* (Phylum Cyanobacteria; Order Nostocales)** • Macroscopically, trichomes form globular, gelatinous masses. Microscopically, trichomes are tapered to a fine point. Note the round, basal heterocysts. Akinetes are absent. *Rivularia* species are found in marine habitats or habitats saltier than freshwater. (Light Micrograph)

11.11 ***Calothrix* (Phylum Cyanobacteria; Order Nostocales)** • The genus *Calothrix* once included species now classified in the genus *Rivularia*, but DNA hybridization did not support their inclusion. *Calothrix* species are freshwater or terrestrial, grow in tapered trichomes, and possess heterocysts. They are facultative heterotrophs that ferment in the dark. This photograph shows *Calothrix* growing in the outfall of a runoff channel from Black Sand Basin into the Firehole River at Yellowstone National Park. The dark brown mats are *Calothrix*. It prefers a neutral to slightly alkaline pH and grows between 30°C and 45°C. (See the steam?)

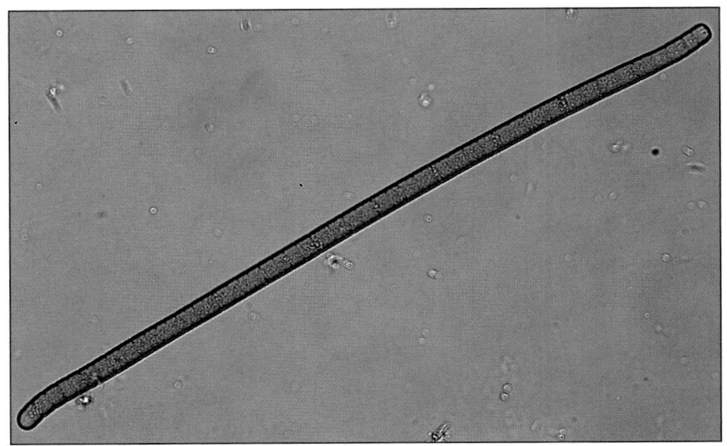

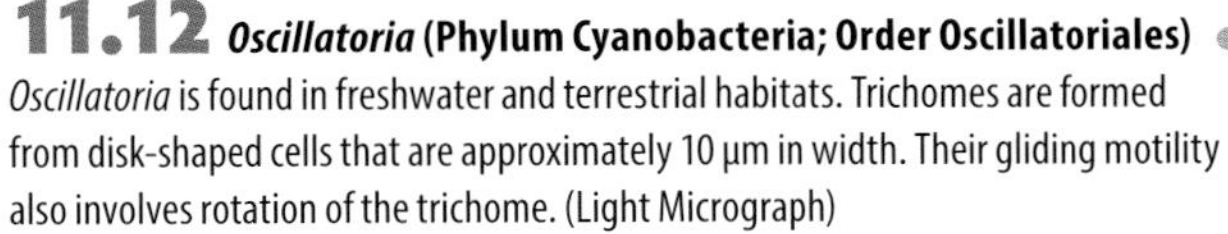

11.12 ***Oscillatoria* (Phylum Cyanobacteria; Order Oscillatoriales)** • *Oscillatoria* is found in freshwater and terrestrial habitats. Trichomes are formed from disk-shaped cells that are approximately 10 µm in width. Their gliding motility also involves rotation of the trichome. (Light Micrograph)

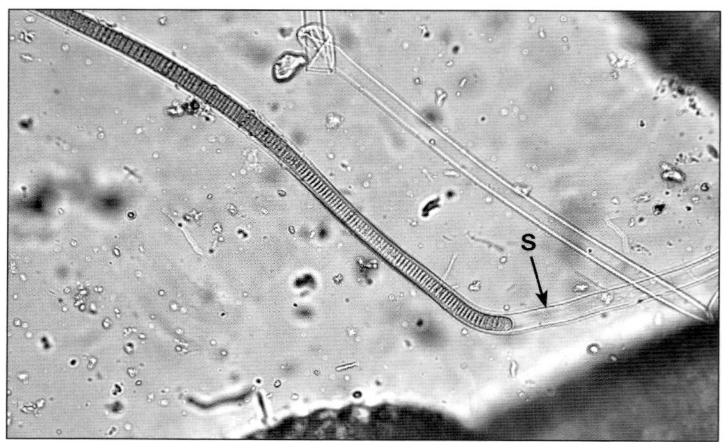

11.13 ***Lyngbya*** **(Phylum Cyanobacteria; Order Oscillatoriales)** • *Lyngbya* trichomes have a sheath (**S**) that distinguish them from *Oscillatoria* (see Fig. 11.12). The trichomes are approximately 20 µm in width. (Light Micrograph)

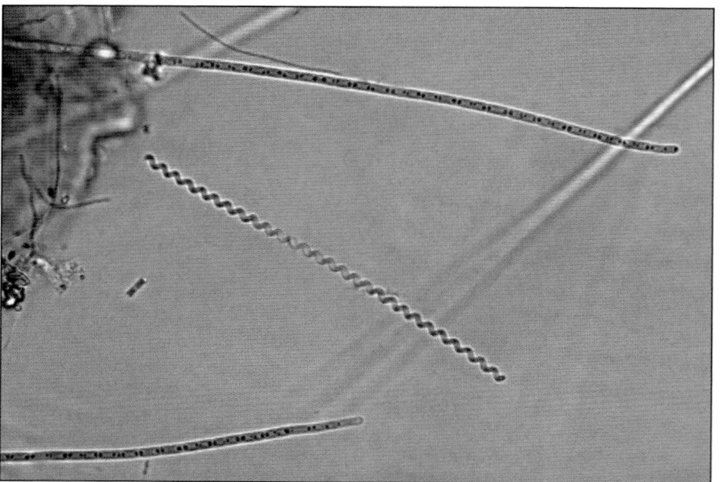

11.15 ***Spirulina*** **(Phylum Cyanobacteria; Order Spirulinales)** • *Spirulina* is perhaps the most distinctive cyanobacterium because of its helical shape. The width of the helix varies, but can obtain sizes up to 12 µm. *Spirulina* species are found in both marine and freshwater. This specimen was in a sample of San Diego Bay water. *Spirulina* is sold in health food stores as a dietary supplement. (Light Micrograph)

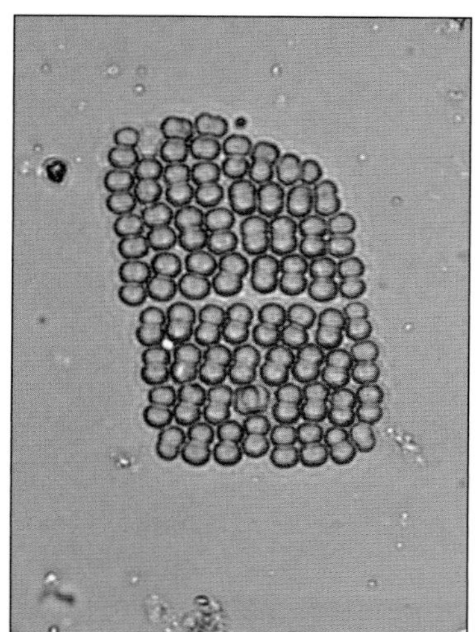

11.17 ***Merismopedia*** **(Phylum Cyanobacteria; Order Synechococcales)** • Cell division in two perpendicular directions produces the planar arrangement characteristic of this genus. The cells are enclosed in a mucilaginous sheath and are approximately 1–2 µm in diameter. Most are freshwater species. (Light Micrograph)

11.14 ***Phormidium*** **(Phylum Cyanobacteria; Order Oscillatoriales)** • *Phormidium* species are freshwater organisms and like other members of the Oscillatoriales, they grow in trichomes composed of disk-shaped cells. The ends of the trichome are generally tapered, and a thin sheath is also present. They grow at neutral pH and between 35°C and 57°C. *Phormidium* is the most obvious member of the orange mat surrounding Grand Prismatic Spring at Yellowstone National Park. The species growing there possesses carotenoid pigments to shield the cells from the harmful effects of solar radiation. The brown mat to the right is *Calothrix*, which has a maximum temperature approximately 10°C less than *Phormidium*. So once again, we see temperature affecting distribution of the organisms living around geothermal features.

11.16 ***Synechococcus*** **(Phylum Cyanobacteria; Order Synechococcales)** • *Synechococcus* strains grow across the full spectrum of habitats—freshwater to seawater to hot springs. The yellow mat in this photo is composed of a thermophilic *Synechococcus* strain growing in a runoff channel from Grand Prismatic Spring (out of frame, but to the right) in Yellowstone National Park. The spring itself is 87°C, but the water in the runoff channel cools as it flows away from its source and provides different habitats for species adapted to them. If you look carefully at the colors on the far "bank," you can see stratification of *Synechococcus* (yellow) in the hottest water, *Phormidium* (orange) in the middle, and *Calothrix* (brown) in the coolest water on top. *Synechococcus* cells are short rods a little more than 1 µm in length, nonmotile, and obligately photoautotrophic.

11.18 Microbialites (A) Fossil stromatolites have been dated to 3.5 billion years before the present. These marine stromatolites are found at Shark Bay, Western Australia. They are approximately 1 m tall. (B) These are microbialites along the shore of Antelope Island in the Great Salt Lake, Utah. Originally domed, they have collapsed, leaving only their basal ring.

Phylum Chlorobi

Chlorobi are Gram-negative, obligate anoxygenic phototrophs, and are informally referred to as the **green sulfur bacteria.** (Their role in the sulfur cycle is discussed on p. 311. Also see Fig. 19.36.) They possess chlorosomes, but unlike the chloroflexi, use H_2S as the electron donor. Carbon fixation occurs by the reverse Krebs cycle.

While found in similar habitats as the purple sulfur bacteria (i.e., aquatic, **anoxic,** H_2S-rich mud), they are able to grow at deeper levels because of the chlorosomes' efficiency at capturing the limited light. *Chlorobium* (order Chlorobiales) is the type genus of the phylum. A Winogradsky column with green sulfur bacterial growth (and likely including *Chlorobium*) is shown in Figure 11.19. Also see pages 305 and 306.

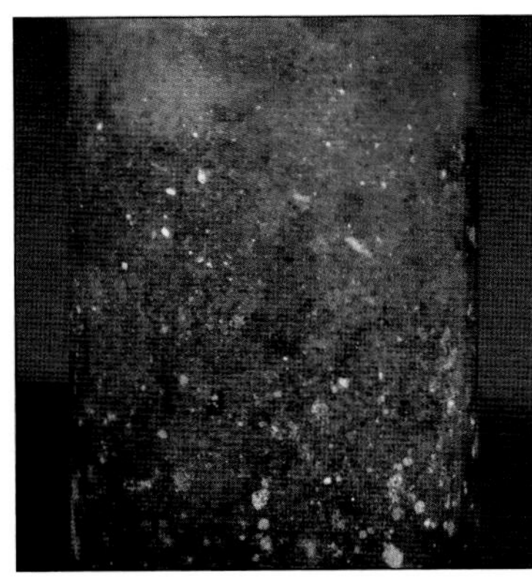

11.19 Green Sulfur Bacteria An accumulation of green sulfur bacteria (GSB) is seen at the left side of this Winogradsky column. *Chlorobium* is a likely inhabitant.

Bergey's Manual of Systematic Bacteriology, 2nd ed., Volume 2

Phylum Proteobacteria

The largest and most diverse Bacterial phylum is the Proteobacteria. All are Gram negative and show relationship based on 16s RNA comparisons. Beyond that, they exhibit the gamut of aerotolerance, energy metabolism, and cell morphology categories. They also comprise the most commonly cultivated Gram-negative organisms of medical, industrial, and general importance. The phylum is split into five classes: Alphaproteobacteria, Betaproteobacteria, Gammaproteobacteria, Deltaproteobacteria, and Epsilonproteobacteria.

Class Gammaproteobacteria (Volume 2B)

Chromatiales

There are three families in the Chromatiales (also known as **purple sulfur bacteria**). Their unifying feature is the ability to photosynthesize using H_2S as the electron donor for CO_2 reduction in photosynthesis. The S^0 resulting from H_2S oxidation is either stored as intracellular or extracellular granules.

Because oxygen is not a product, these photoautotrophs are said to perform **anoxygenic photosynthesis** (as opposed to the oxygenic photosynthesis of cyanobacteria). The photosynthetic pigments bacteriochlorophyll *a* and carotenoids are embedded in the membranes of vesicles formed from infoldings of the cytoplasmic membrane.

Purple sulfur bacteria are found in freshwater and marine mud and sediments illuminated by sunlight but lacking oxygen. The genera *Chromatium* and *Allochromatium* (Fig. 11.20) belong to this group. For more information about the sulfur cycle, see pages 309–312 in Section 19.

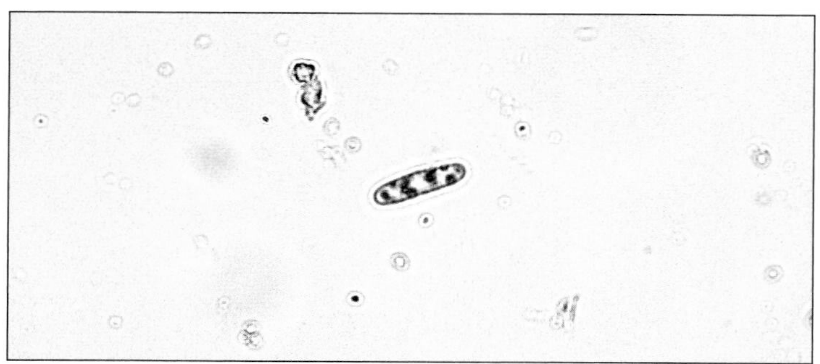

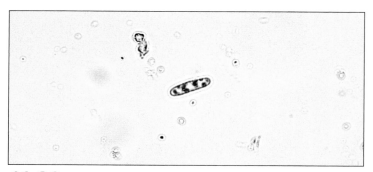

11.20 ***Allochromatium*** **(Phylum Proteobacteria; Class γ-Proteobacteria; Order Chromatiales)** • This sulfur-oxidizing bacterium was provisionally identified as *Allochromatium*, a purple sulfur bacterium, based on its morphology, size, and the evenly distributed sulfur granules in the cytosol. (Light Micrograph)

Thiotrichales

There are three families within the order and we will cover one: Thiotrichaceae. Metabolic diversity abounds in the family, because it is composed of chemolithotrophic and chemoorganotrophic species. The genus *Beggiatoa* will serve as our example.

All *Beggiatoa* species are microaerophilic and have a respiratory metabolism, using oxygen or at times, nitrate, as the final electron acceptor. They oxidize H_2S and store the resulting S^0 as granules. Their cell wall stains Gram negative, but has a unique structure with extra layers external to the typical "outer" membrane. Absence of a sheath surrounding their filaments differentiates them from other Thiotrichaceae.

Beggiatoa species grow in mats associated with marine and freshwater sulfide-rich sediments, and use gliding motility in response to environmental gradients to position themselves favorably within the mat. There are metabolic and morphological differences between species occupying the two environments.

Freshwater *Beggiatoa* species form long, thin (<5 nm) filaments (Figs. 11.21 and 19.37). They are also chemoorganotrophic, using acetate or other small organic compounds as both a carbon and energy source. Marine species' filaments are generally thicker (up to 200 µm.), though some species form thin filaments. The former are obligate chemolithoautotrophs, whereas the latter are facultative chemolithoautotrophs. Much still remains to be learned about details of their metabolism, however.

Beggiatoa species demonstrate negative tropic responses to light, oxygen, and H_2S. Within their mat, there is an oxygen and light gradient diminishing downward from above, and an H_2S gradient diminishing upward from below.

During the daytime when photosynthesis occurs, they migrate downward due to the light and oxygen produced by oxygenic photosynthesizers. At night, they migrate upward to reach their optimum oxygen level (as oxygen is being removed by aerobic respiration, but not replaced by photosynthesis).

Thick marine species and freshwater species utilize stored nitrate and elemental sulfur as final electron acceptors when they find themselves in an anaerobic environment.

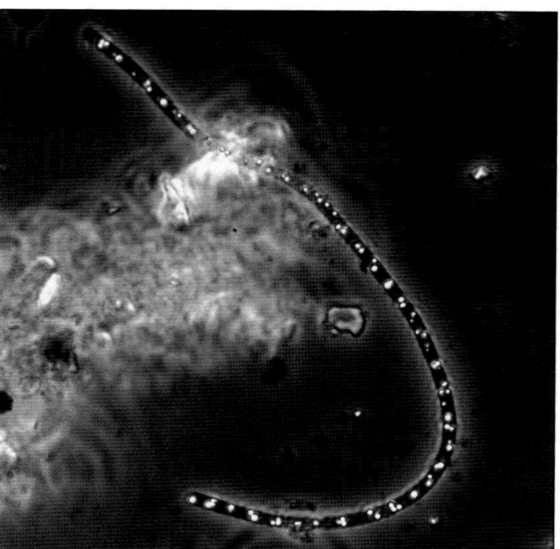

11.21 ***Beggiatoa*** **(Phylum Proteobacteria; Class γ-Proteobacteria; Order Thiotrichales)** • *Beggiatoa* is a common gliding bacterium of anaerobic muds containing H_2S. The circles in the cytosol are elemental sulfur granules produced by H_2S oxidation, the mechanism by which *Beggiatoa* gets its energy. (Phase Contrast)

Pseudomonadales

Members of the Pseudomonadales are respiratory, aerobic chemoorganotrophs and are nonmotile or motile by polar or peritrichous flagella. Two pseudomonad families are treated here: Pseudomonadaceae (type genus *Pseudomonas*) and Moraxellaceae (type genus *Moraxella*).

The Pseudomonadaceae holds eight genera, including *Pseudomonas* and *Azotobacter*. *Pseudomonas* species (Fig. 11.22) are rod shaped and up to 5 µm in length. They are found in a variety of environments and some are opportunistic pathogens (see Fig. 12.52).

Some species are fluorescent and others, such as *P. aeruginosa*, produce distinctive pigments (see Fig. 3.23). Fluorescent species have been shown to benefit plants by reducing plant root pathogens through antibiotic and HCN production and sequestering iron. They also produce plant hormones that promote root growth.

Azotobacter (Fig. 11.23A) is another member of the Pseudomonadaceae. Cells are rods with rounded ends. Resistant **cysts** (Fig. 11.23B) are formed in response to environmental cues, but aren't the equivalent of endospores. *Azotobacter* species are free-living soil bacteria (in contrast to *Rhizobium* with its root nodules) that perform nitrogen fixation. For more on the nitrogen cycle, see pages 307–309 and Figure 19.31.

There are three genera in the Moraxellaceae, including the type genus *Moraxella*. *Moraxella* species usually grow in pairs or short chains of short rods or cocci, though *M. catarrhalis* grows as tetrads (Fig. 11.24). Flagella are absent, but some strains demonstrate pilus-mediated **twitching motility** on surfaces. They are obligate aerobes and are typically catalase and oxidase positive. Acids are not produced from carbohydrate catabolism.

11.22 ***Pseudomonas putida*** **(Phylum Proteobacteria; Class γ-Proteobacteria; Order Pseudomonadales)** • Of the more than 50 *Pseudomonas* species, most are obligate aerobes, but a few (e.g., *P. aeruginosa* and *P. fluorescens*) can also use nitrate as a final electron acceptor and respire anaerobically. *P. putida* belongs with the former group. Its cells are about 5 µm long by slightly less than 1 µm in diameter and it is motile by more than one polar flagellum. In addition, it produces pyoverdine, a yellow-green fluorescent pigment, that sequesters iron from the environment. (Gram Stain)

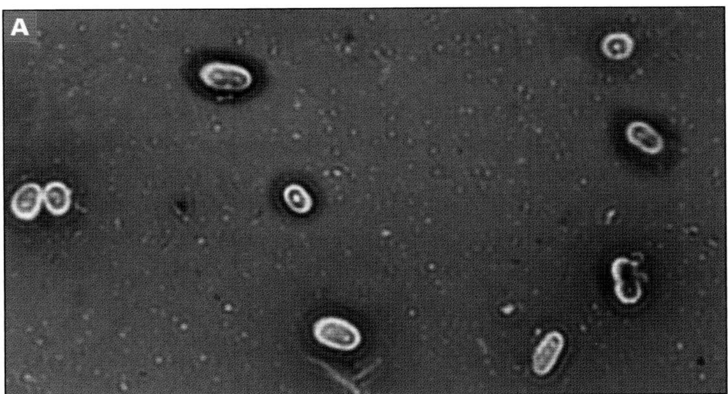

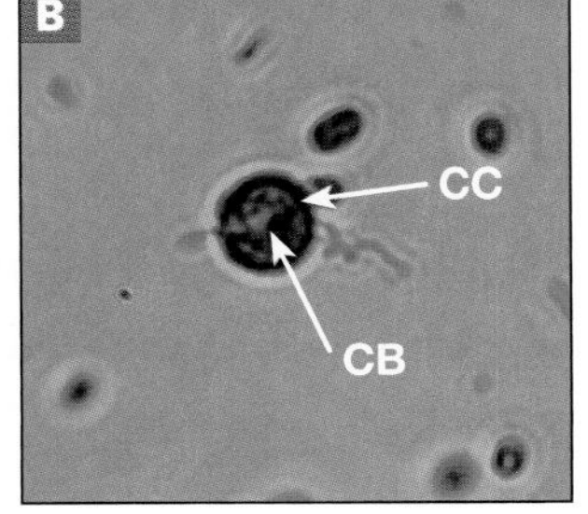

11.23 ***Azotobacter vinelandi*** **(Phylum Proteobacteria; Class γ-Proteobacteria; Order Pseudomonadales)** • (**A**) *Azotobacter vinelandi* is a soil organism that fixes nitrogen, but also reduces nitrate to nitrite (not all *Azotobacter* species can do this). Cells are plump rods with blunt or rounded ends, peritrichous flagella, and a capsule. These were grown in nitrogen-free broth, an enrichment medium used to isolate nitrogen-fixing bacteria. (Capsule Stain) (**B**) *Azotobacter* species can produce cysts that are resistant to certain chemicals and desiccation. They are a resting stage, but are not as metabolically inert as bacterial endospores. In the micrograph, the cyst coat (**CC**) is seen surrounding the central body (**CB**), which is basically, the resting cell. (*Azotobacter* Cyst Stain)

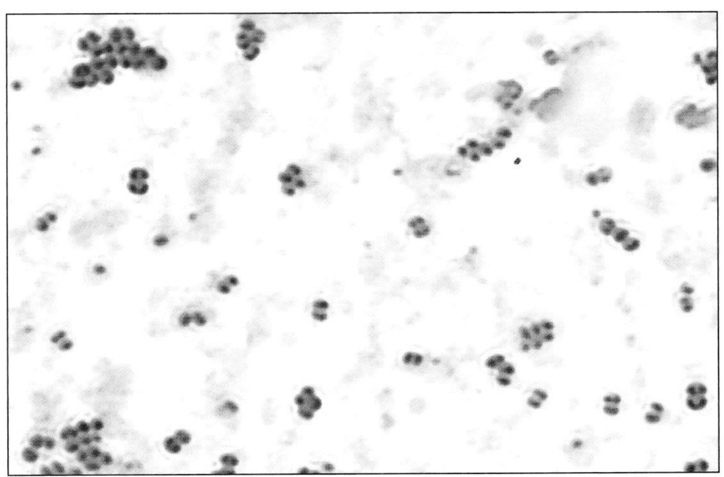

11.24 ***Moraxella*** **(Phylum Proteobacteria; Class γ-Proteobacteria; Order Pseudomonadales)** • *Moraxella catarrhalis* is primarily isolated from the nasopharynx and is unusual among Gram-negative bacteria for its resistance to penicillin. Cells are found as singles, pairs, and tetrads, with a diameter close to 1 µm. (Gram Stain)

Vibrionales

Vibrio species are straight to curved rods that are motile by polar flagella. All are facultative anaerobes, capable of aerobic respiration and fermentation (Fig. 11.25). Most are found in aquatic habitats and some are bioluminescent (see Fig. 19.27). Important genera are *Vibrio* with more than 40 species, and *Photobacterium*. Some *Vibrio* species are pathogenic (e.g., *V. cholerae*; see Figs. 12.73 and 12.74).

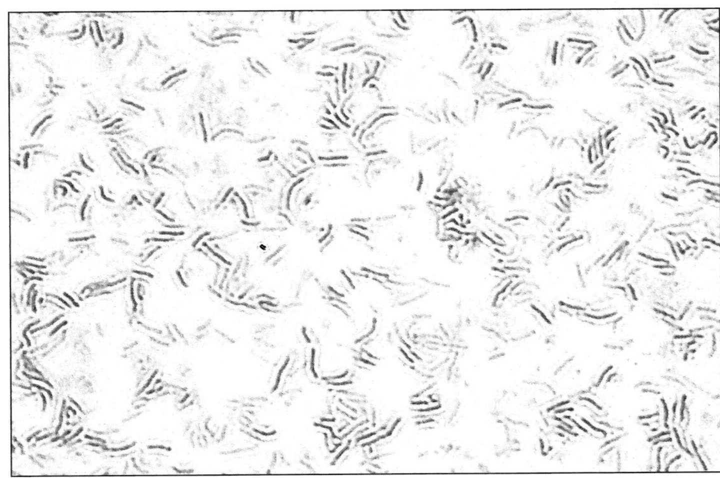

11.25 ***Vibrio anguillarum*** **(Phylum Proteobacteria; Class γ-Proteobacteria; Order Vibrionales)** • *Vibrio* species are facultative anaerobes whose cells range from short straight rods to curved rods. They are motile by polar flagella surrounded by a sheath that is continuous with the outer membrane. *V. anguillarum* is a marine species of fairly long, thin, and slightly curved rods. Its name is descriptive: *anguilla* is Latin for "eel." (Gram Stain)

Enterobacteriales

This order comprises 42 genera, all of which are in the family Enterobacteriaceae. Cells are straight rods and are motile by peritrichous flagella or nonmotile. They are chemoorganotrophs, capable of both aerobic respiration and fermentation, in which acid (and often gas) is produced from glucose and often many other sugars.

Most are catalase positive (see p. 79), reduce NO_3 (see p. 100), and are oxidase negative (see p. 105). They also share a unique antigen (enterobacterial common antigen) found in the outer leaflet of the outer membrane. Although the name Enterobacteriaceae literally means "gut bacteria," their habitats also include soil, water, and plant material.

Important genera include the type genus *Enterobacter* (Fig. 11.26), *Photorhabdus* (Fig. 11.27), *Citrobacter* (see Fig. 12.17), *Erwinia*, *Escherichia* (see Fig. 12.25), *Klebsiella* (see Fig. 12.33), *Proteus* (see Fig. 12.50), *Salmonella* (see Figs. 12.55 and 12.56), *Shigella* (see Fig. 12.59), and *Yersinia* (see Fig. 12.75). Figures from Section 12 are pathogenic species.

11.26 ***Enterobacter cloacae* (Phylum Proteobacteria; Class γ-Proteobacteria; Order Enterobacteriales)** • Like the majority of *Enterobacter* species, *E. cloacae* has an IMViC formula of indole (–), methyl red (–), Voges-Proskauer (+), and citrate (+). *E. cloacae* rods are approximately 2–3 µm long by 1 µm in diameter, which is fairly typical for Enterobacteriaceae. A few *E. cloacae* strains growing in association with rice roots have been shown to fix nitrogen. (Gram Stain)

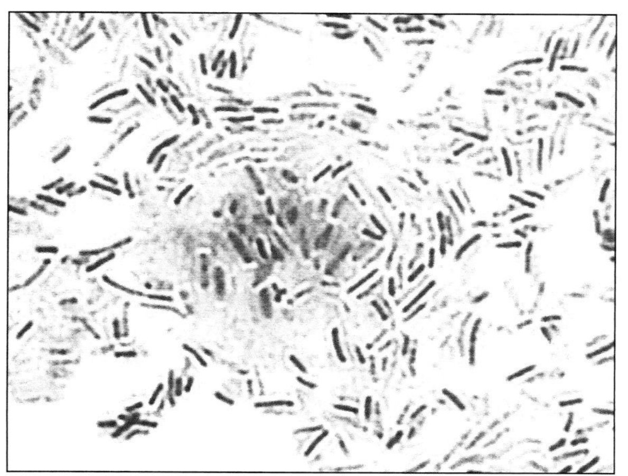

11.27 ***Photorhabdus* (Phylum Proteobacteria; Class γ-Proteobacteria; Order Enterobacteriales)** • The three species of *Photorhabdus* occupy the gut of juvenile stages of the nematode worm *Heterorhabdus*. When the juvenile infects an insect host, *Photorhabdus* enters the insect, multiplies, and kills it. At the same time, the nematodes reproduce, and when they leave the insect carcass, they take *Photorhabdus* with them to repeat the cycle. Although *Photorhabdus* DNA results in only 4% hybridization with *E. coli* DNA, it possesses the enterobacterial common antigen, a key genetic marker for the Enterobacteriaceae. So, for the time being it remains in the family. (Gram Stain)

Class Alphaproteobacteria (Volume 2C)

Rhodospirillales

Rhodospirillaceae (type genus *Rhodospirillum*) and Acetobacteraceae (type genus: *Acetobacter*) are the two families of this order.

Members of Rhodospirillaceae belong to a physiological bacterial group called the **purple nonsulfur bacteria (PNSB)**. *Rhodospirillum* species are metabolically versatile, freshwater chemotrophs that ferment when the environment is dark and anoxic (oxygen deficient). Anaerobically and in the light, they are photoheterotrophs.

Their photosynthetic machinery is carried in multiple membrane-bound photosynthetic vesicles derived from the cytoplasmic membrane. Oxidation of hydrogen and sulfide (in low concentrations) provide electrons. They are also capable of nitrogen fixation. Cells are curved rods to spirals (Fig. 11.28) and polar flagella provide motility.

Acetobacter (Fig. 11.29) and *Gluconobacter* (Fig. 11.30) are two genera of **acetic acid bacteria** within the Acetobacteraceae. They are obligate aerobic respirers that oxidize ethanol to acetic acid. The former then oxidizes acetic acid to CO_2 and H_2O, whereas the latter does not. Both use the **pentose phosphate pathway** (see Appendix, Fig. A.4) for sugar catabolism in place of glycolysis because they lack phosphofructokinase (see Appendix, Fig. A.2). Both are Gram-negative rods and if motile, have peritrichous flagella.

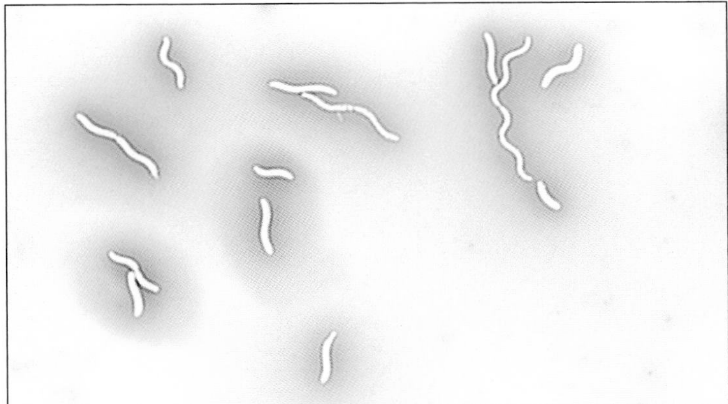

11.28 ***Rhodospirillum rubrum*** **(Phylum Proteobacteria; Class α-Proteobacteria; Order Rhodospirillales)** • *R. rubrum* is a freshwater mesophile and grows as curved or spiral-shaped cells, obtaining lengths up to 10 µm. Under appropriate conditions, it can grow as a photoheterotroph, photoautotroph, or a chemotroph. (Congo Red Negative Stain)

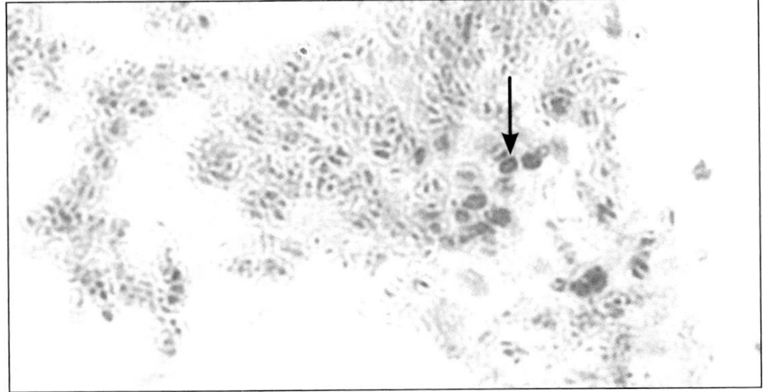

11.29 ***Acetobacter aceti*** **(Phylum Proteobacteria; Class α-Proteobacteria; Order Rhodospirillales)** • *Acetobacter* species are obligate aerobes capable of oxidizing ethanol to acetic acid (acetate) and further oxidizing the latter to CO_2 and H_2O during aerobic respiration. In older cultures, involution forms can be observed. Some of these irregularly shaped cells are visible in this micrograph (**arrow**). Cells are short and elliptical or straight rods obtaining a length of 4 µm. They may be arranged as singles, pairs, or rarely chains. They grow in association with plants. (Gram Stain)

Rickettsiales

Members of this order are obligate intracellular parasites that include a vertebrate and an arthropod host in their life cycles. They are small, Gram-negative rods and many are pathogenic. Important genera are *Rickettsia* (family Rickettsiaceae) and *Ehrlichia* (family Anaplasmataceae).

Rickettsia species have small genomes and lack the ability to make enzymes of carbohydrate metabolism and lipid, amino acid, and nucleotide synthesis, but they do have a complete citric acid cycle. They also have ATP/ADP translocase genes, which enable them to exchange ADP for host cytoplasm ATP. (Think about that!) *Rickettsia rickettsii* causes Rocky Mountain Spotted Fever and is covered in more detail in Section 12 (see Fig. 12.54).

Ehrlichia species are Gram negative and pleomorphic. Transmitted by ticks, they are intracellular parasites of leukocytes and their precursors, among other cells. They have a larger genome than rickettsias and have more metabolic abilities. Infection eventually results in host cell death. In humans, they cause ehrlichiosis, a disease with flu-like symptoms.

Sphingomonadales

Sphingomonadaceae is the only family in the order. Cells stain Gram negative, but have an atypical cell wall, with **sphingolipids** in the outer membrane instead of an LPS layer. The type genus is *Sphingomonas* (Fig. 11.31) and its species grow as straight, curved, or pleomorphic rods.

All are aerobic respirers and many produce an extracellular capsule. They have been isolated from a variety of natural habitats and some (e.g., *S. paucimobilis*) are opportunistic pathogens of humans. Some species are capable of degrading environmental pollutants, such as dioxin and aromatic hydrocarbons, among others.

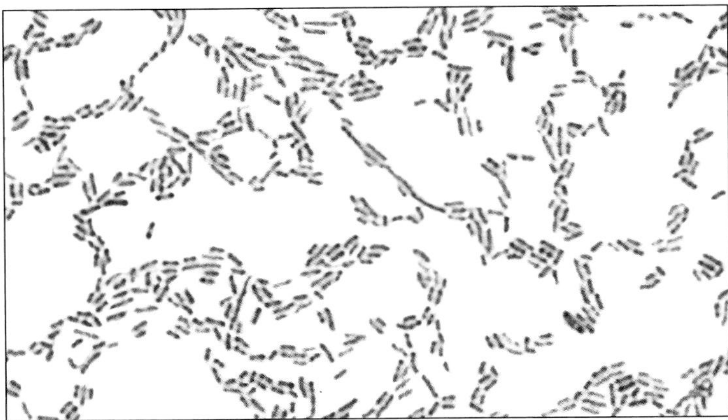

11.30 ***Gluconobacter oxydans*** **(Phylum Proteobacteria; Class α-Proteobacteria; Order Rhodospirillales)** • *Gluconobacter* cells resemble *Acetobacter* in size, shape, and arrangement, but chains are rare. Like *Acetobacter, Gluconobacter* is capable of oxidizing ethanol, but it can't complete the oxidation to CO_2 and H_2O. They typically grow associated with fruits and flowers, which are sugar-rich environments. (Gram Stain)

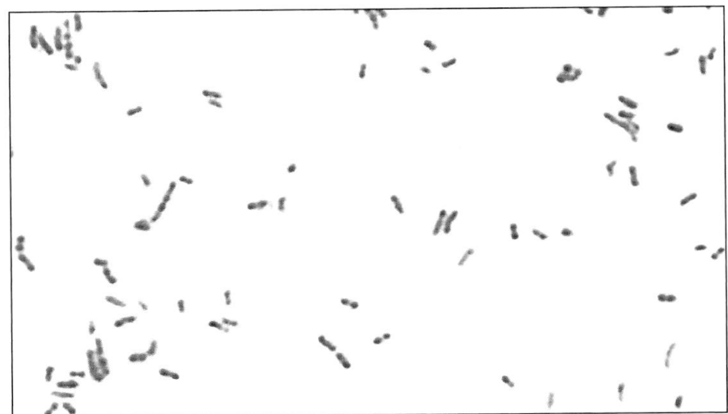

11.31 ***Sphingomonas capsulata*** **(Phylum Proteobacteria; Class α-Proteobacteria; Order Sphingomonadales)** • *Sphingomonas capsulata* is a nonmotile, obligate aerobe. Rods are up to 4 µm in length and 1.4 µm in diameter. Appropriately named in 1962, it no longer produces capsules in spite of its specific epithet. What's in a name? (Gram Stain)

Caulobacterales

Caulobacter (Fig. 11.32) is the type genus of the single family Caulobacteraceae. Cells are Gram negative and vibrioid or elongated with tapered ends. An adhesive **stalk**, made from the cell wall and cytoplasmic membrane, emerges from one end.

During division, the cell produces a daughter cell from the unattached end, which develops a flagellum and swims away after the cells separate. It eventually loses the flagellum, develops a stalk, and attaches to a surface. Caulobacters are obligate aerobic respirers found in freshwater, sewage, and bottled water!

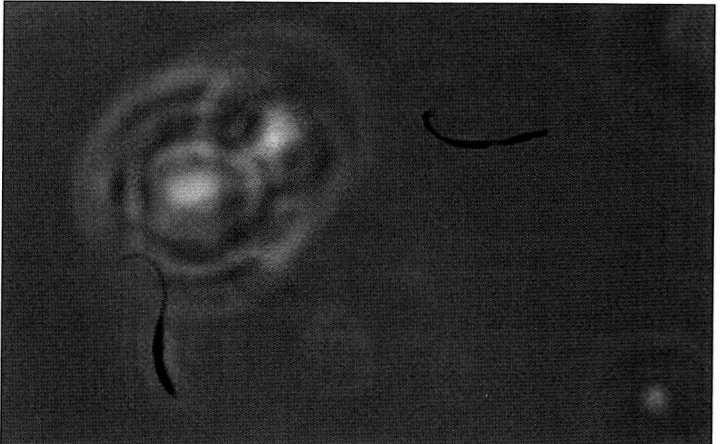

11.32 ***Caulobacter*** **(Phylum Proteobacteria; Class α-Proteobacteria; Order Caulobacteriales)** • Shown are two *Caulobacters* in the process of dividing. The daughter cell of each will develop its flagellum and swim away after the cells separate. The four species of *Caulobacter* are obligate aerobes and are found in freshwater habitats. (Phase Contrast)

Rhizobiales

Rhizobiales includes phenotypically diverse species that share similarities of 16S ribosomal RNA sequences. All are Gram negative. Two genera within the family Rhizobiaceae are considered here: *Rhizobium* and *Agrobacterium*.

Both are soil organisms and many strains have the ability to infect plants and induce tumor formation. They are aerobic respirers and when cultured with a carbohydrate source, produce an extracellular slime. In addition, one genus, *Nitrobacter*, within the family Bradyrhizobiaceae is covered.

Rhizobium (Fig. 11.33) species are able to utilize a variety of carbon and nitrogen sources but do not perform glycolysis. Rather, glucose is metabolized using the pentose phosphate and Entner-Doudoroff pathways (see Appendix, Figs. A.3 and A.4). Up to 6 peritrichous flagella produce motility.

R. leguminosarum is capable of infecting **leguminous** plant roots and inducing **root nodule** formation. Once established, the cells are capable of nitrogen fixation, which provides reduced nitrogen to themselves and the host plant. For more information on the nitrogen cycle, see page 307.

Agrobacterium (Fig. 11.34) species possessing the **Ti** (tumor inducing) **plasmid** have the ability to infect a wide variety of dicot plants and induce tumor formation. A specific plasmid region (T-DNA) is able to integrate in the host's DNA and cause the cell to undergo mitosis, which produces the tumor and results in **crown gall disease**. Genetic engineers have exploited this infectious ability and use the Ti plasmid as a vector to produce **transgenic plants**.

Nitrobacter species are chemolithoautotrophs that play an important role in the nitrogen cycle. Oxidation of NO_2^- to NO_3^- provides them energy and CO_2 serves as their carbon source. They are also capable of chemoorganotrophic metabolism (aerobic and anaerobic respiration). Their cells are rods of various shapes (**pleomorphic**).

The electron microscope reveals intracytoplasmic membranes organized at one pole of the cell called **carboxysomes** and are specialized regions for carbon-fixation. Reproduction occurs via binary fission or budding.

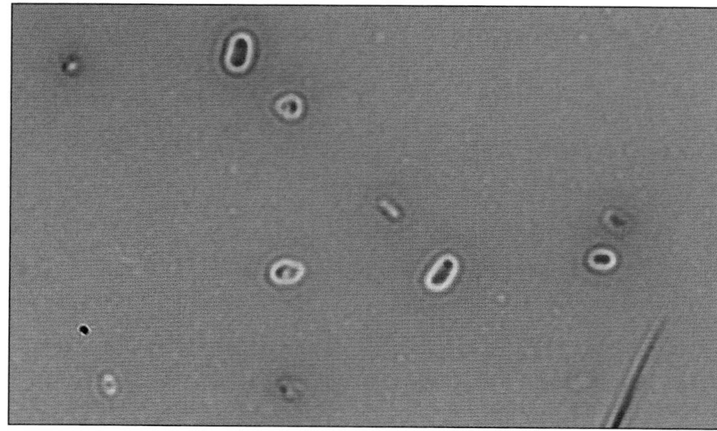

11.33 ***Rhizobium*** **(Phylum Proteobacteria; Class α-Proteobacteria; Order Rhizobiales)** • *Rhizobium* species are obligate aerobes that live in soil. They are capable of entering leguminous plant roots and inducing the formation of root nodules (see Figs. 19.34 and 19.35). Within the root nodule, *Rhizobium* cells fix nitrogen (reduce N_2 to NH_4, a form more easily incorporated into biochemicals) for themselves and their hosts. These cells were grown in culture and produced a small capsule. The longest cells are approximately 3 μm in length. (Capsule Stain)

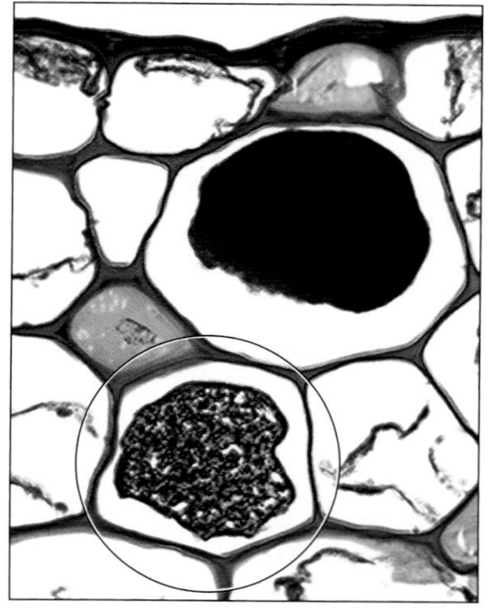

11.34 ***Agrobacterium tumefaciens*** **(Phylum Proteobacteria; Class α-Proteobacteria; Order Rhizobiales)** • Shown is a cross-section of a *Sedum* (stonecrop) stem. During preparation, the plant cell protoplasts have shrunk away from the cell walls and look like debris, but the protoplast of the circled cell is infected with *Agrobacterium* and retained its shape and most of its size. The dark object filling one cell is an amyloplast (starch). (Quadruple Stain)

Class Betaproteobacteria (Volume 2C)

This is a very diverse class that contains four orders, but that appear to be related based on 16S rRNA sequences. Three of the orders are discussed below.

Burkholderiales

The Alcaligenaceae is one of four families in the order. Species in the family are obligate aerobes and chemoorganotrophic, utilizing organic acids and amino acids as carbon sources. *Alcaligenes* (Fig. 11.35) and *Bordetella* (see Fig. 12.10) are two genera in the family. The former exists as single rods or coccobacilli, is motile, and catalase (see p. 79) and oxidase (see p. 105) positive.

Some species are capable of nitrate reduction or nitrite reduction, but usually not both. *Alcaligenes faecalis* is found in a variety of environments, including water, soil, and vertebrates. It is rarely an opportunistic human pathogen.

Bordetella species are minute coccobacilli (<0.5 μm wide by <1 μm long) and fastidious, requiring nicotinamide (for NAD and NADP), cysteine (or other organic sulfur source), and amino acids (for nitrogen) from their environment. *Bordetella pertussis* is the causative agent of whooping cough.

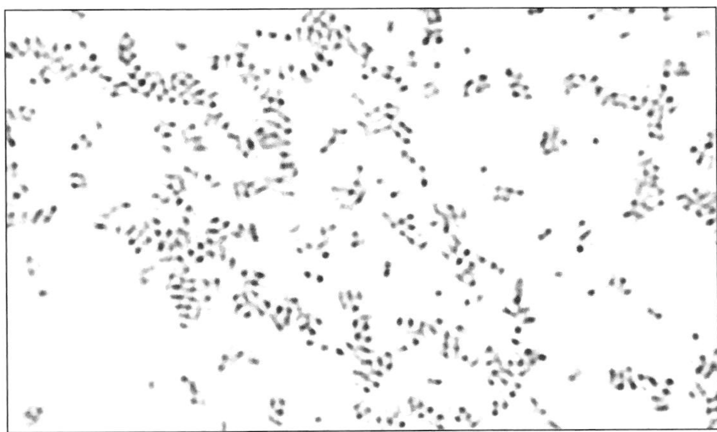

11.35 ***Alcaligenes faecalis*** **(Phylum Proteobacteria; Class β-Proteobacteria; Order Burkholderiales)** • In this field, rods and coccobacilli are both present. Most *Alcaligenes* species are obligate aerobes, but *A. faecalis* is also capable of anaerobic respiration with nitrite as the final electron acceptor. It is found in soil, water, and various clinical specimens, but is rarely an opportunistic pathogen. (Gram Stain)

Neisseriales

Neisseriales is organized into a single family, Neisseriaceae. Cells are typically single cocci, diplococci (often with adjacent sides flattened), or tetrads, (Fig. 11.36), though some species produce rods and spirals. Their wall is structurally Gram negative, but they may be difficult to decolorize and stain Gram positive.

They are chemoorganotrophic, oxidase positive (see p. 105), nonmotile, and aerobic or facultatively anaerobic. The type genus is *Neisseria* with two pathogenic species: *N. gonorrhoeae* (see Fig. 12.43) and *N. meningitidis* (see Fig. 12.45) that cause gonorrhea and meningitis, respectively. For more information, see page 206.

Aquaspirillum and *Chromobacterium* are examples of noncoccus species in the family. *Aquaspirillum* species are typically spiral shaped with polar flagella. They respire aerobically and may be obligate aerobes or microaerophiles. Organic acids or amino acids serve as a carbon source, but typically carbohydrates are not utilized.

Chromobacterium violaceum is a chemoorganotrophic, facultatively anaerobic, motile rod and is the only species in the genus. It is characterized by production of a purple pigment, **violacein** (see Fig. 11.37 and Figs. 3.5B, and 3.28).

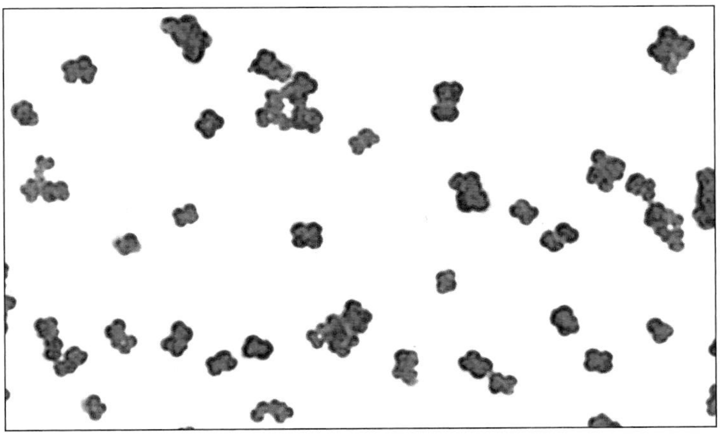

11.36 ***Neisseria sicca*** **(Phylum Proteobacteria; Class β-Proteobacteria; Order Neisseriales)** • *Neisseria sicca* typically grows in tetrads, with cells larger than 1 μm. It colonizes the human nasopharynx, sputum, and saliva, and is an opportunistic pathogen. (Gram Stain)

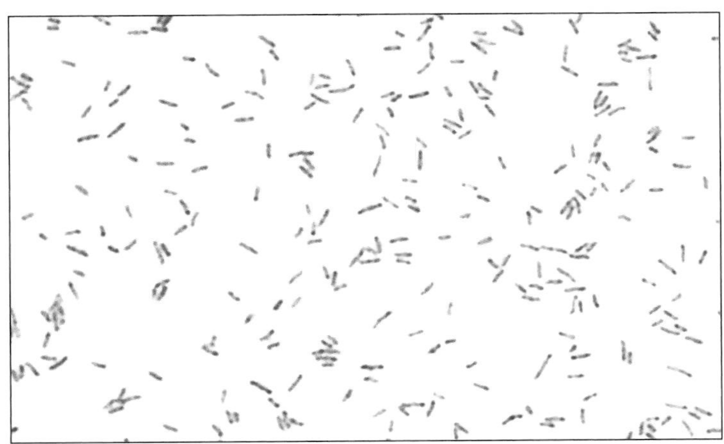

11.37 ***Chromobacterium violaceum*** **(Phylum Proteobacteria; Class β-Proteobacteria; Order Neisseriales)** • *Chromobacterium violaceum* is the only species in the genus. *C. violaceum* is pretty easy to recognize on a mixed culture plate because of its dark purple color produced by the pigment violacein. Its cells are straight rods between 1.5 and 3 μm. A single polar flagellum, often accompanied by additional subpolar or lateral flagella, produce motility. (Gram Stain)

Nitrosomonadales

This order is physiologically, morphologically, and ecologically diverse. There are three families, two of which are covered here. The Nitrosomonadaceae consists of four chemolithoautotrophic genera, all with the ability to oxidize ammonia to nitrite and fix CO_2.

Nitrosomonas (Fig. 11.38) cells range from single to short chains of cocci or rods. The electron microscope reveals flattened membranous cytoplasmic vesicles that are involved in ATP production from ammonia oxidation. Some species can oxidize urea, producing CO_2 for autotrophic growth and NH_3 for ATP production—highly efficient! *Nitrosomonas* is most frequently isolated from aquatic habitats rich in ammonia, and occasionally from soil.

The Spirillaceae comprises a single genus with a single species, *Spirillum volutans* (see Fig. 6.37). Its cells are large spirilla, gaining lengths up to 60 µm. Unlike most motile bacteria its flagella are visible with the light microscope, but only because they're in bundles of up to 75 individual flagella at each end of the cell. *S. volutans* lives in stagnant freshwater and is microaerophilic, but performs aerobic respiration and is oxidase positive (p. 105), but catalase negative (p. 79).

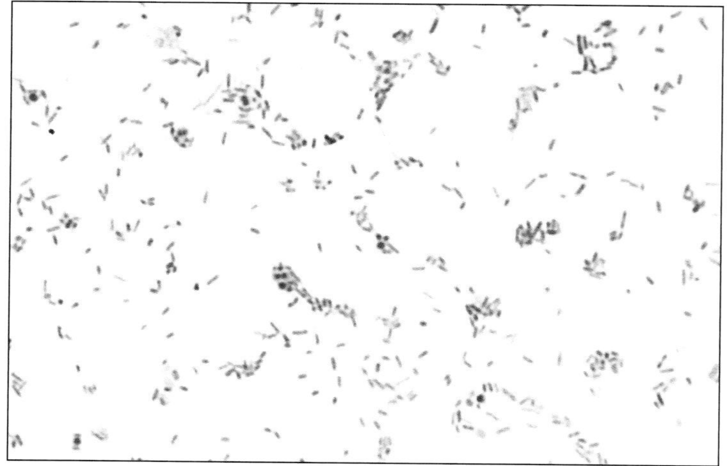

11.38 ***Nitrosomonas*** **(Phylum Proteobacteria; Class β-Proteobacteria; Order Nitrosomonadales)** • *Nitrosomonas* is a genus of Gram-negative nitrifying bacteria found in seawater, brackish water, and freshwater, as well as soils. The cells are straight rods 0.7–1.5 µm wide by 1.5–2.4 µm long. They are aerobic chemoautotrophs that obtain energy by oxidizing ammonia to nitrite. This specimen was obtained from soil grown on an enrichment medium containing ammonia as the only nitrogen source. (Gram Stain)

Class Deltaproteobacteria (Volume 2C)

Deltaproteobacteria is a fairly large, and morphologically and metabolically diverse class based on 16S rRNA similarities. There are currently eight orders. *Desulfovibrio* species (order Desulfovibrionales) serve as metabolic representatives of the anaerobes. They are chemoorganotrophs that respire anaerobically, reducing sulfite or sulfate to H_2S in the ETS.

In nature, they occupy anoxic muds of freshwater and marine habitats, as well as animal intestinal tracts. In a Winogradsky column, they occupy and produce the lowermost black region (see Figs. 19.28 and 19.30, p. 306). Its cells are curved (sometimes straight), motile rods. Figure 19.39 shows a community of sulfur reducing bacteria recovered from anoxic mud, some of which are undoubtedly *Desulfovibrio*.

Aerobic Deltaproteobacteria include *Bdellovibrio* (order Bdellovibrionales). *Bdellovibrio* species are motile, curved rods that respire aerobically. They are found in soil, sewage, and various aquatic habitats. Remarkably, they are predators of Gram-negative bacteria.

The **attack phase** of their life cycle is performed by the motile cells described above. Once a motile cell collides with its prey and attaches, the flagellum continues to rotate, pushing it through the cell wall that has been enzymatically weakened. Development continues in the periplasm, and the prey is converted into a **bdelloplast**.

Feeding off the prey, growth in length, and then division into progeny attack phase cells follows. Each new cell develops a flagellum and swims away. Depending on the temperature and size of the prey cell, this process can be completed in as little as three hours and produce up to six progeny.

Class Epsilonproteobacteria (Volume 2C)

Campylobacterales

The species of this order are Gram-negative, chemoorganotrophic, microaerophiles. Cell morphologies range from curved rods to S-shaped or helical rods. Most are motile and exhibit a corkscrew motion. Carbohydrates are not used as an energy source.

Campylobacter is the type genus of the Campylobacteriaceae and includes two pathogenic species: *C. jejuni* (see Figs. 12.15 and 12.16) and *C. fetus*. The former causes gastroenteritis and abortion; the latter causes abortion in cattle and sheep.

Helicobacter is the type genus of the family Helicobacteraceae and its cells are rods of varying shapes, including spirilla. They are chemoorganotrophic and oxidase positive (see p. 105). *H. pylori* is a human pathogen (see p. 199 and Fig. 12.32).

Bergey's Manual of Systematic Bacteriology, 2nd ed., Volume 3

Phylum Firmicutes

Members of this phylum are diverse and generally have a low G + C% (<50%). Most are Gram positive, many of these with teichoic acids, but some genera are Gram negative. Cell morphology ranges from cocci to rods of various shapes. Some produce endospores. The phylum includes aerobes, facultative anaerobes, and obligate anaerobes. Most are chemoorganotrophic mesophiles and neutrophiles. The phylum is divided into three classes: Bacilli, Clostridia, and Erysipelotrichia, of which we discuss the first two.

Class Bacilli

Bacillales

Bacillaceae is the largest family (of the more than 10) within the order Bacillales, and *Bacillus* is by far the largest genus within the family. *Bacillus* species are rod-shaped and generally stain Gram positive, at least when young. They form endospores (see Fig. 6.27) and are either aerobic or facultatively anaerobic.

Most are soil organisms, though *B. anthracis* is a pathogen of cattle and humans (see Figs. 12.3 through 12.5) and *B. cereus* causes food poisoning (see Figs. 12.6 and 12.7). *B. mycoides* (see Fig. 3.24C) produces very distinctive colonies.

Bacillus thuringiensis (Fig. 11.39) is closely related to *B. cereus* and produces the cereulide toxin responsible for emetic food poisoning (see *Bacillus cereus*, p. 185). In addition, *B. thuringiensis* produces **crystal** (**cry**) and **cytolytic** (**cyt**) toxins, collectively known as **δ-enterotoxins** or **Bt toxin.**

Their production coincides with sporulation, and are visible in cells as **parasporal crystals.** Because of their insecticidal properties, Bt toxin has been employed in agriculture. It has also been shown to be toxic to nematode worms and some cancer cells.

In the last couple of decades 16S rRNA sequence comparisons have resulted in the reclassification of many *Bacillus* species and their placement into new families and genera, still within the order Bacillales. In addition, previously unrecognized relationships have led to inclusion of genera within the order.

These include, with family and distinctive features: *Geobacillus* (Bacillaceae; thermophilic); *Brevibacillus* (Fig. 11.40) and *Paenibacillus* (Paenibacillaceae; soil saprophytes with endospores that distend the mother cell); and *Alicyclobacillus* (Alicyclobacillaceae; obligate aerobes).

The relationship of *Bacillus* to *Sporosarcina* (Planococcaceae; see Fig. 6.28), a genus dating to the early 20th century, was also identified and some former *Bacillus* species were placed into it. Also, *Kurthia* species (Fig. 11.41) were placed in the Planococcaceae. They are found in animal feces (pigs and chickens), milk, and meat, and are unusual within the order in not producing endospores.

Listeriaceae comprises two non-spore forming genera, *Listeria* and *Brochothrix*, with the former being the larger of the two. *Listeria* cells are short rods with blunt ends. They are aerobic or facultatively anaerobic, motile, but do not form endospores, and are found in a wide variety of habitats. *L. monocytogenes* is the causative agent of listeriosis (see Fig. 12.36).

Staphylococcus (see Figs. 5.26 and 12.61 through 12.64) is the largest genus of the four in Staphylococcaceae.

11.39 ***Bacillus thuringiensis*** **Parasporal Crystals (Phylum Firmicutes; Order Bacillales)** • *Bacillus thuringiensis* produces insecticidal toxins that are stored as parasporal crystals (**P**) located toward the ends of cells. Endospores (**E**) and released spores (**S**) are also visible. (Parasporal Crystal Stain.)

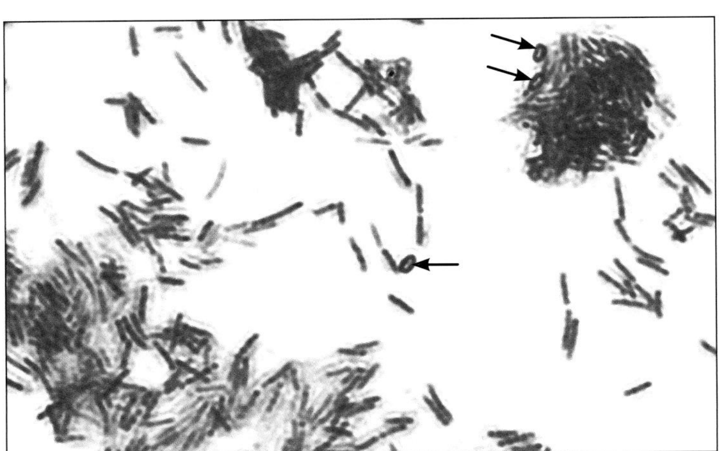

11.40 ***Brevibacillus brevis*** **(Phylum Firmicutes; Order Bacillales)** • *Brevibacillus brevis* cells are straight, motile rods up to 5 µm in length and about 1 µm in diameter. This species used to be *Bacillus brevis*, but was put in its new genus in 1996, but like *Bacillus*, it is a spore-former (**arrows**) that sometimes gives a Gram-variable result. (Crystal Violet Stain)

Staphylococci typically grow in grapelike clusters of spherical cells. No endospores are formed and they are usually catalase positive (see p. 79) and oxidase negative (see p. 105).

Metabolically, they can ferment and aerobically respire, and are facultative anaerobes. *S. aureus* (see p. 214) and *S. epidermidis* (see p. 215) are normal inhabitants of the human body, but both are opportunistic pathogens.

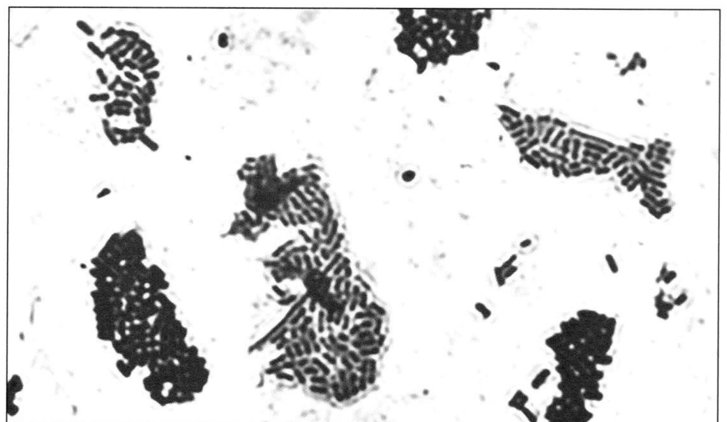

11.41 ***Kurthia zopfii*** **(Phylum Firmicutes; Order Bacillales)** • *Kurthia* is an obligate aerobe that, if motile, has peritrichous flagella. It is seen here as single rods with rounded ends reaching a length of 5 µm, but it also forms chains. *Kurthia zopfii* is named after H. Kurth, a German bacteriologist and W. Zopf, a German botanist—a double eponym. (Crystal Violet Stain)

Lactobacillales

This order comprises six families, of which only three will be addressed: Lactobacillaceae, Enterococcaceae, and Streptococcaceae.

The type genus of the Lactobacillaceae is *Lactobacillus* (Fig. 11.42), whose cells range from long, thin rods to coccobacilli. They are Gram positive, do not produce endospores, and are nutritionally fastidious, aerotolerant anaerobes. Lacking a cytochrome system, their metabolism is strictly fermentative with lactate being the primary end product.

Some species are **homofermentative**, meaning lactate is the sole end-product of fermentation; others are **heterofermentative** and produce multiple end-products. They are found in the oral cavity, intestines, and vagina of mammals and are not considered to be pathogenic to healthy individuals. Some species are useful in making dairy products (e.g., cheese and yogurt), whereas others are active in food spoilage.

The family Streptococcaceae is composed of species whose cells are spherical to ovoid and usually occur in pairs or chains. They are catalase negative (see p. 79) and produce lactic acid as the end-product of fermentation. Often, elevated CO_2 in the environment is required for growth.

Hemolytic reaction on blood agar (see p. 76) and serological testing for cell wall carbohydrates and lipoteichoic acids (for placement into one of the Lancefield groups, first established by Rebecca Lancefield in 1933, see p. 136) remain important characteristics in identifying *Streptococcus* species. *S. pneumoniae* (see p. 218) and *S. pyogenes* (see p. 219) are important human pathogens, but the majority of species are commensals of mucous membranes in warm-blooded animals.

Lactococcus species were once included in the genus *Streptococcus* because of morphological and metabolic similarities. However, because of many biochemical differences and the presence of the Lancefield Group N antigen, they were placed in their own genus.

The family Enterococcaceae comprises four genera, but the type genus (and largest genus) is *Enterococcus* (Fig. 11.43). Members of this genus are ovoid Gram-positive cocci usually arranged in pairs or short chains. Most perform homofermentative metabolism with lactic acid being the primary end-product. Catalase (see p. 79) is not produced.

Most are resistant to 40% bile and hydrolyze esculin (see p. 75). At one time they were classified as Lancefield Group D *Streptococcus*, but more recently identified *Enterococcus* species do not demonstrate this antigen. Of the 40 + species, two are common human commensals and opportunistic pathogens, mostly acquired in health-care

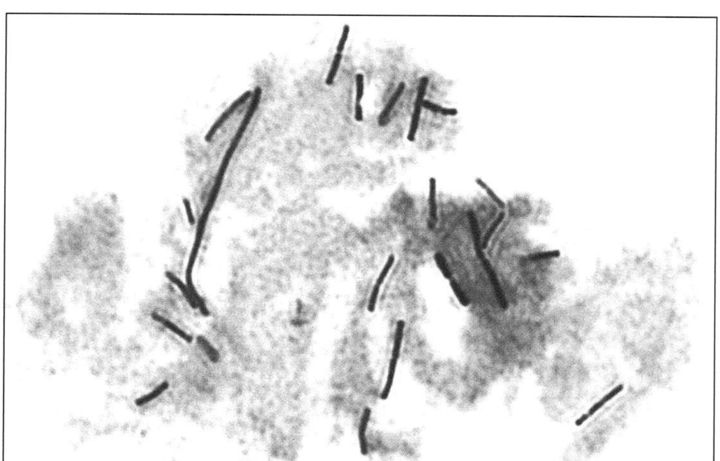

11.42 ***Lactobacillus delbrueckii*** **(Phylum Firmicutes; Order Lactobacillales)** • *Lactobacillus* cells are Gram-positive, non-sporing rods of varying sizes. Metabolism is fermentative, with lactate comprising the majority of end-products. *Lactobacillus* species, including *L. delbrueckii*, are often used to make yogurt or are included in probiotics. These were grown in culture. (Gram Stain)

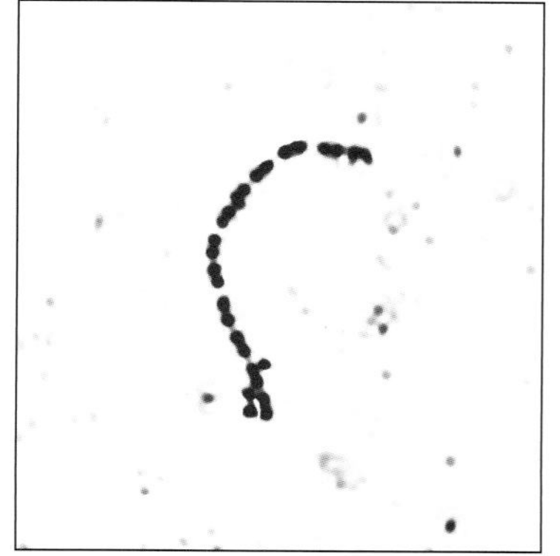

11.43 ***Enterococcus faecium*** **(Phylum Firmicutes; Order Lactobacillales)** • Short chains of ovoid cocci are typical of the genus *Enterococcus*. Cells generally fall in the 1 to 2 µm range. *Enterococcus* species are also homofermentative, which means they (usually) produce a single fermentation product, lactic acid. (Gram Stain)

settings: *E. faecalis* and *E. faecium*. Enterococcal diseases include urinary tract infections, bacteremia, and endocarditis. Resistance to vancomycin (VRE strains-vancomycin resistant enterococci) and penicillin is common.

Class Clostridia

Clostridiales

This order is large and diverse. We will focus on the type genus, *Clostridium* (Fig. 11.44), which itself has more than 180 species. As a rule, clostridia are Gram-positive, obligately anaerobic, catalase-negative (see p. 79), endospore-forming rods. They lack an electron transport chain and are thus fermentative.

The fermentable substrate is often either carbohydrate(s) or amino acids, and the end-products represent a wide range of organic compounds. Most are found in anaerobic regions of soil and mammalian digestive tracts. Five pathogenic species are covered beginning on page 192.

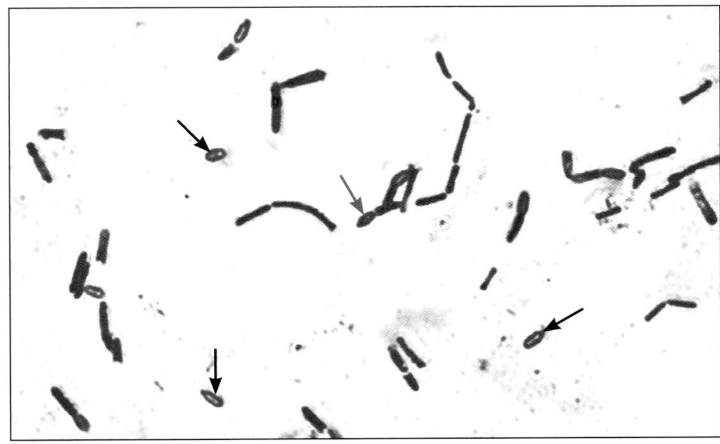

11.44 ***Clostridium beijerinckii*** **(Phylum Firmicutes; Order Clostridiales)** Members of this genus are anaerobic, Gram-positive, endospore-forming rods. *Clostridium* spores often distend the cell, making the region look swollen (**red arrow**), which is an important diagnostic feature for species differentiation. Some free spores are also visible (**black arrows**). Cells range in size from 0.5–1.7 µm in diameter to 1.7–8.0 µm in length. Fermentation products are mainly butyric and acetic acid. (Gram Stain)

Bergey's Manual of Systematic Bacteriology, 2nd ed., Volume 4

Phylum Bacteroidetes

Unique molecular similarities strongly suggest that the phyla Bacteroidetes and Chlorobi (see p. 169) are closely related, even though the former are chemoorganotrophs. Cells are nonmotile, obligately anaerobic, Gram-negative rods. *Bacteroides* species (family Bacteroidaceae) ferment sugars, but rarely catabolize protein. Most are able to grow in the presence of bile and hydrolyze esculin (see p. 75).

B. fragilis constitutes the most common anaerobe isolated from the human colon. It is an opportunistic pathogen (see p. 186 and Figs. 12.8 and 12.9). *Salinibacter ruber* (Fig. 11.45) is an extreme halophile that requires high salt concentrations (>15%) in order to grow. It is a motile, obligate aerobe, that produces red colonies.

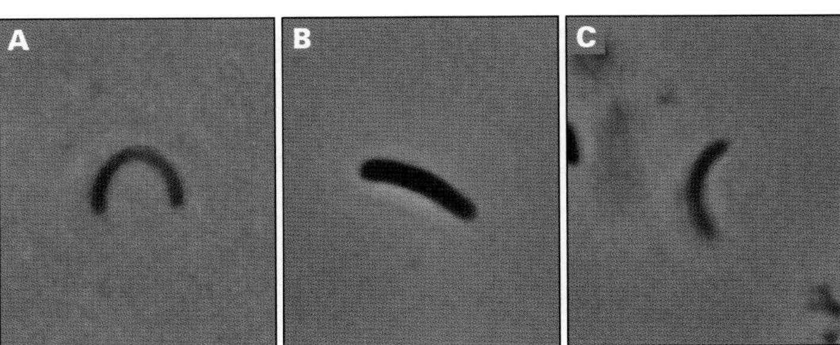

11.45 ***Salinibacter ruber*** **(Phylum Bacteroidetes; Order Bacteroidales)** *Salinibacter ruber* is the only species in the genus and is an extreme halophile related to *Bacteroides fragilis* (see Fig. 12.8), an opportunistic pathogen. (**A**) The specimen shown here came from the North Arm of the Great Salt Lake in Utah, where the salinity exceeds 25%. (For comparison, seawater is approximately 3.5% NaCl.) (**B** and **C**) The specimens shown were identified based on their distinctive curvature. Being extreme halophiles, they don't just tolerate the salty environment, they actually *require* it. (Phase Contrast)

Phylum Spirochaetes

Spirochaetales is the only order in the phylum. Ecologically, they are a diverse group comprising free living, symbiotic, or parasitic chemoheterotrophic species. The Gram-negative cells are flexible, tightly coiled rods with the cell wall, cell membrane, and cytoplasm constituting the **protoplasmic cylinder**. The outer membrane or **outer sheath** surrounds the protoplasmic cylinder.

Between two and 100 **periplasmic flagella** are wound along the spiral cell in the space between the sheath and the protoplasmic cylinder. One end of each flagellum is anchored to a pole of the protoplasmic cylinder, with an equal number attached at each end. The flagella can propel the cell forward, cause it to rotate on its axis, or flex.

Aerobic, facultatively anaerobic, and anaerobic species are represented. Important genera include *Spirochaeta* (Fig. 11.46), *Borrelia* (see p. 188 and Fig. 12.12), *Treponema* (see p. 220 and Fig. 12.72), and *Leptospira*.

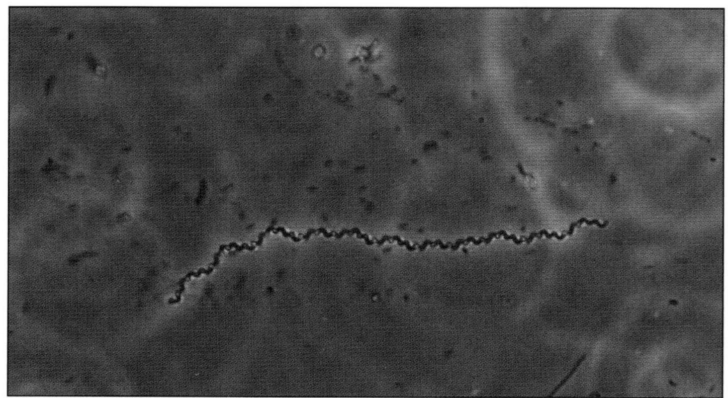

11.46 ***Spirochaeta* (Phylum Spirochaetes; Order Spirochaetales)** • *Spirochaeta* species are chemoorganotrophic, relying primarily on carbohydrates as their carbon source. They are also either obligate or facultative anaerobes. The cells are helical, often tightly so (as in this specimen), flexible rods reaching a length of up to 250 µm. Motility is provided by two periplasmic flagella in most species. This specimen was collected from a seawater sample. (Phase Contrast)

Bergey's Manual of Systematic Bacteriology, 2nd ed., Volume 5

Phylum Actinobacteria

Actinobacteria is a large and diverse phylum with a single class of the same name. Cells are typically Gram-positive rods with a high G + C% and most are aerobes. They are found in soil and plant material, though there are some pathogens.

Five orders will be covered here, with families in parentheses: Bifidobacteriales (Bifidobacteriaceae); Corynebacteriales (Corynebacteriaceae, Mycobacteriaceae, Nocardiaceae); Micrococcales (Micrococcaceae, Cellulomonadaceae); Propionibacteriales (Propionibacteriaceae); and Streptomycetales (Streptomycetaceae).

Bifidobacteriales

Bifidobacteriales and its single family were formed as a result of 16S rRNA data. The type genus *Bifidobacterium* (with 30 + species) is one of several genera and will serve as an example. These obligate anaerobes ferment glucose into acetic and lactic acids, but no gas. Its rods are nonmotile and pleomorphic, often with branching or protrusions of various shapes (Fig. 11.47).

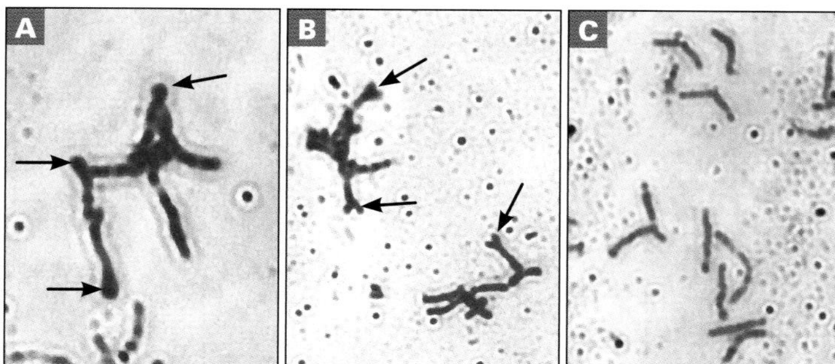

11.47 ***Bifidobacterium* (Phylum Actinobacteria; Order Bifidobacteriales)** • As seen in these examples, *Bifidobacterium* cells are varied in shape. (**A**) Cells with swollen ends are seen (**arrows**). (**B**) Cells with bifurcated ends are seen (**arrows**), as are more typically shaped rods. (**C**) The more common thin cells, some curved, are seen. *Bifidobacterium* species are obligate anaerobes that ferment sugars to lactic and acetic acid end-products. They are often included in commercial probiotic products, one of which was used to isolate these specimens. (Gram Stains)

Natural habitats include human and animal intestines, though individual *Bifidobacterium* species are typically not found in both. They are also a common ingredient in probiotics. Read the label.

Corynebacteriales

Cells of the family Corynebacteriaceae are irregular rods that sometimes form "V" shapes as a result of **snapping division**, where the adjoining wall between cells incompletely separates (Fig. 11.48). They are also sometimes club-shaped (see Fig. 6.34).

Metabolism is fermentative with acid, but no gas, being the end-product. They are facultative anaerobes and are

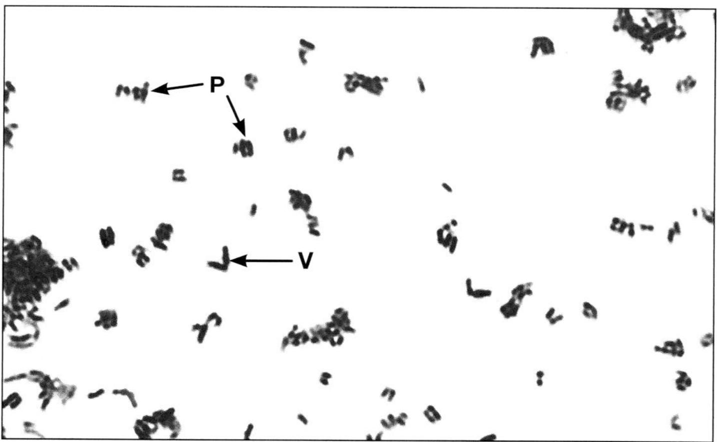

11.48 ***Corynebacterium pseudodiphtheriticum* (Phylum Actinobacteria; Order Corynebacteriales)** • Cells of *Corynebacterium* species are rods of varying lengths, shapes, and arrangements. Two distinctive arrangements are shown in this micrograph: "V" shape (**V**) and palisade arrangement (**P**), in which the rods are stacked side-to-side. Both are due to "snapping division" that results in incomplete wall separation at the time of division. Species are chemoorganotrophic and most are facultatively anaerobic, with the remainder being aerobic. *C. pseudodiphtheriticum* cells are relatively short rods (about 2 µm). It is part of the nasal mucosa microbiome. (Gram Stain)

catalase positive (see p. 79). *Corynebacterium* species are found on mammalian mucous membranes. *C. diphtheriae* is a human pathogen (see p. 195 and Fig. 12.24).

A defining characteristic of the Mycobacteriaceae is that they are **acid-fast** at some point in their life cycle (see p. 63 and Figs. 6.19 and 6.20) due to an abundance of waxy **mycolic acid** in their walls. As a result, they stain weakly Gram positive.

Mycobacterium species (Fig. 11.49) are nonmotile and are nutritionally non-fastidious, though *M. leprae* cannot be cultivated in standard bacteriological media because it is an intracellular parasite. Doubling times range from days to weeks and differentiation between *Mycobacterium* species relies, in part, on whether the isolate is a **slow grower** or a **fast grower**.

Differentiation also relies on whether the isolate produces a carotenoid pigment in the light (**photochromogenic**), in the light or dark (**scotochromogenic**) or not at all (**nonchromogenic**). Two important pathogens are *M. leprae* (see p. 203 and Fig. 12.38) and *M. tuberculosis* (see p. 204 and Figs. 12.39 and 12.40).

Species of Nocardiaceae produce branched filaments (**hyphae**) that cover and penetrate the growth surface. **Aerial hyphae** are also produced. Hyphae may fragment into rod and coccus-shaped cells. Nocardias are chemoorganotrophic and are able to oxidize a variety of carbohydrates. They are aerobic, weakly Gram positive or Gram variable, and partially acid-fast.

Natural habitats include soil and water. Some *Nocardia* species (see Figs. 12.47–12.49) are human pathogens. *Rhodococcus* species (Fig. 11.50) have a morphology similar to *Nocardia* and are metabolically versatile. As a result, they may become very useful in **bioremediation**.

Micrococcales

Our focus is on the Micrococcaceae and Cellulomonadaceae, two of more than a dozen families in the order.

Micrococcaceae species are obligate aerobes with nonmotile cells ranging from cocci to pleomorphic rods. They are typically catalase and oxidase positive (see pp. 79 and 105, respectively).

Micrococcus is the type genus for the family Micrococcaceae. Its cells are cocci, ranging in diameter from 0.5–2 µm, and are often arranged into tetrads or irregular clusters. They are found on human skin and in the soil.

In 1995, three species were removed from the genus *Micrococcus* and placed in a new genus, *Kocuria* (Fig. 11.51), based on differences in cell wall structure and menaquinone and fatty acid variations. They are human commensals, though some may be opportunistic pathogens, especially in immunocompromised patients.

Arthrobacter is also in this family (Fig. 11.52). Its cells are rods during exponential growth, but fragment to form cocci in the stationary phase. They, too, have been characterized by

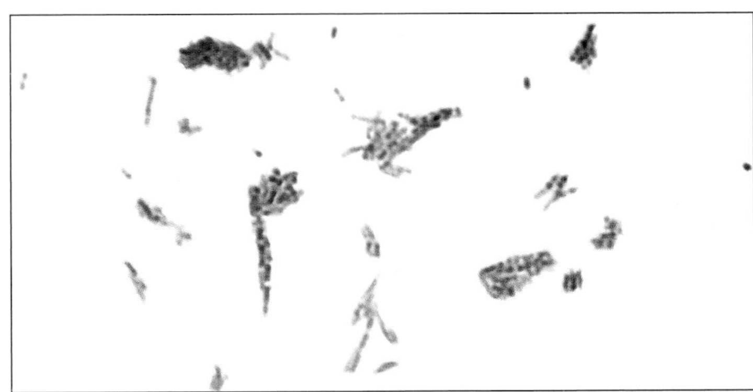

11.49 ***Mycobacterium nonchromogenicum* (Phylum Actinobacteria; Order Corynebacteriales)** • There are more than 100 species in the genus *Mycobacterium*. Some are pathogens (see pp. 203–205), but most are free-living in soil and water. Most are also obligate aerobes, but some are microaerophilic. All are chemoorganotrophs, but they still require CO_2 for growth. Cells are acid-fast rods that often have a beaded appearance, as in this micrograph of *M. nonchromogenicum*. Many species form pigmented colonies, but as the specific epithet implies, *M. nonchromogenicum* is not one of them. (Acid-Fast Stain)

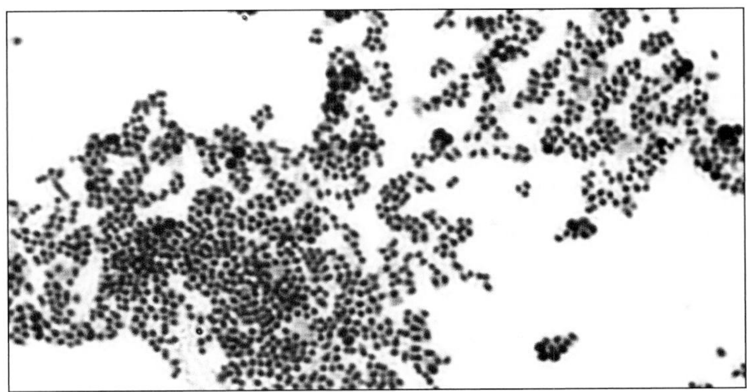

11.50 ***Rhodococcus* (Phylum Actinobacteria; Order Corynebacteriales)** • *Rhodococcus* species are nonmotile, obligate aerobes, that stain either Gram positive or Gram variable (notice the pink cells in the micrograph) and are partially acid-fast. Younger cells are single rods (as in this micrograph), but form branched mycelia on an agar surface with age. They are chemoorganotrophic and obligate aerobes. (Gram Stain)

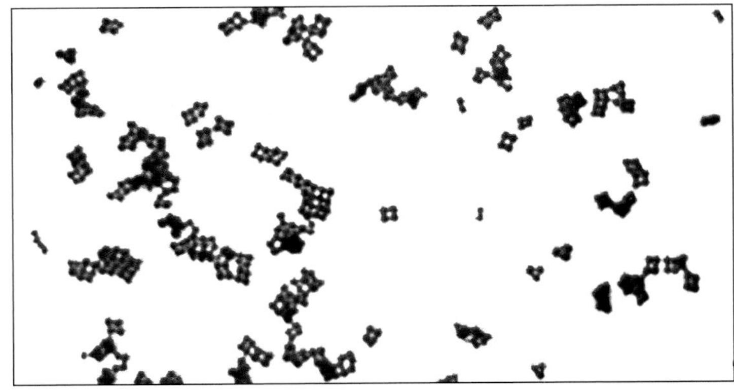

11.51 ***Kocuria rosea* (Phylum Actinobacteria; Order Micrococcales)** • Formerly, *Kocuria rosea* was known as *Micrococcus roseus*, but in 1995 it (along with two other *Micrococcus* species) was placed in the newly formed genus, *Kocuria*. Cells in the genus are Gram-positive, nonmotile, and unencapsulated cocci. They respire aerobically and are chemoorganotrophic. Colonies are pinkish (*rosea* means "rose colored"). They are part of the normal microbiome of skin and mucous membranes. *K. rosea* cells range in diameter from 1–1.5 µm and form pairs, tetrads, and irregular clusters. (Gram Stain)

fatty acids, quinones, and peptide interbridge compositions, though there is enough variation among these to form subgroups within the genus. They are mainly soil organisms.

Cellulomonas (Fig. 11.53) species, in the family Cellulomonadaceae, are composed of chemoorganotrophs that respire aerobically or ferment. Some species are motile. A distinguishing feature of the family is that cells produced by a particular species show great morphological variability. Most are soil organisms that are cellulolytic (digest cellulose), and so are also found associated with decaying wood.

Propionibacteriales

Propionibacterium is the type genus of the family Propionibacteriaceae. Cells are Gram positive, pleomorphic rods that are often arranged in "V" or "Y" configurations. Metabolism is fermentative, with large amounts of propionic acid as the end-product. They are aerotolerant to varying degrees. *Propionibacterium* is found in dairy products and is the primary fermenter in Swiss cheese production. Some, such as *P. acnes* (Fig. 11.54), also inhabit human skin.

Streptomycetales

Superficially, members of the family Streptomycetaceae have much in common with fungi, but they aren't considered fungi because they are prokaryotic. However, due to these similarities (and for historical reasons), fungal terminology is applied to them.

Streptomycetes produce branching **hyphae** that intertwine to form a **mycelium**. As the mycelium ages, aerial hyphae emerge and produce reproductive spores at their tips. The type genus *Streptomyces* (Fig. 11.55) comprises 500+ species.

They are Gram-positive obligate aerobes with the ability to metabolize diverse carbon compounds from their environment, which is frequently the soil. Many produce **geosmin**, a chemical responsible for the "earthy" fragrance of soil. The majority of *Streptomyces* species also produce antibiotics.

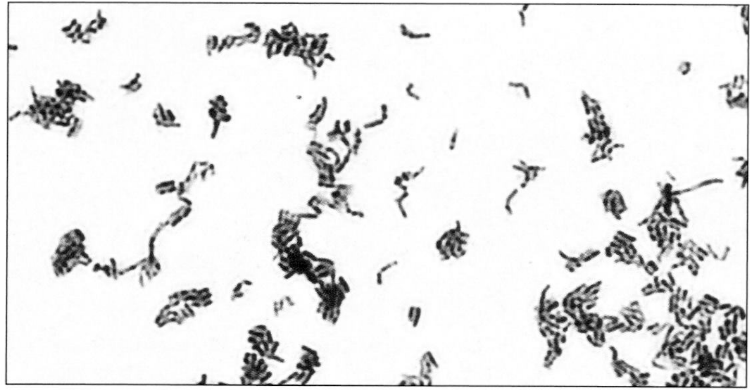

11.52 ***Arthrobacter*** **(Phylum Actinobacteria; Order Micrococcales)** • Cells in this preparation made from *Arthrobacter* grown in culture demonstrate the angular arrangement from which their name is derived (*arthro* means "joint"). The slide was made during exponential growth of the culture, as evidenced by most cells being rod-shaped. As a culture enters stationary phase, the rods fragment and form coccid cells. (Gram Stain)

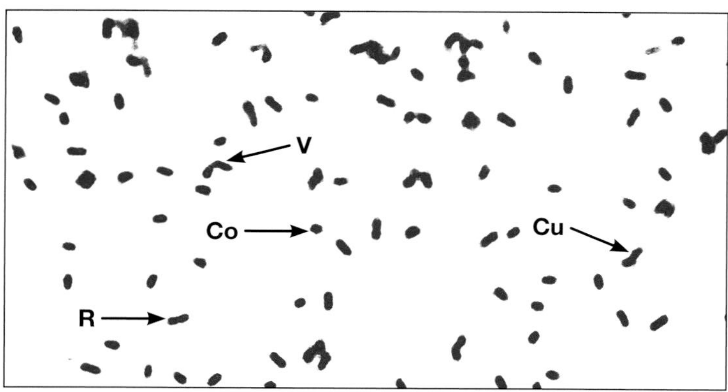

11.53 ***Cellulomonas*** **(Phylum Actinobacteria; Order Micrococcales)** • *Cellulomonas* cells in young cultures assume a variety of shapes: coccus (Co), curved (Cu), rod (R), and V-shaped (V). With age, most cells are rod-shaped. The rods in this micrograph are 2–3 µm in length. They respire aerobically, but also ferment with acid as the end-product. (Gram Stain)

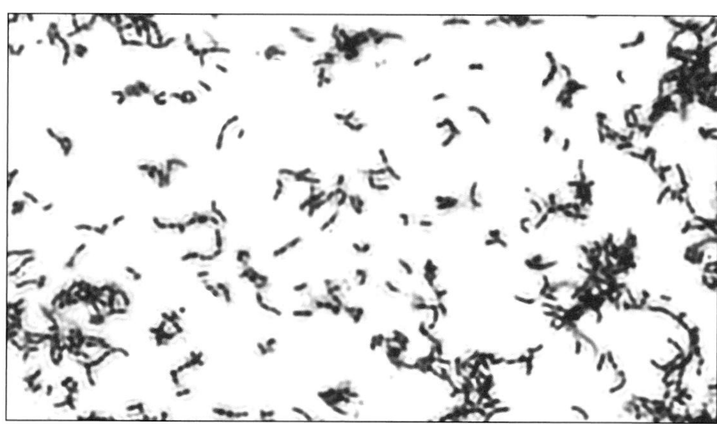

11.54 ***Propionibacterium*** **(Phylum Actinobacteria; Order Propionibacteriales** • This micrograph is of *Propionibacterium acnes* grown in culture. Note the angular pleomorphic rods. *P. acnes* is a commensal that lives in sebaceous glands (associated with hair follicles) and gets its energy from digesting the oily secretion. If the pore becomes blocked and ruptures, an infection can occur and produce a pimple. (Gram Stain)

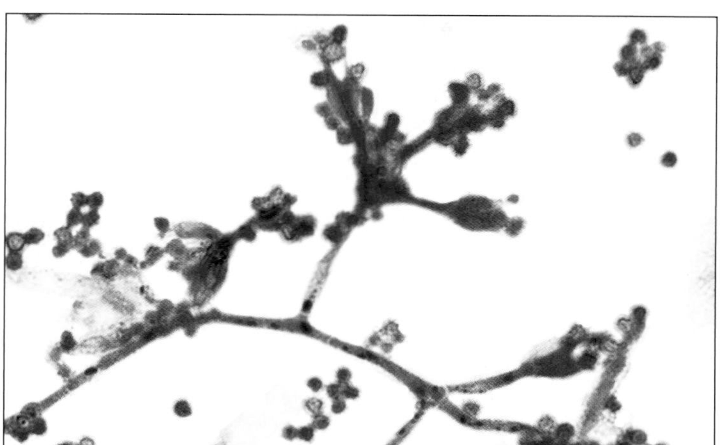

11.55 ***Streptomyces*** **(Phylum Actinobacteria; Order Streptomycetales)** • This is a micrograph of an unidentified *Streptomyces* species grown in culture, but its natural habitat is soil. Note the branched hyphae and spores. The spores are approximately 1 µm in length. (Methylene Blue Stain)

Bacterial Pathogens

SECTION 12

Unlike Section 11, in which Bacteria were organized taxonomically (which was a theme in that section), we have organized them alphabetically in this section to make it easier for the reader to find a pathogen of interest. (Note that cross references for serological, nucleic acid, and other "modern" tests are provided. They are not provided for more traditional biochemical tests. These are mostly found in Section 7.)

Aeromonas spp.

There are currently 36 species in the genus *Aeromonas* (class Gammaproteobacteria), but most infections are due to two species: *Aeromonas hydrophila* and *A. sobria*, which we consider together as *Aeromonas* species. They are leech gut symbionts found in aquatic habitats worldwide, and also may be carried by medicinal leeches. Portals of entry include leech bite wounds, open trauma wounds, or ingestion of contaminated food or water.

When ingested, they adhere to the intestinal mucous membrane and produce a **cholera-like enterotoxin** that causes acute watery diarrhea similar to cholera. When introduced through a superficial wound, the organism may be contained locally or spread throughout surrounding tissue in a condition called **cellulitis**. Rarely, when introduced through an open wound (especially deep penetrating wounds), it can produce a gas gangrene-like infection called ***A. hydrophila*** **myonecrosis**.

Both cellulitis and *A. hydrophila* myonecrosis can quickly lead to septicemia in immunocompromised adults and infants. Mortality rates are high among leukemia, lymphoma, and otherwise immunosuppressed patients following sepsis. *Aeromonas* spp. are not part of the normal human microbiota.

Antibiotic treatment may not be needed for gastroenteritis infections; fluid replacement is often sufficient. Various antibiotics can be used for treatment of superficial wounds and septicemia. Some resistance due to β-lactamases has been identified.

Differential Characteristics

Aeromonas species (Figs. 12.1 and 12.2) are facultatively anaerobic, Gram-negative, slightly curved rods with polar flagella. They ferment sucrose and mannitol, and are positive for ONPG, though lactose is variable.

A positive oxidase test result helps differentiate them from enterics and O-F test results differentiate them from *Pseudomonas* species: OF vs. O, respectively, as do indole results (*Aeromonas* is positive). Other diagnostic procedures include Gram stain, aerobic and anaerobic blood culture, stool culture, wound culture, api 20 E (see p. 121), Vitek 2 (see p. 125), and MALDI-TOF MS (see p. 150).

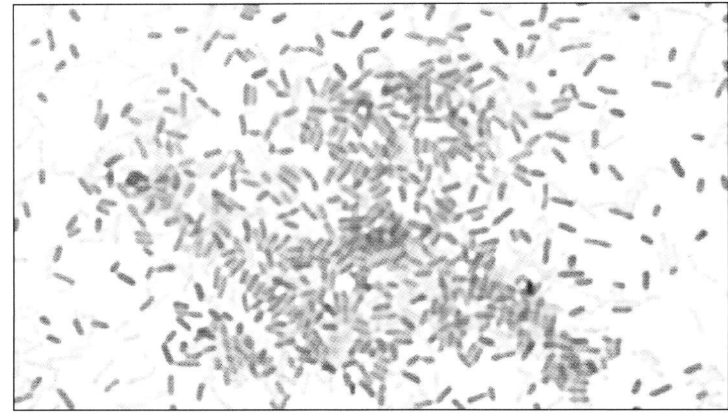

12.1 ***Aeromonas hydrophila*** **Stock Culture** • Rods can range in size from 1 µm or less in width and 1.0–3.5 µm in length with rounded ends. *A. hydrophila* is a motile, facultative anaerobe that produces a cholera-like enterotoxin. (Gram Stain)

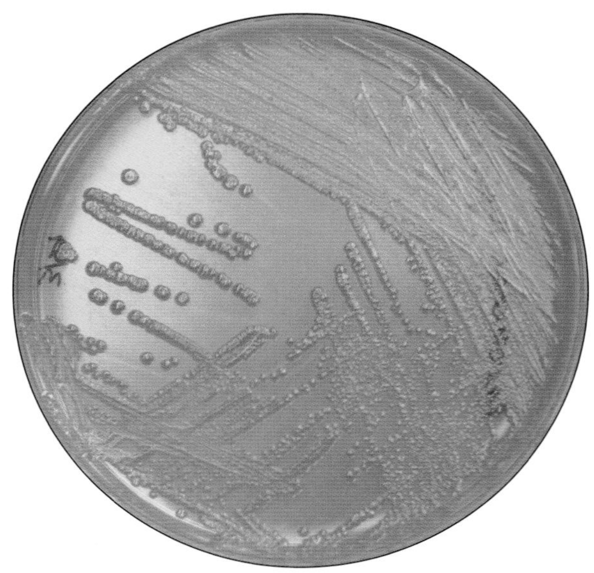

12.2 ***Aeromonas hydrophila*** **Growing on MacConkey Agar** MacConkey agar is typically used to identify coliforms—Gram-negative organisms that discolor the medium and produce pink colonies as a result of acid production from lactose fermentation. *A. hydrophila* doesn't ferment lactose and can be distinguished from coliforms by its lack of pink color. The larger colonies on this plate are about 2–3 mm in diameter.

Bacillus anthracis

Bacillus anthracis (phylum Firmicutes) causes **anthrax** in animals (mostly herbivores) and, more rarely, in humans. The intestinal tract of asymptomatic large animals is its principal habitat; however, *B. anthracis* spores have been known to remain viable in the soils of wet and warm climates for years.

Four types of anthrax occur in humans: cutaneous anthrax, inhalation anthrax (wool sorter's disease), injectional anthrax, and gastrointestinal anthrax. The cutaneous and inhalation forms are most frequently the result of contact with spores carried in animal products (wool, leather hides, etc.), infected animals, or contaminated soils.

Injectional anthrax is commonly associated with intravenous drug users. Gastrointestinal disease usually results from the consumption of contaminated meat. In 2018, only one case of anthrax was reported in the United States.

Two virulence factors account for the organism's pathogenicity: (1) a capsule that enables it to survive phagocytosis and disseminate via the bloodstream, and (2) secretion of three exotoxins—**edema factor** (EF), **lethal factor** (LF), and **protective antigen** (PA). Two plasmids—pχ01 and pχ02, carry the genes for toxin and capsule production, respectively. In addition, another gene product of pχ01 helps regulate capsule production by pχ02.

Protective antigen binds to receptors on phagocytic cells and is enzymatically split into two subunits, one of which is discarded. The other forms pores in the phagocyte's membrane with other PA fragments. Each fragment acts as a receptor for either EF or LF and promotes their entry into the cell via endocytosis. Once in the cell, EF and LF are free to exert their toxic effects.

Edema factor is largely responsible for the necrotic lesions (eschars) surrounded by large areas of swelling characteristic of cutaneous anthrax. Lethal factor is a protease that interferes with cytokine production, neutrophil chemotaxis, and macrophage apoptosis, among other effects—all of which compromise the immune system and allow rapid multiplication of the pathogen. This is followed by entry into the blood, producing bacteremia, shock, and meningitis.

Penicillin has been the traditional treatment for anthrax, although *B. anthracis* has demonstrated susceptibility to a broad variety of antibiotics. However, the strains of *B. anthracis* isolated from victims of the bioterrorist attacks of 2001, although susceptible to penicillin and ampicillin, demonstrated in vitro resistance to several previously effective antibiotics.

Even more significantly, the organisms were found to contain constitutive and inducible β-lactamases, which could lead to penicillin resistance (see β-lactamase test on p. 74). As a result, the CDC has advised against the use of penicillin or ampicillin. Today the recommended treatment is a combination of intravenous ciprofloxacin with one or two additional antibiotics for 7 to 10 days. In addition, a vaccine is available for researchers and animal product workers.

Differential Characteristics

B. anthracis (Figs. 12.3 and 12.4) is a facultatively anaerobic, nonmotile, encapsulated, endospore-forming, Gram-positive rod. It will not grow on phenylethyl alcohol blood agar, is catalase-positive, oxidase-negative, gelatinase-negative, and nonhemolytic.

Diagnostic procedures include Gram stain, endospore stain, and aerobic culture for colony morphology (Fig.12.5). Also available are api CHB (see p. 121), indirect hemagglutination (see p. 134), and ELISA (see p. 137) tests.

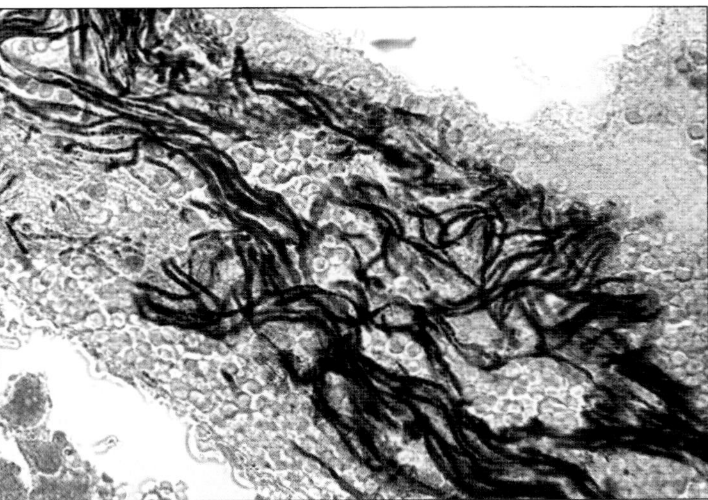

12.3 ***Bacillus anthracis*** **in a Mouse Liver Sinusoid** ● *B. anthracis* is a facultative anaerobe. Its cells are approximately 3.0–5.0 µm long by 1 µm wide. Most of these are in chains. The pink ovals are red blood cells. (Stained Section)

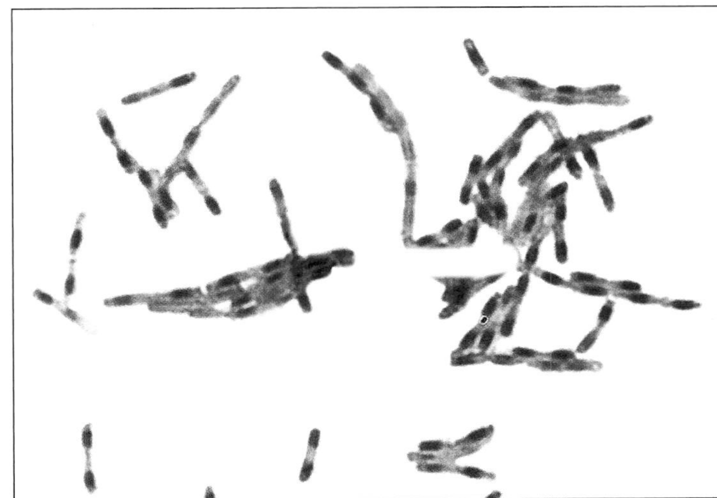

12.4 ***Bacillus anthracis*** **Stock Culture** ● *Bacillus anthracis* is encapsulated, but they don't show in this spore stain. *B. anthracis* resembles *B. cereus*, but is nonmotile and nonhemolytic. Note the elliptical, central to subterminal endospores, and chains of cells. It is the causative agent of anthrax. (Endospore Stain)

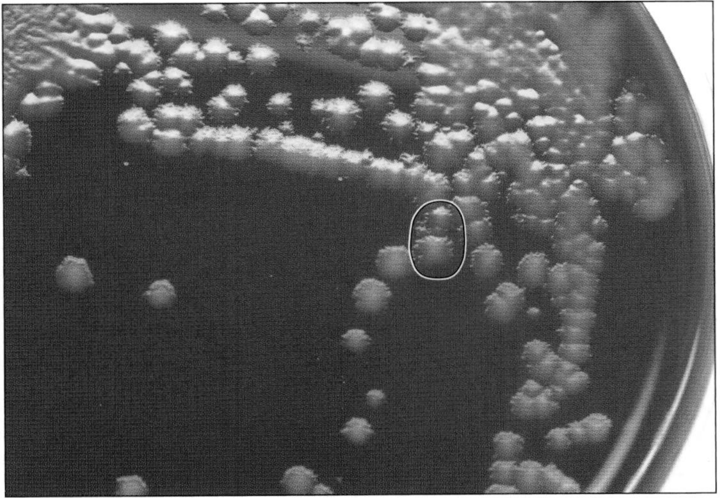

12.5 ***Bacillus anthracis*** **Growing on Sheep Blood Agar** ● *B. anthracis* typically produces large, circular or irregular, gray, finely granular colonies on sheep blood agar. Extensions from the colonies (two are circled) give the appearance of a medusa head (of Greek mythology). Given enough room on the plate, the extensions can get much longer. It is nonhemolytic.

Bacillus cereus

Bacillus cereus (phylum Firmicutes) is one of several opportunistic pathogens within the genus that comprise the *B. cereus* group. They are considered together because of similarities in mechanisms and symptoms of infection, as well sharing great similarities in DNA sequences. Included are *B. anthracis* (see p. 184) and *B. thuringiensis* (see p. 177), which produces Bt toxin that is used in some insecticides, and a few other species.

B. cereus group species are soil organisms found across the globe. Except for *B. anthracis*, they are opportunistic pathogens that can produce localized and systemic infections of various sorts, but they most frequently cause food-borne illnesses. There are two types of *Bacillus* food poisoning: emetic and diarrheal, and both are caused by toxins.

The emetic form is caused by **cereulide**, a heat and acid stable exotoxin. Symptoms appear within a few hours after ingestion and typically last less than 24 hours. A typical sequence of events leading to illness involves contaminated cooked rice, though other foods, such as pasta and milk products are also culprits. Spores that survive the initial cooking germinate and multiply rapidly, releasing cereulide that is not destroyed by further cooking. Ingestion of the toxin causes vomiting.

The diarrheal form is caused by several heat-labile **enterotoxins** that damage intestinal epithelial cells by creating membrane pores, cause hemolysis of RBCs, and damage immune cells by inducing apoptosis. Unlike emetic food poisoning, diarrheal food poisoning requires ingestion of the organism (usually endospores). Symptoms occur within 16 hours after exposure and typically last 24 to 36 hours. Foods often associated with the diarrheal form are meat, vegetables, and sauces.

In addition to food poisoning, *B. cereus* group species can cause a severe form of eye infection, usually resulting from penetrating eye trauma. If untreated, blindness can occur within a day or so. The infection may also spread to the brain and central nervous system with serious consequences.

Gastrointestinal disorders are not typically treated with antibiotics. *B. cereus* is often resistant to β-lactam drugs, such as penicillin and cephalosporin. The recommended treatment for ocular and other infections is vancomycin by itself or in combination with an aminoglycoside such as gentamicin, neomycin, or streptomycin.

Differential Characteristics

Bacillus cereus (Figs. 12.6 and 12.7) is a motile, facultatively anaerobic, endospore-forming, Gram-positive rod. It produces acid from glucose, maltose, fructose, and salicin fermentation, and is lecithinase- and gelatinase-positive. Diagnostic procedures include Gram stain, identification of endospores, aerobic culture, and serologic methods.

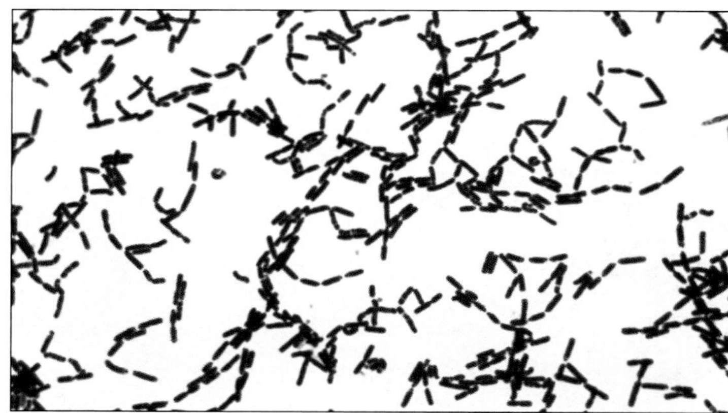

12.6 ***Bacillus cereus*** **Stock Culture** • *Bacillus cereus* rods are motile, encapsulated, and approximately 3–5 µm long by 1 µm wide. Chains are common and endospores, when produced (but not in this specimen—see Fig. 6.25) are elliptical and generally subterminal. (Gram Stain)

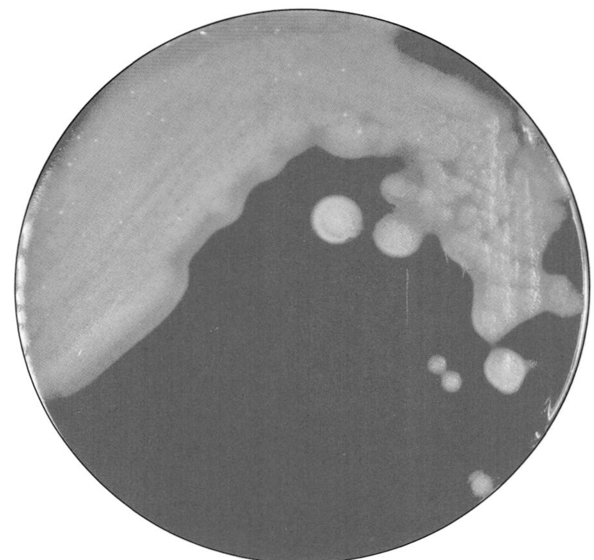

12.7 ***Bacillus cereus*** **Growing on Sheep Blood Agar** • Note the white and finely granular colony morphology compared to *B. anthracis* colonies in Figure 12.5. The larger colonies are approximately 14 mm in diameter. Also notice the β-hemolysis around the colony edges. *B. cereus* produces two kinds of food poisoning: emetic and diarrheal.

Bacteroides fragilis

Several *Bacteroides* species (phylum Bacteroidetes) reside in the human gastrointestinal tract and, to a lesser extent, in the female genital tract. Normal stool contains 10^{11} cells per gram, 1,000 times greater than the facultative anaerobes! Normally, this genus is beneficial, playing essential roles within the gut microbiome of healthy individuals.

Bacteroides fragilis is the most-frequently isolated Gram-negative obligate anaerobe from human infection sites. It is responsible for the majority of intra-abdominal infections, frequently abscesses (peritonitis), and also more rarely causes lung abscesses.

Intra-abdominal abscesses usually result after spillage of intestinal contents into the abdominal cavity from a ruptured appendix or a penetrating wound. Lung abscesses are usually a consequence of aspiration pneumonia and typically consist of a mixture of pathogens (a polymicrobial infection) sometimes including *B. fragilis* and oral streptococci.

Differential Characteristics

B. fragilis cells are nonmotile, encapsulated, Gram-negative, pleomorphic rods with rounded ends, though cell morphology is greatly affected by growth medium and culture age (Figs. 12.8 and 12.9). Sizes vary from 1.5 to 9 µm long.

Provisional identification as *B. fragilis* (or *B. fragilis* group) can be made of an obligately anaerobic, Gram-negative rod that grows on *Bacteroides* bile esculin agar (BBE, see p. 11), is resistant to the antibiotics kanamycin, vancomycin, and colistin, and is catalase positive (unusual for an anaerobe).

Other useful information is provided by colony morphology on blood agar plates and specific fermentation test results: positive for cellobiose and negative for arabinose, salicin, and trehalose. It is also negative for nitrate reduction and H_2S production.

Rifampin and metronidazole are among the most useful antibiotics for treatment. Resistance to other antibiotics is increasing.

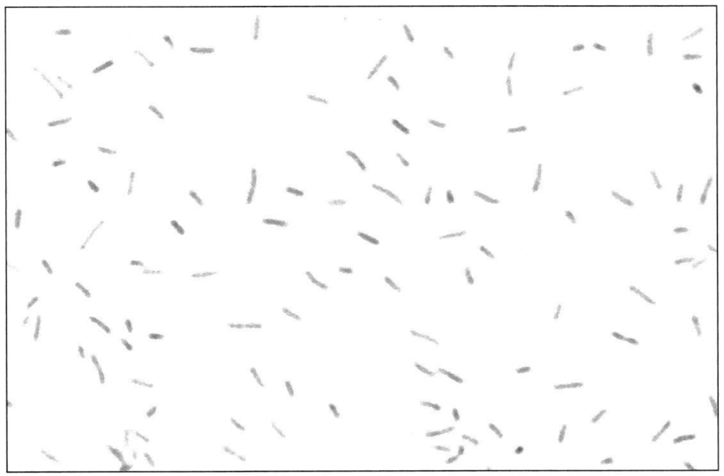

12.8 ***Bacteroides fragilis*** **Stock Culture** • Note the variable lengths (typically between 1.5 and 6 µm) and round ends of these cells. *B. fragilis* is an anaerobe and is responsible for a variety of body cavity abscesses. (Gram Stain)

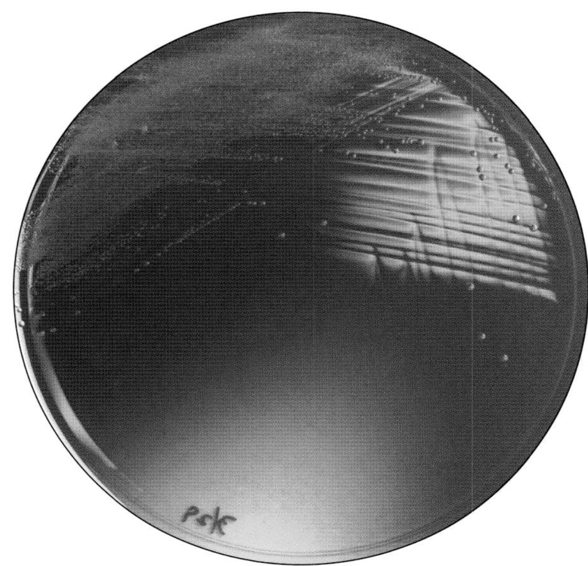

12.9 ***Bacteroides fragilis*** **Growing on** ***Bacteroides*** **Bile Esculin Agar** • Colonies on this plate are less than 1 mm in diameter. Note the darkening of the medium due to the reaction of esculin breakdown products with ferric ammonium citrate. For more information on bile esculin agar, refer to Section 2.

Bordetella pertussis

Bordetella pertussis (class Betaproteobacteria) is a strict pathogen and is not normally found in the microbiome of healthy individuals. It causes the disease formally known as **pertussis,** or less formally as **whooping cough.** Violent coughing, vomiting, and gasping for breath characterize the disease.

With approximately 10 million cases and 400,000 deaths reported annually worldwide, it is still a disease of significance despite vaccine availability. *B. pertussis* can infect anyone with no immunity or diminished immunity, but is most severe and communicable among infants less than 1 year of age. Humans are the exclusive host of this organism, and asymptomatic or unrecognized symptomatic adults are the likely reservoirs.

This highly communicable organism infects greater than 90% of unimmunized people exposed to it. The bacteria enter the mouth or nasopharynx as aerosols and attach themselves to respiratory cilia via pili and other adhesins, and begin releasing **pertussis toxin (PT).** Pertussis toxin is an **A-B toxin**[1] that inhibits chemotaxis and other functions of neutrophils and macrophages by damaging internal signal pathways. In addition, **tracheal cytotoxin** paralyzes cilia and kills ciliated epithelial cells that line the airways, leading to the cough. It is derived from peptidoglycan.

The best treatment is immunization with five doses of pertussis vaccine—the "P" of DTaP (for children) and Tdap (for adolescents and adults)—one at 2, 4, and 6 months of age, again at 15–18 months, and a final dose prior to entering school. Vaccination does not confer lifelong immunity. Boosters with Tdap are recommended every 10 years. Two vaccines are available: whole cell and acellular.

[1] A-B toxins are proteins composed of two functional fragments. The "A" portion ("A" is for "active") is involved in interacting with the host to produce the toxin's damage. It is often an enzyme. The "B" portion ("B" is for "binding") is involved in attaching the toxin to the target cell's receptor and may be involved in transporting the "A" portion into the cell.

The whole cell vaccine contains inactivated *B. pertussis* cells. The acellular vaccine minimally contains inactivated pertussis toxin and may include other *B. pertussis* antigens. It is generally more effective and produces fewer side effects. If infection occurs, then antibiotic treatment with erythromycin is recommended. Resistance to erythromycin is uncommon.

Differential Characteristics

B. pertussis (Figs. 12.10 and 12.11) is a small, nonmotile, obligately aerobic, Gram-negative coccobacillus. To grow *B. pertussis*, a special medium containing 15–25% blood and/or charcoal is used. Though it is oxidase positive, urease negative, and nitrate negative, it is inert in most biochemical tests.

Therefore, diagnostic procedures include direct fluorescent antibody (DFA, see p. 140), nasopharyngeal culture, and enzyme-linked immunosorbent assay (ELISA, see p. 137). Polymerase chain reaction (PCR, see p. 144) is also of use, however it is not appropriate if the specimen was sampled using a calcium alginate fiber and aluminum swab, as these inhibit PCR.

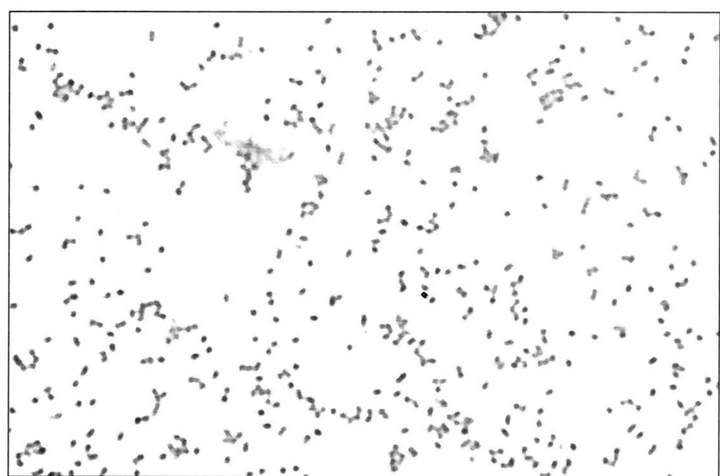

12.10 ***Bordetella pertussis*** **Stock Culture** • These tiny coccobacilli are usually arranged singly or in pairs and exhibit bipolar staining. They can range in size from 0.2–0.5 µm wide by 0.5–1.0 µm in length. *B. pertussis* is an obligate aerobe and is the causative agent of whooping cough. (Gram Stain)

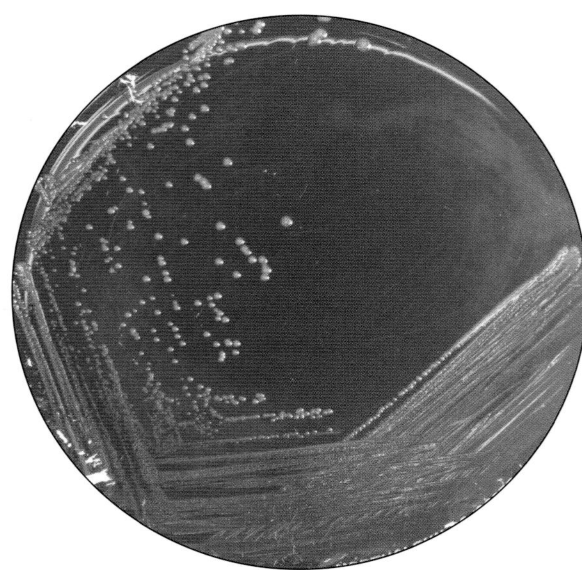

12.11 ***Bordetella pertussis*** **Growing on Regan-Lowe Agar** • Regan-Lowe agar is a selective medium designed to isolate *B. pertussis* and *B. parapertussis* from clinical specimens. Colonies are smooth, round, white, and pearly. These measure approximately 1–2 mm in diameter.

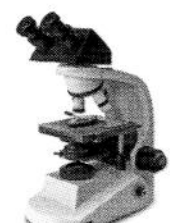

Borrelia burgdorferi

Borrelia burgdorferi (phylum Spirochaetes) causes **Lyme borreliosis,** more commonly called **Lyme disease** because the first reported case occurred in 1977 in Lyme, Connecticut. *B. burgdorferi* was isolated in 1982 and thought to be the sole infective agent in the disease. Subsequent studies, however, have implicated at least nine other *Borrelia* species.

Worldwide occurrence of Lyme disease is dependent on the availability of the hard-bodied tick vector (genus *Ixodes*). In North America *Borrelia* is transmitted by two species—*I. scapularis* and *I. pacificus*, with distribution in eastern and western temperate regions, respectively.

It is estimated that from 30% to 75% of *I. scapularis* carry the organism, which suggests why Lyme disease is the most common vector-borne illness in the United States. Disease-carrying ticks reside most commonly in coastal wooded areas inhabited by white-footed mice and white-tailed deer. Six states (California, Connecticut, Maryland, Minnesota, Oregon, and Massachusetts) have endemic rates of Lyme disease.

Lyme borreliosis is characterized by three distinct stages: Stage 1: a localized, but expanding skin lesion (**erythema migrans**) at the bite site accompanied by headache, fatigue, and malaise; Stage 2: weeks to months later, the inflammation and pain become generalized due to spreading of the spirochete with the possible development of arthritis, meningoencephalitis, or myocarditis; Stage 3: after months or even years of latency the infection becomes chronic, producing severe headaches, muscle and joint pain, and secondary skin lesions.

Ideally, administration of a single oral dose of doxycycline within 72 hours of the bite is used in treatment. Other antibiotics are used if the spirochete has disseminated.

Preventing tick bites is better than treatment. The CDC recommends avoiding areas where ticks are likely found, or wearing protective clothing and applying insect repellant if venturing into such an area. Afterward, checking clothing, equipment, and body for ticks is recommended.

Differential Characteristics

Borrelia burgdorferi (Fig. 12.12) is a motile, microaerophilic, Gram-negative, loosely coiled spirochete. The *B. burgdorferi* population in infected sites is low, so lab cultures and microscopic examination often produce false negative results. In their place, the CDC recommends serological tests targeting patient antibodies to the spirochete.

Initially, either an indirect ELISA (see p. 138) or an indirect immunofluorescent-antibody assay, (see p. 140) is used to detect patient IgM and IgG. A Western blot test has higher specificity and is used to confirm positive ELISA or IFA tests. False negative ELISAs may result from a late rise in IgG in about half the patients and from the effect of antibiotic treatment. PCR (see p. 144) may also be used, but results can be confounded by the scarcity of spirochetes in patient samples and contamination by other species with similar nucleotide sequences.

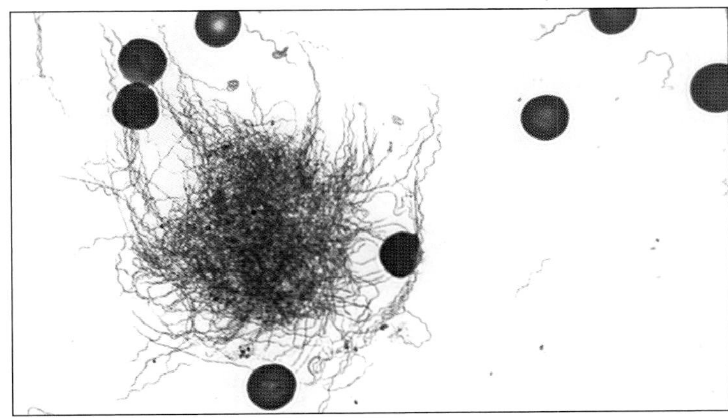

12.12 ***Borrelia burgdorferi* in a Blood Smear** • *B. burgdorferi* is the causative agent of Lyme disease. It is a spirochete, but note the loose coils. Typical dimensions are 0.2–0.3 µm wide by 4–30 µm long. *B. burgdorferi* cells are too thin to be seen using bright field microscopy without staining, but are visible with dark field or phase contrast. The circles are erythrocytes. (Giemsa Stain)

Brucella melitensis

Although most *Brucella* species (class Alphaproteobacteria) cause **brucellosis** (sometimes called **undulant fever**), *B. melitensis* is responsible for the most severe symptoms. Each of the infective agents of brucellosis has a domestic animal reservoir: *B. melitensis* (goats and sheep), *B. abortus* (cattle), *B. suis* (swine), and *B. canis* (dogs).

In spite of their different reservoirs, all infect humans in three principal ways: ingestion, direct contact through abraded skin, or inhalation. It is also a common laboratory acquired infection. In the United States, ingestion of contaminated unpasteurized milk or other dairy products is the most common source of infection.

Once within the body, *B. melitensis* cells enter the blood and are engulfed by neutrophils, which in turn are engulfed by macrophages in organs of the reticuloendothelial system (e.g., spleen and bone marrow). *B. melitensis* survives within phagocytic cells because it inhibits phagosome-lysosome fusion, allowing it to multiply. Early symptoms are nonspecific and often lead to misdiagnosis. These include fever, sweats, muscle and joint pain, and appetite and weight loss. Later symptoms include granuloma formation and tissue destruction in the lymph nodes, bone marrow, kidneys, liver, and spleen, but the mortality rate is low.

Eventually, cell mediated immunity clears the intracellular brucellae, whereas humoral immunity clears the extracellular ones. Antibiotics are limited to those that are able to enter macrophages and the treatment is longer than most, lasting weeks. In 2018, 138 cases of brucellosis were reported in the United States.

Differential Characteristics

Brucella species are small intracellular parasites They are also nonmotile, aerobic, Gram-negative coccobacilli (Fig. 12.13). Species can be presumptively identified from clinical specimens by colony morphology on blood or chocolate

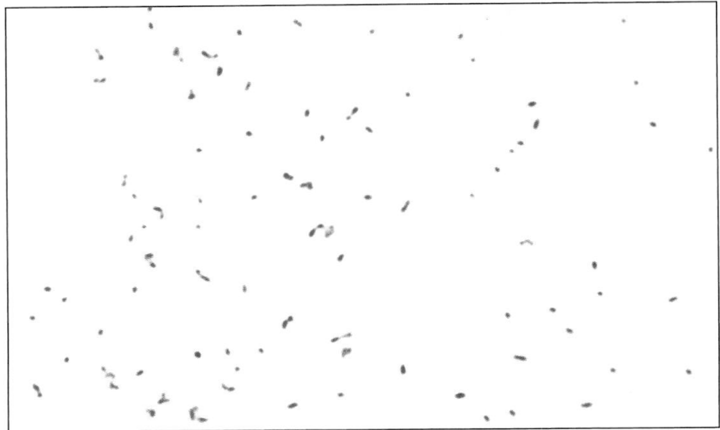

12.13 ***Brucella melitensis* Stock Culture** • Cells are coccobacilli 0.5 µm wide by 0.6–1.5 µm long. Usually arranged singly, these poorly staining cells were counterstained for 5 minutes to darken them for photographic purposes. (Gram Stain)

agar (Fig. 12.14), followed by microscopic examination of smooth, glistening colonies to demonstrate Gram-negative coccobacilli.

A serum agglutination test (see p. 135) using *B. melitensis*-specific serum to detect patient antibodies is used for confirmation, as is an indirect ELISA (see p. 139). Definitive identification utilizes further biochemical testing and PCR (see p. 144) or MALDI-TOF MS procedures (see p. 150).

12.14 ***Brucella melitensis*** **Growing on Sheep Blood Agar** B. *melitensis* is the causative agent of brucellosis. These colonies are round, smooth, convex, with an entire edge, and are up to 4 mm in diameter.

Campylobacter jejuni

Campylobacter jejuni (class Epsilonproteobacteria) is a common, worldwide human pathogen ubiquitously found in domesticated animals such as chickens, cats, and dogs. It is not part of the normal human microbiome and is estimated to cause more than 1.3 million cases of gastroenteritis each year in the United States. To put this number in perspective, *C. jejuni*-caused gastroenteritis is more common than that caused by *Salmonella* and *Shigella* combined.

Transmission of the disease is usually by ingestion of raw milk, undercooked poultry or meat, or contaminated water. Once inside the small intestine or colon, the organism multiplies in the mucous layer and causes watery diarrhea within 1 to 3 days.

During the course of infection an A-B exotoxin called **cytolethal distending toxin (CDT)** enters epithelial cells, destroys their DNA, and causes apoptosis. Inflammation and sometimes bleeding occur, which produces a bloody diarrhea. In most cases the infection is self-limiting and the diarrhea ends after a few days.

Dissemination by the bloodstream may result in salmonellosis-like enteritis and extraintestinal infection. In recent years, this organism has been implicated in a variety of infections including septic arthritis, meningitis, spontaneous abortion, and Guillain-Barré syndrome, a degenerative nerve disorder.

The infection typically resolves on its own, but antibiotics may be used if necessary. Resistance to erythromycin is high in some parts of the world, and resistance to fluoroquinolone and macrolides also has been reported.

Differential Characteristics

C. jejuni is a motile, microaerophilic, capnophilic (requires 5% to 10% CO_2), thermophilic (growth up to 43°C), nonfermentative, Gram-negative helical or curved rod. Cells sometimes appear as "S" or gull-wing shapes (Fig. 12.15).

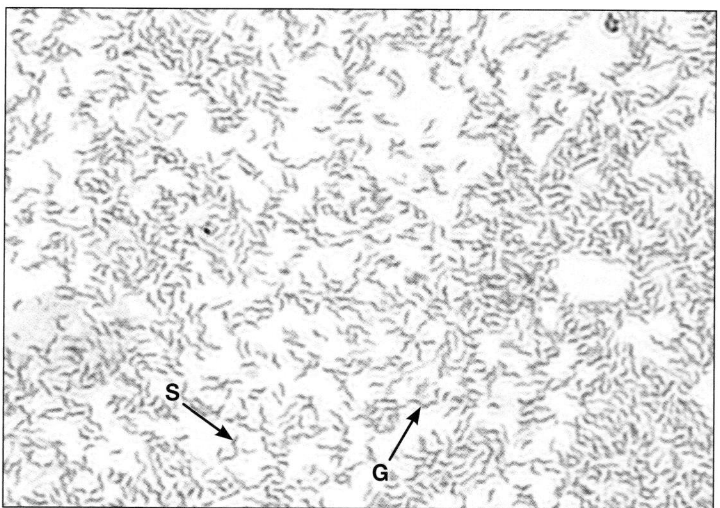

12.15 ***Campylobacter jejuni*** **Stock Culture** Note the characteristic "S" (**S**) and "gull wing" (**G**) shapes of the paired organisms. Cells are less than 1 μm wide and up to 5 μm long. *C. jejuni* is microaerophilic and swims in a corkscrew motion. It causes gastroenteritis. (Gram Stain)

Isolation and presumptive identification from stool samples can be achieved utilizing selective media containing a cocktail of antibiotics mixed with a sheep or horse blood agar base and incubated at 42°C in a microaerophilic environment enriched with CO_2 (Fig. 12.16).

Positive catalase, cytochrome oxidase, and hippurate hydrolysis results corroborate the initial identification. Once isolated, other identification methods can be used, such as API Campy (bioMéreieux) and MALDI-TOF MS (see pp. 121 and 150, respectively).

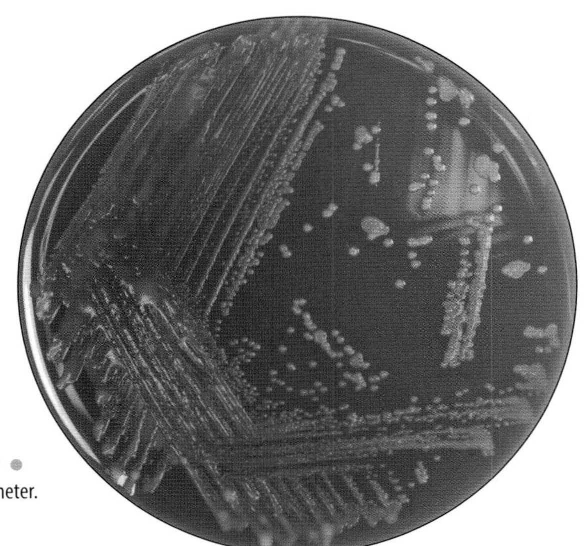

12.16 ***Campylobacter jejuni*** **Growing on Sheep Blood Agar** • These colonies are smooth, round, convex and between 1 and 2 mm in diameter.

Citrobacter spp.

Species in the genus *Citrobacter* (class Gammaproteobacteria) live almost exclusively in the colon of healthy humans. Unlike other members of the family *Enterobacteriaceae*, many of whom live freely outside a human host, presence of *Citrobacter* in the environment almost certainly suggests fecal contamination.

They rarely cause disease in healthy people, but three *Citrobacter* species—*C. freundii*, *C. koseri*, and *C. braakii*—are potent opportunistic pathogens and are responsible for the majority of hospital infections, especially in immunocompromised and neonatal hosts. These include respiratory and urinary nosocomial infections, hospital-acquired bacteremia, and neonatal meningitis. Transmission is usually person-to-person.

Susceptibility to tetracycline, nitrofurantoin, polymyxin B and colistin is common, but *Citrobacter* is resistant to ampicillin, amoxicillin, piperacillin, and cephalosporin.

Differential Characteristics

Citrobacter spp. are motile, facultatively anaerobic, straight Gram-negative rods (Figs. 12.17 and 12.18). They ferment glucose and utilize citrate, are methyl red, catalase, ONPG, and PYR positive, and are Voges-Proskauer, oxidase, and lysine decarboxylase negative.

Diagnostic procedures include aerobic culture, commercialized miniature kits such as api 20 E (bioMéreieux) (see p. 121), and MALDI-TOF MS (see p. 150) for *C. koseri*.

12.17 ***Citrobacter koseri*** **Stock Culture** • The straight rods of *C. koseri* usually appear singly and are about 1 µm wide by 3 µm long. *C. koseri* is a motile, facultative anaerobe. It, and other *Citrobacter* species, are commonly the cause of nosocomial infections. (Gram Stain)

12.18 ***Citrobacter koseri*** **Growing on Sheep Blood Agar** • These low, smooth, white colonies are about 3 mm in diameter. *C. koseri* cultures often smell like feces.

Clostridioides (Clostridium) difficile

Clostridioides difficile (phylum Firmicutes) is a common pathogen found in water, soil, a variety of animal intestines, and the intestinal tracts of a small percentage of healthy humans. (However, it has been found in up to 30% of asymptomatic adult hospital patients who had no recent history of antibiotic treatment).

It produces highly resilient spores that are resistant to antibiotics and disinfectants and are virtually impossible to eliminate from the environment. It has been isolated from health-care workers, asymptomatic hospital patients, hospital bedding, sinks, toilet seats, and endoscopy equipment.

C. difficile causes **antibiotic associated diarrhea (AAD)** and **antibiotic associated colitis (AAC)**, which collectively are referred to as ***C. difficile* infection (CDI)**. It may be a normal member of the patient's gut microbiota or it may be introduced after predisposing factors have allowed it to colonize the colon.

A key factor is that antibiotic treatment may reduce the number of normal bacteria living in the colon, which opens up the opportunity for *C. difficile* to flourish (note the first word in AAD and AAC). Pathogenic strains carry the genes for two toxins: **TcdA** (an enterotoxin) and **TcdB** (a cytopathic toxin), which cause mucosal cell death and fluid accumulation in the colon.

If the resulting diarrhea is self-limiting and stops when antibiotic therapy is discontinued, the patient has suffered from AAD. More severe symptoms and a persistent, sometimes bloody, diarrhea are characterized as AAC. **Pseudomembranous colitis (PMC)** may also result and is potentially fatal.

PMC occurs as dead tissue, neutrophils, and inflammatory cytokines contribute to formation of coalescing plaques over the colon's inner surface. In the most severe cases, the entire mucosa is necrotic and covered with pseudomembrane. Perforation of the colon resulting in peritonitis and septic shock may ensue, both of which can be fatal.

Vancomycin or fidaxomicin are recommended for first-line therapy. Other antibiotics may lead to recurrent CDI if the normal microbiome is reduced and *C. difficile* survives.

Differential Characteristics

C. difficile is a motile, anaerobic, endospore-forming, Gram-positive rod (Fig. 12.19). It is nonhemolytic; gelatinase (may be slow), proline-aminopeptidase, and esculin positive; catalase, lipase, indole, and nitrate negative.

Diagnostic procedures include *Clostridium difficile* toxin assay, *Clostridium difficile* selective agar, fecal leukocyte stain, ELISA to identify the toxins (see p. 137), and PCR (see p. 144).

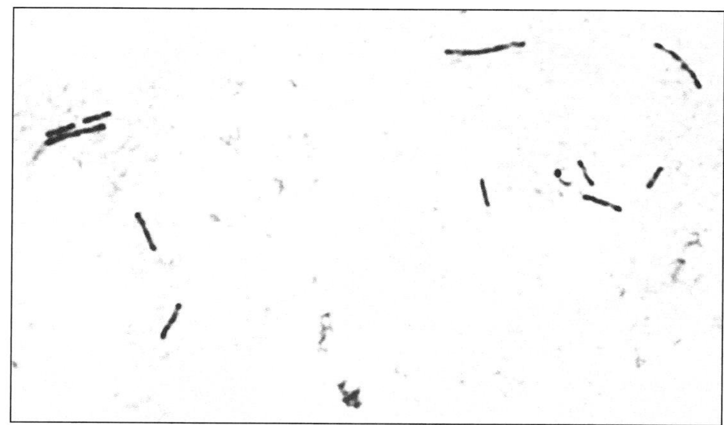

12.19 ***Clostridioides (Clostridium) difficile* Stock Culture** Although not seen in this micrograph, *C. difficile* does produce oval subterminal endospores. Cells can range in size from 0.5–1.9 µm wide by 3.0–17 µm long. It is the causative agent of *C. difficile* infection (CDI), a combination of diarrhea and colitis. (Gram Stain)

Clostridium botulinum

Clostridium botulinum (phylum Firmicutes) is a spore forming anaerobe typically found in soil. Further, the spores are widely distributed in the environment. There are seven known strains—A, B, C, D, E, F, and G—each of which produces an antigenically distinct **botulinum toxin (BoNT)**. Neurotoxins A, B, E, and sometimes F[2] cause botulism in humans.

There are three types of botulism known to occur in humans—foodborne, infant intestinal colonization, and wound. Adult intestinal colonization botulism and iatrogenic botulism are also known, but occur infrequently. The former is similar to the infant form and the latter results from an overdose in cosmetic botulinum treatment.

Although the word *botulism* has become almost synonymous with the food-borne illness, infant intestinal colonization botulism actually occurs more frequently in the United States. (In 2018, there were 157 cases of infant botulism and 17 cases of food-borne botulism in the United States.) Toxin

[2] Type C produces bird disease, and type D infects other mammals. Other clostridial species have been found to produce BoNT. *C. butyricum* and *C. baratii* produce types E and F, and *C. argentinense* (formerly *C. botulinum* Type G) produces type G.

production occurs during spore germination and vegetative growth but remains within the cell until it lyses.

Food-borne botulism most frequently occurs from ingesting botulinum toxin from insufficiently heated home-canned foods. Spores that remain viable in the undercooked canned food germinate and the resulting bacterial population flourishes in the anaerobic, nutrient-rich environment.

Within a few hours to 2 days after ingestion the toxin travels, by way of the bloodstream, to cholinergic synapses where it irreversibly binds to presynaptic motor neurons and blocks release of acetylcholine (ACh) by interfering with its vesicular transport to the membrane.

The result is bilateral flaccid paralysis (extreme weakness), usually beginning with cranial nerves (controlling facial muscles) and descending to the pharynx, neck, arms, and respiratory muscles, frequently resulting in respiratory failure and death. It should be noted that food-borne botulism is easily preventable. Botulinum toxins (but not spores) are heat-labile and destroyed when boiled for at least 20 minutes.[3]

In infant intestinal colonization botulism and wound botulism, infection occurs and *then* toxin is released. Once released, the toxin is absorbed into the bloodstream with progression and presentation of the disease similar to food-borne illness except that constipation is almost always an early sign of infant botulism. Fortunately, with proper treatment, infant botulism mortality is below 3%.

Because food botulism is caused by a rapidly acting toxin, not infection, antibiotics are not used in treatment. Instead, a heptavalent antitoxin (A, B, C, D, E, F, and G) that neutralizes the toxin must be administered quickly to patients suspected of botulism intoxication. Stomach pumping may also be of use.

[3] While this is a biological fact, it's much safer NOT TO EAT THE FOOD.

A pentavalent toxoid vaccine was administered to persons with an occupational risk of exposure to BoNT from 1965 to late 2011, at which time the CDC quit offering it because it was losing its effectiveness. New vaccines using DNA coding for BoNT epitopes delivered to humans via plasmid or viral vectors are being developed.

Differential Characteristics

C. botulinum is a motile, anaerobic, endospore-forming, Gram-positive rod. Its spores are oval, subterminal, and distend the cell (Fig. 12.20). It is usually β-hemolytic; gelatin, lipase, and esculin positive; and catalase, indole, and nitrate negative.

Diagnostic procedures to confirm the clinical diagnosis rely on identification of BoNT in patient samples. Examples are the mouse BoNT bioassay, mouse BoNT neutralization test, and MALDI-TOF MS (see p. 150).

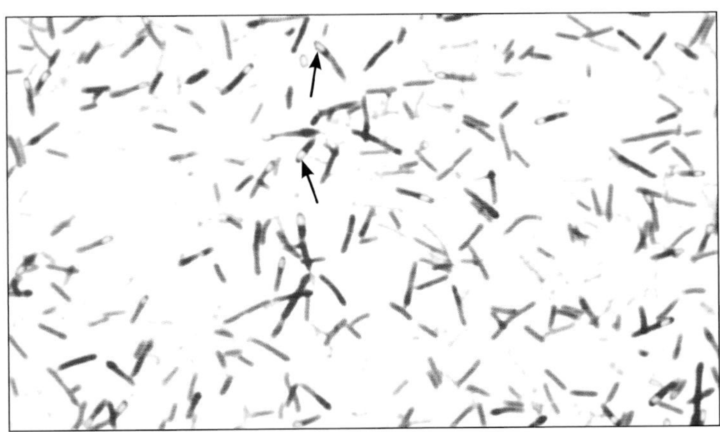

12.20 ***Clostridium botulinum*** **Stock Culture** • Note the unstained, oval, subterminal endospores, which distend some cells (**arrows**). Cell sizes are highly variable between strains, ranging from about 0.5–2.0 µm wide by 3–20+ µm long! *C. botulinum* is the causative agent of botulism. (Gram Stain)

Clostridium perfringens

Clostridium perfringens (phylum Firmicutes) is a very common, obligately anaerobic, spore-forming soil organism. It also inhabits the human colon and is the most frequently isolated *Clostridium* species in human feces. The species has been divided into five types (A–E) based on the toxin (or combinations of) they produce.

Type A produces alpha toxin, a **lecithinase** that breaks down membrane phospholipids resulting in the severe tissue damage associated with **myonecrosis** (gas gangrene) and **necrotizing fasciitis**. The alpha toxin also activates platelets to form clots (thrombi) in smaller blood vessels near the wound, promoting anaerobic conditions more suitable for the pathogen.

Gas gangrene, common in World War I because of the widespread soil contamination of battlefield wounds, is usually associated with traumatic wounds and crushing injuries, but also sometimes occurs after colon resections and septic abortions. The result is typically increased vascular permeability, hypotension, and shock. Death may occur within 2 days of onset of symptoms.

Types B and C produce **beta toxin** that causes **enteritis necroticans** (pig-bel). It is a severe form of enteritis seen in New Guinea following feasts where large amounts of contaminated sweet potato and pork are eaten. Sweet potatoes contain protease inhibitors that prevent beta toxin degradation. The result is intense abdominal pain and bloody diarrhea sometimes accompanied by intestinal perforation. It may also occur in diabetics and malnourished individuals.

Types A, C, and D produce ***C. perfringens*** **enterotoxin (CPE)** associated with the milder form of diarrheal food poisoning. This heat-labile toxin is commonly found in contaminated meat or poultry and their products, such as gravies or stews. Similar to *Bacillus cereus* food poisoning (see p. 185), *C. perfringens* spores that survive the heat during cooking germinate and produce enterotoxin when the conditions become favorable.

The toxin damages the intestinal epithelium (most severely in the ileum), leading to fluid leaking into the lumen. *C. perfringens* has become the second most common bacterial agent causing food-borne illness in the United States.

Various antibiotics may be used to treat gas gangrene, but prompt administration of antitoxin is vital. Surgical removal of necrotic tissue and hyperbaric oxygen therapy are also necessary. Fluid replacement is recommended for diarrheal food poisoning.

Differential Characteristics

Clostridium perfringens is a nonmotile, anaerobic (sometimes aerotolerant), endospore-forming, Gram-positive rod (Fig. 12.21) Although spores are produced by this organism, they are rarely seen in stained preparations. It is gelatinase and lecithinase positive and lipase and indole negative.

Hemolytic reactions differ depending on the strain and source of blood, but most produce some degree of β-hemolysis with an outer ring of α-hemolysis.

Diagnostic procedures include Gram stain of tissue showing Gram-positive rods with the absence of leukocytes, spore size and position (if seen), skin and muscle biopsy using direct or indirect fluorescent antibody (see p. 140), and immunoassays for CPE in stool samples.

12.21 ***Clostridium perfringens*** **Stock Culture** • *C. perfringens* causes gas gangrene and necrotizing fasciitis. Note the blunt ends of the cells and the absence of visible endospores because these cells were grown in culture. Cells can range in size from 0.6–2.4 µm wide by 1.3–19.0 µm long. (Gram Stain)

Clostridium septicum

Clostridium septicum (phylum Firmicutes) is commonly associated with the normal human microbiome, mostly residing in the appendix, but is rarely isolated from healthy stool samples. *C. septicum* bacteremia is a serious condition and often accompanied by some other disease, such as colon cancer, diabetes, or coronary artery disease.

Damage to the colon's mucosa (due to malignancy or *C. septicum*'s α-toxin) provides the opportunity for entry into the blood and subsequent spread to other parts of the body. Once at its new sites, severe tissue damage occurs and leads to death within 48 hours in roughly two-thirds of patients. Treatment is with β-lactam drugs.

Differential Characteristics

C. septicum is a motile, anaerobic, swarming, endospore-forming, Gram-positive rod (Fig. 12.22). It is gelatinase and esculin positive; lipase and indole negative. Diagnostic procedures include Gram stain, and growth and hemolysis patterns on CDC Anaerobe blood agar or phenylethyl alcohol blood agar (see p. 19).

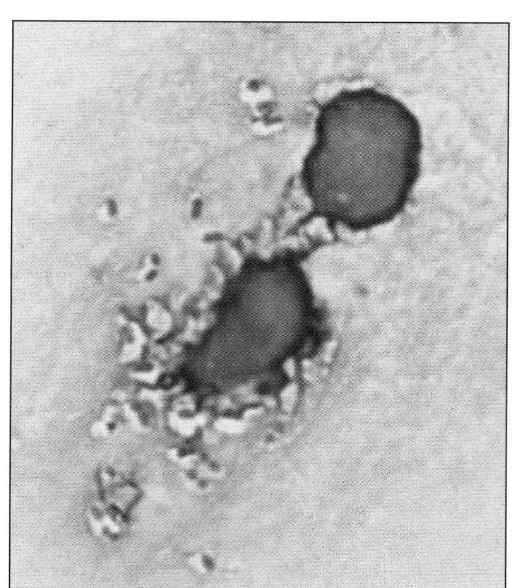

12.22 ***Clostridium septicum*** **in a Human Blood Smear** • *C. septicum* cells can range in size from 0.6–1.9 µm wide by 1.9–35 µm long. It is part of the normal microbiome of the appendix, but can enter the blood and spread to other parts of the body where it produces frequently fatal damage. (Giemsa Stain)

Clostridium tetani

Clostridium tetani (phylum Firmicutes) is an endospore-forming organism found in soil and animal (including human) intestinal tracts. It is the causative agent of **tetanus**, more commonly called "lockjaw," because of the difficulty chewing and swallowing characteristic of the disease's onset. Despite the continued prevalence of disease in developing countries, tetanus has nearly been eradicated in the United States by tetanus toxoid vaccine (part of DTaP and Tdap series).

Tetanus is considered a strictly **toxigenic disease** because the local infection at the site of colonization is typically mild while the effect of released toxin is devastating. Transmission is typically by entry of spores into a traumatic or puncture wound where they germinate, grow, and release the neurotoxin, **tetanospasmin** (TeNT).

TeNT is absorbed and transmitted by motor neurons to the central nervous system where it permanently binds to neurons and blocks the release of the inhibitory neurotransmitter **γ-aminobutyric acid.** The result is a descending severe muscle spasm that begins with facial muscles, neck muscles, and eventually chest muscles, which results in respiratory failure.

Note the comparison between the closely related *C. tetani* and *C. botulinum*. Botulinum toxin travels by way of the bloodstream to peripheral nerve synapses where it blocks release of acetylcholine, resulting in inability to contract muscles (flaccid paralysis); tetanospasmin travels within the nerve cells to central nervous system synapses where it blocks release of inhibitory γ-aminobutyric acid, resulting in the uninhibited or spastic contractions of tetanus.

Vaccination is available as part of the DTaP (administered to children younger than 7 years) and Tdap (administered as a booster at age 11 years) series, followed by booster shots every 10 years. Tetanus antitoxin (immunoglobulin) is used in treatment.

Differential Characteristics

C. tetani is a motile, anaerobic, endospore-forming, Gram-positive rod (Fig. 12.23). It is gelatinase positive and lipase, esculin, and nitrate negative, though diagnostic tests are typically not done. Patient immunization history is of greatest concern in suspected cases.

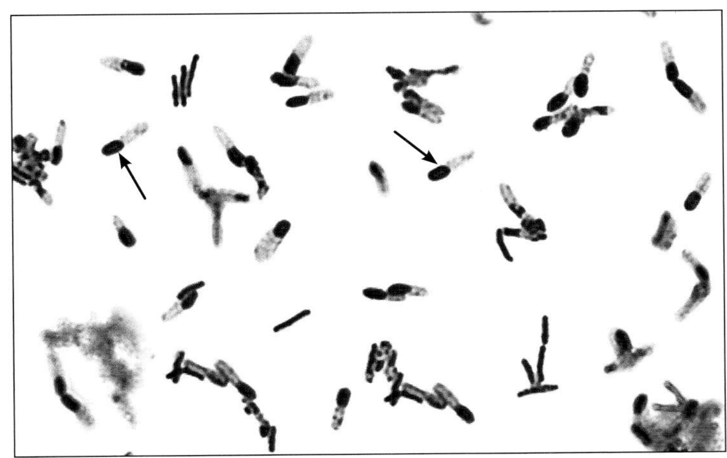

12.23 ***Clostridium tetani*** **Stock Culture** • Note the oval, terminal endospores (**arrows**). Occasionally the spores are subterminal. Cells can range in size from 0.5–1.7 µm wide by 2.1–18.1 µm long. *C. tetani* is the causative agent of tetanus. (Gram Stain)

Corynebacterium diphtheriae

Corynebacterium diphtheriae (phylum Actinobacteria) is the toxigenic agent of **diphtheria** and **cutaneous diphtheria.** Although it can live for months in the environment, *C. diphtheriae* is most often transmitted from person to person in aerosol droplets.

Once a leading cause of childhood mortality, it is now a rare disease in the United States due to routine vaccinations with **diphtheria toxoid.** However, it still exists in developing countries where vaccine is not available in sufficient quantity. In recent years, diphtheria has shifted from a primarily childhood disease to one affecting all ages, including disproportionately high numbers of unimmunized members of low-income communities and intravenous drug users.

C. diphtheriae infects the nasopharynx and is responsible for a local respiratory infection and systemic poisoning from absorption of the cytotoxin produced at the local infection site. In the former, *C. diphtheriae* produces a thick **pseudomembrane** composed of bacteria, fibrin, immune cells, and dead cells, which obstructs the upper respiratory airway. In the latter, cytotoxin is dispersed throughout the body, frequently resulting in myocarditis and congestive heart failure. It can also affect kidney, liver, and nervous system functions.

A third form is cutaneous diphtheria, which is a result of bacterial entry through an open wound causing a local infection with similar systemic effects as the inhalation disease.

The exotoxin (*tox*) gene is introduced into *C. diphtheriae* by a lysogenic phage.[4] The toxin itself has three functional parts carried on two subunits: Fragment B (see footnote 1, p. 187) binds to heparin receptors on host nerve and heart cells. It also has a translocation domain that moves the bound toxin to the inside of the cell. Fragment A is an enzyme that inactivates peptide elongation during protein synthesis.

Vaccination is available as part of the DTaP and Tdap series (in each, the "D" stands for "diphtheria"). Antitoxin to neutralize diphtheria toxin along with penicillin or erythromycin may be used in treatment.

Differential Characteristics

C. diphtheriae is a nonmotile, non-spore-forming, facultatively anaerobic (sometimes strictly aerobic), pleomorphic Gram-positive rod (Fig. 12.24). It produces acid from glucose and maltose fermentation, and no acid from sucrose. It is negative for urease, esculin, pyrazinamidase, and alkaline phosphatase.

[4] A lysogenic phage is a virus that inserts its DNA genome into the bacterial host's DNA and becomes part of its genome.

Diagnostic tests include throat culture, api CoryneStrip, toxin production tests, PCR (see p. 144), and MALDI-TOF MS (see p. 150).

12.24 ***Corynebacterium diphtheriae* Stock Culture** *C. diphtheriae* is frequently seen in a "V-shaped" (**V**) or palisades (**P**) cellular arrangements. Cells can range in size from 1 µm or less wide by 1.0–8.0 µm long. It is the causative agent of diphtheria. (Gram Stain)

Escherichia coli

Escherichia coli (class Gammaproteobacteria) is a member of the large family Enterobacteriaceae, the "Enterics." It inhabits the intestinal tract and female genital tract of healthy humans in addition to other animal hosts. Generally, each host harbors a different strain of the organism.

The strains most important to humans are those that inhabit us and those that live in the intestines of cattle. It is estimated there are 186 genotypic types of *E. coli* due to plasmid diversity and transposable elements (pieces of DNA that can move between and within the genome and plasmids).

E. coli is an **opportunistic pathogen** and in the right place, at the right time may cause disease in most any human tissue. Other *E. coli* strains are true pathogens that cause various forms of **enteritis** or **gastroenteritis**. They are named based on their method of causing diarrhea. They are:

- Enterotoxigenic *E. coli* (ETEC) produces enterotoxins that lead to water entering the intestinal lumen;
- Enteropathogenic *E. coli* (EPEC) attaches to the epithelium and produce lesions;
- Enteroinvasive *E. coli* (EIEC) enters the epithelial cells of the colon (just as *Shigella* does);
- Enteroaggregative *E. coli* (EAEC) attaches to the epithelium in characteristic groups and produces enterotoxins similar to ETEC; and
- Enterohemorrhagic *E. coli* (EHEC) produces enterohemolysins (similar to Shiga toxins) and produces EPEC-like lesions; also known as Shiga-like toxin-producing *E. coli* (STEC).

In most cases, strain identification is not necessary because the condition resolves itself or is managed by antidiarrheal medications. However, EHEC infections are more severe and may be life-threatening. In addition to those listed above, some *E. coli* strains produce extraintestinal infections of the urinary tract, meninges, and blood.

Serotyping is based on three antigens: the polysaccharide of LPS is the O-antigen, flagellin is the H antigen, and capsule compounds comprise the K antigens. Variants of these antigens in different combinations are used to identify different *E. coli* serotypes. However, usually only the O and H antigens are used (e.g., *E. coli* O157:H7). These strains are seen primarily in developing countries, but EHEC O157:H7 has been responsible for outbreaks of hemorrhagic colitis in the United States.

This disease is almost always caused by ingestion of contaminated foods or nosocomial infections; however, person-to-person transmission has been reported. High levels of multidrug resistance are prevalent due to ease of gene transfer.

Differential Characteristics

All members of *Enterobacteriaceae* are oxidase-negative, Gram-negative, facultatively anaerobic rods that produce acid from glucose fermentation (Figs. 12.25 and 12.26). Several commercial multiple test systems are available for the differentiation and identification of the individual species, including EnteroPluri-*Test* (see p. 123), api 20 E (see p. 121), and BBL Crystal E/NF (see p. 122).

Diagnostic tests for identification of individual strains include EIA for Shiga toxin (see p. 137), O157 and H7 serum agglutination tests (see p. 133), PCR (see p. 144), microarray probes (see p. 149), and MALDI-TOF MS (see p. 150).

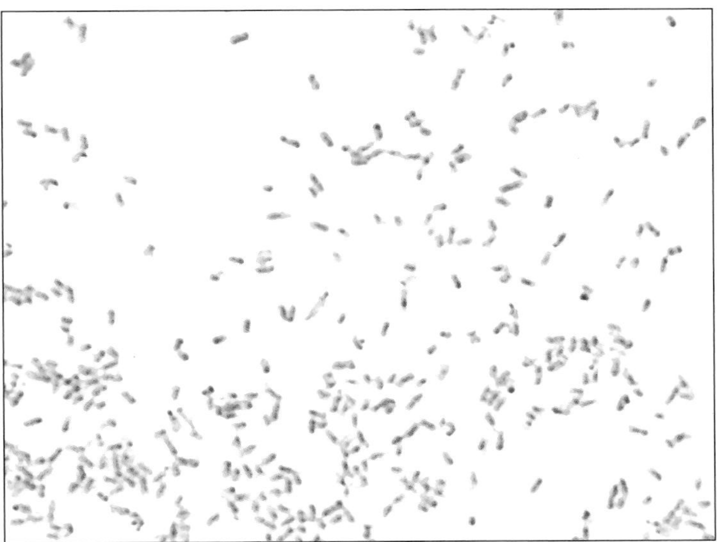

12.25 ***Escherichia coli*** **Stock Culture** • The straight rods are usually arranged singly or in pairs. Cell sizes range from 1–1.5 μm wide to 2–6 μm long. Some *E. coli* strains cause enteritis or gastroenteritis. (Gram Stain)

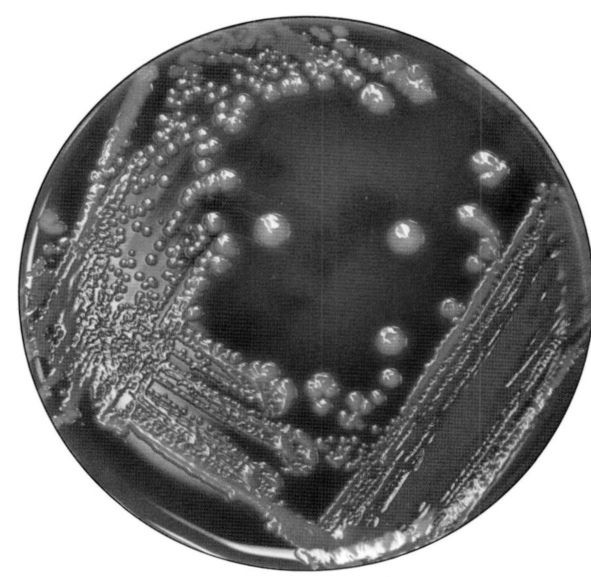

12.26 ***Escherichia coli*** **Growing on Sheep Blood Agar** • The largest of these round, convex, shiny, cream-colored *E. coli* colonies are 6 mm in diameter.

Fusobacterium nucleatum and *Fusobacterium necrophorum*

Fusobacterium nucleatum and *F. necrophorum*, (phylum Fusobacteria) are members of the normal human oral, gastrointestinal, and genital microbiota, but are generally considered pathogens in clinical specimens.

They are involved in a broad spectrum of anaerobic infections (usually with a mixture of organisms). These include periodontal disease, skin ulcers, ulcerative colitis, and are very common agents in upper respiratory infections, including chronic sinusitis, aspiration pneumonia, lung abscesses, and brain abscesses.

They seem to thrive when intestines are unsteady, such as during ulcerative colitis and colon cancer. They have also been shown to cause severe systemic infections in cancer patients following chemotherapy.

The virulence of fusobacteria is largely due to endotoxins that evoke a vigorous immune reaction that can lead to toxic shock with widespread systemic collapse. Some strains produce a β-lactamase, so β-lactam antibiotics are generally avoided in treatment.

Differential Characteristics

Fusobacterium nucleatum (Fig. 12.27) and *F. necrophorum* (Fig. 12.28) are nonmotile, anaerobic, non-sporing, usually spindle-shaped or pleomorphic, Gram-negative rods. In general, they are indole positive, catalase negative, do not grow on bile agar, and may or may not ferment glucose, but don't ferment mannitol, lactose, or rhamnose, and are negative for many of the standard macromolecular hydrolysis tests (e.g., DNase). Butyrate is the major metabolic end-product, which distinguishes them from most *Bacteroides* species.

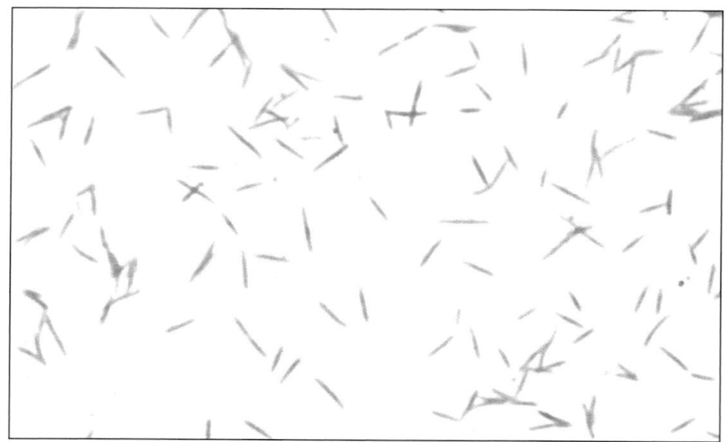

12.27 ***Fusobacterium nucleatum* Stock Culture** ● Note the long, thin, spindle-shaped cells, ranging in length from 5–10 μm. *Fusobacterium* species are anaerobic and cause a variety of infections. (Gram Stain)

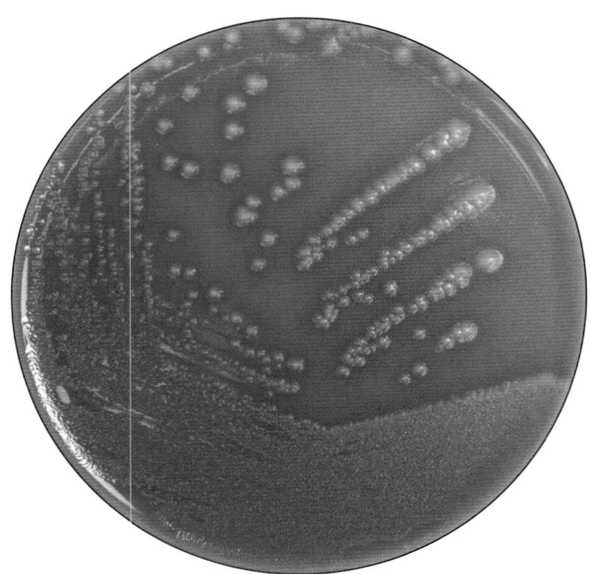

12.28 ***Fusobacterium necrophorum* Growing on Chocolate Agar** ● Note the flat, grayish-brown colonies. The largest colonies are approximately 5 mm in diameter.

Haemophilus influenzae

Haemophilus influenzae (class Gammaproteobacteria) was given the name *Haemophilus* (*haemo* means "blood" and *philos* means "loving") because it requires specific blood factors (**V**, hemin, and **X**, NAD and NADP) to be grown in vitro (Figs. 12.29 and 12.30).

Unencapsulated, commensal strains of *H. influenzae* can be found in the nasopharynx of virtually everyone over the age of 3 months, and may be opportunistic pathogens. Virulent (encapsulated) strains also inhabit the upper respiratory tract, but to a much lesser degree.

In either case, the human upper respiratory tract is its reservoir and *H. influenzae* transmission is from person to person by aerosol droplets. In the United States, 5,573 cases of *Haemophilus influenzae* infection were reported for 2018 (all ages and serotypes).

Of the six known capsular types (a–f), **type b** (**Hib**) is the most pathogenic. *Haemophilus influenzae* b was once the most common cause of bacterial meningitis in children; however, its epidemiologic impact has dramatically changed since the development and widespread use of vaccine in the 1990s. Still, it is the most common cause of bacterial meningitis in unvaccinated children.

Encapsulated forms of *H. influenzae* also cause epiglottitis, cellulitis, otitis media, pneumonia, bronchitis and chronic obstructive pulmonary disease (COPD), septicemia, and endocarditis.

Several vaccines targeting the type b polysaccharide are available and have dramatically reduced the incidence of bacterial meningitis in children. They are administered to infants in multiple doses, often with a booster in the second year.

If treatment is necessary, cephalosporins, such as ceftriaxone or cefotaxime, are used. Ampicillin used to be the antibiotic of choice, but *H. influenzae* b strains developed resistance due to acquisition of a plasmid encoded β-lactamase (see p. 74).

Differential Characteristics

Haemophilus influenzae is a nonmotile, facultatively anaerobic, short or pleomorphic, Gram-negative rod (Fig. 12.31). The requirement for V and X factors is strong, though not absolute, evidence of the genus *Haemophilus*.

H. influenzae is divided into six serogroups (a–f) based on differences in capsular components identifiable by slide agglutination tests (see p. 133). Clinical samples suspected of containing *H. influenzae* are tested for Hib, because it is the most virulent.

The capsules of all strains but Hib have hexose sugars; Hib has pentose sugars, which is the differential basis for the agglutination test. Each serotype demonstrates a unique combination of positive and negative results. PCR (see p. 144) is also an option to identify capsule type. ●

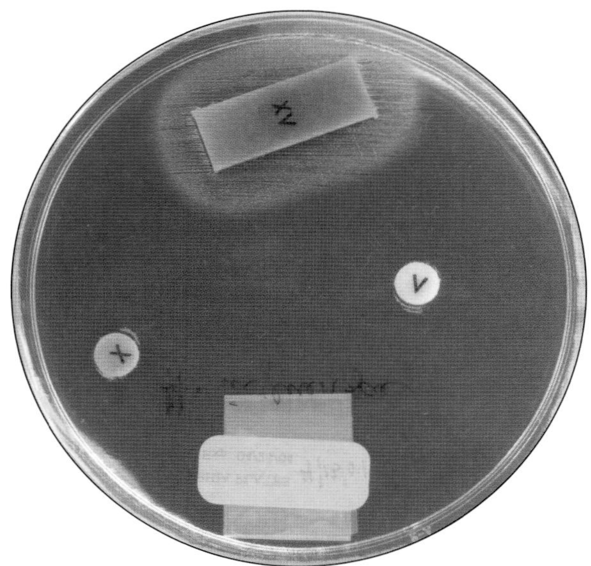

12.29 ***Haemophilus influenzae*** **Demonstrating Characteristic Requirement of Both X and V Blood Factors** • Note the growth surrounding the strip containing both X and V factors, and the absence of growth around the disks containing only X or V factor. The test differentiates between *Haemophilus* species. X factor is hemin and V factor is NAD. Compare with Figure 12.30.

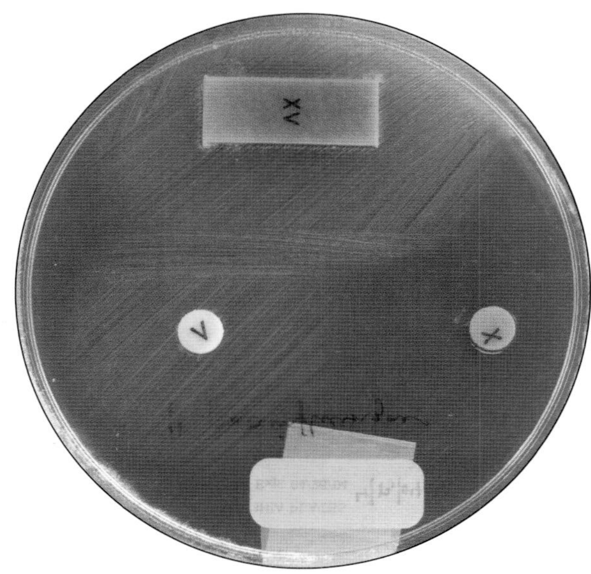

12.30 ***Haemophilus parainfluenzae*** **Demonstrating Characteristic Growth in the X and V Test** • *H. parainfluenzae* can be differentiated from *H. influenzae* because it requires only V factor to survive. Note the growth surrounding both the V disk and X-V strip. *H. ducreyi* (not shown) requires only X factor to grow. Compare to Figure 12.29.

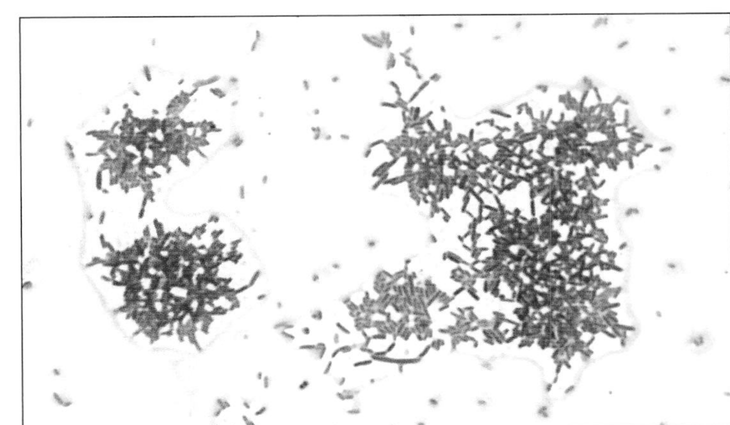

12.31 ***Haemophilus influenzae*** **Stock Culture** • Note the various shapes and sizes of the cells (pleomorphism). Cells can range in size from about 0.5 µm wide by 0.5–3.0 µm long. *H. influenzae* b is encapsulated and is the most virulent strain. It causes meningitis, pneumonia, and chronic obstructive pulmonary disease (COPD). (Gram Stain)

Helicobacter pylori

Discovered in 1982, *Helicobacter pylori* (class Gammaproteobacteria) is associated with gastritis, gastric and duodenal ulcers, and gastric carcinoma (*H. pylori* is the only known bacterial carcinogen). More than half the adults in the world test positive for *H. pylori* antibodies, but the mode of transmission is unclear, with fecal-oral, oral-oral, and common source transmission as viable possibilities.

Infection involves several stages. First, upon entry into the stomach, urease activity allows *H. pylori* to survive the stomach's low pH while it swims through the mucus lining down toward the epithelium. Various adhesins promote attachment to the epithelium and infection has been established. Interaction between *H. pylori* toxins and host neutrophils and macrophages results in a **subclinical gastritis.** Continued urease activity damages epithelial cells because of ammonia's toxicity.

Beyond this, the interaction between *H. pylori* virulence factors and the host's gastric epithelium and immune system is complex. Simply put, two main cytotoxins, VacA and CagA, damage epithelial cells, which leads to further inflammation and potentially ulceration. Host immune response is responsible in part for the severity of the symptoms.

Treatment involving administration of multiple medications (antibiotics and a proton pump inhibitor) is common. Due to *H. pylori*'s ability to embed itself within the stomach mucus lining, infections are typically long lasting and difficult to fully clear.

Differential Characteristics

Helicobacter pylori is a motile, microaerophilic, curved, spiral, or straight and slightly plump, Gram-negative rod (Fig. 12.32). Three key biochemical test results are: positives for catalase and oxidase, and strongly positive for urease. Several different approaches for evaluating urease activity are available. One involves incubating tissue from gastric biopsy in a urea medium similar to the urease test (see p. 118).

A noninvasive method is the urease breath test, in which urea radiolabeled with ^{13}C is swallowed, resulting in exhalation of $^{13}CO_2$, if positive. Another involves swallowing urea radiolabeled with ^{15}N, resulting in urine containing $^{15}NH_3$, if positive. EIA tests for *H. pylori* stool antigen in feces or IgG and IgA antibodies in serum are also available (see p. 137). PCR (see p. 144) is also used on various clinical samples.

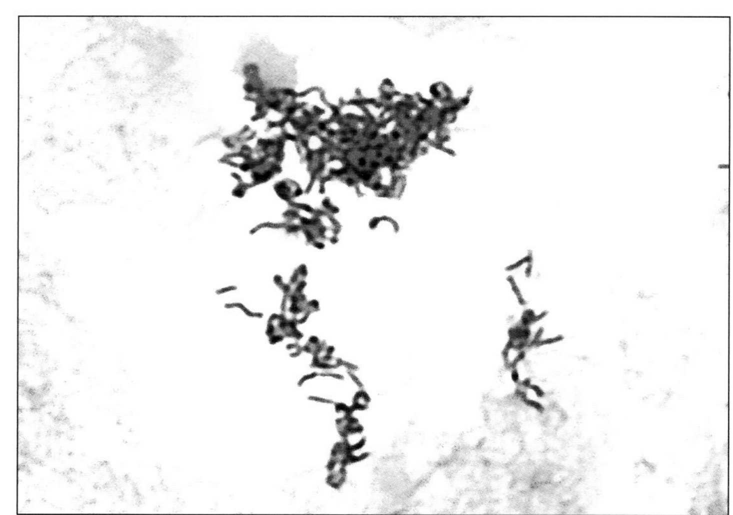

12.32 ***Helicobacter pylori*** **Stock Culture** • *H. pylori* is Gram negative, but young cultures of *H. pylori* grown in vitro frequently stain Gram positive. Note the curved cells. Cells range in size from 0.5 µm wide by 2.5–5 µm long. *H. pylori* causes gastritis, gastric and duodenal ulcers, and gastric carcinoma. (Gram Stain)

Klebsiella pneumoniae

Klebsiella pneumoniae (class Gammaproteobacteria) is found in soil, water, grain, fruits, vegetables, and the intestinal tracts of a variety of animals, including humans. It is also harbored in the nasopharynx. Transmission is via aerosol droplets or by the fecal-oral route.

K. pneumoniae is a very common nosocomial pathogen, but also causes **community-acquired primary lobar pneumonia**—a severe (and frequently fatal) necrotizing infection. Nosocomial infections include pneumonia, urinary tract infections, bronchitis, liver abscesses, surgical wound infections, biliary tract infections, and hospital associated bacteremia. The organism owes its virulence to endotoxin production and its ability to form a protective polysaccharide capsule.

In 2001, a carbapenemase-resistant *K. pneumoniae* strain was reported and in the last two decades resistance has spread through *Klebsiella* species, as well as other members of the Enterobacteriaceae via plasmid transfer. It is a serious enough problem that the designation **"CRE"** (**carbapenemase resistant Enterobacteriaceae**) was coined.

In many instances, carbapenemase is the only remaining effective antibiotic against multidrug-resistant pathogens. Further, it is able to confer resistance to the majority of β-lactam antibiotics. Clearly, treatment options are limited, and physicians may prescribe several antibiotics that show some degree of effectiveness in the hopes that in combination they will be effective. These are known as **drugs of last resort**.

Differential Characteristics

K. pneumoniae is a nonmotile, encapsulated, facultatively anaerobic, Gram-negative rod (Figs. 6.22 and 12.33). It produces large, mucoid colonies (Fig. 12.34) on most isolation media. *K. pneumoniae* is negative for indole, methyl red, and arginine and ornithine decarboxylase, and positive for ONPG, lysine decarboxylase, VP, and citrate.

Diagnostic tests include RapID SS/u System (Remel) for urinary tract infections, api 20 E (see p. 121), and MALDI-TOF MS (see p. 150).

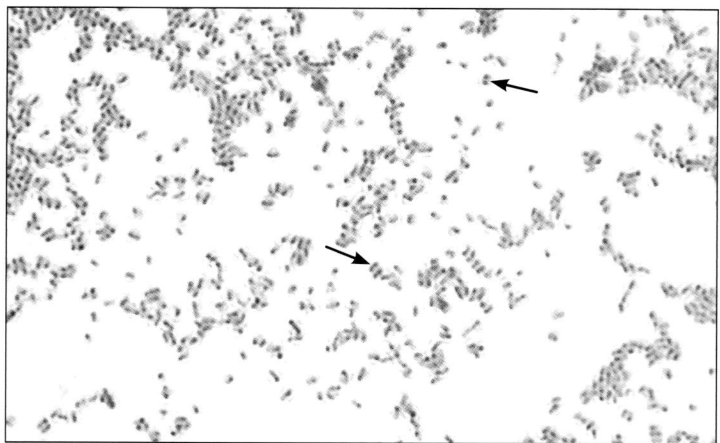

12.33 ***Klebsiella pneumoniae*** **Stock Culture** • *K. pneumoniae* cells range in size from 0.3–1.0 µm wide by 0.6–6.0 µm long. Note the bipolar staining (**arrows**). There is also a hint of a capsule around some cells. (Gram Stain)

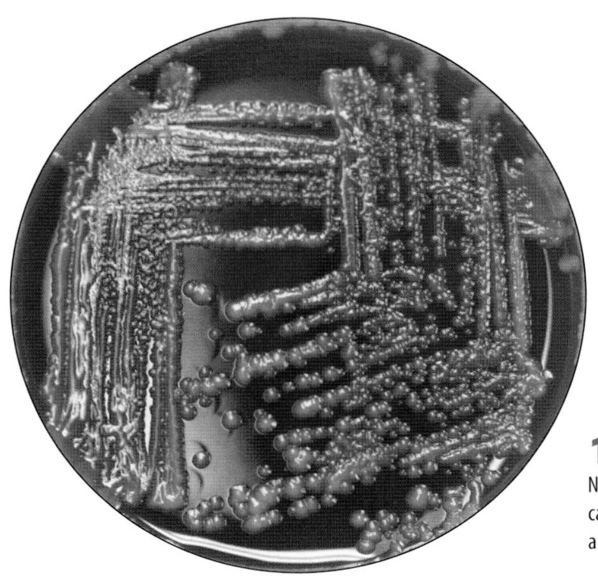

12.34 ***Klebsiella pneumoniae*** **Growing on Sheep Blood Agar** Note the mucoid appearance to the colonies due to *K. pneumoniae's* polysaccharide capsule. The largest colonies are approximately 3 mm in diameter. *K. pneumoniae* is a common nosocomial pathogen.

Legionella pneumophila

Legionella pneumophila (Class Gammaproteobacteria) was discovered following the outbreak of Legionnaires' disease in 1976. It is ubiquitous in nature, typically found in abundance in any aquatic habitat worldwide, including chlorinated hot tubs. In the hotels and hospitals where outbreaks have occurred, the organism has been isolated from freshwater sources, various plumbing fixtures, and cooling systems. In its natural environment, the organism parasitizes amoebae living in the water.

Transmission to humans is by inhalation of environmental aerosols; person-to-person transmission is not known to occur. **Legionellosis** is the general heading used to include all of the diseases caused by this organism (i.e., Legionnaires' disease—a pneumonia-like disease; Pontiac fever—a milder, short-term, flu-like illness; and extrapulmonary spreading most common in immunocompromised patients).

Although capable of surviving extracellularly, *L. pneumophila* is classified as a facultative intracellular pathogen because of its ability to survive and multiply inside phagosomes of pulmonary macrophages. *L. pneumophila* has the ability to induce macrophages to engulf them and inhibit lysosome-phagosome fusion. Instead, the phagosome "recruits" mitochondria and endoplasmic reticulum vesicles to form a ***Legionella*-containing vacuole** (LCV), in which the bacterial cells replicate.

As nutrients become depleted, *L. pneumophila* cells differentiate from a replicative form into an infectious form. These are released to the environment and infect new host macrophages. Antibiotics, such as fluoroquinolones and macrolides, are used for treatment. In 2018, 9,933 *Legionella* associated infections were reported in the United States.

Differential Characteristics

L. pneumophila is a motile, non-sporing, aerobic, fastidious, thin, Gram-negative rod (Fig. 12.35). It derives its carbon and energy from proteins rather than carbohydrates, making it relatively inert in most biochemical tests.

Diagnostic procedures include culture, *Legionella* antigen urine test, direct fluorescent antibody (see p. 140), and PCR (see p. 144) of respiratory samples.

12.35 ***Legionella pneumophila*** **Stock Culture** *L. pneumophila* cells can range in size from less than 1 µm wide by 2–20 µm long or more. It is the causative agent of legionellosis. (Gram Stain)

Listeria monocytogenes

Listeria monocytogenes (phylum Firmicutes) is a very common soil and vegetable matter saprophyte, as well as an intestinal commensal and pathogen of animals, including humans. It is also a common contaminant of meat, raw milk, and cheeses, and enters the body as a food-borne contaminant or through openings in the skin. After attaching to host epithelial cells, it induces phagocytosis.

Once in the phagosome, it produces **listeriolysin O** and phospholipases, resulting in phagosome lysis and release of the cells into the cytoplasm. The protein **ActA** on the cells' surface induces host cell actin to push them to the surface, where they enter protrusions called **filopodia** that allow cell to cell transfer.

Filopodia are subsequently phagocytized by adjacent epithelial cells and macrophages, and the process repeats in these new cells. This spreading mechanism protects *Listeria* from defenses such as complement and antibodies. Most *L. monocytogenes* infections are mild or subclinical, but pregnant women and their fetuses, immunocompromised individuals, and the elderly are most at risk. **Listeriosis** is the general heading used for all diseases caused by the organism.

In immunocompromised and elderly patients, meningitis is the most common form of listeriosis, but bacteremia is also seen. Infection of pregnant women usually leads to mild symptoms resembling influenza, but subsequently may lead to transplacental fetal infection. Fetal infection may lead to spontaneous abortion or premature delivery. In many cases **granulomatosis infantisepticum**, development of skin lesions and granulomas in the brain, lungs, kidneys, skin, and other organs, occurs and is often fatal.

In other neonates, the disease takes one of two forms—early-onset and late-onset. Early-onset disease is acquired in utero and symptoms appear within 48 hours of delivery. These include sepsis, pneumonia, and meningitis (in order of decreasing frequency). Late-onset infections occur between 5 and 14 days after birth and most often manifest as neonatal meningitis. In 2018, 864 listeriosis cases were reported in the United States. Antibiotics are available for treatment.

Differential Characteristics

L. monocytogenes is a small, motile, β-hemolytic, non-sporing, facultatively anaerobic, Gram-positive rod (Fig. 12.36). Key test results in identification of *L. monocytogenes* in clinical samples (blood or cerebrospinal fluid) are Gram reaction, tumbling motility, and positive for catalase, acid from glucose fermentation, VP, esculin hydrolysis (Fig. 12.37), and a narrow zone of β-hemolysis.

Multi-test systems, such as API Coryne (includes *Listeria*) (see p. 121), Vitek 2 (see p. 125), BBL Crystal Gram-Pos ID (see p. 122), and many others. Most molecular methods, such as DNA hybridization (see p. 149), are currently limited to examination of dairy products.

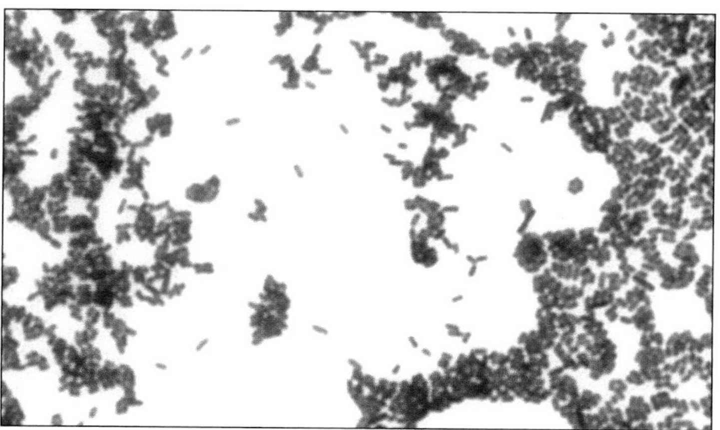

12.36 ***Listeria monocytogenes*** **Stock Culture** • *L. monocytogenes* typically appears as single cells, diplobacilli, chains, or clusters of parallel cells with blunt ends, most of which are represented in this micrograph. Sizes range from 0.5 µm wide to 1–2 µm long. Most cases of listeriosis are mild or subclinical, but the elderly, immunocompromised, and pregnant women are most at risk. (Gram Stain)

12.37 ***Listeria monocytogenes*** **Growing on Oxford Medium** • *L. monocytogenes* colonies are typically about 1 mm in diameter, but these are considerably smaller at about 0.5 mm. Oxford medium is selective for *Listeria* spp. and is also differential for esculin hydrolysis, seen as darkening of the medium. See bile esculin agar on page 75.

Mycobacterium leprae

Mycobacterium leprae (phylum Actinobacteria) is the causative agent of **leprosy (Hansen's disease**[5]**)**—now rare in the United States, but still a serious health concern in parts of South America, Asia, and Africa. Although it can be cultivated in a few laboratory animals, it is believed to occur naturally only in humans and in the nine-banded armadillo of Texas and Louisiana.

Like *Legionella* and *Listeria*, *M. leprae* is an intracellular parasite. Once ingested by a macrophage it survives to multiply within these cells by chemically suppressing the cell's defensive activity.

Two distinctive forms of leprosy are known to occur—**lepromatous leprosy** and **tuberculoid leprosy**[6]. The more severe form of the disease, lepromatous leprosy, occurs in people with compromised cell-mediated immunity. It is a systemic disease that can involve the eyes, nasal passages, and some internal organs, but especially the skin of the face and appendages, and peripheral neurons.

Infection is combatted by macrophages, but they are incapable of killing the bacilli, which then multiply within them. Passage through lymphatic vessels allows entry into the blood, resulting in spreading of the pathogen throughout the body. Formation of bulbous skin lesions hosting large numbers of the pathogen (**multibacillary leprosy**) is the most obvious symptom.

M. leprae also has the ability to infect **Schwann cells** (responsible for myelin production around axons), leading to nerve damage that affects both motor and sensory functions, leading to muscle atrophy and injuries associated with an inability to sense pain. Severe disfigurement occurs due to loss of bone tissue. Because *M. leprae* has a long doubling time (roughly 2 weeks), patients may not see symptoms for years and live for decades with the disease.

Tuberculoid leprosy begins the same way as the lepromatous form. However, a complex, but normal, activation of the immune system occurs and limits the pathogen's growth and in turn, symptoms. Skin lesions lack the abundant bacilli (**paucibacillary leprosy**) of the lepromatous form and spontaneous recovery is likely.

Transmission of both forms is thought to occur by inhalation of nasal mucus carrying *M. leprae*. To minimize selection for resistance, prolonged multidrug therapy is used.

Differential Characteristics

M. leprae is a nonmotile, aerobic, acid-fast, non-sporing, weakly Gram-positive, thin rod (Fig. 12.38). It is uncultivable in vitro, therefore, standard biochemical tests are not useful in identifying it. Clinical diagnoses are based on characteristics of the disease, Gram and acid-fast staining of skin lesion and nasal secretions, and PCR (see p. 144).

A skin test using **lepromin**, an antigen extracted from inactivated *M. leprae*, is also available for tuberculoid leprosy, but is ineffective at identifying lepromatous leprosy cases. This is because the test relies on a cell-mediated response to the antigen, which is absent in these infections.

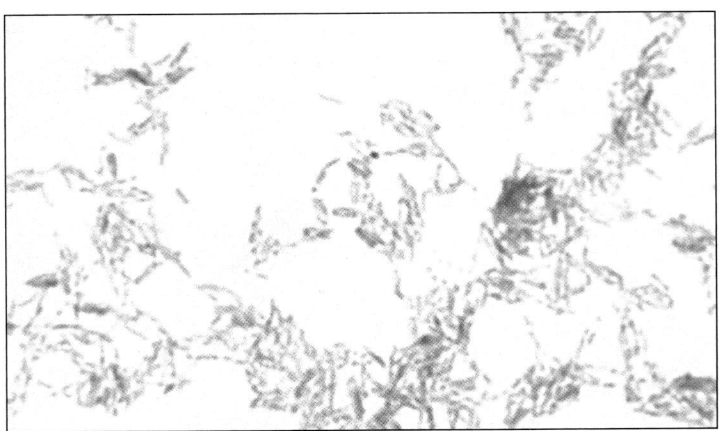

12.38 ***Mycobacterium leprae* Stock Culture** • *M. leprae* rods can be straight or curved and are typically found in clusters. They range in length from 1 to 8 µm by about 0.5 µm in width. *M. leprae* is the causative agent of leprosy. (Acid-Fast Stain)

[5] Named after Norwegian physician Gerhard Henrik Armauer Hansen (1841–1912) who discovered *M. leprae* in 1873.

[6] Several intermediate forms are also known, and are the result of differing bacterial loads.

Mycobacterium tuberculosis

Mycobacterium tuberculosis (phylum Actinobacteria) is the primary pathogen responsible for **tuberculosis (TB)**[7]. Humans are its principal host and reservoir, although it has been isolated from other warm-blooded animals. The disease has afflicted humans for millennia and was the subject of an East Indian hymn dating back to 2500 BC.

The World Health Organization estimates that there were 10 million new TB cases worldwide in 2018, with approximately 10,000 in the United States. They also estimate that one-quarter of the world's population have latent TB infections and that two-thirds of all TB cases are in Southeast Asia and Africa. Worldwide, TB accounted for 1.2 million deaths of HIV-negative patients and 251,000 deaths of HIV-positive patients in 2018.

Transmission is through inhalation of contaminated airborne droplet nuclei that may remain suspended in the air for hours. *M. tuberculosis* also remains infective on surfaces for months. The **ID_{50} (infectious dose)**, that is the number of cells required to infect 50% of a sample group, is <10 cells.

Two major manifestations of the disease exist: **primary tuberculosis** and **secondary tuberculosis**[8]. Primary tuberculosis, the condition produced upon initial exposure to the bacillus, is no more than a mild, flu-like illness for most healthy individuals, and some individuals exhibit no symptoms at all.

In this initial stage, the bacteria enter the lung's alveoli and are ingested by resident macrophages, where they multiply. Death to the macrophage follows, but is itself engulfed by another macrophage, which becomes infected. Macrophages migrate to lymph nodes at the root of the lung, where an immune response is activated.

Eventually, the host immune response kills most of the bacteria, but some remain alive inside small granulomas called **tubercles**, which are visible on chest X-rays. In otherwise healthy individuals, these tubercles usually remain intact for a lifetime, holding the bacteria in check. At this point, the patient has a **latent TB infection**, shows no symptoms, and is not infectious to others.

In some cases, the immune response in not strong enough to limit the initial infection and the bacteria multiply and spread through the body, producing **miliary tuberculosis**, so named because the small lesions are reminiscent of millet grains. This most frequently occurs in children, elderly, and immunocompromised patients. Spreading to the central nervous system often leads to meningitis.

In about 10% to 15% of patients with latent tuberculosis, reactivation of the infection occurs. This is secondary tuberculosis and is the result of a compromised immune system due to age or other factors (e.g., HIV coinfection). Progressive, necrotic lung inflammation results in cavitation of the lungs and bronchi. This increases the patient's infectivity because of the expanded exit routes from the lung, which is further aided by the symptomatic cough. This is the form of tuberculosis people associate with the disease.

Treatment of latent tuberculosis usually involves administration of one or two antibiotics. Active infections may require four or more if the strain is resistant. The **bacilli Calmette-Guérin (BCG)** vaccine, made from an attenuated strain of *Mycobacterium bovis*, has been used since 1921 and is most effective for infants and young children, but its effectiveness for adults is variable. It is not typically administered in the United States because of the low risk of TB and its potential of causing false positive results in the tuberculin skin test.

Differential Characteristics

M. tuberculosis is a nonmotile, aerobic, acid-fast, non-sporing, weakly Gram-positive, thin rod (Fig. 12.39). On solid media, colonies are rough, flat, and dry with an irregular shape (Fig. 12.40). Diagnostic procedures begin with acid-fast stain and culture of sputum, gastric lavage, biopsy, or body fluid samples to verify mycobacteria are present.

Subsequent testing includes the **Mantoux tuberculin skin test** that detects existing cell-mediated immunity to

12.39 ***Mycobacterium tuberculosis* in Mammalian Lung** • *M. tuberculosis* is the causative agent of tuberculosis and its cells are approximately 0.4 µm wide by 1 to 4 µm long. Here, several cells are seen in the mucus of a lung airway. (Acid-Fast Stain)

[7] In addition to *M. tuberculosis*, there are at least nine other *Mycobacterium* species capable of causing tuberculosis. Collectively, these are referred to as the *M. tuberculosis* complex.

[8] These categories may, in fact, be too simplistic. More likely, there is a continuum of disease states possible depending on the degree to which the pathogen or the patient has the upper hand.

M. tuberculosis antigens; the interferon gamma release assay (IGRA), a blood test that measures release of IFN-γ from sensitized WBCs; ELISA for cell wall components (see p. 137); and MALDI-TOF MS (see p. 150). To rapidly identify *M. tuberculosis* in patient samples, PCR (see p. 144) is most commonly used (Fig. 12.41).

Mycobacterium tuberculosis is slow-growing and historically would take 3 to 4 weeks to grow on a standard medium, such as Lowenstein-Jensen agar, thus delaying identification and antibiotic susceptibility testing. Breakthroughs in technology and molecular biology have resulted in faster identification and determination of antibiotic susceptibility.

An example is the **BACTEC MGIT 960** growth chamber (Fig. 12.42) that reduces identification time to 8–14 days. MGIT stands for **Mycobacteria Growth Indicator Tube.** Each tube contains modified Middlebrook 7H9 broth (growth is more rapid in broth) supplemented with a nutrient mixture necessary for *M. tuberculosis* growth and antibiotics to minimize contamination.

Growth is monitored automatically by a fluorescent chemical indicator that only fluoresces in UV light in the absence of oxygen. The chamber is equipped with a photometer that monitors fluorescence automatically and the degree of fluorescence for each sample is recorded. If *M. tuberculosis* is present and growing in a tube, it consumes the oxygen and the chemical fluoresces, which is a positive result.

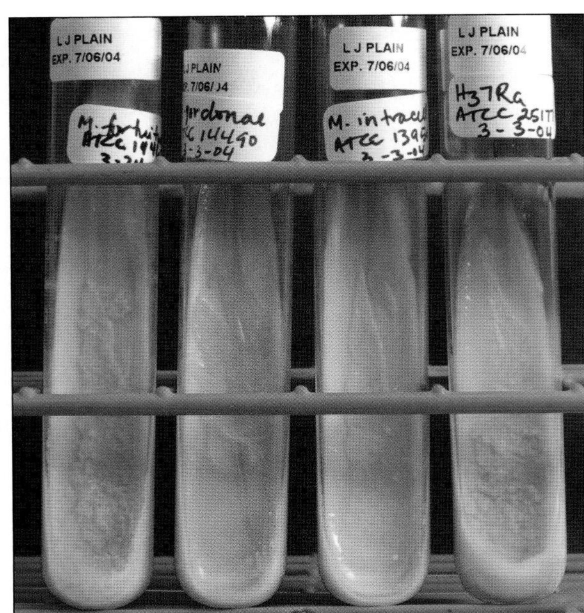

12.40 Collection of *Mycobacterium* Cultures Growing on Lowenstein-Jensen Agar Slants • From left to right, *M. fortuitum*, *M. gordonae*, *M. intracellulare*, and a strain of *M. tuberculosis* (H37Ra). Note the friable (crumbly) growth texture of the *M. fortuitum* and *M. tuberculosis* cultures.

Drug susceptibility testing can also be done, but begins with two tubes inoculated with a patient's isolate. One tube has the antibiotic, and the other does not and acts as a control. If, after incubation, fluorescence is absent in the antibiotic tube but is seen in the control, then the antibiotic is effective. If both tubes fluoresce, the antibiotic is ineffective against the patient's isolate. Antibiotics typically tested are streptomycin, isoniazid, rifampin, and ethambutol.

12.41 GeneXpert (Cepheid) • This automated apparatus performs PCR on patient samples loaded into cartridges (the blue container) to determine the presence of *Mycobacterium tuberculosis*. It also screens for rifampin mutations that are associated with resistance to the antibiotic. Results are returned in less than 2 hours, and this model has the capability of running 16 samples simultaneously. Other cartridges are available for screening MRSA, influenza, Ebola, and Norovirus, among others.

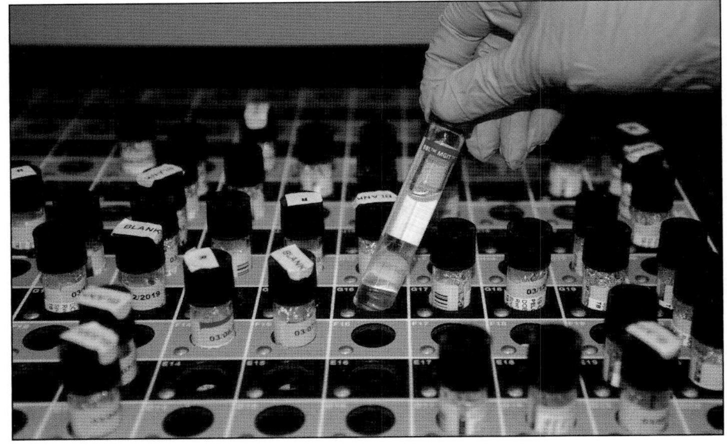

12.42 BD BACTEC MGIT Growth Chamber • MGIT stands for Mycobacteria Growth Indicator Tube, the basic unit of this growth chamber. It is another automated system for determining the presence of *M. tuberculosis* in a patient's sample, as well as establishing antibiotic susceptibility or resistance, if present. See text for details.

Neisseria gonorrhoeae

Neisseria gonorrhoeae (class Betaproteobacteria) causes the exclusively human **sexually transmitted disease** (**STD**), **gonorrhea,** and if found in a patient's sample it is always considered a pathogen. It attaches to urethral or vaginal epithelial cells by **Type IV pili** and other surface proteins. These surface antigens undergo a high rate of genetic variation making it difficult for the host's immune system to keep up.

Asymptomatic infections are more common in women than in men and both serve as the primary reservoir of the pathogen. Acute urethritis is the most common symptom in men, whereas infection of the endocervix in women is most common and is associated with vaginal discharge, bleeding, and difficulty in urination.

Other infection sites produce endometritis, epididymitis, pelvic inflammatory disease (PID), proctitis, pharyngitis, peritonitis, perihepatitis, and conjunctivitis (leading to blindness if not treated). The eyes of neonates may become infected during delivery, but this is prophylactically treated with silver nitrate or a topical antibiotic.

Rarely, the gonococcus enters the blood and produces a **disseminated gonococcal infection** (**DGI**)**,** indicated by **dermatitis-arthritis-tenosynovitis syndrome.** In 2018, 582,248 cases of gonorrhea were reported in the United States.

Antibiotics are used in treatment, but increasing antibiotic resistance is a serious issue. Type IV pili make the cells "**competent**," that is, able to pick up DNA from the environment, which provides a mechanism for **horizontal transfer** of antibiotic resistance genes (rather than random mutation being the sole source of resistance).

Differential Characteristics

N. gonorrhoeae (Figs. 12.43 and 12.44) is an aerobic, Gram-negative diplococcus that sometimes demonstrates "**twitching motility**" (again, due to their Type IV pili). It produces acid only from glucose in the carbohydrate acidification test, and is oxidase and catalase positive.

Diagnostic procedures include Gram stain ID of gonococcus with neutrophils (PMNs) in urethral swabs, nucleic acid probes (see p. 149), PCR (see p. 144), and MALDI-TOF MS (see p. 150).

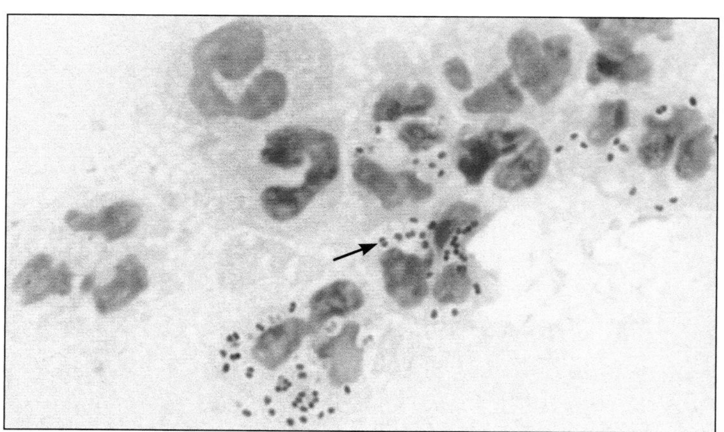

12.43 ***Neisseria gonorrhoeae*** **Inside Polymorphonuclear Leukocytes (Urethral Swab)** • Polymorphonuclear leukocytes (PMNs) are usually neutrophils and are phagocytic. A urethral swab of an individual with gonorrhea will typically show *N. gonorrhoeae* cells that have been engulfed by the PMNs, as in this micrograph. Engulfed or not, *N. gonorrhoeae* cells are typically seen as diplococci with adjacent sides flattened (**arrow**). Individual cells range in size from 0.6–1.9 µm in diameter. (Gram Stain)

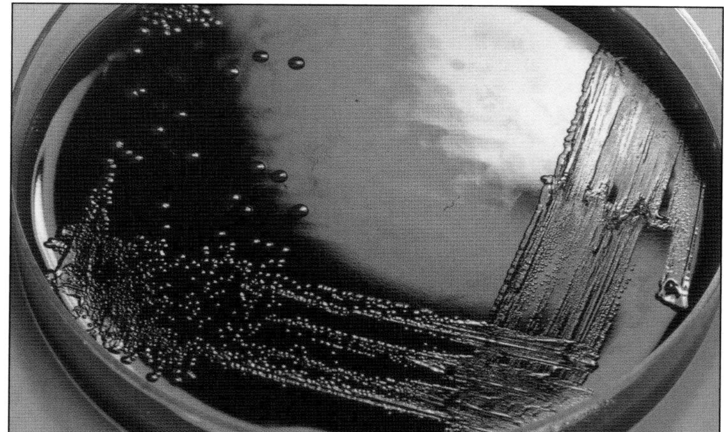

12.44 ***Neisseria gonorrhoeae*** **Growing on Chocolate Agar After 48 Hours Incubation** • *N. gonorrhoeae* is an obligate aerobe and typically forms smaller colonies than other members of the genus. The largest colonies on this plate were approximately 1.5 mm in diameter. Compare colony size with *N. meningitidis* in Figure 12.46.

Neisseria meningitidis

Neisseria meningitidis (class Betaproteobacteria) is the second of two human pathogens in the genus. It resides on mucous membranes of the nasopharynx, oropharynx, and the anogenital region as a commensal and an opportunistic pathogen. It does not remain viable for long outside the human body and must be transferred via droplets or by direct contact with infected respiratory secretions. In most healthy individuals the organism produces a localized infection or no symptoms at all.

In the absence of an early antibody response, it causes **meningococcemia** and the accompanying **meningococcal meningitis,** collectively referred to as **invasive meningococcal disease (IMD),** a devastating and often fatal disease primarily of children and young adults. Symptoms present as a stiff neck, light sensitivity, confusion, and headache, but there is variability.

Antigenic variation of capsular polysaccharides has led to the identification of eight pathogenic serogroups, with types B, C, and Y accounting for the majority of infections in the United States.

The organism's virulence can be attributed to at least five factors: Type IV pili, which help it attach to host mucous membranes, especially in the nasopharynx; a heavy capsule, which helps it evade phagocytosis; an endotoxin (a lipooligosaccharide found in the outer membrane), which facilitates the destruction of red blood cells (RBCs) and activates host responses leading to septic shock; surface components in some strains similar to those of RBCs that fail to stimulate a serum antibody response; and production of an IgA protease that digests IgA, an antibody class found in mucus and other secretions.

Patients with *N. meningitidis* infection are hospitalized and given antibiotics immediately. Antibiotic resistance is less common than in *N. gonorrhoeae*. Vaccines targeting serogroups A, B, C, W-135, and Y are available and confer immunity in at least 85% of recipients. Because their effectiveness lasts only a few years, teenagers and other people at high risk are advised to revaccinate periodically.

Differential Characteristics

N. meningitidis (Figs. 12.45 and 12.46) is an aerobic, Gram-negative diplococcus that sometimes demonstrates "twitching motility." It produces acid from glucose and maltose in the carbohydrate acidification test and is oxidase positive.

Diagnostic tests include Gram stain and culture of cerebrospinal fluid (CSF); latex agglutination tests for some, but not all, serotypes (see p. 133); and MALDI TOF MS (see p. 150).

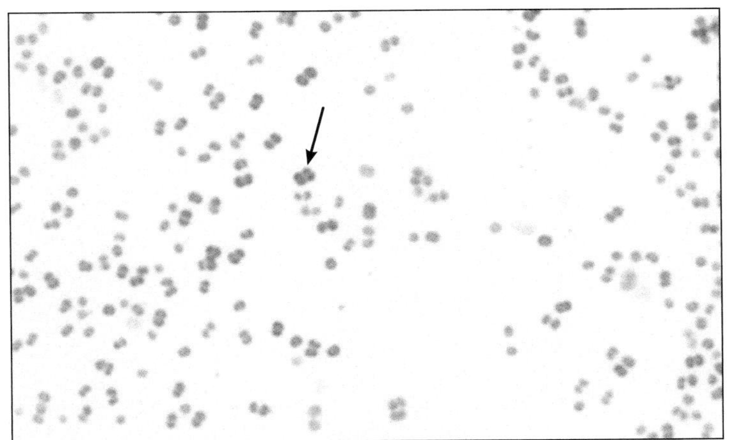

12.45 ***Neisseria meningitidis* Stock Culture** • Meningococci are usually seen as diplococci with adjacent sides flattened, but in young cultures they may form tetrads (**arrow**). Cells are about 1 μm in diameter. *N. meningitidis* causes meningitis. (Gram Stain)

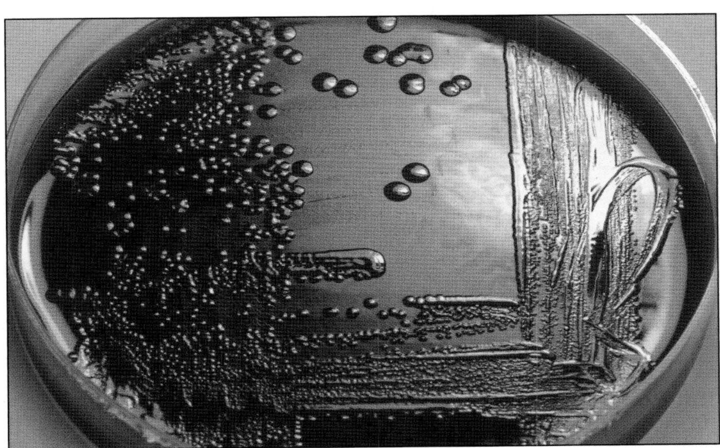

12.46 ***Neisseria meningitidis* on Chocolate Agar After 48 Hours Incubation** • The largest colonies on this plate are approximately 3.5 mm in diameter. Compare colony size with *N. gonorrhoeae* in Figure 12.44.

Nocardia spp.

Nocardia species (phylum Actinobacteria) were once believed to be fungi because of their ability to produce **branching vegetative hyphae**[9] (Figs. 12.47 and 12.48). However, their **mycelia** fragment into rod and coccus-like elements that contain no membrane-bound organelles (and fungi are eukaryotic). They are widespread in nature and are soil saprophytes of vegetation.

N. asteroides was thought to be the specific causative agent of **nocardiosis**, but molecular testing of pathogenic isolates has shown them to be a number of other species. In fact, it is likely that *N. asteroides* is not a pathogen! For convenience, we'll refer to those species that are capable of causing nocardiosis as **nocardiae.**

Nocardiae occasionally infect healthy individuals and produce skin diseases, such as abscesses, cellulitis with lymphocutaneous nodules, and mycetoma. However, they are more commonly opportunistic pathogens of

[9] Despite not being a fungus, fungal terminology is still applied to *Nocardia* species. A hypha (pl. hyphae) is a chain of cells. A mycelium (pl. mycelia) is a mass of hyphae.

immunosuppressed patients following organ transplant or people whose immune systems have otherwise been compromised by AIDS, lymphoma, or corticosteroids.

In the majority of cases, transmission is by inhalation of aerosol droplets leading to **pulmonary nocardiosis** (chronic pneumonia). Dissemination of the organism by the bloodstream typically leads to **central nervous system nocardiosis** (brain abscesses) and infection of virtually all organ systems, and is usually fatal.

Differential Characteristics

Identification of *Nocardia* to the species level is time-consuming because of their slow growth (up to 2 weeks) and is complicated by the absence of a single method capable of identifying all species. Nocardiae (Fig. 12.49) are strictly aerobic, nonmotile, weakly Gram-positive to Gram-variable, partially acid-fast rods, frequently with a beaded appearance when stained. They are lysozyme-resistant and urease-positive.

Diagnostic procedures include Gram stain, acid-fast stain, aerotolerance, and culture of the infected site on tap water agar with the identification of aerial hyphae arising from substrate hyphae.

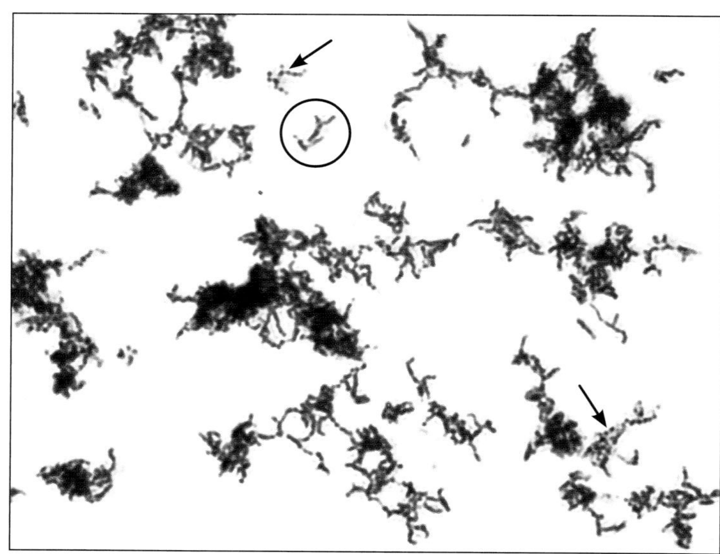

12.48 ***Nocardia asteroides*** **Stock Culture** • The Gram reaction is also weak, as evidenced by the pink cells, due to the lipid mycolic acid in *Nocardia* cell walls. Note the beaded appearance of some cells (**arrows**) and branched filaments (**circle**). Nocardiae are mostly opportunistic pathogens of immunosuppressed people and cause a chronic pneumonia. (Gram Stain)

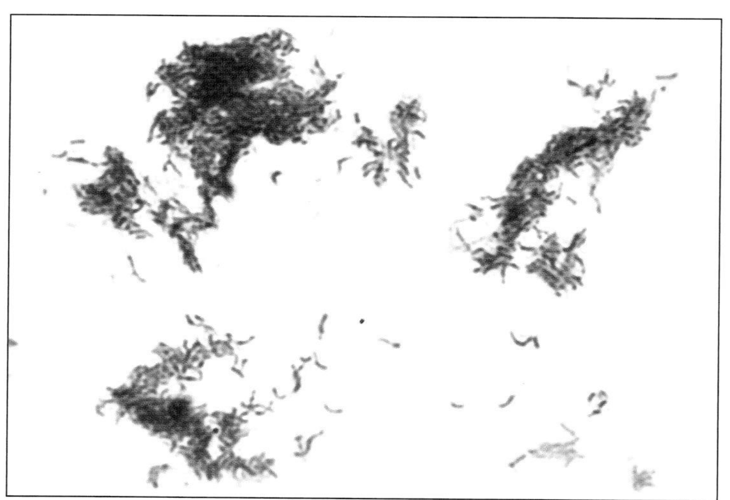

12.47 ***Nocardia asteroides*** **Stock Culture** • Because *Nocardia* species are weakly acid-fast, decolorization is done with a lower concentration of acid-alcohol than is normally used (1% vs. 3%). Notice that some cells still decolorized in this preparation. (Acid-Fast Stain)

12.49 ***Nocardia asteroides*** **Growing on Sabouraud Dextrose Agar** • Note the fungus-like appearance due to aerial elements (best seen along the glass on the back side of the bottle) of the orange-pigmented mycelium.

Proteus mirabilis

Proteus mirabilis (class Betaproteobacteria) is a normal inhabitant of the human intestinal tract as well as a variety of other animals, and is also found in soil and aquatic environments. In clinical settings, it is most frequently isolated from urine samples.

P. mirabilis, like all *Proteus* species, has the ability to periodically migrate on the surface of plated agar media. This cycle of alternating migration and consolidation is due to a characteristic called "swarming motility" (Fig. 12.50) and produces a series of visible concentric rings with a distinct burnt chocolate odor (though smelling clinical cultures is not recommended). Swarming, if seen, is very useful in the identification process.

P. mirabilis is a common nosocomial pathogen isolated from septic wounds and urinary tract infections. Transmission is by direct contact with a carrier or other contaminated source. It is of particular importance as a urinary tract pathogen because of its ability to produce urease (see p. 118).

Urease splits urea, thus creating an alkaline environment and promoting the formation of kidney stones. *Proteus* septicemia is a potential complication of urinary tract infections and can be fatal in weakened individuals.

Most *P. mirabilis* strains are susceptible to ampicillin and cephalosporins. However, resistance due to extended-spectrum β-lactamases (ESBLs) and Amp-C cephalosporinases (AMPCs) is becoming an issue. The genes for these are plasmid borne, not chromosomal, and so are more likely to spread by horizontal gene transfer.

Differential Characteristics

P. mirabilis is a straight, facultatively anaerobic, highly motile (swarming), Gram-negative rod (Fig. 12.51). In a clinical setting, *P. mirabilis* identification can be made based on these results: isolate has flat colonies with tapered edges on MacConkey agar, swarms on blood agar, and is Gram negative, oxidase negative, indole positive, strongly urease positive, and ampicillin susceptible. Alternative options include api 20 E (see p. 121), MALDI-TOF MS (see p. 150), and PCR (see p. 144).

12.50 Three Colonies of *Proteus mirabilis* Growing on Sheep Blood Agar Note the characteristic swarming growth pattern. This occurs as a result of individual cells of two different developmental stages: "swimmers" and "swarmers." For their story, see Figure 6.39.

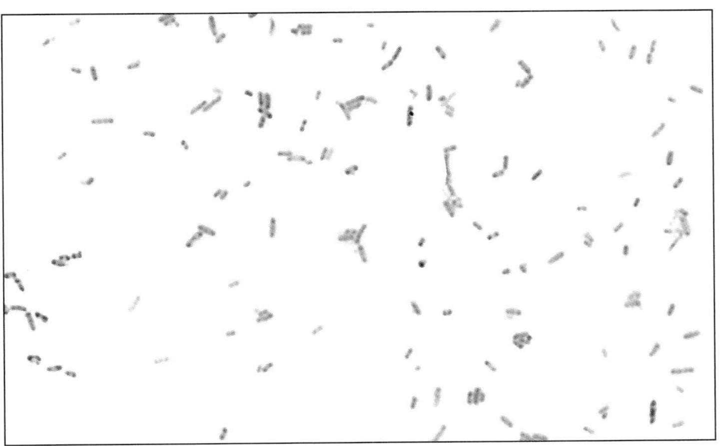

12.51 *Proteus mirabilis* Stock Culture *P. mirabilis* cells are motile, straight rods ranging in size from 0.4–0.8 μm wide by 1.0–3.0 μm long. It is a common nosocomial pathogen that causes urinary tract infections. (Gram Stain)

Pseudomonas aeruginosa

Pseudomonas aeruginosa (class Gammaproteobacteria) is an opportunistic pathogen ubiquitous in soil, water, and living or decaying plant material. It has also been isolated from hospital sinks and tubs, dialysis equipment, contact lens solution, aerators, irrigation fluids, hot tubs, ointments, insoles of shoes, and even in soaps and cleaning solutions. **Biofilm formation** is common and an important virulence factor.

There are 12 clinically important *Pseudomonas* species divided into those that produce fluorescent pigments and those that don't. *P. aeruginosa* belongs to the former group and produces the yellow (in visible light) fluorescent pigment **pyoverdin**, which also acts as a siderophore, an iron scavenger. **Pyocyanin**, a blue, nonfluorescent pigment, may also be secreted and produces a distinctive greenish color around and in the growth when combined with pyoverdin (see Fig. 3.23).

P. aeruginosa is the most significant opportunistic pathogen of its genus and possesses an impressive array of virulence factors, including, pili for attachment to host cells and the aforementioned siderophore (iron is often a limiting factor for growth). Secreted products that affect the host include **exotoxin A** that inhibits protein synthesis, exoenzymes that damage the cytoskeleton, and multiple proteases that damage tissues.

A Type III secretory system is used to inject these into eukaryotic host cells. Some strains secrete a polysaccharide (alginate) that protects them from phagocytosis, antibiotics, and dehydration, as well as participating in biofilm formation, both attached to surfaces such as medical tubing, and also nonattached **microcolonies** in the lungs and sputum of cystic fibrosis patients.

Transmission can be by ingestion, inhalation, or most frequently, through openings in the skin (burn patients are particularly susceptible to this). While healthy individuals are generally unaffected by the organism, immunosuppressed patients are susceptible to a variety of serious infections.

Among the nosocomial infections caused by this organism are pneumonia, wound sepsis, bacteremia, and urinary tract infections. Other infections include: corneal ulcers, swimmer's ear, folliculitis (from contaminated swimming pools or hot tubs), and osteomyelitis of the calcaneus in children due to puncture wounds through the shoe. In cystic fibrosis patients, it is primarily responsible for respiratory tract infections.

A β-lactam with antipseudomonal activity coupled with an aminoglycoside can be used for treatment of planktonic *P. aeruginosa*, but are ineffective against those in a biofilm because of their reduced ability to diffuse through it.

Antibiotic resistance is common and may either be intrinsic or occur as a result of mutation or horizontal transfer via plasmids or transposons. Resistance mechanisms include possession of efflux pumps that remove the antibiotic from within the cell, receptor modification that renders the drug ineffective, β-lactamase production, or a porin size too small for the antibiotic to pass through the outer membrane.

Differential Characteristics

P. aeruginosa is a facultatively anaerobic, motile, straight or slightly curved, Gram-negative rod that produces a green diffusible pigment when grown on solid media (Figs. 12.52 and 12.53). It utilizes glucose oxidatively, but does not ferment carbohydrates, and does not utilize lactose or esculin. It reduces nitrate in anaerobic respiration, is positive for oxidase, catalase, and arginine decarboxylase, and negative for ONPG, urease, and lysine decarboxylase.

Diagnostic procedures include Gram stain, appropriate-site culture, identification of the fluorescent pigment pyoverdin (under UV illumination), presence of the diffusible pigment pyocyanin on a solid medium, PCR (see p. 144), and MALDI-TOF MS (see p. 150).

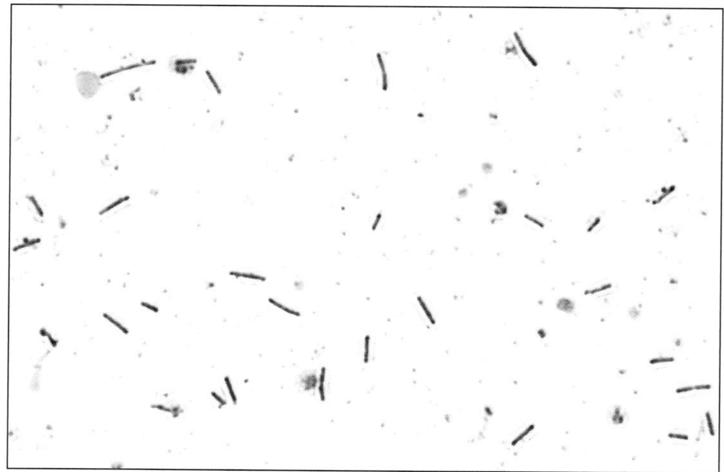

12.52 ***Pseudomonas aeruginosa*** **Stock Culture** • *Pseudomonas aeruginosa* cells are motile, narrow, straight rods approximately 1 μm or less wide by 2–5 μm long. (Gram Stain)

12.53 ***Pseudomonas aeruginosa*** **Growing on Sheep Blood Agar** • Note the distinctive greenish pigment produced by *P. aeruginosa*, especially in the regions of heaviest growth. The largest colonies on this plate are approximately 6 mm in diameter. It is an opportunistic pathogen, with immunosuppressed people being the most at risk.

Rickettsia rickettsii

Rickettsia rickettsii (class Alphaproteobacteria) is a small, Gram-negative, obligately intracellular parasite (Fig. 12.54). The basic biology of *R. rickettsii* is covered in Section 11 (see p. 173). Here, we focus on its medical importance.

R. rickettsii is the causative agent of **Rocky Mountain Spotted Fever** (**RMSF**) and most commonly occurs in midwestern and central southern states of the United States where its tick host is found. Part of its life cycle is spent in the gut of a few hard-bodied tick species and is transmitted to a human or animal host during a blood meal.

Once in the blood, *R. rickettsii* uses outer membrane proteins to bind to **endothelial cell receptors**, which triggers **induced phagocytosis** and the pathogen is engulfed. Normally at this point, the phagosome would fuse with a lysosome and the contents would be digested, but *R. rickettsii* escapes from the phagosome before that happens through the membranolytic action of a phospholipase and hemolysin.

While endothelial cells are the main targets, *R. rickettsii* can also infect macrophages and vascular smooth muscle. Intracellular division occurs and the pathogen spreads to neighboring cells. On a cellular level, infection leads to cytoskeletal and organelle damage. On a larger scale, damage to blood vessels results in fluid leaking into surrounding tissues, and because the infected endothelial cells can potentially be in any part of the body, the organ systems involved vary, as do the symptoms.

However, a rash usually forms a few days after infection, and this is followed a few days later by malaise, fever, headache, muscle ache, nausea, vomiting, and loss of appetite. In severe cases where the brain or lungs are involved, pulmonary edema, encephalitis, and/or coma may result and the infection may become fatal.

Because it possesses so few enzymes, *R. rickettsii* provides few targets for antibiotics to attack and is thus naturally resistant to most of them. Further, those that are effective tend to be bacteriostatic, rather than bactericidal. Doxycycline is the antibiotic of choice.

Differential Characteristics

Because RMSF infection may be fatal within two weeks, rapid lab identification is essential. Unfortunately, as obligate intracellular parasites, *Rickettsia* species have undergone significant genome reduction and possess few enzymes that lend themselves to traditional bacterial tests (e.g., fermentation substrates).

Although it is an intracellular parasite, *R. rickettsii* culture is possible using the yolk sac of embryonated chicken eggs or tissue culture, but these methods run the risk of laboratory-acquired infections and *R. rickettsii* is a BSL-3 organism. Nucleic acid and serological tests are available, including PCR (see p. 144) and serological tests, such as fluorescent antibody staining (see p. 140), ELISA (see p. 137), and latex agglutination (see p. 133) of tissue samples.

However, due to the potential severity of the infection, preliminary diagnosis relies on symptoms and a history of potential contact with ticks, and antibiotic treatment begins before test results are available.

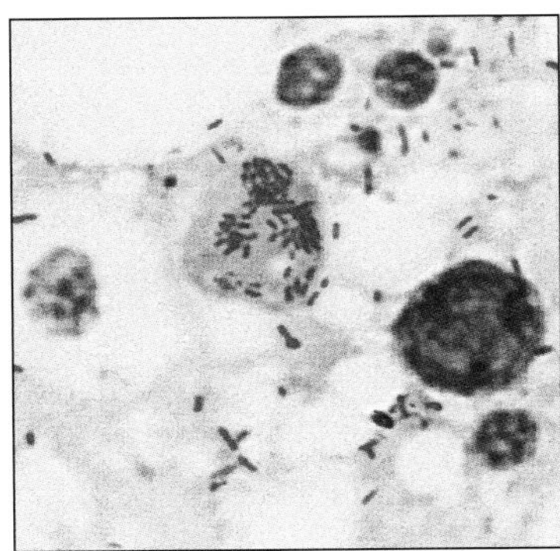

12.54 ***Rickettsia rickettsii* in Yolk Sac** • *R. rickettsii* causes Rocky Mountain Spotted Fever. It can't be grown on standard bacterial media because it's an obligate, intracellular parasite and requires living host cells, such as the yolk sac cells in this photo. *R. rickettsii* cells are Gram-negative and range in size from 0.2–0.3 µm in width by 0.5–2.0 µm in length. (Giemenez Stain)

Salmonella enterica subsp. *enterica*

The genus *Salmonella* (class Gammaproteobacteria) is composed of only two species. These are: *S. bongori*, which infects cold-blooded hosts; and *S. enterica*, which infects warm-blooded hosts. In spite of there being only two species, *Salmonella* nomenclature is challenging, at best.

Together, those two species have been divided into more than 2,500 serotypes based on unique combinations of O (the O polysaccharide of the outer membrane, see Fig. 6.3), H (flagellar), and sometimes Vi (capsular) antigen variants. Further, *S. enterica* has been divided into six subspecies.

Our focus will be on *Salmonella enterica* subsp. *enterica* and four of its serotypes: Enteritidis, Typhimurium, Typhi, and Paratyphi. These four comprise more than half of all *Salmonella* serovars and cause more than 99% of human infections. Because the complete names are cumbersome, they are frequently shortened to the generic name and the serovar name. For instance, *Salmonella enterica* subsp. *enterica*, serotype Enteritidis, becomes *Salmonella* Enteritidis.

S. Enteritidis and *S.* Typhimurium (and other **nontyphoidal *Salmonella*–NTS**) typically reside in the gastrointestinal tract of poultry, rodents, reptiles, and wild birds, and are responsible for **gastroenteritis** in humans. NTS infection is the number-one cause of food-borne illness in the United States. Transmission usually occurs by ingestion of fecal-contaminated food or water, frequently eggs, dairy products, undercooked poultry, and fresh fruits and vegetables, but direct contact with infected animals and sometimes infected people may be responsible.

Salmonella species are susceptible to acidic conditions and have a relatively high infectious dose (between 10^4 to 10^8 cells) to ensure enough survive passage through the stomach (pH<2). Once in the ileum and colon, the survivors attach to, and gain entry into, epithelial cells by inducing phagocytosis. They survive within the phagosome by altering it to prevent fusion with lysosomes and begin to multiply.

Production of other compounds lead to an inflammatory response and fluid loss, resulting in non-bloody diarrhea, abdominal cramps, and fever within less than a day. The infection is usually self-limiting and symptoms generally last for 4 to 7 days. In some instances, entry into the blood can lead to systemic infection, which can be fatal.

Antibiotic treatment is used only in severe cases. In uncomplicated gastroenteritis, fluid replacement is essential, but antibiotic treatment is not recommended due to its contribution to resistant strain development.

Humans are the only reservoir of *Salmonella* Typhi and *S.* Paratyphi. These serotypes are responsible for **typhoid fever** and **paratyphoid fever**, respectively, but the symptoms are so similar that the more general term **enteric fever** is used for both. Transmission occurs by ingesting food or water contaminated with human feces. Incubation may take a week or more and early symptoms include malaise, low grade fever, headache, vomiting, and myalgia.

Because no animal model is available, the mechanisms behind enteric fever are unclear. However, it is thought that the **Vi antigen** (absent in NTS) and toxins are responsible for inhibiting host defenses (inflammation) and promoting spread through the blood, producing acute bacteremia, and host lymphatics, resulting in infections of the liver, spleen, bone marrow, and eventually the kidney and gallbladder.

Damage to the intestinal wall (including perforation) leads to a bloody diarrhea. This phase, accompanied by high fever and sometimes diarrhea, is long lasting and continuous, up to 8 weeks in untreated cases. In a small percentage of patients ("carriers"), the organism is harbored asymptomatically in the gallbladder and sloughed in the feces for up to a year or more. Treatment is with antibiotics, but multidrug resistant *Salmonella* strains are becoming more common.

Differential Characteristics and Diagnostic Tests

Salmonella Enteritidis and *Salmonella* Typhi are straight, motile, facultatively anaerobic, non-sporing, Gram-negative rods (Figs. 12.55 and 12.56), but the latter is also encapsulated. Both are negative for lactose fermentation (Figs. 12.57 and 12.58). *S.* Enteritidis is indole (–), methyl red (+), VP (–), and citrate (+), whereas *S.* Typhi is indole (–), methyl red (+), VP (–), and citrate (–). *S.* Typhi is also negative for gas from glucose.

Because the vast number of strains differ primarily in antigenic structure, serotyping by a reference laboratory is necessary for final identification. Diagnostic procedures include aerobic and anaerobic blood culture, bone marrow culture, stool culture, and PCR (see p. 144).

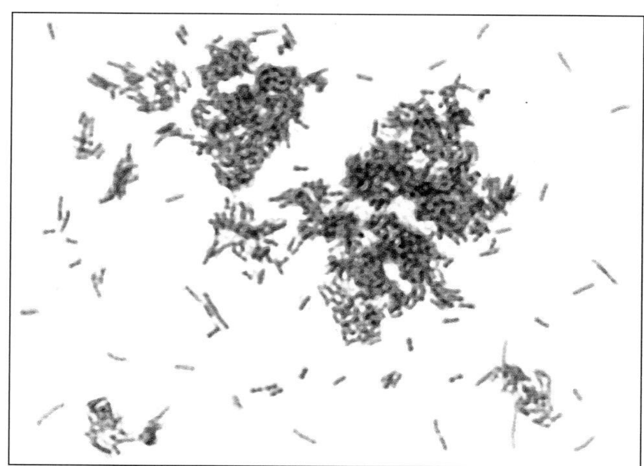

12.55 ***Salmonella*** **Enteritidis Stock Culture** • *S.* Enteritidis cells are straight rods ranging in size from 0.7–1.5 µm wide by 2.0–5.0 µm long. It causes gastroenteritis. (Gram Stain)

12.56 ***Salmonella*** **Typhi Stock Culture** • *S.* Typhi cells look the same as *S.* Enteritidis cells, but cause typhoid fever. (Gram Stain)

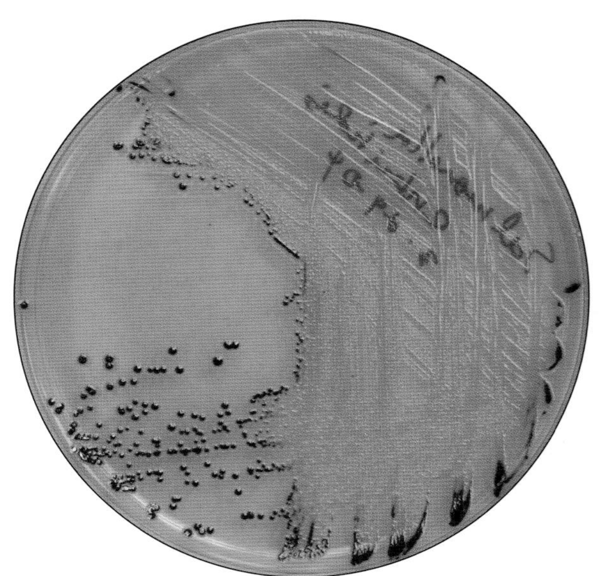

12.57 ***Salmonella*** **Enteritidis Growing on** ***Salmonella-Shigella*** **(SS) Agar** • *S.* Enteritidis is a noncoliform enteric and does not ferment lactose to acid end-products (which would make the growth pink), but note the black colonies due to reaction of H_2S (from sulfur reduction) and ferric citrate in the medium. The largest colonies on this plate are slightly more than 1 mm in diameter. Compare to *Salmonella* Typhi (Fig. 12.56) and *Shigella flexneri* (Fig. 12.60). Also see page 21 for more information about SS Agar.

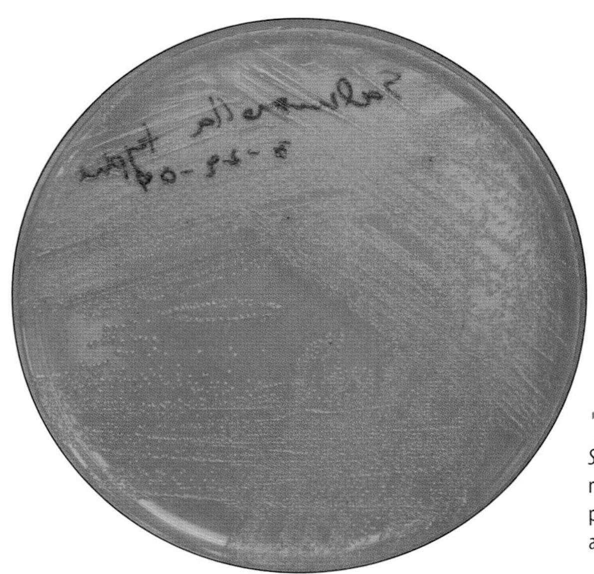

12.58 *Salmonella* **Typhi Growing on** *Salmonella-Shigella* **(SS) Agar** *S.* Typhi is a noncoliform enteric and does not ferment lactose to acid end-products, nor does it reduce sulfur, both evidenced by the colorless colonies. The colonies on this plate are less than 1 mm in diameter. Compare to *Salmonella* Enteritidis (Fig. 12.57) and *Shigella flexneri* (Fig. 12.60).

Shigella dysenteriae

Shigella dysenteriae (class Gammaproteobacteria) is one of four *Shigella* species—*S. dysenteriae* (Fig. 12.59), *S. flexneri* (Fig. 12.60), *S. boydii*, and *S. sonnei*—that are responsible for **bacillary dysentery** (**shigellosis**) in humans, which are the primary hosts. *S. dysenteriae* is endemic in Africa, Asia, and Latin America; *S. flexneri* and *S. sonnei* are found primarily in developed areas, including the United States; and *S. boydii* is mostly restricted to India.

Molecular genetic analysis has called into question the traditional organization of *Shigella* species based on serological groups. It also has uncovered the remarkably close relationship between *Shigella* and *E. coli* strains that cause diarrhea, including enteroinvasive *E. coli* (EIEC; see p. 196), which is more closely related to *Shigella* than other *E. coli* strains.

The majority of shigellosis cases occur in children under 10 years of age. Transmission is by direct person-to-person contact (fecal-oral) or ingestion of food or water contaminated by human feces. It is highly communicable and virulent; it can cause illness with as few as 100 organisms. Although all species of *Shigella* cause the disease, *S. dysenteriae* serotype 1 alone produces the cell-killing **Shiga exotoxin** and is, therefore, responsible for the most severe symptoms.

Unlike *Salmonella*, *Shigella* spp. are resistant to the stomach's acidic environment, which accounts, in part, for the low infectious dose. Once in the intestine, they induce phagocytosis by host epithelial cells in which they multiply and then spread in a process that kills the cells and forms mucosal ulcerations. This process, combined with an acute immune response, is responsible for the purulent bloody diarrhea characteristic of the disease.

A plasmid (pINV) carries most of the virulence genes, which are responsible for entry and spreading, as well as genes for their regulation. Enterotoxins and cytotoxins, both encoded by chromosomal genes, are responsible for fluid loss. *S. dysenteriae* serotype 1 carries a prophage that encodes the Shiga toxin.

There were more than 16,000 shigellosis cases reported in the United States in 2018, but not all cases are reported and the estimates suggest a number closer to 500,000. Worldwide there are typically more than 160 million cases, with 1.1 million deaths.

Oral rehydration and treatment of symptoms is typically adequate, and symptoms disappear within a week. However, without antibiotic treatment these patients may continue to be a reservoir of the pathogen for several weeks after symptoms subside. Antibiotics are strongly recommended for immunocompromised patients and children, or those with severe diarrhea.

Differential Characteristics

S. dysenteriae is a straight, nonmotile, facultatively anaerobic, Gram-negative rod. Key characteristics that distinguish it from *E. coli* are that it is nonmotile, doesn't ferment lactose or produce gas from glucose, and is lysine and ornithine decarboxylase negative. Diagnostic procedures include fecal leukocyte stain, stool culture, and PCR (see p. 144).

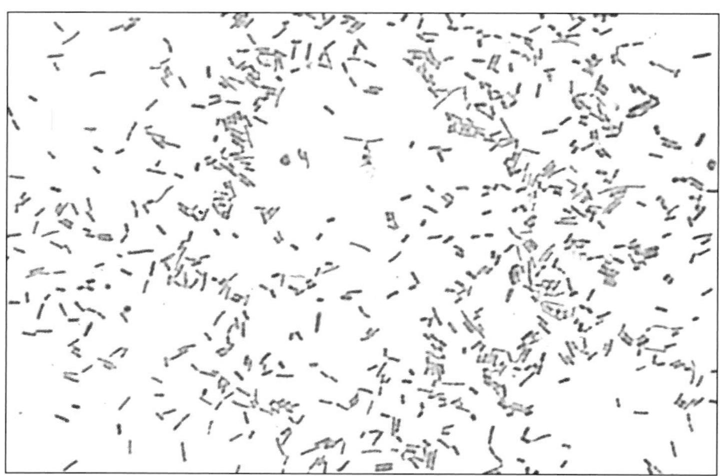

12.59 ***Shigella dysenteriae*** **Stock Culture** • *S. dysenteriae* cells are straight rods ranging in size from 0.7–1.0 μm wide by 1.0–3.0 μm long. *S. dysenteriae* causes bacillary dysentery (shigellosis). (Gram Stain)

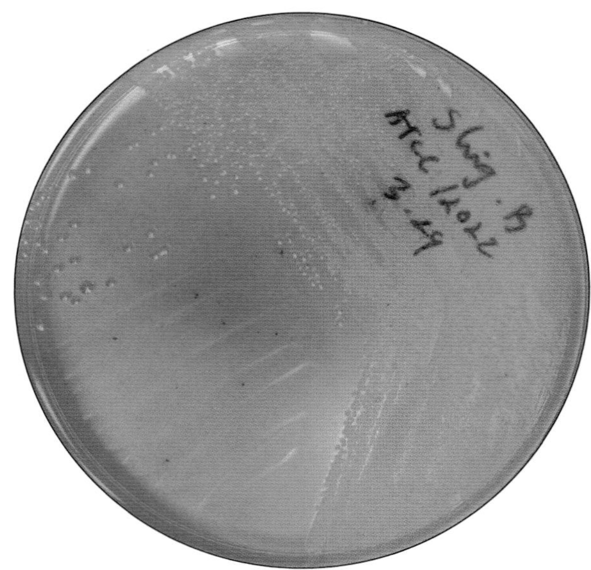

12.60 ***Shigella flexneri*** **Growing on** ***Salmonella-Shigella*** **(SS) Agar** • *S. flexneri* is a noncoliform enteric and does not ferment lactose to acid end-products, nor does it reduce sulfur, both evidenced by the colorless colonies. Compare to *Salmonella* Enteritidis (Fig. 12.57) and *Salmonella* Typhi (Fig. 12.58). *S. flexneri* also causes shigellosis.

Staphylococcus aureus

Staphylococcus aureus (phylum Firmicutes) is a normal human inhabitant, most commonly found in the nasal cavity, but also known to inhabit the skin and vagina. It is responsible for nosocomial as well as community-acquired infections. Examples include impetigo, cellulitis, abscesses, toxic shock syndrome, food poisoning, and scalded skin syndrome. Bacteremia may follow localized infections and result in infections most anywhere in the body.

Factors that increase its virulence include: attachment factors (capsular and biofilm polysaccharides, and teichoic acids); factors that interfere with opsonization and phagocytosis (**Protein A** and **coagulase**); and spreading factors (fibrinolysin, lipases, and hyaluronidase).

Enterotoxins induce vomiting and diarrhea in staphylococcal food poisoning. **Exfoliatins** and **epidermolysins** are exotoxins that break epidermal intercellular junctions and damage that critical microbial barrier, as well as producing a burn-like effect in scalded skin syndrome. An **alpha toxin** produces pores in erythrocytes and other cells, and a **beta toxin** hydrolyzes membrane phospholipids. Both toxins lead to cell death.

Pyrogenic exotoxins cause fever, enhance the activity of endotoxins, and act as **superantigens**. Superantigens nonspecifically activate T-cells, leading to hypersecretion of cytokines that causes an extreme inflammatory response and shock (as in toxic shock syndrome). Staphylococcal **peptidoglycan** also stimulates nonspecific defenses, such as interleukin-1 and complement, that result in swelling and tissue damage.

Factors that make potential hosts more vulnerable to severe *S. aureus* infection include immune deficiencies, breaks in the skin, medical foreign bodies (e.g., sutures or catheters), and existing infections.

S. aureus is transmitted by direct human-to-human contact (often by health-care workers) and fomites (inanimate objects). Aerosols and environmental factors may also play a role. It is a robust organism that resists cleaning solutions and antimicrobial agents, and can survive for weeks in the environment.

Although a fairly large number of antimicrobial agents are effective against *S. aureus*, it, along with other staphylococci, has the ability to acquire resistance to them. Therefore, clinical isolates should be tested for susceptibility as a means of selecting an appropriate drug for each patient.

Methicillin is one of several penicillinase-resistant penicillins, yet resistance to it is at the point of being a crisis. The basis for resistance is a novel **transpeptidase (PBP-2a)** used during peptidoglycan synthesis that is not found in susceptible strains. Because PBP-2a doesn't bind to methicillin (or other β-lactam drugs), the cell is resistant to the drug and is able to build a normal wall. Further, the transpeptidase is encoded by a gene on a transposon, which makes lateral gene transfer relatively common and accounts for its spread.

These resistant strains are known as methicillin-resistant *Staphylococcus aureus*, or **MRSA**. MRSA strains have been designated as **health-care-associated methicillin-resistant *Staphylococcus aureus*** (**HA-MRSA**), which at its peak comprised more than 50% of isolates in some regions of the United States, and **community-acquired methicillin-resistant *Staphylococcus aureus*** (**CA-MRSA**).

HA-MRSA tends to infect the elderly with underlying conditions and most frequently causes bacteremia and pneumonia. It also is generally resistant to multiple antibiotics. CA-MRSA has been more likely to infect a diverse cross-section of the population and cause soft tissue diseases, but it is now being found with increasing frequency in health-care facilities. It is less likely to be resistant to multiple antibiotics than HA-MRSA. In 2000, a third strain was identified: livestock-acquired MRSA (LA-MRSA), with pigs being the primary carrier.

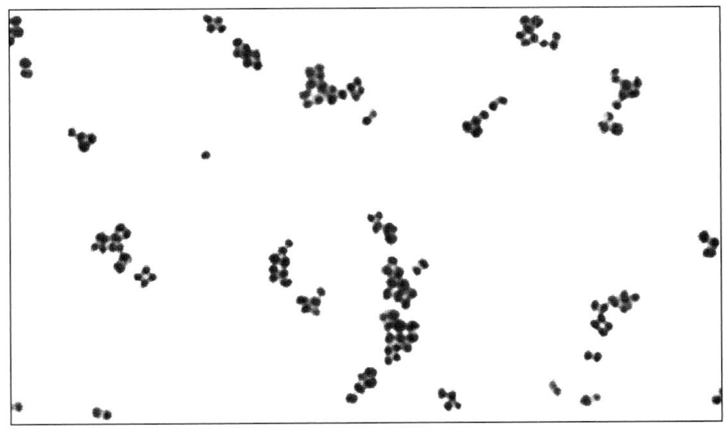

12.61 ***Staphylococcus aureus* Stock Culture** ● Note the grape-like cell clusters characteristic of the genus *Staphylococcus*. Cells are approximately 0.5–1.0 μm in diameter. *S. aureus* is an opportunistic pathogen that can cause a host of diseases. See the text for more details. (Gram Stain)

Differential Characteristics

S. aureus is a nonmotile, facultatively anaerobic, β-hemolytic, catalase positive, Gram-positive coccus (Figs. 12.61 and 12.62). Key characteristics that differentiate *S. aureus* from *S. epidermidis* are colony morphology (large, smooth, opaque, creamy yellow, and β-hemolytic), positive slide coagulase test (bound coagulase; clumping factor), tube coagulase test (free coagulase), and acid production from mannitol and trehalose fermentation. Further, its ability to grow in media containing up to 10% NaCl, produce DNase, and reduce potassium tellurite are useful for differentiating it from other Gram-positive cocci.

Diagnostic procedures include Gram stain to identify characteristic grape-like cell clusters, appropriate-site aerobic culture, biochemical testing, teichoic acid antibody test, and PCR (see p. 144). MALDI-TOF MS (see p. 150) is used to detect MRSA. ●

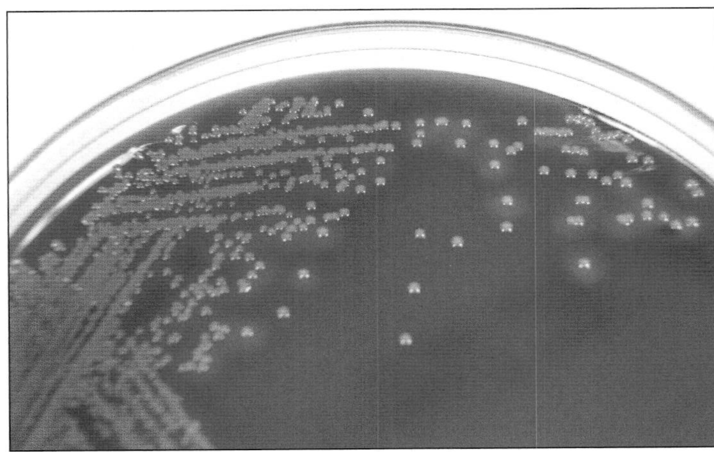

12.62 ***Staphylococcus aureus* Growing on Sheep Blood Agar** ● Note the β-hemolysis beginning around some of the colonies. Note also the absence of yellow color in the colonies, typical of *S. aureus* growth on nutrient agar. The largest colonies on this plate are approximately 1.5 mm in diameter. Compare with *S. epidermidis* colonies in Figure 12.64.

Staphylococcus epidermidis

Staphylococcus epidermidis (phylum Firmicutes) is a normal inhabitant of human skin and mucous membranes. It is the most common **coagulase-negative** *Staphylococcus* (CoNS) encountered clinically, and is considered a "medium pathogenicity" staphylococcus because infections are usually opportunistic and nosocomial in patients with predisposing factors. Infections are frequently associated with medical implants and prosthetic devices.

Septicemia and endocarditis are often the result of infection, but the specifics depend on where it is introduced into the body. Production of a slime layer (or biofilm) enables it to attach to those items, thereby gaining entry to the body. Once inside, the same slime layer retards phagocytosis. It also produces a **delta toxin** that lyses erythrocytes and damages membranes of other cells.

Due to multiple antibiotic resistance and the generally weakened condition of a convalescing patient, disseminated *S. epidermidis* infection can be quite severe and is frequently fatal. Most clinical isolates are resistant to penicillin and methicillin. Susceptibility testing (see p. 291) is recommended for individual isolates. Vancomycin is often the antibiotic of choice.

Differential Characteristics

S. epidermidis is a nonmotile, facultatively anaerobic, nonhemolytic, catalase positive, Gram-positive coccus (Figs. 12.63 and 12.64). Key characteristics that differentiate it from *S. aureus* are colony morphology (small to medium colonies, gray-white, and nonhemolytic), negative slide coagulase test (bound coagulase; clumping factor), tube coagulase test (free coagulase), and negative for acid production from mannitol and trehalose fermentation.

Its ability to grow in media containing up to 10% NaCl is useful for differentiating it from other Gram-positive cocci. Diagnostic procedures include Gram stain, appropriate-site aerobic culture, and standard biochemical and physiological tests.

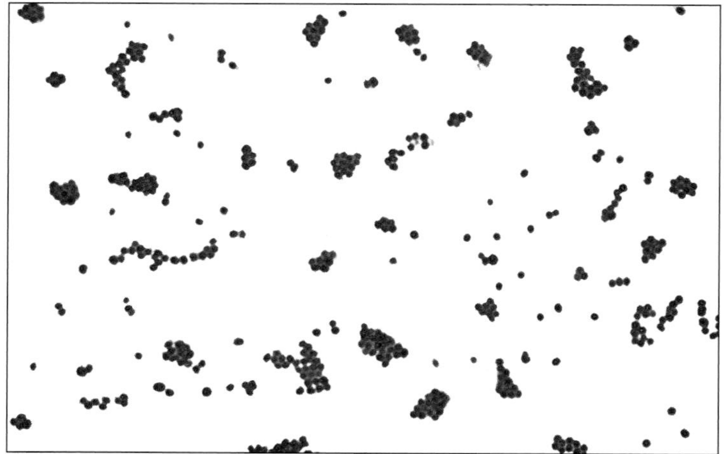

12.63 ***Staphylococcus epidermidis*** **Stock Culture** • Like *S. aureus*, *S epidermidis* also grows in irregular clusters and the cells are of similar size: 0.8–1.0 µm in diameter. It is an opportunistic pathogen, and is responsible for many nosocomial infections, especially those associated with implants or prosthetic devices. (Gram Stain)

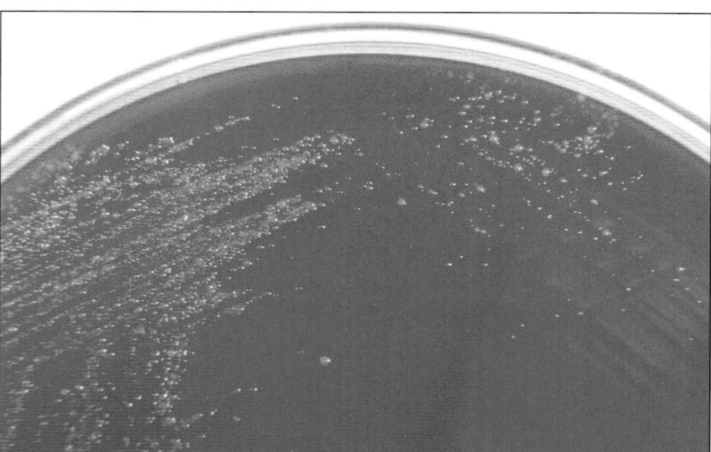

12.64 ***Staphylococcus epidermidis*** **Growing on Sheep Blood Agar** • Note the small white colonies and absence of hemolysis. These colonies are approximately 1 mm in diameter. Compare with *S. aureus* colonies in Figure 12.62.

Streptococcus agalactiae

Streptococcus agalactiae (phylum Firmicutes) belongs to the classic β-**hemolytic streptococcus group**, but a more appropriate group name is **pyogenic streptococci**, because not all species are β-hemolytic. It is also known as (Lancefield) **group B streptococcus**. It is found in the gastrointestinal and urogenital tracts of humans and animals and is a leading cause of neonatal infections.

Transmission can be from an infected mother through tears in fetal membranes, which lead to still births. But more commonly, infection occurs during delivery by an infected mother and results in two distinct disease states. In one (early-onset), sepsis and pneumonia occur in the neonate within the first week of delivery. In the other (late-onset), sepsis and meningitis occur in the baby within the first three months and exposure to the pathogen occurs through handling by the mother or health-care workers.

Its virulence is attributable to a polysaccharide capsule, which allows it to survive phagocytosis, multiply, and eventually spread by way of the bloodstream. Disseminated disease can also cause meningitis, pneumonia, endocarditis, and septic shock in adults, especially in elderly and immunocompromised populations.

Testing the mother for *S. agalactiae* between weeks 35 and 37 of pregnancy is recommended. If positive, then prophylactic use of penicillin G has a high success rate of preventing neonatal infection. Penicillin G is also used to treat infected nonpregnant adults.

Differential Characteristics

S. agalactiae is a β-hemolytic (with a small zone relative to colony size) or nonhemolytic, nonmotile, encapsulated, facultatively anaerobic, Gram-positive coccus (Figs. 12.65 and 12.66). It is positive for sodium hippurate hydrolysis; negative for Voges-Proskauer, catalase, and PYR; and is bacitracin and SXT-resistant.

Diagnostic procedures include appropriate-site aerobic culture, Gram stain, group B *Streptococcus* antigen (Lancefield) test (see p. 136), CAMP test (positive), and PCR (see p. 144).

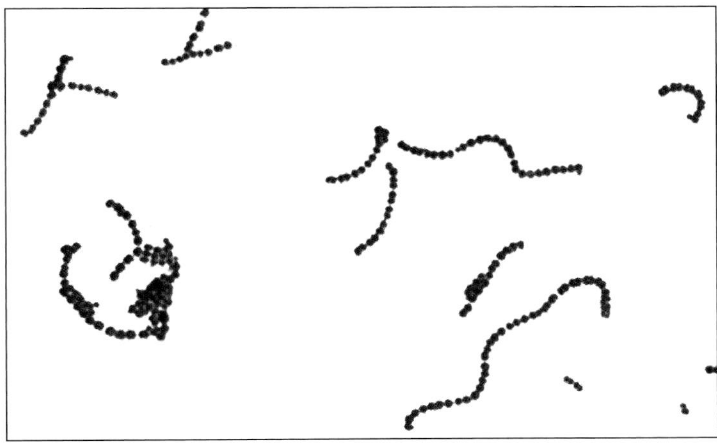

12.65 ***Streptococcus agalactiae*** **(Group B Streptococcus) Stock Culture** • *S. agalactiae* cells are spherical or ovoid and 0.6–1.2 µm in diameter. They are usually seen as pairs or chains, especially when grown in broth as these were. It is a leading cause of neonatal infections and belongs to Group B streptococci. (Gram Stain)

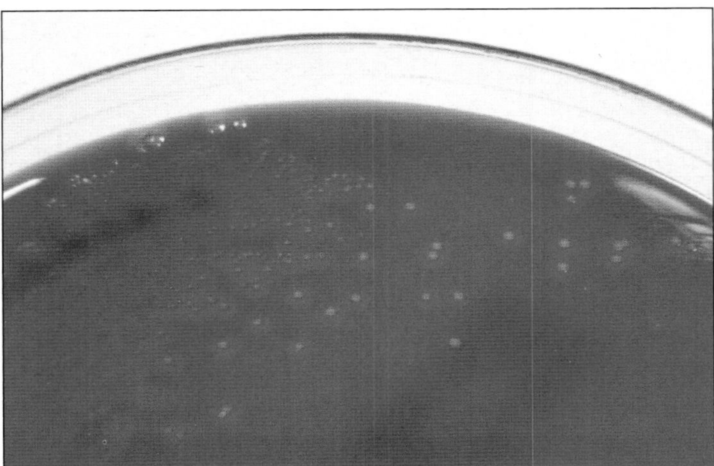

12.66 ***Streptococcus agalactiae*** **(Group B Streptococcus) Growing on Sheep Blood Agar** • Note the small (<1 mm in diameter), white colonies showing the start of β-hemolysis. Some strains are nonhemolytic.

Streptococcus mutans

Streptococcus mutans (phylum Firmicutes) is one member of the streptococcal group known as the **mutans group**. The mutans group is one of five subgroups in the **viridans group**, which also includes the **anginosus group**, **bovis group**, **mitis group**, and **salivarius group**. All viridans streptococci are either α-hemolytic or nonhemolytic, Gram-positive cocci typically found in the mouth, upper respiratory tract, and urogenital tract of humans.

S. mutans is one of the two (the other being *S. sobrinis*) most commonly isolated species of the mutans group. It is responsible for **bacteremia** following dental or urogenital invasive procedures, and in immunosuppressed patients undergoing chemotherapy or bone marrow transplantation. It also can cause **endocarditis**, especially in patients with existing heart valve problems or prosthetic heart valves.

Clinically, the most common encounter with *S. mutans* is in the dentist's chair. It is capable of producing a sticky, extracellular polysaccharide from sucrose that becomes **dental plaque** (a biofilm). The microbial community within dental plaque ferments various sugars to acid end-products that erode tooth enamel and results in dental caries (see Snyder Test, p. 295).

Based on molecular evidence, transmission is from mother to infant and permanent colonization occurs between the second and third years. For serious infections, penicillin G or ceftriaxone are used, but vancomycin can be substituted in cases of resistance.

Differential Characteristics

S. mutans is an α-hemolytic or nonhemolytic, nonmotile, facultatively anaerobic, catalase negative, Gram-positive coccus (Fig. 12.67) and is resistant to optochin (see p. 104). Species identification is usually not clinically necessary for the α-hemolytic and nonhemolytic streptococci.

The various groups can be differentiated from other groups based on their reactions in seven biochemical tests: arginine hydrolysis, esculin hydrolysis, urease, PYR, Voges-Proskauer, and acid production from mannitol and sorbitol fermentation. Diagnostic procedures include Gram stain and appropriate-site aerobic culture.

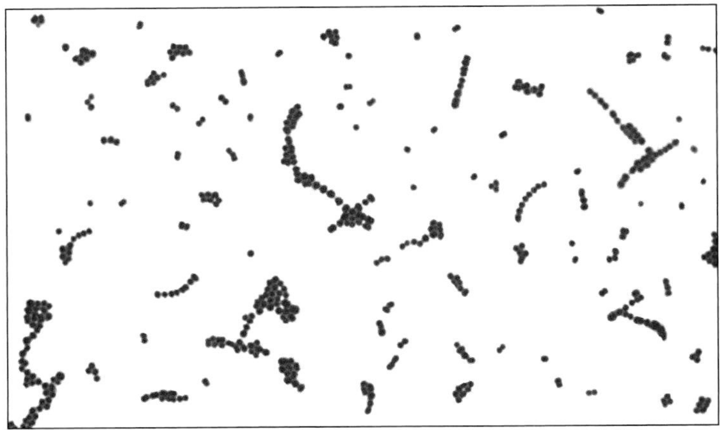

12.67 ***Streptococcus mutans*** **Stock Culture** • *S. mutans* has ovoid cells and usually appears in short chains. Cells range in diameter from 0.5–0.75 µm. *S. mutans* grows in the human oral cavity and is responsible for starting the biofilm that produces dental caries. It does not possess a Lancefield antigen and is ungroupable in that system. (Gram Stain)

Streptococcus pneumoniae

Streptococcus pneumoniae (phylum Firmicutes), commonly called the **pneumococcus,** is estimated to be carried asymptomatically by up to 50% of the human population. Children and adults with children are the principal carriers. It is the leading cause of community-acquired bacterial pneumonia and meningitis in adults, but **otitis media** is the most common result of *S. pneumoniae* infection.

The organism typically colonizes the nasopharynx where it either is eliminated from the body, spreads to the lungs and develops into pneumonia, or is harbored asymptomatically for up to several months. Transmission is usually by direct contact with a carrier or contaminated aerosols.

S. pneumoniae produces a variety of virulence factors. One is a **protease** that breaks down **secretory IgA (sIgA)**, an antibody found in the mucus of mucous membranes. At least 90 different serotypes of *S. pneumoniae* exist and are defined antigenically by their **capsules,** which are their primary virulence factor. Some serotypes are more virulent than others due to their ability to avoid phagocytosis by host cells and the degree to which they stimulate antibody production.

Killed pneumococcal cells release two cell wall virulence factors: **phosphorylcholine** and **pneumolysin.** Phosphorylcholine binds to platelet-activating receptors on platelets, endothelial cells, leukocytes, and other cells, thus interfering with clotting and allowing the pathogen to spread. Pneumolysin activates the complement system and also causes blood vessel damage, both of which lead to fluid accumulation, especially in the lungs, resulting in pneumonia.

In the majority of infections, the invading organisms are cleared with no long-term effects. However, if complicated by bacteremia, meningitis and other secondary infections are the likely result.

As with the staphylococci, drug resistance is an issue with *Streptococcus* and requires isolate susceptibility testing (see p. 291). Penicillin is the antibiotic of choice only if susceptibility has been established. Resistance is due to structurally altered penicillin-binding proteins. Fluoroquinolones showed promise in cases of penicillin resistance, but resistance to these is also on the rise due to mutations in their targets: enzymes associated with modifying DNA structure.

A vaccine containing a mixture of the most common serotypes is available for children through their fifth year. Adults older than 65 should receive the same vaccine given to children, with a follow-up vaccine at least 1 year later.

Differential Characteristics

S. pneumoniae is an α-hemolytic, nonmotile, encapsulated, facultatively anaerobic, Gram-positive diplococcus (or sometimes single cells or short chains) made of football-shaped cells (Figs. 12.68 and 12.69). It is negative for catalase, arginine hydrolysis, esculin hydrolysis, and acid production from mannitol and sorbitol. It is also urease and Voges-Proskauer negative and susceptible to optochin (see p. 104).

Key features differentiating it from other streptococci include the absence of a Lancefield antigen (see p. 136), α-hemolysis, and susceptibility to optochin. Diagnostic procedures include Gram stain, appropriate-site aerobic culture, rapid *Streptococcus* antigen test, PCR (see p. 144), and MALDI-TOF MS (see p. 150).

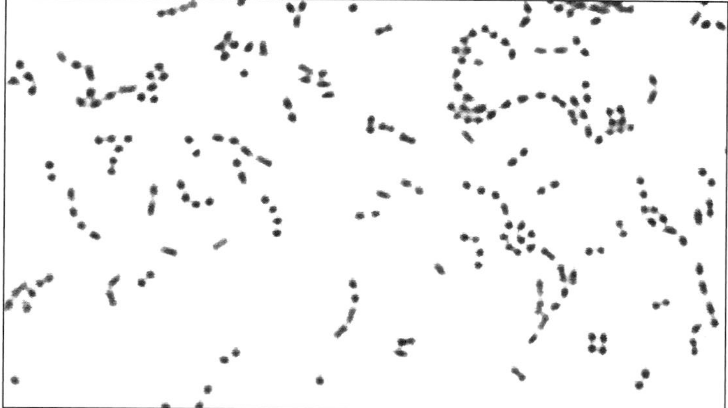

12.68 ***Streptococcus pneumoniae* Stock Culture** • *S. pneumoniae* specimens from sputum samples are typically seen as singles or pairs and the cells are often elongated (lancet-shaped). They range in diameter from 0.5–1.25 µm. *S. pneumoniae* does not possess a Lancefield antigen and is ungroupable in that system. (Gram Stain)

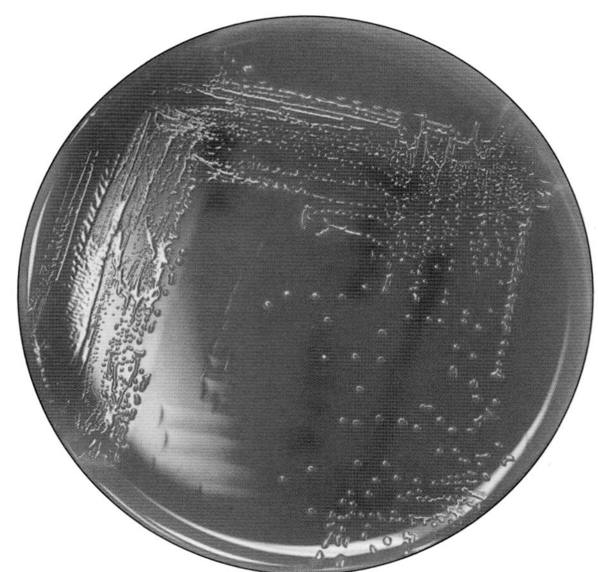

12.69 ***Streptococcus pneumoniae* Growing on Sheep Blood Agar** • These *S. pneumoniae* colonies are less than 1 mm in diameter. Note the greenish color characteristic of α-hemolysis.

Streptococcus pyogenes

Streptococcus pyogenes (phylum Firmicutes), the principal member of group A streptococci, is responsible for a variety of severe infections. It causes **streptococcal pharyngitis** ("strep throat"), impetigo, middle ear infections, mastoiditis, necrotizing fasciitis, puerperal fever, as well as others mentioned in the following text.

The human nose, throat, and skin are reservoirs for *S. pyogenes*, but it is not regarded as a member of the normal microbiota, and thus its presence in patient samples is clinically significant. It is transmitted by direct person-to-person contact or by contaminated aerosols. It possesses an impressive arsenal of virulence factors. Among them are:

- **Streptolysin S** and **Streptolysin O**. Both are hemolysins that attack erythrocytes, leukocytes, and platelets, and are responsible for the β-hemolytic reaction seen by *S. pyogenes* on blood agar plates.
- **Protein F**. This is an **adhesin** that binds to fibronectin in vertebrate cytoplasmic membranes and is responsible for the ability of *S. pyogenes* to attach to host tissues.
- **M protein**. M protein production is unique to *S. pyogenes* among the streptococci. Its presence is associated with poststreptococcal diseases, such as **acute rheumatic fever** (**ARF**) and **glomerulonephritis**. The former is caused by cross-reaction of M-protein antibodies with cardiac tissue. The latter is caused by the accumulation of antigen-antibody complexes in kidney glomeruli, resulting in damage due to complement activation. Further, M protein interferes with the complement system's ability to produce **opsonins** that improve the ability of phagocytic cells to attach to their targets.
- **Streptococcal pyrogenic exotoxins** (**SPEs**). These toxins are **pyrogenic** (fever-causing) and **erythrogenic** (causing redness due to capillary damage). Their erythrogenicity is responsible for the rash of **scarlet fever**. Some act as superantigens and activate T-cells nonspecifically, resulting in the release of cytokines that cause inflammation and shock, as in **streptococcal toxic shock syndrome** (**STTS**). Further, they break down immunoglobins (antibodies) and interfere with the complement system, which affects inflammation and phagocytosis.

Penicillin G is used for treatment of most *S. pyogenes* infections because of its high susceptibility. In cases of penicillin allergies, macrolides may be substituted, though resistance to these is increasing.

Differential Characteristics

S. pyogenes is a β-hemolytic, nonmotile, encapsulated, facultatively anaerobic, catalase negative, Gram-positive coccus (Figs. 12.70 and 12.71). Key features differentiating it from other streptococci include: β-hemolysis (with a large zone relative to colony size), Lancefield group A antigen (see p. 136), bacitracin susceptibility, and positive PYR test (see p. 110).

Diagnostic procedures include throat culture for β-hemolysis, Gram stain, appropriate-site aerobic culture, antibody tests for serum anti-DNase B and serum antistreptolysin-O, streptozyme test, ELISA (see p. 137), PCR (see p. 144), and MALDI-TOF MS (see p. 150).

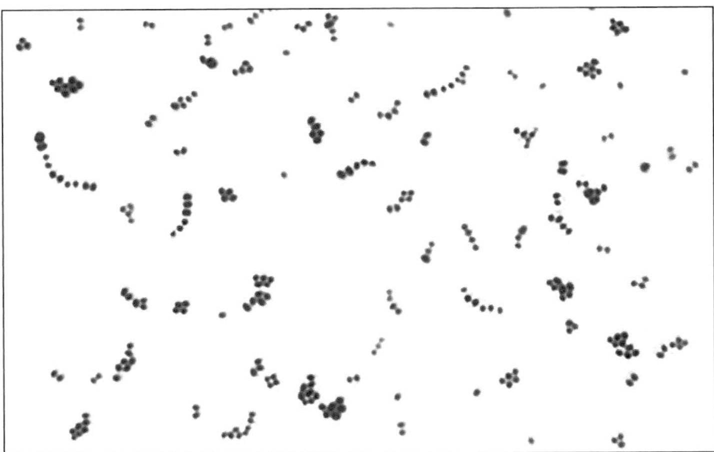

12.70 ***Streptococcus pyogenes*** **(Group A Streptococcus) Stock Culture** Specimens isolated from patients are usually seen in pairs and short chains. *S. pyogenes* belongs to Lancefield Group A and is associated with streptococcal pharyngitis (strep throat), otitis media, and a number of other diseases. See text for more detail. (Gram Stain)

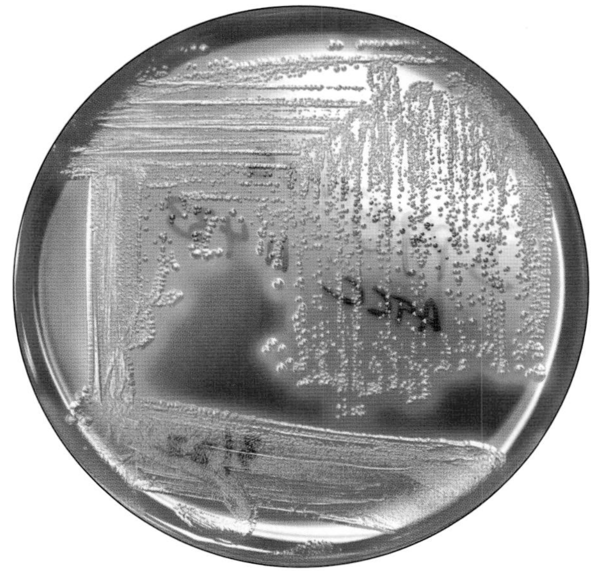

12.71 ***Streptococcus pyogenes*** **(Group A Streptococcus) Growing on Sheep Blood Agar** *S. pneumoniae* colonies are smooth, dome shaped, and mucoid due to cells producing a polysaccharide capsule. The largest colonies on this plate are 2 mm in diameter. Note the extensive β-hemolysis.

Treponema pallidum

Treponema pallidum (phylum Spirochaetes) is an exclusively human pathogen and the infective agent of **syphilis**. It is primarily a sexually transmitted disease (STD) and produces **venereal syphilis**, but vertical transfer from mother to fetus also can occur in **congenital syphilis**.

T. pallidum is "enzymatically challenged," in that glycolysis is its sole mechanism of ATP production and that it is incapable of synthesizing most amino acids, fatty acids, nucleotides, and coenzymes. It relies on its host to provide those compounds and as such, has an abundance and variety of carrier molecules in its periplasm.

Infection begins with entry into the body through mucous membranes, abrasions, or fissures in the epithelia. Some of the organisms attach to host epithelial cells and multiply, while others are carried to lymph nodes where they enter the bloodstream and disseminate throughout the body. Syphilis develops in distinct stages: **incubation, primary, secondary, early latent, late latent**, and **tertiary syphilis**.

Reproduction of the spirochete at the infection site results in an inflammatory response that produces a **chancre**, which is highly infectious. This is the defining symptom of primary syphilis, which lasts about 2 to 6 weeks, after which it heals spontaneously. Hematogenic dissemination to all body regions (including the CNS) takes place and symptoms of secondary syphilis appear 4 to 10 weeks after healing of the chancre.

Symptoms of secondary syphilis are most commonly lesions on the skin and mucous membranes, with rashes on the palms of the hands and soles of the feet. Lesions may also form in the liver, lymph nodes, and muscles. Other nonspecific symptoms include fever and malaise. Again, the patient is highly infectious, but the lesions clear-up within 4 months. Symptoms may reappear during the early latent period, which lasts no more than a year. The late syphilis period follows and the patient has no symptoms and is noninfectious.

About one-third of patients experience tertiary syphilis, the final stage of the disease that occurs 10–25 years later. It is characterized by the destruction of neural tissue (**neurosyphilis**) and/or damage to cardiovascular tissue (**cardiovascular syphilis**) and the formation of large, granulomatous, necrotizing lesions called **gummas** that may form throughout the body.

Congenital syphilis is transmitted from mother to fetus, resulting in an asymptomatic or symptomatic infection with developmental abnormalities.

Penicillin G and ceftriaxone are the treatments of choice. Prevention by safe sex is better. In 2018 the CDC reported more than 35,000 cases of primary and secondary syphilis, and more than 1,300 cases of congenital syphilis in the United States.

Differential Characteristics

T. pallidum is a motile, microaerophilic, Gram-negative spirochete (Fig. 12.72). Initially, darkfield or phase-contrast microscopic examination of the patient's sample is useful. Diagnostic tests include the rapid plasma reagin (Fig. 8.15) and Venereal Disease Research Laboratory (VDRL) tests, which are nontreponemal tests for preliminary screening.

Fluorescent antibody tests (see p. 140) and enzyme immunoassays assays (see p. 137) have proven to be effective methods of determining the presence of *T. pallidum*, as has PCR (see p. 144). Biochemical and physiological tests are not done because the organism does not grow well outside the host.

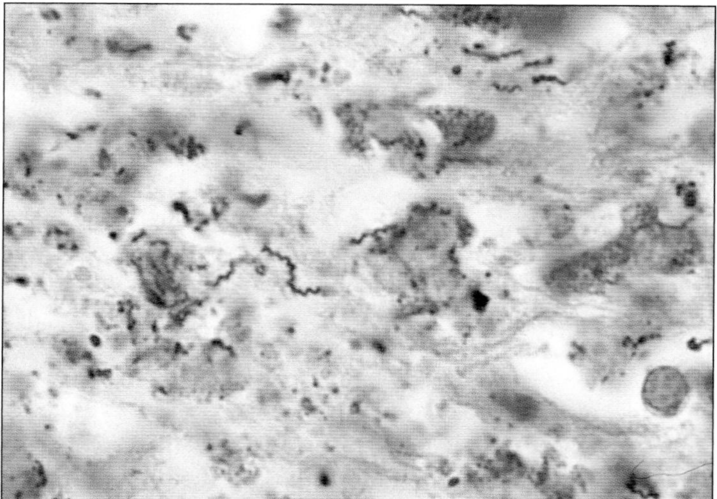

12.72 ***Treponema pallidum*** **in Animal Tissue** • *T. pallidum* is too thin to see with the light microscope without staining. It stains black with a silver stain, and the thin, corkscrew-shaped cells stand out against the gold background. This spirochete demonstrates "corkscrew motility" by means of periplasmic flagella. Cells range in size from 0.2 µm wide by 5–15 µm long. (Silver Stain)

Vibrio cholerae

Vibrio cholerae (class Gammaproteobacteria) strains are typed according to their cell wall composition using the O antigen. Although there are greater than 200 serotypes of this species, only two, the **O1** and **O139 serogroups** (first seen in 1992, but so far only in Asia), have been responsible for the seven cholera pandemics since 1817. The strains have been somewhat arbitrarily divided into the O1 and O139 *V. cholerae,* and the non-O1 non-O139 *V. cholerae* to distinguish the cholera-causing organisms from the rest.

Although immunity to O1 strains does not confer immunity to O139, the diseases caused are indistinguishable. *V. cholerae* O1 has been divided into two biotypes: classical and El Tor. In the early 2000s, new variants started being recovered in Africa and Asia. These hybrids possess features of the classical and El Tor biotypes and are likely the products of lateral gene transfer and genetic recombination.

Most cholera infections are asymptomatic, but the infected carriers act as a reservoir of the pathogen.[10] The organism enters the body by the fecal-oral route, frequently by way of undercooked contaminated seafood. A large inoculum (10^8 cells in water) is required to produce the disease in healthy individuals because sufficient numbers of bacteria must survive the stomach's acidity and reach the small intestine. The infectious dose is considerably smaller if ingested with food.

Virulence factors include the **cholera toxin** (**CT**), encoded by a prophage, and a **toxin coregulated pilus** responsible for attachment to the intestinal epithelium. Once in the small intestine, the organism swims to the mucosal surface and secretes a **mucinase** that "clears" the mucus covering the epithelium, making access easier. Attachment to the mucosal layer by their pili is followed by secretion cholera toxin.

[10] The pathogen also is apparently able to survive in the environment by entering a dormant state characterized as "viable, but not culturable" (VBNC).

The toxin begins a cascade of reactions that ultimately alter electrolyte levels and stimulate a vigorous outpouring of fluids into the intestinal lumen. The result is the characteristic noninflammatory "watery" or "secretory diarrhea," which is frequently fatal within a few hours, due primarily to dehydration and shock.

There were 14 reported cholera cases in the United States in 2018. Because the infection is self-limiting, fluid and electrolyte replacement is standard treatment, but administration of antibiotics minimizes the severity and shortens the duration of infection. A vaccine has been stockpiled by the World Health Organization for use in regions where the disease is endemic, but it is not available in the United States.

Differential Characteristics

V. cholerae is a motile, facultatively anaerobic, salt-tolerant, Gram-negative, straight or curved rod (Fig. 12.73). Key features that distinguish it from *Enterobacteriaceae* include curved rods, cytochrome *c* oxidase positive, sodium required for growth or enhances growth, and growth on TCBS agar (Fig.12.74).

Diagnostic procedures include fecal leukocyte stain, stool culture, commercial multi-test systems (e.g., BBL Crystal and api 20 E), and MALDI-TOF MS (see p. 150). CT can be identified in stool samples using various ELISA (see p. 137) techniques or latex agglutination tests (see p. 133).

12.73 *Vibrio cholerae* Stock Culture • Note the curvature of these *Vibrio cholerae* rods. Rods range in length from 2–4 µm. *V. cholerae* is the causative agent of cholera. (Gram Stain)

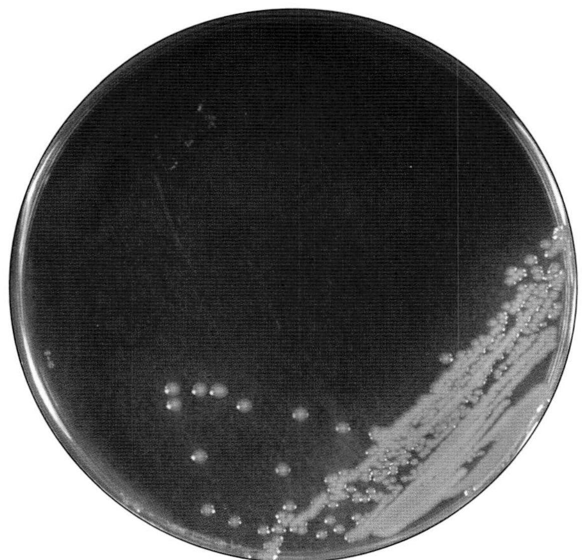

12.74 *Vibrio cholerae* Growing on Thiosulfate Citrate Bile Sucrose (TCBS) Agar • TCBS agar is a selective medium used for isolation of *V. cholerae* and *V. parahaemolyticus* from clinical and environmental specimens (usually fecal-contaminated water). The colonies are yellow due to acid produced by sucrose fermentation. The largest ones are approximately 3.5 mm in diameter. Refer to page 21 for more information on TCBS agar.

Yersinia pestis

Yersinia pestis (class Gammaproteobacteria) has been responsible for dozens of plague epidemics and pandemics over the last several hundred years, one of which took the lives of 25 million Europeans in the fourteenth century.

Existing on every continent except Australia, its habitat is any of a variety of small animals, including rats (for **urban plague**), and rats, ground squirrels, rabbits, mice, and prairie dogs (for **sylvatic plague**). Infected pet cats are also a reservoir and may infect humans directly through scratches, bites, or aerosols, as well as indirectly by harboring infected fleas.

Y. pestis is most commonly transmitted by fleas, but can be inhaled as droplet nuclei, resulting in **pneumonic plague**. Fleas ingest the organism when feeding on infected animal blood and pass it to another animal during a subsequent blood meal. *Y. pestis* produces a coagulase that clots the blood inside the flea's stomach. When the flea attempts to feed again, it regurgitates the coagulated material and contaminated blood back into the bite wound. If the bacteria are deposited directly into the bloodstream by the flea (usually in children), **septicemic plague** is the likely outcome, which results in fever, chills, and possibly leads to shock.

The most common form of the disease is **bubonic plague**, with an incubation period of 2 to 6 days. It is characterized by rapid onset of symptoms, such as severe fever, headache, and inflammation and hemorrhagic necrosis of the inguinal or axillary lymph nodes called **buboes**. This is an extremely serious disease with a very high mortality rate (approaching 50%) when untreated. Pneumonic plague has a shorter incubation time (1 to 3 days), and also begins with fever, headache, and weakness, but these are followed quickly by pneumonia, often with a bloody mucus. Untreated, mortality rate is 100% within 24 hours.

Three plasmid gene products are virulence factors. **Murine toxin**, a phospholipase, is only produced at temperatures consistent with the flea host. It protects *Y. pestis* while in the flea's gut. **Fraction 1 antigen (F1)**, is produced when in the mammalian host. It produces an "envelope" that protects against phagocytosis.

Plasminogen activator (PLA) is a protease that does what its name says: it activates plasminogen, the precursor of **plasmin,** which is responsible for breaking down clots. This enhances the pathogen's ability to spread through the host's body. PLA also breaks down complement, which compromises the host's defenses, especially inflammation and phagocytosis.

Streptomycin, tetracycline, or chloramphenicol are used in treatment. Susceptibility testing is done only in reference laboratories due to the virulence of *Y. pestis* and the fact that resistance is not commonly found in its strains.

Differential Characteristics

Y. pestis is a nonmotile, facultatively anaerobic, Gram-negative rod or coccobacillus (Figs. 12.75 and 12.76). It forms an envelope (not a capsule) when grown at 37°C. Like other enterics, *Y. pestis* is catalase positive, oxidase negative, and ferments glucose to acid, but not gas, end-products. It also ferments mannitol, arabinose, trehalose, and mannose, but not lactose, or sucrose. Most are MR positive, but negative for lysine and ornithine decarboxylase, and arginine dihydrolase.

Diagnostic procedures include aerobic and anaerobic blood culture, appropriate-site aerobic culture, direct fluorescent antibody test (see p. 140) and ELISA for F1 antigen (see p. 137)[11], PCR targeting the PLA or F1 genes (see p. 144), and MALDI-TOF MS (see p. 150).

[11] The lag time between infection (with its rapid onset) and development of measurable patient antibodies make these tests most useful for retrospective studies rather than for diagnosis.

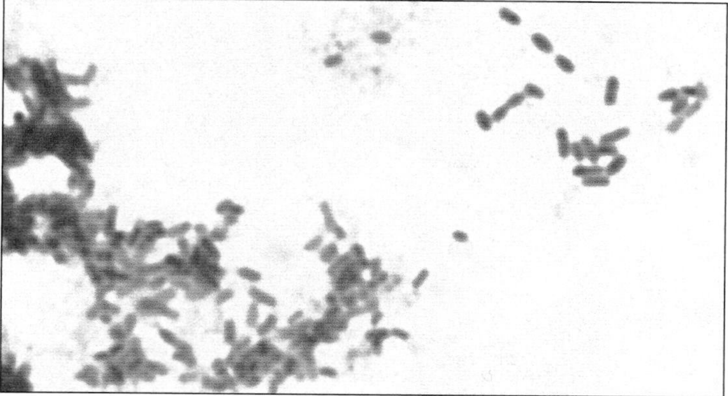

12.75 ***Yersinia pestis* Stock Culture** • *Y. pestis* is typically seen as nonmotile, unencapsulated, straight rods or coccobacilli ranging in size from 0.5–0.8 µm wide by 1–3 µm long. It is the causative agent of plague. (Gram Stain)

12.76 ***Yersinia pestis* on Sheep Blood Agar** • Note the characteristic "fried egg" appearance of these *Y. pestis* colonies growing on sheep blood agar.

SECTION

13

Domain Archaea

At one time, Archaea were classified in the Kingdom Monera along with Bacteria. Molecular evidence obtained by Carl Woese using comparisons of small subunit ribosomal RNA, however, led to the realization that Archaea and Bacteria have very little in common beyond being small prokaryotes.

This initially resulted in their separation within Monera as Archaebacteria and Eubacteria, followed by their removal from Monera and placement into different domains: Archaea and Bacteria, respectively[1]. At the same time, the eukaryotes were placed into the third domain, Eukarya. Refer to Table 1-1, page 4, for a comparison of the three domains.

Many of the earliest archaea discovered and studied were **extremophiles**, meaning that they live in environments that are very hot, very acidic, or very salty. These environments resemble what scientists believe existed on the early Earth, hence the names Archaebacteria and Archaea (*archae* means "primitive").

Microbiologists have subsequently learned that archaea are also commonly found in less extreme environments, such as the ocean and soil, and that they are more abundant than their extremophilic relatives. At the present, five phyla are recognized: **Crenarchaeota**, **Euryarchaeota**, **Nanoarchaeota**, **Korarchaeota** and **Thaumarchaeota.** However, research into Archaea evolutionary relationships is an on-going effort and will undoubtedly lead to revisions of our current interpretations.

[1] Consider the implications: Placing Archaea and Bacteria into different domains means they have less in common with each other than earthworms, amoebas, sunflowers, mushrooms, humans, and all eukaryotes do.

Crenarchaeota

Members of Crenarchaeota are morphologically diverse (including unusual disk-shaped cells), Gram-negative hyperthermophiles, and are distributed worldwide, including hydrothermal vents associated with seafloor spreading. They are metabolically diverse with chemolithotrophic (a majority metabolize sulfur) and chemoheterotrophic species. (See footnote 1 on p. 163 in Section 11 for an explanation of these terms).

Aerotolerance groups include aerobes and facultative anaerobes, but most are obligate anaerobes that perform anaerobic respiration with elemental sulfur (S^0) or nitrate (NO_3) as the final electron acceptors. Fermentation is uncommon. In spite of this metabolic diversity within the phylum, nucleotide sequence comparisons indicate that this is a natural, therefore valid, grouping. Figure 13.1 shows the habitat of *Sulfolobus*, a genus in one of the phylum's three orders.

As a group, *Sulfolobus* species are obligately aerobic hyperthermophiles that live in sulfurous hot springs (called solfataras) with optimum temperatures between 65° and 85°C. They get their energy by oxidizing various forms of reduced sulfur (e.g., H_2S) and producing sulfuric acid (H_2SO_4) as an end-product, which makes their environment highly acidic, ranging from a pH of 1 to 5.

Their cells are variable in shape and are frequently lobed, which explains "lobus" in their name (Fig. 13.2). Glycoprotein subunits in a regular arrangement comprise the wall, known as an **S-layer** (Figs. 13.2 and 13.3). As with Bacteria, motility is produced by rotating flagella, but there appears to be no homology between them. In fact, they are more closely related to Type IV bacterial pili than bacterial flagella. Further, the flagella of Archaea are not hollow and their basal bodies are simpler in construction (Fig. 13.3). For these reasons, there is a trend toward calling the structure an **archaellum.**

13.1 ***Sulfolobus* in Sulfur Cauldron (Phylum Crenarchaeota)** ● This geologic feature of Yellowstone National Park produces copious amounts of H_2S gas. *Sulfolobus* uses the H_2S as an energy source and oxidizes it to sulfuric acid (H_2SO_4), lowering the pH of its environment to 2 or less. The high acidity breaks down rock and soil to produce the muddy conditions seen in the photo. The steam you see rising from the cauldron is due to its high temperature. *Sulfolobus* species grow between 65°C and 85°C. Although *Sulfolobus* is primarily a chemolithoautotroph, it also can grow heterotrophically.

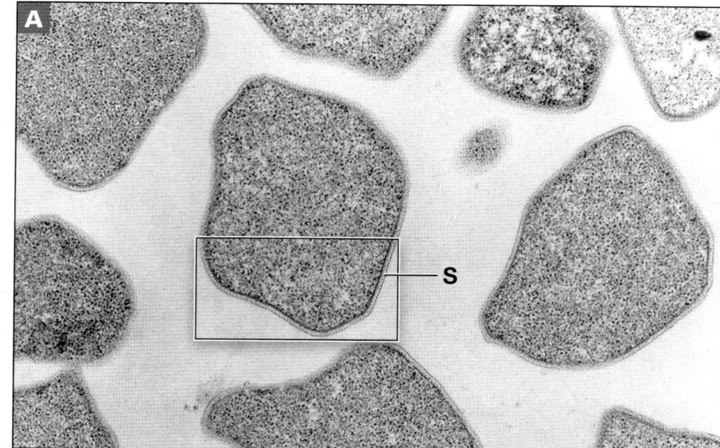

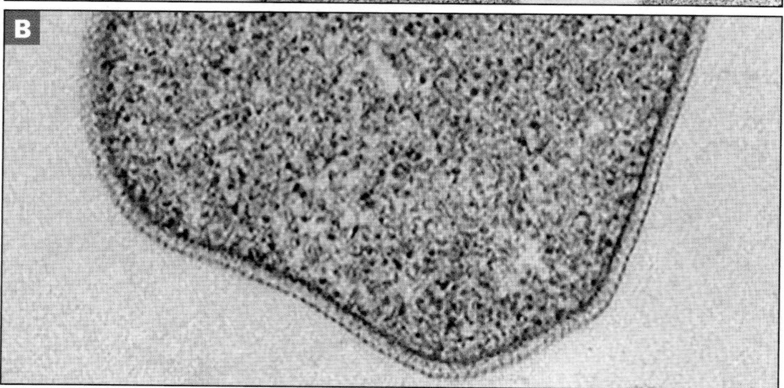

13.2 ***Sulfolobus acidocaldarius* Cell Morphology (Phylum Crenarchaeota)** ● (A) These *Sulfolobus acidocaldarius* cells were originally collected from Yellowstone National Park, WY. They were grown in culture (pH 2, 75°C) and were harvested in mid-log phase for this micrograph. Cells are 1.0–1.5 µm in diameter. The S-layer (**S**), composed of a single layer of glycoprotein anchored to the cytoplasmic membrane, is visible on the surface of each. Also visible, even though this is a thin section, is the lobed morphology of these cells, making their generic name very descriptive. These are *not* spherical cocci. (B) Shown is an enlargement of the rectangular region in Fig. 13.2A. (TEMs)

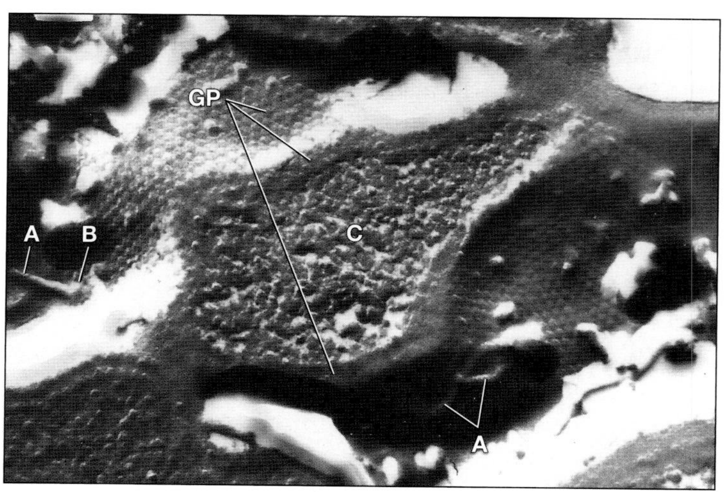

13.3 ***Sulfolobus acidocaldarius* Cell Wall and Archaellum (Phylum Crenarchaeota)** ● Shown is a cross-fractured preparation of the *Sulfolobus* S-layer (surface and cross-sections). Note the regular arrangement of the glycoproteins (**GP**) comprising it. Also visible is cytoplasm (**C**), archaella (**A**), and a basal body (**B**). (Freeze Fracture TEM)

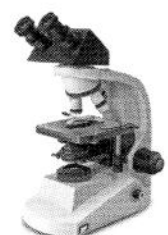

Euryarchaeota

Seven classes comprise Euryarchaeota (Table 13-1). Our emphasis will be on the methanogens, halophiles, and thermoplasmas.

Most microbiologists recognize five orders of methanogens, all of which obtain energy from the oxidation of H_2 gas, formate (CHCOO–), or a few other simple organic compounds and use the electrons and hydrogens from the oxidation to reduce CO_2 to methane (CH_4) in anaerobic respiration. The ability to make methane is the basis for their name: methanogens.

Their cell morphologies range from familiar cocci, rods, and spirals to more unique shapes, such as flat triangles and squares. Their cell walls may possess a compound similar to peptidoglycan called **pseudopeptidoglycan** (or pseudomurein), which has **N-acetyltalosaminuronic** acid in place of *N*-acetylmuramic acid in the glycan chains; a protein **S-layer** ("surface" layer); both; or neither.

Some are motile, usually with tufts of polar archaella, and others are not. Habitats include anaerobic mud of freshwater environments (where they produce "swamp gas"), the rumen of cattle (Fig. 13.4), and sludge digesters (Figs. 13.5 and 13.6). Clearly, methanogens are a diverse group.

Diversity also characterizes halophilic Archaea (Haloarchaea). Representatives vary in aerotolerance (obligate aerobes and facultative anaerobes), motility (motile by polar archaellar tufts and nonmotile forms), and cell morphologies (rods, cocci, flat triangles and disks, and sheets of flat squares, Fig. 13.7). All require salt concentrations of at least 1.5 M NaCl, but most require salt in the 2–5.5 M range.

Survival at such high environmental osmotic pressures requires special adaptations. Among these is the ability to concentrate KCl intracellularly to achieve osmotic balance. A second adaptation involves stability of cellular components in the saline environment. While enzymes of nonhalophiles are denatured by high intracellular salinity, those of halophiles are actually more stable at higher osmotic pressures and are denatured at lower ones. The same is true of their ribosomes.

Halobacterium is a genus of extreme halophiles that is still saddled with its misnomer, because they aren't bacteria. In spite of lacking an outer membrane, they stain Gram negative (Fig. 13.8). Pigments play a major role in *Halobacterium* survival (Fig. 13.9).

Metabolism is chemoheterotrophic (aerobic respiration), but in the absence of oxygen some species can become phototrophic because of the membrane-bound pigment, **bacteriorhodopsin**. Absorption of light by this pigment generates a proton gradient—the pigment acts as light receptor *and* proton pump—and the resulting proton motive force is used to phosphorylate ADP to ATP. It is unclear if these organisms can live photoautotrophically, but they certainly are capable of photoheterotrophic growth.

Halorhodopsin, a second membrane-bound pigment, is used by probably all halophilic cells to pump chloride ions inward (to increase KCl, as mentioned previously). Other pigments are involved in phototactic responses.

13.5 Sludge Digesters • Shown is a row of sludge digesters at the Point Loma Wastewater Treatment Plant in San Diego, CA. Microbial sludge digestion is the final step in sewage treatment, but there are variations on how that treatment occurs. At this plant, these 3-million-gallon tanks contain what remains of the sewage after solids have been screened out and chemical treatments have occurred. Residence time in each digester is approximately 15 days, during which microbial action breaks down organic materials in the anaerobic environment. Methanogens are active in this process. The methane they produce is used to generate electricity at the plant and also supplements the electrical grid. Excess methane is purified and sold as CNG.

13.4 Methanogens in Cow Rumens • Methanogens are found in the anaerobic digestive tracts of ruminants, such as cows. According to the EPA, the gut microbiota of ruminants is responsible for the production of 80 million metric tons of methane (a greenhouse gas) globally. This cow is a resident of the Van Ommering Dairy Farm in Lakeside, CA. The Van Ommering farm uses methane produced by its herd to generate electricity that supplies the farm's energy needs, thereby reducing greenhouse emissions and providing them with their own energy source—a win-win situation.

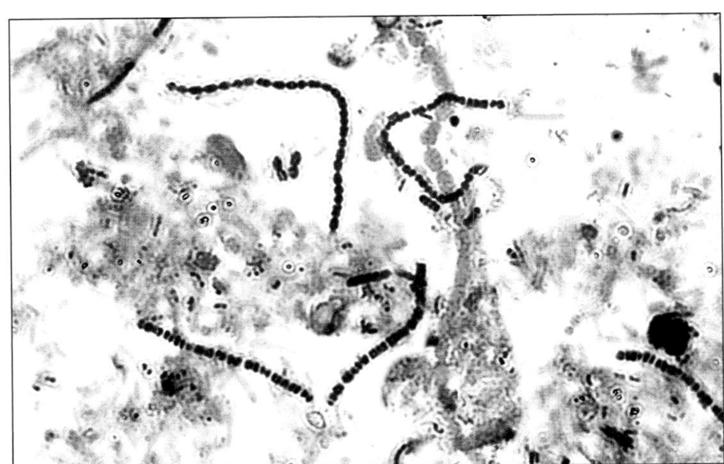

13.6 Composite Gram Stain of Methanogens (Phylum Euryarchaeota; Classes Methanobacteria and Methanococci) and Other Microbial Inhabitants of Digested Sludge • This micrograph shows Gram-stained elements from two micrographs (to illustrate greater cell diversity) of digested sludge collected from one of the digesters shown in Figure 13.5. At this point, sludge is mostly microbes. While it is not possible to identify specific cells in the micrograph based solely on their morphology, methanogens are in there, both Gram positive and Gram negative. (Gram Stain)

Some species, including *H. salinarum*, produce **gas vesicles** made of protein (not membrane), which may coalesce into **gas vacuoles** (Fig. 13.8). It has been suggested that gas vacuoles increase buoyancy in the cells, which makes it easier for them to float/swim to the surface of the water column where oxygen (for aerobic respiration) and light (to operate the proton pump) are more plentiful. Figure 13.10 shows *H. salinarum* growing at the surface of a broth culture.

Thermoplasma cell morphologies range from cocci to filamentous rods. They are facultatively anaerobic heterotrophs that can use elemental sulfur as the final electron acceptor in anaerobic respiration. All are thermophilic (optimum temperature of 60°C) and acidophilic (optimum pH 2). Motility is provided by a single archaellum. Their most distinctive feature is that they lack a cell wall but possess a triple-layer cytoplasmic membrane (Fig. 13.11).

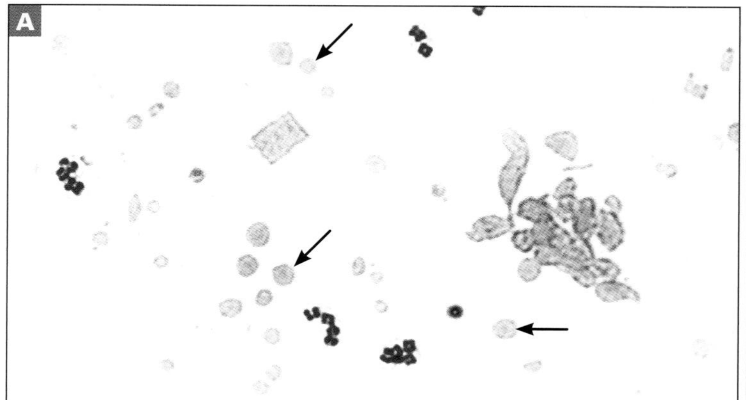

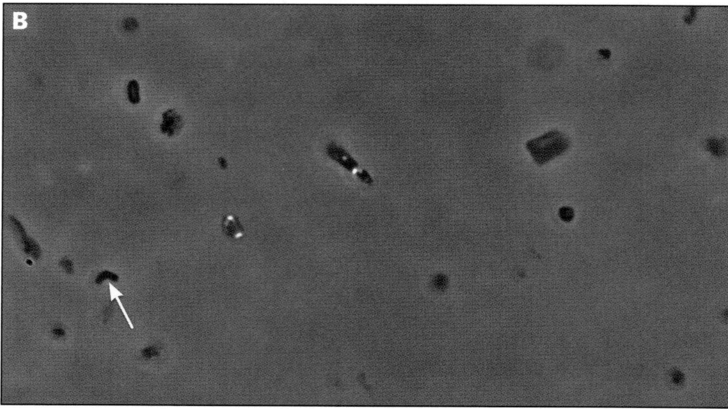

13.7 Haloarchaea Cell Morphologies (Phylum Euryarchaeota; Class Halobacteria) • The samples in these micrographs were collected from the North Arm of the Great Salt Lake in Utah. The North Arm has a salinity exceeding 25% (seawater is approximately 3.5% NaCl, so these are extreme halophiles). Identifying Bacteria and Archaea based on cell morphology alone is difficult, and to complicate this task, many Haloarchaea are pleomorphic. Nevertheless, considering that the extreme environmental conditions limit the possible candidates and that some are very unique in appearance, some provisional identifications are provided. (**A**) Notice the diversity of unusual shapes. Scattered throughout are disk-shaped cells ranging in size from 2–3 µm in diameter (**arrows**) and a rectangular cell, provisionally identified as *Haloquadratum*. It is about 5 µm by 3 µm. The faint, white dots within *Haloquadratum* are likely gas vesicles. (Gram Stain) (**B**) In this phase contrast micrograph, two unidentified cells are seen (probably different species) with gas vesicles (white dots), and another *Haloquadratum* cell is to the right. At the left is a bacterial extreme halophile (**arrow**), *Salinibacter ruber* (see p. 179).

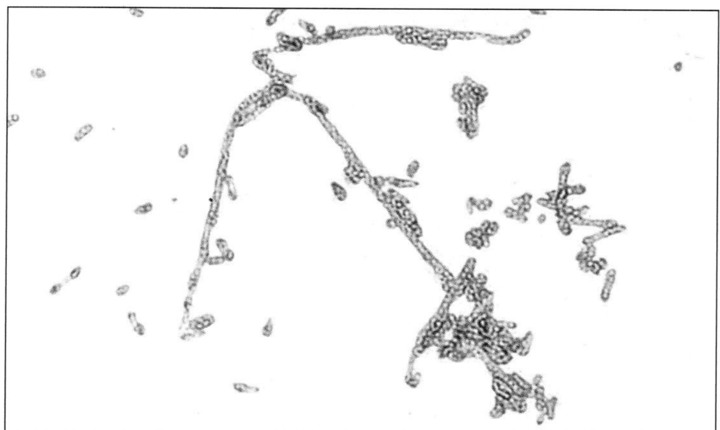

13.8 *Halobacterium salinarum* Grown in Culture (Phylum Euryarchaeota; Class Halobacteria) • *H. salinarum* cells are pleomorphic rods that vary in different media and temperatures. Note the circular, white gas vacuoles in the cells. These provide buoyancy, allowing the cells to float or swim more easily to the water's surface where light and oxygen are more plentiful. (Gram Stain)

13.9 Salterns in San Diego Bay • Salterns are low pools of saltwater used in the harvesting of salt. As water evaporates, the saltwater becomes saltier and saltier, until only salt remains, which can then be sold. The colors in the pools are the result of differently pigmented communities of halophilic microorganisms that are associated with different salinities as the pools dry out.

13.10 ***Halobacterium* Broth Culture (Phylum Euryarchaeota; Class Halobacteria)** ● The pink layer at the top is where *Halobacterium salinarum* is growing. Pigments in *Halobacterium* are used to capture light energy, which can be used to produce a proton gradient for ATP synthesis and operate chloride pumps. Others are involved in phototaxis (movement toward light).

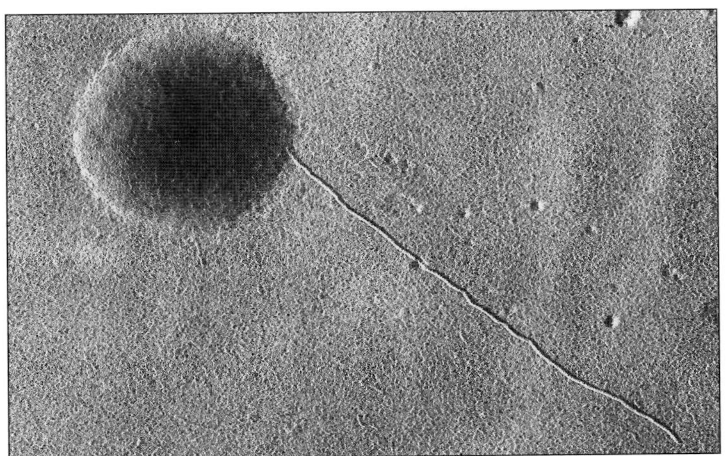

13.11 ***Thermoplasma*, an Archaeon Without a Wall (Phylum Euryarchaeota)** ● *Thermoplasma* cells range from 0.5–5.0 μm in diameter. These thermophilic acidophiles lack a cell wall and are facultatively anaerobic chemoheterotrophs. They swim using their single archaellum. This cell looks flattened due to the preparation process. (TEM, Platinum Shadow Cast)

TABLE **13-1** **Brief Characterizations of the Seven Classes of Phylum Euryarchaeota**

Class	Characteristics
Methanobacteria	Gram-positive cocci to long rods with pseudopeptidoglycan wall; obligate anaerobes that oxidize H_2, using CO_2 as the electron acceptor to form CH_4; do not catabolize carbohydrates, protein, or most other organic compounds (exceptions are methanol, secondary alcohols, formate, and CO)
Methanococci	Obligately anaerobic cocci or rods with proteinaceous cell wall (S-layer); most oxidize H_2, formate, or alcohols with concurrent reduction of CO_2 to CH_4
Halobacteria	Pleomorphic cells; extreme halophiles requiring a minimum salinity of 1.5 M NaCl; chemoorganotrophic
Thermoplasmata	Pleomorphic cells; aerobic (or facultative), thermoacidophilic chemoheterotrophs; cell wall is absent
Thermococci	Obligately anaerobic, hyperthermophilic heterotrophs; reduce S^0 to H_2S in anaerobic (sulfur) respiration
Archaeoglobi	Cocci to irregular cocci; obligately anaerobic, hyperthermophilic, neutrophilic heterotrophs; S^0 inhibits growth; sulfate, sulfite, thiosulfate, and nitrate are used as final electron acceptors in anaerobic respiration
Methanopyri	Gram-positive rods with pseudopeptidoglycan; obligately anaerobic; chemoautotrophic by forming CH_4 from H_2 and CO_2; optimum temperature is 98° (won't grow below 80°C)

Nanoarchaeota, Korarchaeota, and Thaumarchaeota

These three phyla were previously unknown because they were uncultivatable in the laboratory. However, advancements in studying 16S ribosomal RNA genes without cultivating the cells led to their identification.

Nanoarchaeota contains a single rare species called *Nanoarchaeum equitans*, which, at 0.4 μm in diameter, is the smallest known organism on the planet. Their extraordinarily small genome limits their metabolic abilities, which is consistent with their obligately parasitic lifestyle. Their host is a hyperthermophilic *Crenarchaeota* species isolated from a hydrothermal vent.

Korarchaeota are thermophiles found both on land and in marine environments. The first identified species was found in a hot spring (Obsidian Pool) in Yellowstone National Park. It is a hyperthermophilic, obligate anaerobe, and a fermenter of proteins.

Thaumarchaeota have, surprisingly, been found worldwide in mesophilic marine and soil environments with low nutrient availability. They are chemolithotrophic bacilli that perform nitrification (oxidizing ammonia to nitrite).

Domain Eukarya: Simple Eukaryotes

SECTION 14

Simple, microbial eukaryotes are currently placed into four major groupings (Fig. 14.1): Excavata, SAR/HA (including subgroups Stramenopila, Alveolata, Rhizaria, and Hacrobia—a new group of uncertain position), Archaeplastida, and Unikonta (including Amoebozoa and Opisthokonta).

Archaeplastida also includes true plants and Unikonta includes animals. Plants and animals will not be covered, because they don't fit within the traditional boundaries of microbiology.

The subgroups listed below the supergroups are not necessarily of the same taxonomic rank. That is, they are not all phyla or classes, etc.

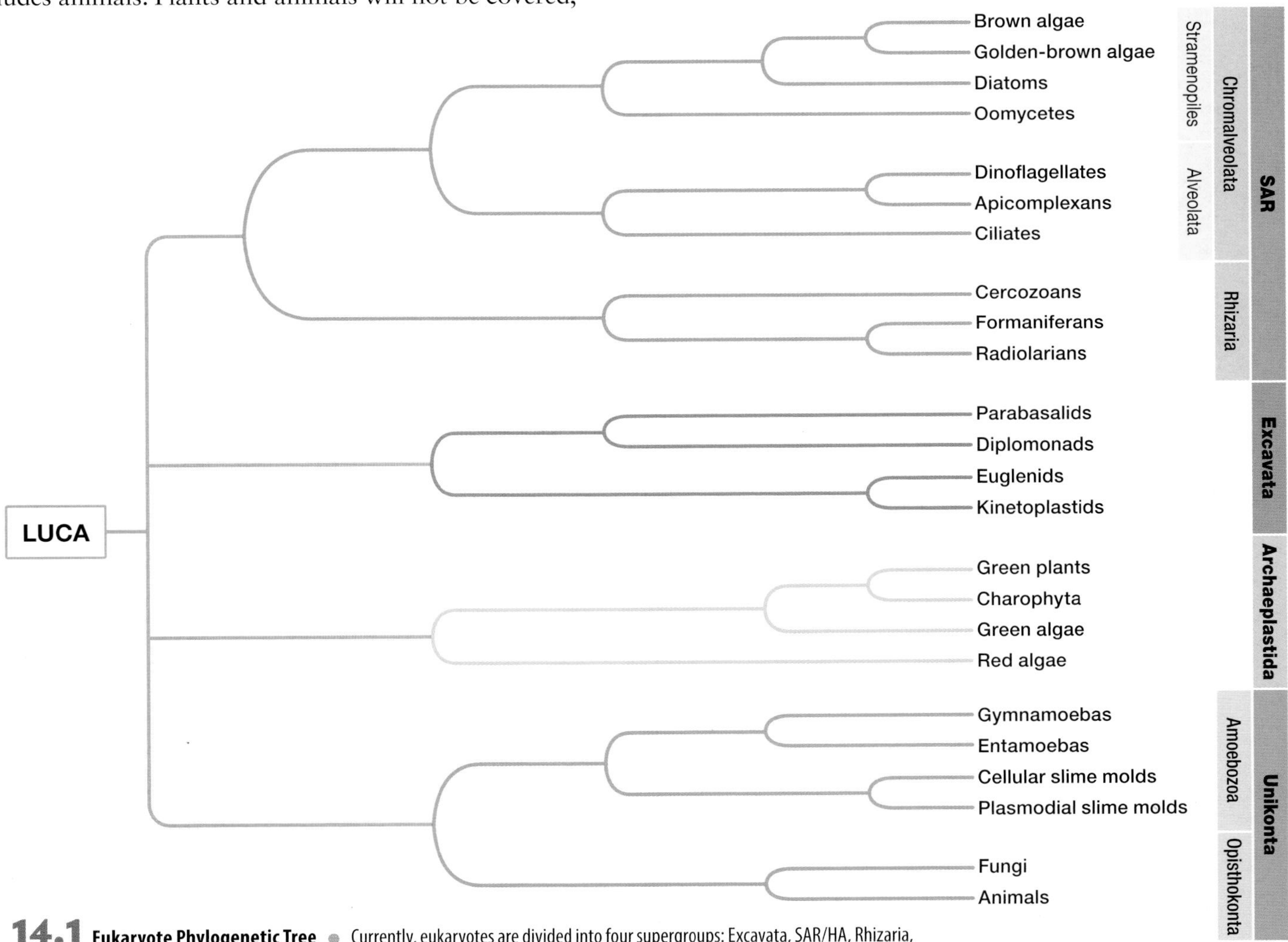

14.1 Eukaryote Phylogenetic Tree • Currently, eukaryotes are divided into four supergroups: Excavata, SAR/HA, Rhizaria, and Unikonta. The names at the end of each branch represent commonly studied eukaryotes, but are not necessarily the same taxonomic rank. Hacrobia is not shown due to its uncertain placement with Stramenopiles or Archaeplastida, or both.

Group Excavata

This group obtains its name from the presence in many members of a feeding groove *excavated* from one side of the cell. Among the major Excavata groups are **parabasalids** and **diplomonads**, both in the phylum Metamonada, and **kinetoplastids** and **euglenids**, both in the phylum Euglenozoa.

- One unifying feature of parabasalids is the presence of a **parabasal body**, a Golgi apparatus associated with the internal flagellar structures. Parabasalids also possess **hydrogenosomes**, organelles that likely evolved multiple times and are thought to be degenerated mitochondria. These organelles earn their name from their ability to make ATP anaerobically with H_2 as a waste product. Other parabasalid features include flagella, a longitudinal aggregation of microtubules called an **axostyle**, and an **undulating membrane** derived from the cytoplasmic membrane associated with a flagellum. *Trichomonas* (Fig. 14.2) and *Dientamoeba* are examples. For more information about these pathogens, see page 255.
- Diplomonads, such as *Giardia* (Fig. 14.3), are unicellular flagellates with two, large nuclei. At one time diplomonads were thought to lack mitochondria, but more recent evidence shows they have degenerated into **mitosomes** that lack the electron transport chain and are thus nonfunctional in generating ATP for the cell. Instead, diplomonads use cytoplasmic biochemical pathways, such as glycolysis, to perform substrate level phosphorylation for ATP synthesis. For more information on *Giardia* as a pathogen, see page 256.
- Euglenids, represented by *Euglena* (Fig. 14.4), are green, photosynthetic protists when light is available, but are capable of heterotrophy when light is not (making them **mixotrophic**). One or two flagella are present and emerge from an invagination of the anterior (forward) cytoplasmic membrane. Flagella possess a **crystalline rod** similar to those of kinetoplastids. A red photoreceptor at the cell's anterior, called an **eyespot,** is another distinctive feature. These are not considered human pathogens.
- Kinetoplastids are placed in the same phylum as euglenids. The presence of a DNA mass within their single mitochondrion, called a **kinetoplast**, is a unifying feature of kinetoplastids. As with euglenids, they have a distinctive **crystalline rod** within their flagella. Free-living and parasitic (pathogenic) species are known. *Leishmania* and *Trypanosoma* (Fig. 14.5) are examples of pathogens. For more information on these pathogens, see pages 257 and 258, respectively.

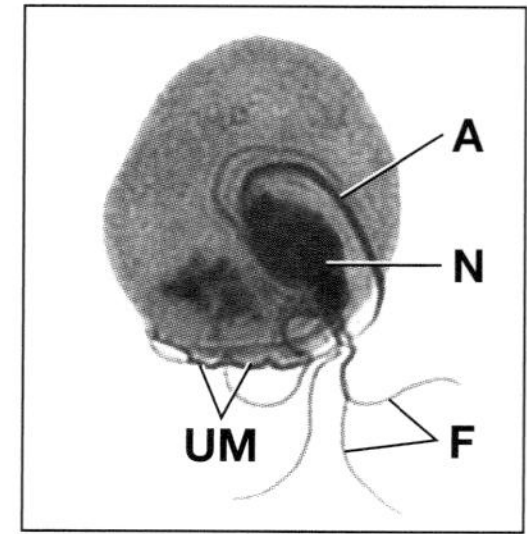

14.2 ***Trichomonas vaginalis*** **(Excavata; Phylum Metamonada)** • *Trichomonas vaginalis* is a parabasalid species and is the causative agent of trichomoniasis, a sexually transmitted disease. Note the prominent nucleus (**N**), four anterior flagella (**F**), undulating membrane (with the fifth anterior flagellum at its free edge), and axostyle made of microtubules (**A**).

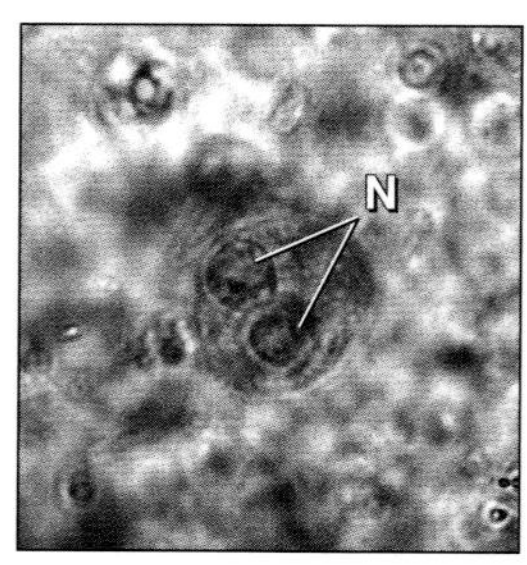

14.3 ***Giardia duodenalis*** **Trophozoite in a Fecal Smear (Excavata; Phylum Metamonada)** • *Giardia* trophozoites attach to the small intestine and feed. This trophozoite is largely recognizable among the debris because of the two nuclei (**N**) typical of diplomonads. *Giardia* cysts have four nuclei and are the more common stage found in feces. (Iron Hematoxylin Stain)

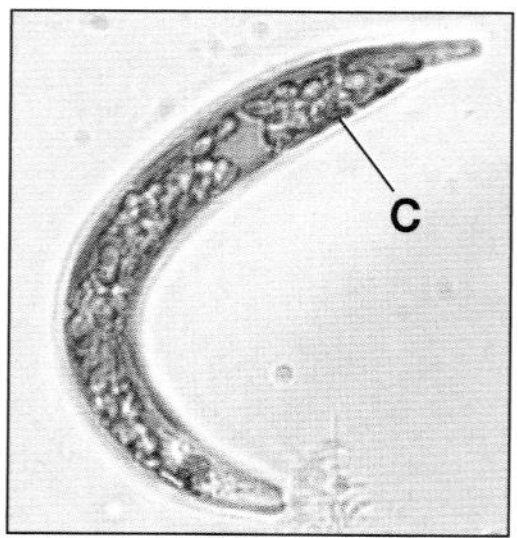

14.4 ***Euglena*** **(Excavata; Division Euglenozoa)** • *Euglena* is a large genus of mixotrophic flagellates. Most species have chloroplasts (**C**), which are discoid in this specimen. A red "eyespot" is located in the colorless anterior of the cell. The single flagellum also emerges from the anterior. (Unstained Wet Mount)

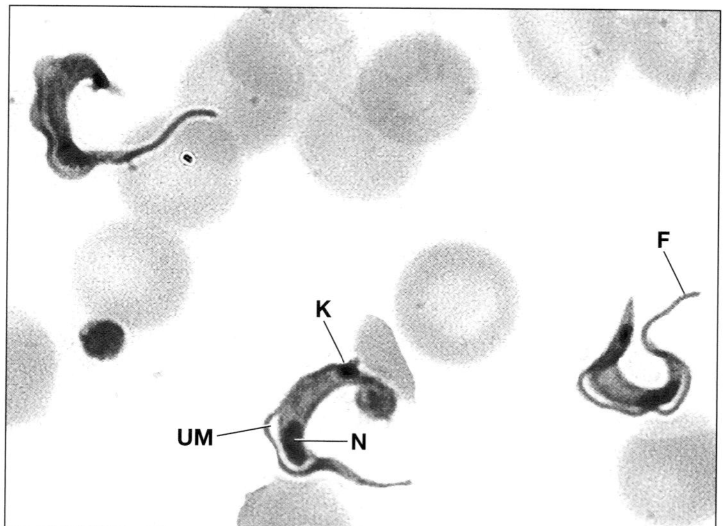

14.5 ***Trypanosoma cruzi*** **in a Blood Smear (Excavata; Division Euglenozoa)** • *Trypanosoma cruzi* is a pathogenic euglenid, but is in the class Kinetoplastida due to the presence of a unique structure called a kinetoplast. The kinetoplast (**K**) is within each cell's single mitochondrion and is composed of DNA. Note these other features: nucleus (**N**), undulating membrane (**UM**), and flagellum (**F**). *T. cruzi* is a blood parasite and causes Chagas disease (American trypanosomiasis). (Wright's Stain)

Group SAR/HA

Subgroup Stramenopila (Heterokontophyta)

Stramenopiles are soil, marine, and freshwater autotrophs and heterotrophs. They are characterized by having a flagellum made of a basal attachment, a hollow shaft, and distinctive glycoprotein filaments ("hairs") split into three parts at the ends. A second smooth flagellum is often present.

There are four groups of stramenopiles: **oomycetes**, which are colorless and heterotrophic, and **diatoms**, the **golden algae**, and the **brown algae**, all of which are photosynthetic. These latter two groups are not covered here.

- Many oomycetes resemble fungi (*myco* means "fungus") in that they grow as cellular filaments and produce sporangia, but these are superficial similarities. Their life cycle and molecular evidence shows that they are not closely related to the fungi. Most are saprophytes (decomposers) but some are plant or fish pathogens. The oomycete genus *Saprolegnia* is shown in Figures 14.6.
- Diatoms, or **bacillariophytes**, are photosynthetic unicellular eukaryotes (Figs. 14.7–14.15). Notice the distinctive golden-brown color from the pigment **fucoxanthin** located in the **chromoplasts**, which may be of variable shape. Oil droplets are also frequently visible. Cell shapes are either round (**centric**) or bilaterally symmetrical and elongated (**pennate**). The cell wall is made of silica embedded in an organic matrix and consists of two halves, with one half overlapping the other in the same way the lid of a Petri dish overlaps its base. The combination of both halves is called a **frustule**. (Often, diatoms can be identified from the frustules of deceased cells.) Some pennate diatoms are motile and have a central, longitudinal line called a **raphe**.

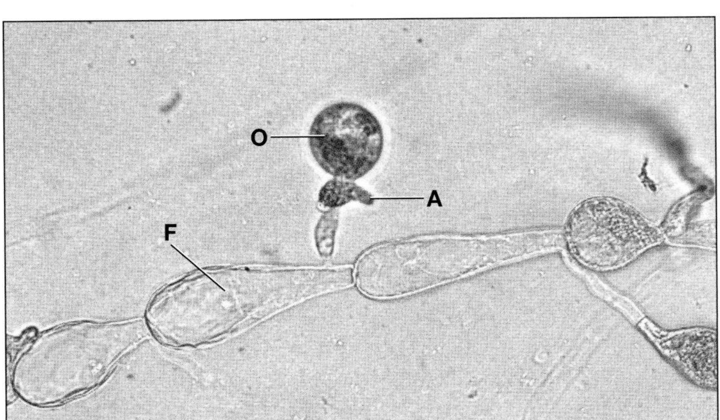

14.6 ***Saprolegnia*** **(Stramenopila; Class Oomycota)** • *Saprolegnia* is an oomycete, or water mold. A diploid filament (**F**) is shown, with the globular **oogonium** (**O**) and the diploid, irregularly shaped **antheridium** (**A**). When an antheridium contacts an oogonium, meiosis occurs in both to produce eggs and haploid male nuclei. The male nuclei migrate into the oogonium where fertilization occurs. *Saprolegnia* is a fish parasite. (Unstained Wet Mount)

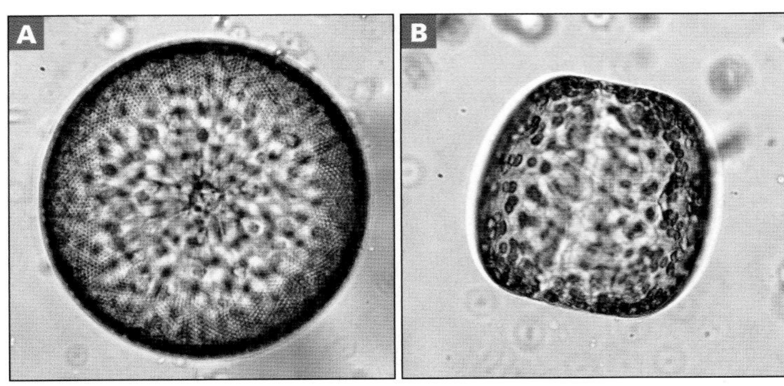

14.7 Centric Diatom—*Melosira* (Stramenopila; Class Bacillariophyceae) • *Melosira* species inhabit marine and freshwater habitats. This specimen was collected from San Diego Bay. Notice the golden-brown chromoplasts and the minute holes in the frustule (cell wall). Also, compare the valve (top/bottom) view (**A**) with the girdle (side) view (**B**). The difference, which is not unusual for diatoms, suggests two different species and this complicates identification. But then, humans don't look the same from above and the side either. The radial symmetry in the valve view makes this a centric diatom. *Melosira* cells divide to form a chain attached by facing valve surfaces. (Unstained Wet Mounts)

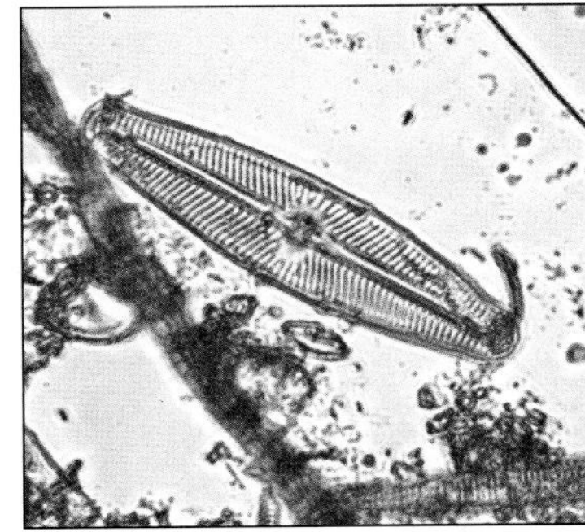

14.8 Pennate Diatom—*Navicula* (Stramenopila; Class Bacillariophyceae) • All of the many *Navicula* species are shaped like a cigar or a boat. Cells are identifiable by prominent transverse lines converging on the central open space. A longitudinal line is also present. The bilateral symmetry of the valve view makes *Navicula* a pennate diatom. (Unstained Wet Mount)

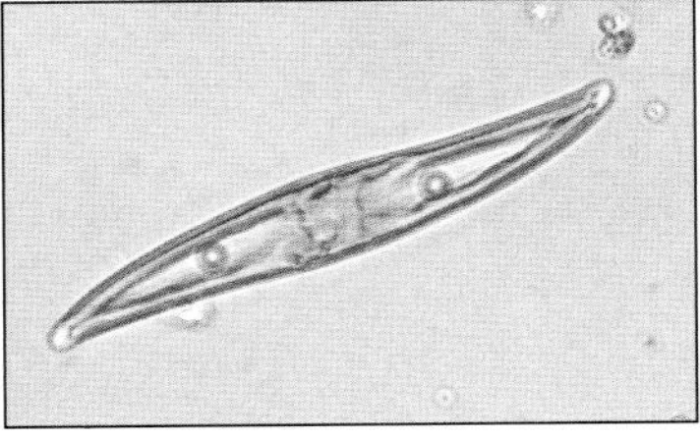

14.9 *Gyrosigma* (Stramenopila; Class Bacillariophyceae) • All species of this genus have sigmoid-shaped cells. Note the golden-brown chromoplasts at the edge and two prominent oil droplets. (Unstained Wet Mount)

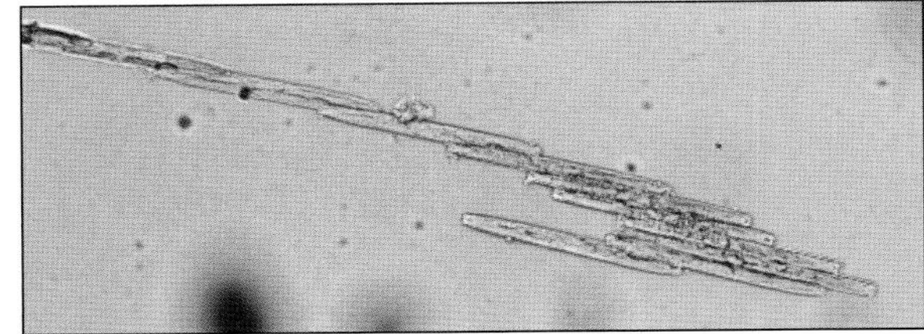

14.10 *Bacillaria* (Stramenopila; Class Bacillariophyceae) • Members of this genus glide with individual cells sliding over one another. At the time this photo was taken, the top cells were in the process of gliding to the right over the lower cells. This specimen was collected from a drainage ditch. (Unstained Wet Mount)

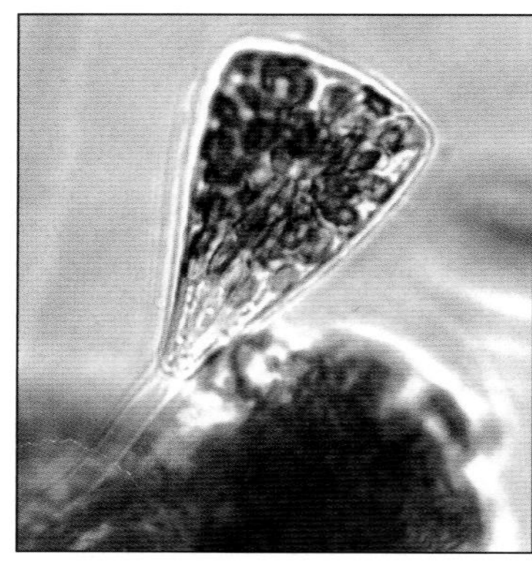

14.11 ***Licomophora* (Stramenopila; Class Fragilariophyceae)** • *Licomophora* is a genus of marine epiphytes. That is, they attach to a surface, frequently another plant, and grow from there without substantially harming the host. You can see the stalk of one, but typically several attach by a common stalk. (Unstained Wet Mount)

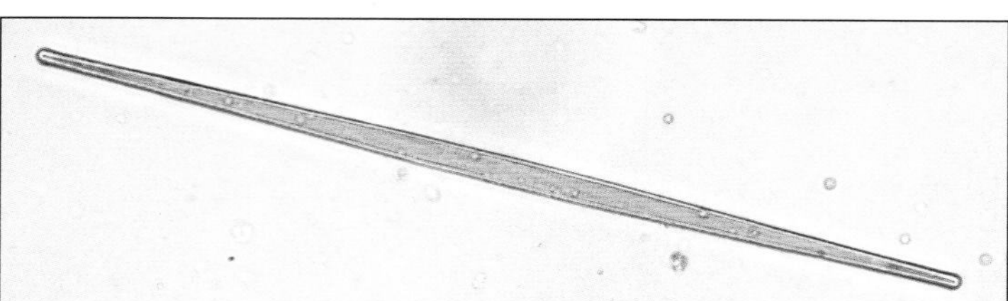

14.12 ***Synedra* (Stramenopila; Class Fragilariophyceae)** • These diatoms are distinctive because of their long, needle-shaped cells. They occur singly or sometimes in groups of cells attached at their ends, radiating outward like spokes of a wheel. (Unstained Wet Mount)

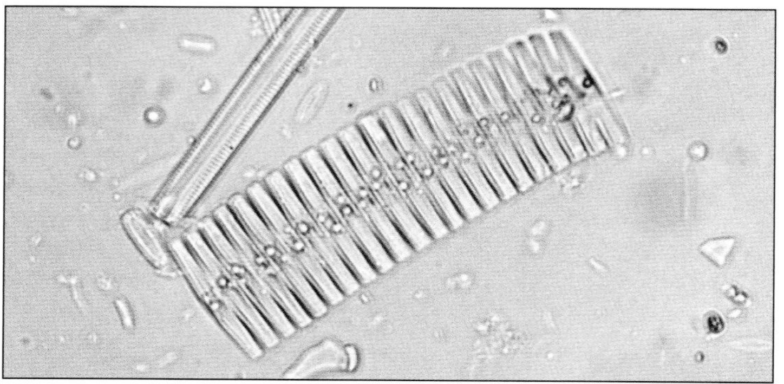

14.13 ***Fragilaria* (Stramenopila; Class Fragilariophyceae)** • *Fragilaria* species are found in marine and freshwater habitats. This specimen was obtained from seawater. *Fragilaria's* rectangular cells form ribbons with cells attached side-by-side. Note the golden-brown chromoplasts and oil droplets. (Unstained Wet Mount)

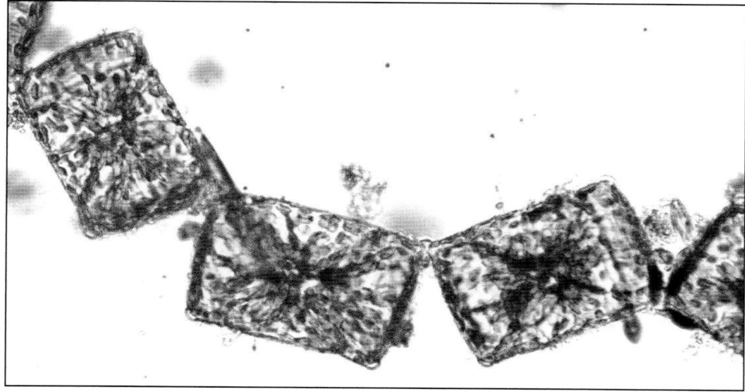

14.14 ***Tabellaria* (Stramenopila; Class Fragilariophyceae)** • *Tabellaria* species form distinctive zigzag colonies. Note the golden-brown chromoplasts. This specimen is a freshwater species collected from a drainage ditch. (Unstained Wet Mount)

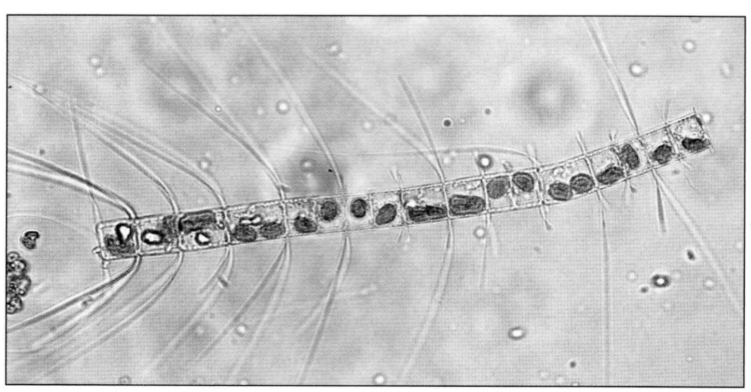

14.15 ***Chaetoceros* (Stramenopila; Class Coscinodiscophyceae)** • Distribution of this large, easily identified marine genus of beautiful diatoms is cosmopolitan. (Unstained Wet Mount)

Subgroup Alveolata

The alveolates possess cytoplasmic membranous sacs (**alveoli**) near the cytoplasmic membrane. It has been speculated that the alveoli are somehow involved in maintaining osmotic balance, but their function is not known for certain. There are three main groups of alveolates: **dinoflagellates**, **apicomplexans**, and **ciliates**.

- Dinoflagellates (Fig. 14.16) are typically unicellular and autotrophic. Most have two flagella: one protruding from the cell and the other positioned in a groove encircling the cell. Most dinoflagellates are **mixotrophic**, though strictly autotrophic and heterotrophic species are known. Some species are responsible for **red tides**, where their pigments color the waters of warm coastal regions. Their excessive growth can be harmful to other species and humans due to oxygen depletion and toxin production.
- Apicomplexans are nonmotile animal parasites and retain a remnant of a plastid. Life cycles are complex, usually involving more than one host. The name "apicomplexa" derives from a complex of organelles at the cell's apex, which are used in penetrating host cells. *Plasmodium* species (Fig. 14.17), the causative agents of malaria, are examples. For more information about *Plasmodium* and malaria, see page 260.
- Ciliates are characterized by their possession of numerous cilia over the cell's entire surface or localized in specific regions. They are marine and freshwater heterotrophs and use the cilia for feeding as well as locomotion. In addition, they have two kinds of nuclei—a **macronucleus** and **micronucleus**. The macronucleus is polyploid and its genes are expressed to carry on cell functions. The micronucleus is responsible for initial formation of the macronucleus and is involved in asexual and sexual reproduction when it undergoes meiosis. (Figs. 14.18–14.20). *Balantidium coli* is a ciliate pathogen. See page 262 for more information.

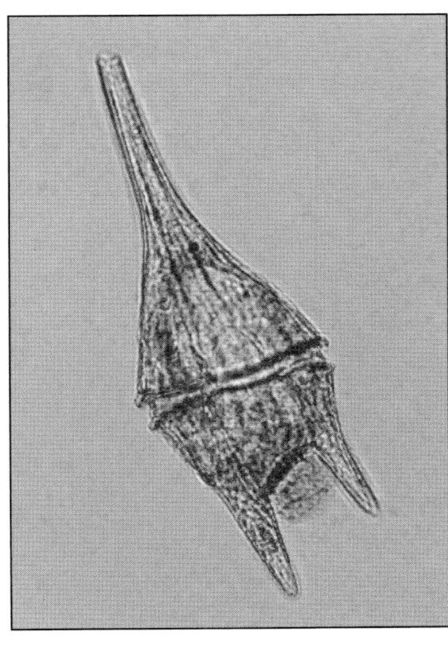

14.16 ***Ceratium*** **(Alveolata; Phylum Dinoflagellata)** • *Ceratium* species are easily identified by the "horns" protruding from the cell wall. Note the groove at the cell's center that encases the circular flagellum. This specimen is a marine species. (Unstained Wet Mount)

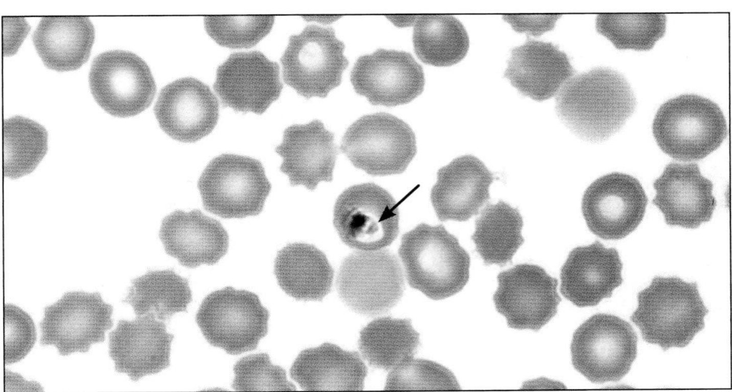

14.17 ***Plasmodium*** **Trophozoite in a Blood Smear (Alveolata; Phylum Apicomplexa)** • A *Plasmodium* **trophozoite** (**arrow**) has infected this red blood cell. Over the course of a *Plasmodium* infection, the parasite assumes many forms. This is the **ring stage**. (Giemsa Stain)

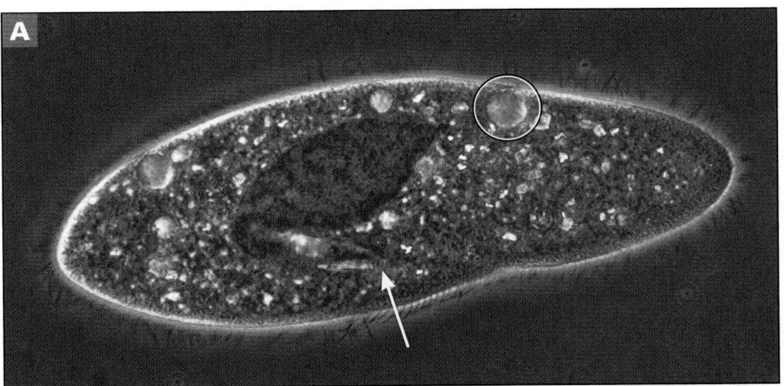

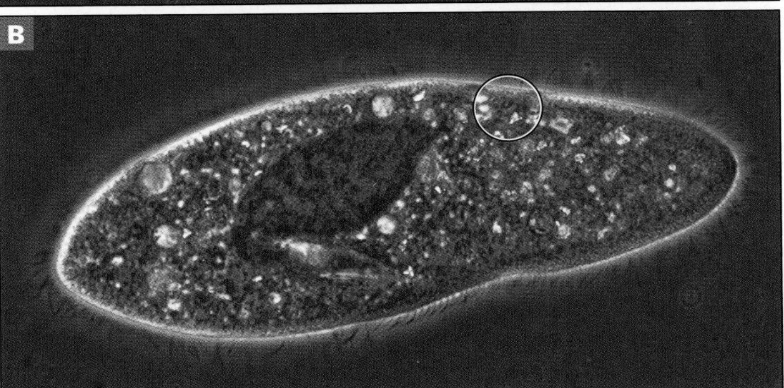

14.18 ***Paramecium*** **(Alveolata; Phylum Ciliophora)** • (**A**) Notice in this phase contrast micrograph the prominent cilia covering the entire cell surface and the oral groove (**arrow**). Also notice the round, light area indicated by the **circle**. This is a contractile vacuole that constantly pumps water out of the cell to keep it from bursting. (**B**) Every few seconds, the contractile vacuole empties into the environment and is no longer a prominent organelle until it fills again. This photo was taken a few seconds after the one in (**A**). (Phase Contrast Wet Mount)

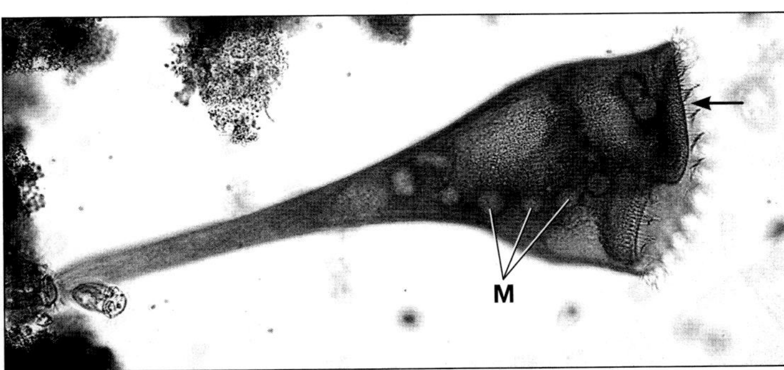

14.19 ***Stentor*** **(Alveolata; Phylum Ciliophora)** • Its trumpet shape, size (up to 2 mm), and beaded macronucleus (**M**) make *Stentor* an easy ciliate to identify. It is covered with cilia, some of which surround its "mouth" (**arrow**) and are used in feeding. This species is naturally green. (Unstained Wet Mount)

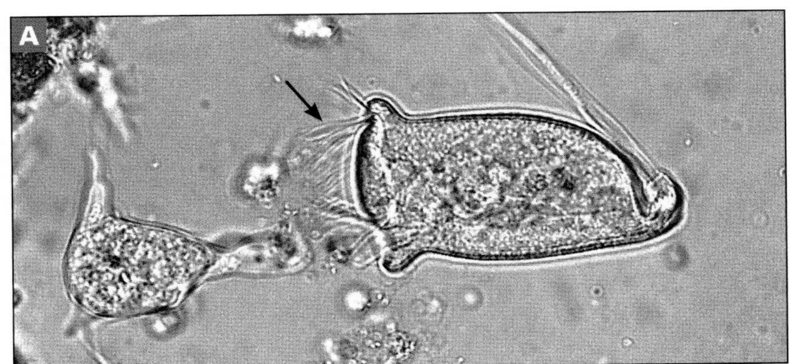

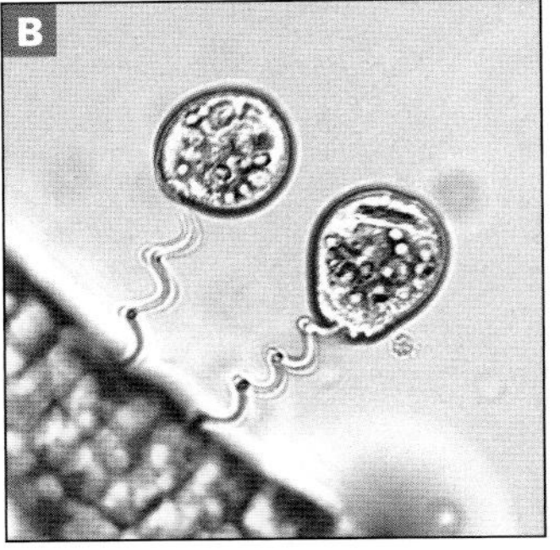

14.20 ***Vorticella*** **(Alveolata; Phylum Ciliophora)** • (**A**) *Vorticella* is a genus of stalked, inverted bell-shaped ciliates, with marine and freshwater representatives. A crown of cilia (**arrow**) surrounds the oral groove and is involved in feeding. This individual seems interested in the amoeba to its left. (**B**) *Vorticella* is capable of retraction when the stalk coils. If you find a *Vorticella*, spend some time watching it until it recoils. It's worth the wait (which shouldn't be more than a minute). (Unstained Wet Mounts)

Subgroup Rhizaria

Rhizarians are amoeboid unicells. Unlike the amoebas described in Amoebozoa (p. 238), whose **pseudopods** (cellular extensions used in moving and feeding) are blunt or lobate, rhizarians have long, thin pseudopods (*pseudo* means "false" and *podos* means "foot"). Most rhizarians are covered with an inorganic shell called a **test**. There are three main rhizarian groups: **foraminiferans, radiolarians**, and **chlorarachniophyta**.

- Foraminiferans have threadlike pseudopods extending from their multichambered, calcium carbonate or chitinous test. All foraminifera are marine and mostly occur in coastal waters. Though heterotrophy is the rule, some forams host endosymbiotic algae and are thus able to live autotrophically. A test from a deceased foraminiferan is shown in Figure 14.21.
- Radiolarians are marine heterotrophs with silica cell walls. Their tests are radially symmetrical with spines (Fig. 14.22).

Subgroup Hacrobia

It is unclear where the haptophytes and cryptomonads belong among or within the eukaryotic supergroups, but current evidence suggests that they are a sister group to SAR. The name Hacrobia has been assigned to that sister group.

Haptophytes are unicellular algae with two smooth flagella. They are distinguished by the presence of a **haptonema** extending from the cytoplasm between the two flagella. It is composed of layers of membranes surrounding seven microtubules, and so is fundamentally different in structure from flagella, which have the 9 + 2 arrangement of microtubules within (nine pairs surrounding two single microtubules).

The haptonema is a feeding structure that gathers food particles and delivers them to a food vacuole at the posterior of the cell. Many haptophytes have calcified scales called **coccoliths**. Figure 14.23 shows the haptophyte *Coccolithophora*.

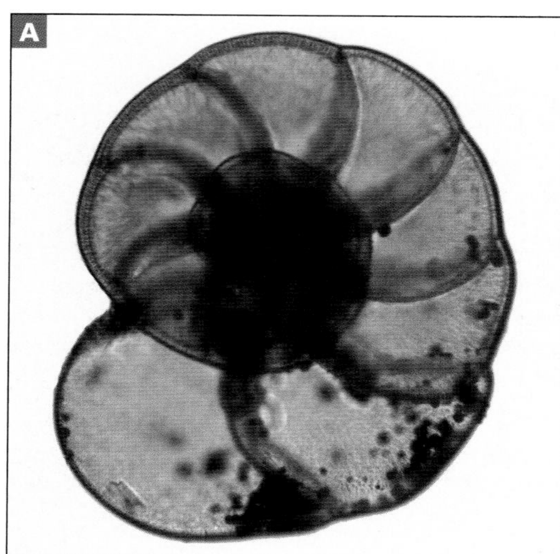

14.21 ***Elphidium*** **Test (Rhizaria; Phylum Foraminifera)** • Shown is a test from one species of the common genus *Elphidium*. (**A**) The entire *Elphidium* test. (**B**) A detail of the same test showing pores through which pseudopodia extend in life. (Unstained Whole Mounts)

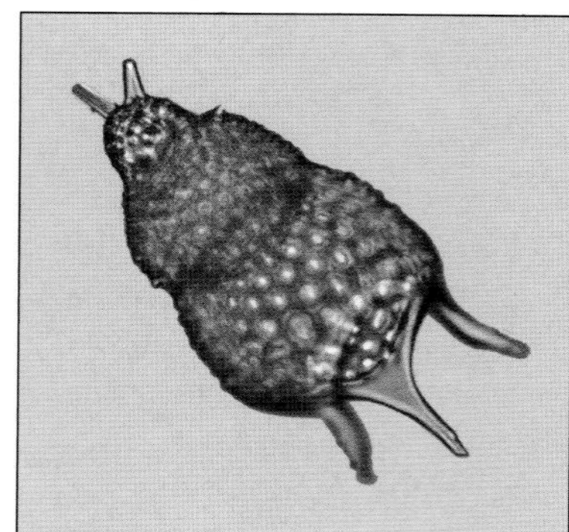

14.22 Radiolarian Test (Rhizaria; Phylum Radiolaria) • This is the radially symmetrical test of an unidentified radiolarian. Like diatoms, radiolarian cell walls are made of silica. (Unstained Whole Mount)

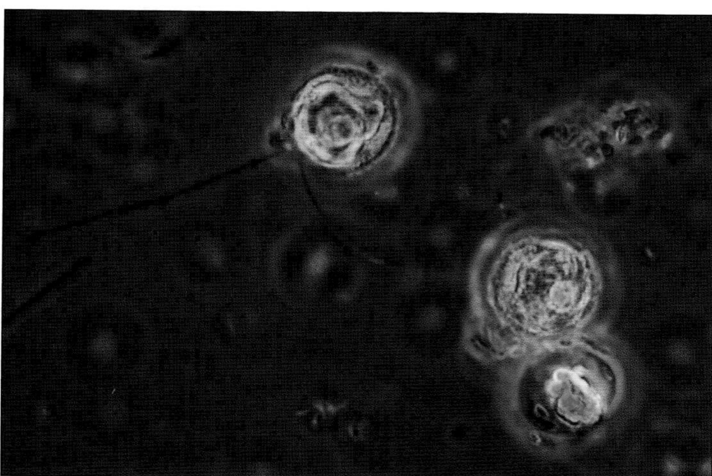

14.23 ***Coccolithophora*** **Grown in Culture (Haptophyta; Class Prymnesiophyceae)** • *Coccolithophora* is a member of the eukaryotic subgroup Hacrobia. Note the two discoid golden chromoplasts, the two smooth flagella, and the small scales on the cell's surface. (Phase Contrast Wet Mount)

Group Archaeplastida

There is strong evidence that the Archaeplastida are descended from a primary endosymbiotic event in which an ancestral cell engulfed a cyanobacterial cell, which eventually evolved into the chloroplasts of **red algae** (division[1] **Rhodophyta**), **green algae** (divisions **Chlorophyta** and **Charophyta**), and land plants.

- Red algae are mostly marine and have complex life cycles. They possess the photosynthetic pigments chlorophylls *a* and *d*, as well as the accessory pigment **phycoerythrin**, which makes them red. All red algae lack flagella. Most rhodophytes are macroalgae and are not in the domain of microbiology, but *Porphyridium* (Fig. 14.24) and *Cyanidioschyzon* (Fig. 14.25) are among the few exceptions. However, the macroalgae *Gelidium* (Fig. 14.26) and *Gracilaria* do contribute to microbiology because they are the primary sources of agar used in making microbiological media.
- Chlorophytes grow as single cells, colonies, and branched or unbranched filaments (Figs. 14.27–14.36). All possess chlorophyll *a* and *b* in chloroplasts of various shapes, which are responsible for their green color. Many also have **pyrenoids** in the chloroplasts where carbon-fixation occurs. If present, flagella are paired and of equal length. Cell walls typically are made of **cellulose.** The majority of chlorophytes live in freshwater, but there are marine and terrestrial species as well. Chlorophyte life cycles routinely involve diploid and haploid stages, often with the zygote being the only diploid cell. Single-celled species, such as *Chlamydomonas* (Fig. 14.27), reproduce sexually and asexually. Haploid cells of opposite mating types (designated " + " and "–") fuse and form a diploid zygote, which undergoes meiosis to produce 4 or 8 vegetative haploid cells. Asexual reproduction occurs when haploid vegetative cells undergo mitosis to produce

[1] "Division" is equivalent to "phylum." It was originally used by botanists when only plant and animal kingdoms were recognized. That terminology has been retained by many taxonomists for groups that once were in the plant kingdom.

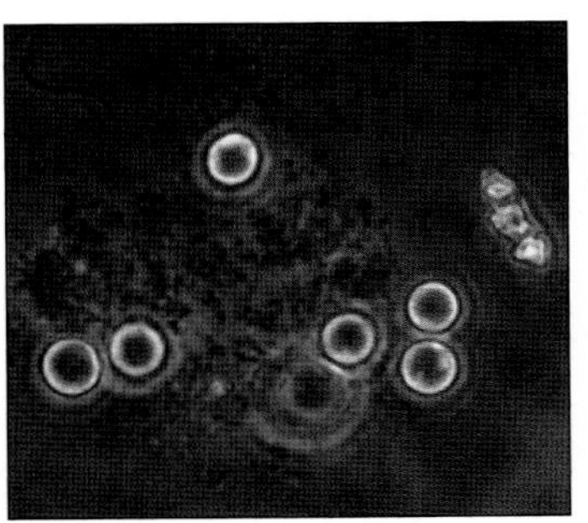

14.24 ***Porphyridium*** **(Archaeplastida; Phylum Rhodophyta)** • Cells of the unicellular red alga *Porphyridium* measure approximately 5 μm in diameter. They are motile and exhibit positive **phototaxis** by secreting mucilage behind themselves that propels them toward the light. (Phase Contrast Wet Mount)

14.25 ***Cyanidioschyzon*** **(Archaeplastida; Division Rhodophyta)** • *Cyanidioschyzon* is a unicellular red alga that is adapted to low pH (1 to 4) and high temperature (40°C to 55°C). Though it is a rhodophyte, it appears green in this runoff channel from Pinwheel Geyser at Yellowstone National Park.

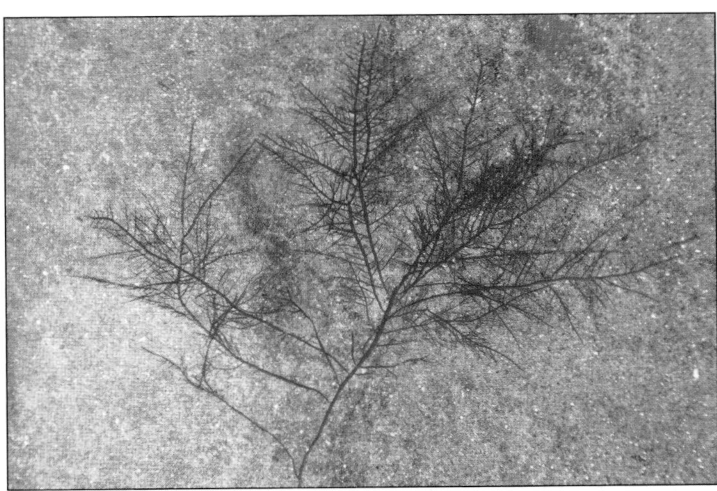

14.26 ***Gelidium*** **(Archaeplastida; Division Rhodophyta)** • Species of the red macroalgae *Gelidium* and others are harvested for the agar they produce in their cell walls. In addition to being used in microbiological media, agar has many other scientific and culinary applications. This specimen was found in the drift on a San Diego beach.

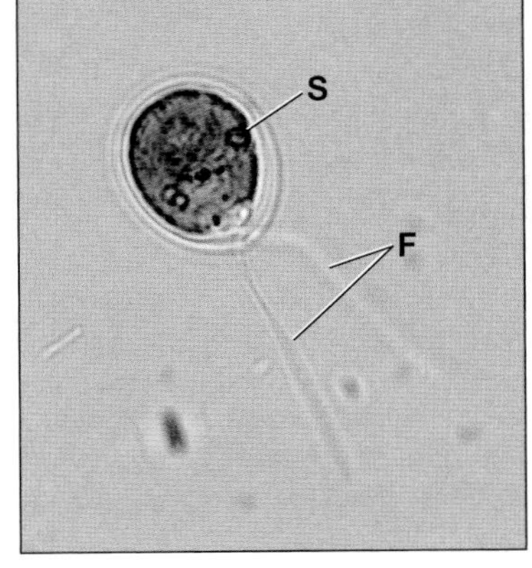

14.27 ***Chlamydomonas*** **(Archaeplastida; Division Chlorophyta)** • This unicellular green alga has two flagella (**F**) and a single, cup-shaped chloroplast. A red-orange stigma (**S**) is found within the chloroplast and is involved in phototaxis. Cells are approximately 30 μm long and 20 μm wide. Gametes and vegetative cells are haploid and structurally indistinguishable. (Unstained Wet Mount)

up to 16 daughter cells within the parent cell's wall, which are then released and develop into vegetative cells. Many filamentous species, such as *Ulothrix* (Fig. 14.35), are composed of haploid cells that have the potential to produce haploid **zoospores** by mitosis that can divide and develop into a new filament. Filament cells also have the potential to produce haploid gametes by mitosis that join to produce a diploid zygote, which undergoes meiosis to produce haploid cells that develop into new filaments. Others, such as *Ulva* (Fig. 14.34), demonstrate **alternation of generations**, with a multicellular diploid **sporophyte** stage that produces haploid spores by meiosis, and a haploid **gametophyte** stage that produces gametes by mitosis. Fertilization produces a diploid sporophyte zygote that develops into the multicellular sporophyte.

- Charophytes (Figs. 14.37–14.40) comprise a second, diverse group that also goes by the common name "green algae." Their distinctions are complex and often not visible to the casual light microscopist. Because of shared characters in flagellar structure, cellulose synthesis, mitosis, and certain enzymes, it is thought charophytes and green plants are derived from a common ancestor that diverged from the chlorophytes. Members of the charophyte class Zygnematophyceae undergo alternation of generations, in which the zygote is the only diploid stage. When haploid cells of opposite mating types ("male" and "female" or " + " and "–") contact, a conjugation tube forms and the cells fuse, as do the nuclei (Fig. 14.39B). The diploid zygote undergoes meiosis to produce four haploid nuclei, three of which disintegrate. This leaves one haploid cell that undergoes cell division to produce a colony of cells characteristic of its species (a filament, semi-cells, etc.).

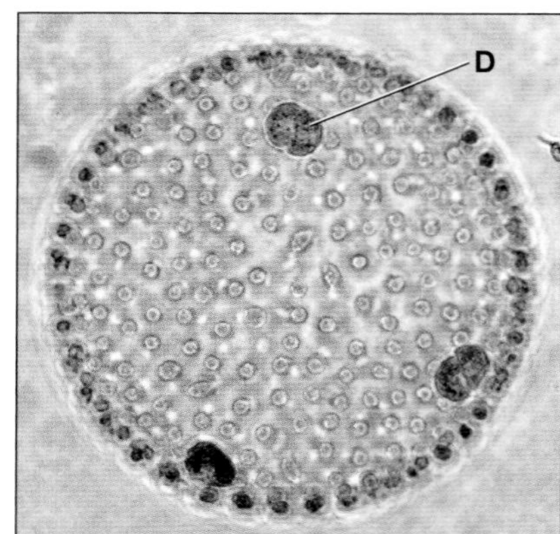

14.29 ***Volvox*** **(Archaeplastida; Division Chlorophyta)** ● *Volvox* is a colonial green alga made of cells similar in shape to *Chlamydomonas*. Daughter colonies (**D**) form asexually from special cells in the parent colony. Release of the daughter colony involves its eversion as it exits through an enzymatically produced pore. Mature colonies can reach a size visible to the naked eye. (Unstained Wet Mount)

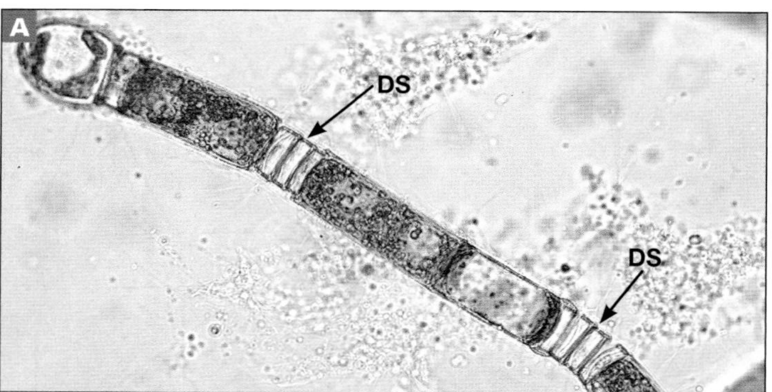

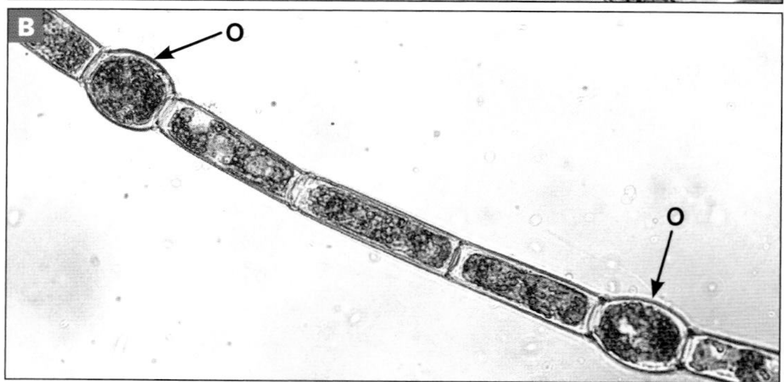

14.30 ***Oedogonium*** **(Archaeplastida; Division Chlorophyta)** ● (A) *Oedogonium* is a large chlorophyte genus composed of freshwater species with unbranched filaments. Division of cells within the filament results in its elongation and produces distinctive division scars (**DS**). (B) The cylindrical cells of this filament are interrupted by two eggs, also called "oogonia" (**O**). (Unstained Wet Mounts)

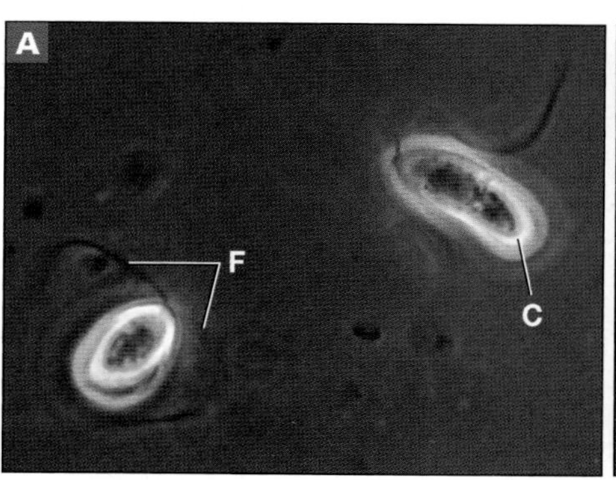

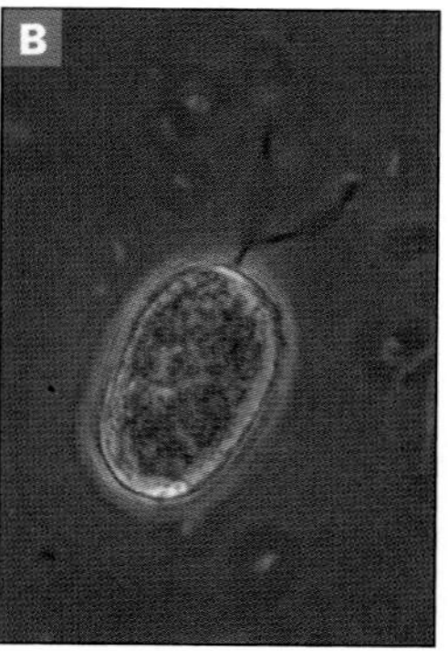

14.28 ***Dunaliella salina*** **(Archaeplastida; Division Chlorophyta)** ● (A) *Dunaliella salina* is a halophilic relative of *Chlamydomonas*. These specimens were taken from the north arm of the Great Salt Lake, where the salinity remains a fairly constant 25% (seawater is about 3.5%). Adaptations to this high salinity include the absence of a cell wall and a high concentration of glycerol, which acts as a compatible solute and maintains osmotic balance. Note the two flagella (**F**) and the single cup-shaped chloroplast (**C**). (B) *D. salina* also has the ability to produce large amounts of the accessory photosynthetic pigment and antioxidant β-carotene, which is responsible for this individual's orange color. It is one of several species used to produce β-carotene commercially. (Phase Contrast Wet Mounts)

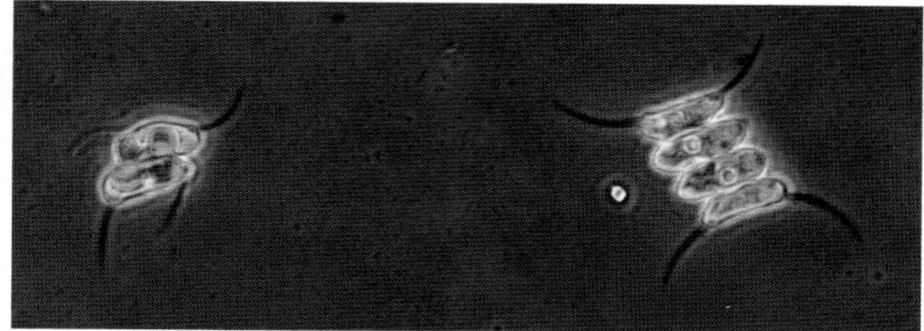

14.31 ***Scenedesmus*** **(Archaeplastida; Division Chlorophyta)** ● This common chlorophyte consists of 4 or 8 cells joined along their edges. The cells on each end have distinctive spines. Asexual reproduction occurs as each cell of the colony divides to produce a new colony within the confines of its cell wall. These are subsequently released as the parental cell wall breaks down. Cells range from 30 µm to 40 µm across. (Phase Contrast Wet Mount)

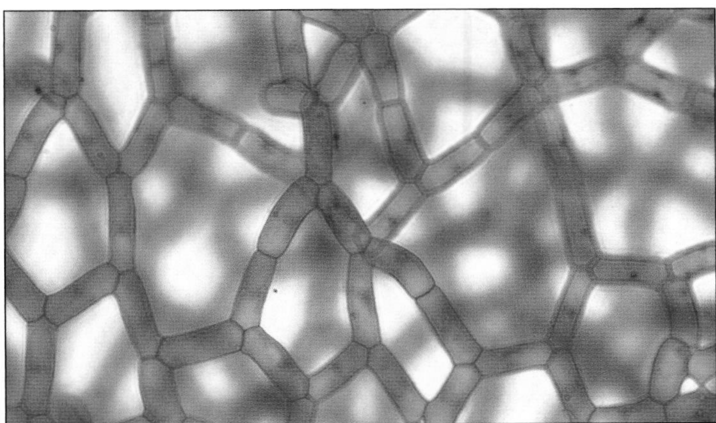

14.32 ***Hydrodictyon*** **(Archaeplastida; Division Chlorophyta)** • The cells of this green alga form open connections with four (sometimes more) of its neighbors to produce a complex network of multinucleate cells. This explains its common name: the "water net." Cells may reach 1 cm in length and each contains a single chloroplast with numerous pyrenoids. This specimen is from a culture, but it is naturally found in freshwater habitats. (Whole Mount)

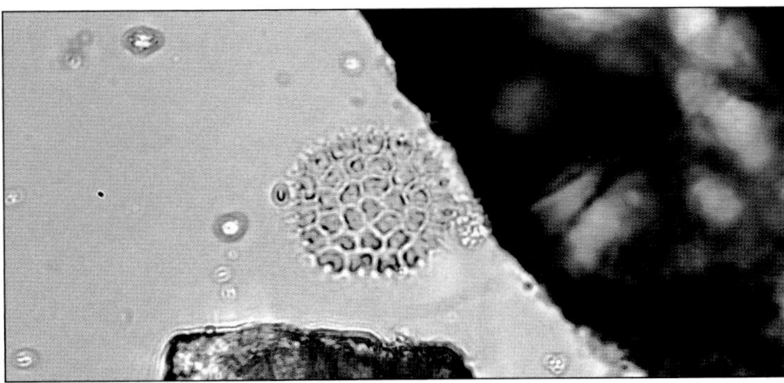

14.33 ***Pediastrum*** **(Archaeplastida; Division Chlorophyta)** • *Pediastrum* is a colonial chlorophyte of freshwater. Its cells are angular and the surface cells often contain bristles that are thought to be used for buoyancy. The cell number and arrangement are constant for each species. Each cell has the potential to produce a colony of the same number and arrangement of cells as the parent colony. (Unstained Wet Mount)

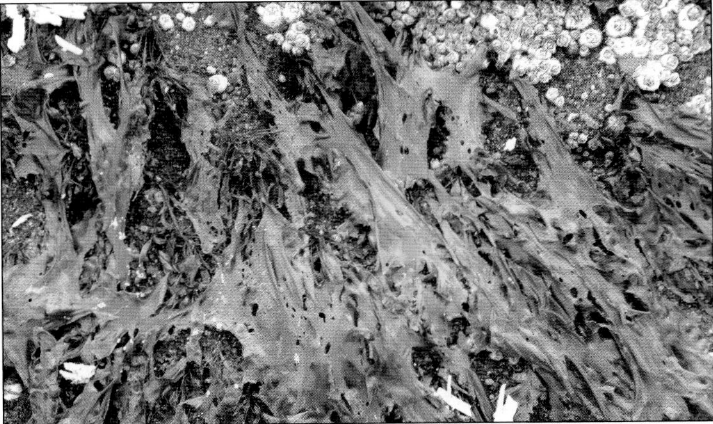

14.34 ***Ulva*** **(Archaeplastida; Division Chlorophyta)** • There are roughly 100 freshwater and marine *Ulva* species. The thallus is sheet-like (giving *Ulva* its common name, "sea lettuce") and only two cells in thickness. *Ulva* species undergo alternation of diploid sporophyte and haploid gametophyte generations that are identical in appearance (**isomorphic**). Shown is *Ulva fenestrata*, which has holes in its thallus (*fenestra* means "window" or "hole"). This specimen was found growing in an Oregon rocky intertidal habitat. The white objects are barnacles.

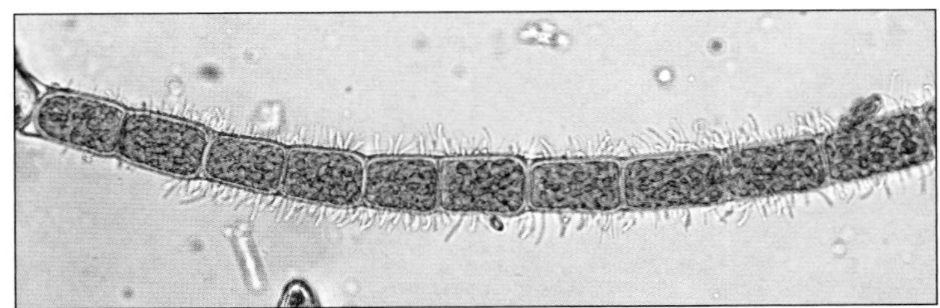

14.35 ***Ulothrix*** **(Archaeplastida; Division Chlorophyta)** • *Ulothrix* is an unbranched, filamentous chlorophyte with both marine and freshwater species. Its chloroplasts are found near the cell wall and form either a complete or incomplete ring around the cytoplasm. The "hairs" on this specimen are epiphytic cyanobacteria. (Unstained Wet Mount)

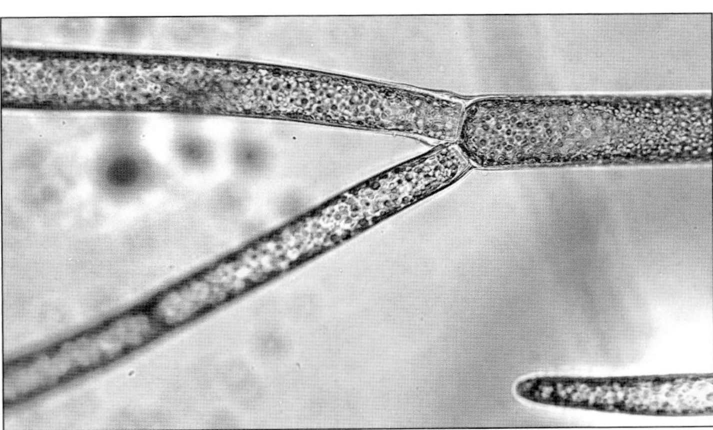

14.36 ***Cladophora*** **(Archaeplastida; Division Chlorophyta)** • If the specimen is green and made of branched filaments with elongated cells, it is most likely a species of *Cladophora*. Most species are freshwater, but this was found in a marine rocky intertidal habitat. The cells are multinucleate and the single chloroplast has numerous pyrenoids that may look like a network of unconnected pieces. This high magnification shows a single branch. Like *Ulva*, *Cladophora* demonstrates isomorphic alternation of generations. (Unstained Wet Mount)

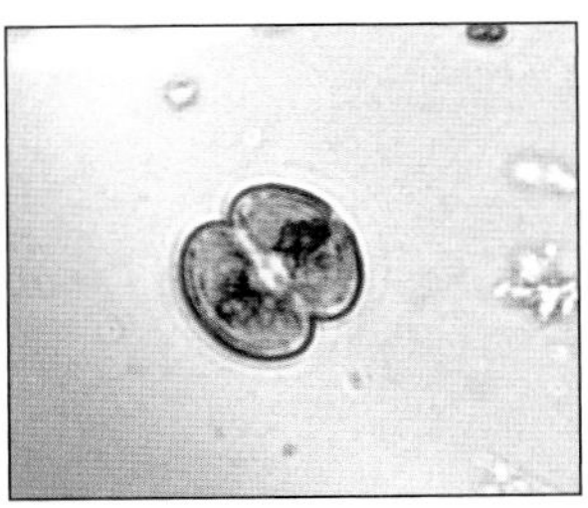

14.37 ***Cosmarium*** **(Archaeplastida; Division Charophyta)** • *Cosmarium* is a freshwater desmid genus with more than 1,000 species. Desmids are characterized by paired haploid "semicells" that can separate, with each forming a new mirror-image semicell. In this species of *Cosmarium* two chloroplasts—one at each end—are present in each semicell. (Unstained Wet Mount)

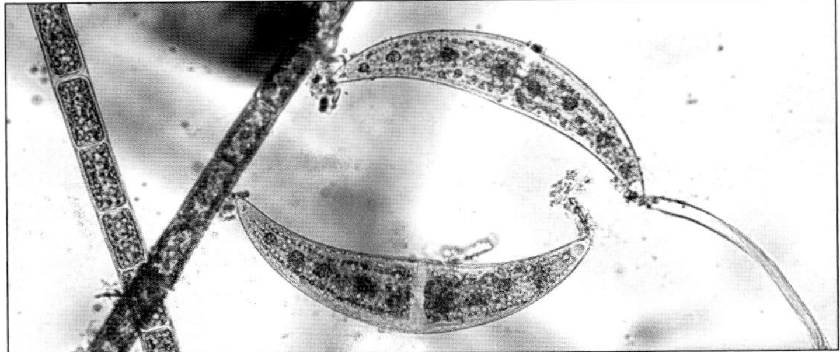

14.38 ***Closterium*** **(Archaeplastida; Division Charophyta)** • Seen here are two *Closterium* individuals from a freshwater sample. *Closterium* is another desmid composed of two semicells, but lacks the distinct constriction typical of most desmids. Vacuoles are present at the outer ends of the cells. Also visible are *Ulothrix* filaments. (Unstained Wet Mount)

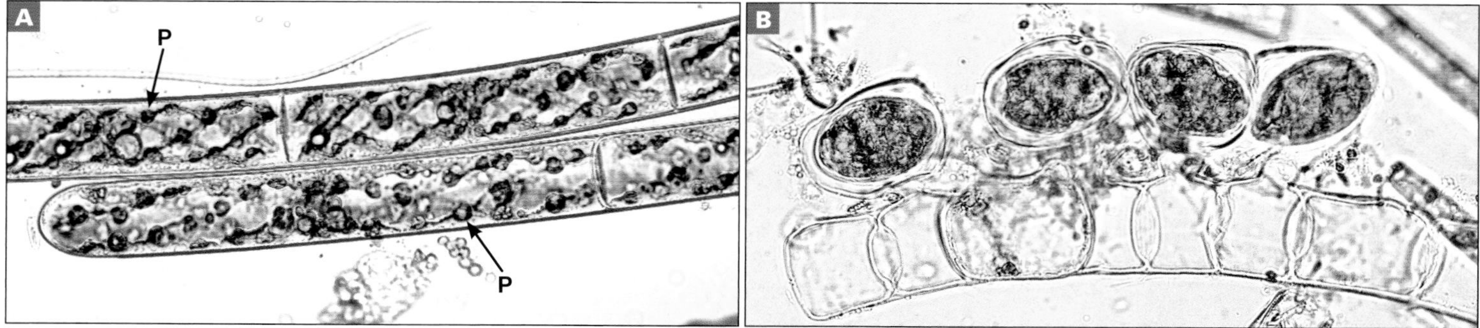

14.39 ***Spirogyra*** **(Archaeplastida; Division Charophyta)** ● (A) *Spirogyra* is a well-known filamentous charophyte because of its distinctive spiral chloroplasts with numerous pyrenoids (**P**). The haploid nucleus is located in the center of the cell and is obscured by the chloroplast. (B) Sexual reproduction occurs as two parallel filaments form conjugation tubes, through which one cell (a "male" gamete) moves to fuse with another cell ("female" gamete) of the other filament, resulting in a diploid zygote. This type of conjugation has happened in four cells in this micrograph. Each zygote undergoes meiosis, with each resulting cell capable of developing into a new haploid filament. (Unstained Wet Mounts)

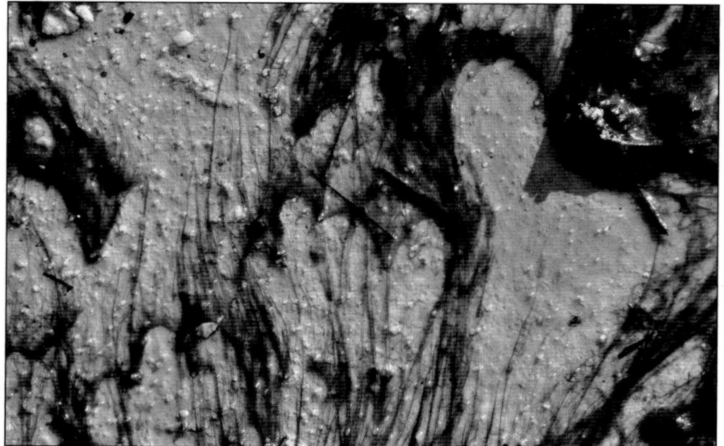

14.40 ***Zygogonium*** **(Archaeplastida; Division Charophyta)** ● This *Zygogonium* species grows as unbranched filaments of cells and is adapted to hot (32°C to 55°C) and acidic (pH 1 to 5) environments. It produces a dark purple, nonphotosynthetic pigment stored in vacuoles that protects against excessive ultraviolet exposure and the toxic effects of excessive iron, which is most effective when the mat is dense (**upper right**). Where growth is not so dense, green strands are visible. This photo was taken at Porcelain Basin in Yellowstone National Park.

Group Unikonta

The group Unikonta is composed of two subgroups: Amoebozoa, which includes amoebas of various types, and Opisthokonta, which includes fungi and animals.

Subgroup Amoebozoa

Amoebozoans are characterized by producing lobe-shaped pseudopodia (as compared to the threadlike pseudopodia of foraminiferans and radiolarians). Pseudopods perform the dual function of feeding and locomotion. Amoebozoans include **gymnamoebas, entamoebas,** and **slime molds.**

- Gymnamoebas include the "classical" amoebas (Fig. 14.41). Most are predatory heterotrophs, and there are marine, freshwater, and soil representatives. The cytoplasm is differentiated into a central portion (**endoplasm**) and the portion just beneath the cytoplasmic membrane (**ectoplasm**). Microfilaments in the ectoplasm produce **cytoplasmic streaming,** which is responsible for pseudopod formation.
- Entamoebas are parasitic heterotrophs (Fig. 14.42). They live in the guts of vertebrates and invertebrates. Important identification features are the presence and location of condensed peripheral and central chromatin. The latter is called the **karyosome.** For more information on *Entamoeba* as a pathogen, see page 263.
- Slime molds are divided into **cellular slime molds** and **plasmodial slime molds.** At some point in the life cycle, all slime molds produce sporangia that liberate reproductive spores. This feature at one time was thought to indicate relationship with fungi, but this has not been borne out by molecular evidence. The vegetative stage of a plasmodial slime mold is usually a brightly colored **plasmodium** (Fig. 14.43) that engages in feeding by phagocytosis as the pseudopodia push their way through soil and decaying organic material. The unique feature of the plasmodium is that it is a single, giant cell easily visible to the naked eye (reaching sizes of several

centimeters). Within the plasmodium are multiple diploid nuclei. Nutrients and oxygen are distributed by cytoplasmic streaming. When nutrients become limiting or the plasmodium begins to dry out, haploid spores are produced. When conditions become favorable again, spores germinate to produce flagellated swarm cells or amoeboid cells which act as gametes in fertilization to reestablish the diploid state. The zygote then undergoes mitosis without cytokinesis to produce a new plasmodium.

- Cellular slime molds (Fig. 14.44) exist as haploid, single-celled feeding amoebas. When food becomes limiting, the individual amoebas begin to aggregate and form a motile, multicellular **slug**. When migration is complete, the slug becomes vertically oriented to form a stalked fruiting body. Cells of the stalk produce cellulose to support the head, which is formed by cells from the rear of the slug that migrate upward. Head cells differentiate into asexual spores that will germinate into amoebas once conditions are favorable again. A sexual cycle occurs when two amoebas fuse to make a diploid zygote, which then begins to consume other amoebas and then encysts within a cellulose wall. Prior to germination, the diploid nucleus undergoes meiosis to produce haploid nuclei, which are distributed to new amoebas that enter the asexual cycle.

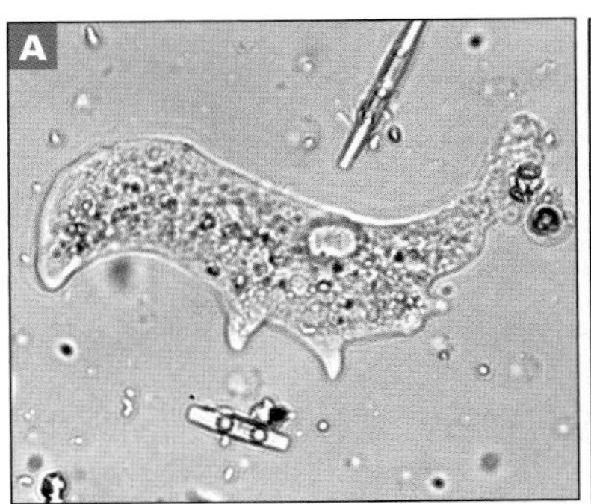

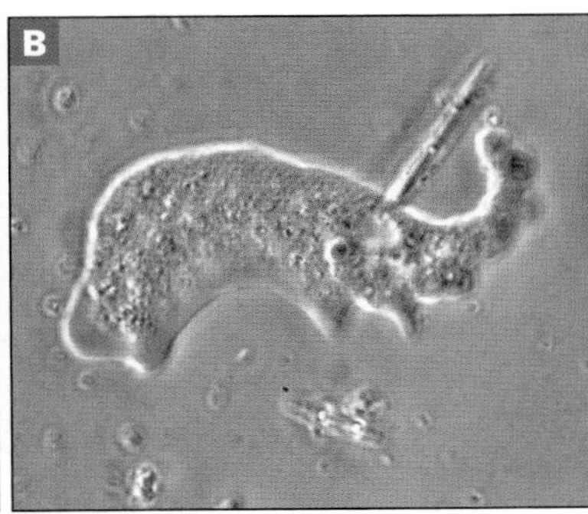

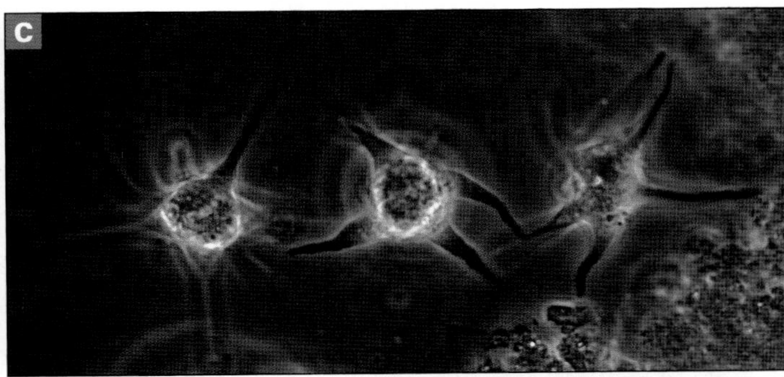

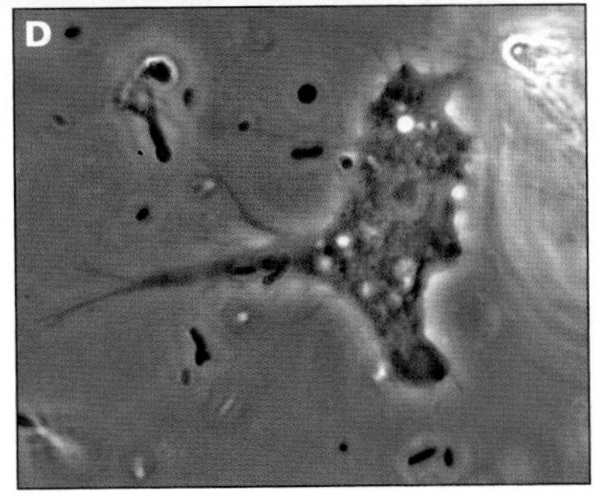

14.41 Gymnamoebas (Phylum Amoebozoa; Subphylum Lobosa) • Cells with flowing cytoplasm and a constantly changing shape are likely to be amoebas. They move and capture prey by extending pseudopods outward. All of these micrographs are unstained wet mounts. (A) This amoeba has multiple lobose pseudopods. In this bright field micrograph, the difference between peripheral cytoplasm (ectoplasm) with that in the interior (endoplasm) is clearly visible. The nucleus is visible, as are numerous food vacuoles (the golden spots—probably an indication of this individual's food preferences.) (Bright Field) (B) This is the same amoeba viewed with phase contrast only one or two seconds after the previous micrograph (notice that the shape is basically the same and the diatom at the top has moved only a short distance). Phase contrast produces a different texture to the image and makes some cell structures more easily seen. Compare the ectoplasm, endoplasm, and nucleus with the bright field image. (C) Astramoebas are unmistakable, with a more-or-less spherical body from which the pseudopods extend. The group shown here goes by the name, "The Three Amoebas." (Phase Contrast) (D) This unidentified amoeba is interesting for its ornate (filose) pseudopods. (Phase Contrast)

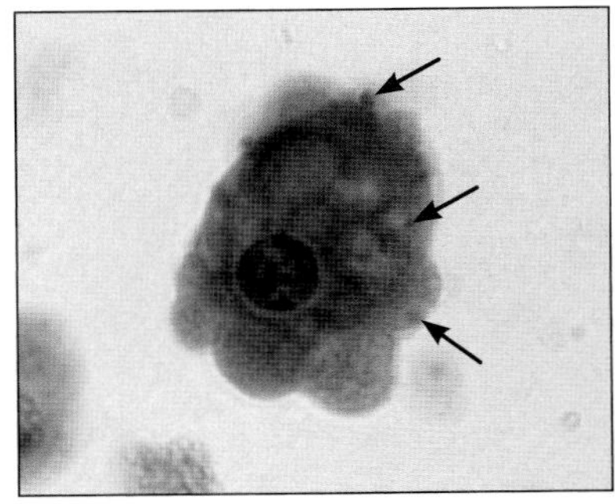

14.42 *Entamoeba* in a Fecal Smear (Phylum Amoebozoa; Subphylum Conosa) • Trophozoites (the active, feeding stage) range in size from 12 µm to 60 µm. In the nucleus, notice the small, dark, central karyosome and the beaded chromatin at the margin. Also notice probable ingested red blood cells (**arrows**) and the finely granular cytoplasm. (Iron Hematoxylin Stain)

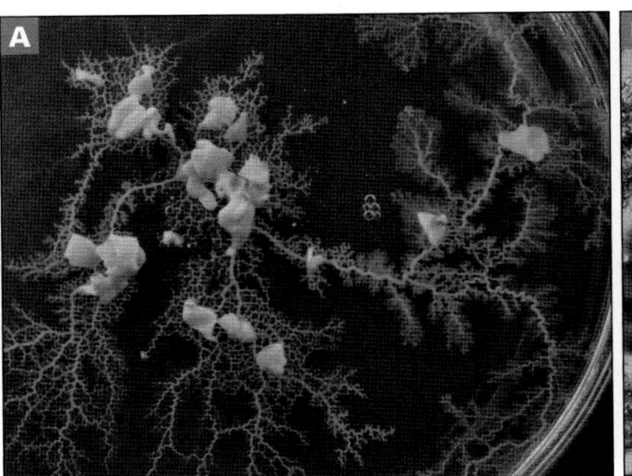

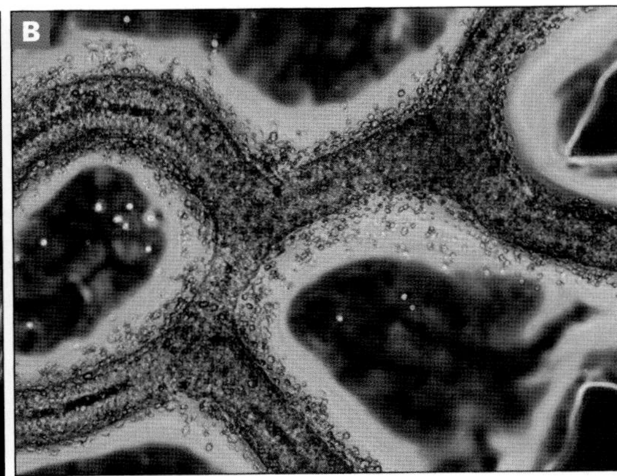

14.43 *Physarium*—Plasmodial Slime Mold (Phylum Amoebozoa; Subphylum Conosa) • (A) Shown is a *Physarium* plasmodium growing on a water agar plate without magnification. The nutrients are supplied by oat flakes (the lumps in the photo) and the plasmodium has spread over the agar surface "in search" of more food. *Physarium* is a typical plasmodial slime mold in that it is brightly colored. (B) This is the same plasmodium viewed under low power. When the photograph was snapped, the central portion of each plasmodial strand was streaming to the left (meaning back toward the older portions). (Unstained Whole Mounts)

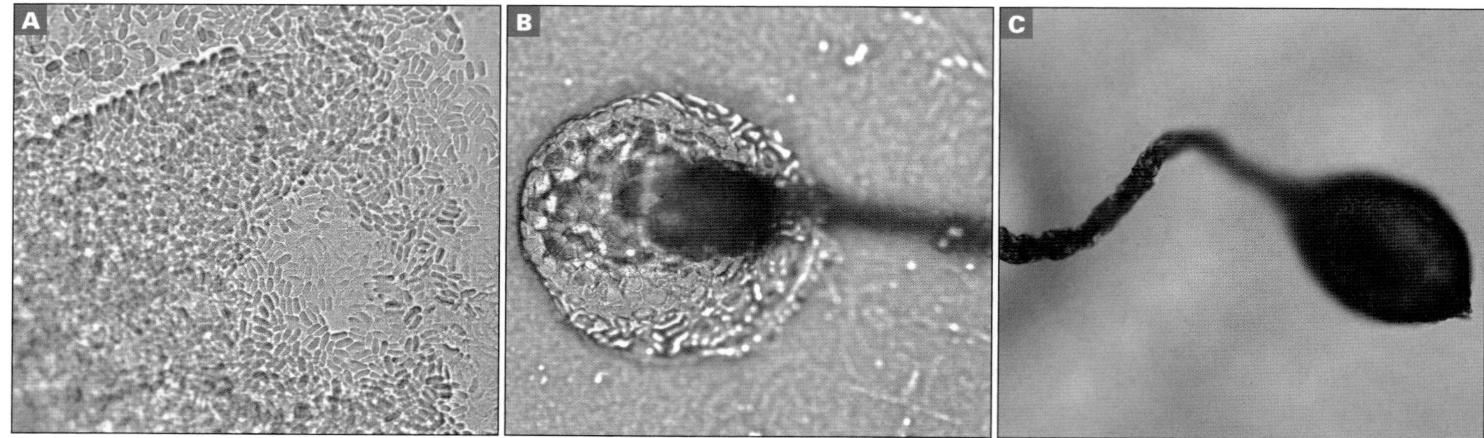

14.44 ***Dictyostelium*—Cellular Slime Mold (Phylum Amoebozoa; Subphylum Conosa)** • (A) Shown is a portion of an aggregating *Dictyostelium* slug. Notice the regular shape of the cells, which is very unlike the irregular shapes of the amoebas when acting as individuals. The blurry part in the lower left is the base of a growing stalk. (B) This is the head of an immature fruiting body that has yet to migrate up the stalk. (C) This mature fruiting body of *Dictyostelium* will eventually release haploid spores that form a new generation of amoeboid cells. (Unstained Whole Mounts)

Subgroup Opisthokonta

Opisthokonts form a second major branch of Unikonta and comprise the fungi, lichens, and the entire animal kingdom. The animal kingdom will not be addressed here.

Fungi are nonmotile eukaryotes and are divided into the divisions Ascomycota, Basidiomycota, Zygomycota, and Blastocladiomycota. They are heterotrophic, but unlike others in this group (that ingest, then digest their food), fungi are **absorptive heterotrophs;** that is, they secrete exoenzymes into the environment, and then absorb the digested nutrients. Their cell wall is usually made of the polysaccharide **chitin,** not cellulose as in plants. Most are **saprophytes** that decompose dead organic matter, but some are **parasites** of plants, animals, or humans.

Fungi are informally divided into unicellular **yeasts** (Fig. 14.45) and filamentous **molds** (Fig. 14.46) based on their overall appearance. **Dimorphic fungi** have both mold and yeast life-cycle stages. Filamentous fungi that produce fleshy reproductive structures—mushrooms, puffballs, and shelf fungi—are referred to as **macrofungi** (Fig. 14.47), even though the majority of the fungus is filamentous and hidden underground or within decaying matter.

Individual fungal filaments are called **hyphae,** and collectively they form a **mycelium.** The hyphae are darkly pigmented in **dematiaceous fungi** and unpigmented in **hyaline** or **moniliaceous fungi** (Fig. 14.48). Hyphae may be **septate,** in which walls separate adjacent cells, or **nonseptate** if walls are absent (Fig. 14.49).

Fungal life cycles usually are complex, involving both sexual and asexual forms of reproduction. Gametes are produced by **gametangia** and spores are produced by a variety of **sporangia.** While gametes and spores are both reproductive cells, they function differently in advancing the life cycle. A spore does so simply by dividing (asexual reproduction), whereas gametes, which are haploid, need to combine by fertilization and produce a zygote with a single diploid nucleus (sexual reproduction).

The familiar situation is that fusion of cytoplasm (**plasmogamy**) and joining of nuclei (**karyogamy**) happen one right after the other. Not so in fungi. Karyogamy is usually delayed following plasmogamy, resulting in a **dikaryotic** cell with two separate haploid nuclei. This cell usually divides repeatedly along with the nuclei to produce a hypha of dikaryotic cells.

At some point karyogamy occurs and a diploid zygote is formed in each cell, which undergoes meiosis to produce haploid sexual spores ("sexual" because they are part of the sexual life cycle). So, unlike animals, the only diploid cell in most fungal life cycles is the zygote, and the spores it produces are characteristic of the specific fungal group.

Many fungi also produce characteristic asexual spores during their life cycle. If they form at the ends of hyphae,

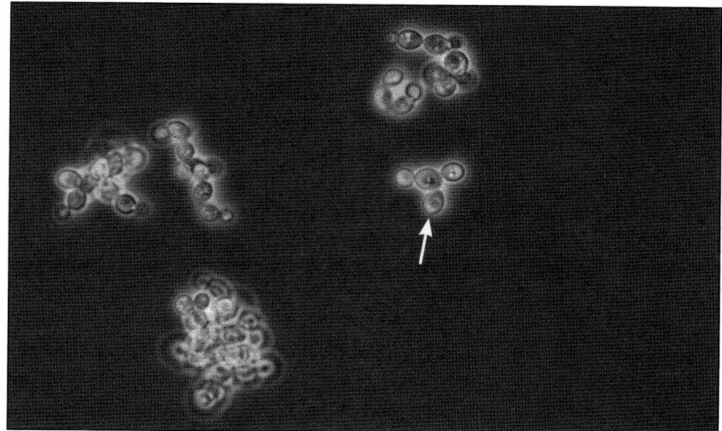

14.45 **Yeast—*Saccharomyces cerevisiae*, the Brewer's Yeast (Opisthokonta; Division Ascomycota)** • Cells of the brewer's yeast *Saccharomyces cerevisiae* are oval with dimensions of 3 μm–8 μm by 5 μm–10 μm. Short chains of cells (pseudohyphae) are visible in this field. (Phase Contrast Wet Mount)

they are called **conidia.** Other asexual spores are **blastospores** (blastoconidia), which are produced by budding, and **arthrospores,** which are produced when a hypha breaks. **Chlamydospores** (**chlamydoconidia**) are formed at the end of some hyphae and are a resting stage.

The fungus kingdom is divided into divisions (phyla) primarily based on the pattern of sexual spore production and the presence of septa in the hyphae. Members of the division **Ascomycota** (formerly class **Ascomycetes**) produce an **ascus** (sac) in which the zygote undergoes meiosis to produce haploid **ascospores** (Fig. 14.50A). Ascomycete hyphae are septate. This group also includes yeasts, which are usually unicellular.

Members of the division **Basidiomycota** (formerly class **Basidiomycetes**) have septate hyphae and during sexual reproduction produce a **basidium** that undergoes meiosis to produce four **basidiospores** attached to its surface (Fig. 14.50B). Both ascomycetes and basidiomycetes produce asexual conidia.

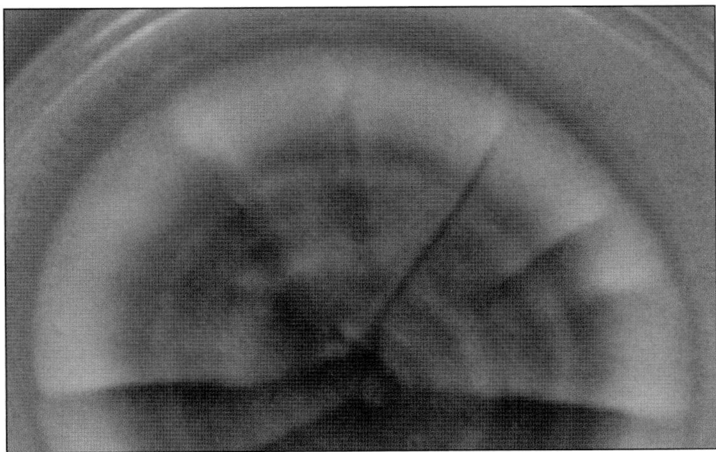

14.46 Mold—*Penicillium italicum* (Opisthokonta; Division Ascomycota) • Molds grow as fuzzy colonies. Their spores are abundant in the environment and frequently show up as contaminants on agar plates. For more information about *Penicillium*, see page 254.

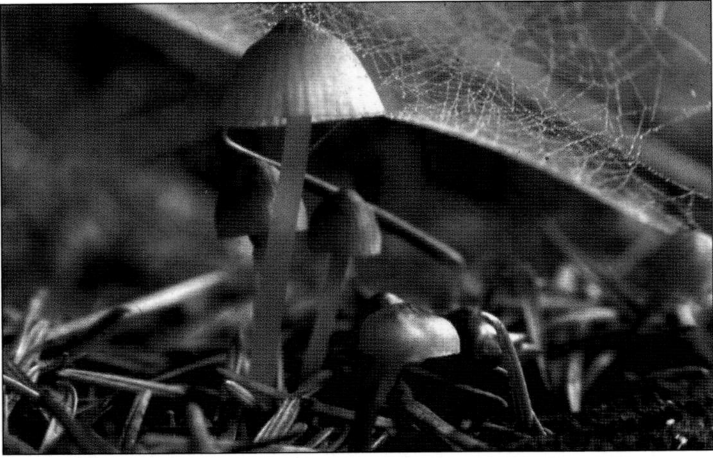

14.47 Macrofungus • This mushroom is the fruiting body from a mycelium that is busily engaged in decomposing the redwood log beneath it.

Members of the division **Zygomycota** are terrestrial and form **zygospores** when the zygote undergoes meiosis. They also produce asexual **sporangiospores.** The role of these spores is described here in the *Rhizopus* life cycle.

There are about a dozen species of the zygomycete *Rhizopus*. They are fast-growing molds that produce white or grayish, cottony growth. The mycelium becomes darker with age as sporangia are produced, giving it a "salt and pepper" appearance (Fig. 14.51).

Microscopically, *Rhizopus* species produce broad (10 μm), hyaline, and usually nonseptate surface and aerial hyphae with irregular diameters. Anchoring **rhizoids** (Fig. 14.52) are produced where the surface hyphae (**stolons**) join the bases of the long, unbranched **sporangiophores.**

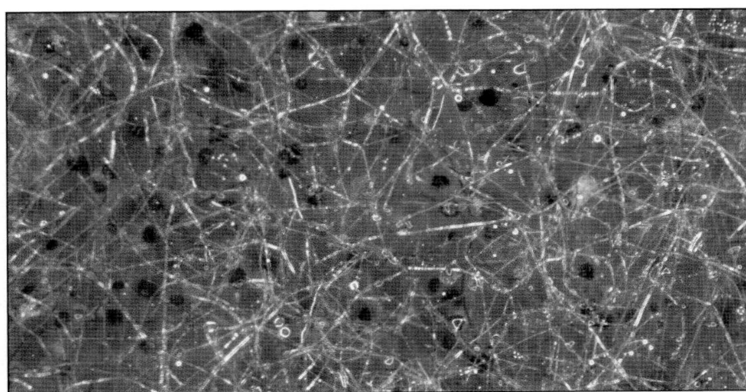

14.48 Hyaline Hyphae • This tangled mass of hyphae belongs to the bread mold, *Rhizopus*. They are hyaline because they lack pigment. The black spots are sporangia.

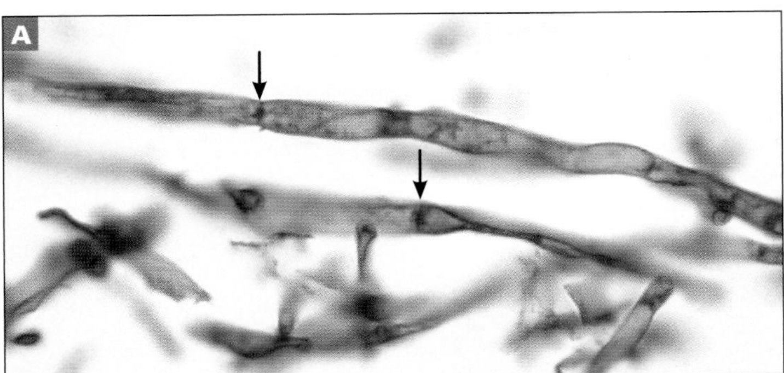

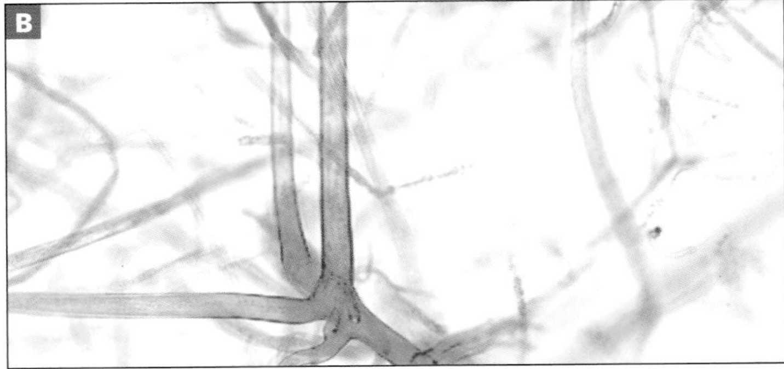

14.49 Septate and Nonseptate Hyphae • (**A**) The hyphae of *Aspergillus* are septate. Note the cross walls (**arrows**) dividing the cells. (**B**) *Rhizopus* provides an example of nonseptate hyphae. (Stained Whole Mounts)

The *Rhizopus* life cycle (Fig. 14.53) has both sexual and asexual phases. Asexual sporangiospores are produced by large, spherical sporangia (Fig. 14.54) borne at the ends of the elevated sporangiophores. A hemispherical **columella** (Fig. 14.55) supports the sporangium. After release, the sporangiospores develop into hyphae identical to those that produced them.

On occasion, sexual reproduction occurs when hyphae of different mating types (designated + and – strains) make contact. Initially, **progametangia** (Fig. 14.56) extend from each hypha and nuclei and cytoplasm flow into them. As a consequence, they enlarge and contact one another. Upon contact, a septum separates the end of each progametangium into a gametangium and a **suspensor** (Fig. 14.57).

The walls between the two gametangia dissolve, and a thick-walled **zygosporangium** develops (Figs. 14.58 and 14.59). Fusion of nuclei (karyogamy) occurs within the zygosporangium, resulting in multiple diploid zygote nuclei. After a dormant period, zygote meiosis occurs and produces haploid zygospores. The zygosporangium germinates and produces a sporangium similar to asexual sporangia. Zygospores are released, develop into new hyphae, and the life cycle is completed.

Rhizopus species are common and are naturally found associated with living and decaying plant material. In the former case they can cause spoilage. *R. oryzae* and *R. microsporus* are the two species most responsible for producing **zygomycosis** (mucormycosis), a condition

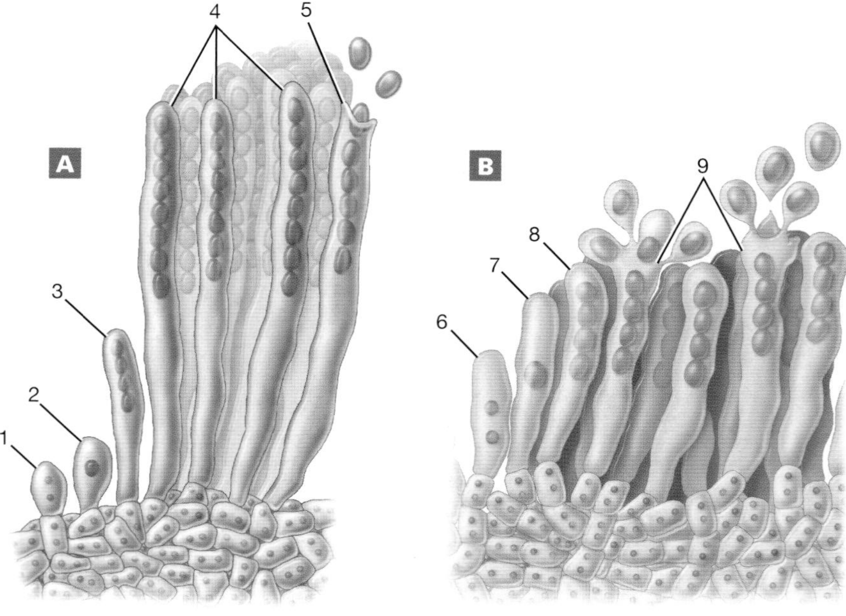

14.50 Sexual Sporangia • Ascomycota and basidiomycota are major fungal divisions distinguished by their sexual sporangia, in which the zygote undergoes meiosis to produce haploid spores. Because crossing over occurs in meiosis, the resulting spores are genetically different from each other, as are the hyphae that develop from them. (A) Shown in this drawing of a magnified ascomycete mycelium are dikaryotic hyphae along the base and asci in various stages of development. Each ascus typically produces four pairs of ascospores, which are released and develop into a new mycelium. Ascomycetes are commonly called sac fungi because their spores develop within the saclike sporangium. Key to Figure Labels: **1** Dikaryotic ascus with two haploid nuclei; **2** Ascus with zygote nucleus following karyogamy; **3** Ascus with four haploid ascospore nuclei following zygotic meiosis; **4** Asci with eight ascospores following mitosis of the four ascospores; **5** Ascus releasing ascospores. (B) Basidiomycetes produce their spores at the ends of elongated basidia that resemble a club, hence their common name, club fungi. This drawing illustrates a portion of the dikaryotic mycelium and several basidia. One could find these along the edges of a mushroom's gills. Each haploid nucleus enters an extension of cytoplasm that buds off and is swept away to develop into a new mycelium. Key to Figure Labels: **6** Dikaryotic basidium; **7** Basidium with zygote nucleus following karyogamy; **8** Basidium with four haploid nuclei following zygotic meiosis; **9** Basidiospores being released.

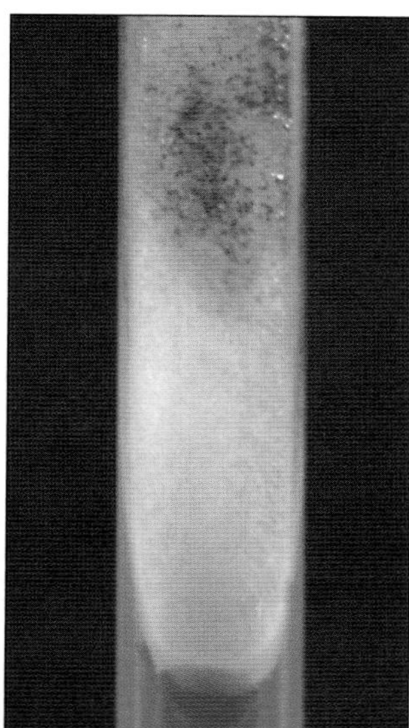

14.51 *Rhizopus* stolonifer Culture (Opisthokonta; Division Zygomycota) • Black asexual sporangia of this bread mold have begun to form, giving the growth a "salt and pepper" appearance. (Sabouraud Agar)

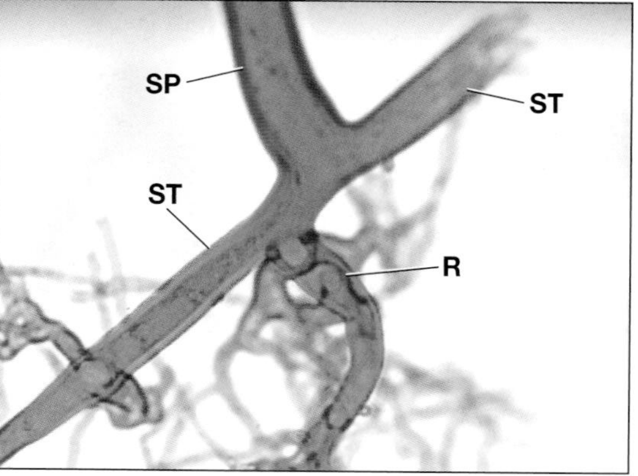

14.52 *Rhizopus* Rhizoids (Opisthokonta; Division Zygomycota) • Anchoring root-like rhizoids (**R**) form at the junction of each sporangiophore (**SP**) and the stolon (**ST**). Note the absence of septa. (Stained Whole Mount)

found most frequently in diabetics, organ transplant recipients, and immunocompromised patients.

Portal of entry and overall health of the patient are critical factors in determining the location and extent of infection. Sites of infection include the nasal cavity, brain, abdominopelvic cavity, skin, and lungs. Entry into the blood leads to rapid spreading of the organism, occlusion of blood vessels, and necrosis of tissues.

Pilolobus is another zygomycete genus, with the unusual ability to aim and shoot its sporangia (Fig. 14.60), which form at the tips of sporangiophores. The **subsporangial vesicle** acts as a lens and orients itself so that the sporangium is aimed at a light source. Using a burst of water from the subsporangial vesicle, spores are released a meter or more toward the light with the "intent" of shooting them toward vegetation that will be ingested by an animal. Spores are distributed in the animal's feces, and the cycle repeats.

Saccharomyces cerevisiae is an ascomycete used in the production of bread, wine, and beer and is an emerging human pathogen. It does not form a mycelium but, rather, produces a colony similar to bacteria (Fig. 14.61). Vegetative **blastospores** are generally oval to round in shape, and asexual reproduction occurs by budding (Fig. 14.62). Short **pseudohyphae** are sometimes produced when the budding cells fail to separate.

In an *S. cerevisiae* population, haploid and diploid cells are indistinguishable from one another and both reproduce asexually by budding. Diploid cells occasionally undergo

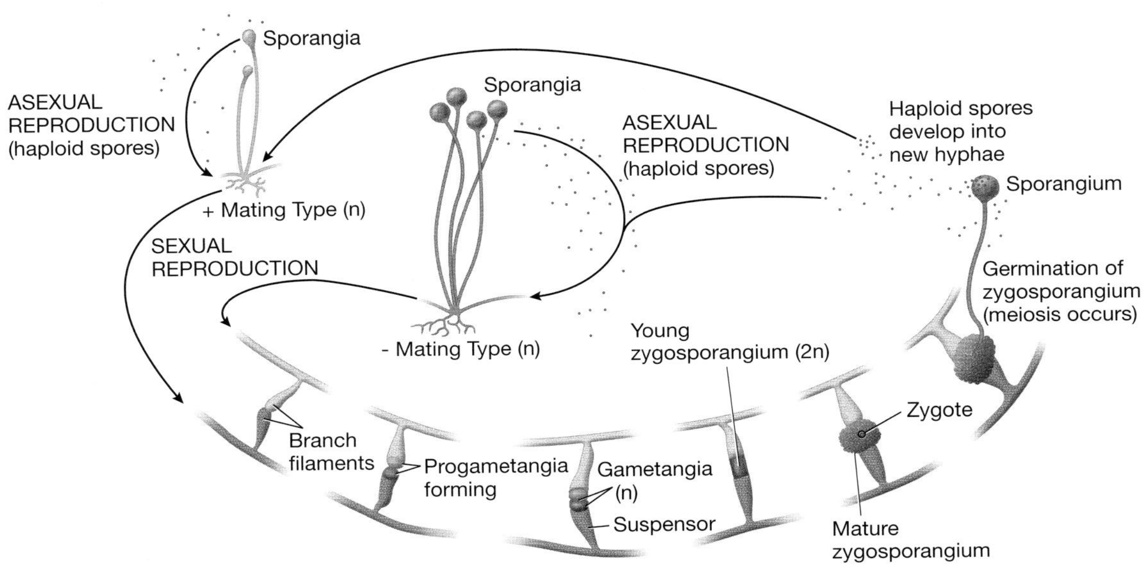

14.53 ***Rhizopus* Life Cycle (Opisthokonta; Division Zygomycota)** • Please refer to the text for details.

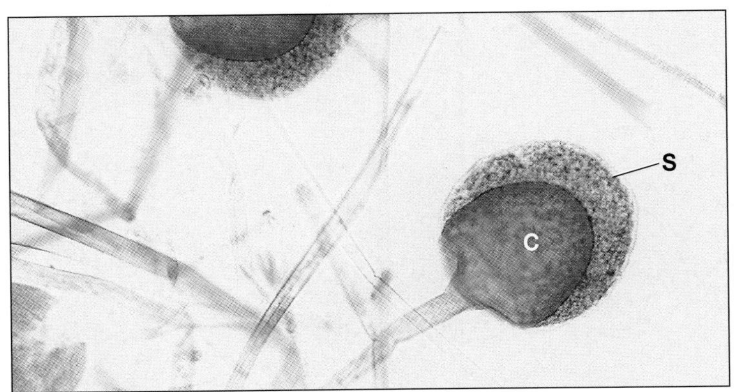

14.54 ***Rhizopus* Sporangiophore (Opisthokonta; Division Zygomycota)** • The sporangium is found at the end of a long, unbranched, and nonseptate sporangiophore. The haploid asexual sporangiospores (**S**) cover the surface of the columella (**C**), which has a flattened base. (Stained Whole Mount)

14.55 ***Rhizopus* Columella (Opisthokonta; Division Zygomycota)** • Note the umbrella-shaped columella that remains after the sporangium has released its spores. (Stained Whole Mount)

meiosis (acting as an ascus) and produce one to four ascospores (Fig. 14.63). When released, ascospores bud to produce more haploid cells. Occasionally, haploid cells of opposite mating types combine to create a new diploid cell, thus acting like gametes.

Microscopic examination, and biochemical and antifungal sensitivity tests are used in identification. Asci stain Gram negative whereas the vegetative cells stain Gram positive. The asci and ascospores stain pink in a Kinyoun acid-fast stain, whereas the vegetative cells stain blue. For information about clinically important ascomycete and basidiomycete pathogens, refer to Section 15.

With fewer than 200 described species, the division **Blastocladiomycota**, represented here by *Allomyces* (Fig. 14.64), is small by comparison to the other fungal divisions. Unlike most fungi, they produce flagellated spores and exhibit alternation of generations in their life cycle.

The diploid **sporothallus** can produce two sporangia types: **mitosporangia** and **meiosporangia.** The former produce motile, diploid **mitospores** by mitosis. These develop and produce more sporothalli. Meiosporangia produce motile, haploid meiospores by meiosis. These develop into haploid **gametothalli,** which produce motile male and female gametes. Upon fertilization, a new diploid sporothallus is produced.

Lichens are a union of two organisms living in a mutualistic symbiosis. One partner, the **mycobiont,** is a fungus. The other partner, the **photobiont,** is either a cyanobacterium or an alga (or sometimes both are present). Most mycobionts are ascomycetes and approximately 20% of fungi worldwide participate in forming lichens. Photobionts are limited to a few dozen genera.

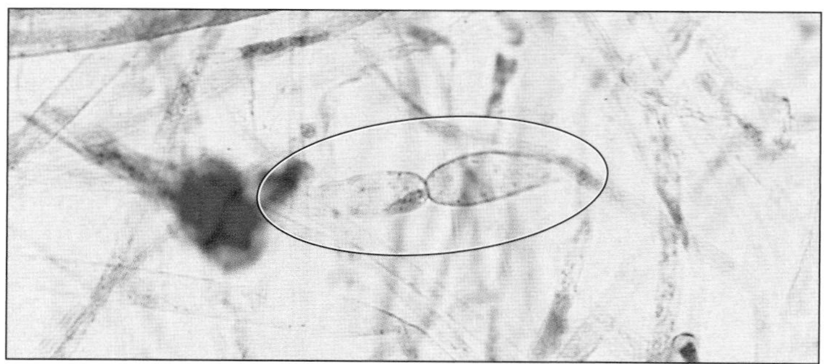

14.56 ***Rhizopus*** **Progametangia (Branch Filaments) (Opisthokonta; Division Zygomycota)** • Progametangia from different hyphae are shown circled in the center of the field. Contact between the progametangia results in each forming a gamete. (Stained Whole Mount)

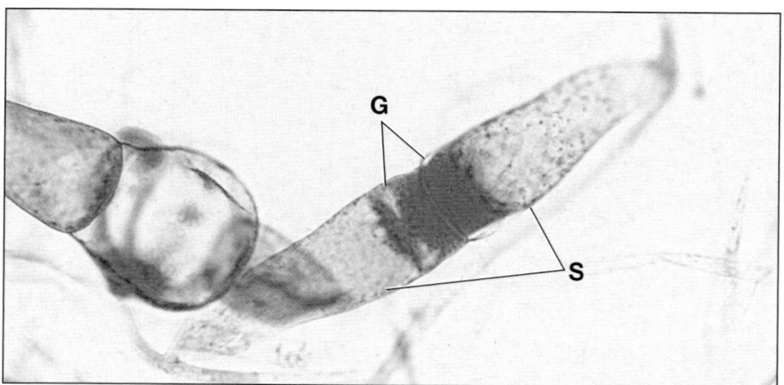

14.57 ***Rhizopus*** **Gametangia and Suspensors (Opisthokonta; Division Zygomycota)** • Gametangia (**G**) and suspensors (**S**) are shown in the center of the field. Gametangia contain haploid nuclei from each mating type. (Stained Whole Mount)

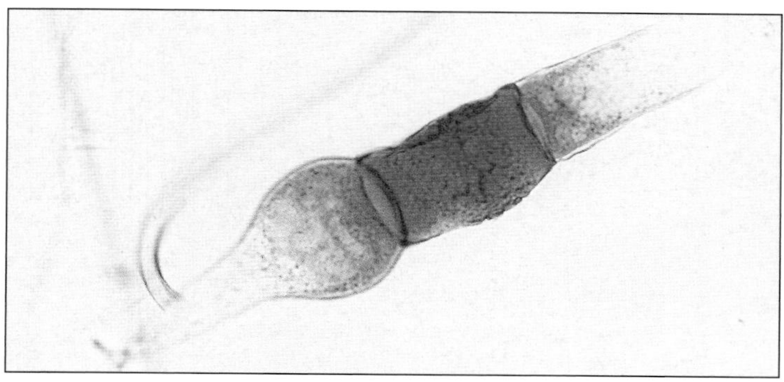

14.58 Young ***Rhizopus*** **Zygospore (Opisthokonta; Division Zygomycota)** • The zygospore forms when the cytoplasm from the two mating strains fuse (**plasmogamy**). (Stained Whole Mount)

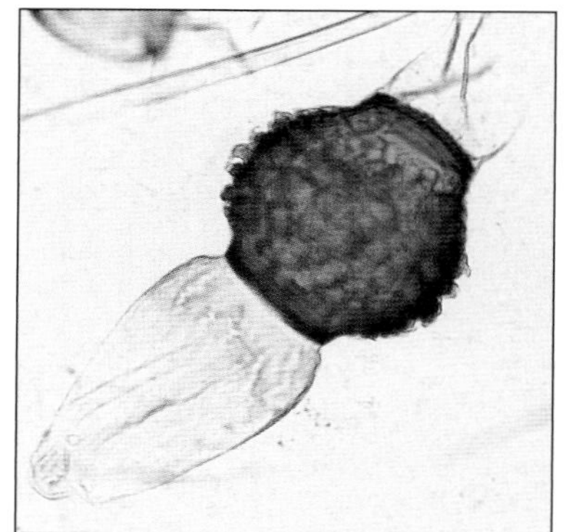

14.59 Mature ***Rhizopus*** **Zygospore (Opisthokonta; Division Zygomycota)** • Haploid nuclei from each strain fuse within the zygospore (**karyogamy**) to produce many diploid nuclei. Meiosis of each occurs to produce numerous haploid spores. (Stained Whole Mount)

14.60 ***Pilolobus*** **Culture (Opisthokonta; Division Zygomycota)** • These *Pilolobus* sporangiophores are oriented vertically because they were incubated in the dark. Had there been a directional light source, they would have been oriented toward it. Spores are in the black cap; the clear, swollen area beneath them is the subsporangial vesicle (**arrows**). The culture was grown on Rabbit Dung Agar; the brown object on the agar surface is a rabbit pellet.

The association between mycobiont and photobiont favors both. In fact, frequently the lichenized fungus can survive in places where the individuals cannot. The photobiont photosynthesizes and produces organic compounds from CO_2 and water. If it is a cyanobacterium, it may also be capable of nitrogen fixation. In turn, the mycobiont benefits from the productivity of its partner and provides it with moisture and protection.

Lichens reproduce asexually by fragmentation of the **thallus** (body) and by producing **propagules** containing both the mycobiont and photobiont. Sexual reproduction also occurs, but the mechanisms are varied and poorly understood. It appears that the sexually reproducing fungus must obtain a new photobiont partner each generation, thus emphasizing the continuity of the fungus and validating the convention of naming lichens after the mycobiont.

The lichen body can usually be categorized into one of three types: **foliose** (leafy), **fruticose** (shrubby), and **crustose** (crusty). Examples are shown in Figures 14.65–14.67. ●

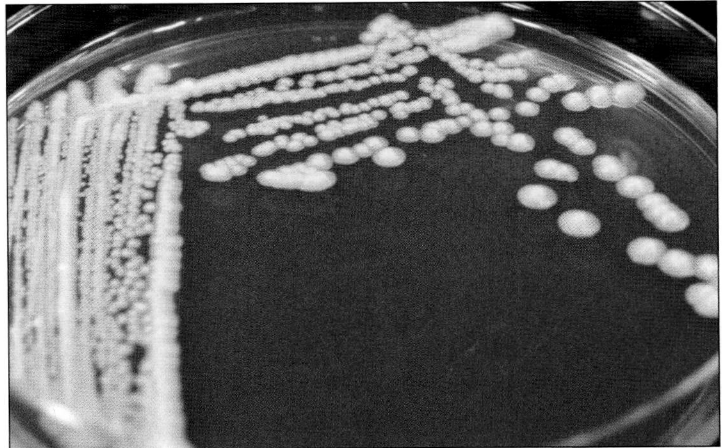

14.61 ***Saccharomyces cerevisiae*** **Colonies (Opisthokonta; Division Ascomycota)** ● Note that *S. cerevisiae* grows in colonies resembling bacterial colonies, and are not fuzzy like mold colonies.

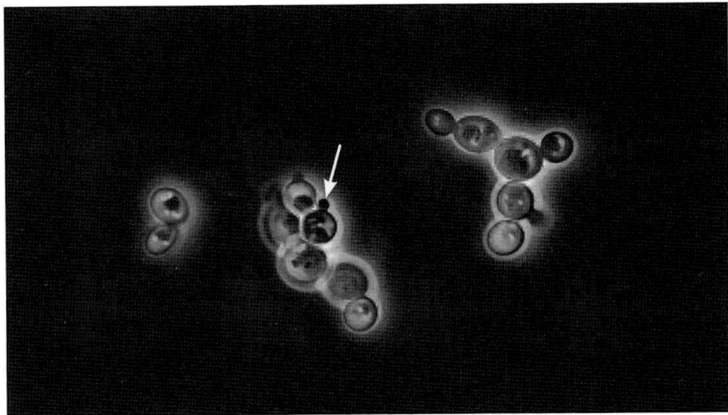

14.62 ***Saccharomyces cerevisiae*** **Vegetative Cells (Opisthokonta; Division Ascomycota)** ● Note the budding cell (blastoconidium) indicated by the **arrow**. (Phase Contrast Wet Mount)

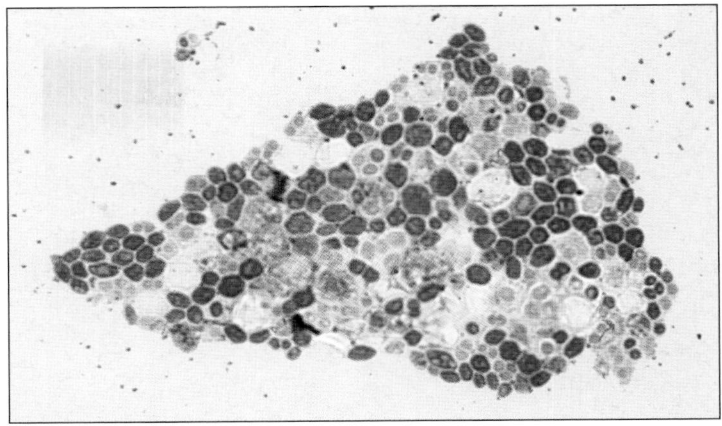

14.63 ***Saccharomyces cerevisiae*** **Ascospores (Opisthokonta; Division Ascomycota)** ● One to four ascospores are produced in the original cell (acting as an ascus) by meiosis. In this specimen, the vegetative cells are blue and the ascospores are red. (Acid-Fast Stain)

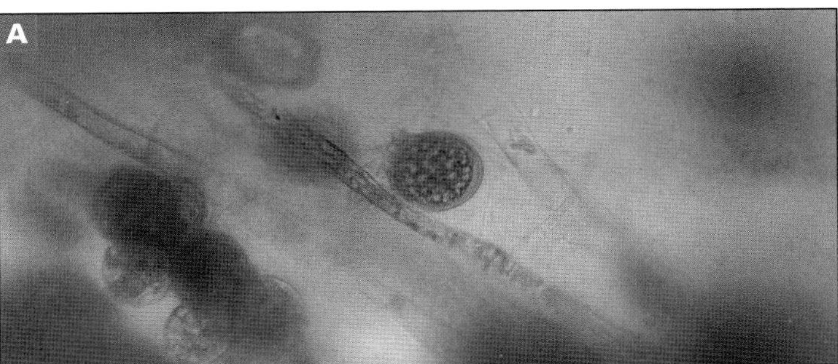

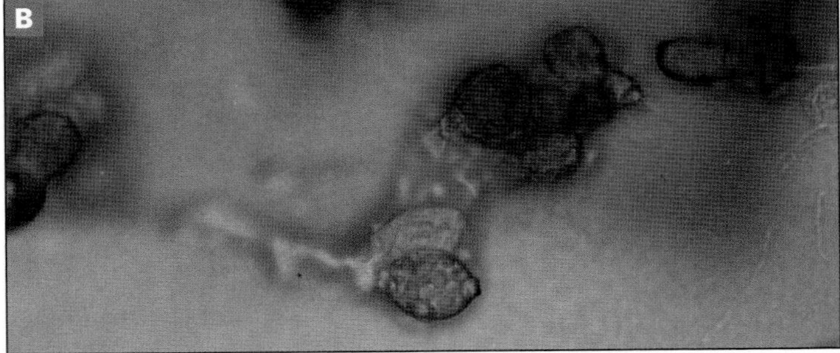

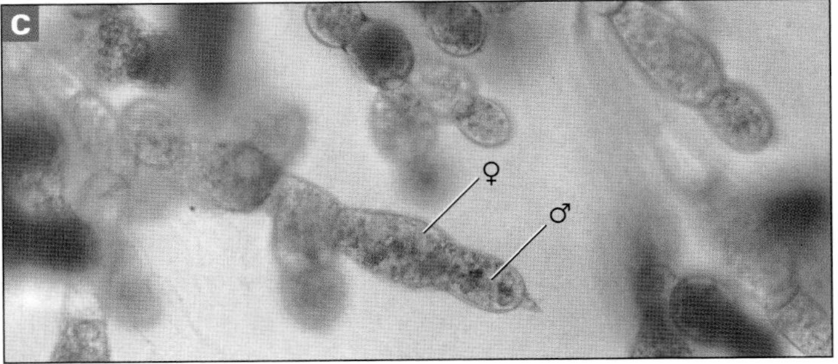

14.64 ***Allomyces*** **Reproductive Structures (Opisthokonta; Division Blastocladiomycota)** ● *Allomyces* is a soil saprophyte (decomposer) that exhibits a life cycle with alternation of diploid and haploid generations. (**A**) Shown is a diploid meiosporangium. It is part of a sporothallus and produces motile, diploid meiosporgania that develop into another sporothallus. (**B**) This is a mitosporangium of a sporothallus. It produces motile, haploid mitospores that develop into haploid gametothalli. (**C**) Shown are male and female gametangia. They produce male and female gametes by mitosis. When gametes fuse, a zygote is produced that develops into a mature sporothallus, completing the life cycle. In *Allomyces*, male gametangia (♂) are golden in color, whereas female gametangia (♀) are colorless. (Unstained Whole Mounts)

14.65 Foliose Lichen ● Note the leafy appearance of this foliose lichen's thallus. It was growing in a public park in Oregon.

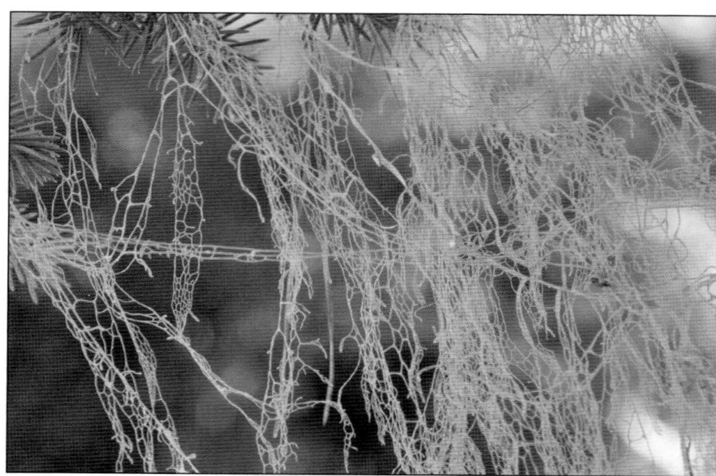

14.66 Fruticose Lichen ● Note the delicate network of thallus strands in this stringy lichen. It was provisionally identified as *Ramalina menziesii* and was photographed in Northern California.

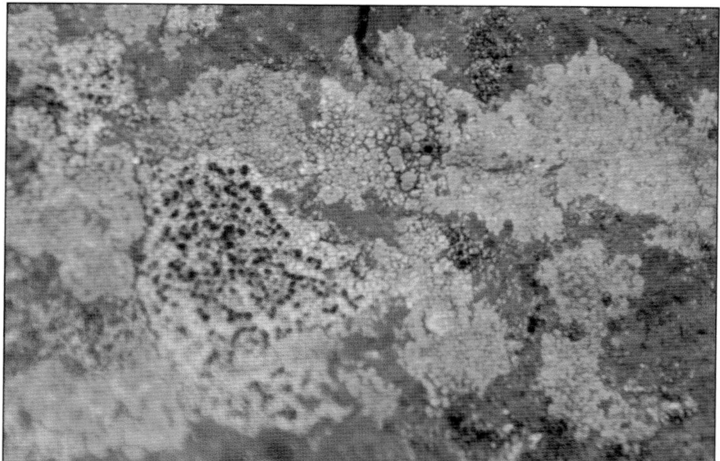

14.67 Crustose Lichen ● This crustose lichen was provisionally identified as *Caloplaca marina*. It was found high in the intertidal zone of a rocky shore in Oregon.

Fungi of Clinical Importance

SECTION 15

This section is a brief treatment of fungal pathogens. For an introduction to fungi in general, please refer to page 240 (Opisthokonts) in Section 14.

Ascomycetous Yeasts

Candida albicans

Candida albicans is one of approximately 200 species in the genus and is part of the normal respiratory, gastrointestinal, and female urogenital tract microbiota, but is also the most common fungal opportunistic pathogen. It is the causative agent of thrush in the oral cavity, vulvovaginitis of the female genitals, and cutaneous candidiasis of the skin.

Systemic candidiasis may follow infection of the lungs, bronchi, or kidneys. Entry into the blood may result in endocarditis. The ability to attach to epithelial surfaces of mucous membranes and subsequently penetrate to deeper tissues are important virulence factors.

Individuals most susceptible to *Candida* infections are diabetics, those with immunodeficiency (e.g., AIDS), catheterized patients, and individuals taking antimicrobial medications. Treatments vary depending on the location of infection and the state of the patient.

Candida albicans produces oval **blastoconidia** (blastospores) 3 µm–6 µm long that reproduce by budding and are the "typical" yeast form (Fig. 15.1). Sometimes the cells separate after budding, but other times they remain attached, forming septate chains of two types: **pseudohyphae**, which are narrower at cell junctions (because they are chains of blastoconidia), and true hyphae, which are not (Fig. 15.2). Both types branch and produce more asexual blastoconidia.

The ability to switch growth pattern between pseudohyphae and true hyphae may increase virulence. Larger, round, thick-walled **chlamydospores** may form at the ends of pseudohyphae and are useful in identification, though they are misnamed as reproductive cells. The sexual life cycle is similar to *S. cerevisiae* (pp. 243–244).

Utilization of selective and differential media allow presumptive identification of *C. albicans* during isolation. For instance, Bismuth Sulfite Glucose Glycine Yeast (BiGGY) Agar (see Fig. 2.6), various chromogenic media, and blood agar or chocolate agar provide useful preliminary information.

- Various chromogenic media are currently available. HardyCHROM *Candida* (Fig. 15.3) is an example, but they all work on the same basic principle: the medium includes chromogenic substrates that produce a unique color when acted upon by species-specific enzymes. All have bacterial inhibitors. Depending on the manufacturer, these media can provide definitive or presumptive identification of species.

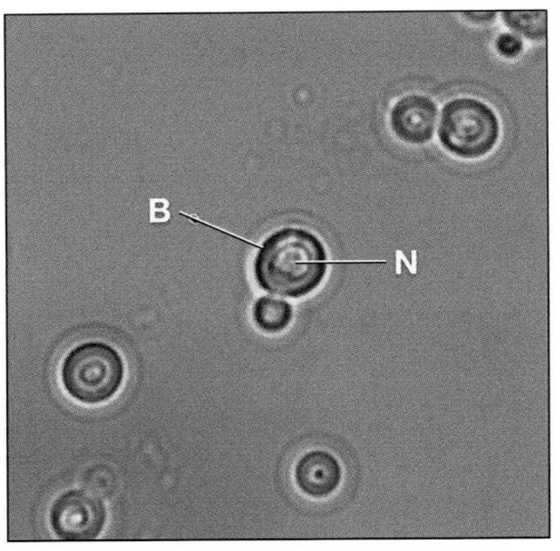

15.1 ***Candida albicans* Blastoconidia (Opisthokonta; Division Ascomycota)** Individual yeast cells are called blastoconidia. They may be formed by budding of single cells (**B**) or by budding from hyphae or pseudohyphae (as in Fig. 15.2). Also note the oval shape and nuclei (**N**). Cells range in length from 3 to 6 µm. (Wet Mount, Lactophenol Cotton Blue Stain)

- Growth of *C. albicans* on blood agar (see p. 76) or chocolate agar (see p. 12) can result in distinctive projections ("feet" or "spikes") extending from the colony margin (Fig. 15.4). Evidence suggests that their presence is a more sensitive indicator of *C. albicans* than germ tube formation. Chocolate agar is similar to blood agar except that the blood is added to molten agar, which causes the cells to lyse and release nutrients located in the cytosol. It is mostly used to grow fastidious bacteria, but obviously also supports *Candida* species. Its name is derived from the brown color produced by lysed RBCs.

Microscopic features of *C. albicans* have been discussed here, but presumptive identification of *C. albicans* can be made by performing a germ tube test, in which the yeast cells are inoculated into serum and observed microscopically after 2 to 3 hours of incubation at 35°C.

A true germ tube (Fig. 15.5) will have no constriction at its base and will develop into a true hypha. If a constriction is present at the base of the protrusion, it is not a germ tube and will develop into a pseudohypha. Other identification methods include nucleic acid probes (p. 149), rarely PCR due to expense (p. 144, api 20X AUX, and MALDI-TOF mass spectrometry (p. 150).

Pneumocystis jirovecii

Pneumocystis jirovecii (formerly *P. carinii*) is an opportunistic pathogen whose taxonomy and natural history has been difficult to work out due to our inability to grow it outside of its mammalian host until 2014[1].

Comparison of ribosomal RNA and genome sequencing has led to the conclusion that *Pneumocystis,* long classified as a protozoan, is more closely related to certain ascomycete yeasts and it is now considered an ascomycete. Its protozoan history is still evident, however, in the terminology associated with it.

Older protozoan terms are shown in parentheses. *Pneumocystis* exists as a **trophic form** (trophozoite), a **sporocyte** (precyst), and a multinucleate **ascus** (cyst) (Fig. 15.6). Asexual reproduction is by fission of the trophic form, not budding as in most yeast. Its sexual phase is still unclear, but it's thought to resemble yeast-like conjugation to form a sporocyte that produces four spores by meiosis. These then undergo mitosis to produce an ascus with eight spores.

Most people become seropositive for *Pneumocystis* in childhood and humans—even immunocompetent individuals—are the apparent reservoirs for the organism. Transmission is through the air, with the primary infection occurring in the lungs. Infection produces pneumocystis pneumonia (PCP) in AIDS patients and immunosuppressed individuals, and interstitial plasma cell pneumonitis in malnourished infants. Identification is made by direct examination of the ascus using immunofluorescence.

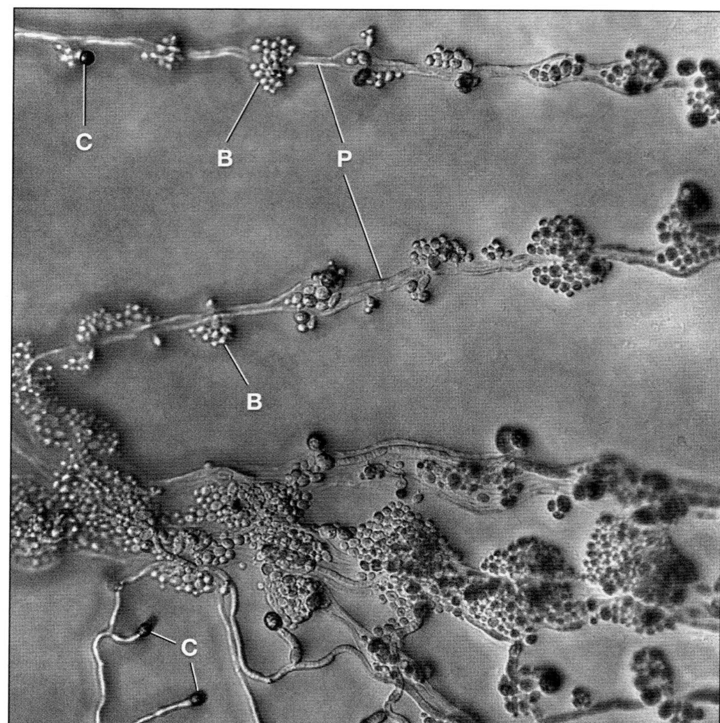

15.2 ***Candida albicans*** **Pseudohyphae (Opisthokonta; Division Ascomycota)** Note the growing clusters of blastoconidia (**B**) at septal junctions on the pseudohyphae (**P**). (The characteristic constrictions are mostly obscured by the blastoconidia.) Thick-walled, terminal chlamydospores (**C**) are also shown. These are distinctive of *C. albicans*. (Unstained Slide Culture)

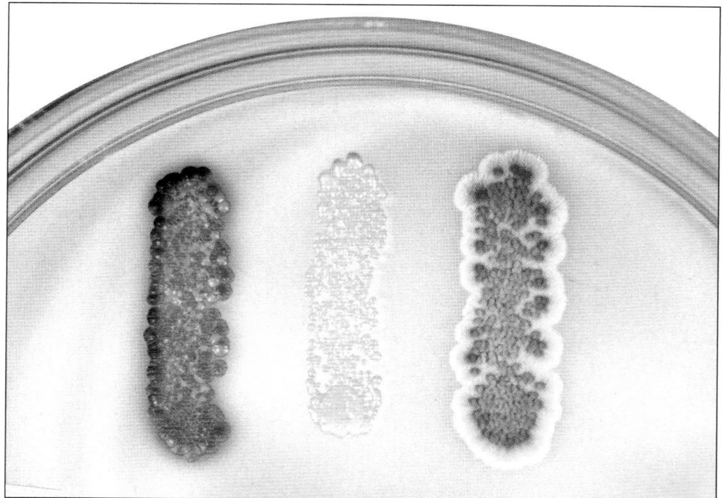

15.3 Colonies on HardyCHROM ***Candida*** **Agar (Opisthokonta; Division Ascomycota)** HardyCHROM *Candida* agar (Hardy Diagnostics, http://hardydiagnostics.com) is a commercial chromogenic medium used for isolation and identification of *Candida albicans*, *C. glabrata* (presumptive ID), *C. krusei*, and *C. tropicalis*. From left to right: *C. albicans*, *Saccharomyces cerevisiae* (for comparison of a non-*Candida* species), and *C. krusei*.

[1] In 2014 Verena Schildgen, et al., reported on successful cultivation of *P. jirovecii* using respiratory epithelial cells. See Verena Schildgen, Stephanie Mai, Soumaya Khalfaoui, Jessica Lüsebrink, Monika Pieper, Romana L. Tillmann, Michael Brockman, Oliver Schildgen. mBio May 2014, 5 (3) e01186-14; DOI: 10.1128/mBio.01186-14.

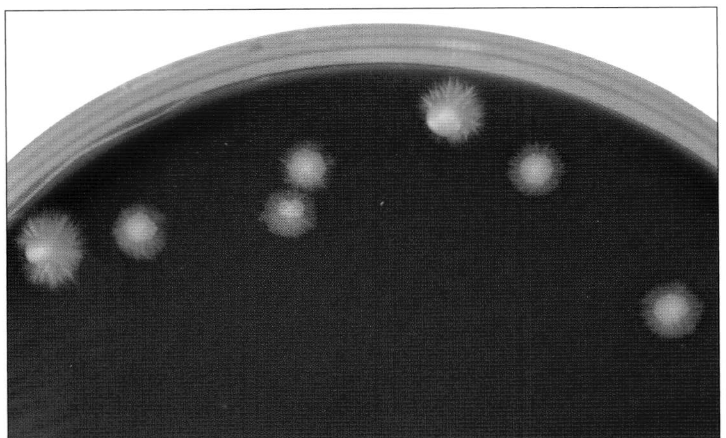

15.4 ***Candida albicans*** **on Chocolate Agar (Opisthokonta; Division Ascomycota)** • *Candida albicans* grown on nutrient-rich media will produce colonies with distinctive "spikes" on their margins. This is a 72-hour culture of *C. albicans* on chocolate agar. Note the brown color of the medium, giving it its name.

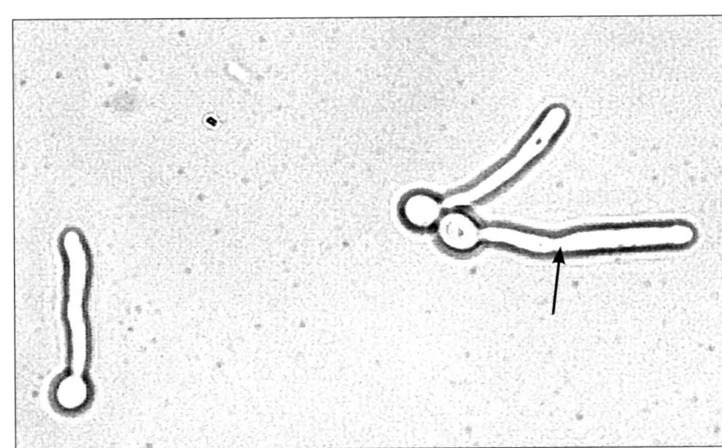

15.5 ***Candida albicans*** **Germ Tube Test (Opisthokonta; Division Ascomycota)** • Presence of germ tubes (**arrow**) allows presumptive identification of *C. albicans*. Germ tubes develop into true hyphae. These were produced by growing *C. albicans* at 35°C for 2 hours, and then observing at 400×. (Unstained Wet Mount)

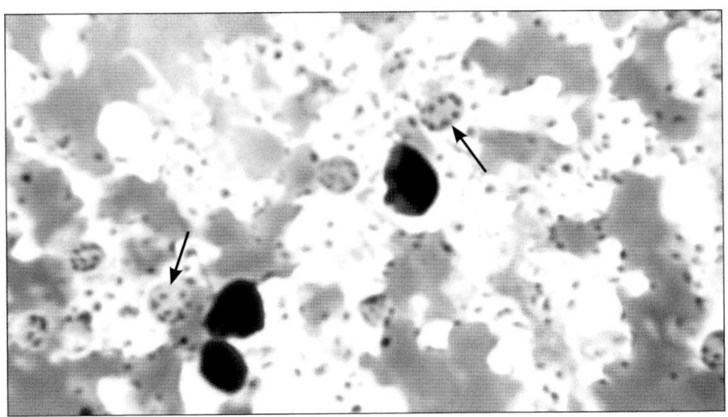

15.6 Section of *Pneumocystis jirovecii* Cyst in an Infected Lung (Opisthokonta; Division Ascomycota) • Zygotic meiosis followed by a mitotic division of each spore results in a mature cyst with eight spores (**arrows**). Once released, the spores become trophic forms (trophozoites). Cysts are between 5 and 8 µm in size. (Periodic Acid-Schiff Stain)

Basidiomycetous Yeasts

Cryptococcus neoformans

Cryptococcus neoformans causes cryptococcosis and is one of only two known serious human pathogens of the genus. (The other, *C. gattii*, was formerly known as *C. neoformans* var gattii.) Five serotypes have been recognized in the two species. *C. neoformans* is an encapsulated, nonfermenting, aerobic yeast (Fig. 15.7) that reproduces by budding, but pseudohyphae are rare.

C. neoformans is often found in soil mixed with accumulated bird (pigeon) droppings that enrich it with nitrogen. Urban sites where pigeons roost may also harbor the organism in the dried feces. Infection by *C. neoformans* from inhalation leads to pneumonia and then meningitis. Cryptococcosis, at one time a disease of poultry workers, is increasing in incidence and is among the more common opportunistic infections of AIDS patients.

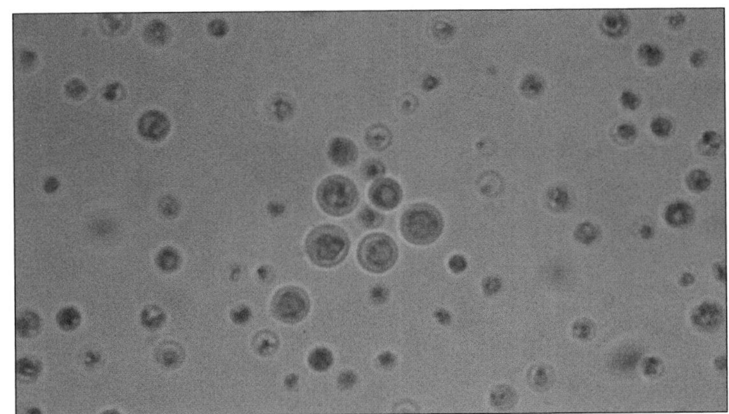

15.7 ***Cryptococcus neoformans*** **(Opisthokonta; Division Basidiomycota)** • *C. neoformans* is characterized by spherical, encapsulated cells that range in size from 3.5–8 µm. The capsule is faintly visible in this preparation. (Wet Mount, Lactophenol Blue Stain)

Due to the capsule, colonies of *C. neoformans* are shiny, cream colored, and mucoid. Identification is made by direct microscopic examination for the presence of a capsule and absence of pseudohyphae. Other diagnostic procedures include agglutination tests (see p. 133) identifying a unique mannan in the patient's serum, urease production (ascomycetous yeasts are urease negative), and api 20C AUX.

Rhodotorula mucilaginosa (R. rubra)

Species in the genus *Rhodotorula* are saprophytes commonly found in environmental samples taken from soil, air, freshwater, and marine habitats. In medical settings they have been isolated from plastic, and often invasive, devices as a contaminant during use. Various foods (packaged and fresh) have also been shown to harbor primarily *R. mucilaginosa*. Prior to 1985, no *Rhodotorula* infections had been reported, but now three species are known opportunistic pathogens (*R. glutinis*, *R. minuta*, and *R. mucilaginosa*), mostly of immunocompromised hosts with the infection resulting in fungemia associated with catheterization. Localized infections have also been reported.

Colonies are various shades of red, orange, or yellow, smooth, soft, and mucoid. (Fig. 15.8). Cells are spherical to oval and may be encapsulated. Hyphae and pseudohyphae are not formed (Fig. 15.9). Identification is done by traditional biochemical testing, colony morphology, and microscopic examination. Nucleic acid sequencing (see p. 142) and MALDI-TOF MS (see p. 150) are also utilized.

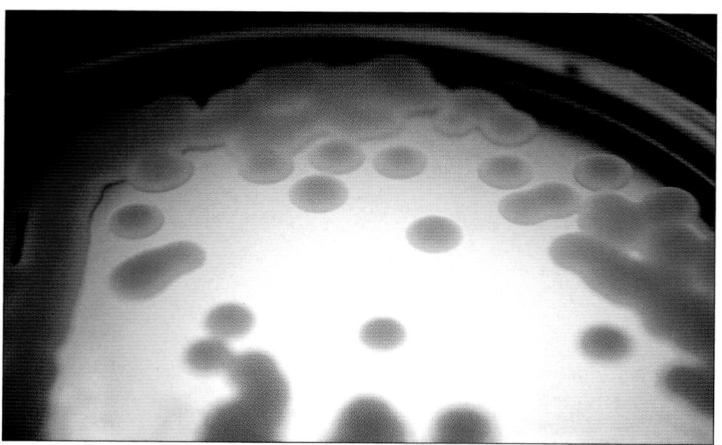

15.8 ***Rhodotorula rubra*** **Culture on Sabouraud Agar (Opisthokonta; Division Basidiomycota)** • *Rhodotorula rubra* is an opportunistic pathogen, primarily in medical settings. It produces colorful colonies, ranging from red to yellow. These colonies grew in about 3 days, but it took another day or two to produce the color.

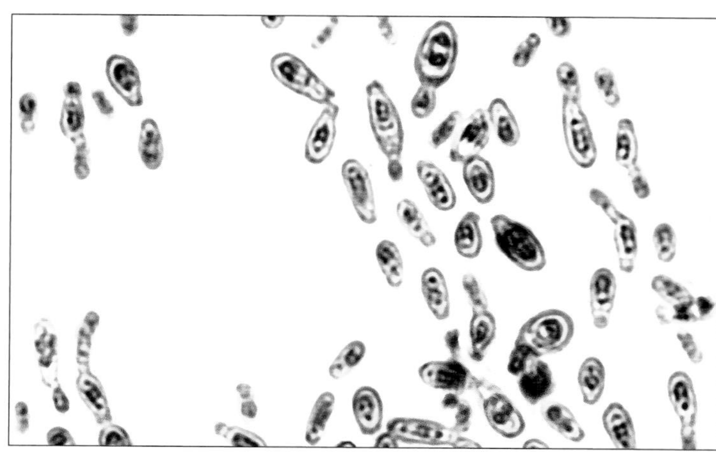

15.9 ***Rhodotorula mucilaginosa*** **(*R. rubra*) Cells (Opisthokonta; Division Basidiomycota)** • *R. mucilaginosa* cells are up to 10 µm long by 5 µm wide and tend not to be as "plump" as *Saccharomyces* cells (see Fig. 14.62). They do not usually form a capsule or pseudohyphae. Some budding cells are visible in this field. (Wet Mount, Safranin Stain)

Monomorphic Molds

Dermatophytes

The dermatophytes include *Epidermophyton, Microsporum, Trichophyton*, and related fungal genera that infect keratinized tissues of mammals—epidermis, nails, and hair. They produce a group of conditions known as "tinea" or "ringworm" (*tinea* means "worm"). The latter name comes from their growth pattern in which the infection spreads outward from the center of initial infection. The actively growing cells are at the edge of the growth, whereas healing tissue is toward the middle, thus giving the impression of a worm at the lesion's periphery.

Various forms of tinea are recognized based on body location: tinea capitis occurs on the head (*caput* means "head"); tinea corporis occurs on the face and trunk (*corpus* means "body"); tinea cruris (also known as "jock itch") occurs on the groin (*cruris* means "of the leg"); tinea unguium occurs on the nails (*unguis* means "nail" or "claw"); and tinea pedis occurs on the feet and causes athlete's foot (*pedis* means "of the foot").

Dermatophytes may infect more than one body region and often produce similar symptoms, so they are considered together here. Inflammation of infected tissue due to fungal antigens is the primary symptom and is often manifested as scaling, discoloration, and itching. Transmission is through contact with infected skin or contaminated items, such as towels, clothing, combs, or bedding.

Identification generally involves consideration of the tissue infected (as some are fairly specific), examination of the infection, and microscopic examination of patient samples and/or cultures. Tissues infected with *Microsporum* fluoresce when illuminated with a **Wood's lamp** (an ultraviolet light source) whereas *Epidermophyton* and *Trichophyton* do not produce fluorescence. Large **macroconidia** and smaller **microconidia** are especially useful microscopic characteristics used in differentiating species.

Epidermophyton

Epidermophyton floccosum is the only pathogen in the genus. It invades the dead, keratinized tissues of skin and nails, but not hair, and is highly contagious. In culture, colonies have an olive-green to khaki central portion with an orange periphery. Older cultures become overgrown with a white aerial mycelium.

Microscopically, characteristic macroconidia can be observed arising from conidiophores either singly or in clusters of two or three. These are septate with one to five cells, and club-shaped with thin, smooth walls. Microconidia are absent.

Microsporum

Terminal, septate macroconidia with thick, rough walls characterize the genus *Microsporum*. Microconidia may be present, but are not common. **Chlamydoconidia** (Fig. 15.10), a resting stage, are produced by some species. Hyphae are septate and either branched or unbranched.

M. audouinii is responsible for scalp ringworm in children, but rarely infects adults. Colonies have a fine, downy surface and are gray to tan with a light-orange reverse (Fig. 15.11). Unlike other species in the genus, *M. audouinii* rarely produces macroconidia or microconidia. Pointed chlamydoconidia may be seen (Fig. 15.10).

M. gypseum infects humans less frequently than *M. audouinii*. Colonies (Fig. 15.11) are variously colored, ranging from tan to reddish brown with a powdery or granular surface and often developing a white border or white center. Thin-walled and rough macroconidia with no more than six cells are abundant (Fig. 15.12). Club-shaped microconidia may also be observed.

Trichophyton spp.

Trichophyton species are the most important dermatophytes causing infection in adults. They are responsible for most cases of tinea pedis and tinea unguium, and occasionally tinea corporis and tinea capitis. Macroconidia are rare, but are located at the ends of hyphae and are club-shaped, smooth, thin-walled, and septate with up to 10 cells. Spherical microconidia are more common. An in vitro hair perforation test is used to distinguish *T. mentagrophytes* from other members of the genus (except *T. terrestre* and some *T. tonsurans*) (Fig. 15.13).

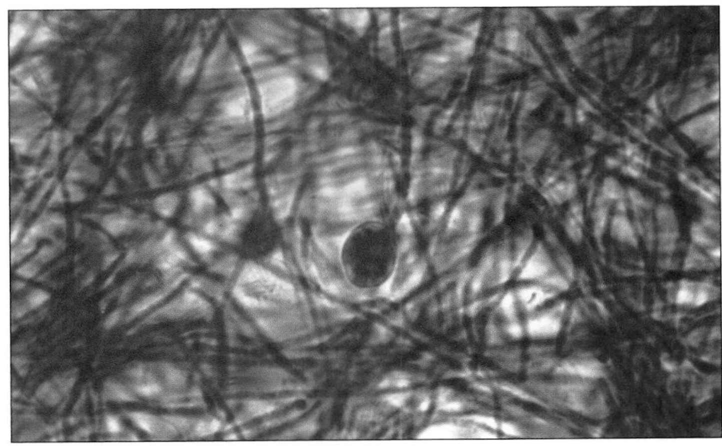

15.10 ***Microsporum audouinii*** **Chlamydoconidia (Opisthokonta; Division Deuteromycota)** • When grown in culture, *M. audouinii* often produces terminal chlamydoconidia, which contain resistant resting cells. (Wet Mount, Lactophenol Cotton Blue Stain)

15.11 ***Microsporum*** **Cultures (Opisthokonta; Division Deuteromycota)** • On the left is *M. gypseum*; *M. audouinii* is on the right.

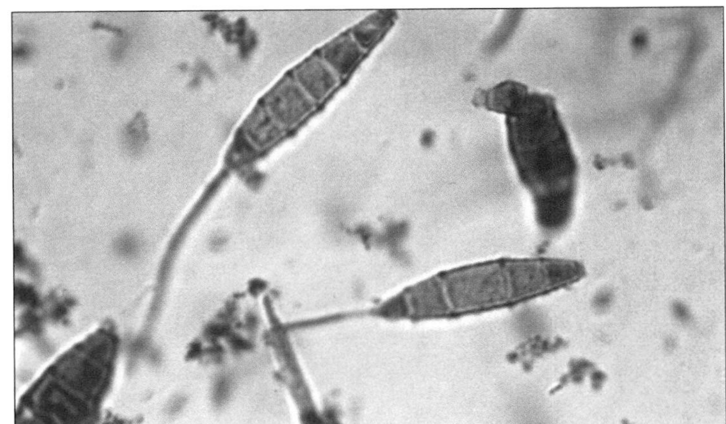

15.12 Macroconidia of ***Microsporum gypseum*** **(Opisthokonta; Division Deuteromycota)** • *M. gypseum* typically has fewer than six cells in each macroconidium. Note the rough surface. Macroconidia range in size from 8–16 µm wide by 22–60 µm long. (Wet Mount, Lactophenol Cotton Blue Stain)

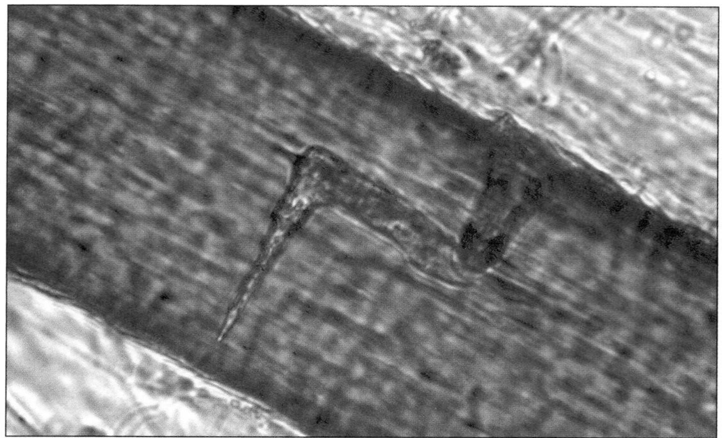

15.13 ***Trichophyton mentagrophytes* Hair Perforation Test (Opisthokonta; Division Ascomycota)** • *Trichophyton mentagrophytes* infection can be diagnosed by a hair perforation test. Note the indentation in the hair. (Wet Mount, Lactophenol Cotton Blue Stain)

Aspergillus

Aspergillus species are commonly found in soil and on plants. More than 200 species have been identified. They are characterized by green to yellow or brown granular colonies with a white edge (Fig. 15.14). One species, *A. niger*, produces distinctive black colonies (Fig. 15.15). Vegetative hyphae are **hyaline** (unpigmented) and septate.

The *Aspergillus* fruiting body is distinctive, with chains of conidia arising from one (uniseriate) or two (biseriate) rows of **phialides** attached to a swollen **vesicle** at the end of an unbranched conidiophore (Fig. 15.16). The conidiophore grows from a foot cell in the vegetative hypha (Fig. 15.17). Fruiting body structure and size, and conidia color are useful in species identification.

Since their discovery it was thought that *Aspergillus* species were incapable of sexual reproduction, but evidence is mounting that under the right environmental conditions viable ascospores can be produced.

A. fumigatus and other species are opportunistic pathogens that cause aspergillosis, an umbrella term covering many diseases. Immunocompromised patients are at highest risk of infection. Invasive aspergillosis is the most severe form and has high mortality. It results in necrotizing pneumonia and may spread to other organs, such as the heart and central nervous system. Antifungal medications can be used in treatment, but identification of the pathogen to species level is necessary because of their different sensitivities.

Allergic aspergillosis may occur in individuals who are in frequent contact with the spores and become sensitized to them. Subsequent contact produces symptoms similar to asthma. Aspergilloma (fungus ball) involves colonization of the paranasal sinuses or lungs, resulting in abscess formation. It is frequently asymptomatic and resolves without treatment, but in serious cases it can be fatal.

Some *Aspergillus* species are of commercial importance. Fermentation by *A. oryzae* and *A. soyae* are used in the production of sake and soy sauce. Aspergilli are also used in commercial production of citric acid, various enzymes, and many other products. Further, *A. niger* and *A. oryzae* have been designated as **GRAS** (generally regarded as safe) by the FDA and WHO, which has opened the door for using them as hosts in biotechnological applications.

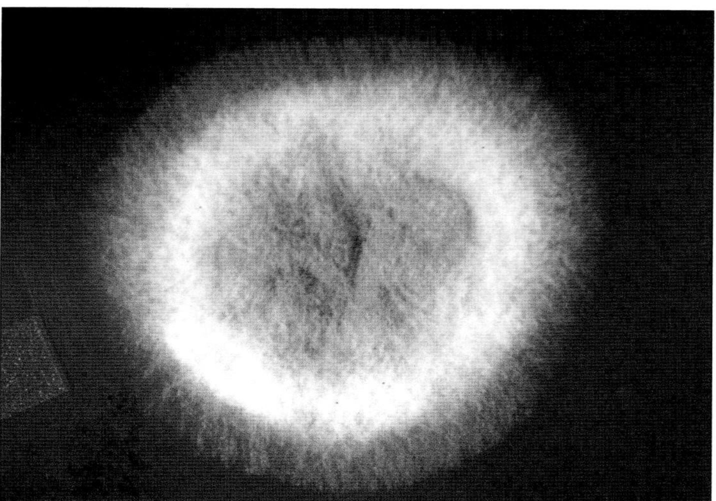

15.14 ***Aspergillus fumigatus* Colony on Sabouraud Dextrose Agar (Opisthokonta; Division Ascomycota)** • Note the rugose topography and green, granular appearance with a white margin of this *Aspergillus fumigatus* colony. The reverse is white. Although a common cause of aspergillosis, normally healthy people are not at great risk from *A. fumigatus* infection.

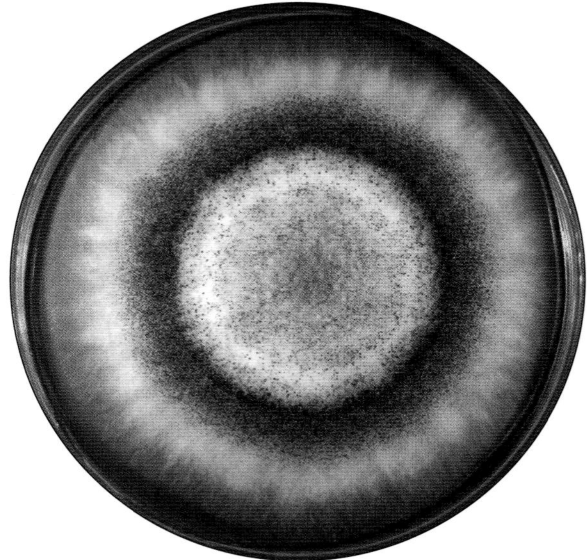

15.15 ***Aspergillus niger* Colony on Sabouraud Dextrose Agar (Opisthokonta; Division Ascomycota)** • *Aspergillus niger* produces distinctive black colonies. This colony was grown 7 days at 25°C and filled the plate.

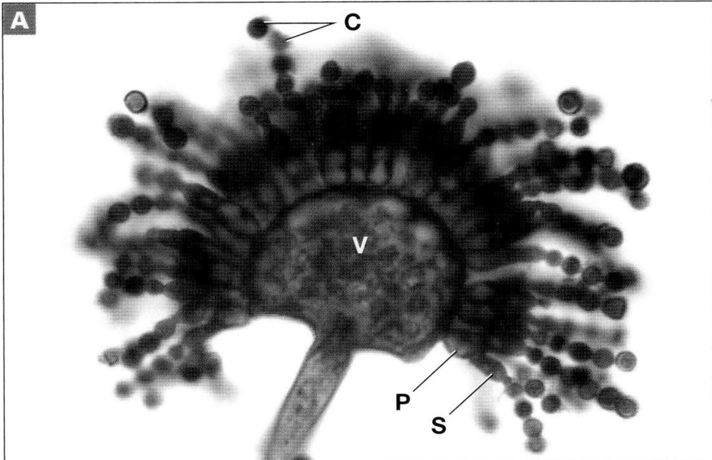

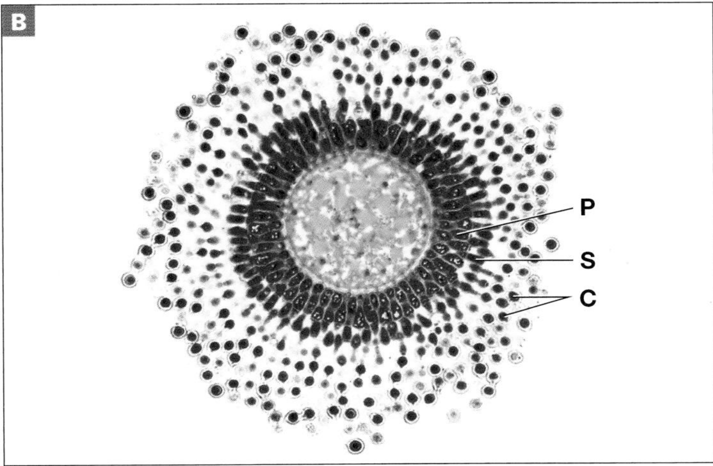

15.16 ***Aspergillus* Conidial Heads (Opisthokonta; Division Ascomycota)** (A) Shown is a longitudinal section of an *Aspergillus niger* conidiophore. The conidia (**C**), primary (**P**) and secondary (**S**) phialides, and vesicle (**V**) are visible. This is a biseriate conidium. (B) This *Aspergillus* conidial head is also biseriate, but is seen in a cross section.

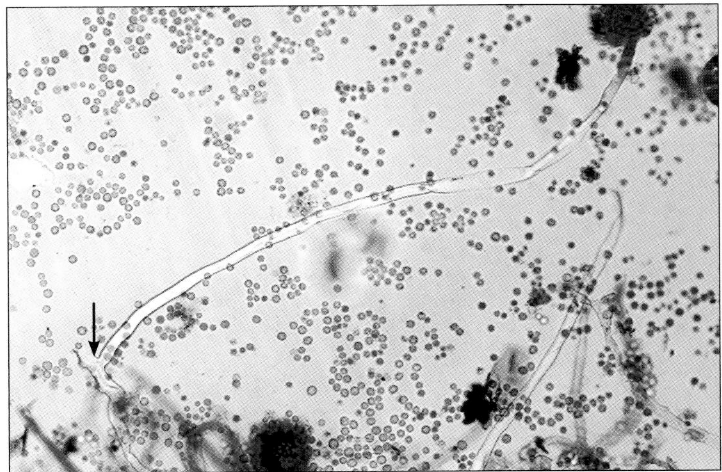

15.17 ***Aspergillus* Foot Cell (Opisthokonta; Division Ascomycota)** Sporangiophores emerge from a foot cell with the shape of an inverted "T" (**arrow**). This is a wet mount of *Aspergillus flavus*. Notice all the conidia in the field. (Wet Mount, Lactophenol Cotton Blue Stain)

Coccidioides immitis

Coccidioides immitis causes coccidioidomycosis ("Valley fever"), a lung disease associated with desert regions of the southwestern United States, northern Mexico, and parts of Central and South America. Sixty percent of infections are asymptomatic and self-limiting, but in some individuals, symptoms are influenza-like and may include hypersensitivity reactions. Rarely, the disease becomes disseminated and then may be lethal. It is the most frequent cause of laboratory-acquired fungal infections, so extreme care must be taken when handling it.

The mold form consists of branched, septate, hyaline hyphae. Barrel-shaped **arthroconidia** develop separated by thin-walled **disjunctor cells.** When the disjunctor cells die, arthroconidia become separated (Fig. 15.18). Infection occurs when these become airborne and are inhaled. Once in the host, arthroconidia develop into thick-walled **spherules** containing endospores (not to be confused with bacterial endospores), which may in turn develop into spherules upon release. In culture, arthroconidia develop into hyphae.

C. immitis colonies are white and wooly, but may develop a variety of colors with age. The yeast form does not grow on standard culture media. Colony morphology, hyphae with alternating barrel-shaped arthroconidia and disjunctor cells, and direct examination of patient samples for the presence of spherules are used in identifying *C. immitis*. DNA probes (p. 149) and serological methods are also used in identification.

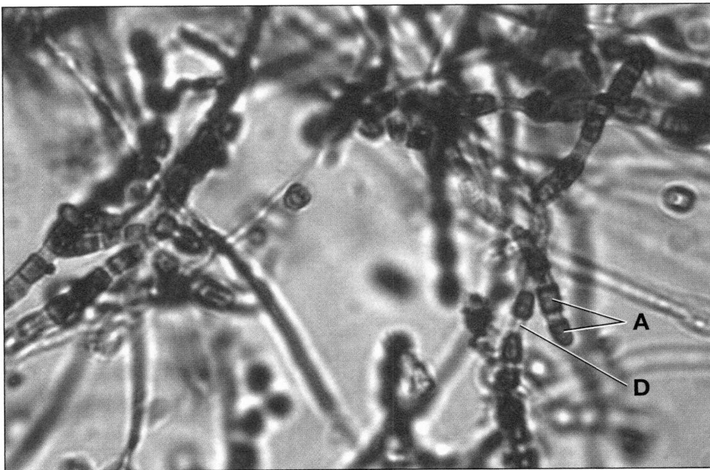

15.18 ***Coccidioides immitis* Arthroconidia (Opisthokonta; Division Ascomycota)** *C. immitis* hyphae have barrel-shaped arthroconidia (**A**) separated by thinner-walled disjunctor cells (**D**). When the disjunctor cells die, infectious arthroconidia are dispersed in the air. Arthroconidia range in size from 3–4 µm wide by 3–6 µm long. (Wet Mount, Lactophenol Cotton Blue Stain)

Fonsecaea

Fonsecaea pedrosoi is a **dematiaceous** (darkly pigmented) fungus that causes chromoblastomycosis, a disease of skin and subcutaneous tissues, particularly in the lower extremities. It has a worldwide distribution, but is more common in tropical and subtropical regions.

Colonies are olive green to black with a velvety texture and a black reverse (Fig. 15.19). The branched and septate hyphae have dark cell walls due to the pigment melanin. The dark conidia are borne on septate conidiophores in four basic patterns. At least two of these must be seen to identify the isolate as *F. pedrosoi*. MALDI-TOF MS is also used in identification (see p. 150).

15.19 ***Fonsecaea pedrosoi* Culture (Opisthokonta; Division Ascomycota)** • This dematiaceous fungus is the primary cause of chromoblastomycosis. Its black color is due to the pigment melanin.

Penicillium

Members of the genus *Penicillium* are ubiquitous, being found in the air, soil, and decaying organic matter worldwide. More than 300 species have been identified. They produce distinctive green, powdery, radially furrowed colonies with a white apron (Fig. 15.20) and light-colored reverse surface. The hyphae are septate and thin.

Distinctive *Penicillium* fruiting bodies, consisting of **metulae**, phialides, and chains of spherical conidia, are located at the ends of branched or unbranched conidiophores (Fig. 15.21). Although not an important feature in laboratory identification, sexual reproduction occurs under the proper conditions and results in the formation of ascospores within an ascus.

Penicillium is best known for its production of the antibiotic penicillin (discovered by Scottish physician-scientist Alexander Fleming in 1928), but it is also a common contaminant. One pathogen, *P. marneffei*, is endemic to Asia and is responsible for disseminated opportunistic infections of the lungs, liver, and skin in immunosuppressed and immunocompromised patients. It is thermally dimorphic, producing a typical velvety colony with a distinctive red pigment at 25°C but converting to a yeast form at 35°C.

Other species are important, because they cause spoilage of fruits and vegetables. *P. chrysogenum* is common in damp, indoor habitats and their spores can cause allergic reactions. Other *Penicillium* species are of commercial importance for fermentations used in cheese production. Examples include *P. roqueforti* (Roquefort cheese) and *P. camemberti* (Camembert and Brie cheeses).

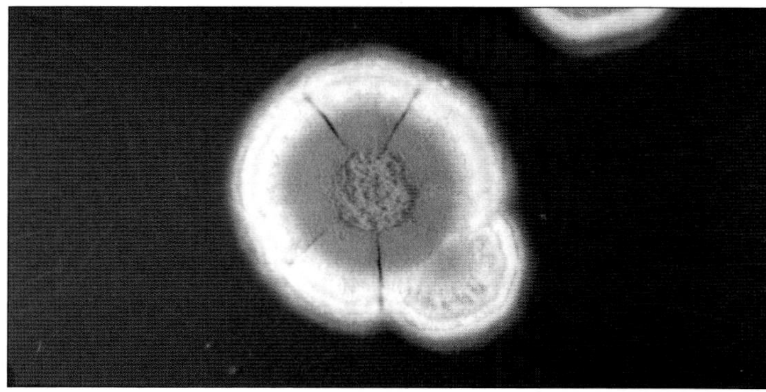

15.20 ***Penicillium notatum* Colony on Sabouraud Dextrose Agar (Opisthokonta; Division Ascomycota)** • The green, granular surface with radial furrows and a white apron are typical of the genus.

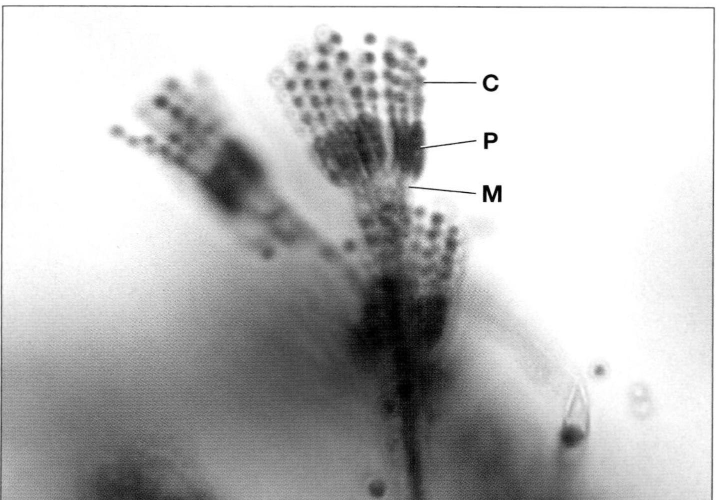

15.21 ***Penicillium* Conidiophore (Opisthokonta; Division Ascomycota)** • *Penicillium* species produce a characteristic brush-shaped conidiophore (*penicillus* means "paintbrush"). Metulae (**M**), phialides (**P**), and chains of spherical conidia (**C**) are visible.

Protozoans of Clinical Importance

SECTION 16

The supergroups of Domain Eukarya were introduced in Section 14 (see Fig. 14.1). These are: Archaeplastida, Excavata, SAR, and Unikonta. Human pathogens are found in the last three and they, along with some commensals found in clinical specimens, are the subject of this section.

Group Excavata

Phylum Metamonada

Dientamoeba fragilis

At one time, *Dientamoeba fragilis* was considered to be an amoeba, but cytological evidence suggests it is better classified as a parabasalid (see p. 230). Until 2014, only the trophozoite (Fig. 16.1) was known, but then a precyst and cyst form were identified.

The trophozoite lives primarily in the cecum where it feeds on the bacterial and yeast microbiota as well as cellular debris. It is found in approximately 4% of humans and is being identified in stool samples more frequently. Typical symptoms include diarrhea, abdominal pain, and fatigue. The mode of transmission is unclear, but there is evidence supporting the idea that *D. fragilis* is transmitted in the eggs of parasitic helminths, such as *Ascaris* (see p. 275) and *Enterobius vermicularis* (see p. 277).

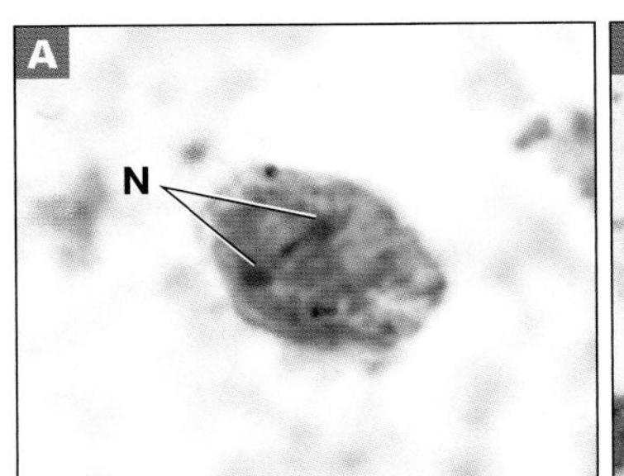

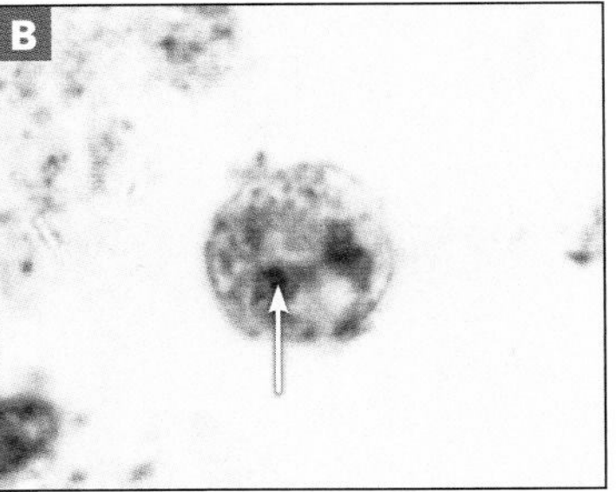

16.1 ***Dientamoeba fragilis* Trophozoites in Fecal Smears (Excavata; Phylum Metamonada)** Trophozoites are 5–12 µm in diameter and are the stage most frequently seen in stool samples. Most cells have two nuclei (**N**), but some have only one. The nucleus contains a single karyosome (dense staining chromatin) and the nuclear envelope is indistinct. The cytoplasm is often vacuolated. (**A**) A typical trophozoite. (Trichrome Stain) (**B**) A trophozoite with a fragmented karyosome (**arrow**). (Iron Hematoxylin Stain)

Trichomonas vaginalis

Trichomonas vaginalis (Fig. 16.2) is a parabasalid. Parabasalid motility comes from a group of anterior flagella (*Trichomonas* has four) and an **undulating membrane**, an extension of the plasma membrane associated with the posterior flagellum. ATP is produced by **hydrogenosomes**, degenerated mitochondria, with H_2 gas as the end-product.

T. vaginalis is the causative agent of trichomoniasis in humans. It causes inflammation of genitourinary mucosal surfaces—typically the vagina, vulva, and cervix in females and the urethra, prostate, and seminal vesicles in males—and infection in females is more common.

Most infections are asymptomatic or mild. Some erosion of surface tissues and a discharge may be associated with infection. The degree of infection is affected by host factors, especially the bacterial microbiota present and the pH of the

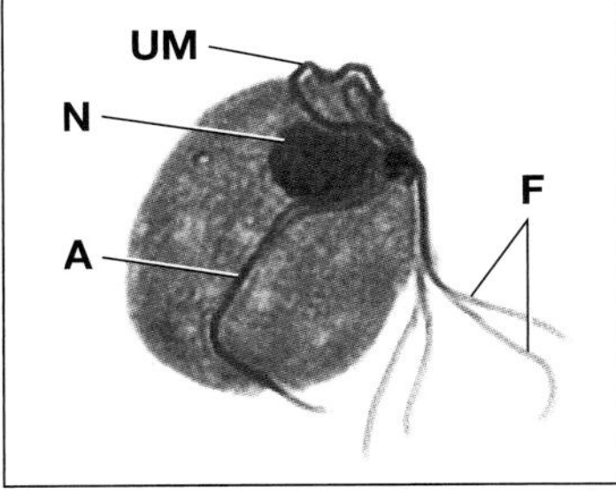

16.2 ***Trichomonas vaginalis* (Excavata; Phylum Metamonada)** The trophozoite is the only stage of the *Trichomonas* life cycle. The four anterior flagella (**F**) are visible, as is the nucleus (**N**), undulating membrane (**UM**), and axostyle (**A**). Cells range in size from 5–15 µm wide by 7–23 µm long.

mucosal surfaces. Transmission typically is by sexual intercourse. Diagnosis is made by microscopic examination of urethral or vaginal discharge.

Giardia duodenalis

Giardia duodenalis (also known as *Giardia intestinalis)* is a diplomonad (see p. 230) with worldwide distribution. Diplomonads possess two nuclei of equal size, multiple flagella, and (apparently) degenerated mitochondria called **mitosomes.** Mitosomes lack a functional electron transport chain, thus diplomonads are anaerobes. Their function is unclear.

G. duodenalis is the causative agent of giardiasis. It most frequently occupies the duodenum where it feeds on nutrients and secretions. The trophozoite (Fig. 16.3) is heart-shaped with four pairs of flagella and a sucking disk that allows it to attach to the mucosal surface and resist gut peristalsis. Trophozoites divide longitudinally and produce two daughter trophozoites as a result.

Multinucleate cysts (Fig. 16.4) lacking flagella are formed as the organism passes through the colon. Cysts are shed in the feces and may produce infection of a new host upon ingestion. Transmission typically involves fecal-contaminated water or food, but direct fecal-oral contact transmission is also possible.

The organism attaches to epithelial cells in the duodenal crypts but does not penetrate to deeper tissues. However, it does damage microvilli and interferes with nutrient absorption by the host. Most infections are asymptomatic and clear up on their own.

Chronic diarrhea, dehydration, abdominal pain, and other symptoms may occur if the infection produces a population large enough to involve a significant surface area of the small intestine. Diagnosis is made by identifying trophozoites or cysts in stool specimens, serological tests for antigens, and nucleic acid tests.

Chilomastix mesnili

Chilomastix mesnili is most commonly found in warmer climates, but can be found most anywhere. It exists as a trophozoite and a cyst (Figs. 16.5 and 16.6). Both may be found in stool samples and are used in identifying infection by this nonpathogen. It typically lives in the cecum and large intestine as a commensal, but may be indicative of infection by other parasites. Infection occurs through ingestion of cysts.

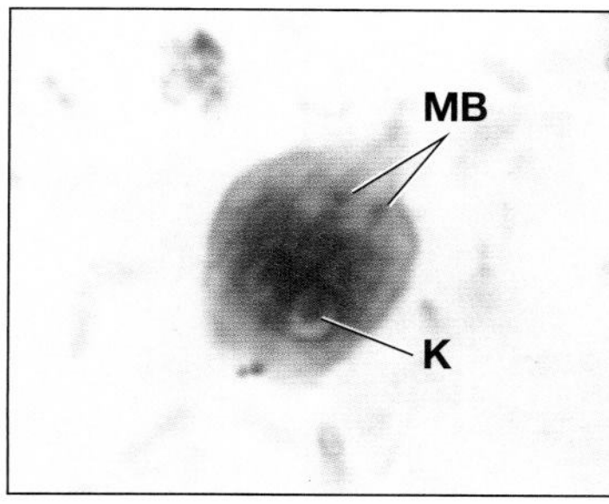

16.3 ***Giardia duodenalis*** **Trophozoite in a Fecal Smear (Excavata; Phylum Metamonada)** • Trophozoites have a long, tapering posterior end and range in size from 9–21 µm by 5–15 µm. There are two nuclei with karyosomes (**K**). Two median bodies (**MB**) are visible, but the four pairs of flagella are not. (Iron Hematoxylin Stain)

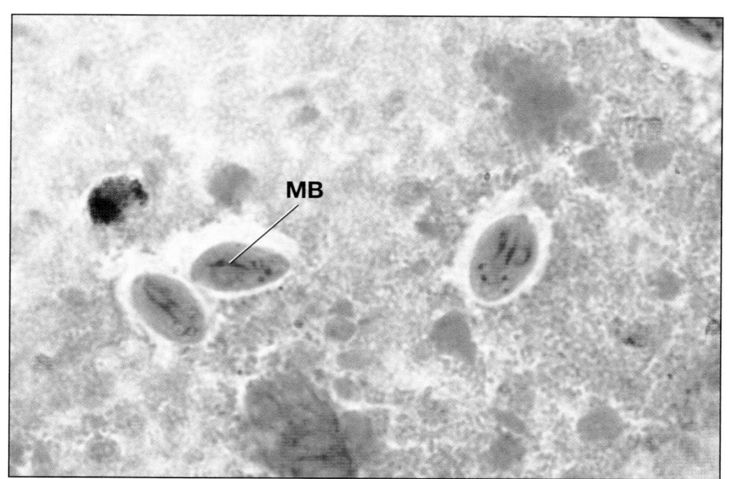

16.4 ***Giardia duodenalis*** **Cyst in a Fecal Smear (Excavata; Phylum Metamonada)** • Giardia cysts are smaller than trophozoites (8–12 µm by 7–10 µm), but the four nuclei with eccentric karyosomes (dark dots, especially obvious in the cell at the right) and the median bodies (**MB**) are still visible. (Trichrome Stain)

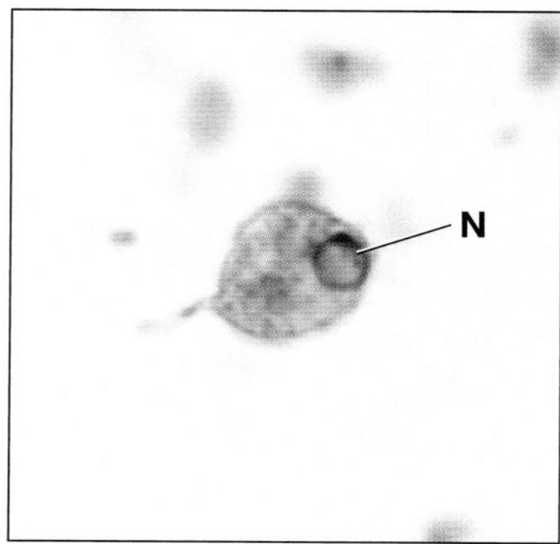

16.5 ***Chilomastix mesnili*** **Trophozoite in a Fecal Smear (Excavata; Phylum Metamonada)** • Trophozoites are elongated with a tapering posterior and a blunt anterior end that holds the nucleus (**N**). The dimensions are 6–20 µm long by 5–7 µm wide. There are four flagella: three at the anterior end (which may be difficult to see) and one associated with the prominent cytostome ("cell mouth"), which is also difficult to see in this preparation. (Iron Hematoxylin Stain)

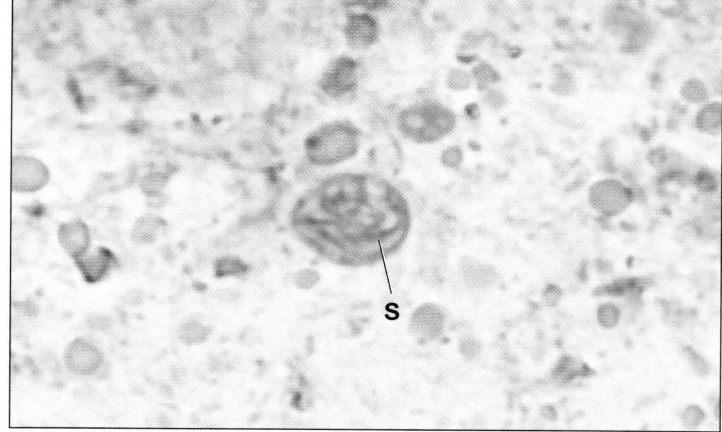

16.6 ***Chilomastix mesnili*** **Cyst in a Fecal Smear (Excavata; Phylum Metamonada)** • Cysts are lemon-shaped, often with an anterior knob, and are 6–10 µm long. The nucleus may be difficult to see. A distinctive "shepherd's crook" (**S**) associated with the cytostome is also visible. (Trichrome Stain)

Phylum Percolozoa

Naegleria fowleri

Naegleria fowleri is a free-living soil and water microbe with an amoeboid stage (Fig. 16.7), a cyst stage, and a flagellated stage. Under the proper conditions, it also is a facultative parasite that causes primary amebic meningoencephalitis (PAM). Infection is most prevalent in children and young adults.

It probably occurs when the individual forces contaminated water containing the cells up the nasal passages (as in diving). The organisms travel up the olfactory nerves into the cranial vault where they multiply and digest the olfactory bulbs and cerebral cortex. Symptoms occur about a week after infection and include fever, severe headache, and coma. Death occurs within about a week from the onset of symptoms.

Phylum Euglenozoa

Leishmania donovani

Leishmania donovani, a kinetoplastid in the phylum Euglenozoa (see p. 230), actually represents a number of geographically separate species and subspecies that are difficult to distinguish morphologically. All produce visceral leishmaniasis, or kala-azar, a disease found in tropical and subtropical regions.

The pathogen exists as a nonflagellated **amastigote** (Fig. 16.8) in the mammalian host (humans, dogs, and rodents) and as an infective, motile **promastigote** in the female sand fly vector (Fig. 16.9). Promastigotes are introduced into the mammalian host by sand fly bites. Distribution of the disease is associated with distribution of the appropriate sand fly vector.

Upon introduction into the host by the sand fly, the organism is phagocytized by macrophages and converts to the amastigote stage. Mitotic divisions result in filling of the macrophage, which bursts and releases the parasites. Phagocytosis by other macrophages follows and the process repeats. In this way, the organism spreads through much of the reticuloendothelial system, including lymph nodes, liver, spleen, and bone marrow. Kala-azar is a progressive disease and is fatal if untreated.

Amastigotes in an infected host may be ingested by a sand fly during a blood meal. Once inside the sand fly, they develop into promastigotes and multiply. They eventually occupy the fly's buccal cavity where they can be transmitted to a new mammalian host during a subsequent blood meal.

Transmission requires the vector and does not occur by direct contact. Where infection is common, successful diagnosis may be based on presence of disease symptoms and signs. Serological tests for antigens and PCR (see p. 144) are also commonly used for rapid diagnosis.

Other *Leishmania* species are responsible for cutaneous and mucocutaneous leishmaniasis, which are infections of skin and oral, nasal and pharyngeal mucous membranes. In all cases, infection involves macrophages in the affected region. Transmission is by direct contact with lesions or by sand fly bites.

Trypanosoma brucei and *Trypanosoma cruzi*

Trypanosomes are heterotrophic members of the Euglenozoa and like *Leishmania* species are kinetoplastids (see p. 230), which are characterized by a crystalline rod within their flagella in addition to the usual 9 + 2 arrangement of microtubules. *Trypanosoma brucei* is divided into three subspecies: *T. brucei brucei* (which is nonpathogenic), and *T. brucei gambiense* and *T. brucei rhodesiense* (Fig. 16.10), which cause African trypanosomiasis, also known as African sleeping sickness.

The organisms are morphologically very similar but differ in geographic range and disease progress. West African trypanosomiasis (caused by *T. brucei gambiense*) is generally a mild, chronic disease that may last for years before the nervous system is affected, whereas East African trypanosomiasis (caused by *T. brucei rhodesiense*) is more acute and results in death within a year. Modern molecular methods that compare proteins, RNA, and DNA are used to differentiate between them.

African trypanosomes have a complex life cycle. One stage of the life cycle, the **epimastigote**, multiplies in an intermediate host, the tsetse fly (genus *Glossina*). The

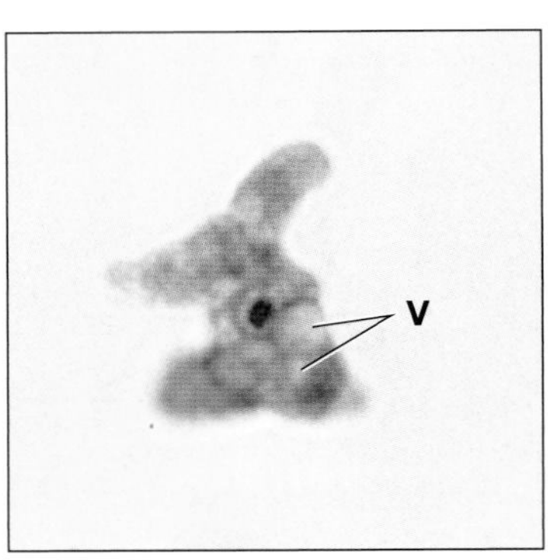

16.7 ***Naegleria fowleri*** **Trophozoite (Excavata; Phylum Percolozoa)** • Trophozoites are between 10–35 µm in size. Notice the large karyosome within the nucleus and the lobed pseudopods. Vacuoles (**V**) are also visible in the cytoplasm. This cell was grown in culture. (Iron Hematoxylin Stain)

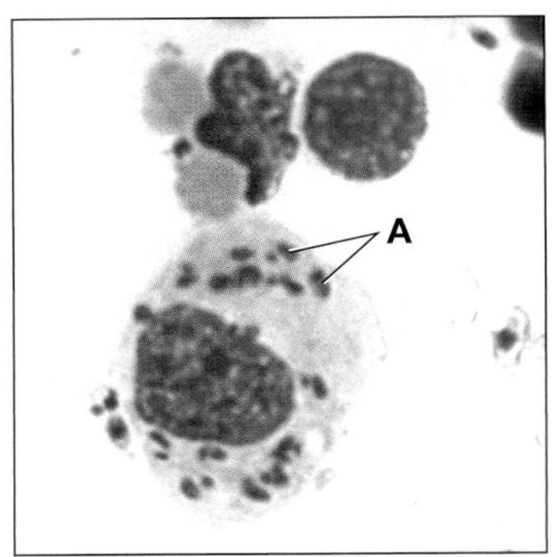

16.8 ***Leishmania donovani*** **Amastigotes in Spleen Tissue (Excavata; Phylum Euglenozoa)** • Shown are amastigotes (**A**) in a spleen cell. They are 3–5 µm in size and multiply within phagocytic and other cells by binary fission. Amastigotes are also known as Leishman-Donovan (L-D) bodies.

infective **trypomastigote** stage is then transmitted to the human host through tsetse fly bites.

Once introduced, trypomastigotes multiply and produce a chancre at the site of the bite. They enter the lymphatic system and spread through the blood, and ultimately to the heart and brain. The infective cycle is complete when an infected individual (humans, cattle, and some wild animals are reservoirs) is bitten by a tsetse fly, which ingests the organism during its blood meal. The fly becomes infective for its lifespan.

Progressive symptoms include headache, fever, and anemia, followed by symptoms characteristic of the infected sites. The symptoms of sleeping sickness—sleepiness, emaciation, and unconsciousness—begin when the central nervous system becomes infected. Depending on the infecting subspecies, the disease may last for months or years, but the mortality rate is high.

Death results from heart failure, meningitis, or severe debility of some other organ(s). As with most diseases, the earlier treatment begins the better the prognosis. A variety of antiprotozoal drugs are used with varying side effects, but those administered later in the disease must cross the blood-brain barrier and are difficult to administer safely.

Immune response to the pathogen is hampered by the trypanosome's ability to change surface antigens faster than the immune system can produce appropriate antibodies. This antigenic variation also makes development of a vaccine a challenge.

Diagnosis is made from clinical symptoms and identification of the trypomastigote in patient specimens (e.g., blood, CSF, and chancre aspirate). An ELISA (see p. 137) and an indirect agglutination test (see p. 133) also have been developed to detect trypanosome antigens in patient samples.

Trypanosoma cruzi (Fig. 16.11) causes American trypanosomiasis (Chagas disease). Cone-nosed ("kissing") bugs are the insect vector. They transmit the infective trypomastigote during a blood meal through their feces. Scratching introduces the organism into the bite wound or conjunctiva. A local lesion (**chagoma**) forms at the entry site and is accompanied by fever.

Spreading occurs via lymphatics (producing lymphadenitis) and trypomastigotes may be found in the blood within a couple of weeks. Trypanosomes then become localized in reticuloendothelial cells of the spleen, liver, and bone marrow where they become **amastigotes** and multiply intracellularly. Infected individuals may infect the cone-nosed bugs during a subsequent blood meal.

American trypanosomiasis occurs in South and Central America. It may be fatal, mild, or asymptomatic in adults. It is especially severe in children, who often introduce the trypanosome through the conjunctiva, leading to edema of the eyelids and face of the affected side. The disease may spread to the central nervous system or to the heart, causing severe myocarditis.

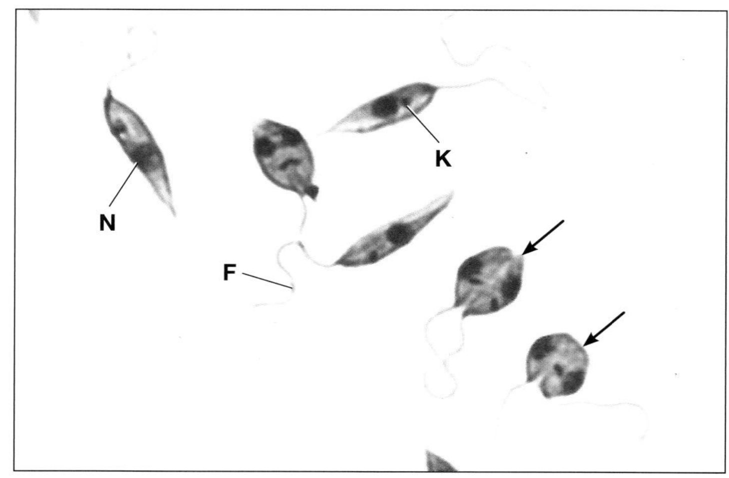

16.9 ***Leishmania donovani*** **Promastigotes (Excavata; Phylum Euglenozoa)** • Promastigotes range in size from 15–20 µm long by 1.5–3.5 µm wide and are the infectious stage in the life cycle. They enter the host during a sand fly bite. Notice the anterior flagellum (**F**), the kinetoplast (**K**), and the nucleus (**N**). The two cells at the lower right are dividing (**arrows**). These cells were grown in culture.

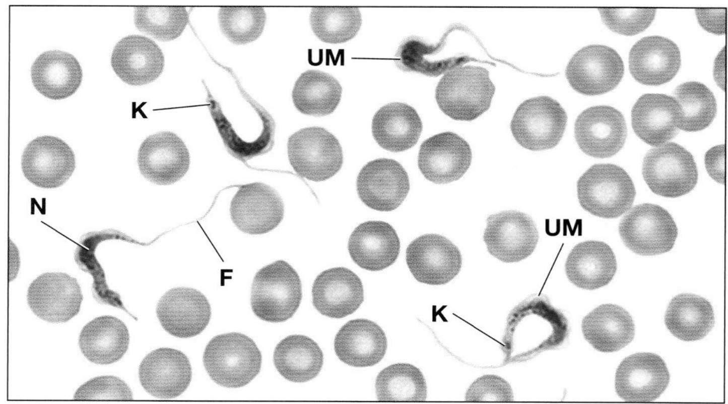

16.10 ***Trypanosoma brucei*** **Trypomastigotes in a Blood Smear (Excavata; Phylum Euglenozoa)** • This species is the causative agent of East African trypanosomiasis, and the trypomastigote is the infectious stage. Trypomastigotes range in size from 14–33 µm long by 1.5–3.5 µm wide. The central nucleus (**N**), posterior kinetoplast (**K**), anterior flagellum (**F**), and undulating membrane (**UM**) are visible. Trypomastigotes are variable in size, but may reach a length of 40 µm. (Giemsa Stain)

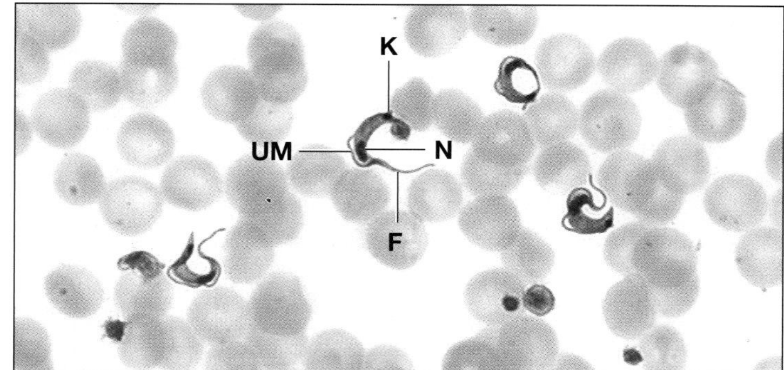

16.11 ***Trypanosoma cruzi*** **Trypomastigotes in a Blood Smear (Excavata; Phylum Euglenozoa)** • *T. cruzi* is the causative agent of American trypanosomiasis (Chagas disease) and its vector is the cone-nosed beetle. Cells in blood smears are about 20 µm long and assume a "C" or "V" shape, as seen here. Visible are the nucleus (**N**), undulating membrane (**UM**), posterior kinetoplast (**K**), and anterior flagellum (**F**). (Wright's Stain)

Group SAR

Subgroup Stramenopila (Heterokontophyta)

Blastocystis spp.

Blastocystis spp. (formerly *Blastocystis hominis*) is a stramenopile (see p. 231) commensal microbe that occupies the large intestine of up to 20% of the population. In most cases, patients carrying *Blastocystis* trophozoites show no symptoms, but in some "it" (actually, one of more than a dozen species) appears responsible for symptoms such as fever, nausea, diarrhea, and abdominal cramps.

Fecal-oral transmission of the cyst is responsible for its spread. Identification is made by finding cysts (the central body form), in a stool sample (Fig. 16.12) and by detecting antibodies with ELISA (see p. 137) or fluorescent antibody tests (see p. 140).

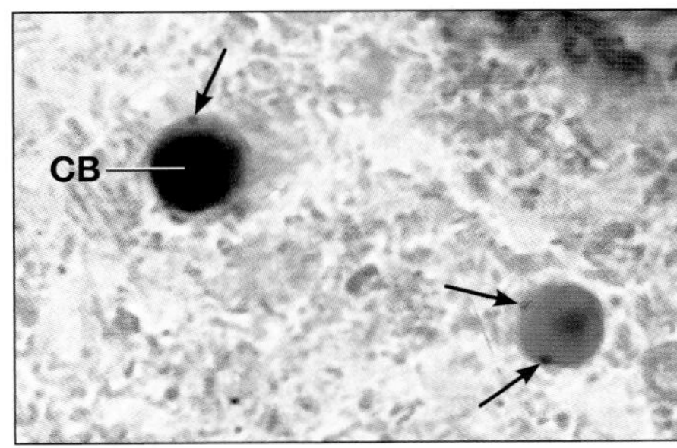

16.12 ***Blastocystis hominis*** **Trophozoites in a Fecal Smear (Stramenopila)** • Trophozoites vary greatly in size over the range of 6–40 µm. A large central body (**CB**) surrounded by several small nuclei is distinctive of the trophozoite (**arrows**). Staining properties also may vary, as shown in these two specimens. (Trichrome Stain)

Subgroup Alveolata

Babesia microti

Babesia microti belongs to a group of animal parasites called apicomplexans (see p. 232). It is a blood parasite found in the northeastern United States and other parts of the world. Mice (genus *Peromyscus*) are the main reservoir and the parasite is introduced into human hosts by a bite from an infected tick (genus *Ixodes)*. The parasites then infect RBCs and resemble young *Plasmodium* trophozoites (Fig. 16.13).

Symptoms of babesiosis, which are not easily distinguished from other diseases, appear in non-splenectomized individuals after about a week and include malaise, fever, and generalized aches and pains. Immunosuppressed or splenectomized hosts infected by *Babesia* may experience a fulminant, life-threatening hemolytic disease. Diagnosis is by detection of parasites in the blood.

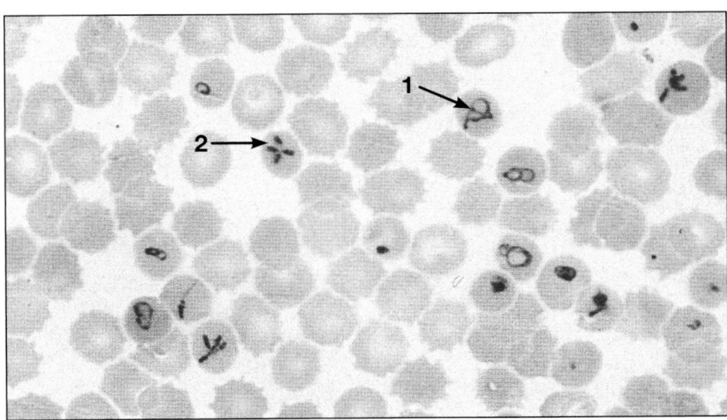

16.13 ***Babesia microti*** **Trophozoites in a Blood Smear (Alveolata; Phylum Apicomplexa)** • *Babesia microti* infects red blood cells at the sporozoite stage. Once inside, they develop into trophozoites (ring forms). The ring forms resemble *Plasmodium falciparum* (**arrow 1**), but are more variable in size, never pigmented, and occasionally form a cross-like tetrad (**arrow 2**). (Giemsa Stain)

Cryptosporidium parvum

Cryptosporidium parvum is an apicomplexan (see p. 232) parasite of intestinal microvilli and causes cryptosporidiosis. Infection of immunocompetent individuals is usually self-limiting, but infection of immunocompromised patients results in a long-term disease characterized by profuse, watery diarrhea.

Infective **oocysts** containing **sporozoites** (Fig. 16.14) in fecal-contaminated water or food are ingested by the host. Sporozoites then undergo multiple developmental stages in the small intestine, eventually producing oocysts, which are passed in the feces and complete the cycle. Infection may also occur through contact with infected animals. Diagnosis is made by finding acid-fast oocysts in feces or fluorescent antibody detection from stool samples (see p. 140).

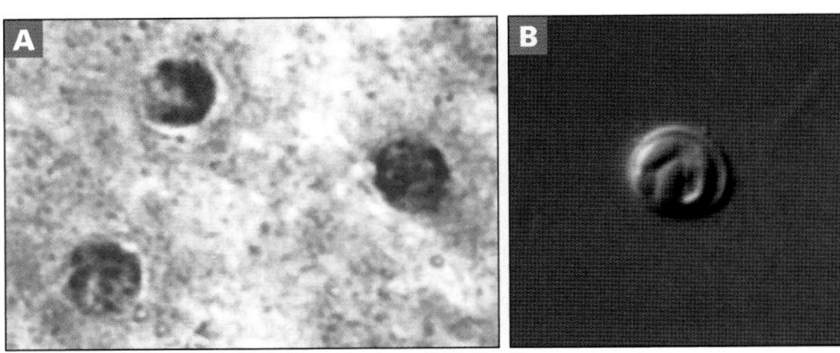

16.14 ***Cryptosporidium parvum*** **Oocysts (Alveolata; Phylum Apicomplexa)** • (**A**) Cryptosporidium oocysts contain sporozoites (visible as denser-staining objects within the cyst), which, after ingestion, are released in the small intestine. These undergo progression through several different developmental stages, but ultimately become oocysts that are passed in the feces. Oocysts are typically about 5 µm in diameter. These were found in a fecal sample. (Modified Acid-Fast Stain) (**B**) Water departments are greatly concerned about fecal contamination in drinking water, so they screen for parasites, such as *Cryptosporidium* and others. This oocyst, with obvious sporozoites, was found in water prior to treatment. (Differential Interference Contrast Microscopy)

Plasmodium spp.

Plasmodia are apicomplexans (see p. 232) with a complex life cycle, part of which is in various vertebrate cells while the other part involves an insect. In humans, liver and red blood cells are infected, and the insect vector is the female *Anopheles* mosquito. However, in some parts of the world transmission may also occur via blood transfusion or use of contaminated needles. A generalized life cycle is shown in Figure 16.15. Representative life cycle stages for the various species are shown in Figures 16.16 to 16.23.

Four species of *Plasmodium* cause different forms of malaria in humans, with the first two accounting for more than 90% of the infections:

1. *P. vivax* (benign tertian malaria)
2. *P. falciparum* (malignant tertian malaria)
3. *P. malariae* (quartan malaria)
4. *P. ovale* (ovale malaria)

The life cycles are similar for each species, as is the progress of the disease, so the following description is a generic one with only certain specific variations added.

The **sporozoite** stage of the pathogen is introduced into a human host's blood during a bite from an infected female *Anopheles* mosquito. Sporozoites then infect liver cells and produce the asexual **merozoite** (trophozoite) stage. Merozoites are released from lysed liver cells, enter the blood, and infect erythrocytes. (Reinfection of the liver occurs at this stage in all except *P. falciparum* infections.) Once in the erythrocytes, merozoites feed on hemoglobin, become vacuolated, and take on a ring shape.

Merozoite reproduction in RBCs goes through a multinucleate **schizont** stage, which undergoes cytokinesis to produce the uninucleate merozoites. (In *P. falciparum* infections, this occurs in visceral capillaries and the schizont is not seen in peripheral blood.) Thus begins a cyclic pattern of reproduction in which more merozoites are produced and released from the red blood cells synchronously.

Merozoite reproductive cycles are tied to the symptoms of malaria. For the first week or so after infection, merozoites are reproducing in the liver and the patient is asymptomatic. Once released from the liver, multiple merozoites infect RBCs, but only one lineage survives and a synchronized periodicity of classic symptoms begins.

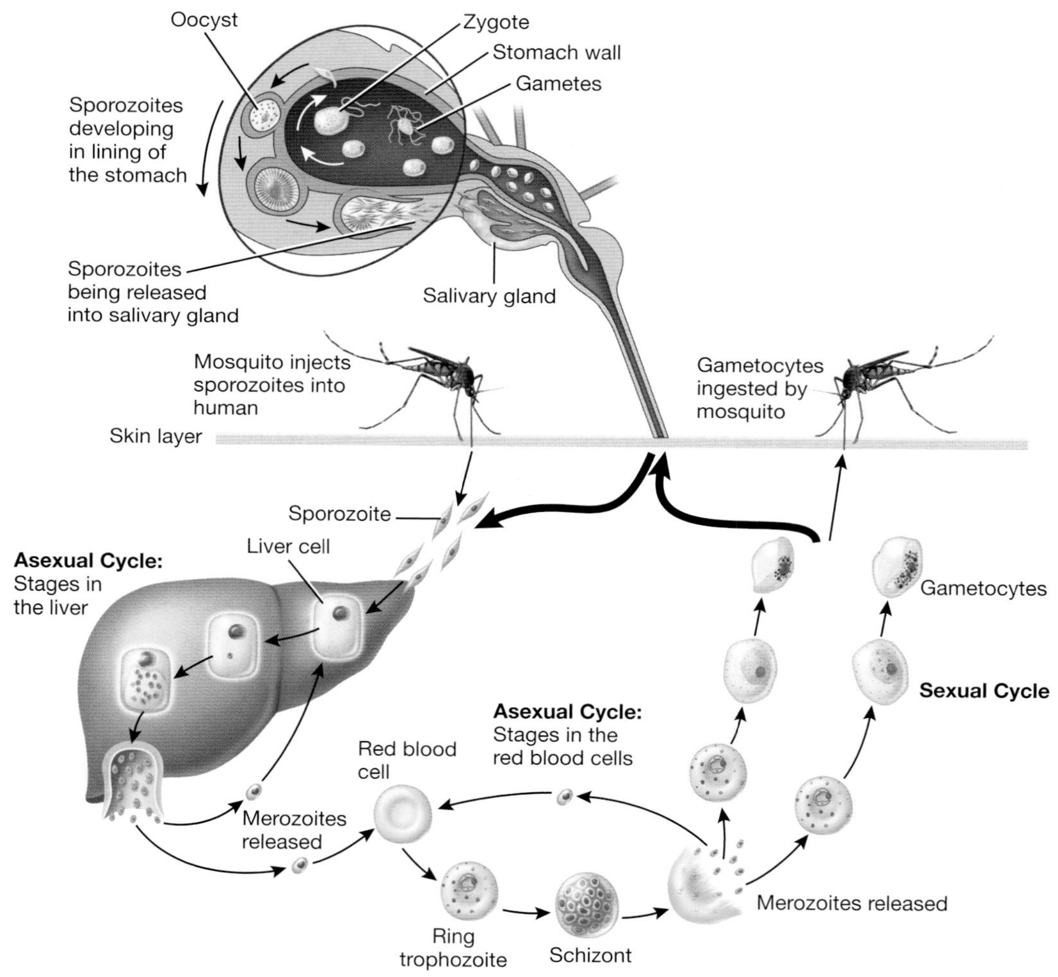

16.15 Generalized *Plasmodium* Life Cycle • See the text for details.

Erythrocyte rupture and release of merozoites and metabolic wastes lead to chills (lasting an hour or so), fever (lasting a few hours), and sweating (lasting a few hours). These are followed by a return to normal or subnormal body temperature during which merozoites re-infect the red blood cells and the cycle repeats.

The sexual phase of the life cycle begins when certain merozoites enter erythrocytes and differentiate into male or female **gametocytes**, which are ingested by a female *Anopheles* mosquito during a blood meal. Fertilization occurs in the mosquito and the zygote eventually develops into a cyst within its gut wall. After many divisions, the cyst releases sporozoites, some of which enter the mosquito's salivary glands ready to be transmitted back to the human host.

Most malarial infections are eventually cleared, but not before the patient has developed anemia and has suffered permanent damage to the spleen and liver. The most severe infections involve *P. falciparum*.

Erythrocytes infected by *P. falciparum* develop abnormal projections that cause them to adhere to the lining of small blood vessels. This can lead to obstruction of the vessels, thrombosis, or local ischemia, which account for many of the fatal complications of this type of malaria—including liver, kidney, and brain damage.

Diagnosis of malaria is made by a combination of appropriate symptoms presented by the patient and a positive identification of the parasite in a blood smear. If speciation is not possible from blood smears, PCR (see p. 144) is available.

Treatment is with antimalarial drugs, but a number of factors come into play before a choice is made. These include: *Plasmodium* species involved, parasite load, drug resistance, and patient travel history, among others. Travelers to parts of the world where they may be exposed to malaria are advised to take doxycycline prophylactically before, during, and for four weeks after their trip.

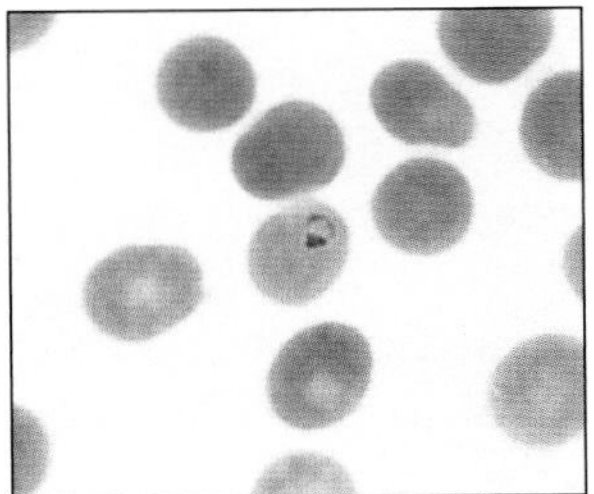

16.16 ***Plasmodium falciparum* Ring Stage in a Blood Smear (Alveolata; Phylum Apicomplexa)** • The ring stage is a young trophozoite (merozoite). Note the chromatin dots in the nucleus. (Wright's Stain)

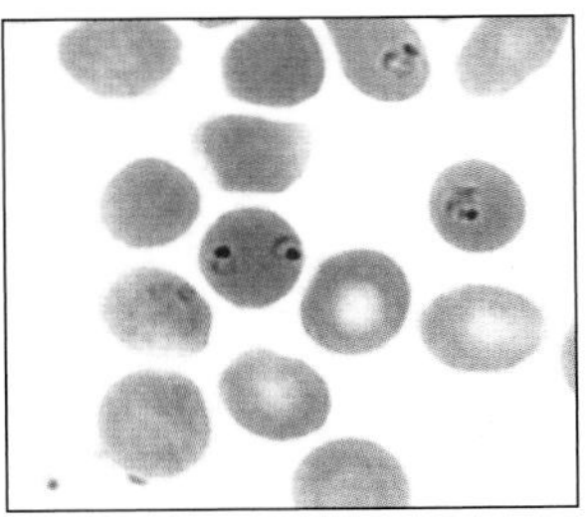

16.17 ***Plasmodium falciparum* Double Infection in a Blood Smear (Alveolata; Phylum Apicomplexa)** • It is not unusual to find RBCs with a double infection with *P. falciparum*. (Wright's Stain)

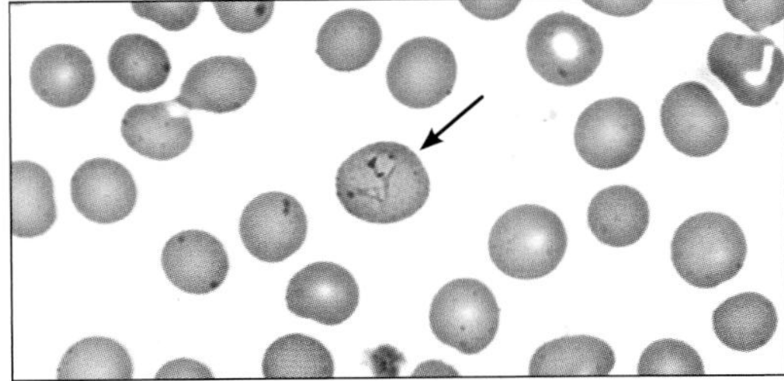

16.18 Erythrocyte Infected with *Plasmodium vivax* in a Blood Smear (Alveolata; Phylum Apicomplexa) • The parasite is in the ring stage, and the RBC (**arrow**) exhibits characteristic cytoplasmic Schüffner's dots—not many, but they're there. Schüffner's dots are also seen in RBCs infected with *P. ovale*. Their presence assists in identification of the infecting species. (Giemsa Stain)

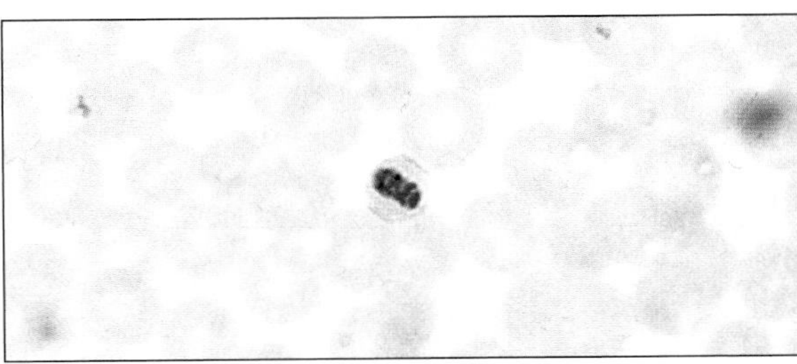

16.19 Band Trophozoite of *Plasmodium malariae* in a Blood Smear (Alveolata; Phylum Apicomplexa) • Older trophozoites of *P. malariae* may elongate to form a band. This is useful in identifying the infecting species. (Giemsa Stain)

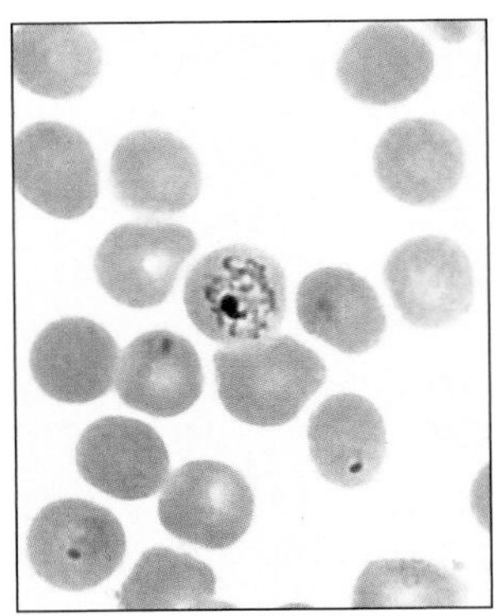

16.20 ***Plasmodium falciparum* Developing Schizont in a Blood Smear (Alveolata; Phylum Apicomplexa)** • Schizonts are a multinucleate stage that develops from the trophozoite (merozoite). Cytokinesis produces a new generation of merozoites (one for each nucleus), which lyse the RBC and are released to produce another round of infection. This process is responsible for the periodicity of symptoms associated with malaria. *P. falciparum* schizonts are not usually seen in peripheral blood smears because schizont-infected cells reside in visceral capillaries. (Wright's Stain)

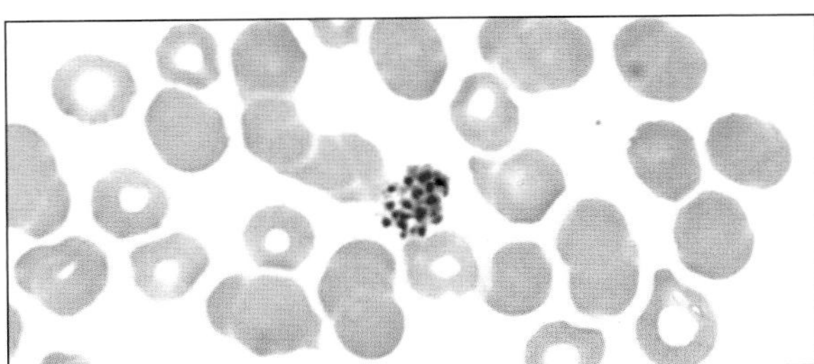

16.21 Mature *Plasmodium vivax* Schizont in a Blood Smear (Alveolata; Phylum Apicomplexa) • This schizont has approximately 16 merozoites. More than 12 merozoites produced by a schizont distinguishes *P. vivax* from *P. malariae* and *P. ovale*, which both typically have eight, but up to 12. A *P. falciparum schizont* may have up to 24 merozoites, but these are not typically seen in peripheral blood smears and so aren't likely to be confused with *P. vivax*. (Wright's Stain)

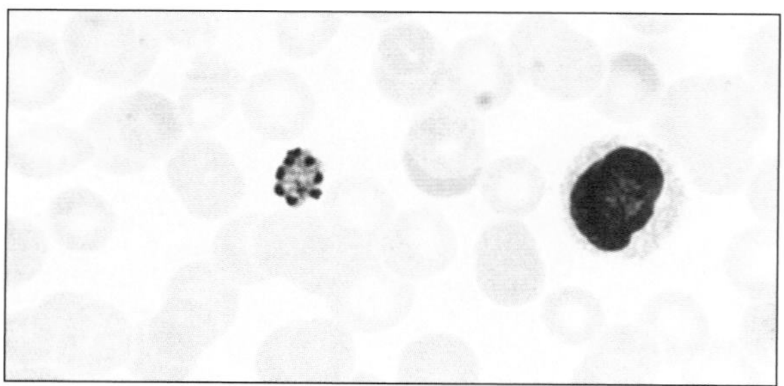

16.22 ***Plasmodium malariae*** **Schizont in a Blood Smear (Alveolata; Phylum Apicomplexa)** ● The *P. malariae* schizont has eight merozoites in a distinctive rosette arrangement. (Giemsa Stain)

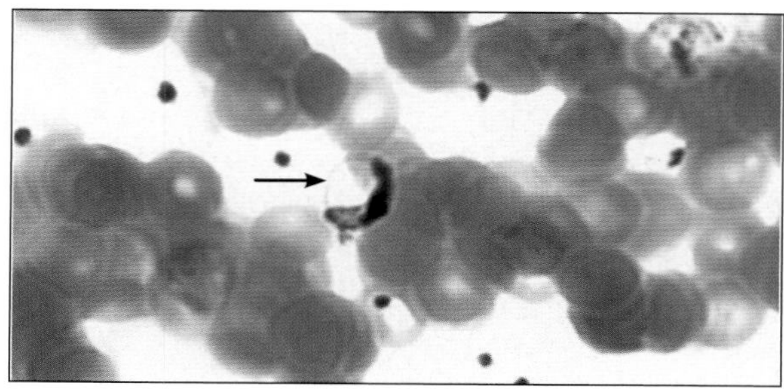

16.23 ***Plasmodium falciparum*** **Gametocyte in a Blood Smear (Alveolata; Phylum Apicomplexa)** ● Gametocytes are the last life cycle stage in a *Plasmodium* infection. If ingested by a female *Anopheles* mosquito during a blood meal, they differentiate into male and female gametes within her gut. Fertilization follows and the zygotes eventually develop into sporozoites that migrate to her salivary gland ready to be deposited into the next host. Differentiation between microgametocytes ("male") and megagametocytes ("female") is difficult in this species. The erythrocyte membrane is visible around the gametocyte (**arrow**). (Wright's Stain)

Toxoplasma gondii

Toxoplasma gondii (Fig. 16.24) is an apicomplexan (see p. 232) with worldwide distribution and is the causative agent of toxoplasmosis. It is an obligate intracellular parasite and cats are the definitive hosts.

Its life cycle has sexual and asexual phases, with the sexual phase only occurring in the lining of cat intestines, where thick-walled, resistant **oocysts** are produced and shed in the feces. Each mature oocyst contains eight **sporozoites** and if ingested by a cat, the sexual cycle may be repeated as the sporozoites produce **gametocytes,** which in turn produce gametes, undergo fertilization, and produce another oocyst generation.

Ingestion of an oocyst by an intermediate host (animal or human) begins the asexual cycle. Oocyst germination occurs in the duodenum and the released sporozoites infect the epithelial cells. After division and differentiation, rapidly dividing, motile **tachyzoites** (trophozoites) are released into the blood and spread the infection to lymph nodes and other parts of the reticuloendothelial system (e.g., monocytes).

As the infection progresses, tachyzoites differentiate into slowly dividing **bradyzoites,** which in turn form **tissue cysts** that may reside in the host's skeletal muscle, liver, heart, or brain for the rest of its life. Tissue cysts ingested by a cat eating an infected animal develop into gametocytes in the cat's intestines. Gametes are formed, fertilization produces an oocyst, and the life cycle is completed. Ingestion of tissue cysts by an intermediate host leads to production of tissue cysts within that host.

Ingestion of oocysts or tissue cysts and transplacental transmission to a fetus are the main routes of human infection. Acute infection of immunocompetent individuals results in mild cold or flu-like symptoms, swollen cervical lymph nodes, or there may be no symptoms at all. Once tissue cysts are formed, spreading of the pathogen usually stops and any symptoms disappear.

The more serious form of the disease involves infection of a fetus across the placenta (especially in the first or second trimester) from an infected mother. This type of infection may result in stillbirth, or liver damage and brain damage. Infection of immunocompromised patients or rupturing of an existing tissue cyst typically leads to CNS infection and subsequent alteration of neurological functioning.

Unlike many of the other pathogens, diagnosis is not routinely made microscopically because *Toxoplasma* is found mostly within cells that are not easily sampled (as opposed to blood pathogens, for example). Instead, diagnosis usually is done by testing for *Toxoplasma* antibodies.

Balantidium coli

Balantidiasis is found worldwide and is caused by the ciliate (see p. 232) pathogen *Balantidium coli*, which exists in two forms: a vegetative trophozoite (Fig. 16.25) and a cyst (Fig. 16.26). Laboratory diagnosis is made by identifying either the cyst or the trophozoite in feces, with the latter being more commonly found. A major risk factor is contact with pigs.

The ciliate trophozoite is highly motile and has a macronucleus and a micronucleus. Cysts in sewage-contaminated water are the infective form and after ingestion become trophozoites in the intestines.

Trophozoites reproduce by longitudinal division and live either in the colon's lumen or attach to its mucosa, generally feeding on bacteria. They may cause mucosal ulcerations, but not to the extent produced by *Entamoeba histolytica*.

Symptoms of acute infection include bloody and mucoid feces, nausea, vomiting, and abdominal tenderness, among others. Diarrhea alternating with constipation may occur in chronic infections. Most infections probably are asymptomatic.

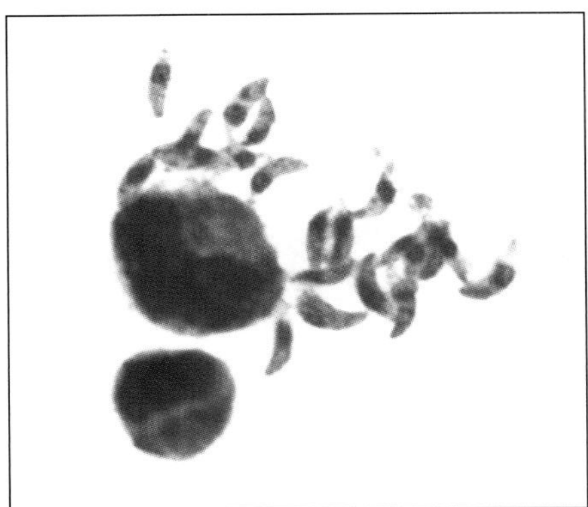

16.24 ***Toxoplasma gondii*** **Tachyzoites (Trophozoites) in a Blood Smear (Alveolata; Phylum Apicomplexa)** • *Toxoplasma* tachyzoites are responsible for asexual reproduction in the *Toxoplasma* life cycle. See the text for details. Tachyzoites are motile, bow-shaped cells with prominent nuclei. They are about 6 µm in length. The larger cells are leukocytes.

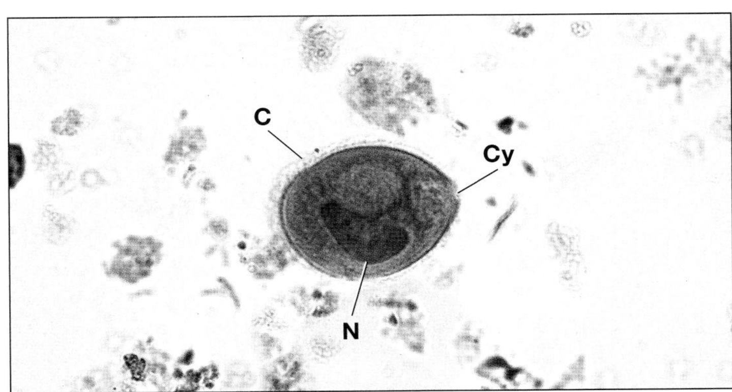

16.25 ***Balantidium coli*** **Trophozoite in a Fecal Smear (Alveolata; Phylum Ciliophora)** • Trophozoites are oval in shape with dimensions of 50–100 µm long by 40–70 µm wide. Cilia (**C**) cover the cell surface. Internally, the macronucleus (**N**) is prominent; the adjacent micronucleus is not. An anterior cytostome (**Cy**) is usually visible in most specimens.

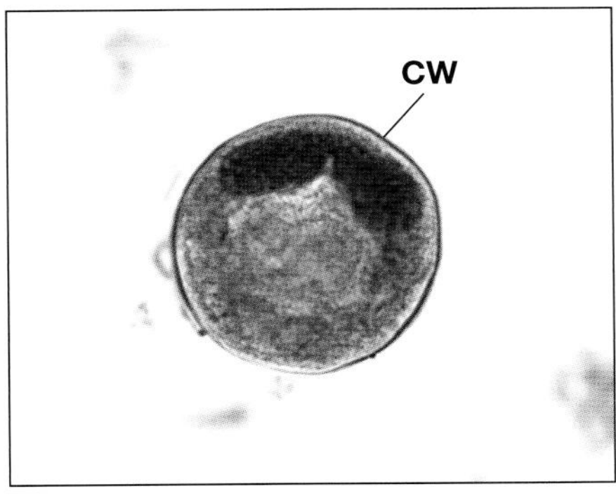

16.26 ***Balantidium coli*** **Cyst in a Fecal Smear (Alveolata; Phylum Ciliophora)** • Cysts are usually spherical and have a diameter in the range of 50–75 µm. There is a cyst wall (**CW**) and the cilia are absent. As in the trophozoite, the macronucleus is prominent, but the micronucleus may not be.

Group Unikonta

Phylum Amoebozoa

Entamoeba histolytica

Entamoeba histolytica belongs to the phylum Amoebozoa (p. 238). It is the causative agent of amebic dysentery (amebiasis), a disease most common in areas with poor sanitation worldwide, but especially regions with warmer climates.

Identification is made by finding either trophozoites (Fig. 16.27) or cysts (Fig. 16.28) in a stool sample, though serological tests and PCR (see p. 144) have been shown to be more sensitive and specific in differentiating *E. histolytica* from *E. dispar*, a nonpathogenic species morphologically indistinguishable from *E. histolytica*.

Infection occurs when a human host ingests cysts, either through direct fecal-oral contact or contaminated food or water. It is the ability of cysts to withstand the stomach's acidic environment that makes them the infective stage; they are able to deliver trophozoites safely to the less harsh environment of the small intestine.

Excystation in the small intestine results in release of one tetranucleate trophozoite in which each nucleus goes through one mitotic division (producing eight nuclei) followed by the cell doing three rounds of cytokinesis to produce eight small, uninucleate trophozoites.

Trophozoites live in the colon and feed on bacteria. As a consequence of relative dehydration (the colon's primary job is to reabsorb water from the undigested, unabsorbed food in it), trophozoites convert into cysts that are passed in the feces. Mature cysts have four nuclei. They are hardy and able to survive several days outside the host, which makes them problematic as contaminants of water and food. If ingested, the life cycle is completed.

Most infections (~90%) are asymptomatic, because the trophozoites form colonies in the mucus lining the colon and feed there. Invasive amebiasis occurs when trophozoites penetrate the mucus layer, attach to a specific epithelial receptor, and cause ulceration of the mucosa leading to acute amebic dysentery, fulminating colitis, or appendicitis.

In the most severe cases, trophozoites enter the blood and infect other organs, especially the liver, lungs, or brain. Infections can be treated with tissue amebicides (e.g., metronidazole) that target trophozoites and/or luminal amebicides that target cysts.

Another member of the genus *Entamoeba* deserves mention here. *Entamoeba coli* is a fairly common, nonpathogenic intestinal commensal that must be differentiated from *E. histolytica* in stool samples. Its characteristic features are given in the captions for Figures 16.29 and 16.30.

Endolimax nana

Endolimax nana is a very common commensal amoebozoan (see p. 238) that resides mainly in the cecum of humans. It exists as a trophozoite and cyst (Figs. 16.31 and 16.32) and its life cycle resembles that of *Entamoeba histolytica.* Infection occurs by ingestion of cysts in fecal-contaminated food or water. Identification is made by finding the trophozoites and/or cysts in stool samples.

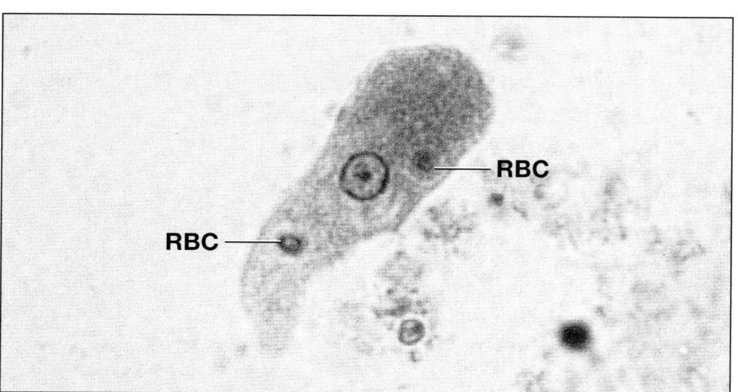

16.27 ***Entamoeba histolytica*** **Trophozoite in a Fecal Smear (Unikonta; Phylum Amoebozoa)** • Trophozoites range in size from 12–60 µm. Notice the small, central karyosome, the beaded chromatin at the nucleus's margin, the ingested red blood cells (**RBC**), and the finely granular cytoplasm. Compare with an *Entamoeba coli* trophozoite in Figure 16.29. (Iron Hematoxylin Stain)

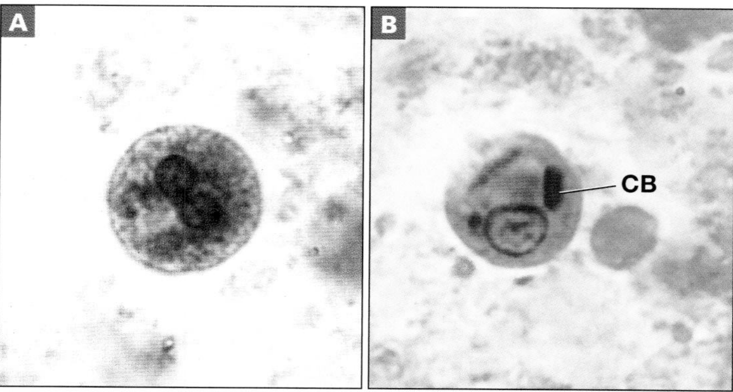

16.28 ***Entamoeba histolytica*** **Cysts in Fecal Smears (Unikonta; Phylum Amoebozoa)** • (A) *E. histolytica* cysts are spherical with a diameter of 10–20 µm. Two of the four nuclei are visible; other nuclear characteristics are as in the trophozoite. Compare with an *Entamoeba coli* cyst in Figure 16.30. (Iron Hematoxylin Stain) (B) This *E. histolytica* cyst has cytoplasmic chromatoidal bars (**CB**). These are found in approximately 10% of the cysts, have blunt ends, and are composed of ribonucleoprotein. (Trichrome Stain)

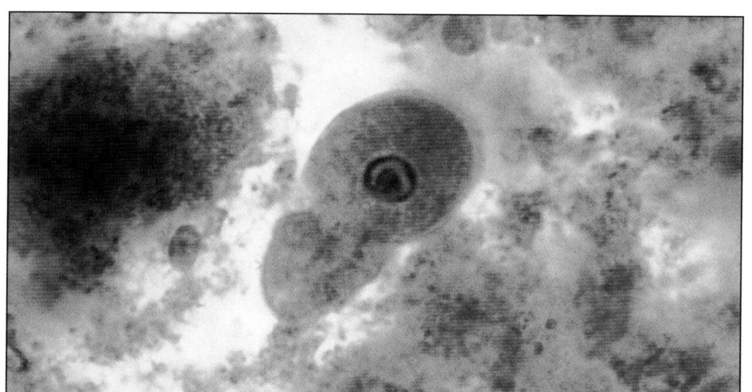

16.29 ***Entamoeba coli*** **Trophozoite in a Fecal Smear (Unikonta; Phylum Amoebozoa)** • Trophozoites have a size range of 15–50 µm. Notice the relatively large and eccentrically positioned karyosome in the nucleus, the unclumped chromatin at the nucleus's periphery, and the vacuolated cytoplasm (lighter regions) lacking ingested RBCs. The usually nonpathogenic *E. coli* must be distinguished from the potentially pathogenic *E. histolytica*, so compare with Figure 16.27. (Trichrome Stain)

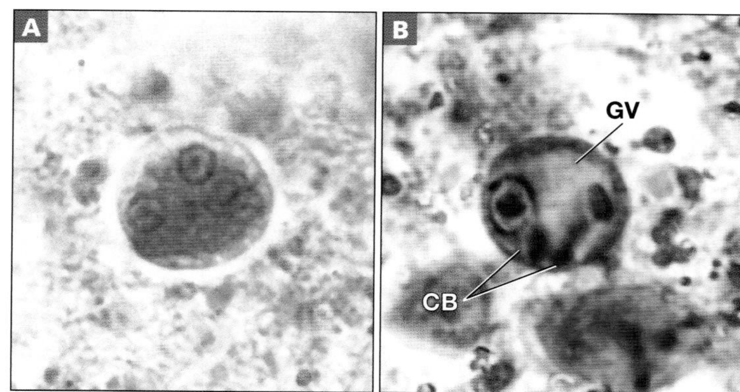

16.30 ***Entamoeba coli*** **Cysts in Fecal Smears (Unikonta; Phylum Amoebozoa)** • Cysts are typically spherical and are between 10–35 µm in diameter. They contain eight, or sometimes 16 nuclei. This makes differentiation from *E. histolytica* cysts simpler, because they never have more than four nuclei. (A) Five nuclei are visible in this specimen. (Trichrome Stain) (B) Chromatoidal bars (**CB**) and a large glycogen vacuole (**GV**) characteristic of immature *E. coli* cysts are visible in this specimen. Compare with Fig. 16.28. (Trichrome Stain)

Iodamoeba bütschlii

Iodamoeba bütschlii is less commonly found in humans than *Entamoeba coli* or *Endolimax nana*, but when present, it lives in the cecum and feeds on the other resident organisms. Transmission is by fecal contamination, but it is nonpathogenic. Laboratory diagnosis is done by identifying trophozoites and/or cysts (Figs. 16.33 and 16.34) in stool specimens. Distinguishing characteristics are given in the captions. ●

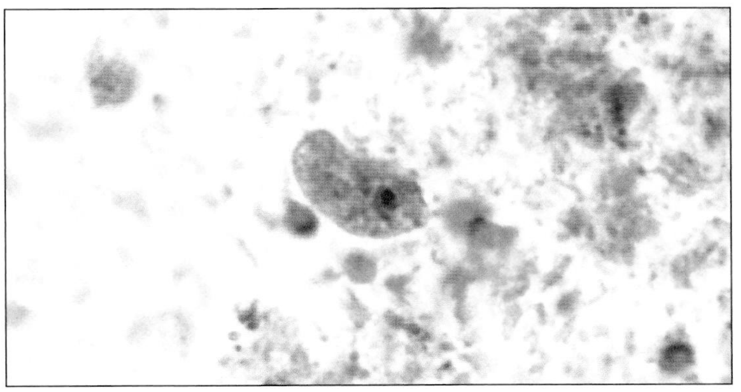

16.31 ***Endolimax nana*** **Trophozoite in a Fecal Smear (Unikonta; Phylum Amoebozoa)** ● Trophozoites are small, with a size range of 6–15 µm. Notice the large karyosome filling the majority of the nuclear space as well as the absence of peripheral chromatin. (Trichrome Stain)

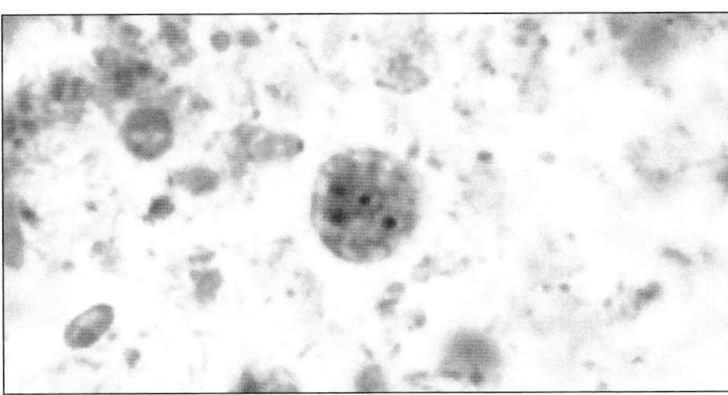

16.32 ***Endolimax nana*** **Cyst in a Fecal Smear (Unikonta; Phylum Amoebozoa)** ● *E. nana* cysts range in size from 5–14 µm. There are typically four nuclei, with each having the distinctive large karyosome. (Trichrome Stain)

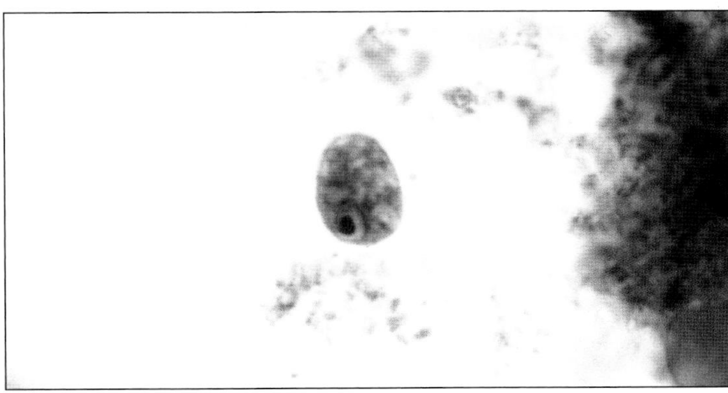

16.33 ***Iodamoeba bütschlii*** **Trophozoite in a Fecal Smear (Unikonta; Phylum Amoebozoa)** ● *I. bütschlii* trophozoites are 6–15 µm in length. The nucleus has a large karyosome, but lacks peripheral chromatin. On occasion, fine karyosome strands may be observed radiating outward from the karyosome to the nuclear membrane. (Iron Hematoxylin Stain)

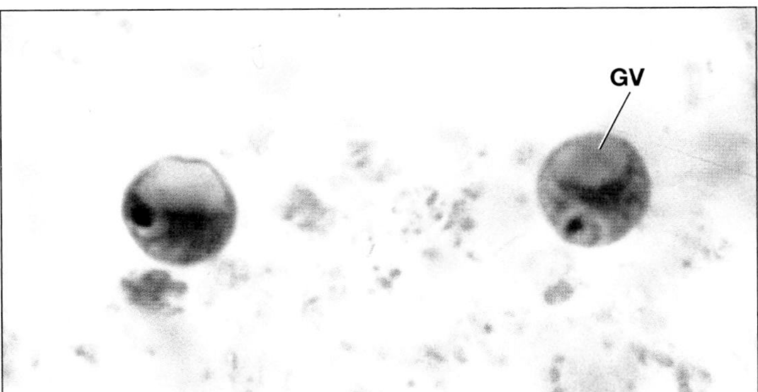

16.34 ***Iodamoeba bütschlii*** **Cysts in a Fecal Smear (Unikonta; Phylum Amoebozoa)** ● Cysts of *I. bütschlii* range in size from 6–15 µm. The single nucleus has a large karyosome and no peripheral chromatin. A glycogen vacuole (**GV**) is typically found in these cysts. (Iron Hematoxylin Stain)

Parasitic Helminths

SECTION 17

A study of helminths is appropriate to the microbiology lab, because clinical specimens may contain microscopic evidence of helminth infection. The three major groups of parasitic worms encountered in lab situations are **trematodes** (flukes), **cestodes** (tapeworms), and **nematodes** (roundworms).

Life cycles of parasitic worms are often complex, sometimes involving several hosts, and their details are beyond the scope of this book. Emphasis here is on a brief background and clinically important diagnostic features of selected worms.

Trematode Parasites Found in Clinical Specimens

Trematode worms belong to the phylum Platyhelminthes, which literally means "flat worm," and is their common name as well as a description of their shape. Flatworms lack a body cavity, true organs, and (usually) have only one opening to their digestive tract.

The class Trematoda is largely composed of parasitic species with life cycles involving two or more hosts (one definitive host—a vertebrate—and one or more intermediate hosts—frequently a mollusk, such as a snail) and multiple developmental stages. The larval stages reproduce asexually, but the adult worms, which live in the definitive host, are generally hermaphroditic and represent the sexual stage of the life cycle.

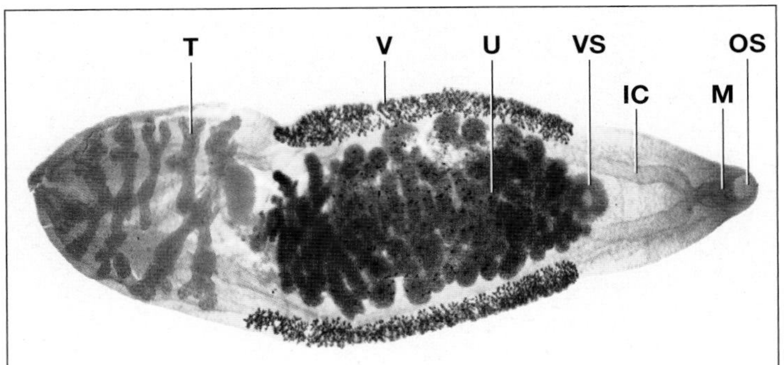

17.1 ***Clonorchis sinensis*** **Adult (Phylum Platyhelminthes; Class Trematoda)** Adults range in size from 1–2.5 cm. Visible in this specimen are the oral sucker (**OS**), mouth (**M**), intestinal ceca (**IC**), ventral sucker (**VS**), uterus (**U**) with eggs, vitellaria–yolk glands (**V**), and testis (**T**). (Stained Whole Mount)

Clonorchis (Opisthorchis) sinensis

Clonorchis sinensis is the Chinese liver fluke (Fig. 17.1), which causes clonorchiasis, a liver disease. It is a common parasite of people living in China, Japan, Korea, Vietnam, and Taiwan, and is becoming more common in the United States. Infection typically occurs when undercooked, infected fish is ingested. The ingested juveniles mature into adults in the duodenum and migrate to the liver bile ducts where they begin producing eggs in approximately one month. Eggs passed in the feces continue larval development within aquatic snails and then within fish to complete the life cycle.

Most infections are asymptomatic, but patients may have a fever, abdominal pain, and jaundice. Damage to the bile duct epithelium, liver, and gallbladder is related to the number of worms and the duration of infection. Diagnosis is made by identifying the characteristic eggs in feces (Fig. 17.2) and anti-helminthic medication is available for treatment. Preventative measures include improved sanitation and thorough cooking of fish and freshwater plants.

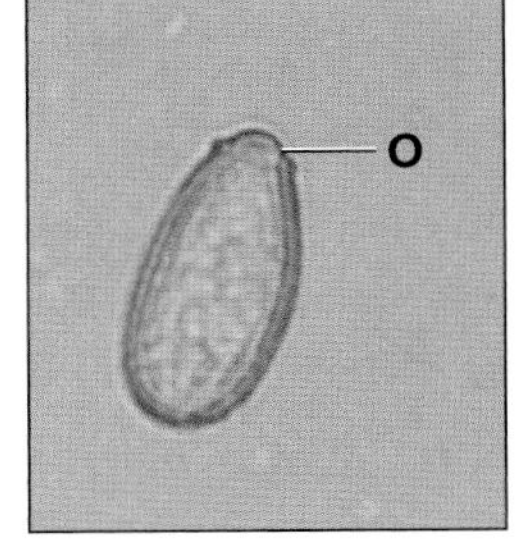

17.2 ***Clonorchis sinensis*** **Egg in a Fecal Specimen (Phylum Platyhelminthes; Class Trematoda)** Eggs are thick shelled and between 27 and 35 µm long. There is a distinctive operculum (**O**) that rests upon a ring ("shoulders") and often a knob at the abopercular end (not visible in this specimen). After passage in the feces and once in the snail host, the operculum detaches and the miracidium within emerges. (Wet Mount; D'Antoni's Iodine Stain)

Fasciola hepatica

Fasciola hepatica (Fig. 17.3) is the common liver fluke and causes fascioliasis. Domestic sheep and cattle are typical mammalian hosts. Human infection results when **metacercaria** attached to aquatic vegetation are ingested. Penetration of the intestine and subsequently the liver by metacercaria leads the juveniles to the bile ducts where they develop into adults.

Migration by the juveniles damages the liver; adults damage the bile ducts, gall bladder, and liver, resulting in cirrhosis, jaundice, or in severe cases, abscesses. Eggs are passed in the feces and aquatic snails are the intermediate host that releases metacercaria. *F. gigantica* is a close relative that also produces fascioliasis. Diagnosis is made by identification of eggs in feces (Fig. 17.4). ELISAs identifying antigenic secretory and excretory products also are in use. Anti-helminthic medication is available for treatment.

Fasciolopsis buski

Fasciolopsis buski (Fig. 17.5) is common in central and Southeast Asia where it infects humans and pigs. It causes fasciolopsiasis and is the largest human fluke parasite. Its life cycle is similar to *Fasciola hepatica*, but differs in that its infection site is the small intestine, not the liver.

Consequences of infection include inflammation of the intestinal wall and obstruction of the gut if the worms are numerous. Chronic infections lead to ulceration, bleeding, diarrhea, and abscesses of the intestinal wall. Metabolites from the worm may sensitize the host, which may cause death. Diagnosis is made by identification of the eggs (up to 25,000 per day per worm!) in fecal samples (Fig. 17.6). Anti-helminthic medication is available for treatment.

Paragonimus westermani

Paragonimus westermani (Fig. 17.7) is a lung fluke and one of several species to cause paragonimiasis, a disease mostly found in Asia, Africa, and South America. *P. westermani* is primarily a parasite of carnivores, but humans (omnivores) may get infected when eating undercooked crabs or crayfish carrying the cysts (Fig. 17.8).

Ingested juveniles excyst in the duodenum, penetrate the intestinal wall and diaphragm, and find their way to the lungs where the adults mature. Eggs (Fig. 17.9) are released into the bronchioles in approximately 2 to 3 months and are

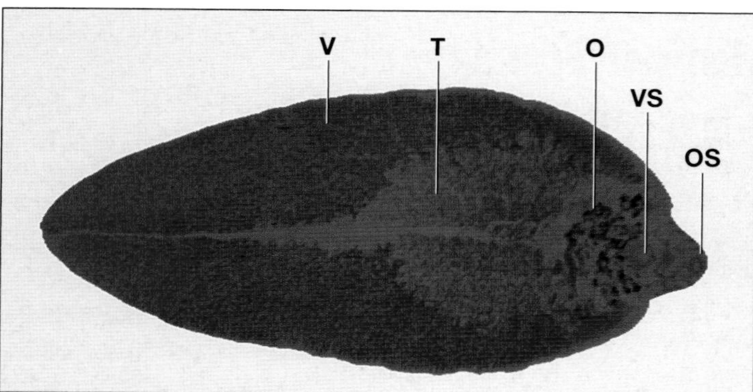

17.3 ***Fasciola hepatica* Adult (Phylum Platyhelminthes; Class Trematoda)** • Adults are leaf shaped with a conical anterior end. They are about 3 cm long by 1 cm in width. In this specimen, the oral sucker (**OS**), ventral sucker (**VS**), ovary (**O**), testis (**T**), and vitellaria (**V**) are visible. (Stained Whole Mount)

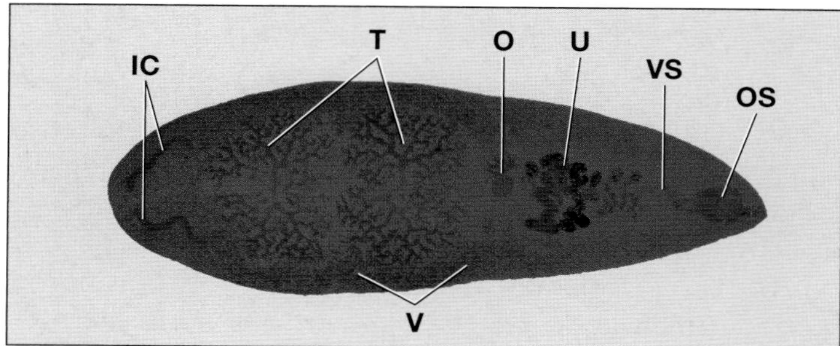

17.5 ***Fasciolopsis buski* Adult (Phylum Platyhelminthes; Class Trematoda)** • Adults are up to 7.5 cm long and 2 cm wide. *F. buski* lacks the anterior conical region of *Fasciola hepatica*. Visible are the oral sucker (**OS**), ventral sucker (**VS**), uterus (**U**) with eggs, ovary (**O**), vitellaria (**V**), highly branched testes (**T**), unbranched intestinal ceca (**IC**). (Stained Whole Mount)

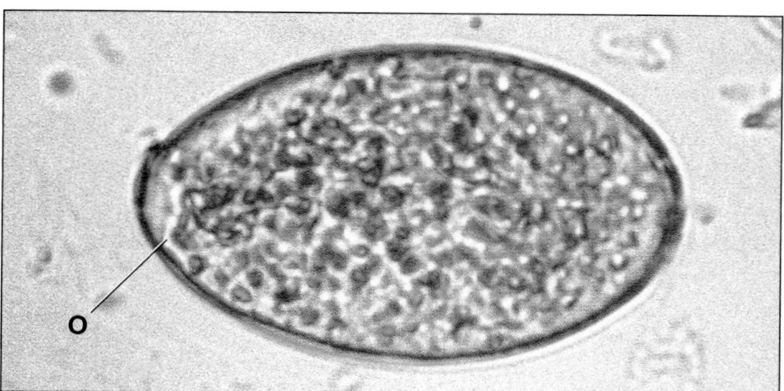

17.4 ***Fasciola hepatica* Egg in a Fecal Specimen (Phylum Platyhelminthes; Class Trematoda)** • The large eggs (130–150 µm long by 60–90 µm wide) are unembryonated in fecal samples and have an inconspicuous operculum (**O**). These eggs are difficult to distinguish from *Fasciolopsis buski*. (Wet Mount; D'Antoni's Iodine Stain)

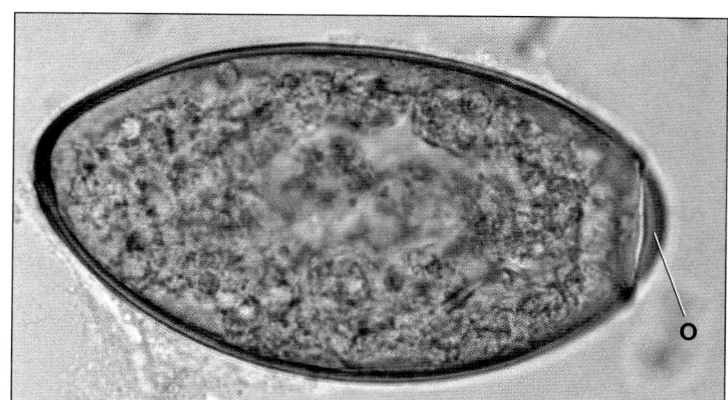

17.6 ***Fasciolopsis buski* Egg in a Fecal Specimen (Phylum Platyhelminthes; Class Trematoda)** • These eggs are unembryonated when passed in the feces and are very similar to those of *Fasciola hepatica*, but have a slightly more prominent operculum (**O**). Their sizes range from 130–140 µm long by 80–85 µm wide. (Wet Mount; D'Antoni's Iodine Stain)

diagnostic of infection. They may be passed in sputum or feces (as a result of coughing and swallowing) and find their way to the first intermediate host, which is a freshwater snail. Development continues in a second intermediate host, crabs or crayfish, which completes the life cycle.

Consequences of lung infection are a local inflammatory response followed by possible ulceration. Symptoms include cough with discolored or bloody sputum and difficulty breathing. These cases are rarely fatal, but may last a couple of decades. If the parasite load is small, the patient may experience no symptoms, though migration of juveniles through the body may cause pain. Occasionally, the wandering juveniles end up in other tissues, such as the brain or spinal cord, which can cause paralysis or death.

Diagnosis is made by finding eggs in feces or sputum or by serological testing for *P. westermani* antibodies in the patient. Anti-helminthic medication is available for treatment, but infection can be avoided by proper cooking of crabs and crayfish and better sanitation.

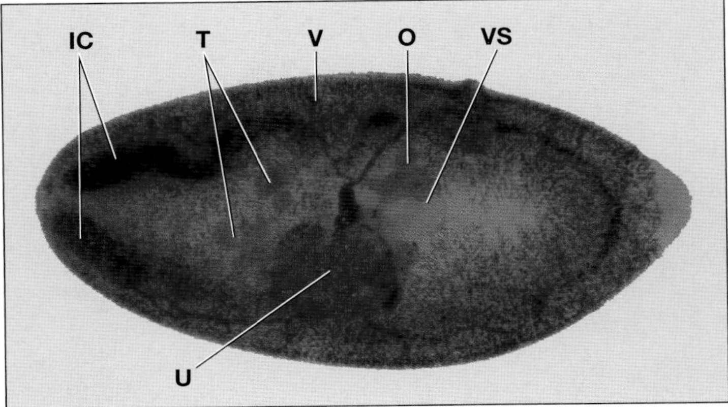

17.7 ***Paragonimus westermani*** **Adult (Phylum Platyhelminthes; Class Trematoda)** • Adults are up to 1.2 cm long, 0.6 cm wide, and 0.5 cm thick. Visible in this specimen are the ventral sucker (**VS**), ovary (**O**), uterus (**U**), testes (**T**), intestinal ceca (**IC**), and vitellaria (**V**). (Stained Whole Mount)

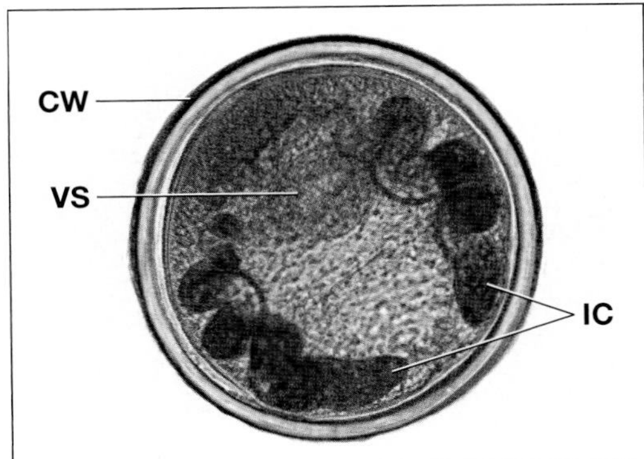

17.8 ***Paragonimus westermani*** **Cyst (Phylum Platyhelminthes; Class Trematoda)** • The cyst stage, which is found in crabs or crayfish, contains a living metacercaria. Once the vertebrate host eats the crab or crayfish, the juvenile fluke is released from the cyst where it begins its migration from the intestines to the lungs. Prominent in this cyst are the ventral sucker (**VS**), intestinal ceca (**IC**), and thick cyst wall (**CW**). (Whole mount)

Schistosoma haematobium

Schistosoma haematobium is one of several blood fluke species that infect humans and cause specific forms of schistosomiasis. All blood flukes have a more rounded body than typical flatworms, an oral and ventral sucker, and separate sexes.

Presence of eggs in patient urine samples is typically used for diagnosis. Serological tests for antischistosomal antibodies are also available, but may give false positive results in patients who have been successfully treated for a prior infection. All forms of schistosomiasis can be treated with anti-helminthic medication.

S. haematobium is an African and Middle Eastern blood fluke that causes urinary schistosomiasis. Infection occurs via contact with contaminated water containing **cercariae**, which penetrate the skin, enter circulation, and continue development in the liver. No metacercaria stage is seen.

After about three weeks in the liver, adult worms, which have separate sexes, colonize the veins associated with the urinary bladder and begin to lay eggs (Fig. 17.10). Some eggs pass through the vein's wall and then the bladder to be passed out in urine, but a majority of them become trapped in the wall and initiate a build-up of fibrous tissue as well as an immune response.

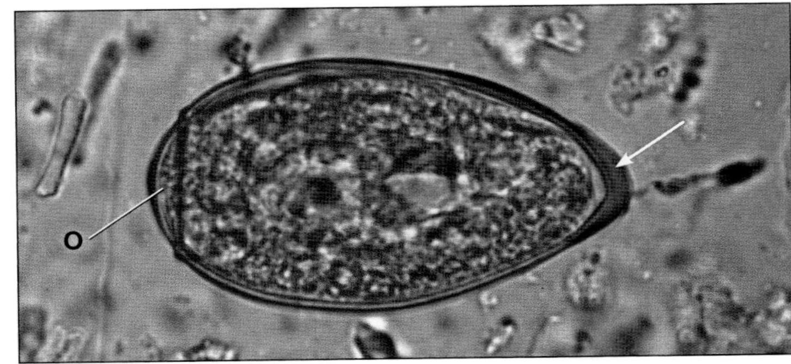

17.9 ***Paragonimus westermani*** **Egg in a Fecal Specimen (Phylum Platyhelminthes; Class Trematoda)** • *P. westermani* eggs are ovoid and range in size from 80–120 µm long by 45–70 µm wide. They have an operculum (**O**) and the shell is especially thick at the abopercular end (**arrow**). They are unembryonated when seen in feces. (Wet Mount; D'Antoni's Iodine Stain)

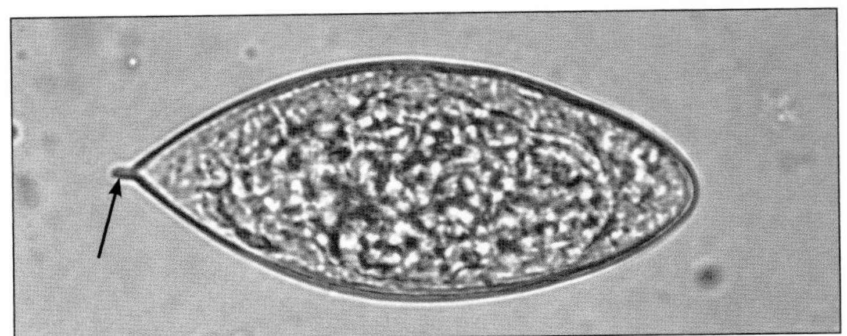

17.10 ***Schistosoma haematobium*** **Egg in a Urine Specimen (Phylum Platyhelminthes; Class Trematoda)** • *S. haematobium* eggs are large (112–170 µm long by 40–70 µm wide), thin-shelled, and lack an operculum. There is a distinctive terminal spine (**arrow**). Each egg contains a miracidium (larva). (Wet Mount)

Symptoms of the disease include hematuria and painful urination. There is also a high probability of developing bladder cancer. If there is a high parasite load and the infection is chronic, other parts of the genitourinary system may become involved. Diagnosis is by finding eggs in urine or less frequently, feces.

Schistosoma japonicum

Schistosoma japonicum is a blood fluke primarily found in Asia. Its life cycle is similar to *S. haematobium*, but the adults reside in veins of the small intestine. Adults produce eggs (Fig. 17.11) that penetrate the intestine and pass out with the feces. Presence of eggs in the feces indicates infection. Some patients are asymptomatic, whereas others have bloody diarrhea, abdominal pain, and lethargy. In some cases, eggs reach the brain and the infection may be fatal.

Schistosoma mansoni

Schistosoma mansoni is a blood fluke found in Brazil, some Caribbean islands, Africa, and parts of the Middle East. Infection occurs via contact with fecal-contaminated water carrying **cercariae** (Fig. 17.12) released from the snail intermediate host. These juveniles penetrate the skin, enter circulation, and continue development in the venules of the large intestine.

Sexes are separate, but females spend most of their time within the **gynecophoric canal** of the male, where mating takes place (Fig. 17.13). Eggs (Figs. 17.14 and 17.15) from adult females penetrate the intestinal wall and are passed out with the feces. Presence of eggs in the feces indicates infection.

Some patients are asymptomatic, whereas others have bloody diarrhea, abdominal pain, and lethargy. Like other forms of schistosomiasis, prevention is problematic because infection occurs through the skin, which requires better water quality for bathing and laundry.

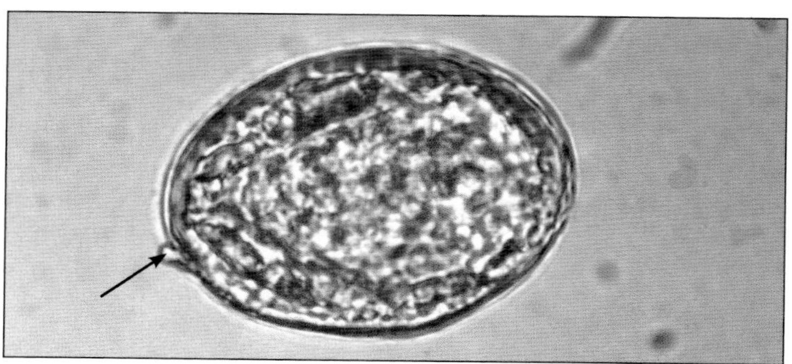

17.11 ***Schistosoma japonicum*** **Egg in a Fecal Specimen (Phylum Platyhelminthes; Class Trematoda)** • *S. japonicum* eggs are thin-shelled and lack an operculum. They range in size from 70–100 µm long by 55–65 µm wide. There may be a small spine (**arrow**) visible near one end. Each egg contains a miracidium (larva). (Wet Mount)

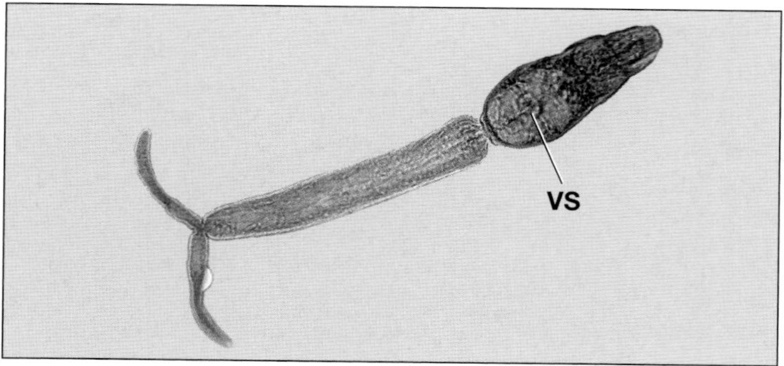

17.12 ***Schistosoma mansoni*** **Cercaria (Phylum Platyhelminthes; Class Trematoda)** • Schistosome cercariae have a forked tail and penetrate the human host directly. Cercariae develop from miracidia within the snail host and are the infectious stage for humans. There is no metacercaria stage, which is unusual among trematodes. The ventral sucker (**VS**) is prominent in this specimen. (Stained Whole Mount)

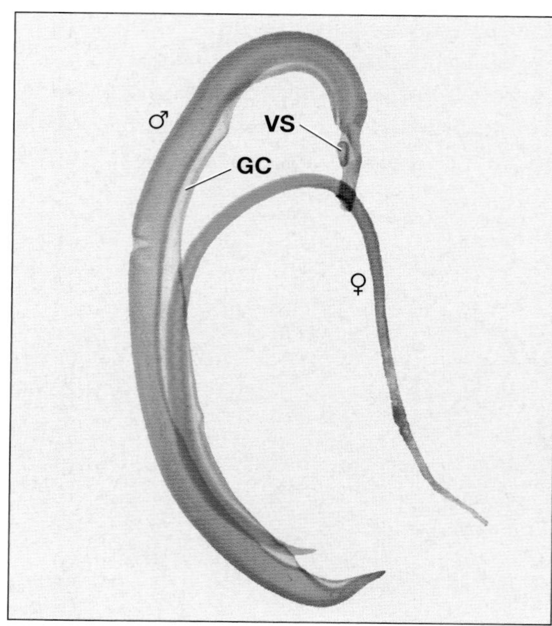

17.13 ***Schistosoma mansoni*** **Adults in Copula (Phylum Platyhelminthes; Class Trematoda)** • The male is larger and has a gynecophoric canal (**GC**) in which the slender female resides during mating. (*Schistosoma* literally means "split body.") The male's ventral sucker (**VS**) is also visible. (Stained Whole Mount)

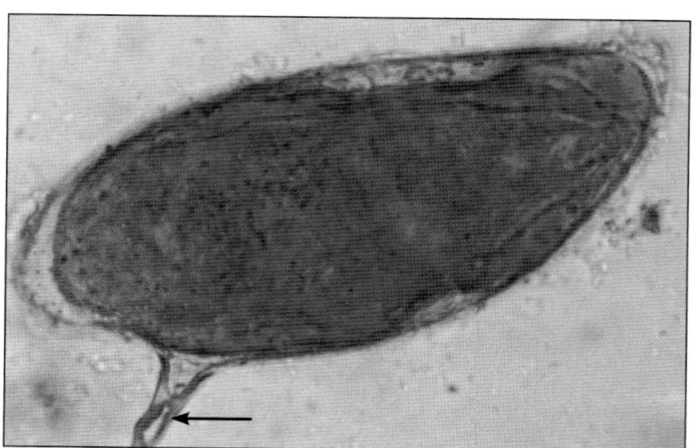

17.14 ***Schistosoma mansoni*** **Egg in a Fecal Specimen (Phylum Platyhelminthes; Class Trematoda)** • *S. mansoni* eggs are large (114–175 µm long by 45–70 µm wide) and contain a miracidium (larva). They are thin-shelled, lack an operculum, and have a distinctive lateral spine (**arrow**). (Wet Mount; D'Antoni's Iodine Stain)

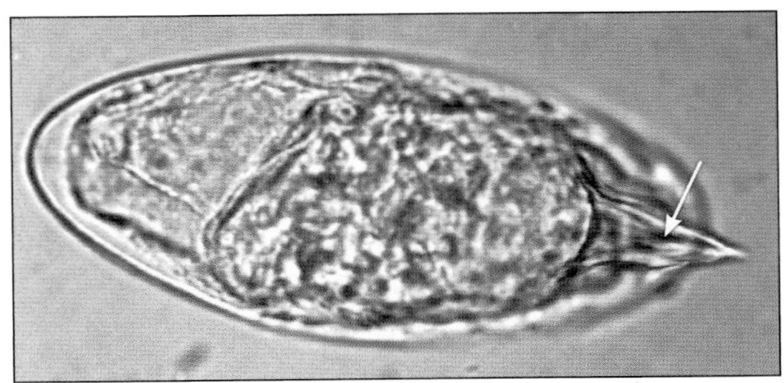

17.15 ***Schistosoma mansoni* Egg in a Fecal Specimen (Phylum Platyhelminthes; Class Cestoda)** ● The lateral spine may be oriented in such a way as to be difficult to see. In this specimen, the spine (**arrow**) is above the egg and the egg might be misidentified as *S. haematobium*. (Wet Mount)

Cestode Parasites Found in Clinical Specimens

Cestodes, commonly called tapeworms, are a class in the phylum Platyhelminthes composed exclusively of vertebrate intestinal parasites. The adult body is divided into segments called **proglottids**, each of which is devoted exclusively to sexual reproduction and carries both egg and sperm producing structures.

The **scolex** (head) is modified with suckers, hooks, or both, for attachment to the intestinal wall where worms may reside for years. The head may also have an evertable **rostellum**. No mouth or digestive tract is necessary because they absorb nutrients from the intestinal contents through their body surface.

Diphyllobothrium latum

Diphyllobothrium latum (*D. lata*) is commonly known as the broad fish tapeworm. It is found in Northern Europe, and the Great Lakes and West Coast regions of North America. The closely related *D. ursi* is responsible for infection in northeastern North America. *D. latum* juveniles infect muscle tissue of fish, which pass on the larvae when they are eaten raw or are undercooked. Adults develop in the intestine and begin egg production between 1 and 2 weeks later.

Infection may result in no symptoms or mild symptoms, such as diarrhea, nausea, abdominal pain, and weakness. In heavy infections, mechanical blockage of the intestine may occur. Rarely, infection results in pernicious anemia due to the worm's uptake of vitamin B_{12}.

Diagnosis is commonly made by finding the eggs (Fig. 17.16) or more rarely, proglottids (Fig. 17.17), in feces. Adults may reach a length of 9 m with more than 4,000 proglottids! Thoroughly cooking fish prior to eating is enough to prevent infection of the consumer.

Dipylidium caninum

Dipylidium caninum (Fig. 17.18) is a common parasite of dogs and cats worldwide and is known as the flea tapeworm. Adult worms reside in dog or cat intestines and release proglottids (Fig. 17.19) containing egg packets that migrate out of the anus. When these dry, they look like rice grains. Larval fleas (the intermediate host) may eat the eggs, each of which contains a juvenile **oncosphere**, and become infected. If the dog, cat, or human—frequently a child due to proximity to the cat or dog pet—ingests one of these fleas, the life cycle is completed in the new host.

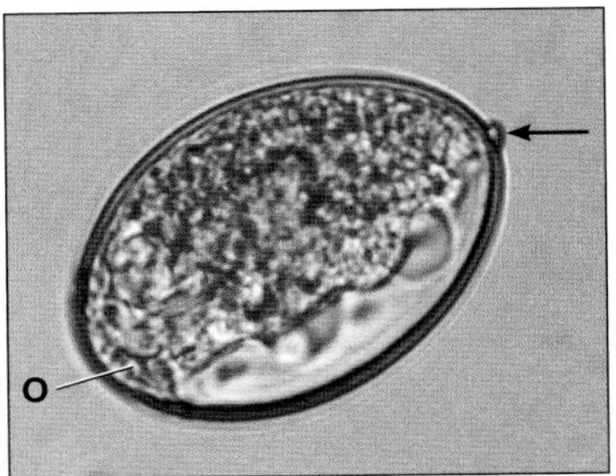

17.16 ***Diphyllobothrium latum* Egg in a Fecal Specimen (Phylum Platyhelminthes; Class Cestoda)** ● The eggs are unembryonated when passed in feces and have an operculum (**O**). A knob (**arrow**) is often present at the abopercular end. Egg dimensions are 58–75 µm long by 44–50 µm wide. (Wet Mount; D'Antoni's Iodine Stain)

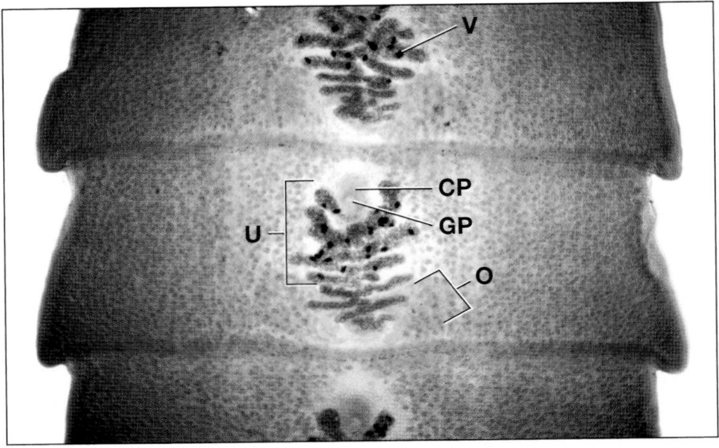

17.17 ***Diphyllobothrium latum* Proglottid (Phylum Platyhelminthes; Class Cestoda)** ● Anterior is at the top of the photo. *D. latum* proglottids are typically wider than long and genital pores are midventral. The distinctive rosette-shaped uterus (**U**), genital pore (**GP**), and the prominent cirrus pouch (**CP**) containing the male copulatory organ are also located midventrally. Two ovaries (**O**) are positioned posteriorly on either side of the uterus. The dark spots throughout most of the proglottid are vitellaria (**V**), and beneath them are the numerous testes (partly visible in the lighter pink regions). (Stained Whole Mount)

Infection may be asymptomatic or produce mild abdominal discomfort, appetite and weight loss, perianal itching, and indigestion. Diagnosis is made by identifying the egg packets (Fig. 17.20). Treatment is usually not necessary because infection is frequently self-limiting and asymptomatic, but anti-helminthic medication may be prescribed for humans and their pets.

Echinococcus granulosus

The definitive host of *Echinococcus granulosus* (Fig. 17.21) is a carnivore, but the life cycle requires an intermediate host, usually an herbivorous mammal. When an herbivore eats the eggs passed in the carnivore's feces, thick-walled **hydatid cysts** (Figs. 17.22 and 17.23) containing **protoscolices** (juvenile worms) develop in its lungs, liver, or other organs. When a carnivore eats an infected intermediate host, the protoscolices develop into adult worms in its intestines and the life cycle is completed.

Humans involved in raising domesticated herbivores (e.g., sheep with their associated dogs) are most susceptible to infection, but are dead-end hosts because the parasite cannot reproduce and humans are not likely to be eaten. Instead, infected humans develop hydatid disease, in which hydatid cysts form in the lung, liver, or other organs, a process that may take many years. Hydatid disease is found in countries around the Mediterranean Sea, South America, and Africa, but only sporadically in the United States.

Symptoms depend on the location and size of the hydatid cyst, which interferes with normal organ function. Due to sensitization by the parasite's antigens, release of fluid from the cyst can result in potentially fatal anaphylactic shock of the host. Diagnosis is made by detection of the cyst by CT scan, ultrasound, or X-ray. Treatment can involve antiparasitic medications, aspiration and sterilization of the cyst's contents, or surgical removal.

Hymenolepis diminuta

Hymenolepis diminuta is a tapeworm that has various arthropods as its intermediate hosts and rodents as the definitive hosts, hence the common name rat tapeworm. Humans may also be infected via ingestion of arthropods infected with larval **cysticercoids** (Fig. 17.24).

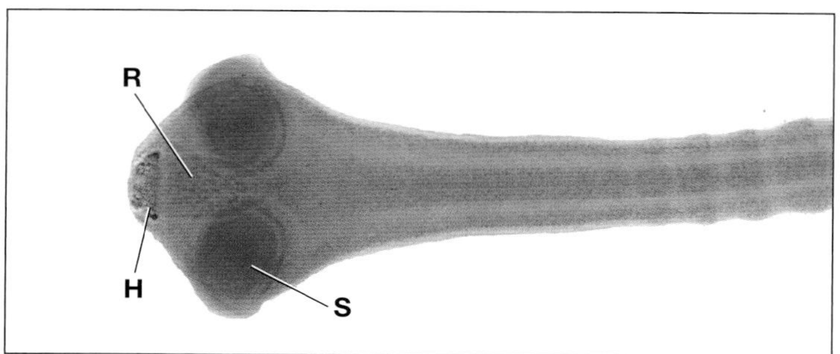

17.18 ***Dipylidium caninum*** **Scolex (Phylum Platyhelminthes; Class Cestoda)** • Adult worms may reach a length of 50 cm with a width of 3 mm. The scolex has four suckers (**S**) and a retractable rostellum (retracted in this photo–**R**) with rows of hooks (**H**). (Stained Whole Mount)

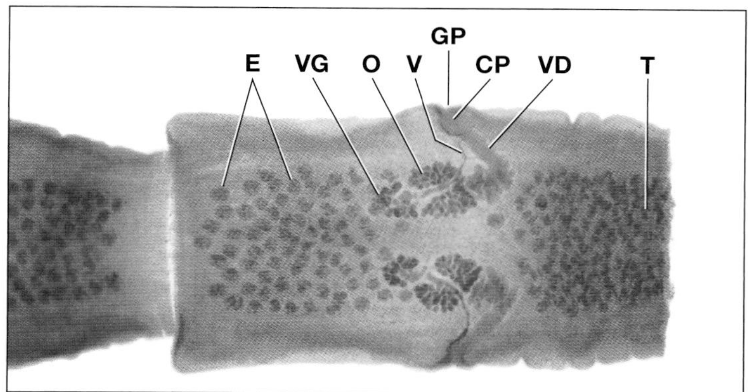

17.19 ***Dipylidium caninum*** **Proglottid (Phylum Platyhelminthes; Class Cestoda)** • Anterior is to the right. *D. caninum* proglottids are distinctive because each has paired reproductive organs, but only one side is labeled. Visible are eggs (**E**), the testes (**T**), vitelline gland (**VG**), ovary (**O**), vagina (**V**), genital pore (**GP**), cirrus pouch (**CP**), and vas deferens (**VD**). Early in the transition from mature to gravid, the uterus breaks up into egg packets (see Fig. 17.20), each with a dozen or more eggs and fill the now gravid proglottid. The genital pores on each side of the proglottid give this worm its common name—the "double-pored tapeworm." (Stained Whole Mount)

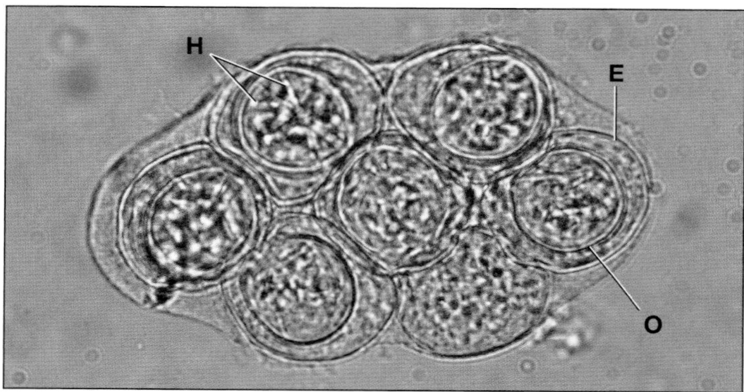

17.20 ***Dipylidium caninum*** **Egg Packets (Phylum Platyhelminthes; Class Cestoda)** • Each *D. caninum* egg packet is composed of a dozen or so eggs (**E**), each with an oncosphere (**O**). Each oncosphere possesses six hooklets (**H**). After ingestion by a flea larva, oncospheres mature into cysticercoids, which are the infective stage for dog, cat, and human hosts. (Stained Whole Mount)

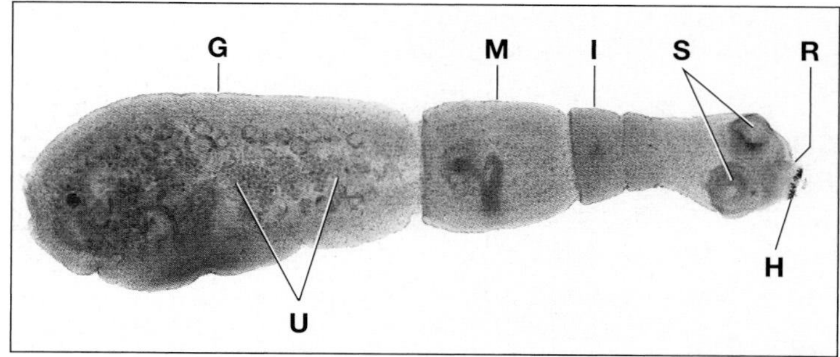

17.21 ***Echinococcus granulosus*** **Adult (Phylum Platyhelminthes; Class Cestoda)** • Adult worms are about 0.5 cm in length. The scolex has a nonretractable rostellum (**R**) armed with a ring of hooks (**H**). There are four suckers (**S**). Behind the scolex is a narrow neck and typically three proglottids in various stages of development: immature (**I**), mature (**M**), and gravid (**G**). The uterine sac (**U**) with eggs is in the gravid segment. Once released, eggs are infectious. (Stained Whole Mount)

Adult worms develop in and attach to the intestinal mucosa, with one host harboring numerous worms. They release eggs (Fig. 17.25), which exit with the feces and are useful in diagnosis. Proglottids are rarely found in feces and are not of diagnostic value. Many cases of hymenolepiasis are asymptomatic, but some result in mild abdominal discomfort and digestive upset. Anti-helminthic medications are available.

Hymenolepis (Vampirolepis) nana

Hymenolepis nana (Figs. 17.26 and 17.27) is the dwarf tapeworm, the most common human cestode parasite in the world. It is found in areas of high population density with poor sanitation and, like *H. diminuta*, causes hymenolepiasis. No intermediate host is required and humans can become infected by ingesting embryonated eggs directly (Fig. 17.28).

Once in the gut, **oncospheres** (larvae retained within an egg) develop into juveniles within the lymphatics of intestinal villi. These juveniles are released into the intestinal lumen within a week, attach to the mucosa, and mature into adults. Infection may involve hundreds of worms, yet symptoms are usually absent or mild: diarrhea, nausea, loss of appetite, or abdominal pain.

Eggs may reinfect the same host (**autoinfection**) or pass out with the feces to infect a new host. Eggs in the feces are used for identification, but proglottids are not because they are rarely passed. Treatment with anti-helminthic medications is available and prevention requires improved sanitation.

Taenia species

Two taeniid worms are important human pathogens. These are *Taenia saginata* (*Taeniarhynchus saginatus*), the beef tapeworm, and *Taenia solium*, the pork tapeworm. Both tapeworms are found worldwide and both cause taeniasis. They can grow to lengths of several meters and survive in human intestines for a couple of decades. *T. solium* may produce 1,000+ proglottids, each with 50,000 eggs, whereas *T. saginata* may produce 2,000+ proglottids, each with 100,000 eggs! Humans are the definitive host for both.

Cattle are the intermediate hosts of *T. saginata* and become infected when they ingest eggs or proglottids in fecal-contaminated soil or water. Juveniles are dispersed throughout the body and infect various organs, but most

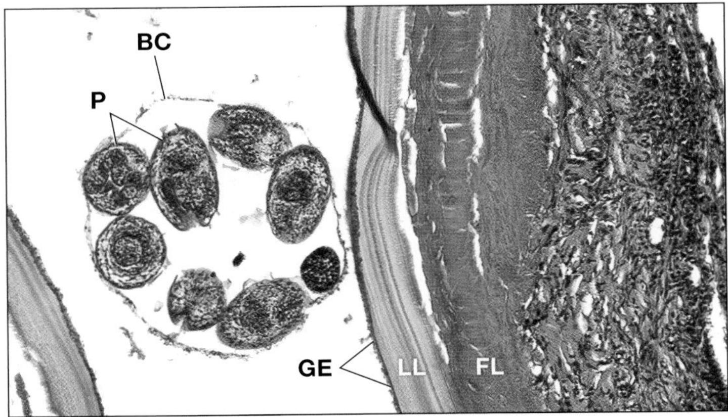

17.22 ***Echinococcus granulosus*** **Hydatid Cyst in Lung Tissue (Phylum Platyhelminthes; Class Cestoda)** • Shown is a small portion of a hydatid cyst in the lung. The cyst wall consists of an internal germinal epithelium (**GE**); an acellular, laminated layer (**LL**); and a fibrous layer of host tissue (**FL**). Brood capsules (**BC**) form from the germinal epithelium and remain attached to it by a stalk (not shown in this section). They are filled with fluid and the epithelium produces many *E. granulosus* protoscolices (**P**) by invagination. This brood capsule is 300 µm in diameter. The material at the far right is the host's lung. (Section; H & E Stain)

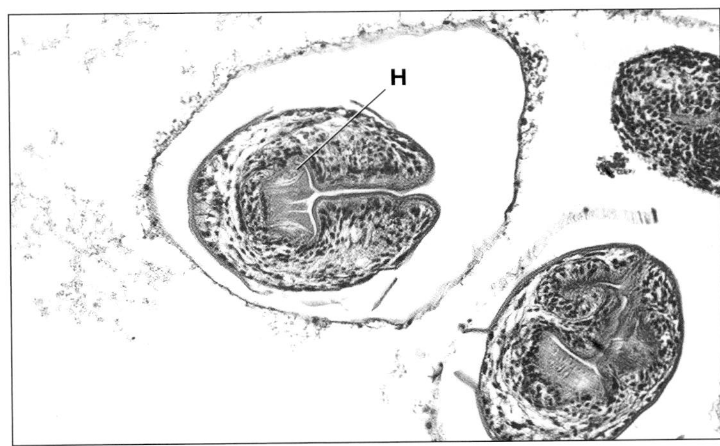

17.23 ***Echinococcus granulosus*** **Protoscolex (Phylum Platyhelminthes; Class Cestoda)** • The protoscolex contains an invaginated scolex with hooks (**H**). Upon ingestion by the host, the protoscolex evaginates and produces an infectious scolex that attaches to the intestinal wall, matures, and produces eggs. (Longitudinal Section; H & E Stain)

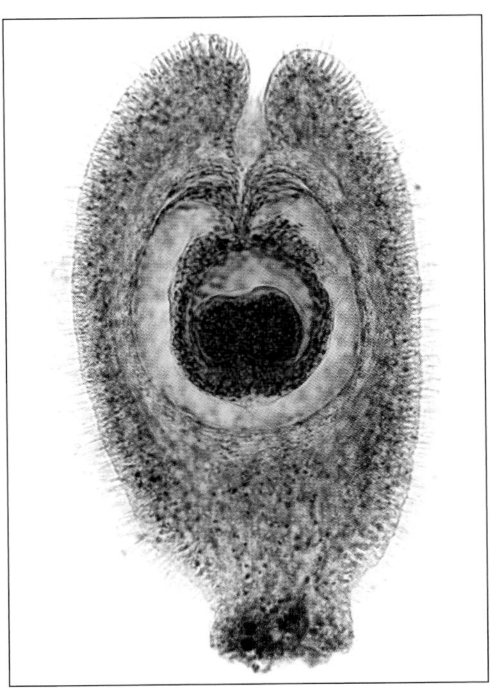

17.24 ***Hymenolepis diminuta*** **Cysticercoid (Phylum Platyhelminthes; Class Cestoda)** • The larval cysticercoid is the infectious stage and is found in arthropods of various sorts. Rodents are the definitive host, but if ingested by humans, the cysticercoid can develop into an adult in the intestines. Eating grain containing infected beetles is a common mode of infection. (Stained Whole Mount)

important to the life cycle, skeletal muscle. As a result, humans who eat undercooked beef containing juvenile **cysticerci** become infected.

In the presence of bile salts, cysticerci develop into adults, attach to the small intestine's wall using hooks on their **scolices**, and begin producing gravid proglottids within a few weeks, which are passed in the feces along with eggs. Symptoms of infection are usually mild nausea, diarrhea, abdominal pain, and headache.

Diagnosis to species is impossible with only the eggs (Fig. 17.29); specific identification requires a scolex or gravid proglottid (Figs. 17.30 and 17.31). Treatment is with anti-helminthic drugs. Finding a scolex in a fecal sample is considered a sure sign of successful treatment.

The *T. solium* life cycle is similar to *T. saginata*, but the host is pork, not beef, so human infection with the adult worm occurs when undercooked pork containing cysticerci (Fig. 17.32) is eaten. Symptoms of infection are generally mild and involve the digestive tract. However, if a human ingests eggs, cysticerci develop and may be found in any tissue, including the eyes, brain, heart, liver, lungs, and skeletal muscle.

Symptoms of cysticercosis depend on the tissue infected, but mostly they are not severe. However, infection of the brain can lead to seizures, and death of a cysticercus can produce a rapidly fatal inflammatory response. As with *T. saginata*, diagnosis to species is impossible with only the eggs; specific identification requires a scolex or gravid proglottid (Figs. 17.33 and 17.34). Treatment is the same as for *T. saginata*.

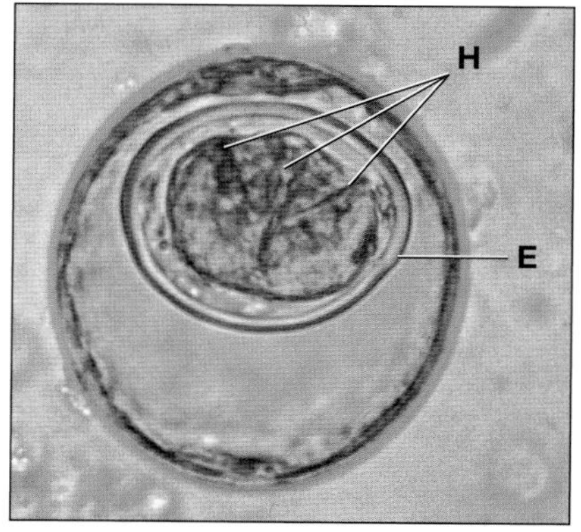

17.25 ***Hymenolepis diminuta*** **Egg in a Fecal Sample (Phylum Platyhelminthes; Class Cestoda)** • The spherical, thick-shelled eggs of *H. diminuta* range in size from 70–85 µm long by 60–80 µm wide. The embryo (**E**) is centrally positioned separate from the wall. Three pairs of hooks (**H**) are also present. (Wet Mount; D'Antoni's Iodine Stain)

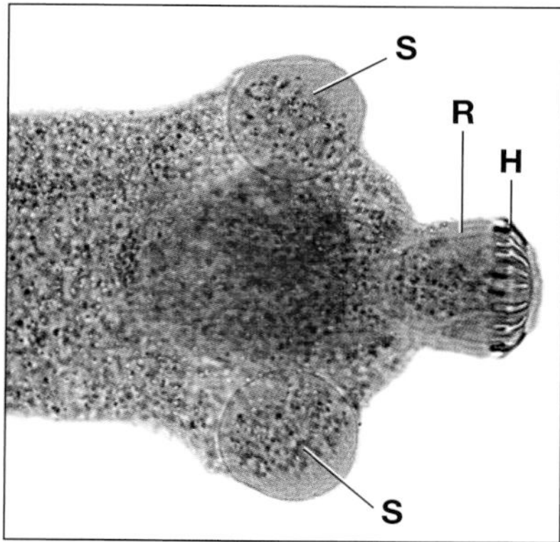

17.26 ***Hymenolepis nana*** **Scolex (Phylum Platyhelminthes; Class Cestoda)** • Adults gain a length of up to 10 cm, but are only 1 mm in width. The retractable rostellum (**R**) is armed with up to 30 hooks (**H**). Four suckers (**S**) are also present. (Stained Whole Mount)

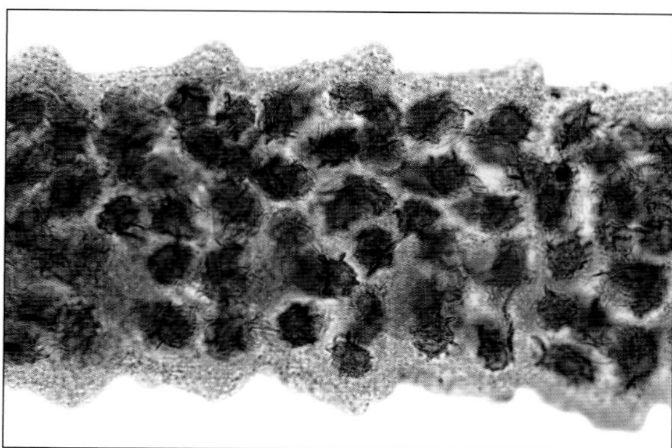

17.27 ***Hymenolepis nana*** **Proglottids (Phylum Platyhelminthes; Class Cestoda)** • Immature *H. nana* proglottids are wider than long, but as they mature, they lengthen. These are gravid proglottids with developing eggs. (Stained Whole Mount)

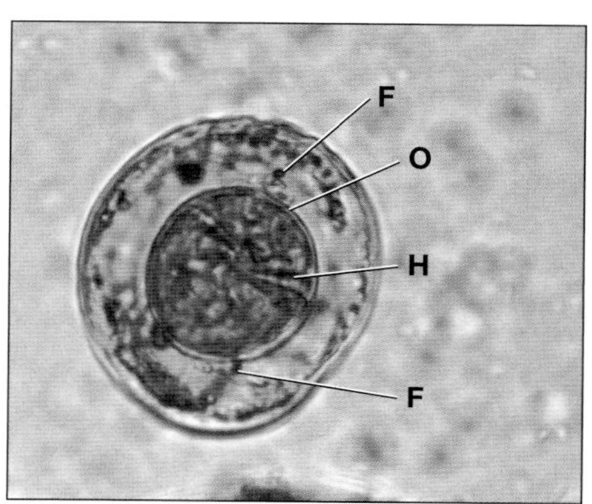

17.28 ***Hymenolepis nana*** **Egg in a Fecal Sample (Phylum Platyhelminthes; Class Cestoda)** • The *H. nana* egg resembles the egg of *H. diminuta*, but it is smaller (30–47 µm in diameter) and has a thinner shell. The oncosphere (**O**) is separated from the shell and contains six hooks (**H**). Another distinguishing feature is the presence of between four and eight filaments (**F**) arising from either end of the oncosphere. (Wet Mount; D'Antoni's Iodine Stain)

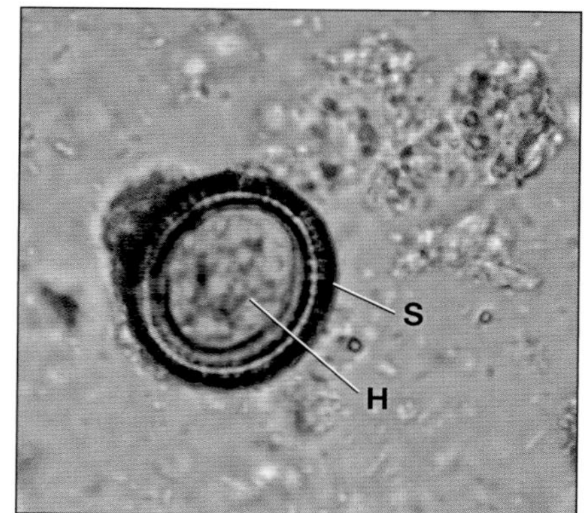

17.29 ***Taenia*** **Egg in Feces (Phylum Platyhelminthes; Class Cestoda)** • Taeniid eggs are distinctive looking enough to identify to genus, but not distinctive enough to speciate. Eggs are spherical and approximately 40 µm in diameter with a striated shell (**S**). The oncosphere contains six hooks (**H**). (Wet Mount; D'Antoni's Iodine Stain)

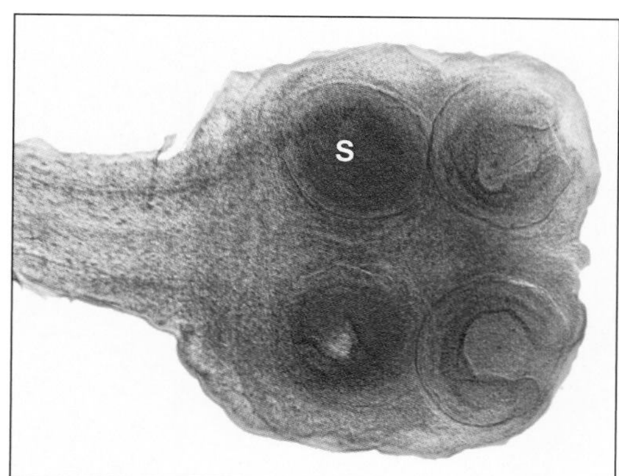

17.30 ***Taenia saginata*** **Scolex (Phylum Platyhelminthes; Class Cestoda)** • The *T. saginata* scolex has four suckers (**S**) and no hooks. (Stained Whole Mount)

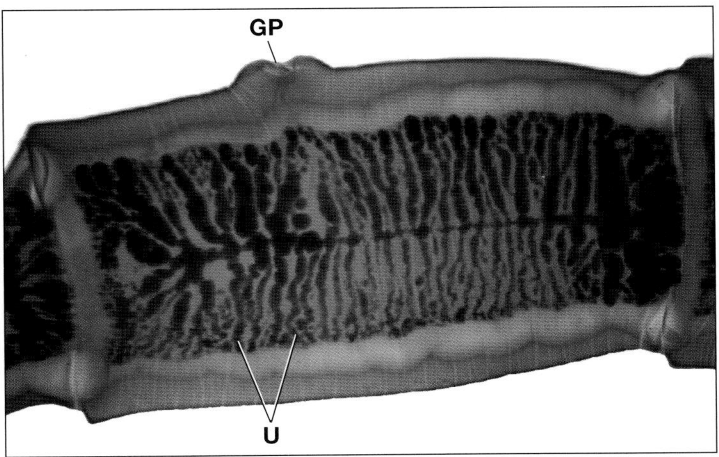

17.31 ***Taenia saginata*** **Proglottid (Phylum Platyhelminthes; Class Cestoda)**
The egg-filled uterus (**U**) of *T. saginata* proglottids consists of a central portion with 15–20 lateral branches that have been injected with India ink to make counting easier. Compare to the *T. solium* uterus in Fig. 17.34. Each proglottid possesses a distinctive-looking genital pore (**GP**). (Stained Whole Mount)

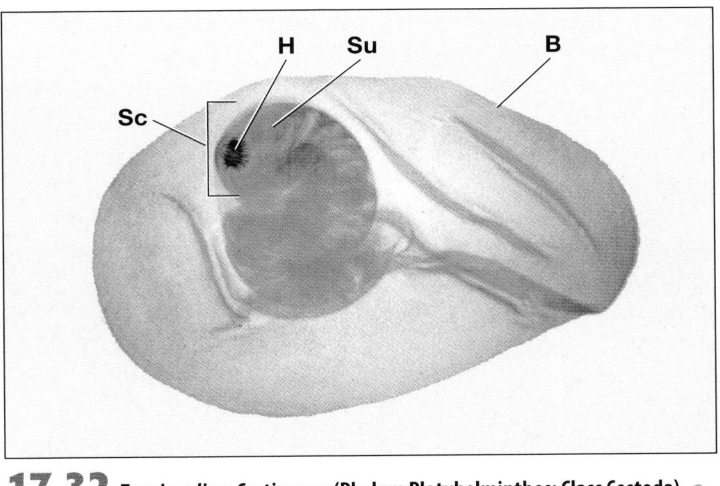

17.32 ***Taenia solium*** **Cysticercus (Phylum Platyhelminthes; Class Cestoda)**
The *T. solium* cysticercus (cysticercus cellulosae) consists of a bladder (**B**) with an invaginated scolex (**Sc**) with its hooks (**H**) and suckers (**Su**). Human infection by a cysticercus occurs after ingesting *T. solium* eggs that may develop into cysticerci subcutaneously or within various organs. (Stained Whole Mount)

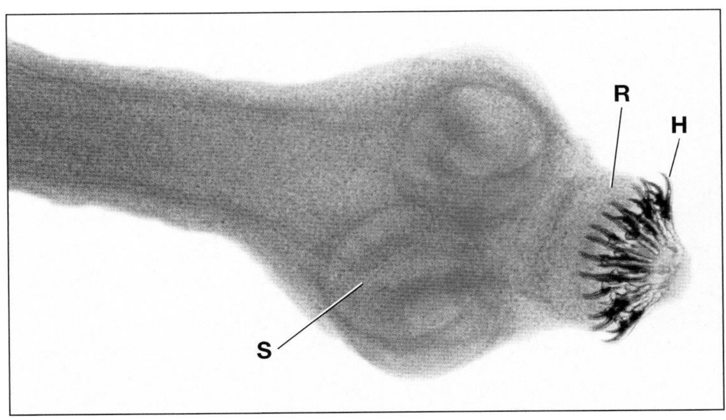

17.33 ***Taenia solium*** **Scolex (Phylum Platyhelminthes; Class Cestoda)**
The *T. solium* scolex has four suckers (**S**) and a rostellum (**R**) with two rings of hooks (**H**). (Stained Whole Mount)

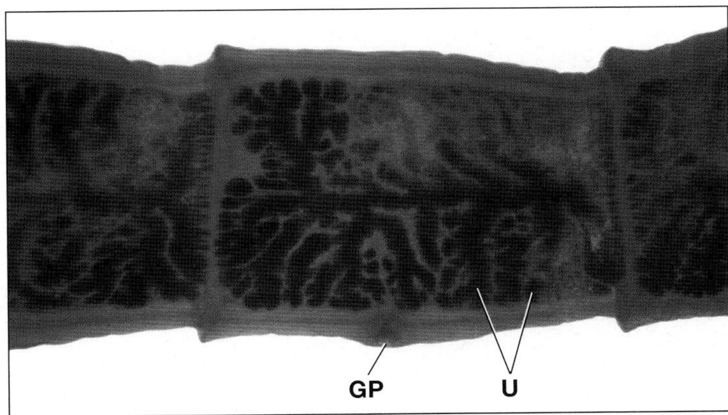

17.34 ***Taenia solium*** **Proglottid (Phylum Platyhelminthes; Class Cestoda)**
The egg-filled uterus (**U**) of *T. solium* proglottids consists of a central portion with 7–13 lateral branches that have been injected with India ink to make counting easier. Compare to the *T. saginata* uterus in Figure 17.31. Each proglottid possesses a distinctive-looking genital pore (**GP**). (Stained Whole Mount)

Nematode Parasites Found in Clinical Specimens

Free-living species in the phylum Nematoda, commonly called roundworms, are abundant in virtually all terrestrial and aquatic ecosystems. Parasitic species, which are our focus, equal free-living species in number with more than 60 species being human parasites. Anatomically, nematodes have a complete digestive tract, a body cavity, reproductive organs in separate sexes, and are unsegmented.

Ascaris lumbricoides

Ascaris lumbricoides (Figs. 17.35 and 17.36) is a large nematode—females may reach a length of 49 cm—and is most commonly found in tropical regions. Human infection occurs when eggs in fecal-contaminated soil or food are ingested. Juveniles emerge in the intestine, pass through its wall, and then migrate to the lungs and other tissues.

After a period of development in the lungs, juveniles move up the respiratory tree to the esophagus and are swallowed again. Adults then reside in the small intestine and produce eggs (Figs. 17.37 and 17.38), which are passed in the feces. There is no intermediate host.

Patients are usually asymptomatic, but may experience varying degrees of intestinal obstruction and malnutrition, depending on the parasite load. Juvenile worms in the lungs can cause *Ascaris* pneumonia as a result of a hypersensitivity reaction. If secondary bacterial infections occur, the pneumonia can be fatal.

Identification of an *Ascaris* infection is made by observing eggs in feces. Several anti-helminthic drugs are available for treatment and improved sanitation and personal hygiene are effective in minimizing transmission.

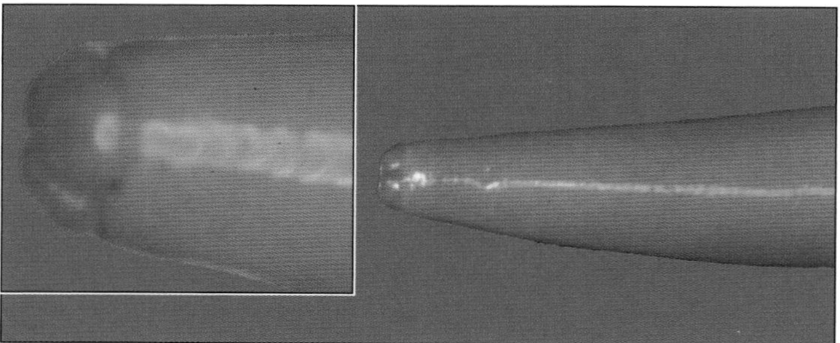

17.35 ***Ascaris lumbricoides* Anterior (Phylum Nematoda)** • *A. lumbricoides* has a cylindrical shape with three prominent mouth parts (see inset). (Preserved Specimen)

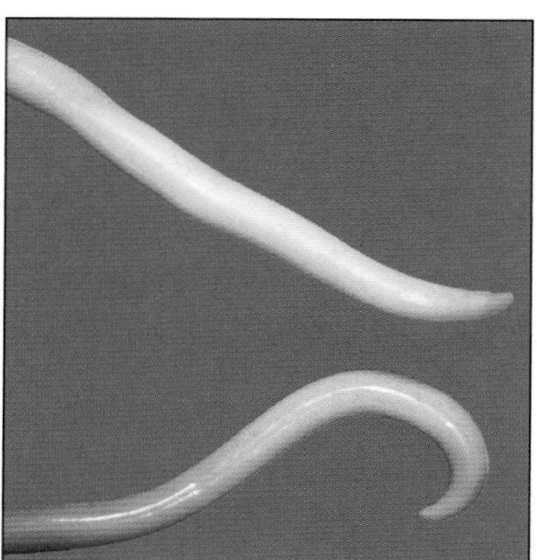

17.36 ***Ascaris lumbricoides* Adult Worms (Phylum Nematoda)** • *A. lumbricoides* males (bottom) are shorter than females (top)—up to 31 cm vs. 35 cm—and have a curved posterior. (Preserved Specimen)

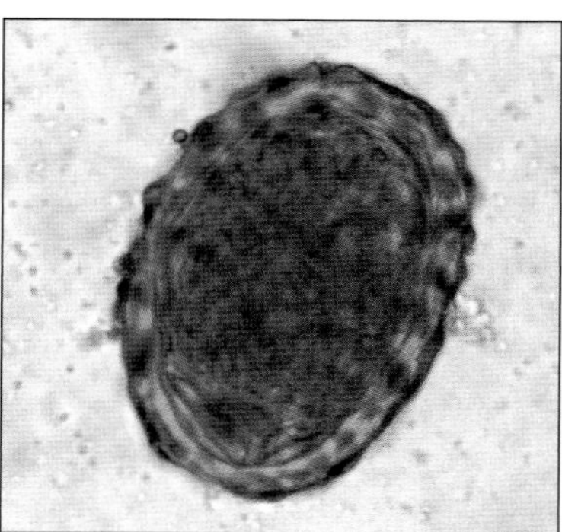

17.37 Fertile *Ascaris lumbricoides* Egg in Feces (Phylum Nematoda) • Fertile *A. lumbricoides* eggs are 55–75 µm long by 35–50 µm wide and are embryonated. Their surface is covered by small bumps called mammillations. (Wet Mount; D'Antoni's Iodine Stain)

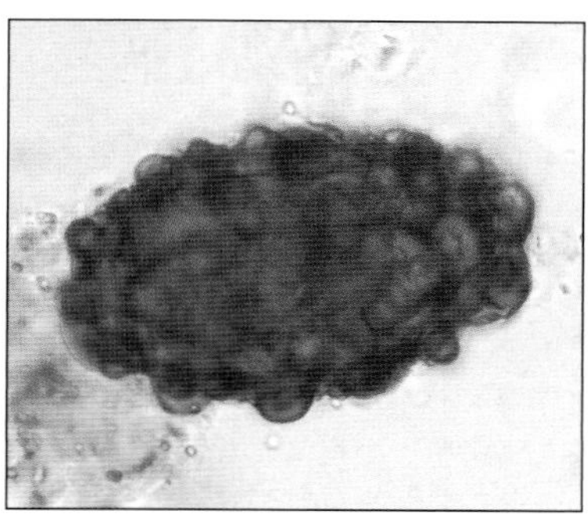

17.38 Infertile *Ascaris lumbricoides* Egg in Feces (Phylum Nematoda) • Infertile eggs may also be passed in the feces and are recognizable because they are longer (up to 90 µm) than fertile eggs and have no embryo inside. (Wet Mount; D'Antoni's Iodine Stain)

Capillaria hepatica

Capillaria hepatica is mostly a rodent parasite, but human infections do occur. The life cycle in rodent and carnivorous hosts begins with ingestion of eggs that develop into adults in the liver where they lay eggs. Transmission of eggs occurs when the host dies and decomposes, or is eaten by a predator that passes the eggs in its feces.

Human infection results when food or soil contaminated with eggs is ingested. Juveniles emerge in the small intestine and migrate to the liver where development into adults occurs. Eggs (Fig. 17.39) are deposited in the liver, but cannot develop there. The main symptom of human capillariasis is hepatitis with eosinophilia, but other symptoms of liver dysfunction may be present.

Identification is made by liver biopsy or postmortem examinations. Anti-helminthic medications are available for treatment. Prevention is possible with improved sanitation and attention to personal hygiene.

Dirofilaria immitis

Dirofilaria immitis, a **filarial** nematode, is the dog heartworm and causes dirofilariasis. It is distributed throughout warm temperate to tropical regions of the world and is also found in most parts of the United States, though it is most common in the gulf and southeastern states. Dogs and other carnivores are the definitive hosts and mosquitoes are the intermediate hosts.

When an infected mosquito takes a blood meal from a susceptible host, infective larvae enter the host's bloodstream and develop into adults in the right ventricle of the heart and pulmonary vessels, a process that can take up to 6 months. Gravid females release **microfilariae** (Fig. 17.40) into the blood, where they are ingested by mosquitoes during a blood meal. Development into infective larvae occurs in the mosquito and the life cycle is complete.

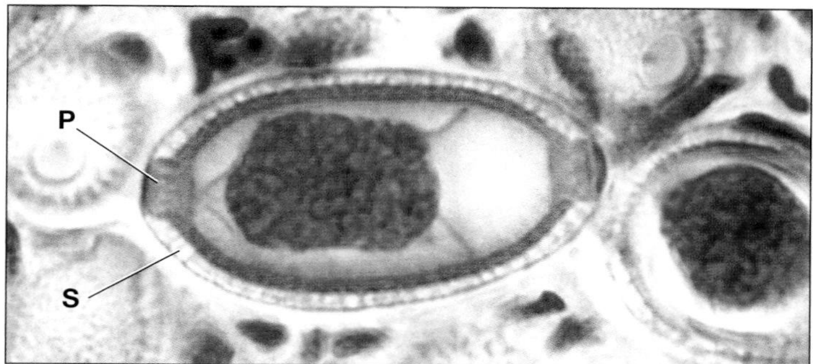

17.39 ***Capillaria hepatica* Eggs in the Liver (Phylum Nematoda)** • *C. hepatica* eggs are 51–67 µm long by 30–35 µm wide and have "plugs" (**P**) at either end and a striated sheath (**S**). Eggs are incapable of embryonation within the liver and are only passed in the feces if an infected liver has been eaten, but the consumer doesn't get infected because the eggs are not embryonated. Once on the soil, they embryonate and await ingestion by a new host that will become infected. While not found in feces of infected individuals, they can be found in feces if an infected liver has been eaten. Therefore, a laboratorian must be able to distinguish them from the similar eggs of *Trichuris trichiura*, which has much more prominent plugs at each end and no striations (see Fig. 17.54). (Longitudinal Section; H & E Stain)

Adult females can reach a length of a little more than one foot, whereas adult males are considerably shorter. An infected dog may have as many as 300 adult worms blocking its pulmonary vessels and interfering with proper heart valve functioning. Infected dogs may be asymptomatic, or present symptoms such as coughing, exercise intolerance, loss of appetite, and weight loss. Diagnosis comes from a blood smear positive for microfilariae. Infection can be fatal; however, treatment is available that targets infective larvae, but not adults.

While humans are not a host for *D. immitis*, infection can occur, but the parasite does not develop into adults. Rather, they die during development within the heart and are carried into pulmonary arteries where they become trapped as the vessels become smaller. Their presence results in pulmonary dirofilariasis, which produces inflammation, restricted blood flow, and the formation of granulomatous nodules resulting from immune cell accumulation.

Approximately half the patients are asymptomatic, but symptoms, such as cough, chest pain, low-grade fever, and malaise may develop. Microfilariae are not found in the blood. However, radiographs can detect the presence of nodules, which may then be sectioned for microscopic examination, though degradation of the worm can make its identification difficult. Treatment is surgical removal of the nodule.

Enterobius vermicularis

Enterobius vermicularis (Fig. 17.41) is the human pinworm. It is found worldwide and is prevalent among people in institutions, because conditions favor fecal-oral transmission of the eggs (Fig. 17.42). Poor sanitary habits of children make them especially prone to infecting others. Transmission may also involve eggs being carried on air currents and then inhaled by a susceptible host.

After ingestion, eggs hatch in the duodenum and mature in the large intestine where the adults reside. Adult females emerge from the anus at night to lay thousands of eggs in the perianal region. About one-third of pinworm infections are asymptomatic; the other two-thirds usually do not produce serious symptoms.

Diagnosis is made by identifying the eggs. Because eggs are laid externally, they are rarely found in feces. Instead of examining fecal samples, samples from the perianal region are taken with cellophane tape and examined microscopically for eggs. Several anti-helminthic drugs are available for treatment and improved personal hygiene is effective in minimizing transmission.

Hookworms *(Ancylostoma duodenale* and *Necator americanus)*

The hookworms *Ancylostoma duodenale* (Figs. 17.43 through 17.45) and *Necator americanus* (Fig. 17.46), have very similar morphologies and life cycles, and the eggs are indistinguishable, so they are considered together here. Both are found in regions where warm, moist soil is present.

Infection occurs when juveniles penetrate the skin, enter the blood, and travel to the lungs. They penetrate the respiratory membrane and are carried up and out of the lungs by ciliary action to the pharynx, where they are swallowed. When they reach the small intestine, they attach and mature into adults that feed on blood and tissues of the host. Adults are rarely seen, because they remain attached to the intestinal mucosa. Eggs (Fig. 17.47) are passed in the feces and are diagnostic of infection.

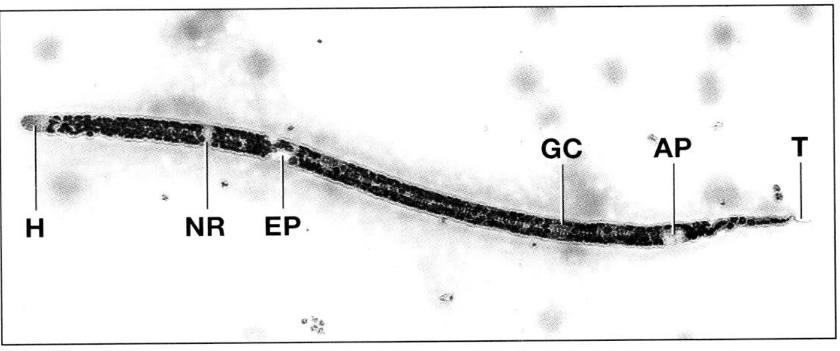

17.40 ***Dirofilaria immitis*** **Microfilaria in a Blood Smear (Phylum Nematoda)** • *D. immitis* microfilariae are released into the blood by mature female worms and are the larval stage transmitted to mosquitoes during a blood meal. They are approximately 300 µm long by 7 µm wide. The purple circles are nuclei. Also visible are the head end (**H**), nerve ring (**NR**), excretory pore (**EP**), genital cell (**G**), anal pore (**AP**), and tail end (**T**). (Giemsa Stain)

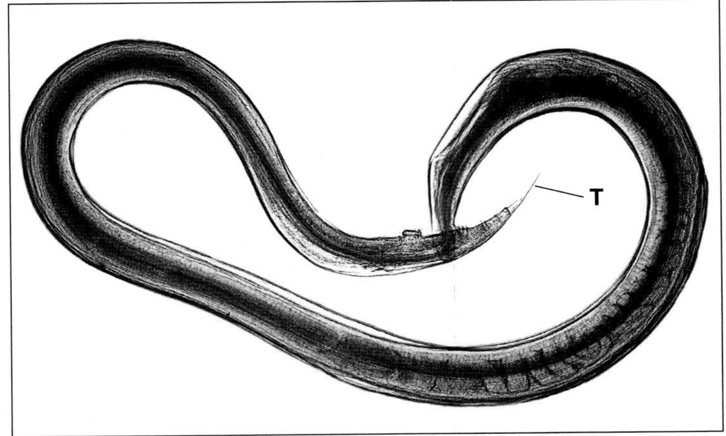

17.41 ***Enterobius vermicularis*** **Adult Female (Phylum Nematoda)** • Female pinworms are about 1 cm long and have a pointed tail (**T**) from which this group derives its common name. Males are about half that size and have a hooked tail. The dark interior is the digestive tract. (Whole Mount)

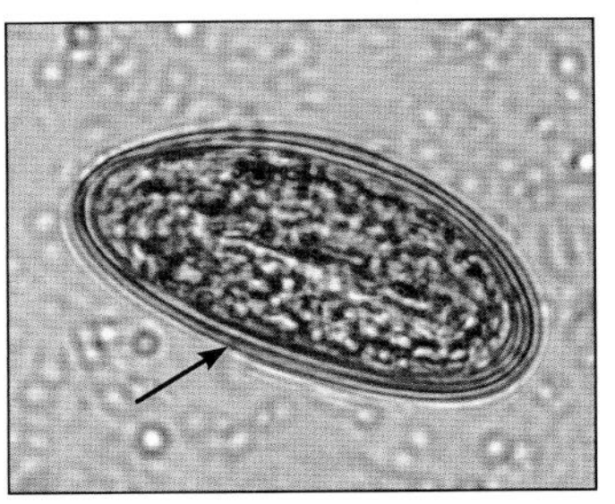

17.42 ***Enterobius vermicularis*** **Egg (Phylum Nematoda)** • The eggs of *E. vermicularis* are 50–60 µm long by 40 µm wide with one side flattened (**arrow**). They are usually embryonated in typical preparations. (Wet Mount; D'Antoni's Iodine Stain)

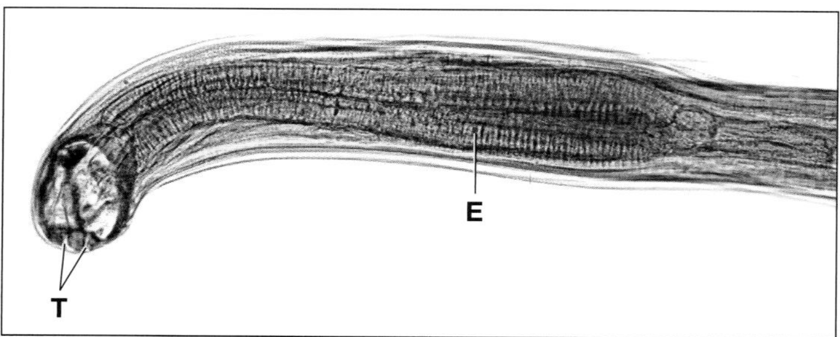

17.43 Anterior of *Ancylostoma duodenale* (Phylum Nematoda) ● This anterior view of *A. duodenale* shows the mouth, two teeth (**T**), and thick-walled esophagus (**E**). The bend in the head gives this group its common name—hookworm. Compare with Figure 17.45. (Unstained Whole Mount)

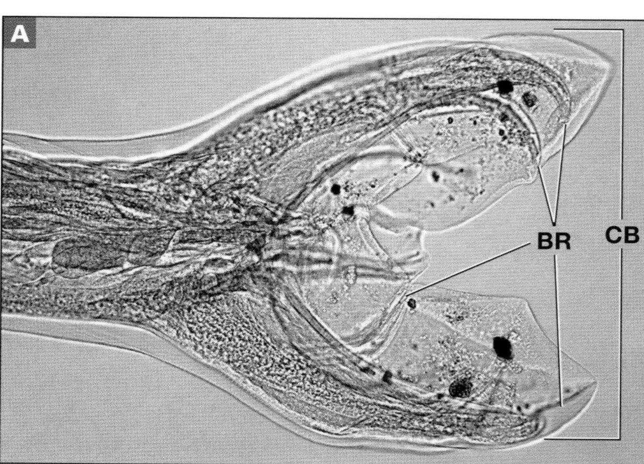

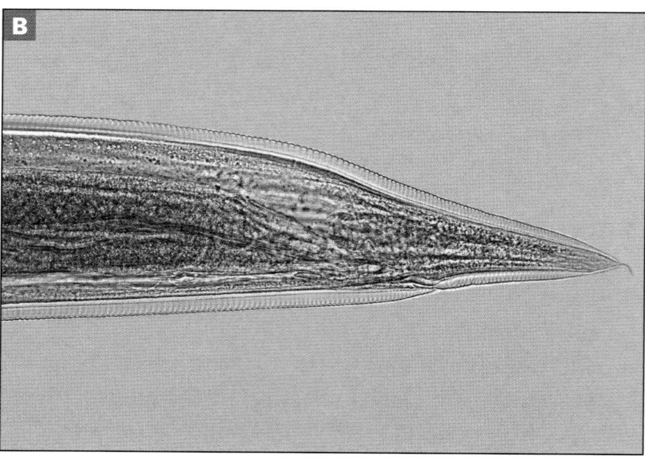

17.44 Sexual Dimorphism in *Ancylostoma duodenale* (Phylum Nematoda) ● Besides internal reproductive organs, (**A**) males and (**B**) females differ in their tail morphologies. Males have a copulatory bursa (**CB**) at the posterior end; females do not. The arrangement of the bursal rays (**BR**) is helpful in identification. (Whole Mounts)

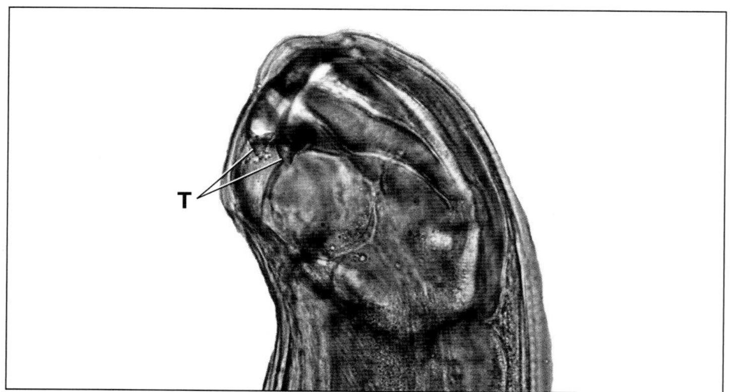

17.45 *Ancylostoma duodenale* Head (Phylum Nematoda) ● *Ancylostoma duodenale* and *Necator americanus* adult worms look very similar, but they can easily be distinguished because the former has teeth (**T**) and the latter has cutting plates (see Fig. 17.46). This micrograph is a composite of two photos taken in different focal planes to show the two teeth clearly. (Whole Mount)

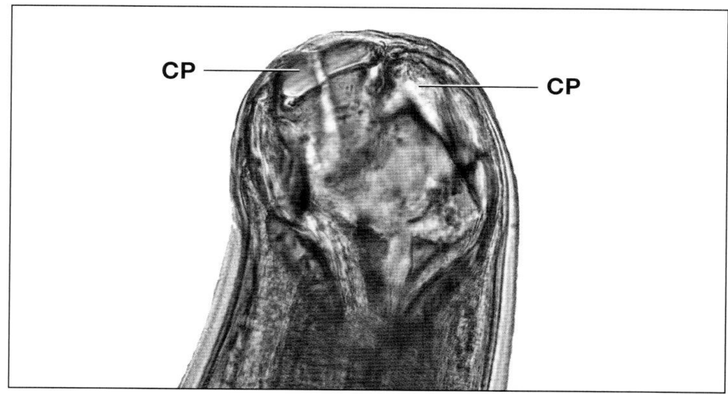

17.46 *Necator americanus* Head (Phylum Nematoda) ● This detail of *N. americanus* shows the cutting plates (**CP**) that help to differentiate it from *Ancylostoma duodenale* (compare with Fig. 17.45). Also notice the hooked head. (Whole Mount)

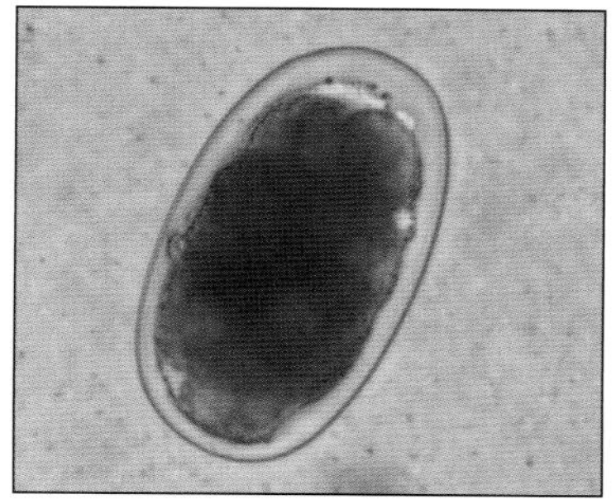

17.47 Hookworm Egg in Feces (Phylum Nematoda) ● Hookworm eggs are 55–75 µm long by 36–40 µm wide. They have a thin shell and contain a developing embryo (seen here at about the 16-cell stage) that is separated from the shell when seen in fecal samples. (Wet Mount; D'Antoni's Iodine Stain)

Hookworms produce immunosuppressant proteins that enhance their ability to survive in the host, but most infections are still asymptomatic. When symptoms appear, their severity is related to the parasite load. Severe symptoms of bloody diarrhea and iron-deficiency anemia are only seen in acute heavy or chronic infections. A rash may develop where larvae have penetrated the skin and this site is susceptible to secondary bacterial infection.

Humans are not the definitive host, but hookworm infection can lead to a condition known as larval migrans, where the larvae wander subcutaneous tissues, but do not mature. Treatment is available with a variety of anti-helminthic drugs. Attention to personal hygiene, covering the skin (e.g., with footwear), and not frequenting areas suspected of soil contamination are successful preventative measures.

Onchocerca volvulus

Onchocerca volvulus is found in Africa, Mexico, and parts of South and Central America. It causes onchocerciasis ("river blindness") and is transmitted through bites of infected black flies (*Simulium* spp.). Filarial juveniles enter the tissues and develop into adults in about a year. Adults then reside in subcutaneous tissues and become surrounded by a collagenous capsule (Fig. 17.48).

Microfilariae (Fig. 17.49) develop and may be picked up by a black fly when feeding. Within the black fly, they complete their life cycle by developing into filarial larvae that migrate to its proboscis, ready to be transferred to a new host during a blood meal. Damage due to the adult worm is negligible, at worst forming a nodule. The microfilariae are more troublesome.

Dead microfilariae may cause dermatitis followed by a thickening, cracking, and depigmentation of the skin. Living microfilariae may infect the eyes, die, and stimulate an immune response. Sclerosing keratitis, which results in blindness, is one consequence of chronic eye inflammation. Diagnosis is by demonstration of microfilariae in skin snips. Annual or semiannual chemotherapy is the treatment of choice.

Strongyloides stercoralis

Strongyloides stercoralis is the intestinal threadworm. Infection occurs when **rhabditiform larvae** in fecal-contaminated soil penetrate the skin. The juveniles are carried by the blood to the lungs where they migrate to the pharynx and are swallowed. In the intestines, they develop into parthenogenetic females (*S. stercoralis* does not produce parasitic adult male worms) that burrow into the intestinal mucosa.

Each day, females release a few dozen eggs that develop into rhabditiform larvae (Fig. 17.50) before they are passed in the feces or reinfect the intestines (**autoinfection**). Once passed, larvae may become infective and complete the life cycle or they may follow a developmental path that produces

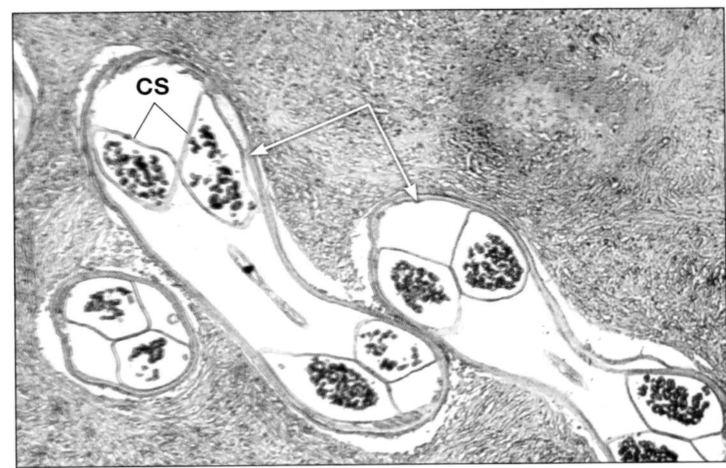

17.48 **Adult *Onchocerca volvulus* in Nodules (Phylum Nematoda)** • The adult worms are highly coiled within fibrous nodules (**arrows**) beneath the skin. Some worm cross-sections are labeled (**CS**). (Section; H & E Stain)

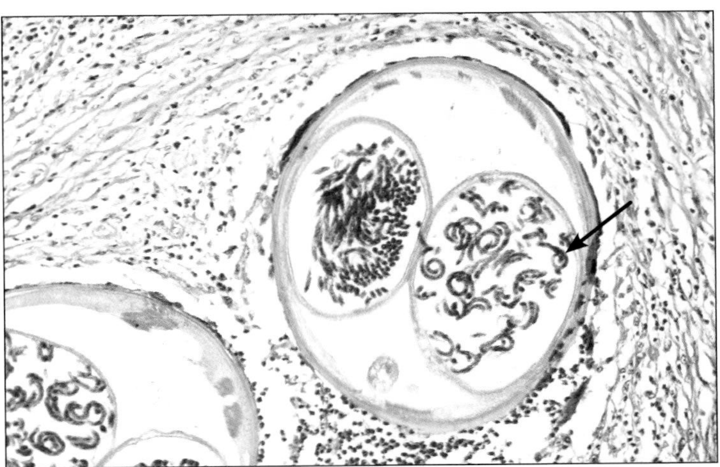

17.49 **Adult *Onchocerca volvulus* Showing Developing Microfilariae (Phylum Nematoda)** • Microfilariae are indicated by the **arrow**. They are 220–360 μm long. (Cross Section; H & E Stain)

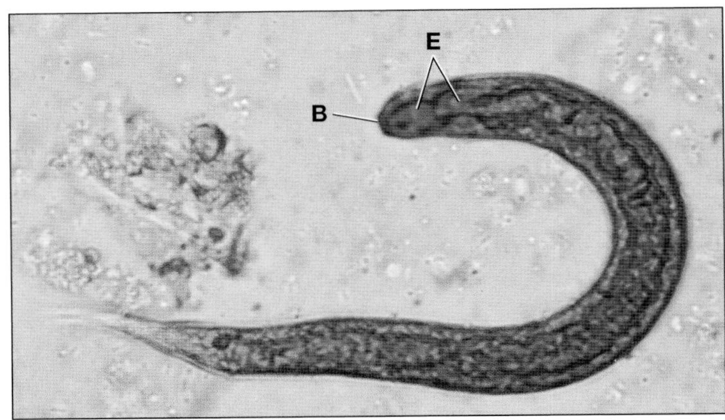

17.50 ***Strongyloides stercoralis* Rhabditiform Larva in a Fecal Sample (Phylum Nematoda)** • These larvae may be distinguished from hookworm larvae (which are rarely in feces) by their short buccal cavity (**B**). The name "rhabditiform" refers to the esophagus (**E**) shape, which has a constriction within it (between the two leader lines). (Wet Mount; D'Antoni's Iodine Stain)

free-living adult males and females. Free-living adults produce infective juveniles and the cycle (with sexual reproduction) is completed.

Symptoms of infection may be itching or secondary bacterial infection at the entry site of infective larvae, a cough and burning of the chest during the pulmonary phase, and abdominal pain and perhaps septicemia during the intestinal phase. Immune-suppressed individuals or those taking corticosteroids are at risk of higher levels of autoinfection leading to disseminated strongyloidiasis, which is fatal in up to 90% of cases.

Diagnosis depends on finding rhabditiform larvae (rarely eggs) in fresh fecal samples. Failing that (because of their relatively small numbers), serological tests (e.g., ELISAs, IFAs—see Section 8) that identify antibodies to *Strongyloides* are also available. Anti-helminthic medications are used for treatment of most infections and may be supplemented with antibiotics to prevent secondary bacterial infections. Attention to good personal hygiene and wearing shoes in potentially contaminated soil are good preventative measures.

Trichinella spiralis

Trichinella spiralis produces trichinosis, a disease of carnivorous mammals. Infection occurs when undercooked meat (e.g., pork) containing infective juveniles in **nurse cells** is eaten. These juveniles emerge from their nurse cells and enter the intestinal mucosa. Between 2 and 3 days later, the juveniles have developed into mature adults that burrow between intestinal epithelial cells.

Juveniles emerge from the females and are distributed throughout the body. When they are in skeletal muscle, they enter the muscle fibers and each induces the formation of a nurse cell (Figs. 17.51 and 17.52). In as little as 4 weeks, these juveniles become infective. Humans, unless they are eaten, are a dead-end host for this parasite.

Symptoms of infection are many and varied because the juveniles migrate throughout the body. Some consequences of infection are pneumonia, meningitis, deafness, and nephritis. Death may occur due to heart, respiratory, or kidney failure, but most infections are subclinical. Diagnosis is by muscle biopsy. Anti-helminthic drugs are available to kill the adults, but once in the muscle cells, larvae may be difficult to remove.

Trichuris trichiura

The whipworm *Trichuris trichiura* is a parasite of the large intestine and is often associated with *Ascaris* infections. Infection occurs through ingestion of eggs in fecal-contaminated soil or on plants. The juveniles then emerge and penetrate the large intestine's mucosa. As they grow, their thicker posterior projects into the intestinal lumen while the thinner anterior remains buried in the mucosa and feeds on cell contents and blood.

Adult females (Fig. 17.53) release up to 20,000 eggs a day (Fig. 17.54), which pass out with the feces and are diagnostic of infection; adult worms are rarely seen. Most infections are asymptomatic. With heavy worm burdens (more than 100), dysentery, anemia, and slowed growth and cognitive development are common. Anti-helminthic drugs are available for treatment.

Wuchereria bancrofti

Wuchereria bancrofti is the most commonly identified filarial worm. It is found in tropical and subtropical regions worldwide and causes lymphatic filariasis. Humans are the definitive host and several mosquito species are the intermediate host. Infection occurs when they transfer infective juveniles during a blood meal.

Once in the host, larvae migrate into the large lymphatic vessels of the lower body where they mature and may reside for years. Adults are found in coiled bunches and the females

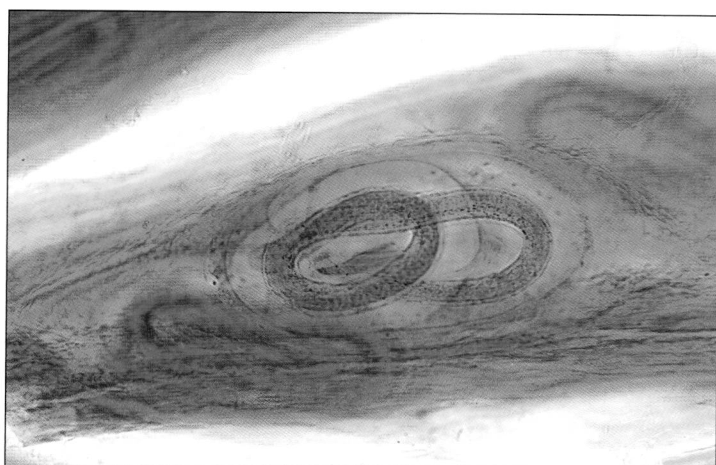

17.51 *Trichinella spiralis* Larva in Skeletal Muscle (Phylum Nematoda) The spiral juvenile and its nurse cell (skeletal muscle) are visible in this preparation. (Stained Whole Mount)

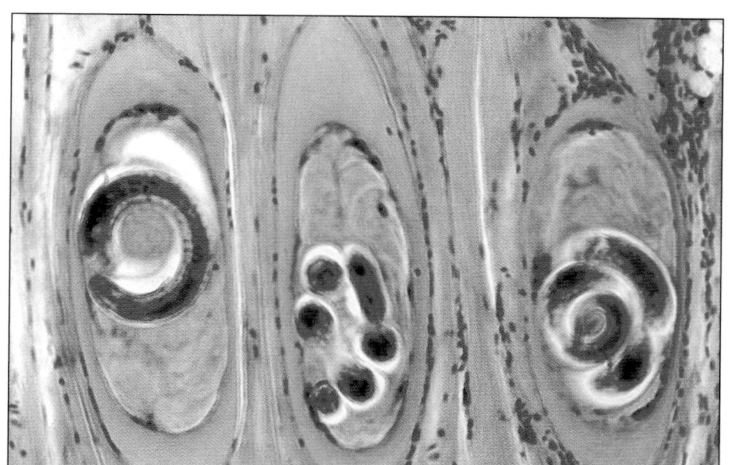

17.52 *Trichinella spiralis* Larvae in Skeletal Muscle (Phylum Nematoda) Each larva has entered a different skeletal muscle cell and converted it into a nurse cell that sustains it with nourishment. (Stained Section)

release microfilariae (Fig. 17.55) by the thousands. Microfilariae enter the blood and circulate there, often with a daily periodicity—most abundant at night when the mosquito vector is active and in lung capillaries during the day when it is hot.

Some infections are asymptomatic, whereas others result in acute inflammation of lymphatics associated with fever, chills, tenderness, and toxemia. In the most serious cases, obstruction of lymphatic vessels occurs and results in elephantiasis, a disease caused by accumulation of lymph fluid in the tissues, an accumulation of fibrous connective tissue, and a thickening of the skin.

Diagnosis of infection is made by identifying microfilariae in blood smears or in some cases, ultrasound. Treatment of active infection is with an anti-helminthic drug. Protecting oneself against mosquito bites is an effective means of prevention, though not always a practical or attainable one.

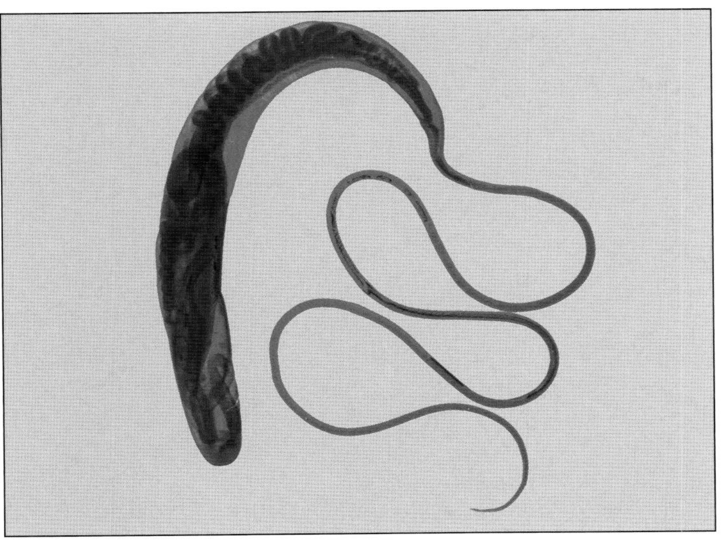

17.53 Adult Female *Trichuris trichiura* (Phylum Nematoda) • Adults are about 5 cm in length, with females being slightly longer than males. The blunt posterior end (seen here) is indicative of females. The long, thin anterior portion remains embedded in the intestinal mucosa and feeds. This specimen was approximately 30 µm wide at the anterior end and 1,000 µm at its widest point. (Stained Whole Mount)

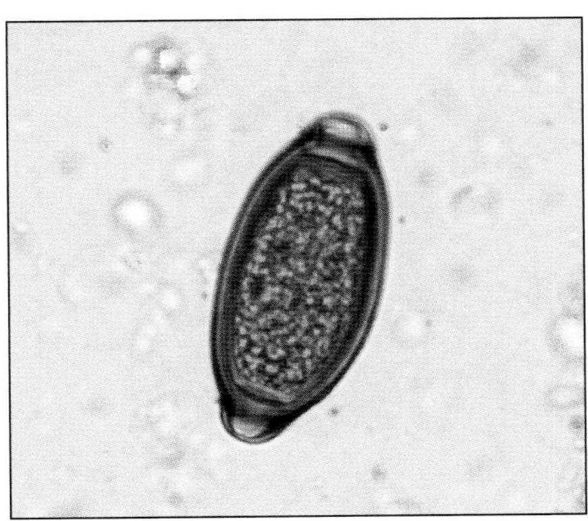

17.54 *Trichuris trichiura* Egg in Fecal Sample (Phylum Nematoda) • The barrel-shaped eggs of *T. trichiura* have distinctive plugs at either end. They are 50–55 µm in length by 22–24 µm wide. They resemble *Capillaria hepatica* eggs, but those are not frequently found in fecal samples, have a striated surface, and less prominent plugs (see Fig. 17.39). (Wet Mount; D'Antoni's Iodine Stain)

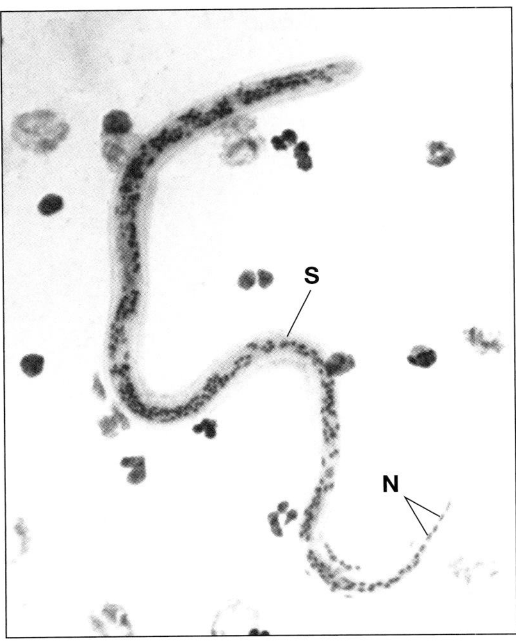

17.55 *Wuchereria bancrofti* Microfilaria in a Blood Smear (Phylum Nematoda) • The microfilariae of *W. bancrofti* can be distinguished from others in the blood by the sheath (**S**) and the single column of nuclei (**N**) not extending to the tip of the tail. (Stained Whole Mount)

Quantitative Microbiology

SECTION **18**

Standard Plate Count (Viable Count)

Purpose

The standard plate count (viable count) is one method of determining the density of a microbial population in a liquid medium. It provides an estimate of actual *living* cells in the sample (hence "viable" count) and is used in the food and dairy industries to monitor contamination. If samples are taken over a period of time, a growth curve of the culture can be produced.

Principle

The standard plate count is a procedure that allows microbiologists to estimate population density in a liquid sample by plating a known volume of a very dilute portion of that sample and counting the number of colonies it produces. The inoculum that is transferred to the plate contains a *known* proportion of the original sample because it is the product of a **serial dilution**.

As shown in Figure 18.1, a serial dilution is simply a series of controlled transfers down a line of dilution blanks (tubes containing a known volume of sterile **diluent**—water, saline, or buffer). The series begins with a sample containing an unknown concentration (density) of cells and ends with a very dilute mixture containing only a few—or no—cells.

Each dilution blank in the series receives a known volume from the mixture in the previous tube and delivers a known volume to the next, typically reducing the cell density to 1/10 or 1/100 at each step. (Greater dilutions in single steps are generally avoided, because they can be accomplished more conveniently and with greater accuracy by simply combining 1/10 or 1/100 dilutions.) To ensure the greatest degree of accuracy, all samples must be mixed well prior to any transfer.

For example, if the original sample in Figure 18.1 contains 1,000,000 cells/mL, following the first transfer the 1/100 dilution (10 μL into 990 μL of diluent) in dilution tube 1 would contain 10,000 cells/mL (1,000,000 cells/mL × 1/100 = 10,000 cells/mL). In the second dilution (tube 2) the 1/100 dilution would reduce it further to 100 cells/mL (10,000 cells/mL × 1/100 = 100 cells/mL).

Because the cell density of the original sample is not known at this time, only the dilutions (without mL units) are recorded on the dilution tubes. By convention, dilutions are expressed in scientific notation. Therefore, a 1/10 dilution is written as 10^{-1} and a 1/100 dilution is written as 10^{-2}.

A known volume (usually 0.1 mL) of appropriate dilutions (depending on the *estimated* cell density of the original sample) is then spread onto agar plates to produce at least one **countable plate**. A countable plate is one that contains between 30 and 300 colonies (Fig. 18.2). A count lower than 30 colonies is considered statistically unreliable and greater than 300 is typically too many to be viewed as individual colonies on a standard 100 mm Petri dish.

In examining the serial dilution diagram (Fig. 18.1), you can see that the first transfer in the series is a simple dilution, but that all successive transfers are compound dilutions. Both dilution types can be calculated using the following formula:

$$V_1D_1 = V_2D_2$$

where V_1 and D_1 are the volume and dilution of the concentrated broth, respectively, while V_2 and D_2 are the volume and dilution of the completed dilution. Undiluted samples are always expressed as 1. (Undiluted sample is 100% sample, which equals 1.) Here's how to calculate the dilution in Tube 1: 10 μL (V_1) of undiluted sample ($D_1 = 1$) is transferred to 990 μL of diluent to make a final volume of 1000 μL (V_2). Solve for D_2 using the permuted formula:

$$D_2 = \frac{V_1D_1}{V_2} = 10\ \mu L \times \frac{1}{1000\ \mu L} = \frac{1}{100} = 10^{-2}$$

Notice that the dilution has no units because the microliters cancel.

Compound dilutions are calculated using the same formula[1]. However, because D_1 in compound dilutions no longer represents undiluted sample, but rather a fraction of the original density, it must be represented as something less than 1 (e.g., 10^{-1}, 10^{-2}, etc.). Here's how to calculate the dilution in Tube 2: 10 μL (V_1) of the 10^{-2} dilution (D_1) from Tube 1 are transferred to 990 μL of diluent, making V_2 1000 μL. Solve for D_2 using the permuted formula:

$$D_2 = \frac{V_1D_1}{V_2} = \frac{10\ \mu L \times 10^{-2}}{1000\ \mu L} \times 10^{-1} \times 10^{-3} = 10^{-4}$$

Spreading a known volume of a known dilution onto an agar plate and counting the colonies that develop provides all the information you need to calculate the original cell density (OCD) because each colony is assumed to have grown from a single cell or cell type. This is the basic formula for this calculation:

$$OCD = \frac{CFU}{D \times V}$$

[1] This formula will work with all necessary dilution calculations. For calculations involving unconventional volumes or dilutions, the formula is essential, but for simple tenfold or hundredfold dilutions like the ones described, the final compounded dilution in a series can be calculated simply by multiplying each of the simple dilutions by each other. For example, a series of three 10^{-1} dilutions would yield a final dilution of 10^{-3} ($10^{-1} \times 10^{-1} \times 10^{-1} = 10^{-3}$). Three 10^{-2} dilutions would yield a final dilution of 10^{-6} ($10^{-2} \times 10^{-2} \times 10^{-2} = 10^{-6}$). We encourage you to use whatever method is better for you. In time you will be doing the calculations in your head.

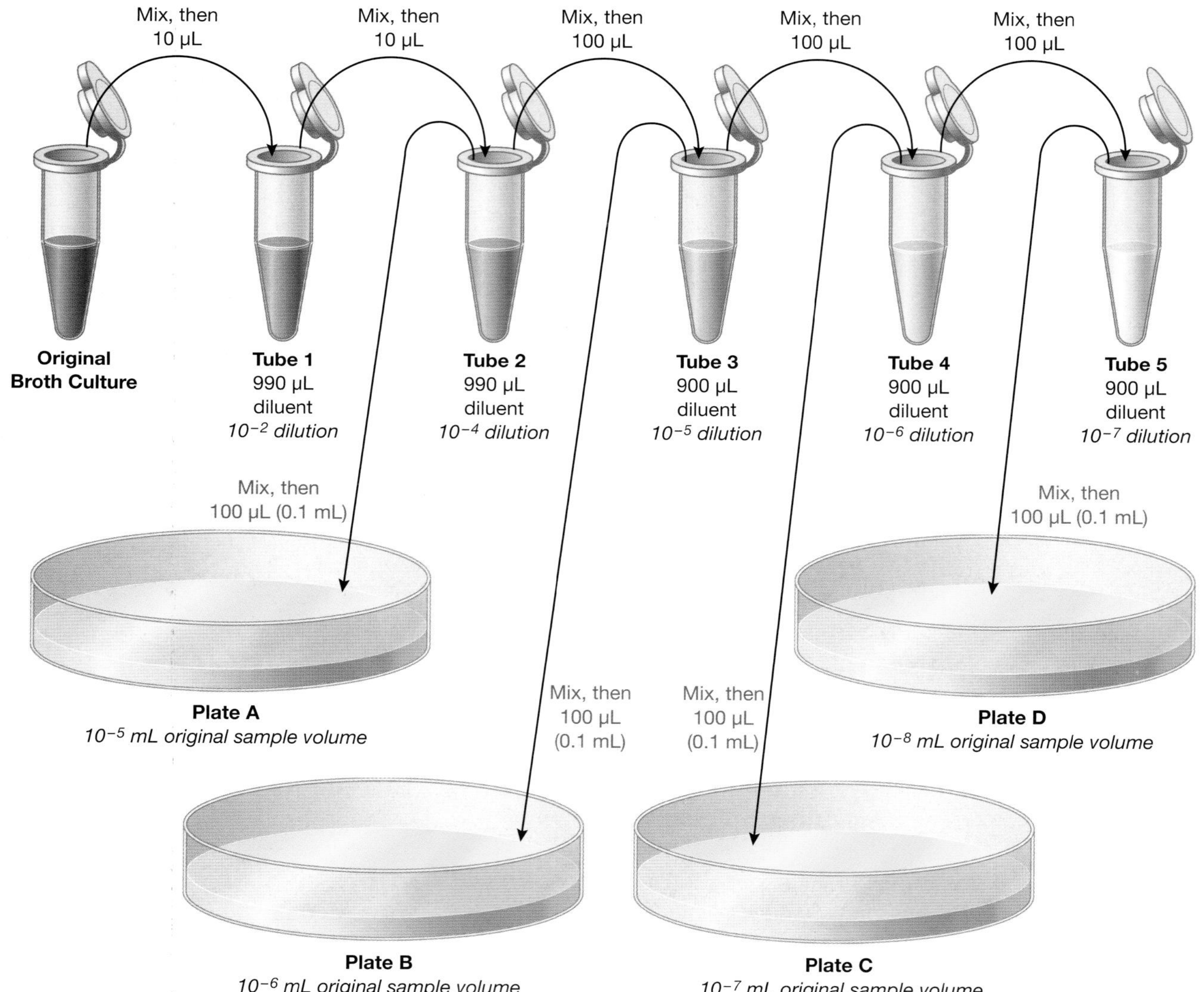

18.1 Serial Dilution • This is a sample dilution scheme. The dilution assigned to each tube (written below the tube) represents the proportion of original sample inside that tube. For example, the dilution in Tube 2 is 10^{-4}, so the proportion of original sample inside it is 1/10,000th of the total volume. Transferring 0.1 mL (=100 μL) of that diluted sample to Plate A results in an original sample volume of $0.1\ mL \times 10^{-4} = 10^{-5}$ mL.

CFU (colony forming units) is actually the number of colonies that develop on the plate. CFU is the preferred term because colonies could develop from single cells or from groups of cells, depending on the typical cellular arrangement of the organism. (We have to admit that we really don't know how many cells started each colony, so we don't overstate our case and equate colonies to CFUs rather than to individual cells.)

D is the dilution as written on the dilution tube from which the inoculum comes. V is the volume transferred to the plate. (Note that this volume must be included in the formula because densities are expressed in CFU/mL. Plating some volume other than 1 mL must be accounted for. For instance, inoculation of 0.1 mL would contain 1/10th as many cells as 1 mL.)

As you can see in the formula, the volume of *original sample* being transferred to a plate is the product of the *volume transferred* and the *dilution* of the tube from which it came. Therefore 0.1 mL transferred from a 10^{-2} dilution contains only 10^{-3} mL of the original sample ($0.1 \text{ mL} \times 10^{-2} = 10^{-3}$ mL). The convention among microbiologists is to condense D and V in the denominator into "**Original sample volume**"[2] (expressed in mL). The formula thus becomes:

$$\text{OCD} = \frac{\text{CFU}}{\text{Original sample volume}}$$

The sample volume is written on the plate at the time of inoculation. Following a period of incubation, the plates are examined, colonies are counted on the countable plate(s), and calculation is a simple division problem. Suppose, for example, you count 37 colonies (Fig. 18.3) on a plate inoculated with 0.1 mL of a 10^{-5} dilution. Knowing that this plate now contains 10^{-6} mL of original sample, calculation would be as follows:

$$\text{OCD} = \frac{\text{CFU}}{\text{Original sample volume}} = \frac{37 \text{ CFU}}{10^{-6} \text{ mL}} = 3.7 \times 10^{7} \frac{\text{CFU}}{\text{mL}}$$

[2] Some microbiologists refer to this as the "plate dilution." We prefer to use OSV in this introduction because it emphasizes what is really happening; that is, OSV is the volume of the original sample deposited on the plate.

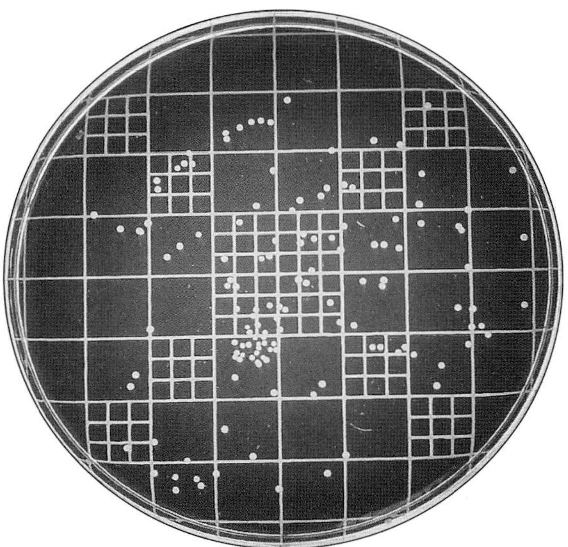

18.2 Countable Plate • A countable plate has between 30 and 300 colonies. Therefore, this plate (shown on a colony counter grid) with approximately 130 colonies is countable and can be used to calculate cell (or CFU) density in the original sample. Plates with fewer than 30 colonies are TFTC ("too few to count"). Plates with more than 300 colonies are TMTC ("too many to count").

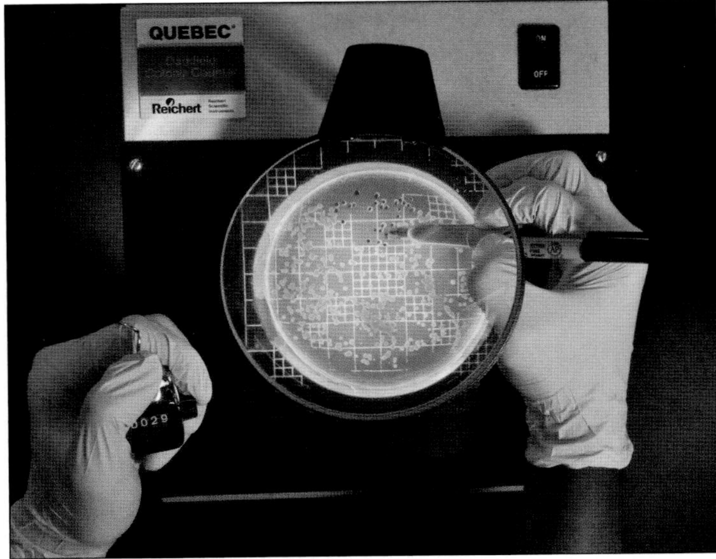

18.3 Colony Counter • The magnifying lens and grid (Fig. 18.2) make colony counting easier. Counted colonies are either punched with an inoculating needle or toothpick or marked on the Petri dish base with a pen (as in the photo) to ensure all colonies are counted and none are counted twice. A hand tally counter (seen in the left hand) is used to ensure that distractions don't cause the microbiologist to lose track of the counted colonies. Other, more sophisticated counting methods are available. One uses an electronic pen that, when touched to the plastic Petri dish below a colony, records a tally. There are also software systems that capture an image of the plate and then colonies are counted by the computer.

Direct Count (Petroff-Hausser Counting Chamber)

Purpose

This direct count method is used to determine cell density in a sample.

Principle

Microbial direct counts, like plate counts, take a small portion of a sample and use the data gathered from it to calculate the overall population cell density. This is made possible with a device called a **Petroff-Hausser counting chamber.** The Petroff-Hausser counting chamber is very much like a microscope slide with a 0.02 mm deep chamber, or "well," in the center containing an etched grid (Fig. 18.4).

The grid is one square millimeter and consists of 25 large squares, each of which contains 16 small squares, making a total of 400 small squares. Figures 18.5 and 18.6 illustrate the counting grid.

Each small square is 1/20 mm by 1/20 mm, making an area of 1/400 mm^2. When the cover glass is put in place it rests 2/100 (0.02) mm above the grid, so the volume above *each small square* is 1/400 $mm^2 \times 2/100$ mm = 2/40,000 $mm^3 = 5 \times 10^{-5}\ mm^3$. One $mm^3 = 10^{-3}$ cm (cubic centimeters), so $5 \times 10^{-5}\ mm^3 = 5 \times 10^{-8}\ cm^3$. Because 1 cm^3 = 1 mL, the volume above each small square is 5×10^{-8} mL. (Phew!)

This may seem like an extremely small volume—and it is—but the space above each small square is large enough to hold about 50,000 average-size cocci. Fortunately, dilution procedures prevent this scenario from occurring and cell counting is easily done using the microscope and a hand counter.

Original cell density (OCD) is determined by counting the cells found in a predetermined group of small squares and dividing by the number of squares counted (to get an average of cells per square) multiplied by the dilution[3] and the volume of sample above one small square.

The following is a standard formula for calculating original cell density in a direct count:

$$\text{OCD} = \frac{\text{Cells counted}}{(\text{Squares})(\text{Dilution})(\text{Volume})}$$

[3] Dilutions are calculated using the following formula:

$$D_2 = \frac{V_1 D_1}{V_2}$$

D_2 is the new dilution to be determined. V_1 is the volume of sample being diluted. D_1 is the dilution of the sample before adding diluent (undiluted samples have a dilution factor of 1). V_2 is the new combined volume of sample and diluent after the dilution is completed.

For most accurate estimates, some experts recommend a minimum overall count of 600 cells in one or more samples taken from a single population. Optimum density for counting is between 5 and 15 cells per small square. Counting may be done manually or by using a computer analysis program to count cells from digital images.

If, for example, 200 cells from a sample with a dilution of 10^{-2} were counted in 16 squares (remembering that the volume above a single small square is 5×10^{-8} mL), the cell density in the original sample would be calculated as follows:

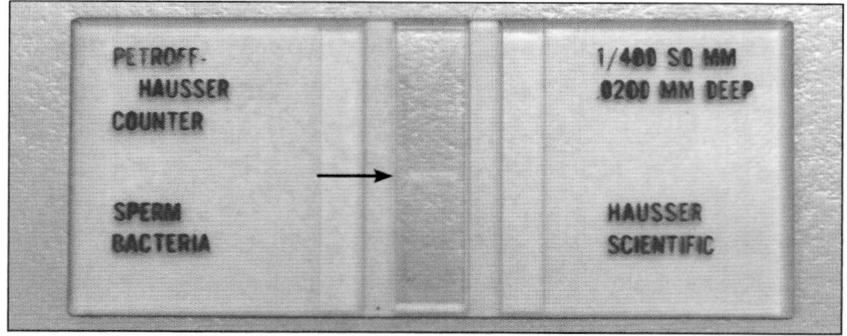

18.4 Petroff-Hausser Counting Chamber • The Petroff-Hausser counting chamber is a precision instrument used for the direct counting of bacterial and other cells. A well-mixed bacterial suspension is drawn by capillary action from a pipette into the chamber enclosed by a coverslip. The cells are then counted against the grid of small squares in the center of the chamber. The horizontal lines of the grid are faintly visible at the **arrow**.

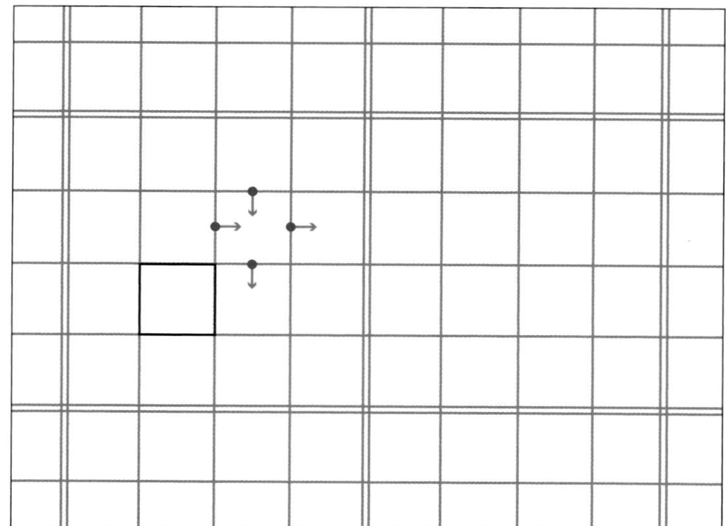

18.5 Petroff-Hausser Counting Chamber Grid • Shown is a drawing of a portion of the Petroff-Hausser counting chamber grid. The smallest squares are the ones referred to in the formula. The volume above a small square is 5×10^{-8} mL. To ensure that all cells are counted only once, include cells on a line with the square below or to the right, as indicated by the cells (**violet**) and arrows (**red**).

$$OCD = \frac{\text{Cells counted}}{(\text{Squares})(\text{Dilution})(\text{Volume})}$$

$$OCD = \frac{200 \text{ cells}}{(16)(10^{-2})(5 \times 10^{-8} \text{ mL})}$$

$$OCD = 2.5 \times 10^{10} \text{ cells/mL}$$

The advantages of direct counting are that it is fast, easy to do, and relatively inexpensive. The major disadvantage is that living as well as dead cells are counted.

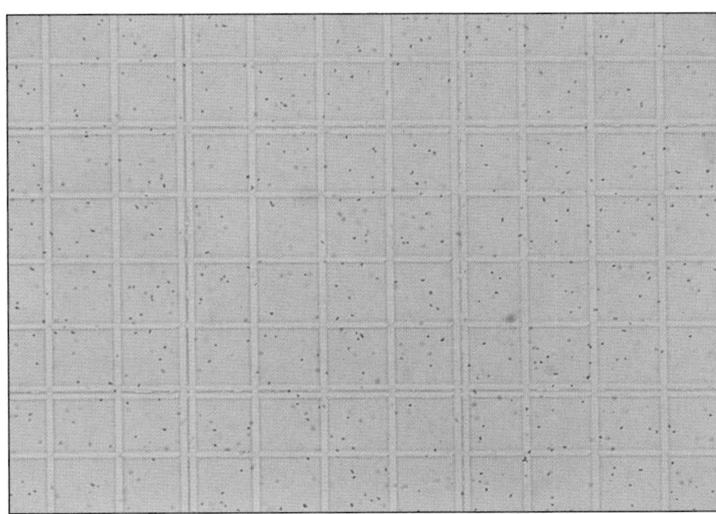

18.6 Bacterial Cells in a Petroff-Hausser Counting Chamber • This is a 10^{-4} dilution of *Vibrio natriegens* on the grid viewed with the high dry lens. Try counting cells in several small squares, take the average, and calculate the original cell density.

Plaque Assay for Determination of Phage Titer

Purpose

This technique is used to determine the concentration of viral particles in a sample. Samples taken over a period of time can be used to construct a viral growth curve.

Principle

Viruses that attack bacteria are called **bacteriophages**, or simply phages (see Figs. 10.3 and 18.7). Some viruses attach to the bacterial cell wall and inject viral DNA into the bacterial cytoplasm. The viral genome then commands the host cell to produce more viral DNA and viral proteins, which are used for the assembly of more phages.

Once assembly is complete, the cell lyses and releases the phage progeny, which then attack other bacterial cells and begin the replicative cycle all over again. This process, called the **lytic cycle**, is shown in Figure 10.4.

Lysis of bacterial cells growing in a lawn on an agar plate produces a clearing that can be viewed with the naked eye. These clearings are called **plaques**. The plaque assay uses this phenomenon as a means of calculating the phage concentration in a given sample.

When a sample of bacteriophage (generally diluted by means of a serial dilution) is added to a plate inoculated with enough bacterial host to produce a lawn of growth[4], the number of plaques formed can be used to calculate the original phage titer, or density.

[4] Gaps in the lawn may be mistaken for plaques or may contain virus particles that didn't have the opportunity to infect host cells, both of which reduce the accuracy of the count.

The plaque assay technique is similar to the standard plate count, in that it employs a serial dilution to produce countable plates needed for later calculations. One key difference is that the plaque assay is done using the **pour-plate technique**, in which bacterial cells and viruses are first added to molten agar and then poured into the plate.

In this procedure, diluted phage is added directly to a small amount of *E. coli* culture and allowed a 10-minute (±5 minutes) adsorption period to attach to and infect the bacterial cells. Then this phage–host mixture is added to a tube of molten soft agar, mixed, and poured onto prepared nutrient agar plates as an agar overlay.

The consistency of the solidified soft agar is sufficient to immobilize the bacteria while allowing the smaller bacteriophages to diffuse short distances and infect surrounding cells when released from lysed host cells. During incubation, the

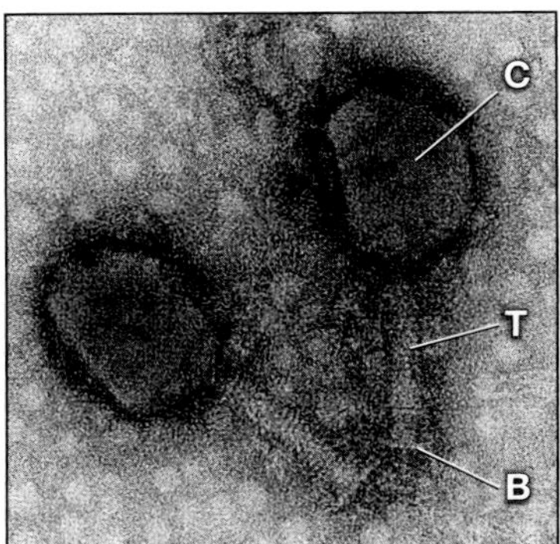

18.7 T4 Coliphage • This is a negative stain of two T4 phage particles. Shown are the capsid (**C**), tail (**T**), and base plate (**B**). Tail fibers are visible only as broken pieces in the background. The length of this phage from base plate to tip of capsid is approximately 180 nm (0.18 μm). (TEM)

phage host produces a lawn of growth on the plate in which plaques appear where contiguous cells have been lysed by the virus (Fig. 18.8).

The procedure for counting plaques is the same as that for the standard plate count. To be statistically reliable and practical, countable plates must have between 30 and 300 plaques. Calculating **phage titer** (original phage density) uses the same formula as other plate counts except that **plaque forming units (PFU)** instead of CFU (colony forming units) becomes the numerator in the equation. Phage titer, therefore, is expressed in PFU/mL and the formula is written as follows:

$$\text{Phage titer} = \frac{\text{PFU}}{\text{Volume plated} \times \text{Dilution}}$$

As with the standard plate count, it is customary to condense *volume plated* and *dilution* into *original sample volume*. The formula then becomes:

$$\text{Phage titer} = \frac{\text{PFU}}{\text{Original sample volume}}$$

The original sample volume is written on the plate at the time of inoculation. Following a period of incubation, the plates are examined, plaques are counted on the countable plate(s), and calculation is a simple division problem.

Suppose, for example, you inoculated a plate with 0.1 mL of a 10^{-4} phage dilution. (Remember, you are calculating the *phage density;* the *E. coli* has nothing to do with the calculations.) This plate now contains 10^{-5} mL of original phage sample. If you subsequently counted 115 plaques on the plate, calculation would be as follows:

$$\text{Original phage density} = \frac{1.15 \times 10^{2}\ \text{PFU}}{10^{-5}\ \text{mL}} = 1.15 \times 10^{7}\ \frac{\text{PFU}}{\text{mL}}$$

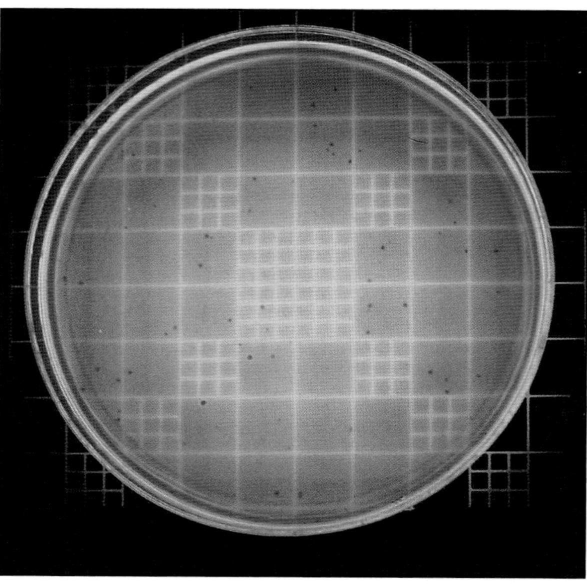

18.8 Plaque Assay Countable Plate • This plaque assay plate has between 30 and 300 plaques; therefore, it is "countable." They look dark because the photograph was shot against the dark background of a colony counter. Without the colony counter and viewed with transmitted light the plaques will look like clear spots in the hazy bacterial lawn.

Urine Streak—Semiquantitative Method

Purpose

Urine culture is a common method of detecting and quantifying urinary tract infections (UTIs). It frequently is combined with media for specific identification of members of *Enterobacteriaceae* (e.g., MacConkey agar) or *Streptococcus* (sheep blood agar).

Principle

Urine culture is a semiquantitative CFU counting method that quickly produces countable plates without a serial dilution. The instrument used in this procedure is a volumetric loop, calibrated to hold 0.001 mL (1 μL) or 0.01 mL (10 μL) of sample.

Urine culture procedures using volumetric loops are useful in situations where a rapid diagnosis is essential and approximations ($\pm 10^{2}$ CFU/mL) are sufficient to choose a course of action. Volumetric loops are useful in situations where population density is not likely to exceed 10^{5} CFU/mL.

In this standard procedure, a loopful of urine (generally from a clean catch midstream sample or a bladder catheter sample) is carefully transferred to a blood agar plate. The initial inoculation is a single streak across the diameter of the agar plate. Then the plate is turned 90° and (without flaming the loop) streaked again, this time across the original line in a zigzag pattern to evenly disperse the bacteria over the entire plate.

The plate is rotated 90° again and the entire surface is streaked a third time parallel to the original streak (Figs. 18.9–18.12). Following a period of incubation, the resulting colonies are counted and population density, usually referred to as "original cell density," or OCD, is calculated.

OCD is recorded in "colony forming units" per milliliter (CFU/mL). CFU/mL is determined by dividing the number of colonies on the plate by the volume of the loop. For example, if 75 colonies are counted on a plate inoculated with a 0.001 mL loop, the calculation would be as follows:

$$OCD = \frac{CFU}{\text{Loop volume}}$$

$$OCD = \frac{75\ CFU}{0.001\ mL}$$

$$OCD = 7.5 \times 10^4\ CFU/mL$$

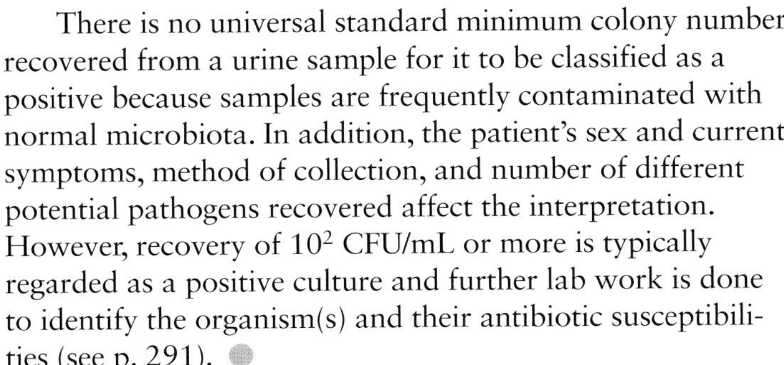

There is no universal standard minimum colony number recovered from a urine sample for it to be classified as a positive because samples are frequently contaminated with normal microbiota. In addition, the patient's sex and current symptoms, method of collection, and number of different potential pathogens recovered affect the interpretation. However, recovery of 10^2 CFU/mL or more is typically regarded as a positive culture and further lab work is done to identify the organism(s) and their antibiotic susceptibilities (see p. 291).

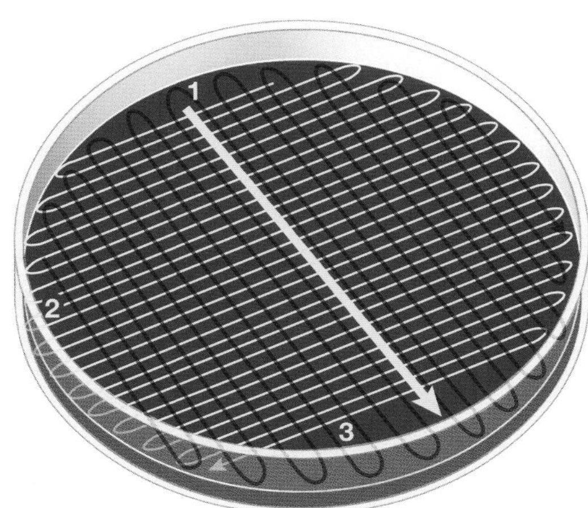

18.9 Semiquantitative Streak Method • Streak 1 is a simple streak line across the diameter of the plate. Streak 2 is a tight streak perpendicular to Streak 1. Streak 3 is like Streak 2, but parallel to Streak 1. Done properly, cells should be evenly spread across the entire plate's surface. The loop is not flamed between streaks. (For starters, it's probably plastic, but that aside, flaming would reduce the number of cells transferred to the plate and affect the results.)

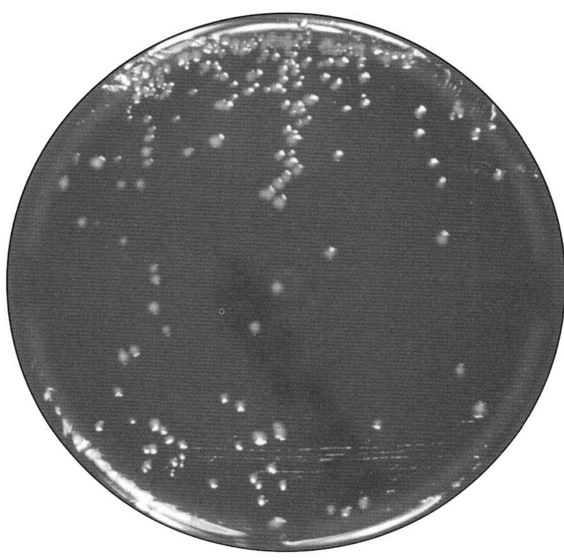

18.10 Urine Streak on Sheep Blood Agar • This plate was inoculated with a 0.01 mL volumetric loop using a fake urine sample spiked with *Escherichia coli* and *E. faecalis*. The cell density (in CFU/mL) can be determined by dividing the number of colonies by 0.01. As a general (but not absolute) rule, more than 10^2 CFU/mL is considered a positive test.

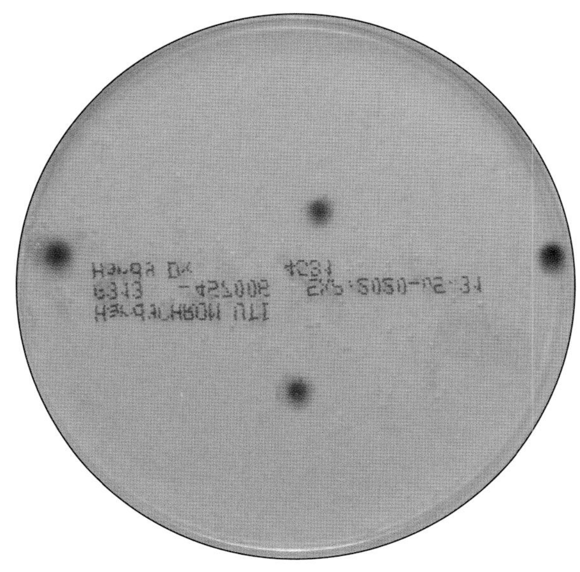

18.11 HardyCHROM UTI Medium–Streak Plate • This medium enables differentiation of bacterial species or genera commonly found in urine samples by color production. Different chromogenic substrates for enzymes unique to each species are included in the medium. Shown is a streak plate made with a 0.01 mL loop and fake urine containing *Escherichia coli* (magenta) and *Proteus mirabilis* (buff). We counted 122 colonies, so the density in the sample was $1.22 \times 10^2/10^{-2} = 1.22 \times 10^4$ CFU/mL

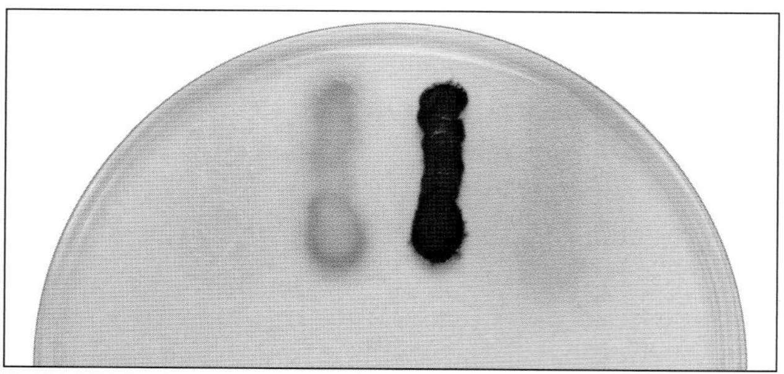

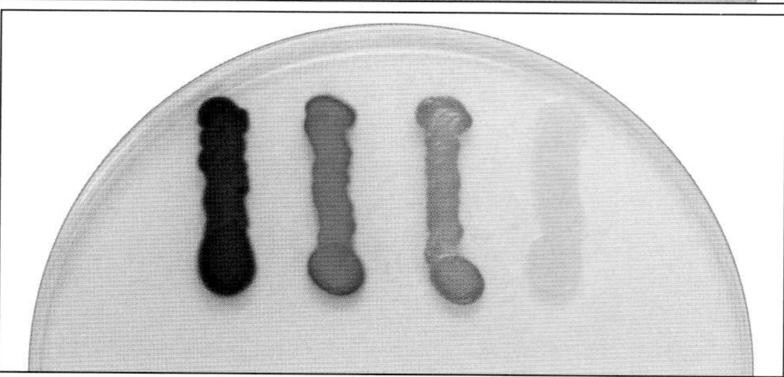

18.12 HardyCHROM UTI Medium—Species Diversity
Here is a sampling of the species diversity this medium can identify. Top (left to right): *Pseudomonas aeruginosa*, *Escherichia coli*, *Klebsiella pneumoniae*, and *Proteus vulgaris*. Bottom (left to right): *Klebsiella (Enterobacter) aerogenes*, *Enterococcus faecalis*, *Staphylococcus saprophyticus*, and *Staphylococcus aureus*. In normal usage, this medium would be streaked as in Figure 18.9.

Medical, Environmental, and Food Microbiology

SECTION 19

Part A: Medical Microbiology

Antimicrobial Susceptibility Tests (Kirby-Bauer Method and E-Test)

Purpose

The disk diffusion test is a standardized method used to measure the effectiveness of antibiotics and other chemotherapeutic agents on pathogenic microorganisms. In many cases, it is an essential tool in prescribing the appropriate treatment for a patient.

Principle

Antibiotics are natural antimicrobial agents produced by microorganisms. One type of penicillin, for example, is produced by the mold *Penicillium notatum*. Today, because many agents used to treat bacterial infections are synthetic, the terms **antimicrobials** or **antimicrobics** are used to describe all substances used for this purpose.

In 1966 Alfred Bauer, William Kirby, and associates first published a standardized disk diffusion method of testing the effectiveness of more than 20 antimicrobial chemicals. Their methodology, which bears their names, has become a valuable tool for establishing the effectiveness of antimicrobics against pathogenic microorganisms in clinical laboratories.

In the test, antimicrobic-impregnated paper disks are placed on an agar plate inoculated to form a bacterial lawn. The plates are incubated to allow growth of the bacteria and time for the agent to diffuse into the agar. As the drug moves through the agar, it establishes a concentration gradient. If the organism is susceptible to it, a clear zone—**zone of inhibition**—will appear around the disk where the concentration is high enough to stop growth (Figs. 19.1 and 19.2).

The junction of the zone of inhibition with growth is critical to interpretation of the test. It is at this junction where the concentration of antimicrobic has become too low to effectively stop growth. This junction represents

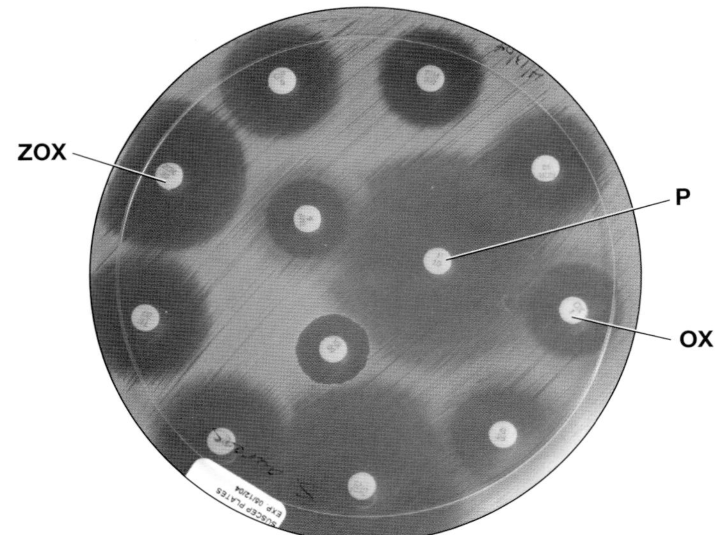

19.1 Disk Diffusion Test of Methicillin-Susceptible *Staphylococcus aureus* This plate illustrates the effect of (clockwise from top outer right) Nitrofurantoin (F/M300), Norfloxacin (NOR 10), Oxacillin (OX 1), Sulfisoxazole (G .25), Ticarcillin (TIC 75), Trimethoprim-Sulfamethoxazole (SXT), Tetracycline (TE 30), Ceftizoxime (ZOX 30), Ciprofloxacin (CIP 5), and (inner circle from right) Penicillin (P 10), Vancomycin (VA 30), and Trimethoprim (TMP 5) on methicillin-susceptible *S. aureus*. Compare the zone sizes with those in Figure 19.2, paying particular attention to ceftizoxime, oxacillin and penicillin. Note that the zones are black because the plate was photographed against a black background.

the **minimum inhibitory concentration** (**MIC**) for that particular strain.

The basic disk diffusion test does not quantify the MIC (though variations are available that do this), but rather utilizes the zone of inhibition's size, which depends upon the sensitivity of the organism to the specific antimicrobial agent (among other factors). Interestingly, the clear zone does not necessarily mean that the bacteria have been killed.

Drugs that kill the organism are said to be **bactericidal**, but other drugs are **bacteriostatic**; that is, they stop the bacteria from dividing, but do not kill them. Some mechanisms of antibiotic action and resistance are given in Table 19-1.

All aspects of the Kirby-Bauer procedure are standardized to ensure reliable results. Therefore, care must be taken to adhere to these standards. Mueller-Hinton agar, which has a pH between 7.2 and 7.4, is poured to a depth of 4 mm in

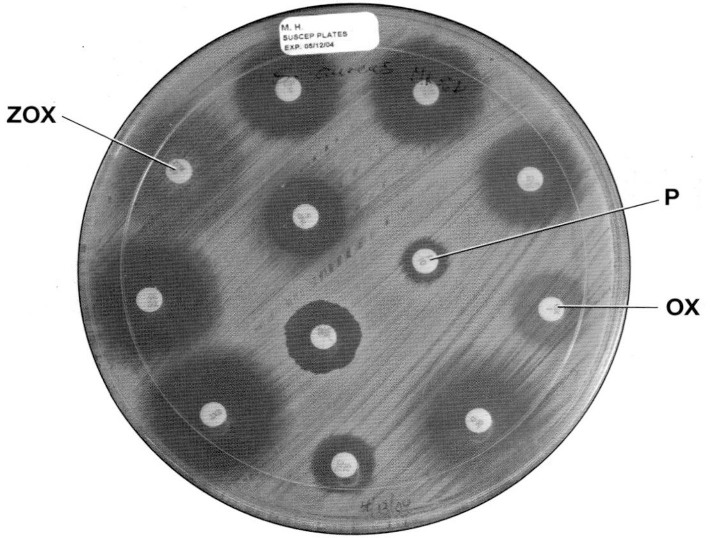

19.2 Disk Diffusion Test of Methicillin-Resistant *Staphylococcus aureus* (MRSA) • The Kirby-Bauer test illustrating the effect of the same antibiotics as in Figure 19.1 on methicillin-resistant *S. aureus*. Compare the zone sizes with those in Figure 19.1 and note the breakthrough growth surrounding ceftizoxime (**ZOX**) and oxacillin (**OX**) and the significantly smaller zone surrounding penicillin (**P**). Note that the zones are black because the plate was photographed against a black background.

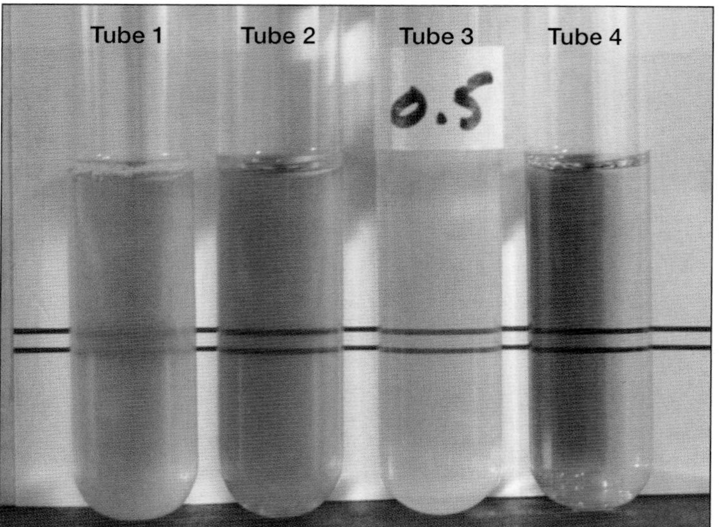

19.3 McFarland Standards • This is a comparison of a 0.5 McFarland turbidity standard (Tube 3) to three broths having varying degrees of turbidity. Each of the 11 McFarland standards (0.5 to 10) contains a specific percentage of precipitated barium sulfate to produce turbidity. In the Kirby-Bauer procedure, the test culture is diluted to match the 0.5 McFarland standard (equivalent to $1.5 \pm 0.5 \times 10^8$ cells per mL) before inoculating the plate. Comparison is made visually by placing a card with sharp black lines behind the tubes. Notice that the turbidity of Tube 2 matches the McFarland standard exactly, whereas Tubes 1 and 4 are too turbid and too clear, respectively. Alternatively, cultures may be standardized with a spectrophotometer. The correct turbidity will have an absorbance reading of 0.08 to 0.10 at 625 nm in a 1 cm cuvette.

TABLE **19-1 Antibiotic Targets and Resistance Mechanisms** • Not all antibiotics affect cells in the same way. Some attack the bacterial cell wall, and others interfere with biosynthesis reactions. Resistance mechanisms can be broken down into four main categories: (a) altered target such that the antibiotic no longer can interact with the cellular process, (b) an alteration in how the drug is taken into the cell, (c) enzymatic destruction of the drug, and (d) development or increased activity of an efflux mechanism.

Antibiotic	Cellular Target	Resistance Mechanism
Chloramphenicol	Prevents peptide bond formation during translation	1. Poor uptake of drug 2. Inactivation of drug
Ciprofloxacin	Interferes with DNA replication	1. Altered target 2. Poor uptake of drug
Penicillin	Inhibits cross-linking of the cell wall's peptidoglycan	One or more of: 1. Altered target 2. Poor uptake of drug 3. Production of β-lactamases
Streptomycin	Blocks initiation complex formation in protein synthesis	1. Altered target
Tetracycline	Blocks attachment of aminoacyl tRNA to A site on ribosome	1. Efflux mechanism
Trimethoprim	Inhibits purine and pyrimidine synthesis	1. Altered target

either 150 mm (as in Figs. 19.1 and 19.2) or 100 mm Petri dishes. The depth is important because of its effect upon the diffusion. Thick agar slows lateral diffusion and thus produces smaller zones than plates held to the 4 mm standard. Incubation time and temperature are also important, as are a multitude of other factors.

Inoculation is made with a broth culture, generally incubated 4 to 6 hours to bring it into exponential phase, and subsequently diluted to match a **0.5 McFarland turbidity standard**, which corresponds to a cell density between 1 and 2×10^8 CFU/mL (Fig. 19.3). Alternatively, a spectrophotometer can be used to bring the broth culture to an absorbance of 0.08 to 0.10 at 625 nm in a 1 cm cuvette, which corresponds to the 0.5 McFarland standard.

The U.S. Food and Drug Administration establishes the disk concentration of each agent, which is printed on the disk along with a code indicating the chemical agent. Disks are then dispensed onto the inoculated plate (Fig. 19.4), which is incubated at 35 ± 2°C for 16 to 18 hours. After incubation, the plates are removed and the zone diameters are measured in millimeters (Fig. 19.5) and compared to a standardized table of results (Table 19-2).

The Clinical Laboratory Standards Institute (CLSI) in Wayne, Pennsylvania, has performed extensive research (and continues to do so) to establish the zone diameter interpretive standards provided in Table 19-2—and there are many more antimicrobial agent/pathogen combinations we have not reproduced.

The zone diameters (break points) established by CLSI for resistance and susceptibility are published in a document titled, *M100 Performance Standards for Antimicrobial*

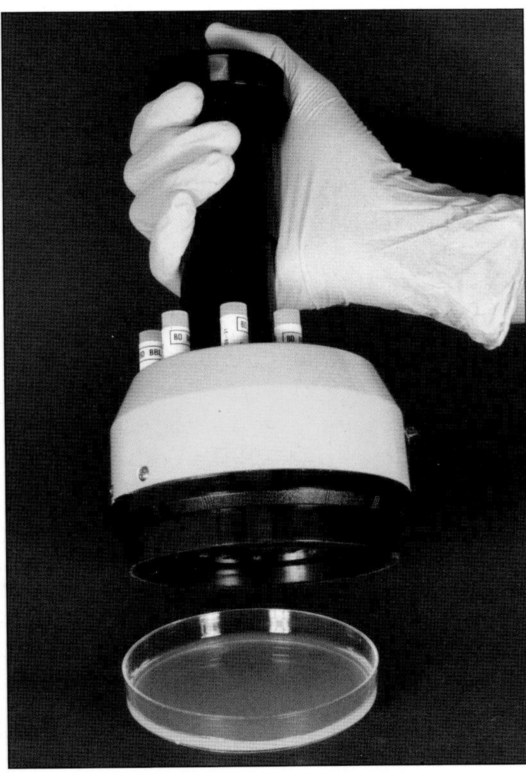

19.4 Disk Dispenser • This antibiotic disk dispenser is used to uniformly deposit disks on a Mueller-Hinton agar plate. Disk cartridges are seen projecting out of the dispenser's top.

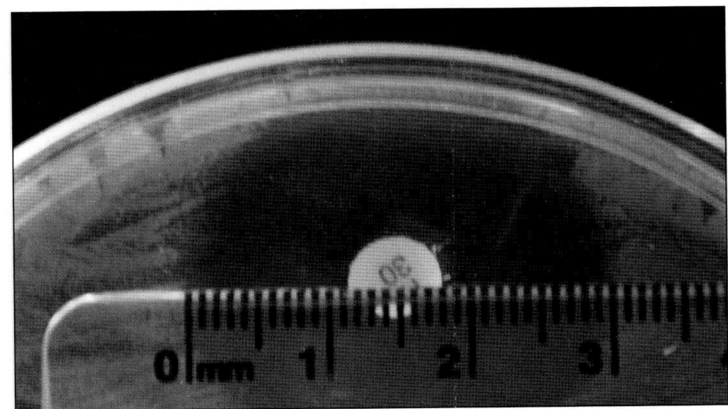

19.5 Measuring the Zones of Inhibition • Using a metric ruler and a dark, nonreflective background, each zone's diameter is measured. Standards for comparison (Table 19-2) are given in millimeters (mm). This zone is 29 mm in diameter.

TABLE **19-2** Zone Diameter Interpretive Chart

Antibiotic	Organism(s)	Code	Disk Potency	Zone Diameter Interpretive Standards (mm)		
				Resistant	Intermediate	Susceptible
Chloramphenicol	*Enterobacteriaceae* and *Staphylococcus*	C 30	30 µg	≤ 12	13–17	≥ 18
Penicillin	*Staphylococcus*	P 10	10 U	≤ 28		≥ 29
Streptomycin	*Enterobacteriaceae*	S 10	10 µg	≤ 11	12–14	≥ 15
Tetracycline	*Enterobacteriaceae* *Staphylococcus*	TE 30 TE 30	30 µg 30 µg	≤ 11 ≤ 14	12–14 15–18	≥ 15 ≥ 19

This table includes a sampling of common antibiotics and contains data provided by the Clinical and Laboratory Standards Institute (CLSI). Permission to use portions (specifically Tables 2A and 2C) of *M100 Performance Standards for Antimicrobial Susceptibility Testing,* 30th edition, (2020) has been granted by CLSI. The interpretive data are valid only if the methodology in CLSI Standard M02 Performance Standards for Antimicrobial Disk Susceptibility Tests, 13th edition, (2018) is followed. CLSI updates the interpretive tables in M100 annually through new editions of the supplement. Users should refer to the most recent editions. The most current editions of both documents may be obtained from CLSI, 950 West Valley Road, Suite 2500, Wayne, PA 19087. Contact also may be made via the website (www.CLSI.org), email (customerservice@clsi.org), and by phone (1.877.447.1888).

Susceptibility Testing, 30th edition (2020). It is updated biennially and is the source of the information in Table 19-2.

Many factors affect the zone diameter size breakpoints, such as molecular weight, mechanism of action, and concentration on the disk, to name a few. Add to this that all *strains* of a species may not be resistant or susceptible, and even resistant strains may produce a zone of inhibition. It's a complex world.

So how are the breakpoints established? Simply put, zone diameter breakpoints for resistance and susceptibility to a particular antimicrobial agent for a species (like those in Table 19-2) are determined by testing a large number of *known* resistant and susceptible *strains* and the MIC and zone diameter is established for each. The zone diameter below which *all known resistant strains* fall is the **resistance breakpoint.** Likewise, the zone diameter above which all *known susceptible strains* fall is the **susceptibility breakpoint.** In some cases, there may be an intermediate range of zone diameters in which some resistant and susceptible strains fall.

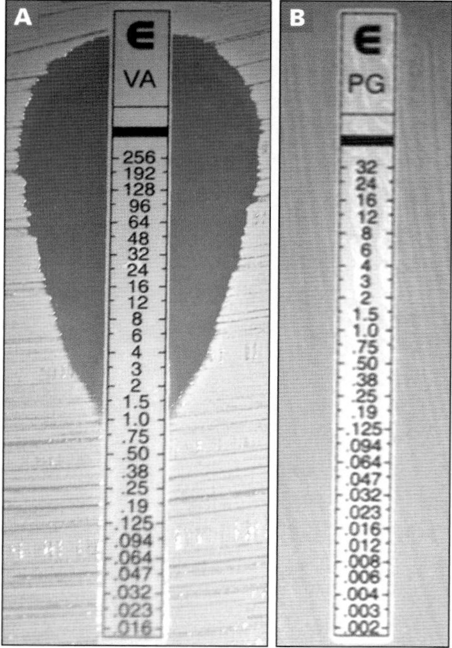

19.6 E-Test ● The E-test is a procedure in which susceptibility to a particular antibiotic can be quantified as a Minimum Inhibitory Concentration (MIC). After incubation, the MIC is determined by where the zone of inhibition intersects the scale printed on the strip. (**A**) Shown is the zone formed by the antibiotic vancomycin when incubated with methicillin-resistant *Staphylococcus aureus* (MRSA). (**B**) The antibiotic penicillin is generally not effective against Gram-negative bacteria. Shown is *Escherichia coli* grown with a penicillin G strip. Note the absence of an inhibition zone.

When a patient's isolate is tested, its zone diameter is compared to the Zone Diameter Interpretive Chart to see if it is susceptible or resistant to that particular antimicrobic based on data gathered from testing multiple known resistant and susceptible strains. The results (because multiple antimicrobics are tested simultaneously) provide information about the strain infecting the patient and the prescribed medication has a high probability of being effective.

The **E-test** system for determining antibiotic sensitivity, illustrated in Figure 19.6, is an alternative to the Kirby-Bauer method and has the added advantage of allowing the MIC to be determined. It consists of a paper strip with a gradient of antibiotic concentrations on one surface and a printed scale on the other.

After an agar plate is inoculated with a lawn of bacteria (as in the Kirby-Bauer method), the strip is placed, antibiotic side down, on the agar surface. During incubation, the antibiotic will diffuse into the agar (higher concentrations traveling farther than lower concentrations) and an elliptical zone of inhibition develops. The point at which the inhibition zone intersects the scale printed on the strip is the MIC.

Normally in a Kirby-Bauer test, the zones around each disk are distinct, separate, and circular. Occasionally, a **synergistic effect** of two antibiotics produces a clearing between the disks extending beyond the perimeters of the otherwise circular zones (Fig. 19.7). In this region, each antibiotic concentration is below its MIC, but is bactericidal or bacteriostatic in combination with the other.

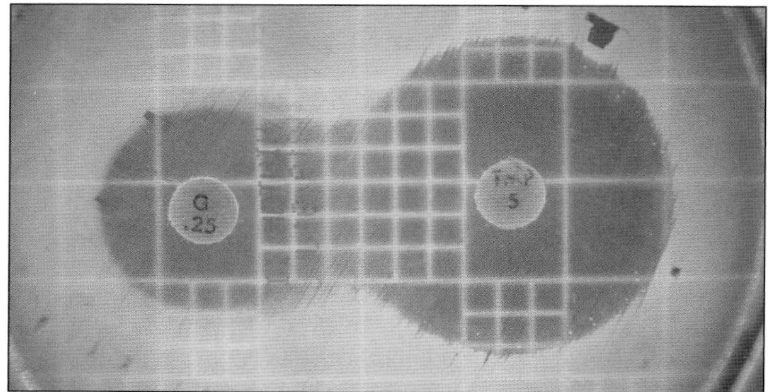

19.7 Antibiotic Synergism ● Shown are sulfisoxazole (**G**) and trimethoprim (**TMP**) disks on an inoculated Mueller-Hinton plate. Ordinarily, zones of inhibition are circular around each disk because diffusion occurs in all directions equally. However, if you complete the circular zones in this photo with a pencil or in your head, you'll see that there is a region of clearing connecting them. This is an example of synergism between the antibiotics, where the clearing comes from the combined action of the two antibiotics in a region that is beyond the MIC of each individually. The numbers on the disks represent micrograms (µg) of antibiotic. Figure 7.89 illustrates how the two antibiotics inhibit enzymes in the same pathway, which makes synergism possible.

Snyder Test

Purpose

The Snyder test is designed to qualitatively indicate susceptibility to dental caries (tooth decay), caused primarily by lactobacilli and oral streptococci.

Principle

The oral cavity presents a variety of habitats for microbial growth, including the tongue, floor of the mouth, and buccal (cheek) and palatal surfaces. These surfaces are covered with stratified squamous epithelium that sheds on a regular basis, thus making heavy microbial colonization difficult (see Fig. 6.8).

Then there are the teeth, the only organs in the body that present hard surfaces to the outside world. Microbial communities are prone to attaching firmly to the teeth in **biofilms**, more commonly known as **dental plaque (Fig. 19.8)**. This firm attachment to teeth makes colonization more permanent and more abundant than on the oral epithelial surfaces.

Tooth biofilms are inhabited by a diverse assemblage (microbiome) of bacteria that become attached to a proteinaceous film covering teeth called a **pellicle.** Initial attachment of bacteria to the tooth surface involves many factors, but a variety of bacterial **adhesins** bind specifically to host protein receptors, which accounts in large part for the presence of microorganisms characteristic of dental plaque. Many of these microbes are fermenters of sugars, such as glucose and sucrose, and produce acid as an end-product. It is the acid that demineralizes tooth enamel and results in **caries** (cavities).

The worst offenders are *Streptococcus mutans* (see Fig. 12.67), other streptococci, *Actinomyces* species, and *Lactobacillus* species. Not surprisingly, the people who are most at risk are those with a high dietary intake of sugars and poor dental hygiene. To make matters worse, these organisms have evolved mechanisms to tolerate low pH conditions (pH of ~5.5) that inhibit growth of other less harmful bacteria.

Snyder test medium (Fig. 19.9) is formulated to favor the growth of oral bacteria and discourage the growth of other bacteria. This is accomplished by lowering the medium's pH to 4.8. Glucose is added as a fermentable carbohydrate, and bromocresol green is the pH indicator.

Lactobacilli and oral streptococci can survive at this low pH, ferment the glucose, and lower the pH even further. Bromocresol green is green at or above pH 4.8 and yellow below. Development of yellow color after incubation, therefore, is evidence of fermentation with acid end-products and, further, is highly suggestive of the presence of dental decay-causing bacteria.

The medium is autoclaved for sterilization, cooled to just over 45°C, and maintained in a warm-water bath until needed. The molten agar then is inoculated with a small amount of saliva, mixed well, and incubated for up to 72 hours. The agar tubes are checked at 24-hour intervals for any change in color.

High susceptibility to dental caries is indicated if the medium turns yellow within 24 hours. Moderate and slight susceptibility are indicated by a change within 48 and 72 hours, respectively. No change by 72 hours is considered a negative result. These results are summarized in Table 19-3.

19.8 Tooth Scraping This sample was taken from the surface of the second molar with a sterile wooden stick several hours after brushing. Note the cell diversity, especially Gram-positive rods and cocci, comprising the community on the tooth's surface. What you see, however, is likely only a fraction of the cells residing on the tooth. The community within a plaque biofilm (if present to a greater degree) is firmly attached to the tooth and not easily scraped (or brushed) away. (Gram Stain)

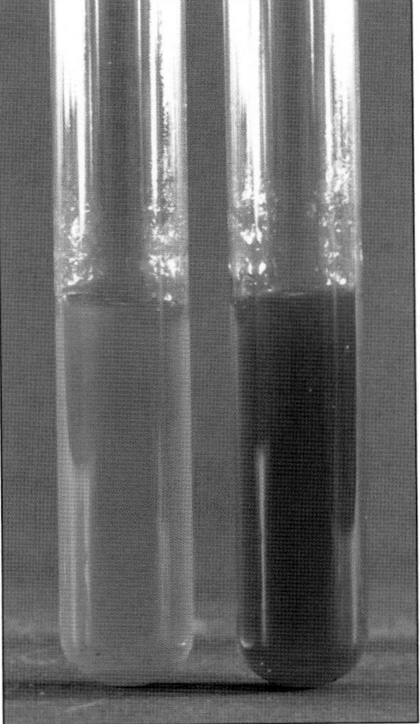

19.9 Snyder Test Results A positive result is on the left and a negative result is on the right. Acid from glucose fermentation has lowered the pH in the positive tube, changing the pH indicator (bromocresol green) from green to yellow. To be of full value, the time it took for the agar to turn positive must be recorded. The test should be read every 24 hours and is complete at 72 hours. See Table 19-3 for an explanation of how time figures into the interpretation.

TABLE **19-3** **Snyder Test Results and Interpretations**

Result	Interpretation
Yellow at 24 hours	High susceptibility to dental caries
Yellow at 48 hours	Moderate susceptibility to dental caries
Yellow at 72 hours	Slight susceptibility to dental caries
Yellow at > 72 hours	Negative

Clinical Sample Collection and Transport

Purpose

Proper collection and transport of patient specimens are crucial to the correct identification of pathogens by a clinical laboratory.

Principle

Sample collection and transport are the first steps in identifying pathogens from patients, and their importance cannot be overstated. Improper collection and transport can make microbial identification by the laboratory more difficult, to the point of impossible if the sample is unusable.

First and foremost, collection of patient specimens for laboratory diagnosis requires that the site sampled has an active infection. Collection of patient specimens may involve tissue removal, collection of sputum, urine or feces, fluid aspiration, venipuncture, or a surface swab. The method of choice is dictated by the body region and suspected pathogen, but in every case care must be taken to prevent contamination by environmental surroundings or the sample taker's own microbiota.

Further, the appropriate sampling instrument and transport medium must be used. Proper training of medical staff in sample collection and transport is imperative, because the laboratory can do little when the sample is contaminated or is not transported properly.

Samples are frequently obtained with swabs. Swabs come in a variety of types—wooden, plastic, or metal shafts with cotton, Dacron, or calcium alginate tips (Fig. 19.10). Plastic swabs are used most often because wooden swabs may harbor toxins that interfere with microbial growth. Flexible wire swabs are recommended for nasopharyngeal and male urethral samples. Cotton tips are useful in collecting non-fastidious organisms but may contain chemicals that are inhibitory to fastidious ones. Dacron tips have the widest application and may even be used for collecting viral samples. Calcium alginate-tipped swabs are best used for *Chlamydia* samples.

Once collected, the sample must be labeled with all relevant information, including patient name and ID number, sample site, date and time of collection, and collector's initials.

Various guidelines have been developed for transporting samples within a hospital or between locations, but are beyond the scope of this book. It stands to reason, though, that the sample be in a leak-proof container and in an environment that is suitable for survival of its contents.

A third consideration is time: The faster the sample gets to the laboratory for processing, the better. Bacterial samples should be transported within 2 hours, if possible; the sample probably will be useless after 24 hours. And fourth, once in the laboratory, the sample should be processed in a timely fashion. In some cases, refrigeration is acceptable for a given amount of time before processing.

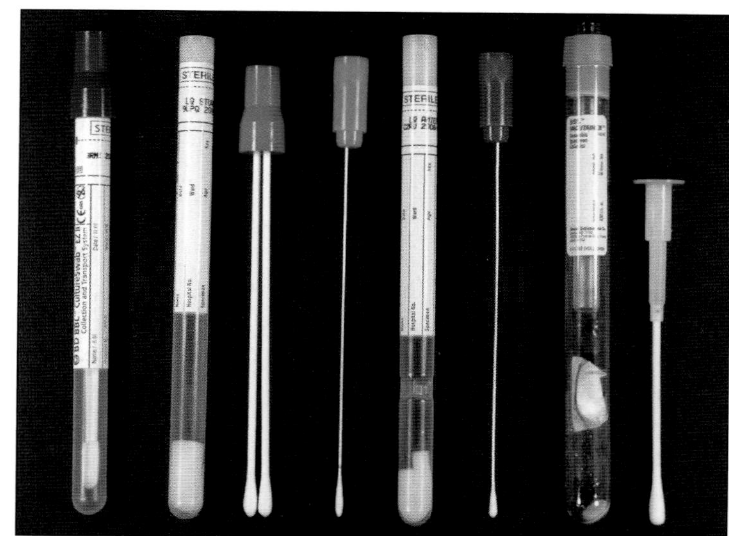

19.10 Collection and Transport Media • Various collection swabs and transport media are shown. The tube with the maroon cap contains a plastic shaft with a polyurethane tip. Because the polyurethane tip is nontoxic, no transport medium is necessary. The red cap double swab has a rayon tip. The tube to the left contains Stuart's liquid transport medium, a buffered medium lacking nutrients that keeps the sample moist but doesn't promote growth. The orange-capped swab has a regular aluminum wire with a Dacron tip. The green-capped swab has a soft aluminum wire with a Dacron tip. Both can be transported in the tube between them, which contains Amie's medium. So far, all tubes are for transport of aerobic organisms. The last system on the right is for transporting anaerobes. See Figure 19.11 for more information.

Various transport media are available depending on the application. Amies, Stuart's, and Cary-Blair are commonly used; we will focus on Amies here. Amies is a defined medium with a variety of chloride salts to maintain osmotic pressure. Phosphate buffers maintain the pH, and thioglycollate produces a reduced environment to minimize oxidative damage to the cells.

Some formulations contain charcoal to neutralize fatty acids and bacterial toxins. Note the absence of a carbon or nitrogen source. This medium is designed to maintain the bacteria, not provide for their growth.

Transport of anaerobes requires a special container that produces anaerobic conditions inside when the swab is inserted. A color indicator is used as a control to assure that the inside is anaerobic (Fig. 19.11).

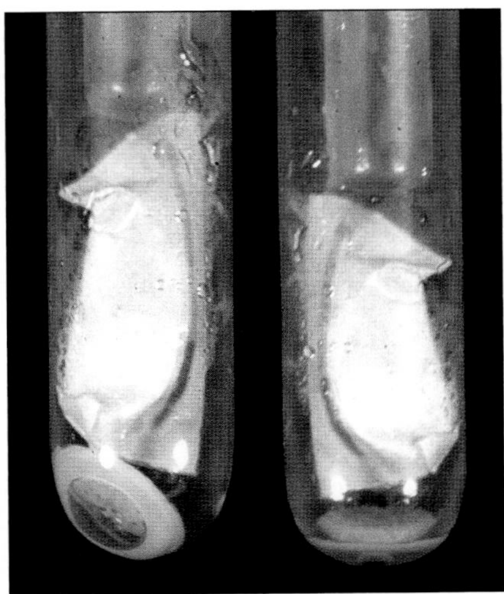

19.11 Transport System for Anaerobes • This is a close-up view of two anaerobic transport tubes. They are the same as the tube shown at the right of Figure 19.10. The sample is taken and inserted into the tube, where it breaks open a vial that produces anaerobic conditions. A color indicator turns pink if oxygen is present, making the sample useless.

Part B: Environmental Microbiology

Environmental Sampling: The RODAC Plate

Purpose

The RODAC plate is used to monitor surface contamination in food preparation, veterinary, pharmaceutical, and medical settings. The plates can also be used to assess the efficiency of decontamination of a surface by taking a sample with different plates before and after treatment.

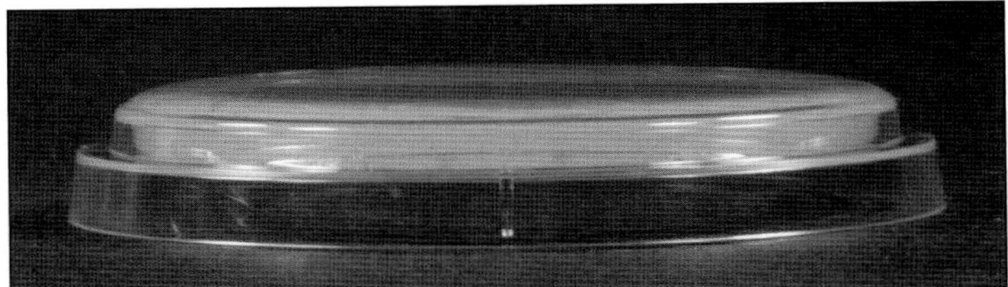

19.12 RODAC Plate • RODAC plates are used for sampling contamination on surfaces. Notice that the agar extends above the edges of the plate to allow contact with the surface to be sampled.

Principle

Monitoring of microbial surface contamination is an important practice in medical, veterinary, pharmaceutical, and food preparation settings. Often, the RODAC (Replicate Organism Detection and Counting) plate is used. It is a specially designed agar plate into which the medium is poured to produce a convex surface extending above the edge of the plate (Fig. 19.12). It is designed to support the lid above the agar without touching it.

The sterile plate is opened and pressed on a surface to be sampled. Most are 65 mm in diameter, which is smaller than standard-sized 100 mm Petri dishes. The smaller size makes it easier to apply uniform pressure across the whole plate when taking the sample. In addition, the base is marked in 16 1-cm squares, allowing an estimate of cell density on the surface (Fig. 19.13).

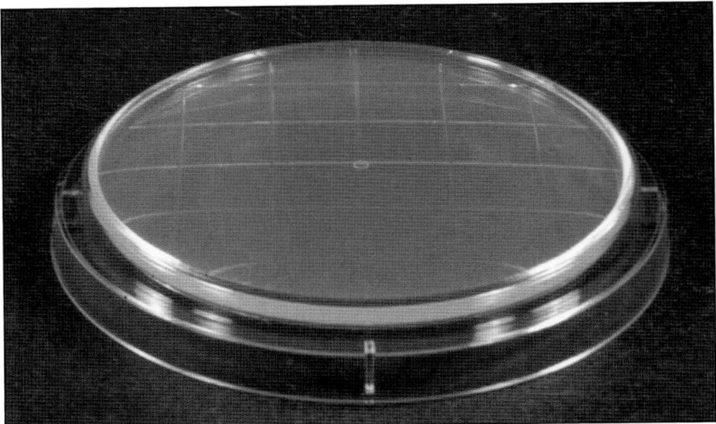

19.13 Grid on the RODAC Plate • Typically, RODAC plates are 65 mm in diameter. There is a grid of 16 squares molded into the base, each with an area of 1 cm^2 (seen here through the agar). Colonies growing in the grid can be counted and an average number of CFU per cm^2 of surface can be determined.

The plate can be filled with a variety of media, the choice of which depends on the surface being sampled and the microbes to be recovered. For instance, tryptic soy agar (TSA), a good, general-purpose growth medium, can be used. TSA can be supplemented with 5% sheep blood to improve recovery of fastidious bacteria.

Fecal coliform density (always a concern) can be checked using m-FC agar or MacConkey agar (see Figs. 2.19 and 2.20). Monitoring yeast and mold contamination often employs Sabouraud dextrose agar (see Fig. 2.25). Because the surface sampled may have been recently disinfected, polysorbate 80 and lecithin are added to counteract the effect of residual disinfectant.

The acceptable amount of growth on a RODAC plate is determined by the surface being sampled. It stands to reason that a surgical area would have a lower limit of acceptability than a food preparation area. Table 19-4 provides some guidelines.

TABLE **19-4** **Interpretation of RODAC Plate Colony Counts (Colonies Per Plate)[1]**

Interpretation	Critical Surfaces[2]	Floors
Good	0–5	0–25
Fair	6–15	26–50
Poor	>16	>50

[1] Adapted from BBL Trypticase Soy Agar with Lecithin and Polysorbate 80 package insert.

[2] Critical surfaces include those in operating rooms, nurseries, table tops, toilet seats, and other nonporous surfaces.

Importance of Water Quality Testing

Water quality is a public health concern and communities monitor their drinking water, recreational water, and wastewater to ensure that it is "safe," that is, free of harmful (toxic or carcinogenic) chemicals and microbial pathogens. This being a microbiology book, we'll allow someone else to address the chemicals.

Microbes of primary concern are pathogens that are transmitted via the **fecal-oral route**, where an infected person passes the pathogen in their feces and another person ingests it by consuming contaminated water or food. Then they get sick and the cycle continues.

So, public health agencies monitor their water for fecal contamination on a regular basis, the frequency of which is based on the use of the water being tested and the degree of its distribution. For instance, drinking water in a large public system may be tested hourly, whereas recreational water may be tested weekly, and well water may be tested annually.

While fecal pathogens are the main biological concern, it would be labor intensive and cost prohibitive to test water samples for every possible fecal pathogen. Instead, water is tested for specific organisms found in feces. These are referred to as **fecal indicator bacteria (FIB)**. They have the qualities of being easily grown, are abundant in feces, and are easily identified. If FIB are found in a water sample, then there is the potential for fecal pathogens to be there, too, and appropriate action is taken. Most of the time, the FIB organisms of choice are **coliforms**.

Coliforms are members of the Enterobacteriaceae (see p. 172), most of which are harmless to humans and many of which are found in the environment. However, others (also mostly harmless) occupy the intestines of warm-blooded animals and serve as a good indicator of fecal contamination.

The EPA sets standards, based on its use, for the number of coliforms that can be present in water to be considered safe. What sets coliforms apart from the other members of Enterobacteriaceae is that they ferment lactose to acid and often gas end-products. As you read through the following four water quality tests, this ability is exploited to identify the presence, and sometimes quantity, of coliforms.

Different water quality tests have different specificities for detecting coliforms. The most general tests are qualitative and simply detect coliform **presence** or **absence**. Quantitative tests detect coliform density in a sample (usually per 100 mL), not just presence or absence. Some tests detect **total coliform** density; that is, the density of all coliforms, including environmental as well as fecal coliforms.

More specific tests detect only **fecal coliform** density (a subset of total coliforms), which indicates fecal contamination *and the potential presence of fecal pathogens*. The most specific tests detect ***Escherichia coli*** density (Fig. 19.14). *E. coli* is generally the most abundant fecal coliform and its presence is also interpreted as fecal contamination. In testing recreational water, *Enterococcus* species are often used as FIB in conjunction with coliforms.

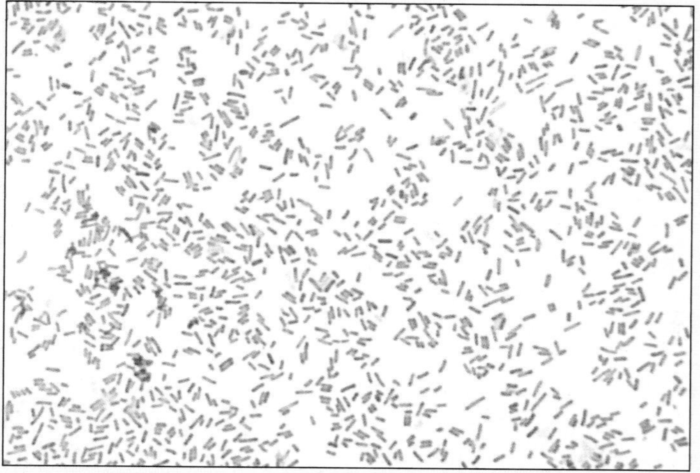

19.14 Gram Stain of *Escherichia Coli* • *E. coli* is a principal coliform detected by the Colilert test system, as well as the other water quality tests covered in this section.

The maximum acceptable FIB density is determined by a governing agency (like the EPA) and is different for different water uses. For instance, the maximum acceptable FIB density in recreational water is higher than in drinking water. The standards for presence/absence tests are usually described as "no more than X% of samples can be positive within a certain time period."

So, what's "safe" water? The answer to that depends on the water's intended use: drinking water, recreation, or wastewater. It also depends somewhat on where you live. Because of these variables, we've avoided printing maximum contaminant levels and suggest you check with your community public utilities department for local standards.

Water Quality Testing: Colilert Test

Purpose

The Colilert[1] test is a commercial preparation that examines drinking water for the presence of total coliforms and *Escherichia coli* (Fig. 19.14), the latter of which is an indicator of fecal contamination. This is a qualitative test (presence or absence), not a quantitative test (density). (For quantitative tests of water quality, see Membrane Filter Technique, Multiple Tube Fermentation Method, and QuantiTray/2000 Kit in this section.)

Principle

Colilert reagent contains nutrients and salts that favor the growth of coliforms and inhibit the growth of noncoliforms. It also contains the indicator nutrients, *o*-Nitrophenyl-β-D-galactopyranoside (**ONPG**) and 4-Methylumbelliferyl-β-D-glucuronide (**MUG**). The test is conducted by adding Colilert reagent to a nonfluorescent bottle containing a 100 mL drinking water sample. The sample is incubated at 35°C for 24–28 hours and then compared to the control.

All coliforms ferment lactose using the enzyme **β-galactosidase.** ONPG is an artificial substrate of the same enzyme and, when hydrolyzed, produces the yellow compound ***o*-Nitrophenol.** Refer to Figure 7.68 for this reaction and Figure 19.15 for the test results. If the test bottle is as yellow or more yellow than the uninoculated control (comparator), the first portion of the test is positive and reported as "positive for total coliforms." Positive samples are then used in the MUG test.

MUG acts as a substrate for the *E. coli* enzyme β-glucuronidase, from which a fluorescent compound is produced. A positive MUG result produces fluorescence when viewed under an ultraviolet lamp (Fig. 19.16). If the fluorescence is equal to or greater than that of the control, the presence of *E. coli* has been confirmed and is reported as "positive for *E. coli*." A sample that is not as yellow and does not fluoresce is considered negative and is reported as "total coliforms absent" and "*E. coli* absent."

[1] Colilert is a registered trademark of IDEXX Laboratories, Inc. https://www.idexx.com/en/water/products/?cy=y_category_252&cx=x_category_259&ts=all

19.15 Colilert ONPG • The Colilert ONPG (*o*-Nitrophenyl-β-D-galactopyranoside) test medium is used to determine presence or absence of total coliform bacteria in water samples. Coliforms are able to ferment lactose to acid end-products using the enzyme β-galactosidase. ONPG is an artificial substrate for the enzyme and coliforms will digest it, producing a yellow color. Any bottle equal to or exceeding the yellow of the control are considered positive for presence of total coliforms. From left to right: Water sample without coliforms (ONPG-negative); *Klebsiella pneumoniae* (ONPG-positive; coliform); *Escherichia coli* (ONPG-positive; coliform); comparator (uninoculated control for color comparison). Note that the positive tests were not from a real water sample. They were inoculated with positive controls for demonstration purposes.

19.16 Colilert MUG • After reading bottles for total coliforms, the positives are checked for *E. coli*. Shown are the same bottles as in Figure 19.15, but illuminated with UV light. Of the organisms of interest in water samples (coliforms), *E. coli* is the only (most likely) one that has the enzyme β-glucuronidase that breaks down MUG (4-Methylumbelliferyl-β-D-glucuronide) to a fluorescent product. From left to right: Water sample negative for coliforms (which wouldn't be used in this test, but we're showing it for the sake of completeness); *K. pneumoniae* (coliform, MUG-negative); *E. coli* (coliform, MUG-positive and therefore *E. coli*); comparator (uninoculated control for color comparison).

Water Quality Testing: Membrane Filter Technique

Purpose

The membrane filter technique has several variations for testing water quality. Described here are three methods that may be used in conjunction, or may be used individually, depending on the specific goal of testing. This is a quantitative technique that provides an estimated number of FIB in the sample per 100 mL based on colony counts.

Principle

In the membrane filter technique (Fig. 19.17A), a vacuum draws a water sample through a porous membrane designed to trap microorganisms larger than 0.45 μm (Fig. 19.17B). After filtering the water sample, the membrane (filter) is applied to the surface of plated **m Endo agar LES**[2] (for determination of **total coliforms**), **m-FC agar** (for **fecal coliforms**), and **m-EI agar** (for determination of ***Enterococcus*** **species**) and incubated for 24 hours at 35°C, 44.5°C, and 41°C, respectively.

m Endo agar LES

m Endo agar LES (Fig. 19.18) is a selective and differential medium that inhibits Gram-positive growth because of the ingredients sodium deoxycholate and sodium lauryl sulfate. It contains lactose for fermentation and basic fuchsin to indicate pH changes. Incubation is at 35.0 ± 0.5°C for 24 ± 2 hours.

Coliforms produce acetaldehyde, which causes colonies to appear red. Subsequent rapid lactose fermentation with acid end-products causes the colonies to become mucoid with a metallic sheen. Noncoliform bacteria (including several dangerous pathogens) tend to be pale pink, colorless, or the color of the medium.

After incubation, all golden or green metallic colonies are counted and are used to calculate "total coliform colonies per 100 mL" using the following formula:

$$\frac{\text{Total coliforms}}{\text{100 mL}} = \frac{\text{Coliform colonies counted} \times 100}{\text{mL of original sample}}$$

A "countable" plate contains between 20 and 80 coliform colonies with a total colony count no larger than 200. To ensure that the colony number falls within this range, it is customary to dilute samples, thereby reducing the number of cells collecting on the membrane. When dilution is necessary, it is important to record only the volume of *original sample* passed through the membrane, and not any added water.

m-Fecal Coliform agar

m-FC agar (Fig. 19.19) is a selective and differential medium used to determine fecal coliforms in water samples. Bile salts

[2] The "m" preceding the media names stands for "membrane."

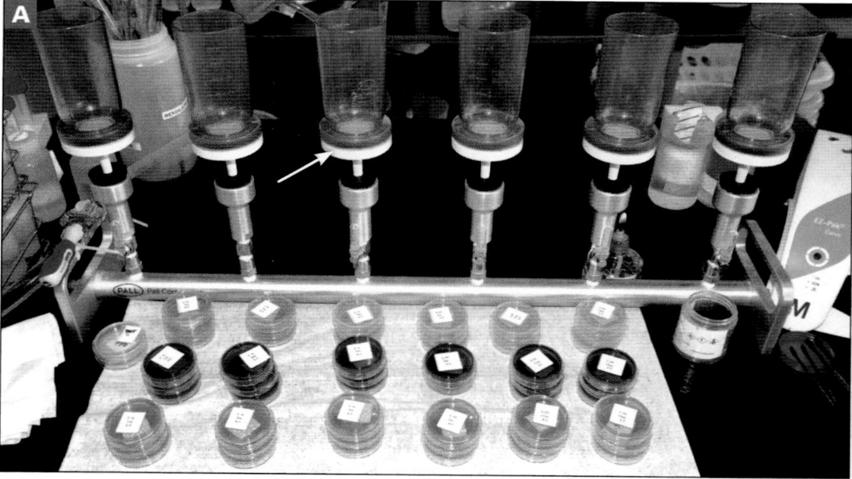

19.17 Membrane Filter Apparatus • (**A**) This apparatus allows six water samples to be run simultaneously. The sterile membrane filter is placed on the white filter base (**arrow**), followed by the orange funnel, which has an opening slightly smaller than the filter's diameter to ensure water passes through the filter and not around it. It is held in place by a magnet that also ensures a tight seal is formed between the filter base and funnel. Well-mixed water samples are added to each of the six assembled funnels and then a vacuum draws the water through the filter and out the silver piping for disposal. Bacteria larger than 0.45μm are trapped on the filters. (*E. coli* is approximately 1 μm by 3 μm.) Filters are then removed and placed on the appropriate medium, using care that the entire filter makes good contact with the medium. The media to be used are (from farthest to closest) m-EI, m-FC, and m Endo LES (see text and Figures 19.18, 19.19, and 19.20). They are in stacks of three because in this particular procedure, water volumes of 0.5 mL, 5.0 ml, and 50 mL are to be run on separate filters for each medium. These plates are 50 mm in diameter. (**B**) Shown is a membrane filter on a 100 mm plate. Note the grid.

inhibit Gram-positive organisms and rosolic acid inhibits most other bacteria except fecal coliforms. Lactose is supplied as a fermentable carbohydrate and aniline blue is a pH indicator that turns blue at low pH.

Incubation is at 44.5 ± 0.2°C for 24 ± 2 hours, at which fecal coliforms can survive, but most environmental coliforms can't. Any fecal coliforms present ferment the lactose to acidic end-products, lower the pH, and aniline blue turns blue, making fecal coliform (positive) colonies blue.

After incubation, "countable" plates with between 20 and 60 blue (coliform) colonies are counted. The calculation for fecal coliforms per 100 mL is the same as for total coliforms.

m-Enterococcus Indoxyl-β-D-Glucoside agar

m-EI agar (Fig. 19.20) is a selective and differential medium used for the recovery of *Enterococcus* species in recreational water samples. Inhibitory ingredients include cycloheximide (fungi) and sodium azide (Gram-negative bacteria). Incubation is at 41 ± 0.5°C for 24 ± 2 hours in an environment with humidity approaching saturation.

All enterococci produce **β-D-glucosidase** and hydrolyze the indoxyl-β-D-glucoside in the medium, which produces a blue end-product. This diffuses into the medium and produces a blue halo around *Enterococcus* colonies. A countable plate has between 20 and 60 colonies and the calculation used is the same as for total coliforms.

Confirmation of *Enterococcus* requires a Gram-stain result of Gram-positive cocci, a positive bile esculin test (see p. 75), and growth on brain-heart infusion agar and in brain-heart infusion broth with 6.5% NaCl.

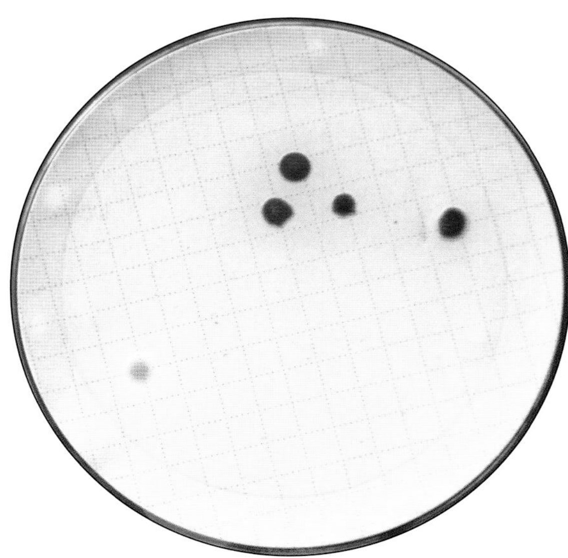

19.19 m-Fecal Coliform (FC) Agar • m-FC agar is also a selective and differential medium for coliforms, but because of its incubation temperature (44.5 ± 0.2°C), fecal coliforms are able to grow, whereas environmental coliforms are not. Blue colonies are fecal coliforms that have produced acid from lactose fermentation. The lowered pH turns the aniline blue indicator blue. There are four fecal coliforms on this plate.

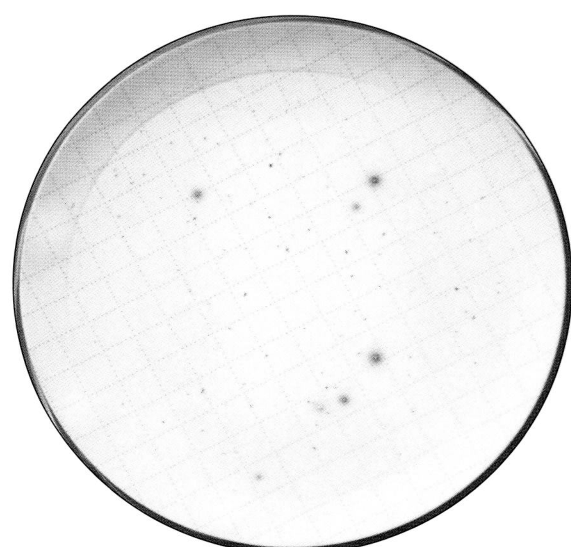

19.20 m-*Enterococcus* Indoxyl-β-D-Glucoside (m-EI) Agar • m-EI agar is a selective and differential medium for *Enterococcus* species. It is used for testing recreational water samples for fecal contamination using enterococci as the indicator organisms. Enterococci produce the enzyme β-D-glucosidase that hydrolyzes indoxyl-β-D-glucoside in the medium. The blue product of this reaction results in a blue halo around colonies that produce the enzyme. There are seven *Enterococcus* colonies on this plate.

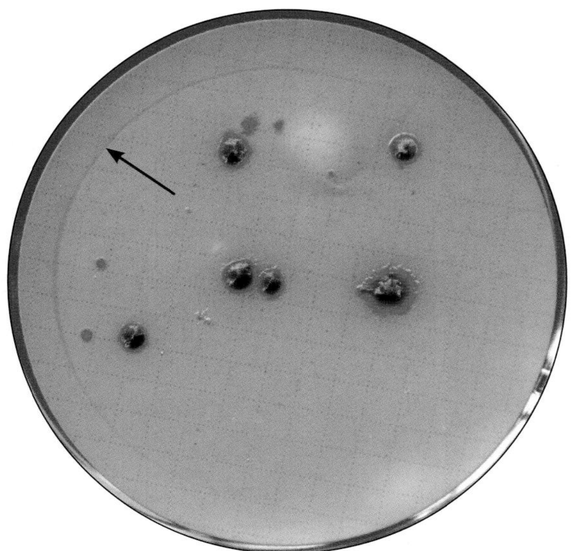

19.18 m Endo Agar LES • Endo agar is a selective and differential medium for coliforms and is used to estimate total coliform density in a water sample. Coliforms ferment lactose in the medium and produce colonies with a green or gold metallic sheen. There are six "total" coliform colonies on this filter. Note the ring (**arrow**) at the filter's edge. This is where the funnel's base rested during suction.

Water Quality Testing: Multiple Tube Fermentation Method (Most Probable Number)

Purpose

This standardized test is used to estimate coliform density (cells/100 mL) in water using statistical methods. It may be used to calculate the density of all coliforms present (total coliforms) or to calculate the density of *Escherichia coli* specifically.

Principle

The **multiple tube fermentation method,** also called **most probable number test,** or **MPN,** is a common means of estimating the number of coliforms present in 100 mL of a water sample using a statistical approach. It is not an actual counting of cells. The same procedure can be used to obtain the MPN for both **total coliforms** and *E. coli*.

The three media used in the procedure are: **lauryl tryptose broth (LTB), brilliant green lactose bile (BGLB)** broth, and **EC** (*E. coli*) broth. All three media contain lactose that coliforms ferment to acid and gas, the latter of which is used to indicate coliform presence. They also contain an ingredient that inhibits the growth of noncoliforms, especially Gram-positive bacteria. In LTB the inhibitory agent is sodium lauryl sulfate. In BGLB and EC broth, the inhibitory agent is bile (oxgall).

Because LTB does not screen out all noncoliforms, it is used to *presumptively* determine the presence or absence of coliforms. BGLB broth is used to *confirm* the presence of coliforms because of its greater inhibitory effect on noncoliforms. EC broth, which includes lactose and bile salts, is selective for *E. coli* when incubated at 45.5°C.

All broths are prepared in 10 mL volumes and contain an inverted Durham tube to trap any gas produced by lactose fermentation. The LTB tubes are arranged in up to 10 groups of five (Fig. 19.21). Each tube in the first set of five receives 1.0 mL of the original sample. Each tube in the second group receives 1.0 mL of a 10^{-1} dilution. Each tube in group three receives 1.0 mL of 10^{-2}, and so on for as many groups of five are used.[3,4]

After inoculation, the LTB tubes are incubated at 35 ± 2°C for up to 48 hours, and then examined for gas production (Fig. 19.22A) or color change due to acid production (Fig. 19.22B). (At this point, the ability of a particular diluted water sample's ability to produce, or not, acid or gas is what's important, which is why we are not concerned with the sample volume being diluted by LTB.) Any positive LTB tubes then are used to inoculate BGLB tubes. Each BGLB receives one or two loopfuls from its respective positive LTB tube (again, volumes are not critical).

[3] The volume of LT broth in the tubes is not part of the dilution factor's calculation. Dilutions of the water sample are made using sterile water prior to inoculating the broths and that is what is used in the calculations. The broth is simply a growth medium; it is not acting as diluent.

[4] The number of groups, number of tubes in each group, dilutions necessary, volume of broth in each tube, and volumes of sample transferred vary significantly depending on the source and expected use of the water being tested.

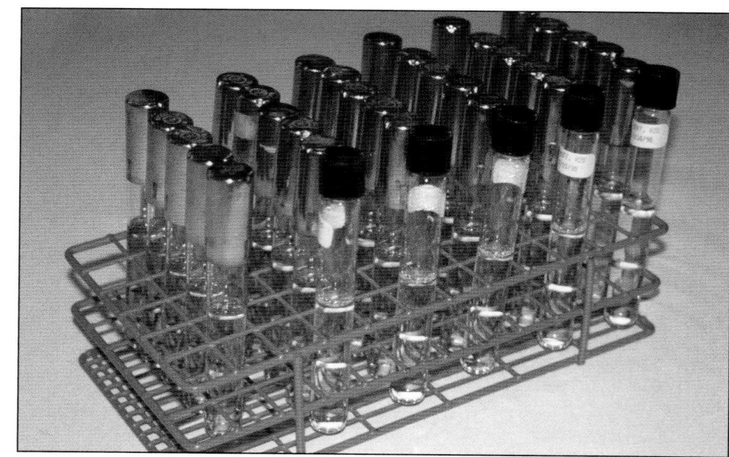

19.21 Multiple Tube Fermentation ● This a multiple tube fermentation test of a seawater sample potentially contaminated by sewage. The test photographed contains six dilution groups (the standard for heavily contaminated samples) rather than three, as described in Table 19-5. The tubes contain lauryl tryptose broth (LTB) and a measured volume of water sample as described in the text. Following incubation, each positive broth (based on gas production) will be used to inoculate a BGLB broth.

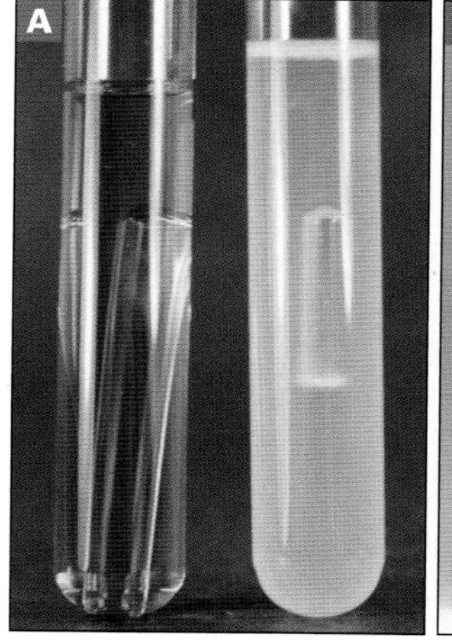

19.22 Lauryl Tryptose Broth (LTB) Results ● (A) The bubble in the Durham tube on the right is *presumptive* evidence of coliform contamination. It would be used to inoculate BGLB and EC broth. The tube on the left is negative. (B) Some protocols call for LTB to include the pH indicator bromocresol purple, that turns yellow with acid production. In these protocols, presumptive evidence of coliform contamination is acid and gas production. From left to right: Presumptive coliform (positive for acid and gas; it would be used to inoculate a BGLB), uninoculated control, and noncoliform.

The BGLB cultures are incubated 48 hours at 35 ± 2°C and examined for gas production (Fig. 19.23). Positive BGLB cultures then are transferred to EC broth and incubated at 45.5°C for 48 hours (Fig. 19.24). After incubation the EC tubes with gas are counted. The same formula is used to calculate total coliform (BGLB) MPN and *E. coli* (EC) MPN, but they are calculated separately only using data from the appropriate tubes. The formula is:

$$\text{MPN per 100 mL} = \frac{100P}{\sqrt{V_n V_a}}$$

Where:

P = total number of positive results (BGLB or EC)

V_n = combined volume of sample in LTB tubes that produced negative results in BGLB or EC

V_a = combined volume of sample in all LTB tubes

It is customary to calculate and report *both* total coliform and *E. coli* densities. Total coliform MPN is calculated using BGLB broth results, and *E. coli* MPN is based on EC broth results.

Using the data from Table 19-5 and the MPN formula previously given, the calculation for total coliform MPN would be as follows:

$$\text{MPN per 100 mL} = \frac{100P}{\sqrt{V_n V_a}}$$

$$\text{MPN per 100 mL} = \frac{100 \times 9}{\sqrt{0.24 \times 5.55}}$$

$$\text{MPN per 100 mL} = 780$$

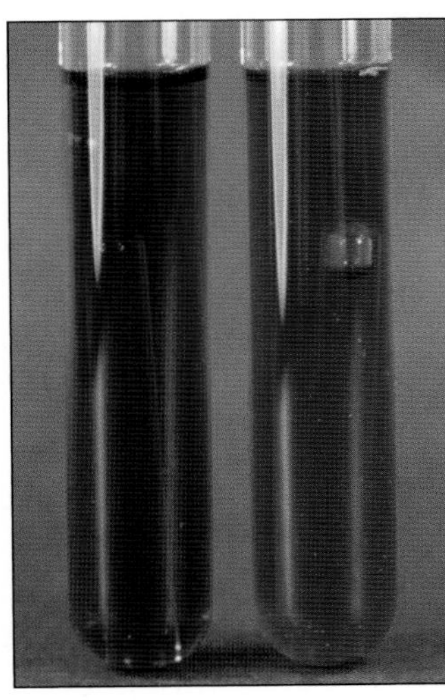

19.23 Brilliant Green Lactose Bile (BGLB) Broth Results • The bubble in the Durham tube on the right is seen as *confirmation* of coliform contamination. It would be used to inoculate an EC broth for potential identification as *E. coli*. The tube on the left is negative.

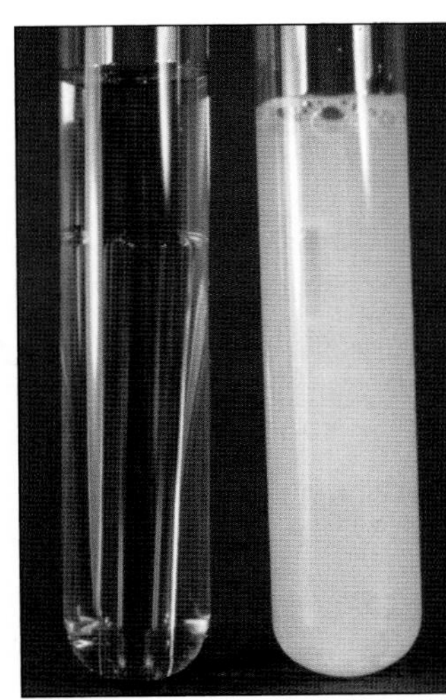

19.24 *E. coli* (EC) Broth Results • EC broth is selective for *E. coli* when incubated at 45.5°C. The bubble in the Durham tube on the right is recorded as confirmation of *Escherichia coli* contamination. The tube on the left is negative.

TABLE **19-5 Example of BGLB Test Results** • The results shown here are of a hypothetical BGLB test using three groups of five tubes (A, B, and C). In this example, the original water sample was diluted to 10^0, 10^{-1}, and 10^{-2}. The three dilutions were used to inoculate the broths in groups A, B, and C, respectively. **Row 1:** The dilution of the inoculum used per group. **Row 2:** The actual amount of original sample that went into each LTB tube. **Row 3:** The number of tubes in each group. **Row 4:** The number of tubes from each group of five that showed evidence of gas production. This total (in red) inserts into the equation as P. **Row 5:** The number of tubes from each group that did *not* show evidence of gas production. **Row 6:** Is used for calculating the "combined volume of sample in negative tubes" and refers to the inoculum that went into the LTB tubes that produced a negative result. This total (in red) inserts into the equation as V_n. **Row 7:** Used for calculating the "combined volume of sample in all tubes" and refers to the total volume of inoculum that went into all LTB tubes. This total (in red) inserts into the equation as V_a. As you can see, the undiluted inoculation (Group A) produced five positive results and zero negative results; the 10^{-1} dilution (Group B) produced three positive results and two negative; and the 10^{-2} dilution (Group C) produced one positive and four negative results. The total volume of original sample that went into LTB tubes was 5.55 mL, 0.24 mL of which produced no gas (shown in red in rows 7 and 6, respectively).

	Group	Group A	Group B	Group C	Totals (A + B + C)
1	Dilution (D)	10^0	10^{-1}	10^{-2}	NA
2	Volume of dilution added to each LTB tube that is original sample (1.0 mL × D)	1.0 mL	0.1 mL	0.01 mL	NA
3	# LTB tubes in group	5	5	5	NA
4	# BGLB positive results	5	3	1	9
5	# BGLB negative results	0	2	4	NA
6	Total volume of *original sample* in all LTB tubes that produced negative BGLB results (D × 1.0 mL × # negative tubes)	0 mL	0.2 mL	0.04 mL	0.24 mL
7	Volume of *original sample* in all LTB tubes inoculated (D × 1.0 mL × # tubes)	5.0 mL	0.5 mL	0.05 mL	5.55 mL

Water Quality Testing: Quanti-Tray/2000 Kit

Purpose

Quanti-Tray/2000[5] is a convenient and rapid method to determine water quality based on total coliform and *E. coli* MPNs (Most Probably Numbers) of water samples. It eliminates the dilutions and multiple inoculation steps (see Multiple Tube Fermentation Test [MPN] on p. 302), and provides results within 24 hours. A similar kit is available for determining *Enterococcus* MPN.

Principle

The Quanti-Tray/2000 determines total coliform and *E. coli* MPNs using a packet with 49 large wells and 48 small wells (Fig. 19.25A). 100 mL of water sample is added to a reagent solution, the mixture is poured into the packet, then sealed and incubated for 24 hours. The reagent solution is a growth medium containing lactose, ONPG, and MUG (see Colilert Test, p. 299).

All coliforms ferment lactose using the enzyme β-galactosidase. ONPG is an artificial substrate of the same enzyme and, when hydrolyzed, produces the yellow compound *o*-**Nitrophenol** (see Figure 7.68). A well that turns yellow after incubation is likely to have coliforms in it. The number of large and small wells that have turned yellow are counted separately (Fig. 19.25B) and a table provided with the kit is used to give the MPN for total coliforms.

Rows in the table list the number of positive large wells (0–49) and columns list the number of positive small wells (0–48). The cell in the table where the appropriate row intersects with the appropriate column contains the total coliform MPN per 100 mL. No cumbersome calculations are necessary, unless the sample was diluted. In that case, the number from the table is simply divided by the dilution factor.

MUG acts as a substrate for the *E. coli* enzyme β-glucuronidase, from which a fluorescent compound is produced. A positive MUG result (meaning *E. coli* is present) produces fluorescence when viewed under an ultraviolet lamp (Fig. 19.25B) After counting the number of positive large and small wells, the table provides the *E. coli* MPN. If the sample has been diluted, then number in the table is divided by the dilution factor to provide the *E. coli* MPN per 100 mL.

[5] Quanti-Tray/2000 is a registered trademark of IDEXX Laboratories, Inc. https://www.idexx.com/en/water/water-products-services/quanti-tray-system/

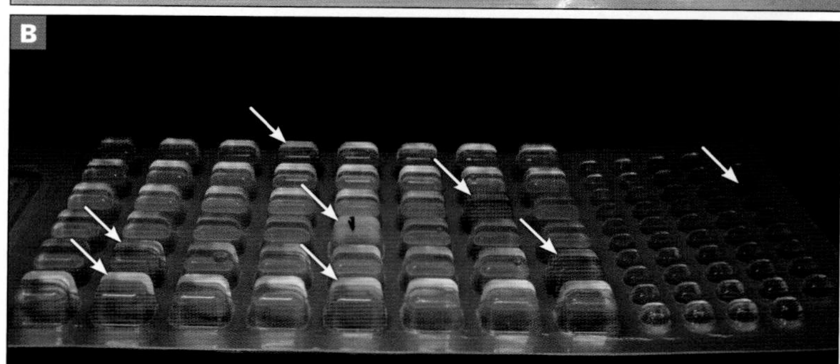

19.25 Quanti-Tray/2000 Most Probable Number (MPN) System (A) Shown is one Quanti-Tray/2000 test. 100 mL of water (in this case, undiluted) is added to the kit's reagent containing ONPG and MUG. Then the mixture is added to the tray to fill the compartments. After sealing, the tray is incubated for (generally) 24 hours at 35°C. This is an unincubated tray. (B) The test indicators are based on the same principles as described in the Colilert test (see p. 299). Large (including the really big one at the left) and small compartments that have turned yellow (due to ONPG digestion) are counted separately and those numbers are used to determine the total coliform MPN. Even though this is being viewed under UV illumination, we can still see that seven large wells are yellow (**arrows**; including the one with the black line) and one small well is yellow (**arrow**). Those numbers are used to read an MPN table provided in the kit. And the answer is: 8.5 total coliforms per 100 mL of sample. Had the sample been diluted, then the number in the table would be divided by the dilution factor. The UV illumination is used to look for fluorescence, which is produced by *E. coli* breaking down the MUG. Only one large well (with the black line) and no small wells show fluorescence. Reading from the same MPN table gives the *E. coli* MPN of 1.0 *E. coli* per 100 mL.

Bioluminescence

Principle

A few marine bacteria from genera *Vibrio* and *Photobacterium* are able to emit light by a process known as **bioluminescence.** Many of these organisms maintain mutualistic relationships with other marine life. For example, *Photobacterium* species living in the flashlight fish receive nutrients from the fish and in return provide a unique device for frightening would-be predators.

Bioluminescent bacteria are able to emit light because of an enzyme called **luciferase** (Fig. 19.26). In the presence of oxygen and a long-chain aldehyde, luciferase catalyzes the oxidation of reduced flavin mononucleotide ($FMNH_2$). In the process, electrons in FMN become excited. Light is emitted when the electronically excited FMN returns to its ground state (Fig. 19.27).

It is estimated that a single *Vibrio* cell burns between 6,000 and 60,000 molecules of ATP per second emitting light. (ATP hydrolysis occurs in conjunction with synthesis of the aldehyde). It also is known that their luminescence occurs only when a certain threshold population size is reached in a phenomenon called **quorum sensing.** This system is controlled by a genetically produced **autoinducer** that must be in sufficient concentration to trigger the reaction.

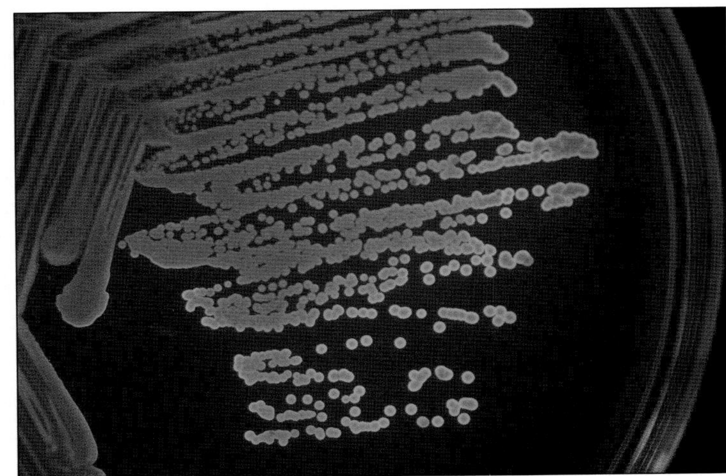

19.27 Bioluminescence on an Agar Plate ● This is an unknown bioluminescent bacterium growing on seawater complete (SWC) agar. This photo is a time exposure. Bioluminescence is not the same as emitting fluorescence as a result of UV light exposure.

$$FMNH_2 + O_2 + R{-}\overset{\overset{\displaystyle O}{\|}}{C}H \xrightarrow{\text{Luciferase}} FMN + H_2O + R{-}\overset{\overset{\displaystyle O}{\|}}{C}{-}OH + \text{Light}$$

19.26 Chemistry of Bioluminescent Bacteria ● The enzyme luciferase catalyzes the oxidation of reduced flavin mononucleotide in the presence of an aldehyde. During the reaction, electrons becomes excited. When they return to their ground state, light is emitted.

Winogradsky Column

Purpose

The Winogradsky column is a method for growing a variety of microbes with uniquely microbial metabolic abilities. Bacterial photoautotrophs, chemolithotrophs, and photoheterotrophs may be found in a mature column. And more "typical" chemoheterotrophs and photoautotrophs also are likely to be found. A mature Winogradsky column is a good source for studying these organisms in the laboratory.

Principle

The Winogradsky column bears the name of its developer, Sergei Winogradsky (1856–1953), a Russian microbiologist and pioneer in microbial ecology. He studied sulfur bacteria because of their ease of handling and cultivation, and then moved on to bacteria associated with the nitrogen cycle.

One of his major discoveries was finding microorganisms (*Beggiatoa* in 1887) capable of the unheard-of type of metabolism that came to be known as chemolithotrophic autotrophy. Until he made his discovery, only photoautotrophs—those performing plant photosynthesis—were known to be autotrophs.

As a result of his work and the work of others, metabolic categories of microorganisms have been identified based on their carbon, energy, and electron sources. Note that in practice, terms are combined to describe the organism more fully.

Winogradsky first used "his" column in the late 19th century. It was (and is) used as a convenient laboratory source to supply for study a variety of **anaerobic, microaerophilic,** and **aerobic** bacteria, including purple nonsulfur bacteria, purple sulfur bacteria, green sulfur bacteria, chemoheterotrophs, and many others (Figs. 19.28–19.30). The basis for the Winogradsky column is threefold.

The first two factors involve opposing gradients that impact the types of organisms that can grow. The first is the oxygen gradient, which gets more and more anaerobic toward the bottom. As a result, obligate aerobes, microaerophiles, facultative anaerobes, and obligate aerobes are found in different locations in the column. The second is the H_2S gradient, which runs opposite in direction to the O_2 gradient.

The third factor is the diffuse light shined upon the column. This promotes growth of phototrophic organisms at levels where they are adapted to the opposing O_2 and H_2S gradients. These layers of phototrophs occur in natural ecosystems but are extremely thin because light does not penetrate mud sediments very far. But with the transparent column, thicker layers develop, which are more easily sampled for cultures.

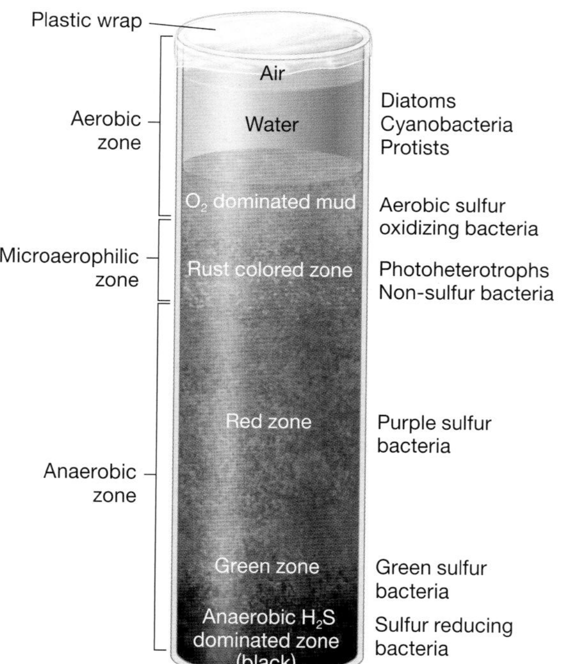

19.28 Artist's Rendition of a Winogradsky Column ● What you put into a Winogradsky column dictates what you grow. Any well-constructed column has an oxygen gradient from top to bottom, with the aerobic zone penetrating perhaps only as much as 20% of the total depth. The remaining portion of the mud column becomes progressively more anaerobic. The different amounts of oxygen lead to layering of microbial communities adapted to that specific environment. This illustration is a generalized picture of the layering that might be seen in a mature column. (A real column often produces intermixed patches rather than distinct layers.) Starting at the top and working downward, the layers are: air, water (containing algae and cyanobacteria), aerobic mud (sulfur oxidizing bacteria), microaerophilic mud (nonsulfur photosynthetic bacteria), red/purple zone (purple sulfur photosynthetic bacteria), green zone (green sulfur photosynthetic bacteria), and black anaerobic zone (sulfur reducing bacteria).

19.29 Freshly Made Winogradsky Column ● The black layer comprising the majority of the column is the unenriched mud. The lighter gray area at the bottom contains mud, $CaCO_3$, $CaSO_4$, and shredded paper mixed into a slurry. Note the absence of air spaces.

19.30 Mature Winogradsky Column at Eight Weeks ● Notice the layers and colors. Also notice that the layers are not as well defined as in the artist's rendition in Figure 19.28. In fact, some look mixed (e.g., the rust and red portions appear mixed in some regions). But the dark, anaerobic zone above the whitish layer at the bottom is well defined. The remainder is—pardon the expression—clear as mud.

Nitrogen Cycle

Biogeochemical cycles, such as the carbon cycle, the nitrogen cycle, and the sulfur cycle, are characterized by **environmental phases**, in which the element is not incorporated into an organism, and an **organismal phase**, in which it is. The nitrogen cycle is especially important as an area of study for microbiologists because it has so many parts in which bacteria participate. It will be helpful to look at Figure 19.31 as you read the following.

Organisms require nitrogen as a component of amino acids, purine and pyrimidine nucleotides, and other compounds. These organic forms of nitrogen do not occur outside cells. Approximately 80% of the air is nitrogen gas (N_2), so it is plentiful in the environment, but it is in an unusable form by all organisms except **nitrogen-fixing bacteria.** Fixation of atmospheric nitrogen occurs through the following reaction:

$$N_2 + 8H^+ + 8e^- + 16ATP \longrightarrow 2NH_3 + H_2 + 16ADP + 16P_i$$

This is a highly endergonic process and is performed by only a few groups of bacteria. Among terrestrial bacteria, *Azotobacter* (a β-proteobacterium; see p. 170 and Fig. 11.23) and *Rhizobium* (an α-proteobacterium; see p. 174 and Fig. 11.33) are usually put forth as examples.

Azotobacter (Fig. 19.32) is a common, free-living, aerobic Gram-negative rod that fixes N_2 in the soil. Oxygen inhibits the **nitrogenase** enzyme responsible for N-fixation, so the ability of *Azotobacter* to fix nitrogen aerobically is attributable to a number of complex, species-specific mechanisms that are beyond the scope of this book.

Azotobacter also has the ability to form resting vegetative cells called **cysts** (see Fig. 11.23B) when environmental

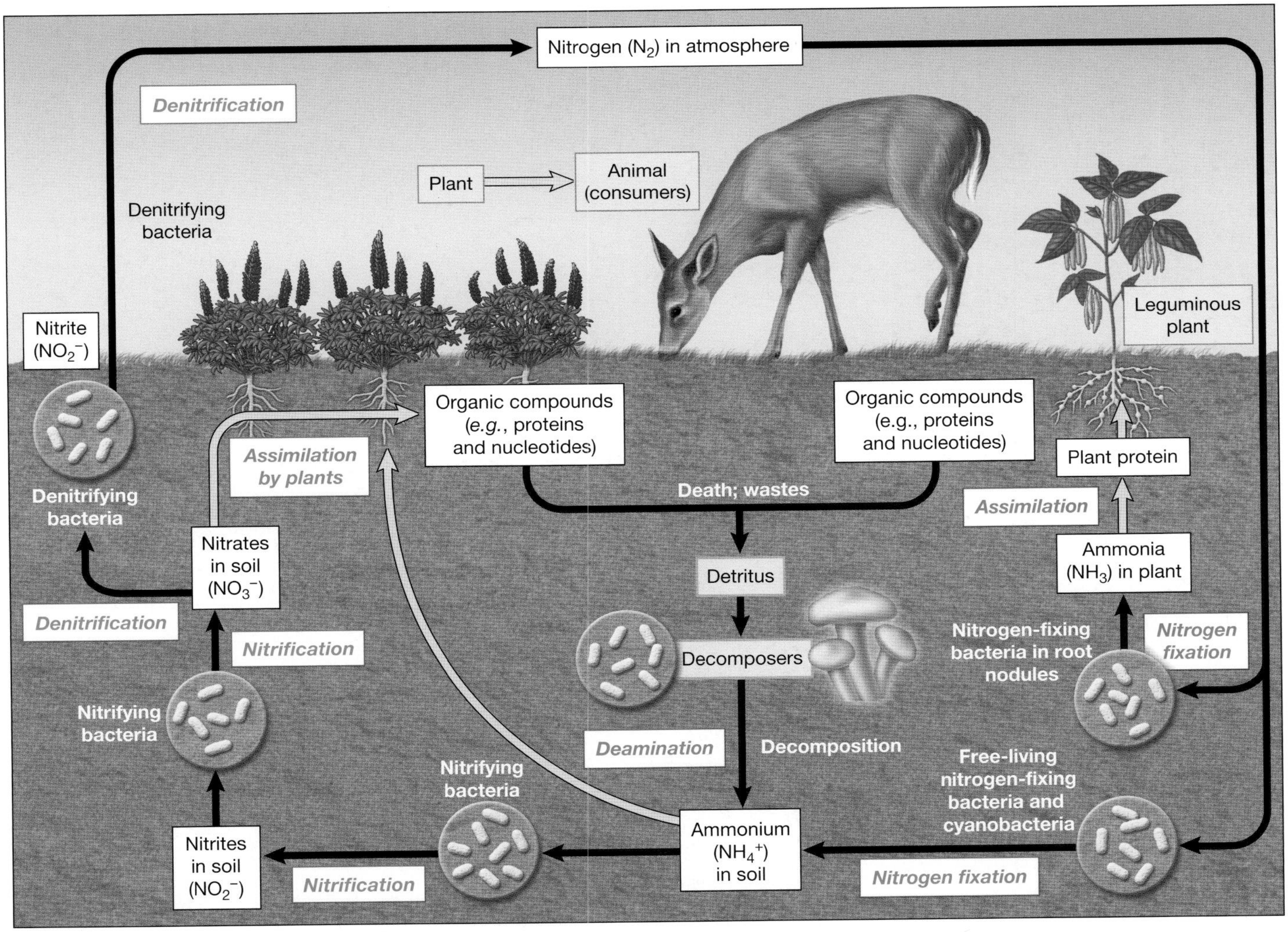

19.31 Nitrogen Cycle • There are three sets of information shown in this nitrogen cycle diagram: the nitrogenous compounds (in white boxes with black type), the process in which they are used (white boxes with blue type), and a general name for the organisms doing each process. Black arrows are microbial processes; gray arrows are macroorganismal. Further (but not labeled), the nitrogen cycle involves chemoheterotrophs (both aerobic and anaerobic respirers) and chemolithotrophs. See text for details.

conditions are not favorable. Although cysts perform somewhat the same function as bacterial endospores, they are different in that they are not as differentiated from the cell that formed them, nor are they as resistant to environmental factors such as desiccation and physical and chemical agents.

Rhizobium (Fig. 19.33) is a mutualistic symbiont of leguminous plants (members of Fabaceae: the Pea family). They rely on a supply of carbon compounds as electron donors from the plant host and, in return, supply the plant with ammonium, which can be assimilated into organic nitrogen (amino acids, nucleotides, etc.). *Rhizobium* and other symbionts are the major N-fixers on land.

Rhizobium is a Gram-negative rod that enters the root and induces the formation of **root nodules** (Figs. 19.34 and 19.35), tumors that are the location of N-fixation. Once in the plant, the cells become irregular in shape and are referred to as **bacteroids.** Their adaptation to fixing nitrogen aerobically is the presence of **leghemoglobin,** which, like our own hemoglobin, binds free oxygen.

Interestingly, leghemoglobin's production is induced through the combined efforts of the host plant and *Rhizobium.* It is so effective at moderating O_2 concentrations in the nodule that the bound form is estimated to outnumber the free form 10,000:1.

In aquatic environments, cyanobacteria are the main nitrogen fixers. An interesting example is the filamentous species *Anabaena* (see Fig. 11.9). Cyanobacteria perform photosynthesis using the same process as green plants. That is, they convert CO_2 and H_2O to carbohydrate and O_2.

Because N-fixation is inhibited by oxygen, *Anabaena* restricts the process to thick-walled cells called **heterocysts,** which lack Photosystem II, the oxygen-producing component of photosynthesis. Further, their thick walls, restrict entry of O_2 by diffusion. Heterocysts rely on physical

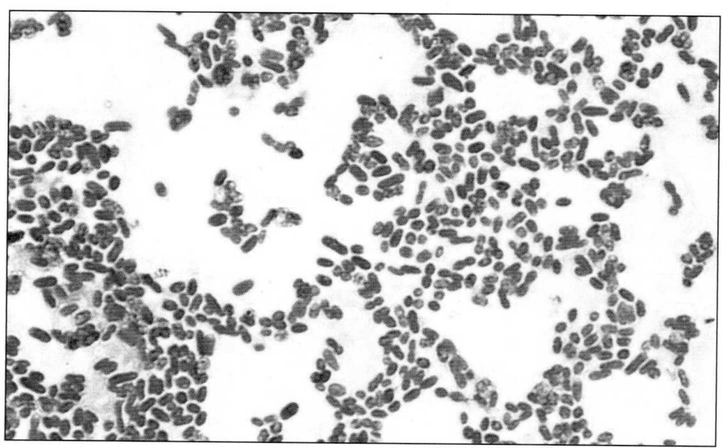

19.32 ***Azotobacter* Gram Stain** • Plump, Gram-negative rods characterize the free-living, nitrogen-fixing *Azotobacter.* These cells were grown in culture.

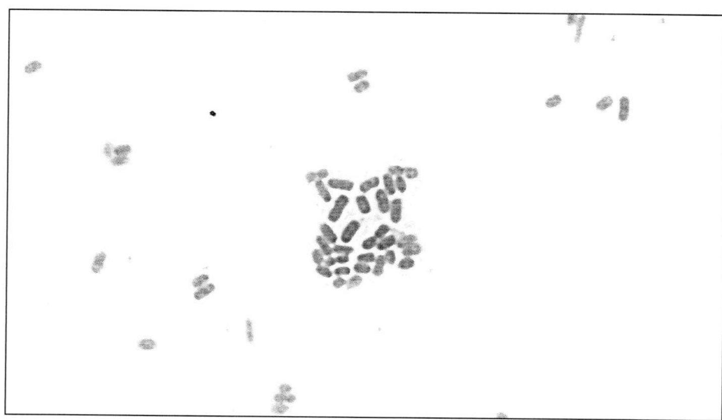

19.33 ***Rhizobium* Gram Stain** • *Rhizobium* occurs naturally as a nitrogen-fixing symbiont of leguminous plants. This Gram stain was made from a culture. *Rhizobium* and other symbionts are the major nitrogen-fixers on land.

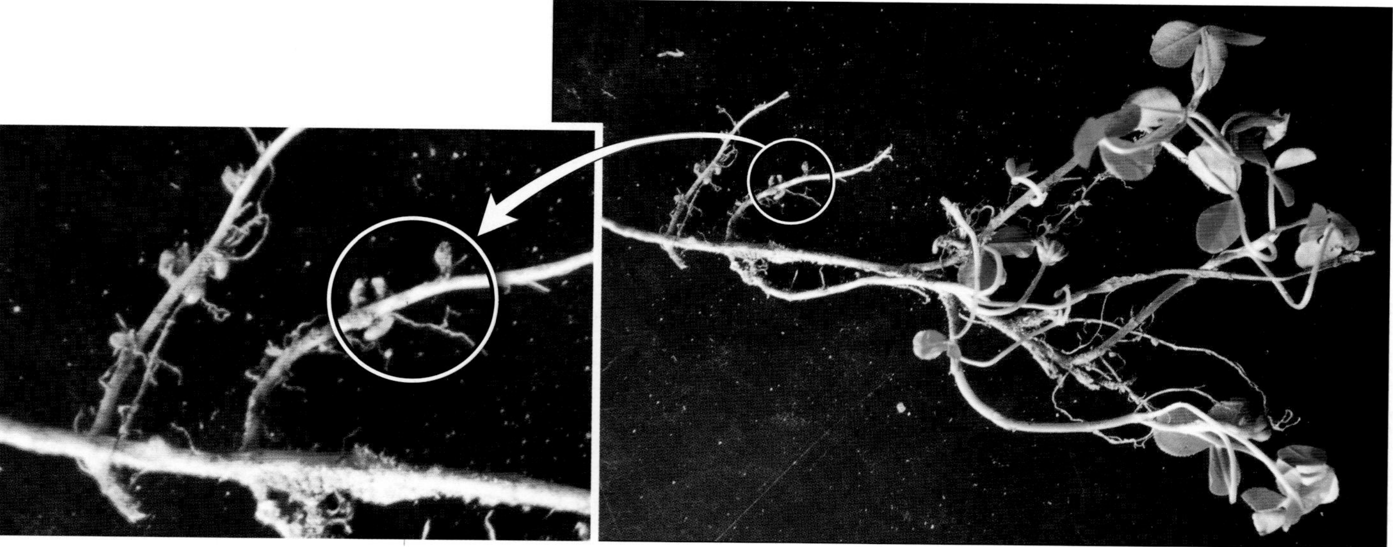

19.34 **Clover Root Nodules** • The roots of this small clover have been infected with *Rhizobium* that causes the formation of tumor-like root nodules (four are circled). The nodule is the site of nitrogen fixation.

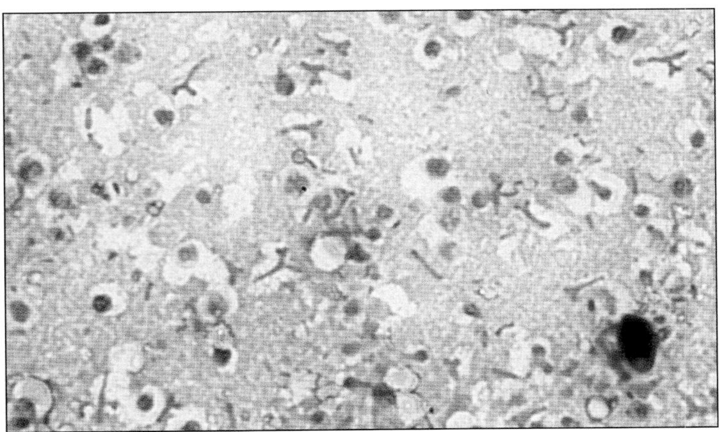

19.35 Root Nodule Section • Symbiotic nitrogen-fixing bacteria, such as *Rhizobium*, induce tumor formation in the roots of certain legumes. Once in the nodule, infected cells become filled with differentiated forms of the bacterium called bacteroids. These are the branched rods in the image.

connection with adjacent cells to supply reducing power for N-fixation and, in return, supply the other cells of the filament with organic nitrogen compounds.

Nitrification is the process of making **nitrate** (NO_3^-) and is strictly a microbial process performed by chemolithotrophic **nitrifying bacteria.** Ammonium ion (NH_4^+) acts as an electron and energy source when it is oxidized to nitrite (NO_2^-), and then to nitrate. This two-step, exergonic process involves two different groups of aerobic chemolithotrophs (which also are autotrophs because they use the energy released to fix carbon dioxide into organic carbon).

First, bacteria, such as *Nitrosomonas*, oxidize ammonium to form nitrite, as follows:

$$NH_4^+ + 1\frac{1}{2}\,O_2 \longrightarrow NO_2^- + H_2O + 2H^+$$

The second step involves different chemolithotrophs, such as *Nitrobacter*, that oxidize nitrite to nitrate, as follows:

$$NO_2^- + \frac{1}{2}\,O_2 \longrightarrow NO_3^-$$

Clay particles in soil bind positively charged ions, whereas negatively charged ions are repelled and are more available for organismal use, because they are freely diffusible in the soil water.

With that fact in mind, nitrification clearly is an important ecological process because it changes the charge on nitrogen from positive (in ammonium ion) to negative (nitrite and nitrate). Nitrate, along with ammonium, is a form of nitrogen that plants can absorb into their roots. Because of their solubility, however, nitrite and nitrate are readily leached from soil, making it less fertile.

In an ecosystem, much of the nitrogen is tied up in organic molecules. No organism lives forever, however, or is immune from producing wastes. **Ammonification** is a consequence of decomposition of organic nitrogen from dead animal and plant protein, and nitrogenous wastes (e.g., urea) from animals. In large part, it is a byproduct of amino acid **deamination** (a step in amino acid catabolism) in cells.

Many bacteria and fungi are capable of ammonification. The ammonia so produced can be recycled through uptake by plants, or it can be oxidized by ammonia oxidizing (nitrifying) bacteria. Deamination of an amino acid occurs as follows:

$$H_2N-\overset{\displaystyle R}{\overset{|}{C}}H-COOH + H_2O \longrightarrow NH_3 + \overset{\displaystyle R}{\overset{|}{C}}H_2-COOH$$

Nitrate reduction is the result of **anaerobic respiration,** in which nitrate is used as the **final electron acceptor.** In some cases, nitrite is the end-product, but in others nitrate is reduced to N_2 gas. Both processes are called **denitrification.** An example of this process is provided by *Pseudomonas denitrificans* and is shown as follows:

$$C_6H_{12}O_6 + 4NO_3 \longrightarrow 6CO_2 + 6H_2O + 2N_2$$

Notice that this reaction is very similar to the summary reaction for aerobic respiration. Because the final electron acceptor is $4NO_3$ and not O_2, however, the products are different. O_2 is not required for—in fact, it often inhibits—anaerobic respiration, so denitrification usually occurs in its absence.

Because nitrate is so soluble, it is leached from soils into aquatic environments and ultimately resides in the oceans. Without denitrification, all the nitrogen would end up in the ocean and be unavailable to terrestrial life.

Sulfur Cycle

Sulfur is one of the most abundant elements on Earth. Having oxidation states from –2 to +6, it is able to form many different compounds usable by living things. Most of the sulfur compounds used by microorganisms are inorganic molecules, used strictly for energy or to be incorporated into organic molecules in biosynthetic processes.

Table 19-6 summarizes some biologically important sulfur compounds and their oxidation states. Figure 19.36 illustrates a simplified version of the sulfur cycle and biogeochemical sulfur transformations. Refer to this figure as you read on.

The sulfur microorganisms are a diverse group and include both Bacteria and Archaea. They live in habitats as diverse as freshwater ponds, lakes, and rivers (especially where there is sewage contamination), water-saturated soils, saltwater lagoons, sulfur solfataras as in Yellowstone National Park, and in and around deep ocean hydrothermal vents. This vast group includes photoautotrophs, photoheterotrophs, chemolithoautotrophs, chemolithoheterotrophs, obligate aerobes, facultative anaerobes, and obligate anaerobes.

Many sulfur oxidizers and reducers live **syntrophically** in mutually dependent communities, in which sulfur is converted back and forth between reduced and oxidized forms. Conversely, sulfur oxidizers, living in and around hydrothermal vents, although still a complex community, have a never-ending source of reduced sulfur flowing up from the vents.

These microbes, receiving no biologically reduced sulfur, thrive in the ecosystem and produce large living mats that cover surrounding surfaces. Figure 19.37 shows a terrestrial microbial mat of sulfur bacteria.

TABLE **19-6** **Sulfur Compounds and Sulfur Organisms That Use Them**

Sulfur Compound	Chemical Formula	Oxidation State	Used By Oxidizers	Used By Reducers
Organic sulfur	R–SH	–2	+	+
Sulfide	H_2S, HS^-, S^{2-}	–2	+	+
Elemental sulfur	S^0	0	+	+
Thiosulfate	$S_2O_3^{2-}$	+2 per S	+	+
Sulfate	SO_4^{2-}	+6		+

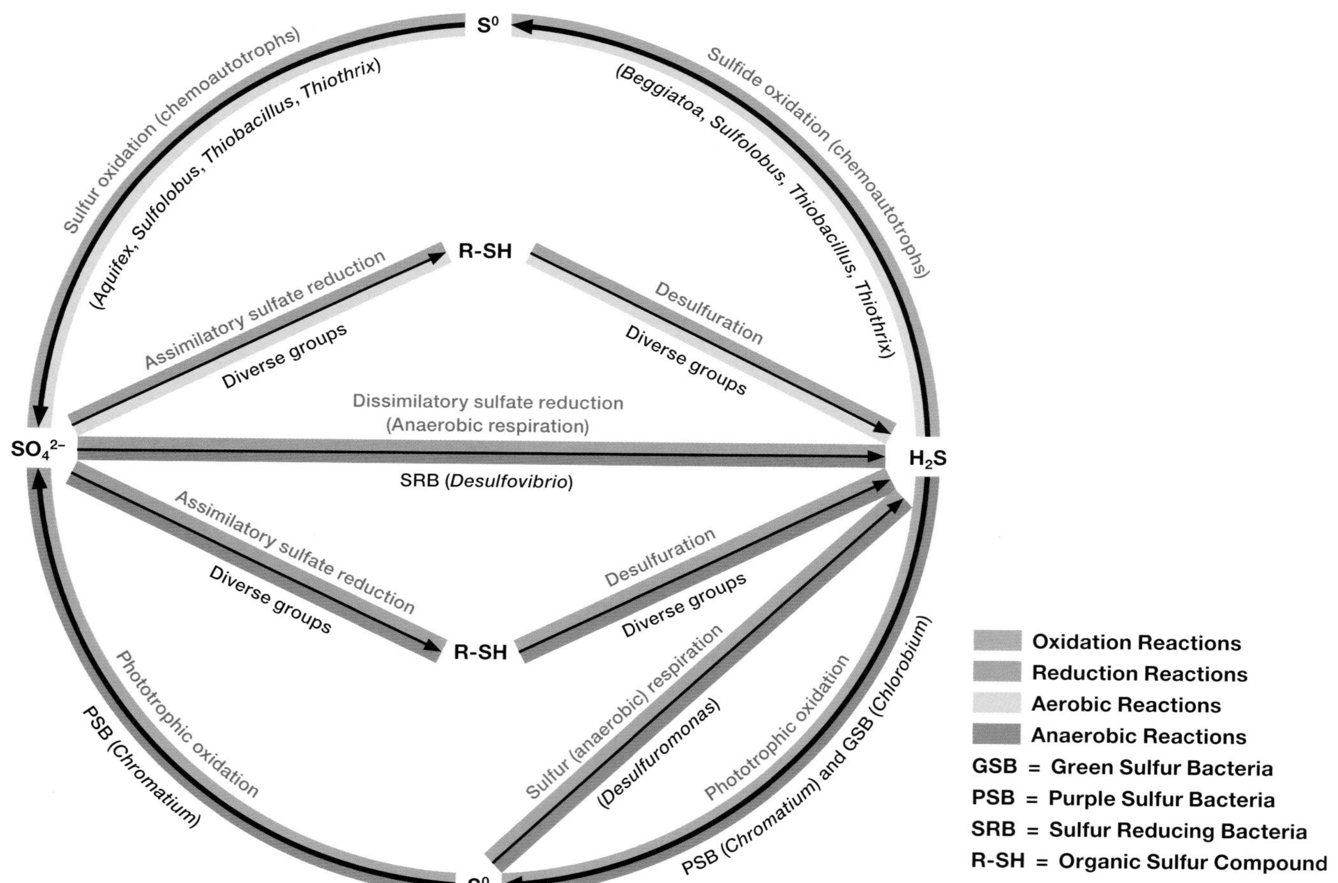

19.36 Biogeochemical Sulfur Transformations ● These are the major sulfur transformations in the sulfur cycle. "R" refers to an organic molecule of variable composition. Refer to Table 19-6 and the text for details.

Sulfur bacteria fall into three major categories—photoautotrophs, chemolithoautotrophs, and the sulfur reducers. The photoautotrophs are **anoxygenic photosynthesizers,** that is, they perform a type of photosynthesis that does not produce oxygen and are represented in Figure 19.28 by **green sulfur** (see p. 169) and **purple sulfur bacteria** (see p. 169), though they are not phylogenetically closely related.

They reside in the anoxic zone of a pond or other aquatic ecosystem close enough to the surface to absorb the sun's energy. The primary photosynthetic pigment is bacteriochlorophyll. It absorbs light of slightly different wavelengths than cyanobacterial and eukaryotic chlorophyll a and is bound to membranes of vesicles or infoldings of the cytoplasmic membrane (recall that bacteria don't have chloroplasts).

Trapped light combined with oxidation of H_2S or elemental sulfur provides the energy and electrons required for CO_2 fixation and carbon reduction, respectively, for the production of carbohydrate. The summary reaction of anoxygenic photosynthesis is analogous to the summary reaction of oxygenic photosynthetic by cyanobacteria and eukaryotes:

photosynthetic eukaryotes:

$$CO_2 + H_2O \longrightarrow [CH_2O] + O_2$$

photosynthetic sulfur bacteria:

$$CO_2 + H_2S \longrightarrow [CH_2O] + S^0$$

In spite of their superficial similarities, there are differences in the details of the two types of photosynthesis, details that are beyond the scope of this book.

Chemolithoautotrophic sulfur oxidizing bacteria get their energy by oxidizing reduced sulfur compounds, most frequently hydrogen sulfide (H_2S), elemental sulfur (S^0), and thiosulfate ($S_2O_3^{2-}$). These organisms differ from photoautotrophs in their energy sources (chemicals rather than light), but they are similar in that they both perform CO_2 fixation and reduction in order make carbohydrates.

These chemolithoautotrophs live in the water and in upper sediment layers of aquatic habitats. As such, they range from strictly aerobic to microaerophilic to facultatively anaerobic. Very common among the sulfur oxidizing bacteria are the so-called **gradient bacteria.** These microaerophiles require both oxygen and H_2S, but only in a very narrow range of concentrations. They typically reside at the level where H_2S rising from sediments and O_2 diffusing down from the air create a critical composition of the two gases.

Gradient bacteria, such as *Beggiatoa* (see p. 170), *Thioplaca*, and *Thiothrix* are facultative chemolithotrophs. They usually obtain their energy from H_2S and fix carbon from CO_2, but are capable of obtaining energy and carbon from organic compounds as well. The chemolithotrophic energy reaction is as follows:

$$H_2S + \frac{1}{2} O_2 \longrightarrow S^0 + H_2O$$

These organisms typically store sulfur granules inside the cells (Fig. 19.38).

Thiobacillus is representative of another group of sulfur oxidizing bacteria. Several of its species perform the same reaction as the gradient bacteria and also store sulfur granules inside the cell. Other, more acidophilic *Thiobacillus* species oxidize sulfur compounds and elemental sulfur all the way to sulfuric acid as follows:

$$S^0 + \frac{1}{2} O_2 + H_2O \longrightarrow H_2SO_4$$

Acidophilic *Thiobacillus* species are obligate chemolithotrophs and derive all of their energy and carbon from reduced sulfur compounds and carbon dioxide, respectively. Not surprisingly, they thrive in environments as acidic as pH 2.

Sulfur-reducing bacteria perform two important types of sulfur reduction, **assimilatory sulfate reduction** and **dissimilatory sulfate reduction.** Assimilatory sulfate reduction is the process of actively transporting sulfate into the cell and converting it to sulfhydryl groups (R–SH) found in the amino acids cysteine and methionine. Because sulfur is

19.37 Microbial Mat of Sulfur Bacteria • The layers of this microbial mat formed at the edge of the Salton Sea in California contain sulfur bacteria. The chemoautotroph *Beggiatoa* is found in the lighter, surface mat. A portion of the surface mat has been removed to reveal the black mud less than 1 cm below in which sulfur reducing bacteria reside. Green sulfur bacteria (photoautotrophs) are found in between.

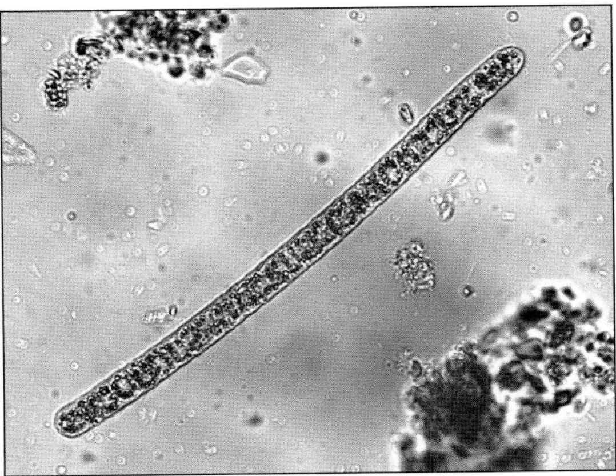

19.38 Chemoautotrophic Sulfur-Oxidizing Bacterium (Bright Field Wet Mount) • This gliding bacterium, provisionally identified as *Beggiatoa*, shows numerous sulfur granules in the cytoplasm. These are the by-product of H_2S oxidation, the method by which *Beggiatoa* obtains energy. This specimen was isolated from the San Diego River estuary.

an essential component needed for biosynthesis, many bacteria, fungi, and even green plants transport it into their cells.

Dissimilatory sulfate reduction is a type of anaerobic respiration done exclusively by sulfate-reducing bacteria (SRB; Fig. 19.39) that use sulfate as the final electron acceptor from the oxidation of an organic (or possibly inorganic) compound.

This process involves a cytochrome system and results in ATP production. Members of the SRB include *Desulfovibrio* (see p. 176) and a number of other genera beginning with *Desulfo-*.

Another type of anaerobic respiration uses elemental sulfur as the final electron acceptor. *Desulfurimonas* is an example and the process is known as **sulfur respiration.**

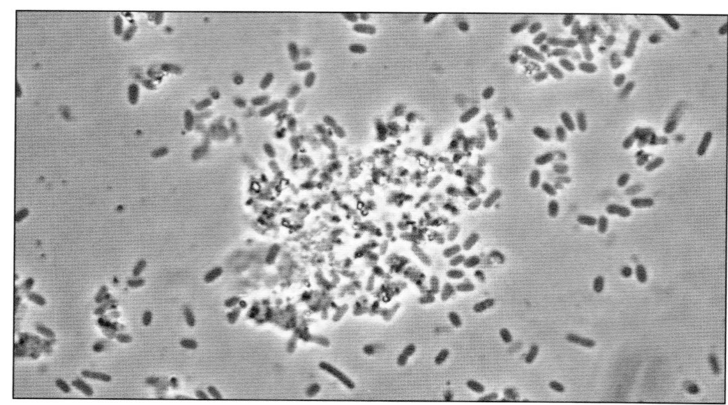

19.39 Sulfur Reducing Bacterium ● This photomicrograph shows a community of unknown sulfur reducers recovered from black (anoxic) pond sediment. (Phase Contrast)

Part C: Food Microbiology

Methylene Blue Reductase Test

Purpose

This procedure is a qualitative method of determining milk quality. It also gives an indication of the contaminant(s) present, with rapid reduction being associated with contamination by high levels of enteric bacteria and *Streptococcus lactis*. It has largely been replaced by quantitative methods, such as the standard plate count (see p. 283) and the Multiple Tube Fermentation test (see p. 302). Still, it's a fun and easy way to assess milk quality.

Principle

Milk is a good source of protein (casein) and carbohydrate (lactose) not only for mammals, but also for microbes—just think back at how many of the media described in Sections 2 and 7 had casein and/or lactose in them.

Because it is such a good nutrient source and it has the potential of becoming contaminated with pathogenic microbes from the cow's skin or udder, or even unsterile machinery, milk quality is always a public health concern. The reduction of methylene blue dye may be used as an indicator of milk quality.

Methylene blue is blue when oxidized and colorless when reduced. It can be reduced enzymatically either aerobically or anaerobically. In the aerobic electron transport system, methylene blue is reduced by cytochromes but immediately is returned to the oxidized state when it subsequently reduces oxygen. Anaerobically, the dye is in the reduced form, and in the absence of an oxidizing substance, remains colorless.

In the methylene blue reductase test, a small quantity of a dilute methylene blue solution is added to a sterilized test tube containing raw milk. The tube then is sealed tightly and incubated in a 35°C water bath. The time it takes the milk to turn from blue to white (because of methylene blue reduction) is a qualitative indicator of the number of microorganisms living in the milk (Fig. 19.40).

The higher the microorganism concentration, the faster the oxygen consumption in the sealed tube, and the faster methylene blue becomes reduced. Good-quality milk takes longer than 6 hours to convert the methylene blue. At the other extreme, very poor-quality milk will convert in less than 30 minutes. See Table 19-7 for more details.

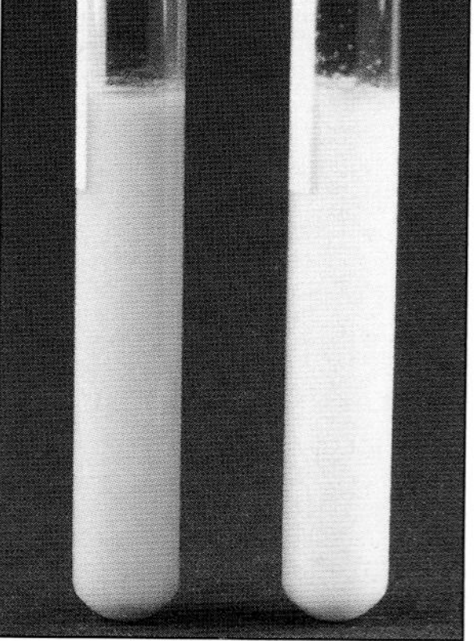

19.40 Methylene Blue Reductase Test Results ● The tube on the left is a control to illustrate the original color of the oxidized medium (milk). The tube on the right indicates bacterial reduction of methylene blue after 20 hours. The speed of reduction corresponds to the concentration of microorganisms present in the milk (Table 19-7).

TABLE **19-7** **Milk Quality Standards for the Methylene Blue Reductase Test (MBRT)**

Reduction Time (t_E)	Milk Quality
Longer than 8 hours	Excellent
Between 6 hours and 8 hours	Good
Between 2.0 hours and 6 hours	Fair
Between 30 minutes and 2.0 hours	Poor
Less than 30 minutes	Very poor

Yogurt

Several species of bacteria are used in the commercial production of yogurt (Fig. 19.41). Most formulations include combinations of two or more species to synergistically enhance growth and to produce the optimum balance of flavor and acidity. One common pairing of organisms in commercial yogurt is that of *Lactobacillus delbrueckii* subsp. *bulgaricus* and *Streptococcus thermophilus*.

Yogurt gets its unique flavor from acetaldehyde, diacetyl, and acetate produced from fermentation of the milk sugar lactose. The proportions of products, and ultimately the flavor, in the yogurt depend upon the types of enzyme systems possessed by the species used. Both species mentioned here contain **constitutive** β-galactosidase systems that break down lactose and convert the glucose to lactate, formate, and acetate via pyruvate in the glycolytic pathway (see pp. 326–331).

As you may remember, lactose is a disaccharide composed of glucose and galactose. *S. thermophilus* does not possess the enzymes needed to metabolize galactose, and *L. delbrueckii* preferentially metabolizes glucose. This results in an accumulation of galactose, which adds sweetness to the yogurt. Acetaldehyde is produced directly from pyruvate by *S. thermophilus* and through the conversion of proteolysis products threonine and glycine by *L. delbrueckii*. Some strains of *S. thermophilus* also produce glucose polymers, which give the yogurt a viscous consistency.

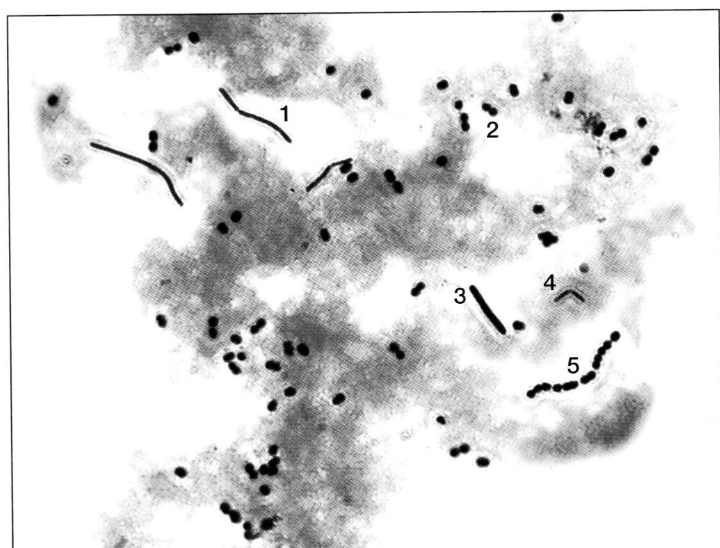

19.41 Microbes in Yogurt • This micrograph is a composite of several micrographs made from one Gram-stained specimen of commercial yogurt. There are at least four (or maybe five—it is not clear if #5 is a chain of #2 or another species) distinct cell morphologies. Rods 1, 3, and 4 are probably three different *Lactobacillus* species.

SECTION 20

Host Defenses

Differential Blood Cell Count

Purpose

A differential blood cell count is done to determine approximate numbers of the various leukocytes in blood. Excess or deficiency of all or a particular group is indicative of certain disease states. Differential counts are automated now, but it is good training to microscopically examine a blood smear in order to be familiar with the cells of interest.

Principle

Leukocytes (white blood cells, or WBCs) are divided into two groups: **granulocytes,** which have prominent cytoplasmic granules, and **agranulocytes**, which lack them. There are three basic types of granulocytes: **neutrophils, basophils,** and **eosinophils**. The two types of agranulocytes are **monocytes** and **lymphocytes**. All leukocytes develop from bone marrow stem cells.

Neutrophils (Fig. 20.1) are 12 µm–15 µm in diameter, about twice the size of an erythrocyte (RBC)[1], and are the most abundant WBCs in blood. Their most prominent cytoplasmic granules are neutral-staining and thus do not have the intense color of other granulocytes when prepared with Wright's or Giemsa stains.

When needed, they leave the blood and enter tissues to phagocytize bacteria, whose recognition requires binding to a variety of different membrane receptors. Once internalized by phagocytosis, bacteria are killed by several mechanisms acting in concert, including oxidation of cell structures by oxygen radicals, enzymatic destruction of the cell wall, and leakage through the cell membrane caused by cationic peptides.

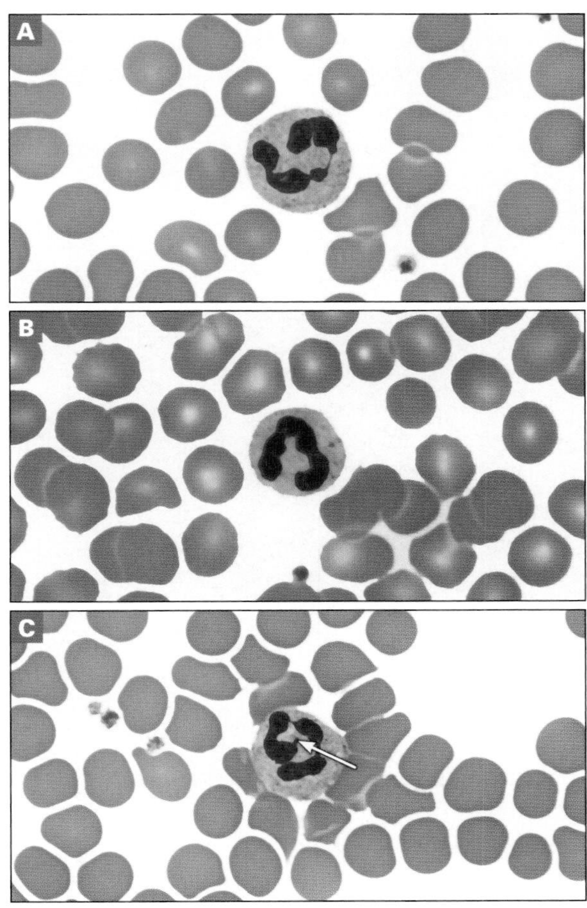

20.1 Neutrophils • Neutrophils are granulocytes with a pinkish to gray cytoplasm. They are the most abundant leukocyte in blood and are not quite twice the size of RBCs. Mature neutrophils have a segmented nucleus (as in micrographs **A** and **C**) with lobes joined by a narrow strand of nuclear material. Immature neutrophils have an unsegmented nucleus and are called band cells (as in micrograph **B**). About 3% of neutrophils in female blood samples demonstrate a "drumstick" protruding from the nucleus, indicated by the **arrow** in micrograph (**C**). This is the region of the inactive X chromosome. (Blood Smears; Wright's Stain)

[1] It is convenient to discuss leukocyte size in terms of erythrocyte size because RBCs are so uniform in diameter. In an isotonic solution, erythrocytes are 7.5 µm in diameter.

Neutrophils typically live only a few days before undergoing **apoptosis.** Host cell debris, dead neutrophils, and dead bacteria form pus (Fig. 20.2).

Mature neutrophils sometimes are referred to as **segs** because their nucleus usually is segmented into two to five lobes. Because of this variation in nuclear appearance, they also are called **polymorphonuclear neutrophils (PMNs).** Immature neutrophils lack this nuclear segmentation and are referred to as **bands** (Fig. 20.1B). This distinction is useful because a patient with an active infection increases neutrophil production, which leads to a higher percentage of immature band cells in the blood.

Another feature seen in 3% of female neutrophils is a "drumstick" extending from the nucleus (Fig. 20.1C). This is the inactive X-chromosome.

Eosinophils (Fig. 20.3) are 12 µm–15 µm in diameter (about twice the size of an RBC), generally have two lobes in their nucleus, and have red staining cytoplasmic granules. Most reside in connective tissues and are sometimes difficult to find in blood smears, where they represent less than 3% of all leukocytes.

Their functions are varied and their numbers increase during allergic reactions, chronic inflammation, and parasitic infections. They phagocytize and digest antigen-antibody complexes, and their secretions include chemicals that control the inflammatory response (e.g., histaminase), and cytotoxic and neurotoxic chemicals that attack parasites.

Basophils (Fig. 20.4) are the least abundant WBCs in normal blood and are difficult to find. They are structurally and functionally similar to tissue mast cells, share a common progenitor in bone marrow (see p. 317), and produce some of the same chemicals (histamine and heparin) as well as bind a class of antibodies called immunoglobin E (IgE). They are 12 µm–15 µm in diameter and their dark-staining cytoplasmic granules usually obscure the one- or two-lobed nucleus.

Agranulocytes include monocytes and lymphocytes. Monocytes (Fig. 20.5) are the blood form of **macrophages** (see p. 318), which phagocytize foreign cells, cell debris, and also act as **antigen-presenting cells** (APCs) during many immune responses. They are the largest of the leukocytes, being two- to three-times the size of RBCs (12 µm–20 µm). Their nucleus is horseshoe-shaped (or at least indented), and the cytoplasm lacks prominent granules, but may appear finely granular.

Lymphocytes (Fig. 20.6) are cells of specific acquired immunity. They are approximately the same size as RBCs, or up to twice their size, and have very little cytoplasm visible around their spherical nucleus.

Two functional lymphocytes types are the **T cell,** involved in **cell-mediated immunity,** and the **B cell,** which converts to a **plasma cell** when activated and produces antibodies in **humoral immunity** (see p. 318).

Natural killer (NK) cells (Fig. 20.6B) comprise a third lymphocyte class that kills foreign or virally infected cells without antigen-antibody interaction. They are larger than B and T lymphocytes, being up to 16 µm, and have large

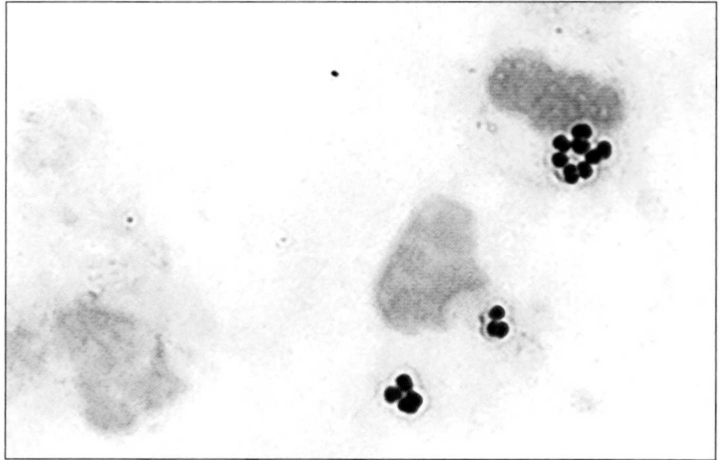

20.2 Pus • Pus occasionally forms at the site of infection. It is composed of neutrophils (living and dead), the invading organism (living and dead), and other cell debris, including damaged host tissue. Here, multiple neutrophils are attacking the Gram-positive cocci that have caused the infection. (Gram Stain)

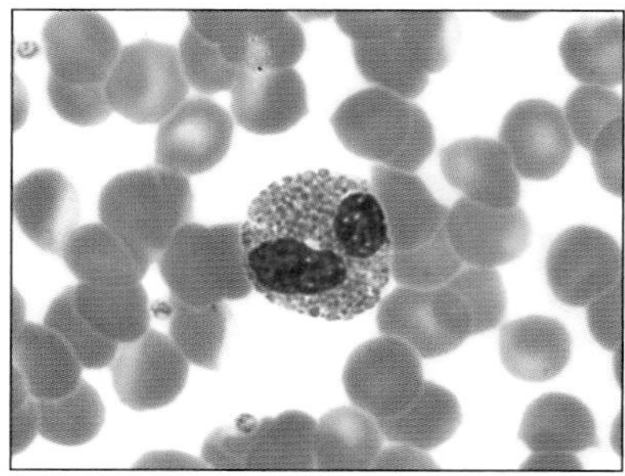

20.3 Eosinophil • These granulocytes are relatively rare in blood and are about twice the size of RBCs. Their cytoplasmic granules stain red, and their nucleus usually has two lobes. (Blood Smear; Wright's Stain)

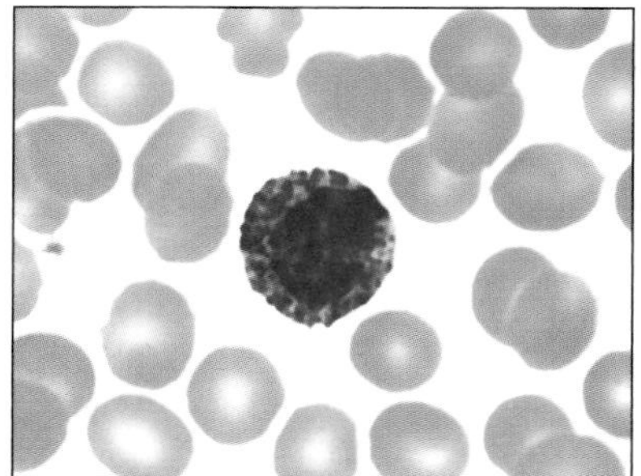

20.4 Basophil • Basophils comprise only about 1% of all WBCs. They are up to twice the size of RBCs and have dark purple cytoplasmic granules that often obscure the nucleus. (Blood Smear; Wright's Stain)

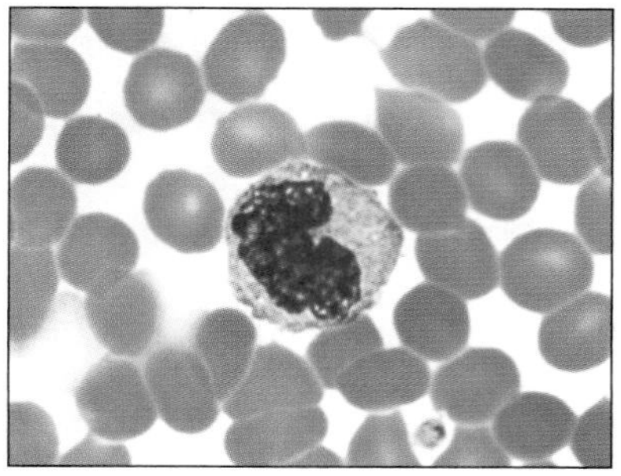

20.5 Monocyte • Monocytes are the blood form of macrophages. They are two- to three-times the size of RBCs and have a round or indented nucleus. Unlike phagocytic neutrophils, monocytes leave the blood and reside in tissues prior to infection. (Blood Smear; Wright's Stain)

cytoplasmic granules, which accounts for their other name: **large granular lymphocytes.**

In a differential white cell count, a sample of blood is observed under the microscope, and at least 100 WBCs are counted and tallied (this task is automated now). Approximate normal percentages for each leukocyte are as follows: neutrophils (mostly segs) 55%–65%, lymphocytes 25%–33%, monocytes 3%–7%, eosinophils 1%–3%, and basophils <0.2%.

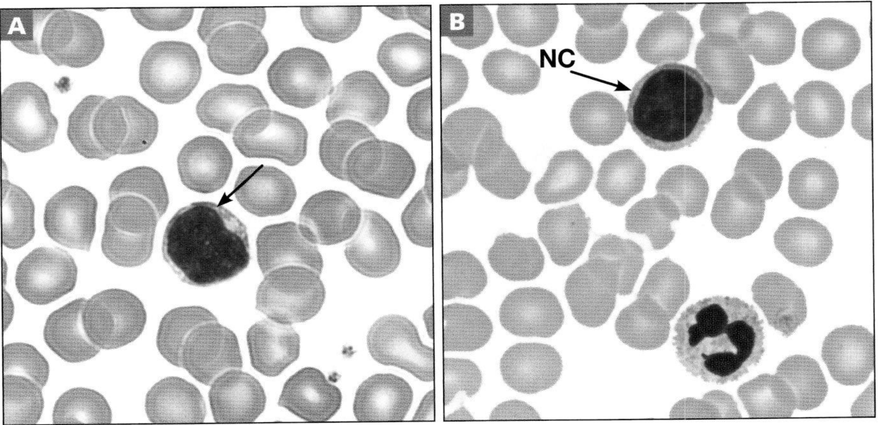

20.6 Lymphocytes • Lymphocytes are common in the blood, comprising 25–33% of all leukocytes. Most are about the size of RBCs and have only a thin halo of cytoplasm encircling their round nucleus. They belong to functional groups called B cells and T cells, which are morphologically indistinguishable. (**A**) Shown is a small lymphocyte. (Blood Smear; Wright's Stain) (**B**) Some lymphocytes are larger than others. These are natural killer cells or some other type of "null" cell (**NC**). A neutrophil is also visible in this field. Note the "drumstick." (Blood Smear; Giemsa Stain)

Immune Cells and Organs

All blood cells are derived from stem cells in bone marrow. However, unlike red blood cells that function within the bloodstream, white blood cells generally function outside the bloodstream within or on the surface of other organs. Discussed here are mast cells, cells of the mononuclear phagocyte system, lymphocytes, and lymphoid organs.

Mast Cells

Like basophils, mast cells (Fig. 20.7) possess prominent granules and secrete many of the same chemicals. Their relationship to each other has confounded physiologists for decades. The evidence now suggests that mast cells develop from their own stem cells, but are related to the stem cells that give rise to RBCs, granulocytes (including basophils), and monocytes. Early in their development, they migrate to mucosal tissues and other sites where they mature into mast cells.

Among numerous other inflammatory mediators, they secrete **histamine** (a vasodilator) and **heparin** (an anticoagulant). They have membrane receptors for **immunoglobin E (IgE)**, which, when bound to their antigen cause the mast cell to degranulate and release their chemical arsenal. An excess

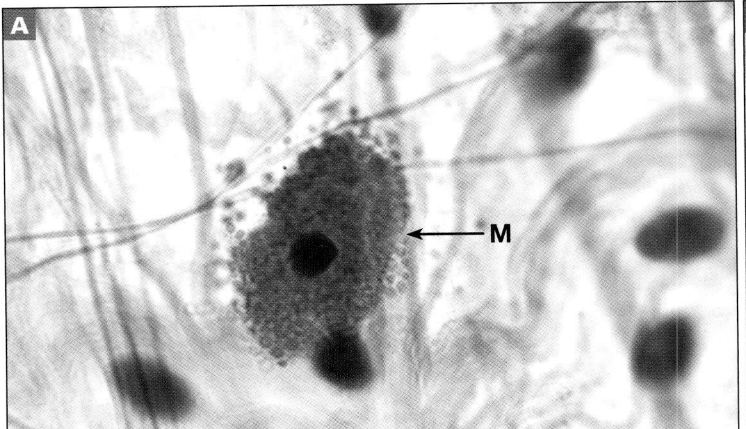

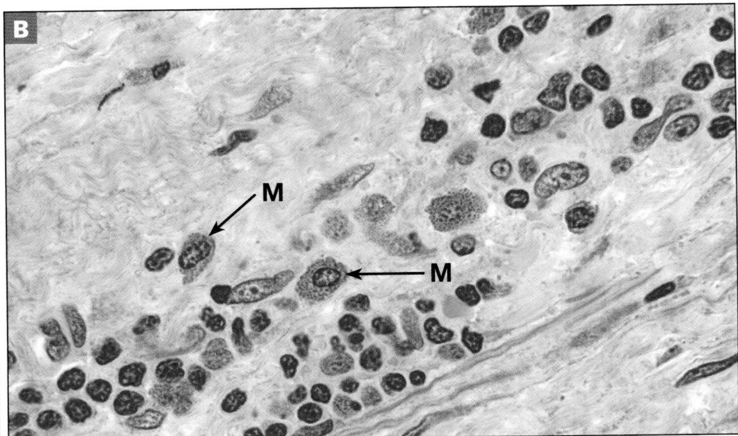

20.7 Mast Cells • Mast cells (**M**) are involved in the inflammatory response by degranulating and releasing histamine and other chemicals. (**A**) A single mast cell is visible in this loose connective tissue spread. When stained properly, mast cells with their granules are unmistakable. The released granules are an artifact of slide preparation. This preparation is a whole mount. (**B**) Several mast cells are visible in this sectioned tonsil specimen. (Hematoxylin-Eosin Stain)

of bound IgE can lead to massive histamine release and produce **Type I hypersensitivity** (allergic) reactions, such as hay fever and anaphylactic shock.

Mononuclear Phagocyte (Reticuloendothelial) System

Monocytes leave the blood and take up residence in tissues, where they differentiate into a variety of tissue-specific phagocytic cells generically referred to as **macrophages**. These include **Kupffer cells** in liver sinusoids, **alveolar macrophages** in the lungs, and **microglia** in the brain (Figs. 20.8A–C).

Not only are these cells important in engulfing invading cells, they also play a role in activating the humoral immune response by "showing" an antigen to the appropriate lymphocyte. In doing so, they are acting in the role of **antigen presenting cells** or **APCs. Dendritic cells** (Fig. 20.8D) are also APCs and are derived from the same stem cells as monocytes. They are found in the skin and mucous membranes.

Lymphocytes and Lymphatic Tissue

Lymphocytes leave the blood and occupy mucosal connective tissue as well as the interior of many organs. Three basic categories of lymphocytes have been identified: **B cells, T cells,** and **null cells.**

When stimulated by a specific antigen, B cells differentiate into **plasma cells** (Fig. 20.9), cells that are responsible for producing antibodies against that antigen, or into **memory B cells,** that reside in lymphatic tissue and await a subsequent exposure to that antigen. Memory cells are responsible for the **secondary (memory) response,** which is more rapid and of greater magnitude than a **primary response.** B cells participate in **humoral immune responses** (*humor* can refer to liquids or moisture, as in *humidity*), so called because the active agents (antibodies) are molecules found in serum and tissue fluid.

Potential T cells migrate to the **thymus,** where they become full-fledged T cells. During maturation, T cells become one of three basic types: **cytotoxic T cells, helper T cells,** and, after exposure to an antigen, **memory T cells.**

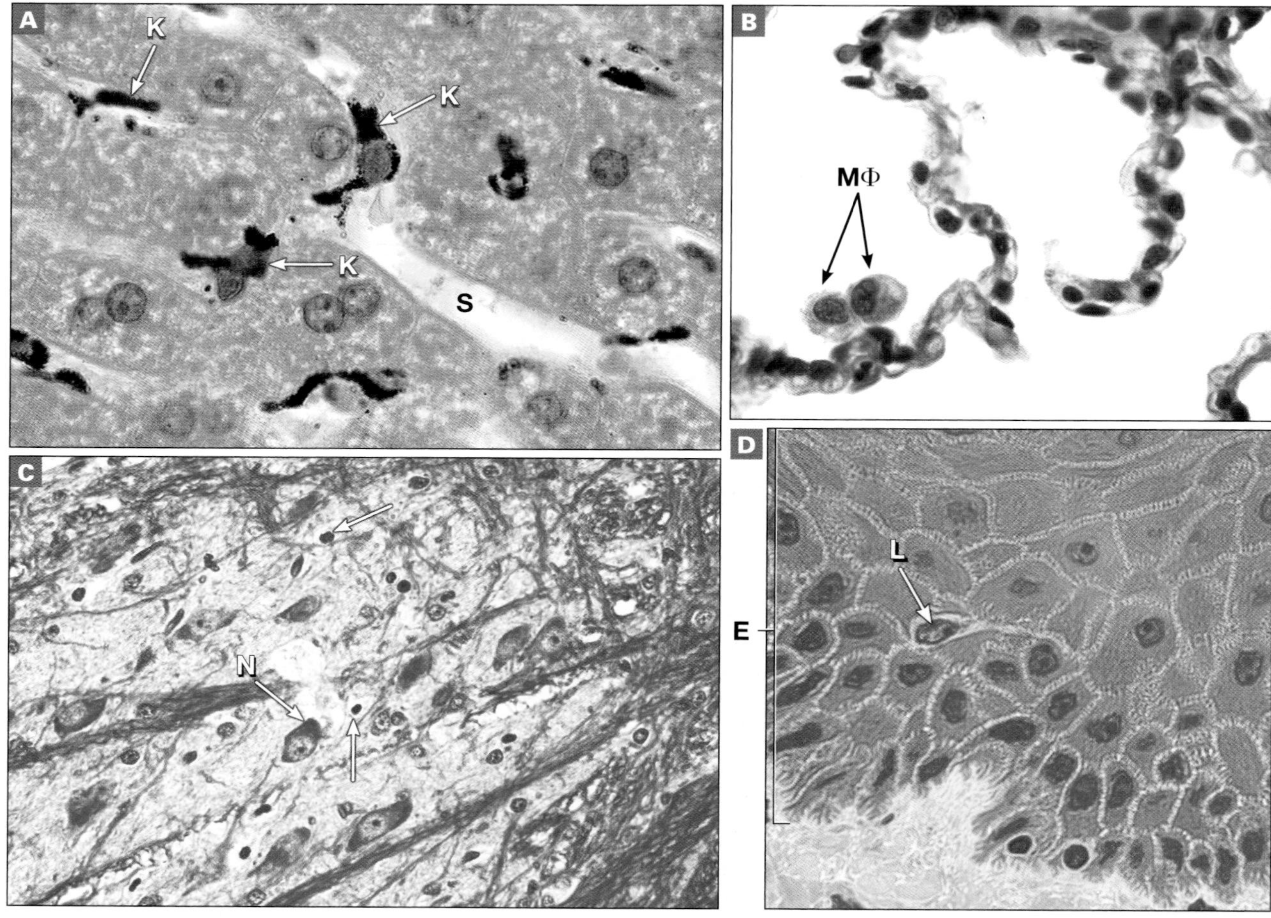

20.8 Macrophages • Macrophages (**MΦ**) are of many types and often are referred to by tissue-specific names. Shown are four examples from sectioned organs. (A) Kupffer cells (**K**) are fixed macrophages that line liver sinusoids (**S**) and remove material from the blood. These macrophages have ingested carbon particles that make them stand out against the stained liver cells. (B) The arrows indicate two alveolar macrophages (**MΦ**) of the lung. The white spaces are lung alveoli. (C) Microglia (**arrows**) are phagocytic cells of neural tissue, but are difficult to identify with a high degree of certainty in routine preparations. They are small cells with a dense and elongated nucleus. The larger cells are neurons (**N**). (D) Dendritic cells are found throughout the body in mucous membrane and the skin. Dendritic cells of the skin are called Langerhans cells (**L**) and are identifiable by their lighter cytoplasm in the deeper epidermal (**E**) layers.

Cytotoxic T cells directly destroy foreign cells or infected host cells. T-helper cells assist B cells and sometimes cytotoxic T cells in their response. Memory T cells fulfill the same role as memory B cells. Cytotoxic T cells participate in **cell-mediated immunity**, so called because the cells (not secreted chemicals) are the active agents. B and T cells are activated when an antigen fits into their unique membrane receptors.

B and T cells cannot be distinguished based on appearance. Rather, they are identified by the presence of specific molecules in their membranes. B cells have B cell markers and T cells have T cell markers, which leaves us with **null cells**, identifiable by the absence of either of those markers. A major subgroup is capable of destroying cells tagged with antibodies and are called **natural killer cells** (Fig. 20.6B).

Lymphoid Organs

Lymph nodes, the **thymus**, and the **spleen** are considered lymphoid *organs* because they are surrounded by a connective tissue capsule that makes them stand apart from surrounding structures. All are packed with lymphocytes.

Lymph nodes (Figs. 20.10 and 20.11) are bean-shaped organs 2 to 10 mm in length located along all but the smallest lymphatic vessels. Lymphatic vessels carry **lymph,** a fluid largely recovered from tissue fluid, and return it to the blood. As the lymph passes through the node's channels (sinuses), antigens it carries contact B cells, T cells, and APCs in the node. If an antigen's shape fits into a B or T cell's receptor, an immune response is initiated.

The thymus (Fig. 20.12) is located in the thoracic cavity between the lungs and is most highly developed at puberty. As an individual ages, the thymus becomes replaced with adipose tissue and its activity diminishes. It is divided into a cortex and medulla. Lymphocytes begin their maturation

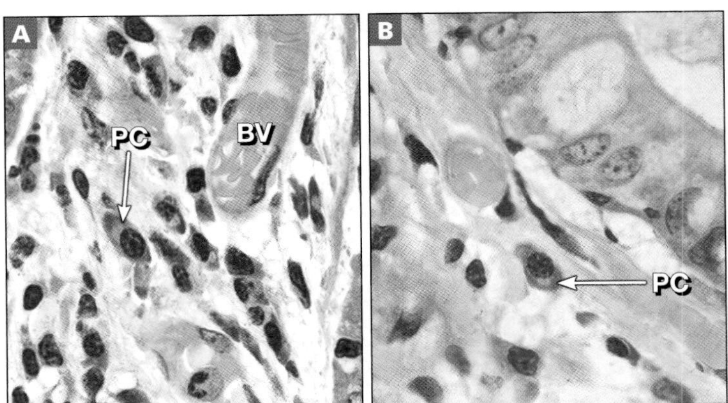

20.9 Plasma Cells • B-lymphocytes develop into plasma cells when stimulated by the appropriate antigen. Plasma cells then secrete protective antibodies against the antigen. (A) Plasma cells (**PC**) are recognizable because of their elongated shape, eccentric nucleus with dark staining spots of chromatin around the periphery (resembling a clock face), and a pale region near the nucleus, which is the site of a Golgi apparatus. Blood vessels (**BV**) are also seen. This specimen is from the colon. (B) Shown is another plasma cell (**arrow**) from the colon. (Sections; Hematoxylin-Eosin Stain)

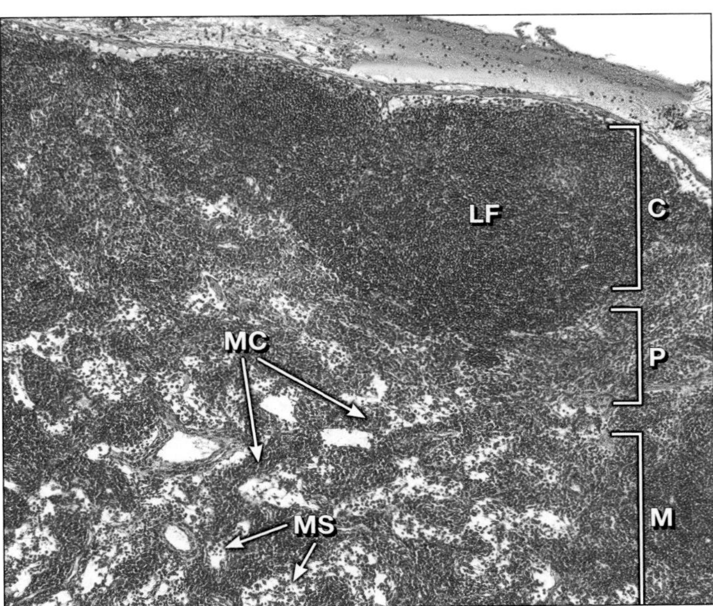

20.10 Lymph Node • Lymph nodes are bean-shaped organs found along the length of lymphatic vessels. Being bean-shaped, one side of the node has an indentation called the hilum. Lymph nodes are generously stocked with lymphocytes, seen as the purple regions in this micrograph. Individual lymphocytes appear as purple dots. Three regions within the node are labeled. From the periphery towards the hilum: the cortex (**C**), composed of spherical aggregations of B cells called lymph follicles (**LF**); the paracortex (**P**) occupied by T cells; and the medulla (**M**), identifiable by its medullary cords (**MC**), made mostly of B cells and plasma cells. In this micrograph, lymph would enter the node at the top, pass through sinuses (some are labeled in the medulla as **MS**), and leave through the lymph vessel at the hilum (not shown) somewhere beyond the lower portion of the photo. Macrophages and dendritic cells are also found in lymph nodes and assist in an immune response. (Section; Hematoxylin-Eosin Stain)

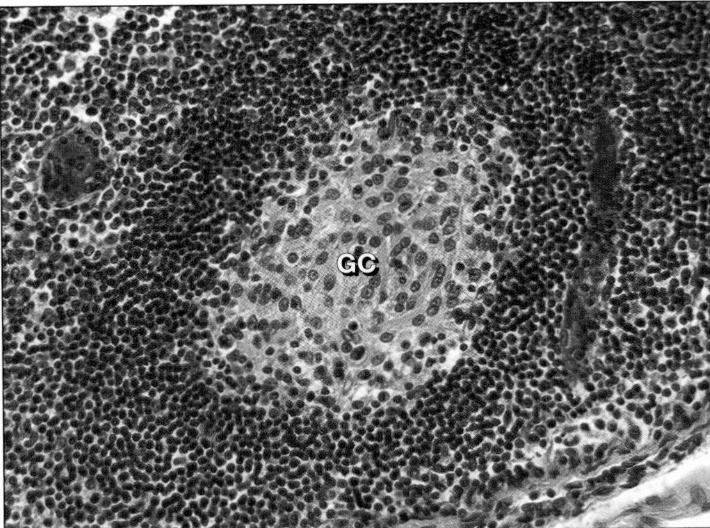

20.11 Single Lymph Follicle • The periphery (cortex) of a node is occupied by lymphatic follicles. Shown is a single follicle that is active in an immune response as evidenced by the lighter central region (germinal center–**GC**) where B cells are multiplying and differentiating to produce plasma cell or memory B cell clones capable of producing antibodies to the antigen that triggered the response. The maturing lymphocytes in the germinal center have more cytoplasm than mature B cells, making the region lighter-staining. Compare the appearance of this follicle to those in Figure 20.10. (Section; Hematoxylin-Eosin Stain)

into T cells in the cortex ("T" stands for thymus) and migrate toward the medulla.

Cells that are not developing properly are removed by macrophages before they reach the medulla. Once in the medulla, **interleukins** promote further maturation. At this point, T cells that have the ability to react with "self" are removed. Those that remain differentiate into cytotoxic T cells or helper T cells, enter the blood, and populate lymphatic tissue throughout the body.

The spleen is a lymphoid organ that occupies the left upper quadrant of the abdominal cavity. It is responsible for immune responses to blood antigens as well as phagocytosis of worn out RBCs and other particulate matter. As such, it is a filter of blood.

Spleen tissue (Fig. 20.13) is divided into regions referred to as **red pulp**, which comprises the majority of the spleen, and **white pulp**. Red pulp is composed of venous sinuses separated by aggregations of macrophages, lymphocytes, plasma cells, RBCs, and other cells. Young, flexible RBCs are able to pass through small openings in the sinus walls and enter the blood. Older, more brittle RBCs break and are removed by macrophages. Released hemoglobin is recycled.

White pulp performs immunological functions and is composed of lymphocyte aggregates. T cells form sheaths around branches of the splenic artery called **central arteries. Splenic nodules**, with B cells in germinal centers, form as expansions of the sheaths and have an eccentric central artery. (They resemble lymphatic follicles, but can be differentiated from them by the central artery.) A lighter zone of active macrophages is found at the periphery of splenic nodules and is the primary site of contact with blood antigens.

Unencapsulated Lymphatic Tissue

Lymphatic tissue is found scattered in various organs. In some cases the lymphocytes are dispersed in connective tissues of organs. In other cases they are organized into lymph follicles within the organ walls.

Lymphocytes associated with mucous membranes are referred to as **Mucosa Associated Lymphatic Tissue (MALT).** Scattered lymphocytes are frequently seen in the mucosal connective tissue and epithelium of various organs. T cells are the primary occupants of this diffuse lymphatic tissue, but some B cells are also present (Fig. 20.14). In other regions, the MALT is more organized, as described in the following text.

Tonsils are probably the most well-known examples of MALT. They are small aggregations of unencapsulated lymphatic tissue located in the oropharynx and nasopharynx. The two **palatine tonsils** (Fig. 20.15) (the ones most frequently removed) are located at the junction of the oral cavity and the oropharynx and are covered by epithelium.

The palatine tonsils consist of lymph follicles composed of B cells, and most have germinal centers. Between the follicles are **interfollicular regions** populated by T cells. Dendritic cells in the epithelium act as APCs. The **pharyngeal tonsil**, located in the nasopharynx, and the **lingual tonsils**, at the base of the tongue, perform similar functions.

Gut Associated Lymphatic Tissue (**GALT**, a subset of MALT) is composed of lymphocytes, sometimes organized into lymph follicles, in the walls of digestive organs. They are most abundant in the **Peyer's Patch** of the ileum (Fig. 20.16).

Special epithelial cells called **M cells** overlie the follicles and transfer antigens to dendritic cells occupying pockets in the basal region of the epithelium. Dendritic cells in turn

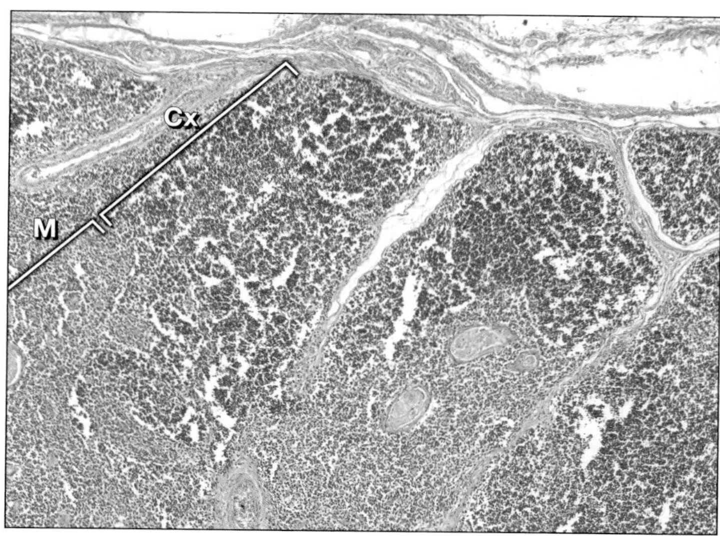

20.12 Thymus • The thymus is divided into a cortex (the darker region–**Cx**) and a medulla (the lighter region–**M**). Lymphocytes begin their maturation into T cells in the cortex and complete the process in the medulla. Along the way, there are check points where cells that are not developing properly or have the ability to react with "self" are removed. Mature T cells enter the blood and populate lymphatic tissue throughout the body. (Section; Hematoxylin-Eosin Stain)

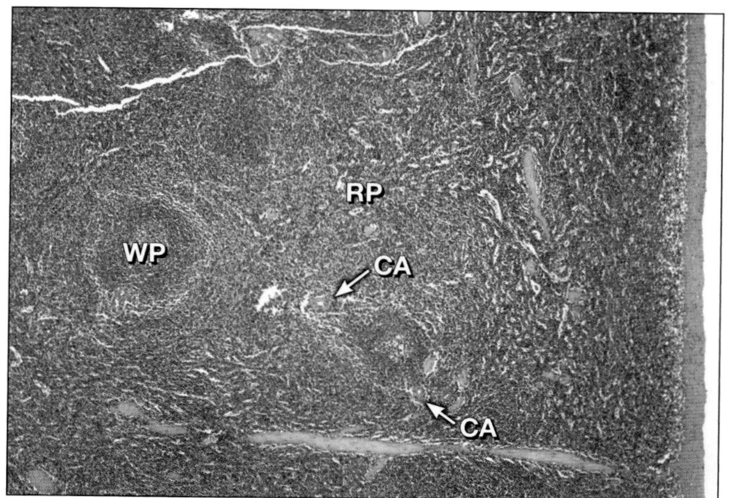

20.13 Spleen • This panoramic view of a spleen section shows regions of red pulp (**RP**) and white pulp (**WP**). Red pulp acts as a blood filter and reservoir. White pulp is largely populated by lymphocytes, but is not white because it has been stained, making it appear the characteristic purple color of lymphatic tissue. Note the eccentric central arteries (**CA**) of the splenic nodules in the white pulp. Lymphocytes of white pulp respond to antigens in the blood. (Section; Hematoxylin-Eosin Stain)

present antigens to T and B cells. GALT is also abundant in the colon and the vermiform appendix, a tubular extension of the cecum.

Bronchus Associated Lymphatic Tissue (BALT) is found in the respiratory tree (Fig. 20.17). It can be diffuse collections of lymphocytes or in the form of lymphatic follicles.

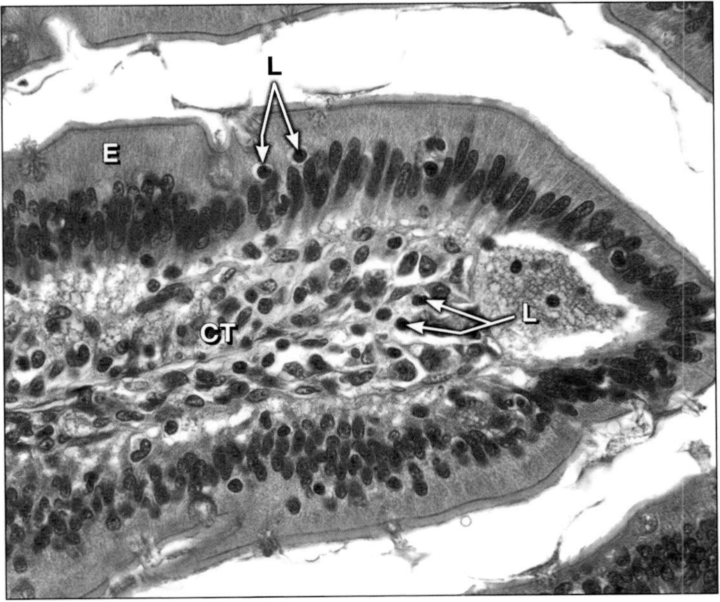

20.14 Diffuse Lymphatic Tissue • Lymphocytes (**arrows**, central **L**) are often found scattered in the connective tissue layer of a mucous membrane, as in this section of a small intestine villus. They are characterized by small, dark-staining nuclei and often form the majority of cells occupying the connective tissue (**CT**). Lymphocytes (**arrows**, upper **L**) also may be seen in the epithelium (**E**). (Section; Hematoxylin-Eosin Stain)

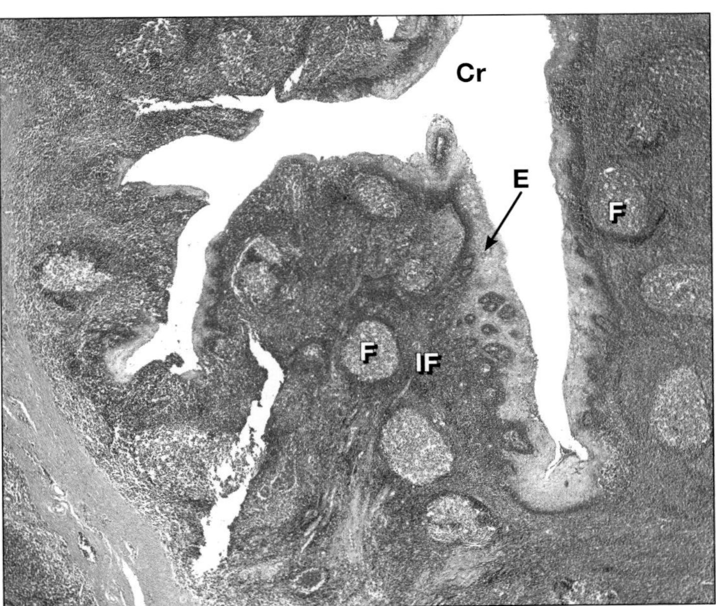

20.15 Palatine Tonsil • The palatine tonsils are covered by epithelium (**E**) that projects down into tonsillar crypts (**Cr**). The bulk of the tonsil is composed of follicles with germinal centers (**F**), with interfollicular (**IF**) regions composed of T cells making up the remainder. (Section; Hematoxylin-Eosin Stain)

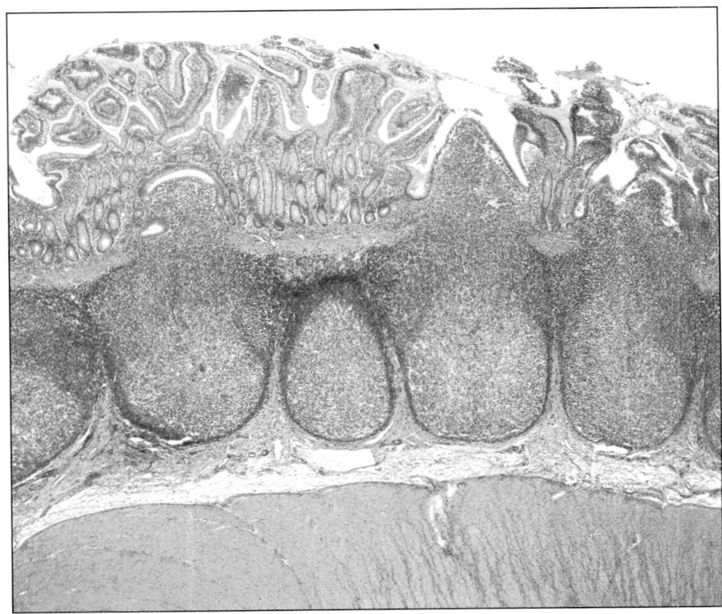

20.16 Peyer's Patch of the Small Intestine • Aggregates of 10–70 submucosal lymph follicles in the ileum are known as Peyer's Patches. The epithelium overlying them contains M cells that capture and transfer antigens to nearby dendritic cells for presentation to B and T cells. Because they are found in the gut, Peyer's Patches are classified as GALT. (Section; Hematoxylin-Eosin Stain)

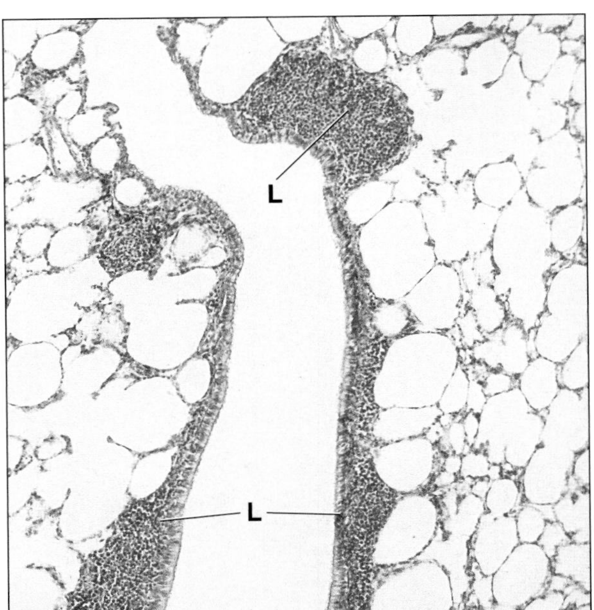

20.17 BALT of a Bronchiole • Lymph follicles and diffuse lymphocytes are also found associated with the respiratory tree, especially at branch points. All those purple dots are lymphocyte (**L**) nuclei within the bronchiole's wall. (Section; Hematoxylin-Eosin Stain)

Biochemical Pathways

APPENDIX

So much of what is done in microbiology relies on an understanding of basic biochemical pathways (Fig. A.1). It is not as important to memorize them (although, with exposure they will become second nature) as it is to understand their importance in metabolism and to interpret diagrams of them when available. The following discussion along with Figure A.1 are provided so you can see how the various biochemical tests presented in this manual fit into the overall scheme of cellular chemistry.

Oxidation of Glucose: Glycolysis, Entner-Doudoroff, and Pentose Phosphate Pathways

Most organisms use glycolysis, also known as the Embden-Meyerhof-Parnas pathway (Fig. A.2), in energy metabolism. It performs the stepwise disassembly of glucose into two pyruvates, releasing some of its energy and electrons in the process. The exergonic (energy-releasing) reactions are associated with ATP synthesis by a process called **substrate-level phosphorylation**. Although a total of four ATPs are produced per glucose in glycolysis, two ATPs are hydrolyzed early in the pathway, leaving a net production of two ATPs per glucose.

In one glycolytic reaction, the loss of an electron pair (oxidation) from a 3-carbon intermediate occurs simultaneously with the reduction of NAD^+ to $NADH + H^+$. The $NADH + H^+$ then may be oxidized in an electron transport chain or a fermentation pathway, depending on the organism and the environmental conditions. The former yields ATP (by **oxidative phosphorylation**) and the latter generally does not. In summary, each glucose completely oxidized in glycolysis yields two pyruvates, $2\ NADH + 2\ H^+$, and a net of 2 ATPs (Table A-1).

Although the intermediates of glycolysis are carbohydrates, many are entry points for amino acid, lipid, and nucleotide catabolism. Many glycolytic intermediates also are a source of carbon skeletons for the synthesis of these other biochemicals. Some of these are shown in Figure A.1.

The Entner-Doudoroff pathway (Fig. A.3) is an alternative means of degrading glucose into two pyruvates. This pathway is found (almost) exclusively among Bacteria (e.g., *Pseudomonas* and *E. coli*, as well as other Gram negatives) and certain Archaea. Some obligate aerobes use this pathway because they lack the enzymes required to convert glucose to glyceraldehyde 3-phosphate in glycolysis.

The pathway also allows utilization of a different category of sugars (aldonic acids) than glycolysis and therefore improves the range of resources available to the organism. It is less efficient than glycolysis because only one ADP is phosphorylated (by substrate-level phosphorylation) and only one NADH is produced. However, it produces one $NADPH + H^+$ that glycolysis does not. Unlike NADH, NADPH is not used as an electron donor in an electron transport chain, so its energy is not used in oxidative phosphorylation. Rather, it is used as reducing power in anabolic reactions. Table A-2 summarizes this pathway.

The pentose phosphate pathway (Fig. A.4) is a complex set of cyclic reactions that provides a mechanism for producing

TABLE **A-1** **Summary of Glycolytic Reactants and Products per Glucose (Maximum Yield)**

Reactant	Product
Glucose ($C_6H_{12}O_6$)	2 Pyruvates ($C_3H_3O_3$)
(2 ATP + 4 ADP) = NET: 2 ADP	(2 ADP + 4 ATP) = NET: 2 ATP
2 NAD^+	2 NADH + 2 H^+

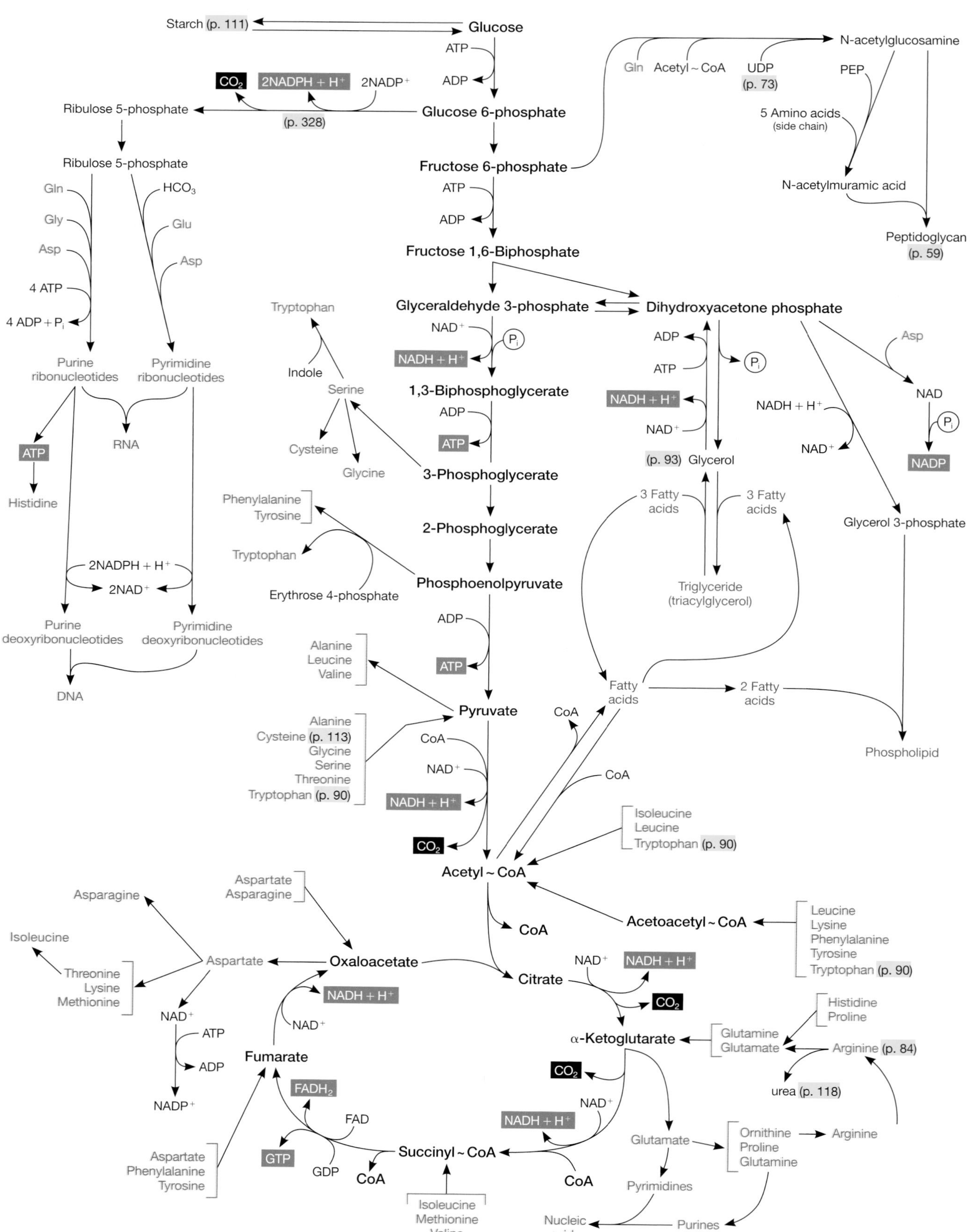

A.1 Integrated Metabolism ● This diagram of metabolism shows basic anabolic and catabolic pathways in bacterial cells along with lab exercises with page references where those reactions are relevant. Remember this diagram is here and check in with it periodically to regain your "forest" perspective if you feel you're getting lost in the "trees." Some details have been left out because of space limitations. A key to this figure is on page 325.

Key to Figure A.1

Black: Carbon intermediates and miscellaneous compounds
CoA = coenzyme A
CO_2 = carbon dioxide
P_i = inorganic phosphate
PEP = phosphoenolpyruvate
Blue: Short term energy carrying molecules
ATP = adenosine triphosphate
ADP = adenosine diphosphate
GTP = guanosine triphosphate
GDP = guanosine diphosphate
UDP = uridine diphosphate
NAD^+/NADH = nicotinamide adenine dinucleotide (oxidized and reduced forms)
$NADP^+$/NADPH = nicotinamide adenine dinucleotide phosphate (oxidized and reduced forms)
FAD/$FADH_2$ = flavin adenine dinucleotide (oxidized and reduced forms)
Green: Amino acids
asp = aspartate
glu = glutamate
gln = glutamine
gly = glycine
Red: Nucleotides and nucleic acids
Violet: Lipids

TABLE **A-2** **Summary of Entner-Doudoroff Reactants and Products per Glucose (Maximum Yield)**

Reactant	Product
Glucose ($C_6H_{12}O_6$)	2 Pyruvates ($C_3H_3O_3$)
(1 ATP + 2 ADP) = NET: 1 ADP	(1 ADP + 2 ATP) = NET: 1 ATP
1 NAD^+	1 NADH + 1 H^+
1 $NADP^+$	1 NADPH + 1 H^+

5-carbon sugars (pentoses) from 6-carbon sugars (hexoses). Pentose sugars produced are used in ribonucleotides and deoxyribonucleotides, as well as being precursors to aromatic amino acids. Further, this pathway produces NADPH, which is used as an electron donor in anabolic pathways.

The pentose phosphate reactants and products are listed in Table A-3. To completely oxidize one hexose to 6 CO_2, a total of six hexoses must enter the cycle as glucose 6-phosphate and follow one of three different routes (notice the symmetry of pathways as drawn). Notice in Figure A.4 that each hexose loses a CO_2 upon entry into the cycle, but at the end five hexoses are produced. Thus, the net reaction is one hexose being oxidized to 6 CO_2.

Notice also the reactions that transfer 2-carbon and 3-carbon fragments between the 5-carbon intermediates. Transketolase catalyzes the 2-carbon transfer, whereas transaldolase catalyzes the 3-carbon transfer. Alternatively, the 5-carbon intermediates can be redirected into pathways for synthesis of aromatic amino acids and nucleotides (not shown).

The decarboxylation of glucose 6-phospate in the pentose phosphate pathway is not as simple as shown in Figure A.4. The preliminary steps include production of 6-phosphogluconate, an intermediate of the Entner-Doudoroff pathway, which allows passage of carbon skeletons between the two pathways. The pentose phosphate pathway also intersects with glycolysis at glyceraldehyde 3-phosphate and fructose 6-phosphate.

TABLE **A-3** **Summary of Pentose Phosphate Reactants and Products per Glucose-6-phosphate (Maximum Yield)**

Reactant	Product
Glucose-6-phosphate (C_6)	$6CO_2$ + 1 P_i
12 $NADP^+$	12 NADPH + 12 H^+

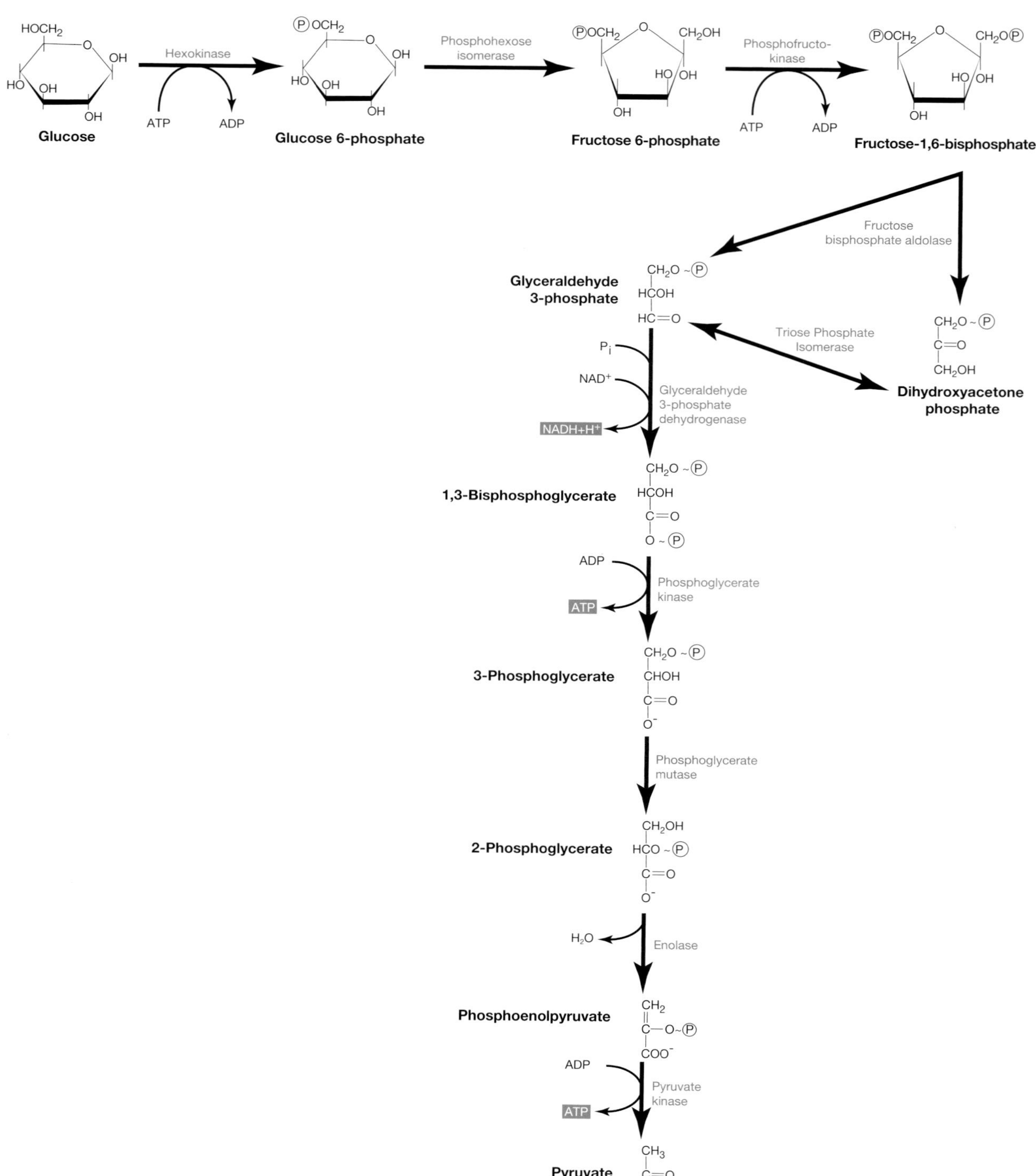

A.2 Glycolysis ● The names of glycolytic intermediates are printed in black ink; the enzyme names are in red. Reducing power (in the form of NADH + H^+) and ATP are highlighted in blue. The major key to getting product yields correct is to recognize that both C_3 compounds (glyceraldehyde 3-phosphate, GAP, and dihydroxyacetone phosphate, DiHAP) produced from splitting fructose 1,6-bisphosphate can pass through the remainder of the pathway because of the triose phosphate isomerase reaction. The conversion of each into pyruvate results in the formation of 2 ATPs and 1 NADH + H+ (Table A-1).

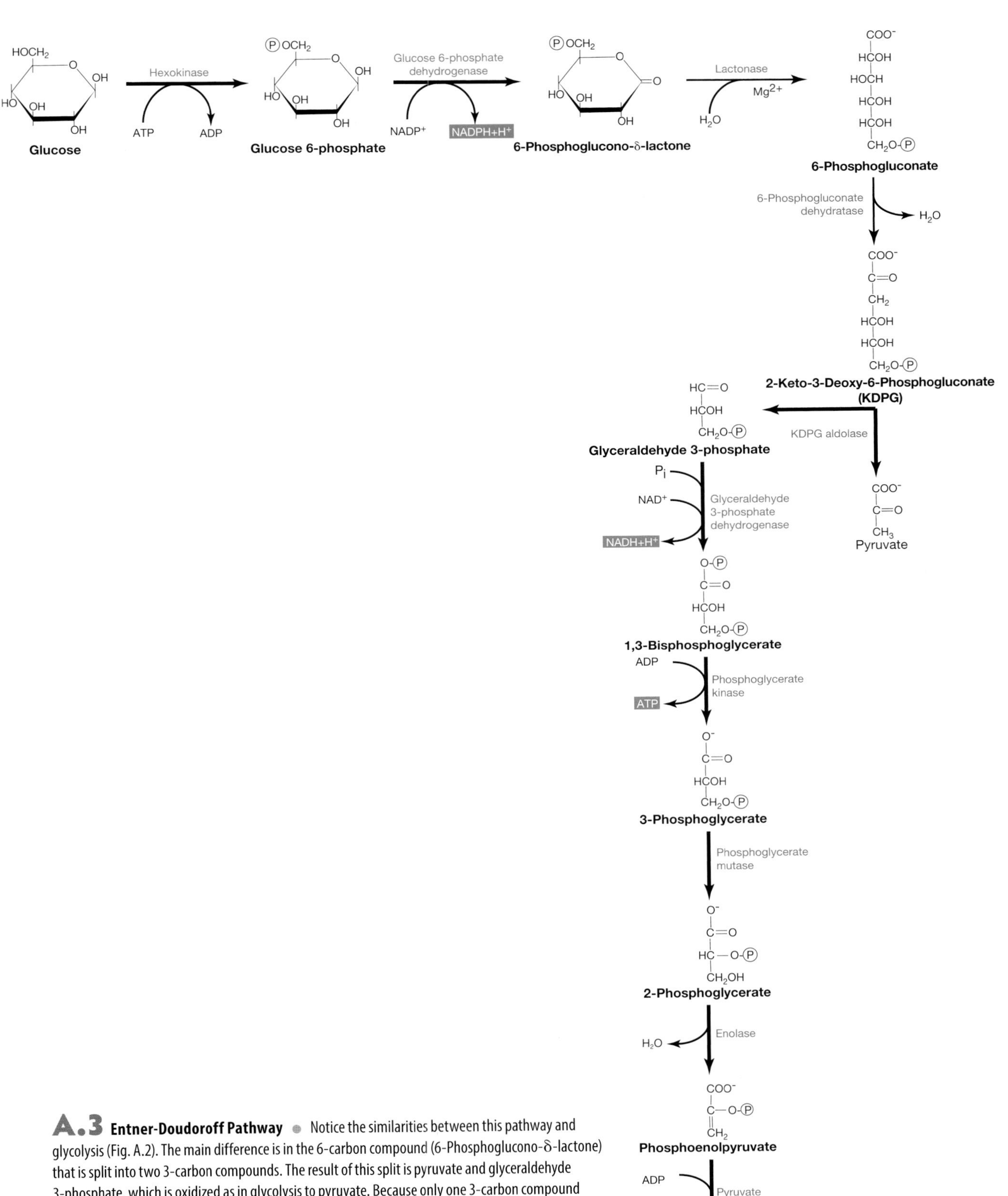

A.3 Entner-Doudoroff Pathway • Notice the similarities between this pathway and glycolysis (Fig. A.2). The main difference is in the 6-carbon compound (6-Phosphoglucono-δ-lactone) that is split into two 3-carbon compounds. The result of this split is pyruvate and glyceraldehyde 3-phosphate, which is oxidized as in glycolysis to pyruvate. Because only one 3-carbon compound goes through the sequence of reactions leading to pyruvate, the ATP and NADH yield is one-half that of glycolysis. However, one NADPH is produced that is not made in glycolysis. (See Table A-2.)

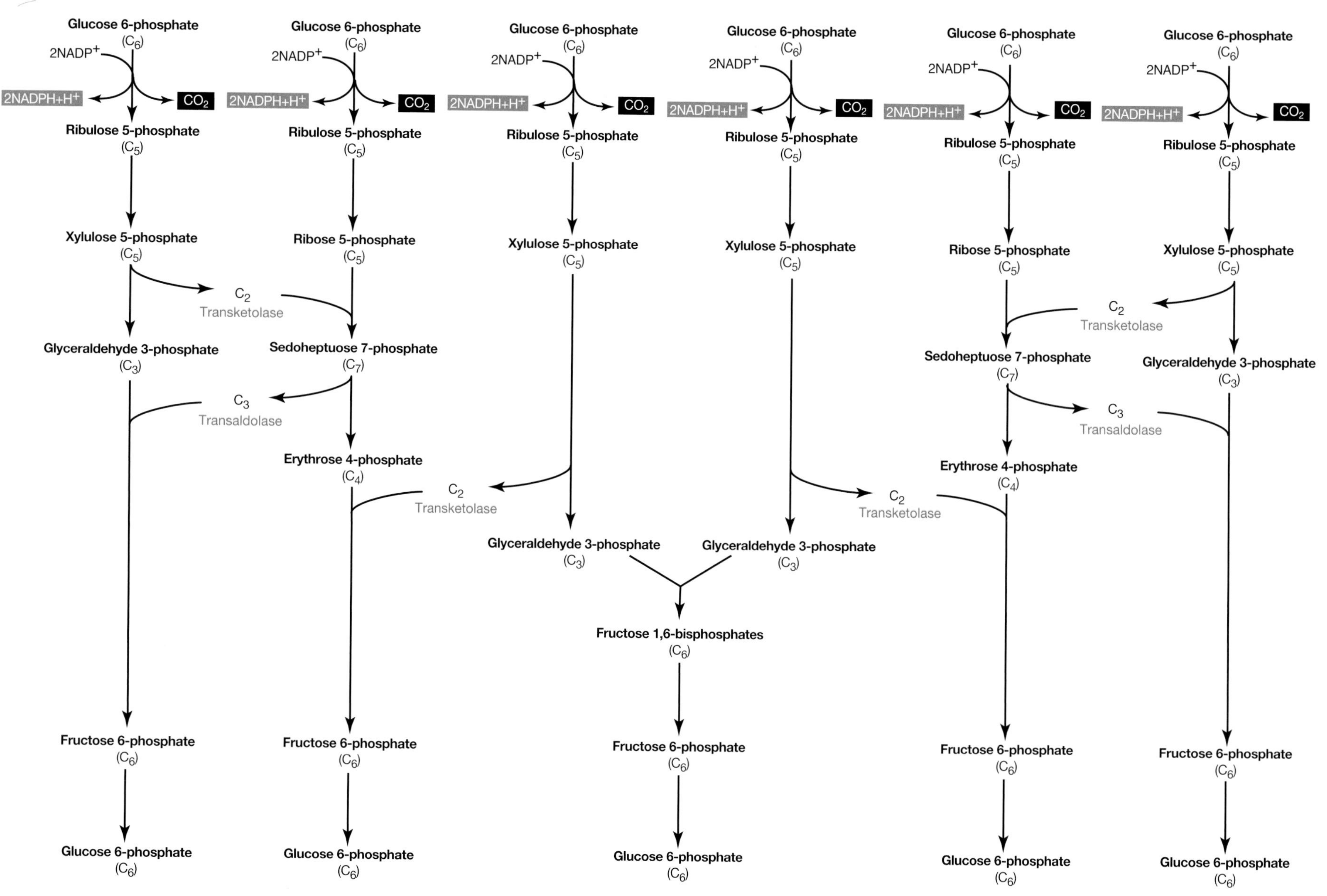

A.4 Pentose Phosphate Cycle ● For every six glucose-6-phosphates that enter and complete the cycle, $6CO_2$ and 12 NADPH + H^+ are produced. Some of the 5-carbon intermediates, however, may be redirected into synthesis of aromatic amino acids and nucleotides. If the cycle is performed as shown, 36 carbons enter as six glucose 6-phosphates ($6 \times C_6 = 36C$). Six CO_2 are immediately lost, leaving a total of 30C to get shuffled around by the remaining reactions to form five glucose 6-phosphates ($5 \times C_6 = 30C$). The enzymes transketolase and transaldolase are responsible for transferring two- and three-carbon fragments, respectively. (See Table A-3.)

Oxidation of Pyruvate: The Citric Acid Cycle and Fermentation

Pyruvate represents a major crossroads in metabolism, and all three pathways discussed so far have the potential to make it, though not in equal quantities. Some organisms are able to further disassemble the pyruvates produced in glycolysis and the other paths and make more ATP and NADH + H^+ in the citric acid cycle. Other organisms simply reduce the pyruvates with electrons from NADH + H^+ without further (or minimal) energy production in fermentation.

The citric acid cycle is a major metabolic pathway used in energy production by organisms that respire aerobically or anaerobically (Fig. A.5). Pyruvate produced in glycolysis or other pathways is first converted to acetyl~coenzyme A during the entry step (also known as the intermediate, or gateway, step). Acetyl~CoA enters the citric acid cycle through a condensation reaction with oxaloacetate.

Products for each pyruvate that enters the cycle via the entry step are: 3 CO_2, 4 NADH + H^+, 1 $FADH_2$, and 1 GTP. (Because two pyruvates are made per glucose, these numbers are doubled in Table A-4, which shows maximum yield per glucose.) The energy released from oxidation of reduced coenzymes (NADH + H^+ and $FADH_2$) in an electron transport chain is then used to make ATP. ATP yields are summarized in Table A-5.

ENTRY STEP
CoA
NAD^+
NADH+H^+
CO_2
Pyruvate
Acetyl ~ CoA
Citrate synthase
H_2O
CoA
Oxaloacetate
Citrate
Aconitase Fe^{2+}
H_2O
Cis-aconitate
H_2O
Aconitase Fe^{2+}
Isocitrate
NAD^+
NADH+H^+
Isocitrate dehydrogenase
Oxalosuccinate
CO_2
Isocitrate dehydrogenase
Mn^{2+}
α-Ketoglutarate
CoA
CO_2
NAD^+
NADH+H^+
α-Ketoglutarate dehydrogenase complex
Succinyl ~ CoA
CoA
Succinyl-CoA ligase
GDP+P_i
GTP
Succinate
Succinate dehydrogenase
FAD
$FADH_2$
Fumarate
H_2O
Fumarase
L-Malate
Malate dehydrogenase
NAD^+
NADH+H^+
CITRIC ACID CYCLE

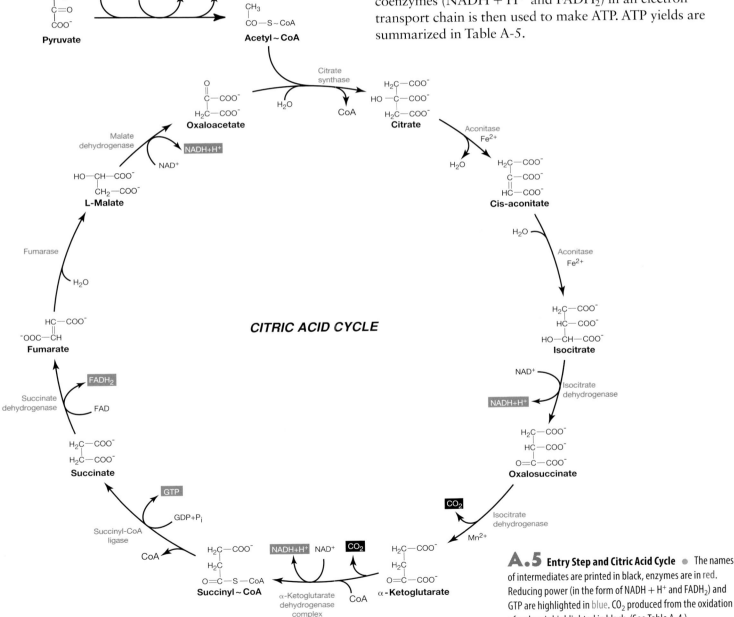

A.5 Entry Step and Citric Acid Cycle • The names of intermediates are printed in black, enzymes are in red. Reducing power (in the form of NADH + H^+ and $FADH_2$) and GTP are highlighted in blue. CO_2 produced from the oxidation of carbon is highlighted in black. (See Table A-4.)

Like glycolysis, many of the citric acid cycle's intermediates are entry points for amino acid, nucleotide, and lipid catabolism, as well as a source of carbon skeletons for synthesis of the same compounds. These pathways are shown in Figure A.1. Single arrows may represent several reactions, and other carbon compounds, not illustrated, may be required to complete a given reaction.

Figure A.6 illustrates some major fermentation pathways exhibited by microbes (though no single organism is capable of all of them). Pyruvate (shown in the blue box) is typically the starting point for each. End-products of fermentation are mostly shown in red boxes. Fermentation allows some cells living under anaerobic conditions to oxidize reduced coenzymes (such as $NADH + H^+$ and shown in blue) generated during glycolysis or other pathways.

Some bacteria (aerotolerant anaerobes) rely solely on fermentation and do not use oxygen even if it is available. Table A-6 summarizes major fermentations and some representative organisms that perform each.

Notice that fermentation end-products typically fall into three categories: gas (black boxes), acid (red boxes), or an organic solvent (an alcohol or a ketone; also red boxes). The specific fermentation performed is the result of the enzymes present in a species and often is used as a basis of classification.

TABLE **A-4** **Summary of Reactants and Products per Glucose in the Entry Step and the Citric Acid Cycle (Maximum Yield)**

Entry Step		Citric Acid Cycle	
Reactant	Product	Reactant	Product
2 Pyruvates + 2 Coenzyme A	2 Acetyl~CoA + 2 CO_2	2 Acetyl~CoA	4 CO_2 + 2 Coenzyme A
2 NAD^+	2 NADH + $2H^+$	6 NAD^+	6 NADH + $6H^+$
		2 GDP + 2 P_i (= 2 ADP + 2 P_i)	2 GTP (= 2 ATP)
		2 FAD	2 $FADH_2$

TABLE **A-5** **ATP Yields from Complete Oxidation of Glucose to CO_2 by a Prokaryote Using Glycolysis, Entry Step, and the Citric Acid Cycle with O_2 as the Final Electron Acceptor**

Compound	Number Produced	ATP Value[1]	Total ATPs per Glucose
$NADH + H^+$	10	3	30
$FADH_2$	2	2	4
ATP (by substrate phosphorylation)	4		4

1 Experimental evidence indicates that the ATP values of 3 per NADH and 2 per $FADH_2$ in the aerobic electron transport chain are probably too high. For mitochondria, values of 2.5 and 1.5, respectively, are closer to reality. However, because multiple electron transport chains have been identified within the domain Bacteria (and even within single species!) no single, universal number for ATP yields per NADH and $FADH_2$ can be assigned. So, we fall back on the comfortable numbers of 3 and 2, recognizing that they are wrong but still can be used to illustrate how ATP yields could be calculated from the various pathways if we only had the correct information.

TABLE **A-6** **Major Fermentations, Their End-Products, and Some Organisms That Perform Them**

Fermentation	Major End-products	Representative Organisms
Alcoholic fermentation	Ethanol and CO_2	*Saccharomyces cerevisiae*
Homofermentation	Lactate	*Streptococcus* and some *Lactobacillus*
Heterofermentation	Lactate, ethanol, and acetate	*Streptococcus*, *Leuconostoc*, and *Lactobacillus*
Mixed acid fermentation	Acetate, formate, succinate, CO_2, H_2, and ethanol	*Escherichia*, *Salmonella*, *Klebsiella*, and *Shigella*
2,3-butanediol fermentation	2,3-butanediol	*Enterobacter*, *Serratia*, and *Erwinia*
Butyrate/butanol fermentation	Butanol, butyrate, acetone, and isopropanol	*Clostridium*, *Butyrivibrio*, and some *Bacillus*
Propionic acid fermentation	Propionate, acetate and CO_2	*Propionibacterium*, *Veillonella*, and some *Clostridium*

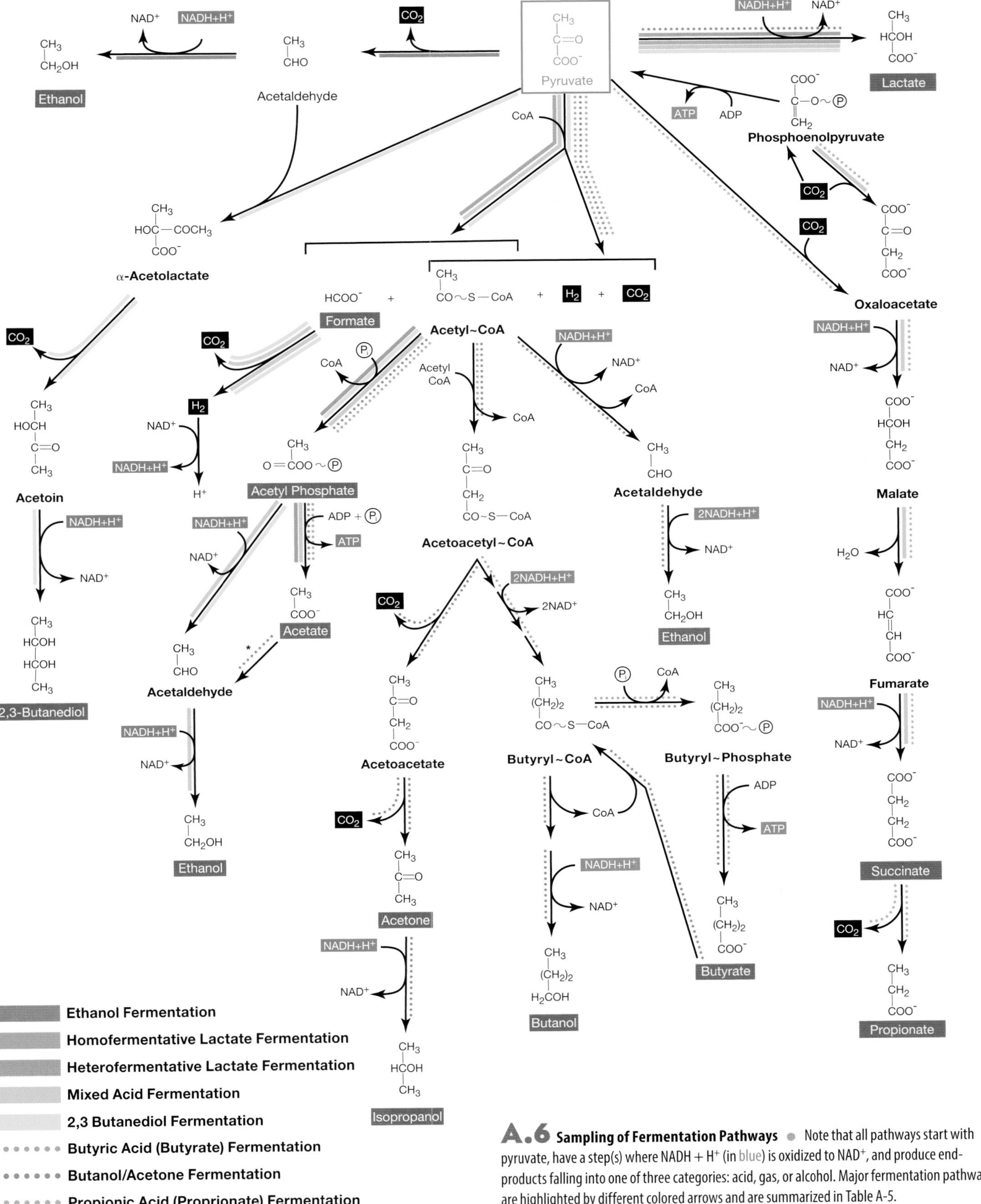

A.6 Sampling of Fermentation Pathways • Note that all pathways start with pyruvate, have a step(s) where NADH + H^+ (in blue) is oxidized to NAD^+, and produce end-products falling into one of three categories: acid, gas, or alcohol. Major fermentation pathways are highlighted by different colored arrows and are summarized in Table A-5.

Suggested General References

Boone, David R. and Richard W. Castenholz, eds. George M. Garrity, editor in chief. *Bergey's Manual of Systematic Bacteriology*, 2nd ed., Vol. One, The *Archaea* and the Deeply Branching and Phototrophic *Bacteria*. New York: Springer, 2001.

Boyd, Eric, Monica Brelsford, Jamie Cornish, John Dore, Jerry Johnson, Susan Kelly, John Peters, Brent Peyton, Heather Rauser, Christine Smith, Suzi Taylor, Jennifer Wirth, and Mark Young. *Living Colors: Microbes of Yellowstone National Park*. Gardiner, MT: Yellowstone Forever, 2013.

Brenner, Don J., Noel R. Krieg, and James T. Staley, eds. George M. Garrity, editor in chief. *Bergey's Manual of Systematic Bacteriology*, 2nd ed., Vol. Two, Part A, The *Proteobacteria*: Introductory Essays. New York: Springer, 2005.

Brenner, Don J., Noel R. Krieg, and James T. Staley, eds. George M. Garrity, editor in chief. *Bergey's Manual of Systematic Bacteriology*, 2nd ed., Vol. Two, Part B, The *Gammaproteobacteria*: Introductory Essays. New York: Springer, 2005.

Brenner, Don J., Noel R. Krieg, and James T. Staley, eds. George M. Garrity, editor in chief. *Bergey's Manual of Systematic Bacteriology*, 2nd ed., Vol. Two, Part C, "The *Alpha-, Beta-, Delta-,* and *Epsilonproteobacteria*: Introductory Essays." New York: Springer, 2005.

Carroll, Karen C. and Michael A. Pfaller, editors in chief. Marie Louise Landry, Alexander J. McAdam, Robin Patel, and Sandra S. Richter, vol. eds. *Manual of Clinical Microbiology*, 12th ed., Volumes 1 and 2. Washington, DC: ASM Press, 2019.

De Vos, Paul, George M. Garrity, Dorothy Jones, Noel R. Krieg, Wolfgang Ludwig, Fred A. Rainey, Karl-Heinz Schleifer and William B. Whitman, eds. William B. Whitman, director of the editorial office, and Aidan C. Parte, managing ed. *Bergey's Manual of Systematic Bacteriology*, 2nd ed., Vol. Three, The *Firmicutes*. New York: Springer, 2009.

Garcia, Lynne S. *Diagnostic Medical Parasitology*, 6th ed. Washington, DC: ASM Press, 2016.

Goodfellow, Michael, Peter Kämpfer, Hans-Jürgen Busse, Martha E. Trujillo, Ken-ichiro Suzuki, Wolfgang Ludwig, and William B. Whitman, eds. William B. Whitman, director of the editorial office, and Aidan C. Parte, managing ed. *Bergey's Manual of Systematic Bacteriology*, 2nd ed., Vol. Five, Parts A and B, The *Actinobacteria*. New York: Springer, 2012.

Krieg, Noel R., James T. Staley, Daniel R. Brown, Brian P. Hedlund, Bruce J. Paster, Naomi L. Ward, Wolfgang Ludwig, and William B. Whitman, eds. William B. Whitman, director of the editorial office, and Aidan C. Parte, managing ed. *Bergey's Manual of Systematic Bacteriology*, 2nd ed., Vol. Four, The *Bacteroidetes, Spirochaetes, Tenericutes (Mollicutes), Acidobacteria, Fibrobacteres Fusobacteria, Dictyoglomi, Gemmatimonadetes, Lentisphaerae, Verrucomicrobia, Chlamydiae,* and *Planctomycetes*. New York: Springer, 2011.

Madigan, Michael T., Kelly S. Bender, Daniel H. Buckley, W. Matthew Sattley, and David A. Stahl. *Brock Biology of Microorganisms*, 15th ed. New York: Pearson, 2018.

Moat, Albert G., John W. Foster, and Michael P. Spector. *Microbial Physiology*, 4th ed. New York, NY: Wiley–Liss, 2002.

Montville, Thomas J., Karl R. Matthews, and Kalmia E. Kniel. *Food Microbiology, An Introduction*, 3rd ed. Washington, DC: ASM Press, 2012.

Procop, Gary W., Deirdre L. Church, Geraldine S. Hall, William M. Janda, Elmer W. Koneman, Paul C. Schreckenberger, and Gail L. Woods. *Koneman's Color Atlas and Textbook of Diagnostic Microbiology*, 7th ed. Philadelphia: Wolters Kluwer, Lippincott Williams & Wilkins, 2017.

Richman, Douglas D., Richard J. Whitley, and Frederick G. Hayden. *Clinical Virology*, 4th ed. Washington, DC: ASM Press, 2017.

Sheehan, Kathy B., David J. Patterson, Brett Leigh Dicks, and Joan M. Henson. *Seen and Unseen, Discovering Microbes of Yellowstone*. Guilford, CT: Falcon Guides, 2005.

Tille, Patricia M. *Bailey & Scott's Diagnostic Microbiology*, 14th ed. St. Louis, MO: Elsevier, 2017.

Urry, Lisa A., Michael L. Cain, Peter V. Minorsky, Steven A. Wasserman, and Jane B. Reece. *Campbell Biology*, 11th ed. New York: Pearson, 2017.

Walsh, Thomas J., Randall T. Hayden, and Davise H. Larone. *Larone's Medically Important Fungi: A Guide to Identification*, 6th ed. Washington, DC: ASM Press, 2018.

White, David, James Drummond, and Clay Fuqua. *The Physiology and Biochemistry of Prokaryotes*, 4th ed. New York: Oxford University Press, 2012.

Yates, Marylynn V., editor in chief, Cindy H. Nakatsu, Robert V. Miller, and Suresh D. Pillai, eds. *Manual of Environmental Microbiology*, 4th ed. Washington DC: ASM Press, 2016.

Zimbro, Mary Jo and David A. Power, eds. *Difco & BBL Manual, Manual of Microbiological Culture Media*. Sparks, MD: Becton, Dickinson and Company, 2003.

Index

➤ Figures and tables are denoted by the page number in *italics* followed by *f* or *t* respectively.